Design, Fabrication, and Characterization of Multifunctional Nanomaterials

Micro & Nano Technologies Series

Design, Fabrication, and Characterization of Multifunctional Nanomaterials

Edited by

Sabu Thomas

Nandakumar Kalarikkal

Ann Rose Abraham

ELSEVIER

Elsevier
Radarweg 29, PO Box 211, 1000 AE Amsterdam, Netherlands
The Boulevard, Langford Lane, Kidlington, Oxford OX5 1GB, United Kingdom
50 Hampshire Street, 5th Floor, Cambridge, MA 02139, United States

Library of Congress Cataloging-in-Publication Data
A catalog record for this book is available from the Library of Congress

British Library Cataloguing-in-Publication Data
A catalogue record for this book is available from the British Library

ISBN: 978-0-12-820558-7

For information on all Elsevier publications visit our website at https://www.elsevier.com/books-and-journals

Publisher: Matthew Deans
Acquisitions Editor: Simon Holt
Editorial Project Manager: Chiara Giglio
Production Project Manager: Sojan P. Pazhayattil
Cover Designer: Christian J. Bilbow

Typeset by TNQ Technologies

Contents

Contributors

Ann Rose Abraham Department of Physics, Sacred Heart College (Autonomous), Kochi, Kerala, India

Shampa Aich Department of Metallurgical & Materials Engineering, Indian Institute of Technology, Kharagpur, West Bengal, India

A.R. Ajitha International and Interuniversity Centre for Nanoscience and Nanotechnology, Mahatma Gandhi University, Kottayam, Kerala, India; Department of Chemistry, Newmann College, Thodupuzha, Idukki, Kerala, India

Nureddin Ashammakhi Center for Minimally Invasive Therapeutics (C-MIT), University of California, Los Angeles, Los Angeles, CA, United States; Department of Bioengineering, Henry Samueli School of Engineering, University of California, Los Angeles, Los Angeles, CA, United States; Department of Radiological Sciences, David Geffen School of Medicine, University of California, Los Angeles, Los Angeles, CA, United States; Department of Biomedical Engineering, College of Engineering, Michigan State University, MI, United States

Sonja Aškrabić Nanostructured Matter Laboratory, Institute of Physics Belgrade, University of Belgrade, Belgrade, Serbia

Pius Augustine Department of Physics, Sacred Heart College (Autonomous), Thevara, Kochi, India; Material Research Centre, Indian Institute of Science, Bangalore, Karnataka, India

Ajit Behera Department of Metallurgical & Materials Engineering, National Institute of Technology, Rourkela, Odisha, India

Patrice Bourson Université de Lorraine, CentraleSupélec, LMOPS, Metz, France

David Chapron Université de Lorraine, CentraleSupélec, LMOPS, Metz, France

Mohammad Ali Darabi Center for Minimally Invasive Therapeutics (C-MIT), University of California, Los Angeles, Los Angeles, CA, United States; Department of Bioengineering, Henry Samueli School of Engineering, University of California, Los Angeles, Los Angeles, CA, United States

Nidhin Divakaran School for Advanced Research in Polymers (SARP) – LARPM, Central Institute of Plastics Engineering & Technology (CIPET) – IPT, Patia, Bhubaneswar, Odisha, India

Dejan M. Djokić Nanostructured Matter Laboratory, Institute of Physics Belgrade, University of Belgrade, Belgrade, Serbia

Zorana D. Dohčević-Mitrović Nanostructured Matter Laboratory, Institute of Physics Belgrade, University of Belgrade, Belgrade, Serbia

Sijo Francis Department of Chemistry, St. Joseph's College, Moolamattom, Kerala, India
Jacek Gapiński Faculty of Physics, Adam Mickiewicz University, Poznań, Poland
Soney C. George Centre for Nanoscience and Technology, Amal Jyothi College of Engineering, Kottayam, Kerala, India
Anwarul Hasan Department of Mechanical and Industrial Engineering, College of Engineering, Qatar University, Doha, Qatar
Garudadhwaj Hota Department of Chemistry, National Institute of Technology, Rourkela, Odisha, India
Jayakumari Isac Centre for Condensed Matter, Department of Physics, CMS College, Kottayam, Kerala, India
K. Jeyasubramanian Mepco Schlenk Engineering College, Sivakasi, Tamil Nadu, India
Neena S. John Centre for Nano and Soft Matter Sciences, Bengaluru, Karnataka, India
Thomasukutty Jose Centre for Nanoscience and Technology, Amal Jyothi College of Engineering, Kottayam, Kerala, India
Nandakumar Kalarikkal School of Pure and Applied Physics, Mahatma Gandhi University, Kottayam, Kerala, India; International and Inter University Centre for Nanoscience and Nanotechnology, Mahatma Gandhi University, Kottayam, Kerala, India; School of Nanoscience and Nanotechnology, Mahatma Gandhi University, Kottayam, Kerala, India
Manoj B. Kale CAS Key Laboratory of Design and Assembly of Functional Nanostructures, and Fujian Provincial Key Laboratory of Nanomaterials, State Key Laboratory of Structural Chemistry, Key Laboratory of Optoelectronic Materials Chemistry and Physics, Fujian Institute of Research on the Structure of Matter, Chinese Academy of Sciences, Fuzhou, Fujian, China
Ebey P. Koshy Department of Chemistry, St. Joseph's College, Moolamattom, Kerala, India
Indradeep Kumar Vels Institute of Science, Technology & Advanced Studies (VISTAS), Chennai, Tamil Nadu, India
Beena Mathew School of Chemical Sciences, Mahatma Gandhi University, Kottayam, Kerala, India
M. Mishra Department of Life Science, National Institute of Technology, Rourkela, Odisha, India
P.M.G. Nambissan Applied Nuclear Physics Division, Saha Institute of Nuclear Physics, Kolkata, India
Fulufhelo Nemavhola Biomechanics Research Group (BRG), Mechanical and Industrial Engineering Department (DMIE), School of Engineering, College of Science, Engineering and Technology (CSET), University of South Africa, Pretoria, South Africa
S. Padhiari Department of Chemistry, National Institute of Technology, Rourkela, Odisha, India
Shabna Patel Department of Mathematics & Science, UGIE, Rourkela, Odisha, India
J. Philip Amal Jyothi College of Engineering, Kanjirappalli, Kerala, India
Mikolaj Pochylski Faculty of Physics, Adam Mickiewicz University, Poznań, Poland
Marc Ponçot Université de Lorraine, CNRS, IJL, Nancy, France

Ippokratis Pountos Academic Department of Trauma and Orthopedics, University of Leeds, Leeds, United Kingdom; Chapel Allerton Hospital, Leeds Teaching Hospitals, Leeds, United Kingdom

K. Priya Madhuri Centre for Nano and Soft Matter Sciences, Bengaluru, Karnataka, India

Dipen Kumar Rajak Sandip Institute of Technology and Research Centre, Nashik, Maharashtra, India

Sreelakshmi Rajeevan Centre for Nanoscience and Technology, Amal Jyothi College of Engineering, Kottayam, Kerala, India; APJ Abdul Kalam Technological University, CET Campus, Thiruvananthapuram, Kerala, India

R. Rakhikrishna Department of Physics, NSS College, Changanassery, Kerala, India

Arunima Reghunadhan Department of Chemistry, Milad-E-Sherif Memorial College, Kayamkulam, Alappuzha, Kerala, India; International and Interuniversity Centre for Nanoscience and Nanotechnology, Mahatma Gandhi University, Kottayam, Kerala, India

Neerish Revaprasadu SARCHI Chair in Nanotechnology, Department of Chemistry, Faculty of Science and Agriculture, University of Zululand, KwaZulu-Natal, Zululand, South Africa

Amirsadegh Rezazadeh Nochehdehi Biomechanics Research Group (BRG), Mechanical and Industrial Engineering Department (DMIE), School of Engineering, College of Science, Engineering and Technology (CSET), University of South Africa, Pretoria, South Africa

Isabelle Royaud Université de Lorraine, CNRS, IJL, Nancy, France

Jitendra Kumar Sahoo Department of Basic Science and Humanities, GIET University, Gunupur, Odisha, India

Harekrushna Sahoo Department of Chemistry, National Institute of Technology, Rourkela, Odisha, India

S.K. Sahoo Department of Chemistry, National Institute of Technology, Rourkela, Odisha, India

Bojan S. Stojadinović Nanostructured Matter Laboratory, Institute of Physics Belgrade, University of Belgrade, Belgrade, Serbia

Nazzar Tellisi Academic Department of Trauma and Orthopedics, University of Leeds, Leeds, United Kingdom; Chapel Allerton Hospital, Leeds Teaching Hospitals, Leeds, United Kingdom

T. Theivasanthi International Research Center, Kalasalingam Academy of Research and Education, Krishnankoil, Tamil Nadu, India

Sabu Thomas International and Inter University Centre for Nanoscience and Nanotechnology, Mahatma Gandhi University, Kottayam, Kerala, India; School of Energy Materials, Mahatma Gandhi University, Kottayam, Kerala, India

M. Tripathy Department of Chemistry, National Institute of Technology, Rourkela, Odisha, India

Rumeysa Tutar Department of Chemistry, Faculty of Engineering, Istanbul University-Cerrahpasa Avcılar, Istanbul, Turkey

Jeena Varghese Faculty of Physics, Adam Mickiewicz University, Poznań, Poland

V.S. Vinila Centre for Condensed Matter, Department of Physics, CMS College, Kottayam, Kerala, India
Runcy Wilson Department of Chemistry, St. Cyril's College, Kilivayal, Kerala, India
Lixin Wu CAS Key Laboratory of Design and Assembly of Functional Nanostructures, Fujian Key Laboratory of Nanomaterials, Fujian Institute of Research on the Structure of Matter, Chinese Academy of Sciences, Fuzhou, Fujian, China

Editors' biographies

Ann Rose Abraham

Ann Rose Abraham, Ph.D. is currently an Assistant Professor at the Department of Physics, Sacred Heart College (Autonomous), Thevara, Kochi, Kerala, India. Dr. Abraham received M.Sc., M.Phil. and Ph.D. degrees in Physics from the School of Pure and Applied Physics, Mahatma Gandhi University, Kerala, India. Her Ph.D. thesis was on the "Development of Hybrid Multiferroic Materials for Tailored Applications." She is an expert in the fields of condensed matter physics, nanomagnetism, multiferroics, and polymeric nanocomposites. She has had research experience at various reputed national institutes such as Bose Institute, Kolkata, India, SAHA Institute of Nuclear Physics, Kolkata, India, and UGC-DAE CSR Centre, Kolkata, India and collaborations with various international laboratories. She is the recipient of a Young Researcher award in the area of physics, and Best Paper Awards—2020 and 2021, a prestigious forum to showcase intellectual capability. She served as assistant professor and examiner at the Department of Basic Sciences, Amal Jyothi College of Engineering, under APJ Abdul Kalam Technological University, Kerala, India. Dr. Abraham is a frequent speaker at national and international conferences. She has authored many book chapters and edited seven books with Taylor and Francis and Elsevier. She has a good number of publications to her credit in many peer-reviewed, high-impact journals of international repute, such as *ACS Journal of Physical Chemistry*, *RSC Physical Chemistry Chemical Physics*, and *New Journal of Chemistry*.

Nandakumar Kalarikkal

Dr. Nandakumar Kalarikkal is an Associate Professor at the School of Pure and Applied Physics and Joint Director of the International and Inter University Centre for Nanoscience and Nanotechnology of Mahatma Gandhi University, Kottayam, Kerala, India. His research activities involve applications of nanostructured materials, laser plasma, and phase transitions. He is the recipient of research fellowships and associateships from prestigious government organizations such as the Department of Science and Technology and Council of Scientific and Industrial Research of the Government of India. He has active collaborations with national and international scientific institutions in India, South Africa, Slovenia, Canada, France, Germany, Malaysia, Australia, and the United States. He has more than 130 publications in peer-reviewed journals. He also co-edited nine books of scientific interest and co-authored many book chapters.

Sabu Thomas

Prof. Sabu Thomas, an outstanding Alumnus of IIT, Kharagpur, is one of India's most renowned scientists in the area of Polymers. After completing his Ph.D. from IIT Kharagpur (1984-1987), he joined MG University as a Lecturer in 1997 and later became its Vice Chancellor. He has taken up a large number of visiting assignments abroad. Under his leadership, the University has been ranked 713th by TIMES, 30th in NIRF and the best University in Kerala. He has supervised 120 Ph.D. students, authored 1,300 publications, and edited 150 books earning him a H-index of 112 and 60,000 citations. He has received Honoris Causa degrees from Russia and France and obtained grants amounting to Rs. 30 crores for research funding from India and abroad. He has been ranked 114th in the list of the world's best scientists and 2nd in India by the Stanford University Ranking in Polymers. He was elected as a Fellow of the European Academy of Sciences. Considering his excellent contributions in teaching, research and administration, Prof. Thomas is the best candidate for the outstanding Alumnus award of IIT KGP.

Contributors' biographies

Jiji Abraham

Dr. Jiji Abraham is working as Assistant Professor in Chemistry in Vimala College (Autonomous) Thrissur. She completed PhD from International and Inter University Center for Nanoscience and nanotechnology Mahatma Gandhi University Kottayam. Here research interests include Polymer nanocomposites, synthesis of nanomaterials, etc. Dr. Abraham has published over 50 research articles which includes 28 journal paper sand 22 book chapters. She has edited one book entitled Rheology and processing of polymer nanocomposites published by Wiley. The H-index of Dr. Abraham is 14 and has more than 600 citations. Dr. Abraham was a visiting research student in Institut Charles Sadron CNRS UPR 22, University de Strasbourg, France. She has presented papers in European Polymer Congress 2015, during June 21–26, 2015, Dresden, Germany; Malaysia Polymer International Conference (MPIC 2017), July 19–20, 2017, Universiti Kebangsaan, Malaysia, International Polymer Characterization Conference POLY-CHAR 2019 (Kathmandu, Nepal) Polymer for Sustainable Development, May 19–23, 2019. Dr. Abraham has delivered many presentations in national/international meetings.

Shampa Aich

Dr. Shampa Aich works in the fields of metallurgical and materials science and engineering, especially on smart materials (SMA/FMSMA), magnetic materials, biomaterials, and surface modifications. Her contribution in those areas include (1) fabrication of NiTi shape memory alloy thin films by magnetron sputtering; (2) Development of microstructure and magnetic properties of magnetic shape memory alloy ribbons in Heusler family by rapid solidification (melt-spinning); (3) development of rare-earth-based as well as rare-earth free permanent magnets; (4) synthesis of wear resistant coating system of TiB (titanium monoboride) whiskers on titanium surfaces by solid state diffusion (Pack boriding); (5) fabrication of macroporous but mechanically tough bioscaffolds on the metal implant/stent surfaces, which can promote cell adhesion and proliferation on metallic implant and can perform controlled on-site drug release and photocatalytic sterilization; (6) deposition of a series of bilayers and multilayers coatings of Ti/TiB2, Ti/TiN, Cr/CrN and Ti/TiN/Cr/CrN on steel substrate by pulsed laser deposition (PLD) technique to improve the tribological properties and cutting efficiency of cutting tool; and (7) to determine the feasibility of growing inert silica shells around the metallic nanoparticles with the help of core-shell nanotechnology to render those nanoparticles stable by preventing their agglomeration and biodegradation to improve their unique properties and potential biomedical applications.

A.R. Ajitha

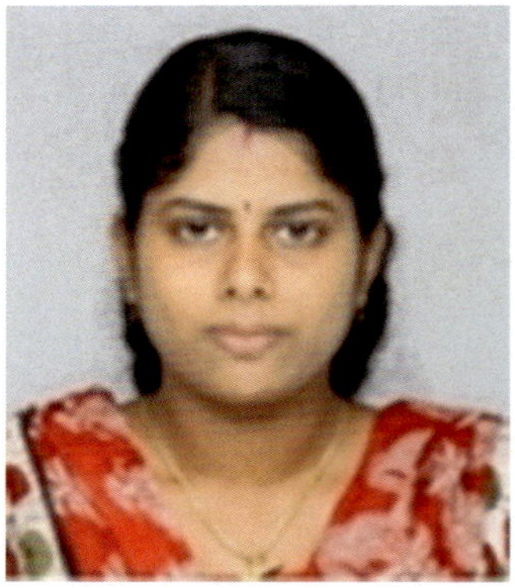

Dr. A.R. Ajitha completed her doctoral research in the area of Polymer Science and is currently working as assistant professor in Newman College, Thodupuzha, Kerala. Her topic of research is MWCNT—PP/PTT blend nanocomposites at IIUCNN, M.G. University, Kottayam with a group under Prof. Sabu Thomas. She received a patent for an invention entitled "A Poly (Trimethylene Terephthalate)/polypropylene blend nanocomposites for effective electromagnetic shielding material for electronic applications" in 2021. She has more than six years of experience in this field. Her research areas include preparation of nanocomposites and their characterization and modification of nanofillers. She has published eight research articles in international journals with a high impact factor. She edited one book related to her research area and has published ten book chapters. In addition, she has attended a good number of conferences and seminars and presented papers in international and national seminars.

Nureddin Ashammakhi

Nureddin Ashammakhi is leading translational research in biomaterials and regenerative therapeutics. He has extensive experience with biodegradable implants, drug release, and nanofiber-based scaffolds. Currently, he is leading research on 3D bioprinting and organ-on-a-chip technology for personalized medicine and regenerative therapy. He was previously a Professor of Biomaterials Technology in Tampere University of Technology, Finland, Chair of Regenerative Medicine in Keele University, UK, Adjunct Professor in Oulu University, Finland, and Visiting Scholar and Adjunct Professor at University of California, Los Angeles, before he joined Michigan State University.

Sonja Aškrabić

Sonja Aškrabić received her PhD degree from Faculty of Physics, University of Belgrade, Serbia, in 2014. Her PhD thesis was focused on phonons and defect states in rare-earth/metal oxide nanomaterials. At present she is a research assistant professor at the Institute of Physics Belgrade, Serbia, in the Nanostructured Matter Laboratory. Her research interests include optical and vibrational properties of oxide nanostructures, hydroxide nanostructures and van der Waals heterostructures of transition metal dichalcogenides; numerical modeling of vibrational spectra; SERS and spectroscopic bioimaging.

Pius Augustine

Dr. Pius Augustine got his PhD in Physics from the Indian Institute of Technology (IIT) Madras, India, and is currently serving as an Assistant Professor in the Department of Physics, Sacred Heart College, Kochi, India. He is also associating with Material Research Center, Indian Institute of Science (IISc) Bangalore, India, through the research award (TARE-Teacher Associate for Research Excellence) by SERB, DST Govt. of India. Dr. Augustine is also a research guide at Mahatma Gandhi University, Kottayam, India. His research interest are ferroelectrics, piezoelectrics, environment friendly piezo systems, 2D semiconductors (MoS_2), contact engineering, valleytronics, etc.

Ajith Behera

Dr. Ajit Behera currently working as Assistant Professor in Metallurgical and Materials Department at National Institute of Technology, Rourkela. He was born in 1987 in Odisha. He has completed his PhD from IIT-Kharagpur in 2016. He received the National "Yuva Rattan Award" in 2020 for his contribution toward society along with his academic carrier. Also received the young faculty award in 2017 and "C.V. Raman Award" in 2019. He has published more than 75 publication including books, book chapters, and journal articles. Currently he is involved with many reputed scientific organization throughout the world. More than 10 PhD students are from his institute/outside the institute, and four foreign exchange students are working on different projects with him. Simultaneously, he is doing a lot to develop the society in terms of device-based application.

Patrice Bourson

Patrice Bourson is a professor at the University of Lorraine in the LMOPS laboratory. He obtained his PhD in 1996 on the study of phase transitions in cyanide-based crystals, both through numerical simulation and temperature Raman studies. After a postdoctorate at the University of Belo Horizonte in Brazil, he specialized in the study of microstructures and defects in materials by Raman spectroscopy, in particular in polymers, by in situ measurements or by coupled experiments.

David Chapron

Dr. David Chapron received a PhD in Physics from the University of Angers (France) in 2006. After 2 years as a postdoctoral researcher at GeorgiaTech Lorraine (Metz, France), he joined as an Assistant Professor at the University of Lorraine (Metz, France)—LMOPS laboratory. His current research interests focus on in situ Raman spectroscopy, optical sensors, semicrystalline polymers, and ferroelectric crystals.

Mohammad Ali Darabi

Dr. Ali Darabi received his PhD degree in the field of Biomedical Engineering from University of Manitoba in Canada in 2019. He completed his postdoc at UCLA and Currently, he is a senior scientist at the Terasaki Institute for Biomedical Innovation and working toward spinning a biotech start up. His research focuses on the design of microcarriers, media development, and bioreactors for cultivated meat production.

Nidhin Divakaran

Dr. Nidhin Divakaran is currently working as a Research Associate at Central Institute of Petrochemicals (CIPET), Government of India, Bhubaneshwar. He pursued a PhD in Materials Physics and Chemistry from Fujian Institute of Research on Structure of Matter, Chinese Academy of Sciences in July 2020. Prior to his PhD, worked in world-class research laboratories at the Indian Space Research Organization (ISRO) and Indian Institute of Technology (IIT). He pursued master's in Nanotechnology at Vellore Institute of Technology (2014), Vellore, and a bachelor's in Instrumentation and Control from Gujarat Technological University (2012). He has several international peer-reviewed publications in the field of polymer nanocomposites and 3D printing. He is also the recipient of the prestigious CAS-TWAS President Fellowship award (2016–20) to pursue his PhD. His research interests include hybrid nanomaterials, synthesis, applications in energy sector, renewable energy, energy storage, fuel cells, batteries, photoelectrocatalysis, H_2 production, environmental, photocatalysis and adsorbents for organic and inorganic pollutant removal, CO_2 conversion, electrocatalysis, design of hybrid electrodes for various applications, nanofluids, chemical, bio, gas, and food sensors, carbon-based materials like CNTs, graphene, and CNFs, polymer nanocomposites and nanofibers for biomedical applications, quantum dots for optoelectronic applications, hydrogen storage materials and bio-nanomaterials, and the additive manufacturing of nanocomposites.

Dejan M. Djokić

Dr. Dejan M. Djokić, born on July 2, 1980 (Valjevo, Serbia), earned a PhD in 2012 at EPFL (Lausanne, Switzerland) in the field of magnetic resonance spectroscopy of novel materials with an emphasis on carbon-based nanomaterials. Previously, he received a master's degree in 2008 at the Faculty of Physics, University of Belgrade, in the subject of infrared spectroscopy in condensed matter physics through engagement within the Center for Solid State Physics and New Materials, Institute of Physics Belgrade. He pursued a postdoctoral career at the University of Geneva from 2013 to 2015. at the Department of Quantum Matter Physics, working on applied high temperature superconducting thin films. In 2016, he started a second postdoctoral engagement at the Institute of Chemical Sciences and

Engineering (EPFL, Lausanne) investigating photoluminescence of carbon nanotubes. Currently, he works at the Institute of Physics Belgrade as scientific collaborator affiliated with the Nanostructured Matter Laboratory, with the aim at studying nanostructured oxide materials and strongly correlated systems based on both Raman and infrared spectroscopy. According to Google Scholar, to date, he has published 19 articles in distinguished international peer-reviewed journals with 135 heterocitations (H-index of 6). His selected fields of expertise include probing electric transport in nanomaterials via noninvasive spectroscopic techniques and investigating phonon anharmonicity due to magnetic degrees of freedom in multiferroic nanoscaled systems.

Zorana Dohčević–Mitrović

Dr. Zorana Dohčević–Mitrović is a Research Professor and Head of the Nanostructured Matter Laboratory, Institute of Physics, Belgrade. Her research interests concern numerical simulations of vibrational and optical properties of nanosized nonoxide ceramics, semiconducting oxide, and multiferroic nanocomposites, and thin films, as well as electrical and magnetic properties of nanostructured systems. From 1991 to 1994, she received a PhD fellowship from the International Center for Theoretical Physics (ICTP, Trieste, Italy) and resided at the CISE Institute, Milano, Italy. In 2001 and 2003, she was given an award for results in physics from the Serbian Ministry of Science and Technological Development. In 2007, her article titled "Temperature-dependent Raman study of Ce0.75Nd0.25O2_d nanocrystals," published in *Applied Physics Letters* (2007), was selected from the American Institute of Physics and the American Physical Society to be published in the December 3, 2007 issue of *Virtual Journal of Nanoscale Science & Technology*, which presents a compilation of 20 articles covering a focused area of frontier research in nanoscience. She was a principal investigator and coordinator of one national project (2011–19), and coordinated several bilateral projects with Italy (2013–15), Switzerland (SCOPES 2009–12), and the Romanian Academy (2008–12). She was engaged and actively participated as WP's coordinator in two FP6 projects and one FP7 project (OPSA, CoMePhS, and NanoCharm). She published more than 200 items, among which are more than 109 papers in reputed journals from the ISI list, with citations more than 2800 times and an H-factor of 32. She is a member of the American Nano Society, American Chemical Society, and Serbian Physical Society. She is a referee for leading international scientific journals and supervises several PhD and master's theses. She is an author for *Optical Properties of Oxide Nanostructures* (Akademska Misao, Belgrade, 2011). She fluently speaks both English and Italian.

Sijo Francis

Dr. Sr. Sijo Francis, Assistant Professor and Head of Department in Chemistry, St. Joseph's College, Moolamattom, Idukki, Kerala, India. She earned her PhD from Mahatma Gandhi University, Kottayam, and finished graduation and postgraduation from Mahatma Gandhi University, Kottayam. She qualified for the CSIR-UGC examination in Chemical Sciences and had a good score in GATE-2010.

Jacek Gapiński

Jacek Gapiński, PhD, Professor at the Faculty of Physics, Adam Mickiewicz University, Poznań, Poland.

Jacek Gapiński obtained his MSc degree in Physics in 1990 on the topic of Dynamic Light Scattering application to determining the size distribution of biological macromolecules. Taking advantage of a five-month stay in the Research Center of Crete, but remaining employed at Adam Mickiewicz University, in 1994 he defended his PhD thesis on the study of supramolecular systems by means of photon correlation spectroscopy. During the following years he continued his research of soft matter by means of different optical methods. To name just a few topics: dynamics of polymers and glass forming molecular liquids (including Brillouin scattering studies), diffusion rate of nanoparticles in complex environments, structure and dynamics of biological macromolecules and micelles, structure and dynamics of strongly interacting colloidal suspensions. More recently, in cooperation with newly founded NanoBioMedical Center at Adam Mickiewicz University, he focused on microscopic techniques, including fluorescence correlation spectroscopy and fluorescence laser scanning microscopy. During his career, he spent almost three years in different research centers, including the Research Center of Crete, Max Planck Institute for Polymer Science in Mainz, Research Center in Juelich (Germany), and ESPCI in Paris. Jacek Gapiński is a coauthor of over 100 scientific articles. In 2010 he obtained habilitation and was employed on a university professor position.

Soney C. George

Soney C. George, PhD, is the Dean Research and Director, Center for Nanoscience and Technology, Amal Jyothi College of Engineering, Kanjirappally, Kerala, India. He is a Fellow of the Royal Society of Chemistry, London, and a recipient of "best researcher of the year" award in 2018 from APJ Abdul Kalam Technological University, Thiruvananthapuram, India. He has also received awards such as best faculty award from the Indian Society for Technical Education, best citation award from the International Journal of Hydrogen Energy, a fast-track award of young scientists by Department of Science and Technology, India, and an Indian Young Scientist Award instituted by the Indian Science Congress Association. He did his post-doctoral studies at the University of Blaise Pascal, France, and Inha University, South Korea. He has published and presented almost 200 publications in journals and at conferences. His major research fields are polymer nanocomposites, polymer membranes, polymer tribology, pervaporation, and supercapcitors. He has guided eight PhD scholars and 102 student projects.

Anwarul Hasan

Dr. Anwarul Hasan is an Associate Professor in the Department of Mechanical and Industrial Engineering, and Biomedical Research Center at Qatar University. Earlier he worked as an Assistant Professor in the Department of Biomedical Engineering and Mechanical Engineering at American University of Beirut, Lebanon. He was also a visiting Assistant Professor and an NSERC Postdoctoral Fellow at the Harvard-MIT Division of Health Sciences and Technology at the Harvard Medical School and Massachusetts Institute of Technology in Boston, USA, from 2014 to 2017 and 2012 to 2013, respectively. Dr. Hasan obtained his PhD from University of Alberta, Canada, in 2010 and worked in industry in Canada from 2010 to 2011. Dr. Hasan has more than 250 peer reviewed publications including over 200 journal articles, and more than 50 conference proceeding papers as well as two edited books on "Tissue Engineering for Artificial Organs." He is a winner of more than sixteen national and international awards. In the latest ranking of the world's top 2% highly cited scientists' list by

Stanford University researchers published in October 2020, Dr. Hasan has been ranked 320 out of 50,331 top biomedical researchers in the world. Dr. Hasan's current research interests involve Biomaterials, Tissue Engineering, 3D Bioprinting, and Organs on chips platforms and microneedle arrays for Diabetic wound healing, cancer biochips, COVID-19 diagnostics, and cardiovascular tissue engineering.

Garudadhwaj Hota

Dr. G. Hota obtained his BSc Chemistry from Sonepur College, Odisha. MSc and MPhil degree in Chemistry from Sambalpur University, Odisha, in the year 1997 and 1998, and then joined as a PhD research fellow in 1999 at IIT Bombay, Powai Mumbai, India. After successfully receiving his PhD degree in 2004, he joined as a Postdoctoral Research fellow at NUS Singapore from the year 2004–2007. In 2007 he joined as an Assistant Professor in the Department of Chemistry, National Institute of Technology (NIT), Rourkela, Odisha, India, and since 2012 he is working as an Associate Professor in the same department. His current research interests include design of functional, hollow, and porous polymer-Inorganic hybrid nanofibers by electrospinning method; construction of graphene based hybrid nanoarchitecture and semiconductor quantum-dot materials for environmental, anitibacterial, photovoltaic, and sensing applications. Development of advanced visible light-active photocatalyst for photocatalytic degradation of toxic pollutants.

Jayakumari Isac

Research Supervisors – Physics, CMS College, Kottayam, Kerala, India.

https://cmscollege.ac.in/research-supervisors-physics.html#

K. Jeyasubramanian

Prof. Jeyasubramanian K. Is a Senior Professor in Department of Chemistry in Mepco Schlenk Engineering College. His research area is on thin film, nanomaterials, and superhydrophobic materials.

Thomasukutty Jose

Dr. Thomasukutty Jose received his master's degree in Chemistry from St. Dominic's College, Kanjirappally, Kerala, India and Ph.D in Polymer chemistry from Bharathiar University, Coimbatore, India. He has 20 publications in peer-reviewed journals and conferences. He received best paper presentation awards in various national and international conferences. He edited one book on "Polymer nanocomposite membranes for pervaporation" which was internationally published by Elsevier. Currently, he is working as Assistant Professor of Chemistry at Sacred Heart College (Autonomous), Thevara, Kerala, India.

Neena S. John

Dr. Neena S. John is a Scientist at the Center for Nano and Soft Matter Sciences (CeNS), Bengaluru, India. Her research interests include inorganic nanomaterials and their diverse morphology with unique properties for applications in molecular sensing, hybrids with carbon nanomaterials as multifunctional materials, nonnoble metal based materials for renewable energy, particularly electrochemical energy and scanning probe microscopy-based investigations of materials. The materials are synthesized in our laboratory to suit the desired applications so as to obtain better performance.

Manoj B. Kale

Manoj B. Kale is a PhD student at the Fujian Institute of Research on the Structure of Matter, University of Chinese Academy of Sciences (P.R. China) in 2020. He has research interests in the synthesis and characterization of nanomaterials and/or polymer nanocomposites for various applications such as conductive thin films, surface coatings, energy conversion, and energy storage.

Ebey P. Koshy

Dr. Ebey P. Koshy had his PhD from Mahatma Gandhi university under the guidance of Prof. VN Rajashekaran pillai. He had a total of 28 years of teaching experience in Department of Chemistry, St. Joseph's College, Moolamattom.

Indradeep Kumar

Indradeep Kumar will receive his Doctorate Degree in Mechanical Engineering (December 2021). He is an Aeronautical Engineer, and to date has published 15-plus research papers in international journals related to nanotechnology. He also applied for a patent, "Water-operated Generator." He is keen about nanotechnology and to date has written three chapters for two different books whose publishers are international (CRC and IGI) and three that are under review by Elsevier.

He is a member of the Royal Society of Chemistry, London (AMRSC), SMIAAEP, FSIESRP (Malaysia), SMISME, Member of Scientific Council (IAESTD), and many others.

He was honored with a doctorate by the World Human Rights Protection Commission. He received the Most Inspiring Youth Icon in Education award. He has also received the Young Scientist, Young Researcher, Research Excellence, and Dedicated Innovative Technologist (Aeronautical Engineering) awards. He was also awarded in International Conference by the Asian Council of Science and Management, Singapore, and got a chance to represent India in Singapore in 2010. In August 2011, he won first prize at a college-level National Quiz Competition.

K. Priya Madhuri

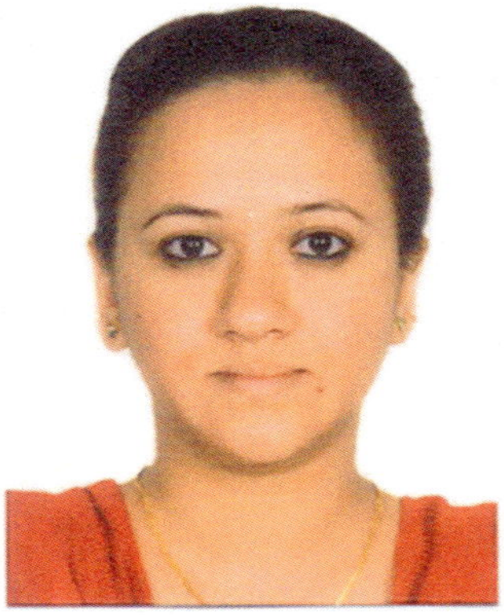

Dr. Priya Madhuri K obtained her MSc in Organic Chemistry from N.M.K.R.V. College for Women, Bengaluru. She obtained her PhD in Chemistry from Centre for Nano and Soft Matter Sciences, Bengaluru, in 2019 on the preparation and properties of thin films of electroactive metal phthalocyanines. She has published her work in various renowned national and international journals. She is presently a National Postdoctoral Fellow at the Department of Inorganic and Physical Chemistry, Indian Institute of Science, Bengaluru under the mentorship of Prof. S. Sampath. Her research interests include probing molecular organization of metal phthalocyanine and related compounds and their nanotechnological applications in sensors and energy.

Beena Mathew

Dr. Beena Mathew received her PhD in Chemistry from the Mahatma Gandhi University, Kerala, India, and the JSPS Postdoctoral fellowship at the Kyushu University, Japan. Presently, she is the Professor of Physical chemistry at the School of Chemical Sciences, Mahatma Gandhi University. She is the former Director of the School of Chemical Sciences. Research areas of interest include nanomaterials, catalysts, sensors, and self-assembled systems.

Monalisa Mishra

Dr. M. Mishra is a faculty of Life Science at NIT Rourkela (a premier academic and research institute in India) since 2014. Before joining at NIT Rourkela, Dr. Mishra had the postdoctoral experiences at multiple universities (University of California, San Diego, USA, Indiana University, Bloomington, USA, and MPI-CBG Dresden, Germany). Dr. Mishra finished her PhD in 2007 at Jacobs University Bremen, Germany, in the group of Prof. VB Mayer-Rochow in the field of Biology. Thereafter, Dr. Mishra experienced in the field of Developmental Biology using advanced microscopic techniques. Currently, her lab is working on developmental biology, nanotoxicology, neurodegenerative and metabolic disorder using Drosophila as a model organism.

Padinharu Madathil Gopalakrishnan Nambissan

Padinharu Madathil Gopalakrishnan Nambissan (P.M.G. Nambissan) is a Senior Professor at Saha Institute of Nuclear Physics, Kolkata, India. He has carried out research using positron annihilation spectroscopy in metals and alloys, quasicrystals, superconductors, polymers, and nanomaterials. Currently his most important areas of interest include semiconductor nanomaterials, nanospinels, and perovskites. He has published 142 research papers in peer-reviewed international journals and presented almost an equal number in national and international conferences. He is a reviewer for several research journals and has contributed chapters in edited volumes on nanoscience and nanotechnology.

Fulufhelo Nemavhola

Fulufhelo Nemavhola is currently the School Director in the College of Science Engineering and Technology. Simultaneously Fulufhelo is an Associate Professor in the Department of Mechanical Engineering at the University of South Africa. In addition to being a registered professional engineer (PrEng) with the Engineering Council of South Africa, Fulufhelo is also a Chartered Engineer (CEng) registered with the Engineering Council of the United Kingdom (ECUK). Fulufhelo graduated with Bachelor of Science (Mechanical Engineering) and Master of Science (Mechanical Engineering) degrees at the University of the Witwatersrand, Johannesburg, South Africa. He held various positions, including those of senior engineer and design engineer in private and parastatal companies before joining the academia. Fulufhelo's interest in academic research then steered him to pursue a doctorate from the Department of Cardiothoracic Surgery of the University of Cape Town in December 2010. The PhD degree was studied under the supervision of Prof Thomas Franz, Prof Neil Davies and Dr Laura Dubuis. His research looked into the remodeling of heart postmyocardial infarctions. His current research focuses in the areas of mechanobiology, soft tissue mechanics, computational biomechanics, and engineering education in open distance e-learning (ODeL). Fulufhelo is involved extensively in research, having authored, coauthored, and presented a number of papers in both national and international forums. Fulufhelo is supervising number of students at doctoral, master's, and honors levels in the areas of areas of experimental mechanics, mechanobiology, soft tissue mechanics, polymer mechanics, and computational biomechanics. Fulufhelo has his heart set on the development of rural villages using technology. As such, Fulufhelo is currently involved in various programs with the aim of developing poor or underprivileged communities.

Amirsadegh Rezazadeh Nochehdehi

Amirsadegh Rezazadeh Nochehdehi is currently part of the academic staff at the University of South Africa (UNISA). He is also a PhD fellow at the Biomechanics Research Group, Department of Mechanical and Industrial Engineering (DMIE), University of South Africa (UNISA), Johannesburg, South Africa. Furthermore, he graduated from Materials and Biomaterials Research Center, Iran (MSc) with a degree in Biomedical Engineering, Division of Biomaterials in 2017. He also graduated from Karaj Branch of Islamic Azad University, Iran (BSc) with a degree in Materials and Metallurgy Engineering, Division of Industrial Metallurgy in 2012. As a research scholar, he has worked in polymer nanocomposites for tissue regeneration applications at International and Interuniversity Center for Nanoscience and Nanotechnology (IIUCNN) in Mahatma Gandhi University (MGU), Kerala, India, in 2018. He also worked with magneto-metallic alloy nanoparticles at Nanotechnology Research Center at University of Zululand, South Africa as a visiting researcher in 2017. In addition, he was quality and safety engineer inspector while working at Tehran Urban and Suburban Railway Operation Company (TUSROC) for a period of five years. He is scientific researcher and also research and development specialist in the field of metallurgy and materials design, advanced materials, hydrogen storage materials, nanoscience and nanotechnology, nanomedicine, nanomaterials, nanocomposites, magnetic nanoparticles, magnetic nanoalloys, magnetic hyperthermia, biomedical science and engineering, biomaterials, biomechanics, mechanics of tissue, and regenerative medicine.

Sandip Padhiari

Mr. Sandip Padhiari received his bachelor's degree from Stewart Science College, Cuttack (Utkal University, Odisha) in 2013, and M.Sc. Chemistry from Ravenshaw University, Cuttack in the year 2015. Then, he worked as a faculty at S.B. Womens' College, Cuttack. Currently he is working as a PhD scholar under the supervision of Prof. Garudadhwaj Hota at Department of Chemistry, National Institute of Technology Rourkela, Odisha. His research area focuses on the development of graphitic carbon nitride based hybrid nanostructured and their photocatalytic and adsorptive application toward pollutant abatements, water splitting, N_2 fixation, and CO_2 reduction. To date he has published five research articles in various international journals.

Shabna Patel

Dr. S. Patel obtained his BSc Chemistry from G. M. Autonomous College, Odisha and MSc. in Chemistry from NIT, Rourkela, Odisha, in 2012. She joined as a PhD research fellow in department of Chemistry NIT Rourkela in 2012 and received her PhD degree in 2018 under the guidance of Dr. G. Hota. Her doctoral works deals with "Functionalized electrospun polyacrylonitrile nanofibers: Synthesis, Characterization and Environmental Applications." She is now working as a lecturer in Chemistry, in U.G.I.E., Rourkela, Odisha. Her research expertise includes electrospun nanofibers, surface functionalization, nanocomposites for antibacterial, energy, and environmental applications.

J. Philip

J. Philip is Dean (Academic), Amal Jyothi College of Engineering, Kanjirappally, Kottayam 686518, Kerala, India; Former Professor, Department of Instrumentation, Cochin University of Science and Technology, Cochin 682022, India. Jacob Philip, PhD, received MSc and PhD degrees in Physics from University of Kerala in 1975 and 1979, respectively, and had postdoctoral training at University of Tennessee, Knoxville, USA. His research has been on diverse areas of condensed matter physics, focused to elastic, thermal, piezoelectric, and optical properties of multifunctional materials. His contributions include techniques to study thermal and elastic properties of solids, such as photoacoustic and photothermal techniques, physical ultrasonics, laser-induced surface acoustic waves, sensors and instrumentation, and so forth. He was a fellow of the Alexander von Humboldt Foundation at University of Heidelberg and a visiting scientist at several institutions in India and abroad. He has authored over 200 journal publications, two books, one edited volume and more than 100 papers in conference proceedings. He has delivered invited lectures and presented papers at several International and National conferences. He has successfully guided 27 researchers to secure a doctoral degree. He is a fellow/member of a good number of science societies, including the Institute of Physics (UK).

Mikolaj Pochylski

Mikolaj Pochylski is an Associate Professor in the Faculty of Physics at Adam Mickiewicz University in Poznan in Poland where he has been since 2007 after receiving his PhD. Until 2014 he worked in the Division of Optics. Since then, he has been associated with Molecular Biophysics Lab.

His research interests revolve around the structure and dynamics of complex systems, in particular, in glass-forming liquids, polymer, and self-organizing amphiphilic liquid mixtures. Recently he started working on high-frequency vibrations in nanostructured materials (colloidal crystals and suspended ultra-thin membranes). Apart from the problems of condensed matter physics, he is interested in developing image analysis methods for objective parametrization and classification of biomedical images.

As main research tools, he uses light scattering spectroscopies (Brillouin, Raman, and photon correlation), polarimetry, and confocal microscopy techniques. He enjoys programming and uses it to automate experimental protocols. He also develops instrumentation for various optical characterizations.

Marc Poncot

Dr. Marc Poncot obtained his PhD in 2009 at the Institut National Polytechnique de Lorraine as part of a CIFRE grant with AcelorMital on the development and characterization of metal/polymer composites for the automotive sector. In 2011, he was appointed Associate Professor at the Ecole Nationale Supérieure des Mines de Nancy, University of Lorraine. He carries out his research at the Institut Jean Lamour within the Physics, Mechanics, and Plasticity team in the field of polymers and composites, elaboration-microstructure-property relations. Since 2019, he has received his Habilitation à Diriger les Recherches (HDR).

Ippokratis Pountos

Mr. Ippokratis Pountos is an orthopedic surgeon who is fellowship-trained in foot and ankle surgery. Following the completion of his orthopedic certification, Mr. Pountos was further fellowship-trained at Leeds University Teaching Hospitals (UK) and in the Atos Clinic in Heidelberg (Germany) on the reconstruction of complex foot and ankle conditions. Mr. Pountos has successfully completed three postgraduate degrees in the United Kingdom. His research interest lies with the biology of bone healing and ways to regenerate bone and cartilage. He holds a senior lecturer position with the University of Leeds (UK). Over the years, Mr. Pountos has published over 55 articles and presented in numerous International meetings. For his research work, Mr. Pountos has competed and won five prestigious prizes including the Best of the Best Paper from the Orthopedic Trauma Association in 2015 and the twice the Best Paper from the British Trauma Society.

Dipen Kumar Rajak

Dr. Dipen Kumar Rajak is currently working as Associate Professor of Department of Mechanical Engineering at G. H. Raisoni Institute of Business Management (GHRIBM), Jalgaon. He has completed his research in the Deformation Behavior of Aluminum Alloy Foam at Indian Institute of Technology (Indian School of Mines) Dhanbad and CSIR-Advanced Materials and Processes Research Institute (AMPRI), Bhopal. He received the Doctor of Philosophy from the Department of Mining Machinery Engineering at Indian Institute of Technology (Indian School of Mines) Dhanbad. He has authored several scientific articles in high impact journals and also written a book chapter called "Aluminium Alloy Foam & An Insight Into Metal Based Foams." His research contribute includes articles in design analysis, numerical validation, artificial neural network, and optimization techniques.

Sreelakshmi Rajeevan

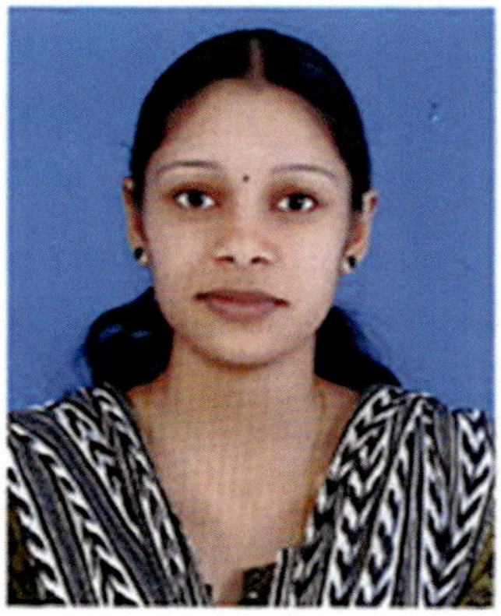

Sreelakshmi Rajeevan, is a research scholar, APJ Abdul Kalam Technological University, Kerala, India, and Center for Nanoscience and Technology, Amal Jyothi College of Engineering, Kanjirappally, Kerala, India. She holds a Master of Technology in Polymer Science and Technology and a Master of Science in Chemistry, both from Mahatma Gandhi University, Kerala. She has two publications in international journals. Her major research fields are polymer nanocomposites, electrochemistry, supercapacitor, energy storage devices.

R. Rakhikrishna

R. Rakhikrishna received MSc degree in Physics from University of Calicut in 2012 and qualified NET. R. Rakhikrishna joined Cochin University of Science and Technology as a research scholar to pursue a PhD in the area of condensed matter physics. She submitted her doctoral thesis in 2020 and has been working on magnetoelectric nanocomposites for her PhD thesis. She has published four journal papers in this area and presented her work at various conferences.

Arunima Reghunadhan

Dr. Arunima Reghunadhan is a researcher in the field of Material Science with a completed master's degree in Analytical Chemistry from University of Kerala with fifth rank and qualified as a UGC-Junior research fellowship in 2011. She started her research career in 2012 at Mahatma Gandhi University (MGU), Kottayam. She is currently working as a guest faculty in Milad-E-Sherif Memorial College, Kayamkulam, Kerala. She completed her doctoral research in 2018 and soon joined Bishop Moore College, Mavelikara, as an FDP substitute teacher. She is interested in nanomaterial synthesis, green synthesis of nanomaterials, characterization, polymer blends, and composites. She has many publications in very reputed international journals, has edited four books, and wrote many book chapters. She was a visiting researcher in the famous Leibniz institute fur Polymerforshung, Dresden eV, Hohe Strasse-6, Germany in the year 2015.

Neerish Revaprasadu

Prof. Neerish Revaprasadu is a Professor of Inorganic Chemistry and South African Research Chair Initiative (SARChI) in Nanotechnology at the University of Zululand. He obtained his BSc (Hons) from University of Natal, South Africa in 1994 and PhD from Imperial College, London, UK, in 2000. He has published 240 research articles and 28 book chapters. He has given more than 80 lectures at national and international conferences. The research focus of the chair is on developing routes for the delivery of high-quality materials. The initial work was on II—VI semi-conductor materials such CdS, CdS, CdTe, PbS, PbTe, and PbSe. The current focus is on earth abundant materials using metal complexes as single source precursors. The materials of interest are Bi, Fe, Ni, Co, and Sb chalcogenides. AACVD and hot injection routes are used to prepare the binary and ternary materials. Work on metal chalcogenide binary, ternary, 2D, and carbon nitride-metal chalcogenide (CN-MC) composite systems for application in energy generation, and storage is also underway. In particular, the materials would be studied for their efficiency in the hydrogen evolution reaction (HER), oxygen evolution reaction (OER), and supercapacitance

Didier Rouxel

Didier Rouxel is a full Professor at the Institut Jean Lamour, Université de Lorraine, France, where he was the former head of the Micro and Nanosystems team and currently leads the piezoelectric polymer group. Prof. Rouxel is graduated from the Ecole Supérieure des Sciences et Techniques de l'Ingénieur de Nancy, Vandoeuvre, France and he completed his PhD in Material Sciences and Engineering from the University of Nancy I, France in 1993.

His major areas of interest include piezoelectric polymers, elastic properties of polymeric materials studied by Brillouin spectroscopy, development of polymer nanocomposite materials and microdevices based on electroactive polymers, piezoelectric nanocrystals, microsensor development for surgery, shape memory alloy-piezoelectric device for energy harvesting, etc.

Prof. Rouxel was expert for the French Agency for Food, Environmental, and Occupational Health and Safety (ANSES) on "Nanomaterials and Health" and "Member of the Year" in 2014 of the French Society of Nanomedicine.

Isabelle Royaud

Professor Isabelle Royaud who is a University lecturer in Polymer physics and Polymer science since 1990 joined the University of Lorraine in 2011 after beginning her career in the University of Lyon. She has obtained her PhD in Polymer Properties (University of Southampton – UK) on "Conformation and crystallinity in Polymers: A study using novel Raman techniques" in 1989 then she obtained her HDR on "Study of the influence of ageing on molecular mobility of polymers" (University of Lyon 1/IMP-Engineering of Polymer Material) in 2000. Since 2011 she has been a Professor at University of Lorraine (EEIGM: European School of Material Science and Engineering, Nancy, France) and a Researcher at Institut Jean Lamour UMR CNRS 7198 (Nancy, France). She is responsible for a research team entitled Physics Mechanics and Plasticity regrouping 11 permanent staff, 8 PhD, postgraduates and masters students. Between 1990 and 2011, she was a University Lecturer in University of Lyon 1 in a Polymer Materials Engineering laboratory (IMP@Lyon1). She is a specialist of relationship architecture/morphology/dielectric, viscoelastic and vibrational properties of polymers. Her scientific production is composed of 76 publications, 2 book chapters, 1 patent, 13 invited lectures in International Conferences, 180 Communications, 89 oral (International Conferences: 50), 91 posters (International Conferences: 62), and 18 invited seminars.

Her h-index (2021) is 20; average citation per item 17.7; sum of time cited:1292 (without self-citation 1226). She has supervised 20 PhD students and participated to 75 PhD defenses. Her main teaching activities (EEIGM) are on polymer physics, processing of composites, functional properties of polymers, vibrational spectroscopy (IR and Raman) of polymers, and the physical aging of polymers. Her research activities deal with the following subjects:

- Polymer physic properties for synthetic polymers and biopolymers.
- Relations between architecture/morphology and mechanical, electrical, and vibrational properties of polymer and multimaterials.
- Influence of aging and morphology on molecular mobility and relaxational phenomena in polymer materials.

- Molecular mobility at the interfaces of nanostructured systems
- in situ monitoring of plastic deformation in polymers at different scales using vibrational techniques, videometric tensile test, synchrotron radiation, and broadband dielectric analysis.

Harekrushna Sahoo

Dr. Harekrushna Sahoo is a faculty of Chemistry at NIT Rourkela (a premier academic institute in India) since 2012. Before joining at NIT Rourkela, Dr. Sahoo had the postdoctoral experiences at multiple universities (UMASS-Amherst [USA], TUD Dresden [Germany], and MBC Dresden [Germany]). Dr. Sahoo finished his PhD in 2007 at Jacobs University Bremen (Germany) in a group advised by Prof. Werner M. Nau in the field of peptide chemistry. Thereafter, Dr. Sahoo experienced in the field of protein and membrane fields using advanced optical spectroscopic techniques. Currently, Dr. Sahoo is working in different research fields such as protein dynamics, extracellular matrix, nanocomposites, and environmental chemistry. To make his research work wider and more visible, he is collaborating with national and international faculties.

Jitendra Kumar Sahoo

Dr. Jitendra Kumar Sahoo completed his PhD from NIT Rourkela in the field of Nano and Environmental Chemistry. Currently he is working as an Assistant Professor in Chemistry at GIET University, Gunupur, Odisha, India. He has established a Nano and Environmental Laboratory at GIET University. He has expertise in the field of wastewater treatment, graphene-based materials, photocatalysis, adsorption of inorganic and organic contaminates, magnetic nanocomposites, and antibacterial activity. He has collaborated with eminent professors from IIT, NIT and CSIR Laboratory of India and foreign professors as well. He has taught various courses to undergraduate and postgraduate students.

Shraban Kumar Sahoo

Mr. Shraban Kumar Sahoo received his MSc degree in Chemistry from National Institute of Technology Rourkela, Odisha, Rourkela in the year 2014. Then, got his PhD degree under the supervision of Prof. Garudadhwaj Hota at Department of Chemistry, National Institute of Technology Rourkela, Odisha. His research area focuses on the development of graphene based hybrid nanostructured and their adsorptive application towards pollutant abatements. He has published 10 research articles in various international repute journals till now.

Bojan Stojadinović

Bojan Stojadinovic was born on May 23, 1988 in Pozarevac, Serbia. He received a BE degree in Computer and Applied Physics Department, Faculty of Physics, Belgrade University in 2011, MSc and PhD degrees in Theoretical and Experimental Physics, Physics of Condensed Matter and Statistical Physics on Faculty of Physics, Belgrade University, in 2012 and 2018, respectively. In 2013, he joined the Physics of nanostructured oxide materials and strongly correlated systems, Institute of Physics Belgrade as research assistant. Since June 2019, he has been at the Laboratory of Physics of nanostructured oxide materials and strongly correlated systems, where he was an assistant research professor at the Institute of Physics Belgrade. Dr. Stojadinovic was a participant in bilateral and COST projects between the Institute of Physics Belgrade and Goethe University Frankfurt, Germany, and TU Vienna, Vienna, Austria. Dr. Stojadinovic is a student and leader of the short-term scientific missions project, entitled Ferroelectric properties of BiFeO3 thin films, as monitored by nanogranular sensor structures prepared by focused electron beam-induced deposition, Goethe University Frankfurt, Germany. His current research interests include multiferroic materials, Raman spectroscopy, physicochemical synthesis of nanomaterials, with 17 publications in that field. Dr. Stojadinovic is the winner of the award for the best doctoral thesis at the University of Belgrade in 2018 awarded by the Chambers of Commerce of Serbia.

Nazzar Tellisi

Nazzar Tellisi MSc, Mmed Sci (Trauma), FRCS, FRCS (Tr and Ortho) is a Consultant Orthopedics and Trauma surgeon with interest in Foot and Ankle at Leeds teaching hospitals NHS Trust. He holds honorary senior clinical lecturer at the University of Leeds and visiting assistant professor in Orthopedics at university of Sharjah. Fellowship trained at the Hospital for Special Surgery, New York, USA, as a clinical fellow with eminent specialists in the field of reconstructive foot and ankle surgery and limb lengthening and reconstruction service with experience in complex foot and ankle reconstruction using Ilizarov techniques.

T. Theivasanthi

Dr. (Ms) T. Theivasanthi PhD works at Kalasalingam University (India) doing research in nanomaterials/nanotechnology and has 18 years of teaching experience and has published many research articles/books and an h-index of 15; delivered lectures (more than 205) in scientific conferences, foreign countries, and renowned institutes of all over India; winner of Publons Global Top Peer Reviewer Award-2018; reviewer in SERB (Dept. of Science & Tech, Govt. of India); life member of scientific bodies; serving as Editorial Board Member/ reviewer for some scientific journals; achieved many awards/honors/recognitions; achieved World Top Reviewer-2018, World Records for nanotechnology invention; other innovative products include low-cost graphene, agricultural nonfertilizer, vegetable powder for diabetes, plants materials for psoriasis, and nanoparticles for treatment of virus infection. ORCID: https://orcid.org/0000-0002-2280-9316 Email: ttheivasanthi@gmail.com

M. Tripathy

Mr. Manamohan Tripathy received his bachelor's degree and M.Sc. in the year 2015 and 2017 from Rajendra Autonomous College, Balanger, Odisha. Then, he received his MPhil degree from Veer Surendra Sai University of Technology, Burla, Odisha. Currently he is working as a Ph.D. Scholar under the supervision of Prof. Garudadhwaj Hota at Department of Chemistry, National Institute of Technology Rourkela, Odisha. His research area focuses on the development of low cost hybrid nanomaterials for adsorptive and photocatalytic removal of contaminants from aqueous medium. He has published five research articles in various international repute journals till now.

Rumeysa Tutar

Dr. Rumeysa Tutar studied chemistry at the Izmir Institute of Technology (IYTE) in Izmir, Turkey. Awarded with The Scientific and Technological Research Council of Turkey (TUBITAK) in 2018, she joined University of California, Los Angeles (UCLA) as a visiting graduate researcher in before getting her PhD Dr. Tutar received her PhD degree in Chemistry from Istanbul University, Turkey in 2020. Currently she is a Research Assistant at Istanbul University-Cerrahpaşa, Turkey. Her research interests are the development of biomaterials for biomedical applications, tissue engineering, and 3D bioprinting.

Jeena Varghese

Ms. Jeena Varghese obtained her MSc degree in Physics from University of Kerala, India in 2016. For her MSc. thesis, she worked on "The resistive switching behavior of PMMA thin films" at the Indian Institute of Space Science and Technology, Trivandrum, Kerala. She was the recipient of INSPIRE scholarship organized by Department of Science and Technology (DST) India, during her BSc and MSc studies. She also obtained an MPhil degree in Physics from Mahatma Gandhi University, India in 2018, and is currently pursuing PhD in Physics under the supervision of Dr. Bartlomiej Graczykowski in Adam Mickiewicz University, Poznan, Poland. She is also working on an FNP funded research project entitled "Photomechanical hybrid membranes nanomechanics (PHNOM)." Her research work focuses on the Brillouin light scattering studies of polymer colloidal crystals. She has worked in two scientific publications, one perspective review article, two book chapters, and one conference publication.

ORCID ID: https://orcid.org/0000-0001-5728-5682.

V.S. Vinila

Dr. V.S. Vinila got her PhD from Mahatma Gandhi University, Kottayam, India. She is the second-rank holder in MSc. Applied Physics of the year 2011, in Mahatma University, Kottayam, India. She authored thirteen papers and coauthored eight papers in international journals. She has previously worked as research associate in a project conducted by Kerala State Council for Science, Technology, and Environment (KSCSTE). Her research areas are nanoceramics, ceramic polymer nanocomposites, fiber-filled polymer composites, superconducting ceramics, and ferroelectric ceramics.

Runcy Wilson

Dr. Runcy Wilson is working as an assistant professor in the Department of Chemistry, St. Cyril's College, Kerala, India. He obtained his PhD in Chemistry from Mahatma Gandhi University, Kottayam, India in 2015. He has written several publications in international journals and book chapters. He has also coedited two books one titled *Transport Properties of Polymeric Membranes* and the other titled *Materials for Potential EMI Shielding Applications: Processing, Properties, and Current Trends* published by Elsevier. Dr. Runcy has almost two years of industrial experience as a junior scientist at the Corporate R&D Center, HLL Lifecare Limited, a Government of India Enterprise, in the area of synthesis of green polymers. His current research interests include polymer nanocomposites for membrane applications, synthesis of biodegradable polymers for medical applications, and development of high-quality EMI shielding material.

Lixin Wu

Dr. Lixin Wu is currently working as a Professor at Haixi Institutes, Fujian Institute of Research on Structure of Matter, Chinese Academy of Sciences. He completed his PhD from Beijing University of Chemical Technology, Beijing, in the year 1988. He pursued his post doc from Pennsylvania State University, USA (1999–2000). He later did post doc at Clemson University, USA, in the year 2000. He later worked in University of California, USA (2000–2002). He worked at Concordia University, Canada, as Associate Researcher from 2002–2007. He also worked as a Materials Scientist at a reputed research company, USA. Dr. Wu has expertise in the field of additive manufacturing, structure and performance of polymer composite materials, polymer nanocomposite materials, preparation and processing of polymer materials, fiber reinforced composite materials, and so forth. He has authored and coauthored several research papers, book chapters, and patents in the field of 3D printing, polymer nanocomposites, materials physics, and chemistry with around 2000 citations. He has supervised and mentored several international PhD and Master students at Chinese Academy of Sciences.

Foreword

It gives me great pleasure and satisfaction to introduce to you the beautiful compilation entitled *Design, Fabrication, and Characterization of Multifunctional Nanomaterials,* edited by Prof. Thomas, Prof. Kalarikkal, and Dr. Abraham. My 10-year-long association with this team of nanomaterial scientists at Mahatma Gandhi University has always fascinated me. I am very much enthralled to find the vast knowledge, palpable love for the subject, and enthusiasm of the editors reflected in many pages of this book.

Nanomaterials represent an important part of the rapidly growing field of nanoscience and nanotechnology. The new generation of nanomaterials features a network of applications in all fields of society and will soon have a tremendous impact on diverse industries, such as medicine, automotive and aeronautics, electronics, chemistry and materials, biotechnology, cosmetics, and environment and ecology. Reducing the size of materials down to the nanoscale can indeed lead to new, sometimes astonishing, properties that hold great promise for a multitude of applications. These inherent properties of nanomaterials make them exhibit one or several functional capabilities, such as magnetism, piezoelectricity, and ferroelectricity, and their combination in multiferroic nanoparticles.

But obtaining these nanostructured objects has required the development of often specific and dedicated elaboration techniques, and also the adaptation of material characterization techniques to this size, if not the development of new analysis techniques. This book proposes a very broad review of these different techniques and approaches for the elaboration and characterization of nanomaterials.

When I searched diligently for a book on fabrication of nanomaterials that meets current needs to recommend to my friends and collaborators worldwide as a solution to address the dilemmas faced in research, I realized that such a book was not yet available, which persuaded me to ask Prof. Thomas and his team to compile one as a pioneering effort in the field of nanomaterials. The book, authored by renowned professors, scientists, and nanomaterial researchers, remarkable as it sounds, covers all aspects of fabrication of nanomaterials, with particular attention to the practical aspects of nanomaterial synthesis technology.

Inspired by the vista of a new world opening up, I am sure this book will be one in which you will learn more about the wonders of nanomaterials and amazing avenues of its synthesis methods, development, and characterization techniques. I hope that this book, which is the output of serious and sustained dedication by Prof. Thomas and his team, will turn out to be an encyclopedia for educators, scientists, and professors, helping students across the world to learn, teach, and practice the art of nanomaterial technology.

Reading about an experience will never replace the experience, but it can give you a taste for it and motivate you to go further. I wish all readers to read and enjoy nanomaterial research!

I can say without a shadow of a doubt that writing this book was a labor of love for the editors and the readers will be able to perceive it too.

Bonne lecture!

Prof. Didier Rouxel
Institut Jean Lamour,
Université de Lorraine,
Nancy Cedex, France

PART I

Characterization techniques of nanomaterials

CHAPTER 1

State-of-the-art technologies for the development of nanoscale materials

Ann Rose Abraham[1], Nandakumar Kalarikkal[2,3,4], Sabu Thomas[3,5]
[1]Department of Physics, Sacred Heart College (Autonomous), Kochi, Kerala, India; [2]School of Pure and Applied Physics, Mahatma Gandhi University, Kottayam, Kerala, India; [3]International and Inter University Centre for Nanoscience and Nanotechnology, Mahatma Gandhi University, Kottayam, Kerala, India; [4]School of Nanoscience and Nanotechnology, Mahatma Gandhi University, Kottayam, Kerala, India; [5]School of Energy Materials, Mahatma Gandhi University, Kottayam, Kerala, India

Chapter Outline

1. Introduction

The production and development of complex functional nanomaterials with controlled size and morphology is possible by the systematic and precise control of experimental conditions. The ongoing research and development in fabrication of nanomaterials have opened novel prospects in production of materials with tailored functionalities, characteristics and properties. Design and fabrication of nanomaterials and exploration of their properties, features, and functionalities is the fundamental step toward research in nanomaterials. Investigation of the structure of nanoparticles and elemental analysis, analysis of morphology, and imperfections or defects studies are the key factors to understand the behavior of nanomaterials and to learn how the various properties are related to their compositions and structure.

The several synthesis methods of functional nanomaterials with desired properties are divided into two categories: the "bottom-up" and the "top-down" approaches. The top-down approach is based on the organized removal or cutting of bulk material to obtain the nanoscale material of required size, geometry, and dimensions. Mechanical methods like grinding, cutting, ball milling, and lithographic techniques like electron beam lithography,

Design, Fabrication, and Characterization of Multifunctional Nanomaterials. https://doi.org/10.1016/B978-0-12-820558-7.00023-6

and photolithography are a few examples. The bottom-up approach involves building-up of materials from atoms or ions to larger entities, starting from gaseous or liquid state material. Physical vapor deposition (PVD), sputtering, laser ablation, chemical techniques such as chemical vapor deposition (CVD), wet-chemistry synthesis like sol-gel method, self-assembly, and so forth are examples. In this chapter we give an overview of various synthesis methods of nanomaterials.

The development of various powerful characterization techniques has offered exceptional prospects for the investigation of nanomaterials on various aspects and has helped to attain a deeper understanding of the properties of such nanomaterials. The X-ray diffraction technique is the most fundamental technique for the analysis of nanomaterials' structure and crystallinity. The crystallite size of nanomaterials can be analyzed using the Debye-Scherrer equation and strain induced in the samples calculated using the Williamson-Hall plot. The Fourier Transform Infrared (FTIR) spectroscopy analysis helps to investigate the functional groups present in the samples. The morphology of the nanomaterials can be investigated effectively using electron microscopy techniques such as scanning electron microscopy (SEM) and transmission electron microscopy (TEM). The high-resolution transmission electron microscopy (HRTEM) and the data of the selected area diffraction (SAED) patterns can be effectively utilized to understand crystalline nature of nanomaterials. Energy dispersive X-ray spectroscopy is widely being used for elemental analysis. Scanning probe microscopy techniques like atomic force microscopy may be employed to investigate the topography of the nanomaterial samples, which can be drop casted into thin-film form. Magnetic force microscopy (MFM) images help to identify the magnetic domain arrangements within the samples. Piezoelectric force microscopy (PFM) is helpful to visualize the ferroelectric or piezoelectric domain arrangements within the samples. The principal techniques used for nanomaterial characterization are summarized in Table 1.1.

The book chapters highlighting the syntheses and characterization techniques of nanomaterials can be introduced briefly as follows. Part I of the book, dealing with Chapters 2–7, give a description of various characterization techniques used for the analyses of nanomaterials. Part II of the book, dealing with Chapters 8–22, mainly focus on various syntheses methods of magnetic, perovskite, carbon nanomaterials, polymer nanocomposites, and other nanosystems. A brief schematic of the techniques described in the chapter is illustrated in Fig. 1.1.

Chapter 2 summarizes the beauty and the power of conventional and micro-Raman spectroscopy, which are the most important nondestructive methods that provide information on the vibrational and electronic properties in confined systems. This chapter presents a brief review on how temperature-dependent Raman spectroscopy comes to be of significant importance for investigating various phenomena in nanostructured materials. Raman spectra of diverse nanomaterials in the form of metal-oxide nanopowders, quantum dots, nanowires, nanoribbons, carbon-based, and multiferroic nanostructures are analyzed

Table 1.1: Principal techniques for nanomaterial characterization.

Characteristics to be analyzed	Characterization techniques
Structural analysis	X-ray diffraction
Three-dimensional geometry and chemical state analysis	X-ray absorption spectroscopy (XAS) [EXAFS, XANES]
Size distribution and shape analysis	Small angle X-ray scattering (SAXS)
Surface area analysis	BET analysis
Structural and functional group analysis	Infrared spectroscopy [1]
Absorbance spectra and band gap analysis	UV-visible absorption spectroscopy
Spectral analysis	Diffuse reflectance spectroscopy (DRS) [2]
Morphological analysis	Scanning electron microscopy (SEM) (resolution of 10 nm or 100 Å) [3]
Morphology and size distribution	Transmission electron microscopy (TEM) [HRTEM, Cryo-TEM, liquid TEM] (resolution of 0.1 nm or 1.0 Å)
Elemental analysis	Energy dispersive X-ray spectroscopy (EDAX) (spatial resolution of 10 nm)
Elemental and composition analysis at surface	X-ray photoelectron spectroscopy (XPS) (resolution of 10 nm) [4]
Topographical analysis	Atomic force microscopy (AFM) (resolution of 0.1 nm)
Topographic and magnetic properties imaging	Magnetic force microscopy (MFM) (resolution of 20 nm)
Topographical and electromechanical analysis	Piezoelectric force microscopy (PFM) (resolution ranges from 10 to 30 nm)
Size, shape, structure and dispersion analysis	Scanning tunneling microscopy (STM) (resolution of 0.1 nm) [5]
Chemical analysis	Raman Spectroscopy [1]
Band structure and light emitting analysis	Photoluminescence spectroscopy (PL) [1,6]
Size and shape analysis	Near field scanning optical microscopy (NSOM)
Particle size distribution	Photon correlation spectroscopy (PCS) or dynamic light scattering (DLS)
Surface charge and stability analysis	Zeta potential [7]
Mass to charge ratio analysis	Mass spectrometry [8]
Defects analysis	Positron annihilation spectroscopy (PAS) [9]
Size, composition, purity and structural analysis	Nuclear magnetic resonance (NMR)
Mass analysis	Thermo-gravimetric analysis (TGA)
Size distribution and dispersion coefficient analysis	Nanoparticle tracking analysis (NTA) [10]
Magnetic hysteresis loop analysis	Vibration sample magnetometer (VSM)
Magnetic properties, oxidation state	Mössbauer spectroscopy
Magnetization saturation, remanence, superparamagnetic blocking temperature	Superconducting quantum interference device (SQUID)
Nanoparticle size, distribution	Ferromagnetic resonance (FMR)
3D atomic structure visualization and analysis	3D electron tomography (resolution of 0.2 nm) [11]

and illustrated. The chapter concludes with Raman spectroscopy at elevated temperatures which is a very convenient method to study the anharmonic interactions through the evolution of optical phonon modes in nanophase materials and also to predict many new thermo-physical, electric and magnetic nanomaterials properties.

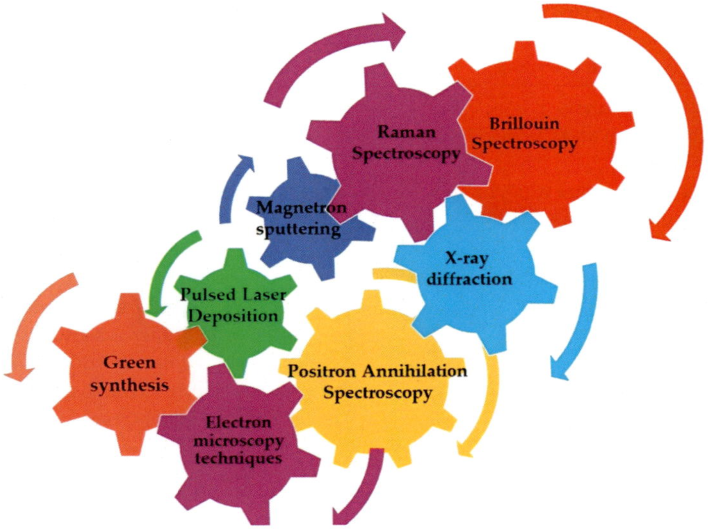

Figure 1.1
Synthesis methods and characterization techniques of nanomaterials.

Chapter 3 is an interesting report on Brillouin Light Scattering which provides a noninvasive, noncontact optical measuring path for the acoustic phonons in hypersonic frequencies. This chapter discusses the development of Brillouin experiments with an insight to its application in the characterization of colloidal nanoparticles.

Chapter 4 elucidates how Raman spectrometry can be coupled in situ to various classical techniques and how it can provide additional and complementary information to understand various systems. It also illustrates that filler addition and its particle size have an influence on the Raman spectroscopy measurement, which is related to the size or volume of defects present in the material and also that a complementary information can be drawn from in situ coupling of experiments to interpret the thermal or dielectric results obtained.

Chapter 5 is a fantastic explanation of defect spectroscopy which has become very important for probing the structural defects in nanomaterials. Positron annihilation spectroscopy is a very sensitive and useful technique for probing the variations of the positron lifetimes, their intensities and the Doppler broadened line-shape parameters, to obtain information regarding the structural transformation taking place in materials. The present chapter contains a brief introduction to the principles of positron annihilation spectroscopy and its potential is exemplified with the help of results obtained from defect studies on calcium-modified magnesium ferrite nanosystems.

Chapter 6 discusses the development and evolution of organ-on-a-chip (OoC) models in testing the safety, efficacy and function of nanomaterials. OoC systems are microscale bioengineered systems capable of recapitulating complex organ functions in health and disease. These are valuable in vitro tool with applications in the study of treatment modalities for an extensive range of conditions ranging from drug discovery to cancer treatment.

Chapter 7 is an amazing description of electrochemical sensors that open up new avenues in electrochemical analysis. Electrochemical biosensors enable on-site sensing applications, which include health care, food quality control, and environmental monitoring in various fields without much investment. An introduction to different electroanalytical techniques and a brief explanation of the electrochemical characterization techniques used particularly for carbon-based nanomaterials, like carbon nanotubes, carbon nanomaterials, graphene, carbon-based quantum dots, and so forth are also described in the chapter.

Chapter 8 discusses the basics of magnetron sputtering process. Various magnetron sputtering techniques such as DC magnetron sputtering processes, RF magnetron sputtering, DC-RF magnetron sputtering, ion beam magnetron sputtering, reactive sputtering, inductively coupled plasma magnetron sputtering, microwave amplified magnetron sputtering, and high power impulse magnetron sputtering have been demonstrated. Formations of smart material thin films and nanocomposites have been discussed in an impressive manner. The advantages and limitations associated with the sputtering process, and the sputtered products and various applications are discussed.

Chapter 9 discusses the basic principles and fabrication strategies of various of self-cleaning materials. The low surface energy materials (silicones, fluorocarbons, organic materials, and inorganic materials) required for superhydrophobic application are discussed extensively with various fabrication techniques such as wet chemical reaction, electrospinning technique, electrochemical deposition, hydrothermal reaction, lithography, layer-by-layer assembly, plasma treatment, and 3D printing techniques. The applications of self-cleaning surface are discussed in terms of automobile parts, solar panel, electronic equipment, surgical equipment, textile materials, and architectural glasses.

Chapter 10 explores the characteristics, behavior, advantages, and disadvantages of conventional Iron sheets blended single-walled carbon nanotubes in electromechanical applications. The nanosheets formed from nano clay have improved electromechanical properties like high strength to weight ratio, negligible rusting ability, damage-tolerance, thermal protection, and control.

Chapter 11 discusses the fascinating role of iron oxide as adsorbent for removal or adsorption of various pollutants from wastewater. The chapter focuses on development of functionalized iron-oxide nanocomposites and its application toward the removal of carcinogenic heavy metals and toxic chemicals from wastewater by adsorption/chemisorption process. This chapter highlights the development of different activated carbon, polymer, amine functionalized, and biochar modified iron-oxide nanoadsorbent toward the removal of heavy metal ions (mainly As, Cd, Cr, Hg, Pb, Ni, Co, and Cu) from aqueous solution. Furthermore, the impact of various parameters such as adsorbent dosage, initial time, temperature, initial concentration of metal ion and ionic strength of solution on removal process have also been summarized.

Chapter 12 is a highly significant chapter that deals with the critical issue of water contamination by highly noxious arsenic species and the promising solution of its adsorption by iron-based nanoadsorbents. The chapter summarizes the synthesis of various iron-based nanomaterials and their characterization techniques. This chapter aims to help the readers to accumulate comprehensive knowledge about the synthesis of iron-based and surface functionalized iron-based nanomaterials for remediation of arsenic species, their characterization, and the possible mechanism for their interaction.

Chapter 13 throws light into the structurally interesting and commercially important Perovskite-based materials, their structure, specifications and energy applications. The use of perovskites in the nanoscale as energy harvesting materials, energy storage devices, fuel cell, solar cell applications and as nano-generators are described in the chapter.

Chapter 14 emphasizes the promising potential of polymer nanocomposites for energy applications like energy storage and electromagnetic (EM) absorption. The synthesis methods, incorporation of nanofillers into polymers and the significant role of polyvinylidene fluoride (PVDF) based nanocomposites in the field of energy storage and energy applications owing to their excellent physicochemical properties, piezo-, pyro-, and ferroelectric properties is highlighted in the chapter.

Chapter 15 gives a concise overview of design and synthesis of $GdBa_2Ca_3Cu_4O_{10.5+\delta}$ superconducting perovskite nanomaterials by solid state reaction technique. Structural studies indicate formation of orthorhombic crystal system. The reduction of lattice defects in the crystal system with increasing particle size and crystallinity is proved. Particle sizes and strains are extracted from Williamson-Hall plots of the samples. Dislocation density and number of particles per unit surface area of GBCCO nanosystems are found to decrease with increase in calcination temperature. Crystallinity and ceramic quality is found to increase and defect density is found to decrease with increase in calcination temperature.

Chapter 16 is a very relevant chapter that describes the development of multiferroic nanoparticles. The synthesis, development and investigation of magnetoelectric properties of four different molar concentrations of sodium potassium lithium niobate-nickel/Cobalt

Ferrite, *xNKLN— (1-x)MFO* nanocomposites. Piezoelectric and magnetoelectric properties indicate that samples with $x = 0.85$ show higher values for both, which proves them better for applications as sensors and actuators. The magnetoelectric coefficients of CFO based samples are lower than the corresponding values for NFO based samples which are interpreted as due to the magnetic hardness of CFO compared to NFO.

Chapter 17 discusses the cheap and economical, easy, green and environment friendly process called Polyol method of synthesis of iron-cobalt (FeCo) and Nickel—Cobalt nano-alloys. The structural, morphological and elemental analyses were done by various characterization techniques. Magnetic studies reveal its potential for medical applications like drug delivery and magnetic Hyperthermia.

Chapter 18 encompasses the "state of the art" research on the "green" synthesis of nanocomposites and their use in photocatalytic applications. The ecofriendly alternatives for synthesis of photocatalytic materials in comparison with chemical or physical techniques are briefed here. A brief report of the various synthesis methods using different natural substances such as microorganisms, enzymes, bacteria, and plant extracts, as an environmentally benign route for green synthesis of nanocomposites and photocatalytic materials is presented in the chapter. The green and biological synthesis methods are promising for the synthesis of ZnO, TiO_2, various metal sulfides, and g-C_3N_4 based materials and their photocatalytic applications are summarized in the chapter.

Chapter 19 focuses on application of 0D carbon nanofillers in the fabrication of polymer nanocomposites. This chapter also discusses the procedure for development of carbon nanofillers reinforced polymer nanocomposites. The chapter, in a nutshell, provides an insight on the role of 0D carbon nanomaterials in the development of polymer nanocomposites and its future prospects toward application in commercial world markets.

Chapter 20 is an insightful description of significance of metal phthalocyanines that form an important class of organic semiconductors and organic electronics. A detailed description of the preparation of MPc-carbon composites is illuminated with examples. The applications of MPc-carbon composites as hole conductor in organic photovoltaics, electro- and photocatalysts in electrochemical hydrogen evolution, oxygen reduction, carbon dioxide reduction, and finally as pseudocapacitors in charge storage devices are discussed in the chapter.

Chapter 21 summarizes the recent development of electrospun composite polymer nanofibers particularly in situ functionalization of nanomaterials in polymer nanofibers matrix. The development of functionalized nanofibrous materials using electrospinning method and their environmental remediation applications is described. The adsorption behavior of electrospun polymer composite and ceramics nanofibers toward removal of inorganic and organic pollutants from water is also emphasized in the chapter.

Chapter 22 highlights the device worthy thin-film growth of PMN-PT using pulsed laser deposition technique. The chapter presents various challenges in the synthesis of PMN-PT ceramic and discusses a novel partial covering combined with modulation and stabilization heating for the realization of device worthy PMN-PT ceramic pellets for laser ablation. The Columbite B-site precursor method used for the synthesis of 0.65 PMN−0.35 PT ceramic and pulsed laser deposition used for thin-film growth is described in the chapter.

2. Conclusion

The nanoparticle characteristics which can be controlled by particle size, shape, distribution, and chemical composition are highly dependent on the nanoparticle syntheses methodologies. Thus, the precise synthesis protocols evolved out as the outcome of continued research have a significant role in production of nanosystems with desired functionalities. The chapters describe the remarkable scientific progresses attained in the development of a wide range of nanomaterials with precise control of shape, size, morphology and properties. Considering the sustainability and environmental impacts, the green syntheses routes of nanomaterials are also explored. Finally, we conclude with a look at the future challenges and prospects for device development through advances in fabrication of nanofibers and thin films.

References

[1] W.J. Salcedo, F.J. Ramirez Fernandez, J.C. Rubimc, Influence of laser excitation on Raman and photoluminescence spectra and FTIR study of porous silicon layers, Braz. J. Phys. 29 (4) (1999) 751−755.
[2] X.H. Tang, X. Wen, S.W. Sun, H.Y. Jiang, New route for synthesis of highly ordered mesoporous silica with very high titanium content, Nanoporous Mater. III (2002) 167.
[3] S. Sagadevan, P. Koteeswari, Analysis of structure, surface morphology, optical and electrical properties of copper nanoparticles, J. Nanomed. Res. 2 (5) (2015) 00040−00048.
[4] S.M. Gamboa, E.R. Rojas, V.V. Martínez, J. Vega-Baudrit, Synthesis and characterization of silver nanoparticles and their application as an antibacterial agent, Int. J. Biosen. Bioelectron 5 (2019) 166−173.
[5] A.J. Shnoudeh, I. Hamad, R.W. Abdo, L. Qadumii, A.Y. Jaber, H.S. Surchi, S.Z. Alkelany, Synthesis, characterization, and applications of metal nanoparticles, in: Biomaterials and Bionanotechnology, Academic Press, 2019, pp. 527−612.
[6] H.F. Liang, C.T.G. Smith, C.A. Mills, S.R.P. Silva, The band structure of graphene oxide examined using photoluminescence spectroscopy, J. Mater. Chem. C 3 (48) (2015) 12484−12491.
[7] P.C. Lin, S. Lin, P.C. Wang, R. Sridhar, Techniques for physicochemical characterization of nanomaterials, Biotechnol. Adv. 32 (4) (2014) 711−726.
[8] J. Griffiths, A brief history of mass spectrometry, Anal. Chem. 80 (15) (2008) 5678−5683.
[9] A. Sarkar, H. Luitel, N. Gogurla, D. Sanyal, Positron annihilation spectroscopic characterization of defects in wide band gap oxide semiconductors, Mater. Res. Express 4 (3) (2017) 035909.
[10] A. Kim, W.B. Ng, W. Bernt, N.J. Cho, Validation of size estimation of nanoparticle tracking analysis on polydisperse macromolecule assembly, Sci. Rep. 9 (1) (2019) 1−14.
[11] S. Mourdikoudis, R.M. Pallares, N.T. Thanh, Characterization techniques for nanoparticles: comparison and complementarity upon studying nanoparticle properties, Nanoscale 10 (27) (2018) 12871−12934.

CHAPTER 2

Temperature-dependent Raman spectroscopy for nanostructured materials characterization

Zorana D. Dohčević-Mitrović, Sonja Aškrabić, Bojan S. Stojadinović, Dejan M. Djokić

Nanostructured Matter Laboratory, Institute of Physics Belgrade, University of Belgrade, Belgrade, Serbia

Chapter Outline

1. Introduction

Raman spectroscopy (RS) is one of the most important nondestructive and noncontact vibrational spectroscopy method providing primarily an information on the vibrational and electronic properties in confined systems [1]. The beauty and the power of RS (conventional and micro-Raman) can be summarized as follows: no sample preparation or damage; high spatial resolution (submicron resolution when combined with a microscopic tool), sensitivity, and easy use. It is powerful and handy analytical tool for the

Design, Fabrication, and Characterization of Multifunctional Nanomaterials. https://doi.org/10.1016/B978-0-12-820558-7.00001-7

characterization of great variety of nanomaterials such as quantum dots, quantum wells, nanowires, nanopowders, nanocomposites and advanced ceramics, thin films, and biomaterials [1].

In nanostructured confined materials, due to the violation of the $q \approx 0$ selection rule, phonons over the entire Brillouin-zone contribute to the Raman line shape causing a shift and asymmetrical broadening of Raman optical or acoustic modes [2,3]. Proper analysis of temperature dependent changes of the phonon mode line shape in nanostructures is of fundamental importance to understand the changes in phonon dynamics. Significant information regarding the phonon-phonon interactions in nanomaterials, i.e., anharmonicity in the lattice potential energy can be obtained. Decay dynamics of the optical/acoustic phonon modes can be complex and dependent not only on anharmonic phonon-phonon decay mechanisms, but on other parameters like particle size and shape, size distribution, type and magnitude of the strain (compressed or tensile) and lattice disorder, structural anisotropy, or surface states. From the temperature dependence of Raman spectra it is possible to deduce about their influence on other nanomaterials properties like thermal expansion, specific heat, or thermal conductivity.

RS, as a complementary technique to X-ray diffraction, has proved to be a useful tool to identify structural stability of nanomaterials. Raman scattering is sensitive to the changes in crystal structure and can be applied for detecting new phases and/or phase transitions in investigated nanomaterials at elevated temperatures.

RS has been extensively employed for investigation the interplay between charge, spin, and lattice degrees of freedom in nanocrystalline materials. Understanding of electron-phonon interaction is of great importance, as this interaction plays a significant role in thermal conduction, charge carrier mobility, and recently discovered superconductivity in nanoparticles [4,5]. Information about the electron-phonon interaction can be obtained from the phonon shift and linewidth of Raman modes, or through the appearance of new modes in heterostructures of 2D materials [6]. The temperature-dependent variation of charge coupled Raman modes enables to deduce about the strength of electron-phonon coupling.

In magnetic nanomaterials, magnetic ordering has a well pronounced effect on the phonon frequencies near magnetic ordering temperature because the exchange coupling between magnetic ions influences Raman mode position, linewidth and integrated intensity. The subtle interplay between the spin and lattice degrees of freedom may be pivotal mechanism for understanding the superconductivity in nanocomposite structures known as nanosuperconductors [7] or can mediate magnetoelectric coupling in type-II multiferroic materials. In composite multiferroic nanostructures spin and phonons can be strongly coupled to the lattice strain as well, and this type of coupling is important aspect in manipulating the magnetic or ferroelectric properties of these materials [8]. An intense spin-phonon interaction manifests as hardening or softening of the phonon frequency or as nontrivial behavior of optical phonon linewidth and integrated intensity near the magnetic ordering temperature. Due to the complexity of magnetic interactions in nanomaterials, the coupling between the lattice and spin degrees of freedom (spin-phonon coupling) can be different for different phonon modes, and the coupling strength may vary even in the case of the same spin-spin interaction. Electron-phonon and spin-phonon coupling strength in low-dimensional materials are size-dependent [9—11], and temperature-dependent RS offers a unique opportunity of probing the temperature evolution of electron-phonon and/or spin-phonon coupling in low-dimensional materials. Since electron-phonon and spin-phonon interactions in nanoscaled materials can be notably modified with respect to bulk materials, it is of great importance to get much better insight into these phenomena having a large impact on the electronic device engineering based on novel nanosized materials [5,6,8].

In this chapter we give a brief overview of the state-of-the-art applications of temperature-dependent RS for analyzing diverse fundamental physical processes in low-dimensional materials. Understanding of these processes enables to predict many novel thermophysical, optical, electric, magnetic, and superconducting nanomaterials features.

2. Anharmonicity in nanostructured materials

Anharmonic effects at elevated temperatures (T) and pressures (p), strongly affect a number of processes in nanostructured materials including thermal and electronic transport properties, structural phase transformation or optical properties. Although theory predicts that heat transport is performed mainly by lower frequency acoustic phonons, the anharmonic decay of optical phonons into acoustic phonons can also contribute to the thermal conductivity [12,13]. Information on phonon scattering processes is therefore important for further improvement of functionality of the nanointegrated devices. RS provides a great deal of information about the phonon vibrations and lattice anharmonicity and the following section will be dedicated to the research of anharmonic interactions in nanomaterials by using temperature-dependent RS.

2.1 Basic theory of phonon-phonon interactions

Anharmonicity in the interatomic potentials (vibrational potential energy), induced by temperature variations, causes the changes of the Raman normal modes frequency (ω), linewidth (Γ), and mode intensity. Because of the anharmonicity of the lattice forces, an optical mode can interchange energy with other lattice modes and decays into phonons of lower energies or is scattered by thermal phonons into modes of different higher energies. The phonon-phonon interactions are characterized by the complex quantity, *phonon self-energy* [14].

$$\Delta(\omega) - i\Gamma(\omega) \tag{2.1}$$

For a phonon of frequency ω and wave vector q in the jth branch, the real part $\Delta_j(\omega, q)$ represents the frequency shift due to anharmonicity, whereas the imaginary part $\Gamma_j(\omega, q)$ corresponds to reciprocal of the phonon lifetime τ ($\tau^{-1} = 2\Gamma$, 2Γ is the full width at half maximum [FWHM] of the Raman peak) [14].

In nonmagnetic, insulating, or semiconductor nanomaterials, the variation of the frequency of the jth normal mode with temperature at constant pressure arises from two contributions: frequency-independent pure-volume contribution ($\Delta_j^E(q)$), emerging from thermal expansion of the crystal and frequency-dependent, pure-temperature contribution ($\Delta_j^A(\omega, q)$), which results from cubic ($\Delta_{j3}(\omega, q)$), and quartic ($\Delta_{j4}(\omega, q)$) anharmonic terms in the vibrational potential energy. The real part of phonon self-energy $\Delta(\omega)$ can be presented as a sum of these two contributions [15].

$$\Delta_j(\omega, q) = \Delta_j^E(q) + \Delta_j^A(\omega, q) = \Delta_j^E(q) + \Delta_{j3}(\omega, q) + \Delta_{j4}(\omega, q) \tag{2.2}$$

Both $\Delta_j^E(q)$ and $\Delta_j^A(\omega, q)$ give rise to temperature-dependent frequency shift from the harmonic mode frequency ω_0. The Raman mode frequency ω_j, as a function of temperature, can be expressed as

$$\omega_j(T) = \omega_0 + \Delta_j^E(q) + \Delta_{j3}(\omega, q) + \Delta_{j4}(\omega, q) \tag{2.3}$$

In most solids $\Delta_j^E(q) < 0$, i.e., lattice dilation results in phonon mode softening. Frequency shift due to thermal expansion can be evaluated from the following equation [16]:

$$\Delta^E = \omega_0\left(\mathrm{e}^{-3\gamma_i \int_0^T \alpha(T')dT'} - 1\right) \tag{2.4}$$

where $\alpha(T)$ is the coefficient of linear thermal expansion and γ_i is the mode-Gruneisen parameter. The multiphonon processes associated with the cubic terms represent the three-phonon decay (scattering) processes. The low-order perturbation calculations show that frequency shift and broadening of the jth Raman mode due to these processes should be linearly dependent on T [17–19].

$$\Delta_{j3}\left(\vec{0}, j; \omega\right) = -\frac{18}{\hbar^2}\sum_{\vec{q_1}, j_1}\sum_{\vec{q_2}, j_2}\left|V\left(\vec{0}, j; \vec{q_1}, j_1; \vec{q_2}, j_2\right)\right|^2 \times \mathscr{P}\left[\frac{n_1+n_2+1}{\omega+\omega_1+\omega_2} - \frac{n_1+n_2+1}{\omega-\omega_1-\omega_2} + \frac{n_1-n_2}{\omega-\omega_1+\omega_2} - \frac{n_1-n_2}{\omega+\omega_1-\omega_2}\right] \tag{2.5}$$

$$\Gamma_{j3}\left(\vec{0}, j; \omega\right) = \frac{18\pi}{\hbar^2}\sum_{\vec{q_1}, j_1}\sum_{\vec{q_2}, j_2}\left|V\left(\vec{0}, j; \vec{q_1}, j_1; \vec{q_2}, j_2\right)\right|^2 \times \{(n_1+n_2+1) \times [\delta(\omega-\omega_1-\omega_2) - \delta(\omega+\omega_1+\omega_2)] + (n_1-n_2)[\delta(\omega+\omega_1-\omega_2) - \delta(\omega-\omega_1+\omega_2)]\} \tag{2.6}$$

The first two terms in Δ_{j3} and Γ_{j3} present the contribution from phonon decay into two phonons of lower energy (*down-conversion* three-phonon processes). The last two terms describe the processes in which a nonequilibrium phonon is destroyed by a thermal phonon and a phonon of higher energy is created (*up-conversion* three-phonon processes). Up-conversion processes contribute at higher temperatures and vanish as T approaches zero, whereas the down-conversion processes always have a finite contribution even at $T = 0$ K [20]. From Eq. (2.5) is obvious that cubic contribution is negative ($\Delta_{j3} < 0$), giving rise to a negative frequency shift. The quartic terms correspond to four-phonon processes and vary linearly, but also quadratically with T [19].

$$\Delta_{j4a}\left(\vec{0},j;\omega\right) = \frac{24}{\hbar}\sum_{\vec{q_1},j_1} V\left(\vec{0},j;\vec{0},j;\vec{q_1},j_1;-\vec{q_1},j_1\right)\left(n_1+\frac{1}{2}\right) \tag{2.7}$$

$$\begin{aligned}\Delta_{j4b}\left(\vec{0},j;\omega\right) &= -\frac{96}{\hbar^2}\sum_{\vec{q_1},j_1}\sum_{\vec{q_2},j_2}\sum_{\vec{q_3},j_3}\left|V\left(\vec{0},j;\vec{q_1},j_1;\vec{q_2},j_2;\vec{q_3},j_3\right)\right|^2 \\ &\times \mathscr{P}\Bigg\{\left[(n_1+1)(n_2+1)(n_3+1)-n_1n_2n_3\right]\times\left[\frac{1}{\omega+\omega_1+\omega_2+\omega_3}-\frac{1}{\omega-\omega_1-\omega_2-\omega_3}\right] \\ &+3\left[n_1(n_2+1)(n_3+1)-(n_1+1)n_2n_3\right]\times\left[\frac{1}{\omega-\omega_1+\omega_2+\omega_3}-\frac{1}{\omega+\omega_1-\omega_2-\omega_3}\right]\Bigg\}\end{aligned} \tag{2.8}$$

$$\begin{aligned}\Delta_{j4c}\left(\vec{0},j;\omega\right) &= -\frac{576}{\hbar^2}\sum_{\vec{q_1},j_1}\sum_{j_2}\sum_{\vec{q_3},j_3} V\left(\vec{0},j;\vec{0},j_1;-\vec{q_1},j_1;\vec{q_1},j_2\right)V\left(\vec{q_1},j_1;-\vec{q_1},j_2;\vec{q_3},j_3;-\vec{q_3},j_3\right) \\ &\times \mathscr{P}\left[\frac{n_1+n_2+1}{\omega_1+\omega_2}-\frac{n_1-n_2}{\omega_1-\omega_2}\right]\left(n_3+\frac{1}{2}\right)\end{aligned} \tag{2.9}$$

$$\begin{aligned}\Gamma^{(4)}\left(\vec{0},j;\omega\right) &= \frac{96\pi}{\hbar^2}\sum_{\vec{q_1},j_1}\sum_{\vec{q_2},j_2}\sum_{\vec{q_3},j_3}\left|V\left(\vec{0},j;\vec{q_1},j_1;\vec{q_2},j_2;\vec{q_3},j_3\right)\right|^2 \\ &\times \{[(n_1+1)(n_2+1)(n_3+1)-n_1n_2n_3][\delta(\omega-\omega_1-\omega_2-\omega_3)-\delta(\omega+\omega_1+\omega_2+\omega_3)] \\ &+3[n_1(n_2+1)(n_3+1)-(n_1+1)n_2n_3]\times[\delta(\omega+\omega_1-\omega_2-\omega_3)-\delta(\omega-\omega_1+\omega_2+\omega_3)]\}\end{aligned} \tag{2.10}$$

In the aforementioned equations $n_j(\omega_j) = \left[\exp\left(\frac{\hbar\omega_j}{k_B T}\right)-1\right]^{-1}$ is the thermal (Bose-Einstein) population factor of the jth phonon mode with wave vector q and frequency ω_j. The quartic terms (Eqs. 2.7–2.9) may be positive or negative. Accordingly, the resultant frequency shift due to phonon-phonon interaction may be either positive or negative depending on the relative magnitudes of the anharmonic terms in the interatomic potential. The *up* and *down-conversion* three (four)-phonon processes are shown diagrammatically in Fig. 2.1A.

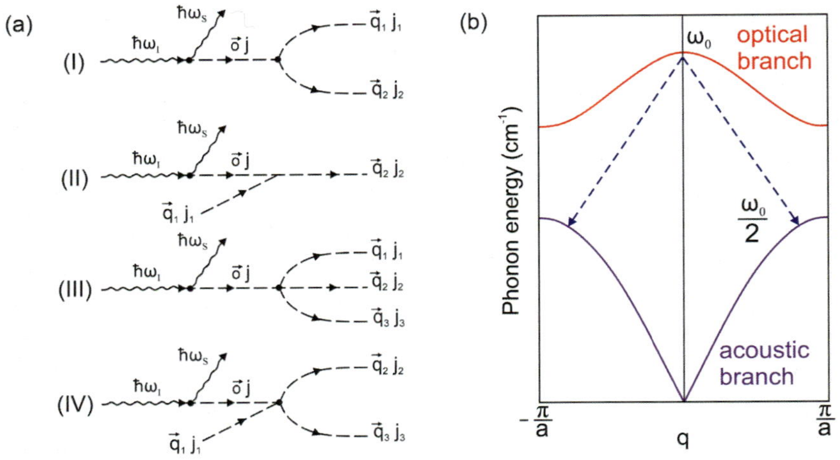

Figure 2.1
Three and four-phonon anharmonic processes (A) (I), (III) down-conversion and (II), (IV) up-conversion processes. (B) three-phonon processes according to Klemens model. *Based on M. Balkanski, R.F. Wallis, E. Haro, Anharmonic effects in light scattering due to optical phonons in silicon. Phys. Rev. B 28 (1983) 1928–1934.*

Following the approach of Klemens, [21], if the optical phonon decays into two or three acoustical phonons of lower mutually equal energies and opposite momenta, the temperature variation of the Raman phonon frequency and linewidth can be simplified:

$$\omega_j(T) = \omega_0 + C\left[1 + \frac{2}{e^x - 1}\right] + D\left[1 + \frac{3}{e^y - 1} + \frac{3}{(e^y - 1)^2}\right] = \omega_0 + \Delta\omega(T) \tag{2.11}$$

$$\Gamma(T) = \Gamma_0 + A\left[1 + \frac{2}{e^x - 1}\right] + B\left[1 + \frac{3}{e^y - 1} + \frac{3}{(e^y - 1)^2}\right] = \Gamma_0 + \Delta\Gamma \tag{2.12}$$

where ω_0 is the harmonic frequency, $x = \frac{\hbar\omega_0}{2k_BT}$, $y = \frac{\hbar\omega_0}{3k_BT}$, A, B, C, D are anharmonic constants and Γ_0 is the intrinsic mode linewidth (temperature-independent broadening). In Fig. 2.1B is presented a three-phonon decay process according to Klemens model. This model, although too simplistic, is generally accepted especially after the work of Hart et al. on Si [22].

2.2 Phonon-phonon interactions in nanomaterials

For many crystalline semiconductor or insulator materials, due to increased temperature, the anharmonic frequency shift and broadening of the Raman modes follow down or up-conversion three-phonon processes, [14,20], but at higher temperatures ($T > 300$ K) four-phonon processes begin to dominate [19]. In nanocrystalline materials, beside the anharmonic effects, temperature induced changes of the Raman line profile depend on several other factors like phonon confinement, strain, disorder/defects. Consequently, anharmonic processes in nanomaterials can show quite distinct behavior from those of bulk counterparts. For more precise determination of influence of these factors on the Raman line profile, the temperature-dependent Raman spectra can be well modeled using phonon confinement model (PCM) [23,24] for spherical nanoparticles in which are incorporated size, inhomogenous strain and anharmonicity [25,26].

$$I(\omega, T) \propto \sum_{i=1}^{n} \int_0^{\infty} \rho(L)dL \int_{BZ} \frac{\exp\left(\frac{-q^2L^2}{8\beta}\right)d^3q}{[\omega - (\omega_i(q) + \Delta\omega)]^2 + \left(\frac{\Gamma(T)}{2}\right)^2} \tag{2.13}$$

In the above equation, the frequency shift $\Delta\omega = \Delta\omega_i(q, L) + \Delta\omega(T)$, where the first term presents strain contribution and the second one originates from anharmonicity; $\rho(L)$ is Gaussian distribution of particle size, q-wave vector expressed in units of $2\pi/a$ (a-lattice cell parameter), L-particle diameter, β-confinement strength and $\Gamma(T)$ is a phonon linewidth, which encompasses broadening due to confinement, strain, and anharmonicity [25]. The integration is performed along all optical dispersion branches within first Brillouin-zone characteristic for certain material. As an example, in Fig. 2.2A are shown

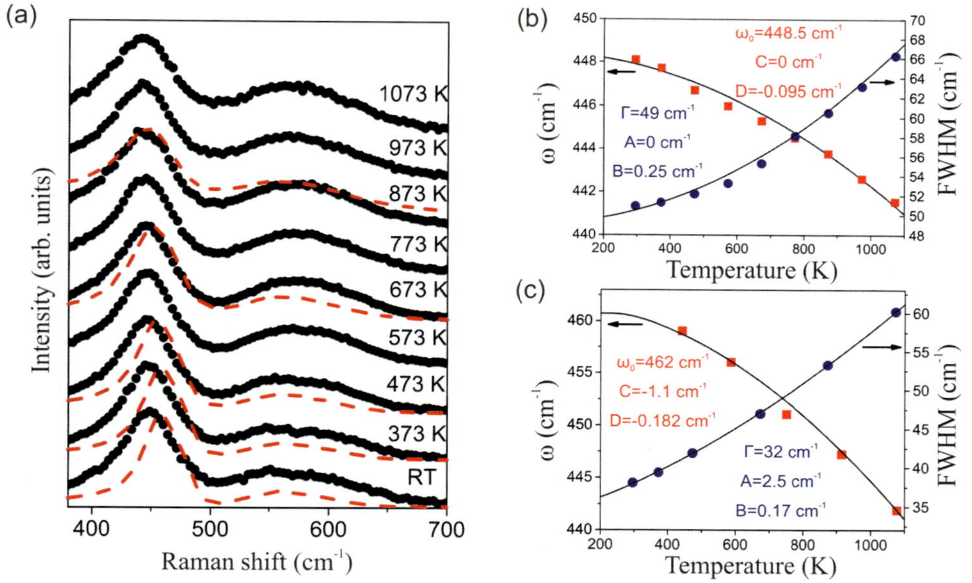

Figure 2.2

(A) Raman spectra of $Ce_{0.75}Nd_{0.25}O_{2-\delta}$ sample upon heating (*circles*) and cooling (*dashed lines*). The variation of frequency and linewidth of F_{2g} mode during (B) heating and (C) cooling. Full lines present calculated values for the peak position and linewidth using the PCM model with Klemens anzats. The best fit anharmonic parameters are given as well. *Reprinted from Z.D. Dohčević-Mitrović, M. Radović, M. Šćepanović, M. Grujić-Brojčin, Z.V. Popović, B. Matović, S. Bošković, Temperature-dependent Raman study of* $Ce_{0.75}Nd_{0.25}O_{2-\delta}$ *nanocrystals, Appl. Phys. Lett. 91 (2007) 203118, with the permission of AIP Publishing.*

Raman spectra of cerium dioxide nanocrystals doped with 25% of Nd ($Ce_{0.75}Nd_{0.25}O_{2-\delta}$) upon heating over the temperature range of 293–1073 K (circles) and gradual cooling down to room temperature (dashed lines) [25]. During the heating, the most intense F_{2g} Raman mode, positioned at ~450 cm^{-1} at room temperature (RT), shifts to lower frequencies, whereas its linewidth increases as presented in Fig. 2.2B. These spectra are fitted using the PCM model (Eq. 2.13) with Klemens anzats (Eqs. 2.11 and 2.12). The best fits of the peak position and linewidth of F_{2g} mode (full lines in Fig. 2.2B) upon heating were obtained including only four-phonon anharmonic processes (D, $B > 0$, $A = C = 0$), contrary to the bulk materials for which three-phonon anharmonic processes are dominant. Such a behavior implies that phonon decay channels are different in nanocrystalline particles. During the cooling process that followed annealing to 1073 K (Fig. 2.2C), anharmonic interactions in $Ce_{0.75}Nd_{0.25}O_{2-\delta}$ nanocrystals are similar to the ones in polycrystalline CeO_2 [25]. Namely, three-phonon anharmonic processes dominated over the four-phonon processes (constants A and C are much higher than B and D), designating that after the heat treatment the particle size increased enough, so that the anharmonic interactions are more similar to bulk materials.

Phonon-phonon interactions were investigated in $Ce_{0.85}Gd_{0.15}O_{2-\delta}$ nanocrystals [27] by following the changes of the F_{2g} Raman mode at elevated temperatures (293–1073 K). High-temperature Raman spectra were analyzed by PCM model (Eq. 2.13) in which are incorporated size, strain and anharmonic phonon decay processes which dominated over the thermal expansion [27]. The temperature variation of F_{2g} Raman mode position and linewidth is presented in Fig. 2.3A and B.

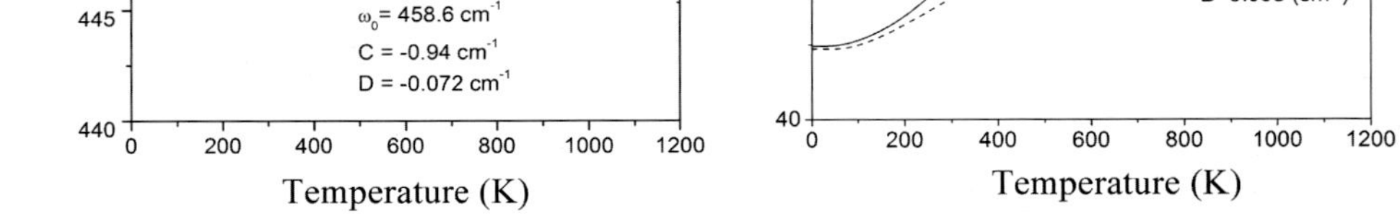

Figure 2.3

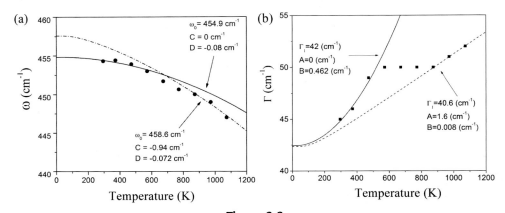

Temperature dependence of (A) frequency and (B) linewidth of the F_{2g} Raman mode in $Ce_{0.85}Gd_{0.15}O_{2-\delta}$ nanopowders. The anharmonic parameters of the best fits are also presented. *Reprinted from S. Askrabić, Z.D. Dohcević-Mitrović, M. Radović, M. Šćepanović, Z.V. Popović, Phonon-phonon interactions in $Ce_{0.85}Gd_{0.15}O_{2-\delta}$ nanocrystals studied by Raman spectroscopy, J. Raman Spectrosc. 40 (2009) 650–655 with the permission of John Wiley & Sons.*

As can be seen from Fig. 2.3A, F_{2g} Raman mode position over the whole temperature range was not possible to fit with a single curve based on three or four-phonon anharmonic processes. The best fit of the experimental values in the 293–573 K temperature range was obtained using only four-phonon anharmonic processes (full line), whereas in the 873–1073 K temperature range three-phonon coupling prevailed (dotted line). It was concluded that anharmonic processes were different in two temperature ranges. The F_{2g} mode linewidth dependency on temperature showed even more distinct behavior, as illustrated in Fig. 2.3B. The $\Gamma(T)$ firstly increased with annealing, then experienced a plateau-like behavior between 573 and 873 K and after that increased again. Such a behavior was ascribed to the effect of increased average particle size which was estimated from the PCM. As the particle size increased with annealing, nanoparticles started to take on the bulk properties and the three-phonon processes became dominant, which was confirmed from the experiment on cooling down the $Ce_{0.85}Gd_{0.15}O_{2-\delta}$ nanocrystals to RT [27]. This study obviously points to subtle interplay between size and anharmonic effects in nanostructured materials, which have to be taken into account.

In the temperature-dependent Raman spectra of ZnO nanopowders [28] it was found that E_2 (low) phonon mode linewidth exhibited anomalous behavior with temperature. In Fig. 2.4A and B are presented frequency (circles) and linewidth (squares) dependence on temperature for two Raman modes, E_2 (high) and E_2 (low) modes. As can be seen from Fig. 2.4A and B, the nonpolar optical phonon (E_2 (high)) has shown expected frequency redshift and increased broadening with increasing temperature (Fig. 2.4A) which is in good accordance with cubic and quartic anharmonicity. On the contrary, the E_2 (low) mode experienced an anomalous linewidth decrease with increasing temperature. This anomaly was ascribed to the increased interference of acoustic and surface modes with E_2 (low) optical phonon mode [28].

Optical phonon modes in most nanostructures experience frequency redshift (softening) with increasing temperature due to the cumulative effect of thermal expansion contribution and three-phonon processes (Eqs. 2.4 and 2.5), which give rise to a negative frequency shift. Contrary to this, the most intense E_g mode (positioned at ~144 cm^{-1} under ambient conditions) of nanophase anatase TiO_2 shifts to higher frequency (blueshift) with temperature rise, whereas the other anatase modes exhibit expected softening [29–32]. Although the blueshift and broadening of this mode with regard to bulk counterpart can be ascribed to the phonon confinement, oxygen nonstoichiometry and surface defects, [29,31], the main contribution originates from the anharmonic interactions that generate positive frequency shift, i.e., four-phonon anharmonic processes (Eq. 2.7) [32].

3. Size/microstrain effects and phase separation

Changes of the wavenumber and the linewidth of the Raman modes at elevated temperatures carry important information about the particle size and shape or about

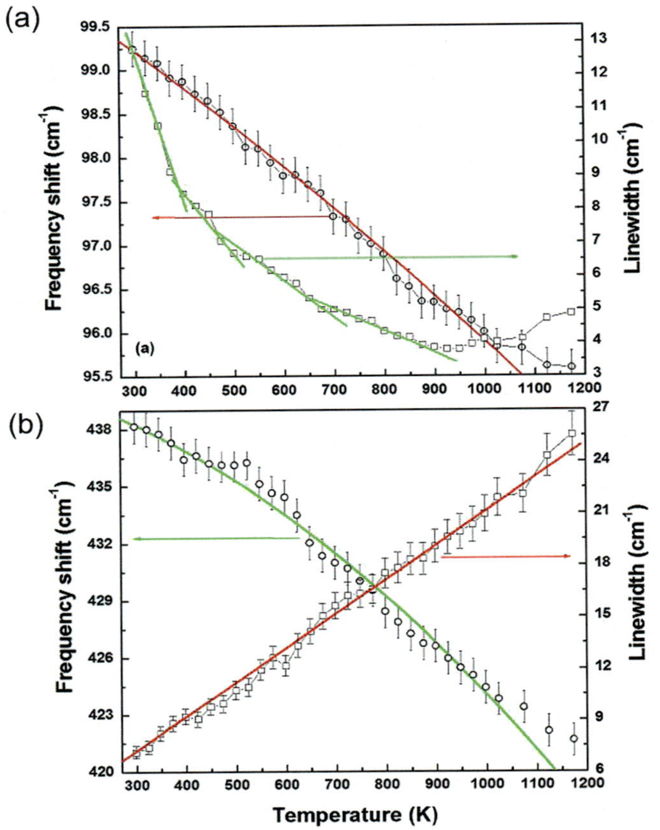

Figure 2.4

Frequency and linewidth dependence on temperature for (A) E_2 (low) and (B) E_2 (high) Raman modes of ZnO nanopowders. The *solid lines* (excluding the linewidth of E_2 (low) mode) are the best theoretical fits using Klemens model. *Reprinted from H.K. Yadav, R.S. Katiyar, V. Gupta, Temperature-dependent dynamics of ZnO nanoparticles probed by Raman scattering: A big divergence in the functional areas of nanoparticles and bulk material. Appl. Phys. Lett. 100 (2012) 051906 with the permission of AIP Publishing.*

structural phase transition which manifests itself through the appearance of new Raman modes. The PCM, described in Section 1.2, is often used to model a Raman mode intensity distribution in nanostructured materials where the size and strain effects lead to significant phonon shift and asymmetric broadening [2,3,23,24]. Using the PCM, effects of particle size, size distribution and microstrain can be disentangled. For example, in related study, oxygen deficient $CeO_{2-\delta}$ nanopowders were annealed in the air in the temperature range (200–500 °C) in order to study structural and vibrational properties of nanophase $CeO_{2-\delta}$ at elevated temperatures [33]. Raman spectra of the annealed samples (S2–S5) are presented in Fig. 2.5A together with as-synthesized $CeO_{2-\delta}$. The most intense F_{2g} mode

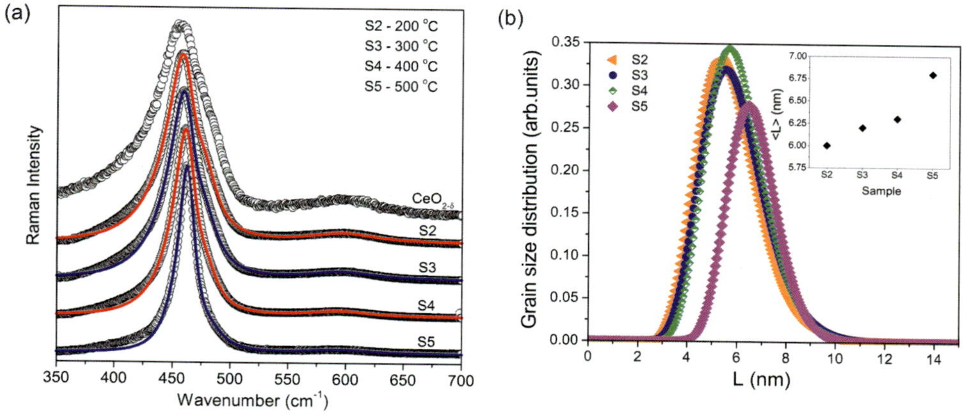

Figure 2.5

(A) Raman spectra of F_{2g} mode for the $CeO_{2-\delta}$ nanopowders at RT and annealed at different temperatures (*circles*), fitted by PCM (*full lines*). (B) Nanocrystalline particle size distribution deduced from the PCM. Samples S2–S5 represent powders annealed at different temperatures 200–500 °C. The inset presents the average particle size increase with annealing, obtained from PCM. *Reprinted from S. Aškrabić, Z. Dohčević-Mitrović, A. Kremenović, N. Lazarević, V. Kahlenberg, Z.V. Popović, Oxygen vacancy-induced microstructural changes of annealed* CeO_{2-x} *nanocrystals, J. Raman Spectrosc. 43 (2012) 76–81 with the permission of John Wiley & Sons.*

(~458 cm^{-1}) was shifted to higher wavenumbers and became more symmetric for the samples annealed at higher temperatures (>300 °C). Another mode which was ascribed to the intrinsic oxygen vacancies defect mode (around 600 cm^{-1}) became less intensive in the samples annealed above 300 °C. This implies that with annealing the concentration of oxygen vacancies decreased. PCM was employed [25,33] to model the changes of F_{2g} mode from Fig. 2.5A. The average lattice parameters were determined by Rietveld analysis of X-ray diffraction spectra [33] and the empirical relation between lattice parameter and particle size, $a = a_0 + k_1 \cdot \exp(-k_2/L)$, was used in PCM, where k_1 and k_2 are fitting parameters. Combined XRD and Raman analysis showed that during the thermal treatment average particle size distribution (Fig. 2.5B) and particle size (inset of Fig. 2.5B) haven't changed much, but the average microstrain was significantly reduced, almost three times in samples annealed at 400 °C and 500 °C. As content of oxygen vacancies decreased too, it was deduced that relaxation of microstrain due to the improved stoichiometry dominantly contributed to the changes seen in the Raman spectra [33].

Another temperature-dependent study of MoO_3 nanoribbons [34] by Raman spectroscopy have shown anomalies in wavenumber and linewidth behavior in the 150–400 °C temperature range. As shown in Fig. 2.6A and B, in $\omega(T)$ and $\Gamma(T)$ dependences (black circles and squares) of 155 cm^{-1} Raman mode, a plateau was observed. The $\omega(T)$ and $\Gamma(T)$ dependences departed from the anharmonic decay model in bulk MoO_3, for which the Raman mode frequency/linewidth monotonously decreased/increased with the

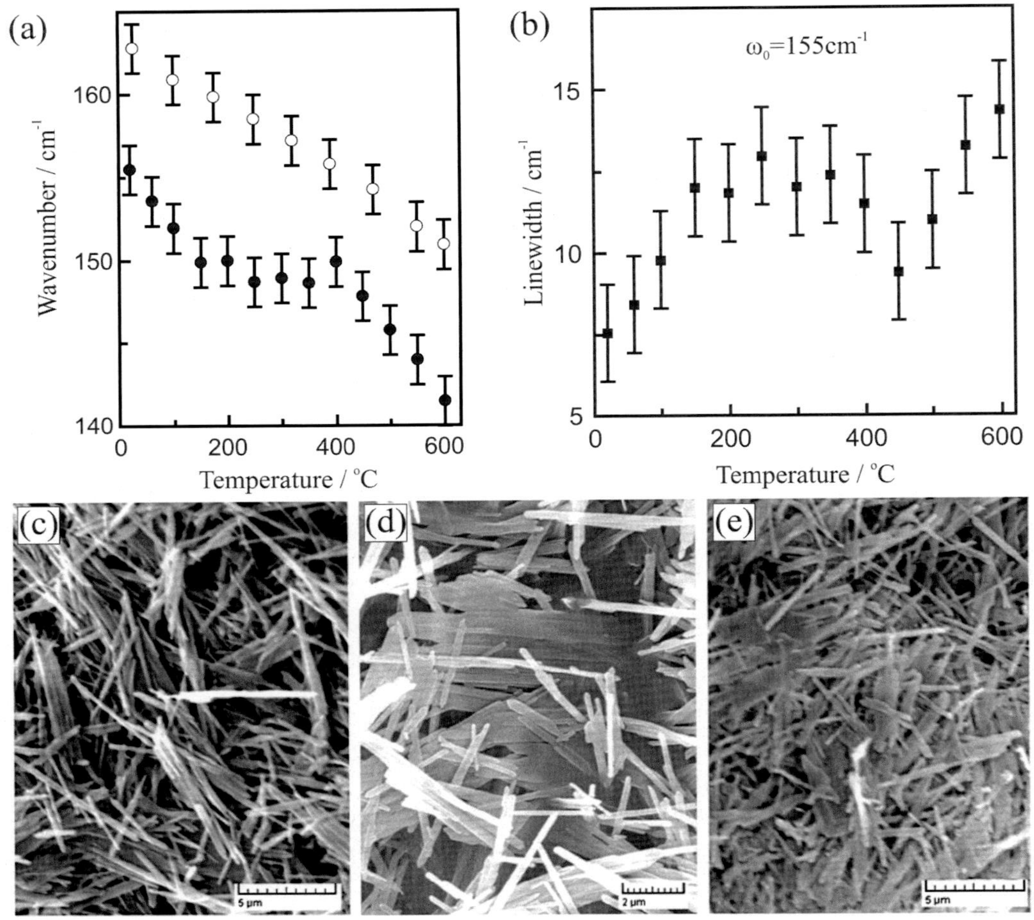

Figure 2.6

Temperature dependence of (A) frequency (*black circles*) and (B) linewidth of 155 cm^{-1} Raman mode of MoO_3 nanoribbons. Open circles present frequency versus temperature plot for bulk counterpart. SEM images at different temperatures: (C) T = 25 °C, (D) T = 150 °C and (E) T = 350 °C. *Reprinted from J.V. Silveira, L.L. Vieira, J.M. Filho, A.J.C. Sampaio, O.L. Alvesc, A.G. Souza Filho, Temperature-dependent Raman spectroscopy study in MoO_3 nanoribbons, J. Raman Spectrosc. 43 (2012) 1407–1412 with the permission of John Wiley & Sons.*

temperature rise. Such a behavior could not be ascribed to a structural phase transition, as MoO_3 nanoribbons are structurally stable up to 650 °C, but can be a consequence of size-induced effects. During the thermal treatment, the crystallite size of MoO_3 nanoribbons increased. In the 150–400 °C temperature range, the anharmonic interactions and the coalescence process are competing processes that lead to the departure from the optical phonon anharmonic decay model and appearance of a plateau [34]. Above 400 °C nanoribbons adopted bulk properties. SEM images of MoO_3 nanoribbons treated at different temperatures (Fig. 2.6C–E) confirmed morphological changes. Namely, in the temperature regime between 150 °C and 400 °C, besides the isolated nanoribbons, large slab structures have been formed due to the coalescence of nanoribbons. At temperatures above 400 °C the nanoribbons were almost totally converted into large slabs behaving more like the bulk counterpart.

A similar appearance of a plateau-like behavior in the F_{2g} mode linewidth dependence on temperature was observed in the case of $Ce_{0.85}Gd_{0.15}O_{2-\delta}$ nanopowders from Fig. 2.3B [27]. From the PCM model it was deduced that anomalous behavior of the F_{2g} mode linewidth in the temperature range 573–873 K can be ascribed to the particle size increase and change of the particle size distribution with annealing when nanomaterials begin to resemble the bulk crystals [27].

RS has a potential to detect the presence of other phases through the appearance of the peaks characteristic for a certain crystalline structure, present even in small concentrations, and particularly in the case when two phases are very similar and cannot be resolved through X-ray diffraction spectra [35,36]. An illustration of phase separation can be found in the Raman spectra of $Ce_{0.85}Gd_{0.15}O_{2-\delta}$ nanopowders which are gradually cooled down from 1073 K to RT [27]. During the cooling process, in the Raman spectra at 473 K a new mode appeared at ~483 cm^{-1}. This new mode presented one of the strongest Raman modes of Gd_2O_3 phase, confirming that phase separation took place [27]. Another example concerns $BaTiO_3$ which has the property of forming four structural phases depending on the temperature [37]. At low temperatures, it has rhombohedral (*R*) structure and the phase transition to orthorombic (*O*), tetragonal (*T*), and cubic (*C*) phase occurs approximately at 180, 275 and 400 K [37]. Temperature-dependent Raman spectra of $BaTiO_3$ nanoparticles, with a 300 nm average particle size, are a good example how to follow the structural phase transitions in this ferroelectric material [37]. In Fig. 2.7A and B are shown Raman spectra of 300 nm $BaTiO_3$ nanoparticles recorded at two different temperatures (83 and 503 K).

The changes in the polar A_1 (TO) mode position with increasing temperatures were found to be a very good indicator of the onset of each of four phases for 300 nm nanoparticles, as presented in Fig. 2.7C. Namely, the redshift of this mode with temperature signifies the presence of one phase, whereas the temperature at which this mode starts to shift to higher wavenumbers (blueshift) points to the onset of a phase transition. In Fig. 2.7C are also indicated the temperatures of the structural phase transitions of the corresponding phases (*R*, *O*, *T*, *C*).

4. Raman thermometry

Less known application of RS concerns its use in thermometry. RS as a noncontact method of high spatial resolution can be used to measure the local temperature of the

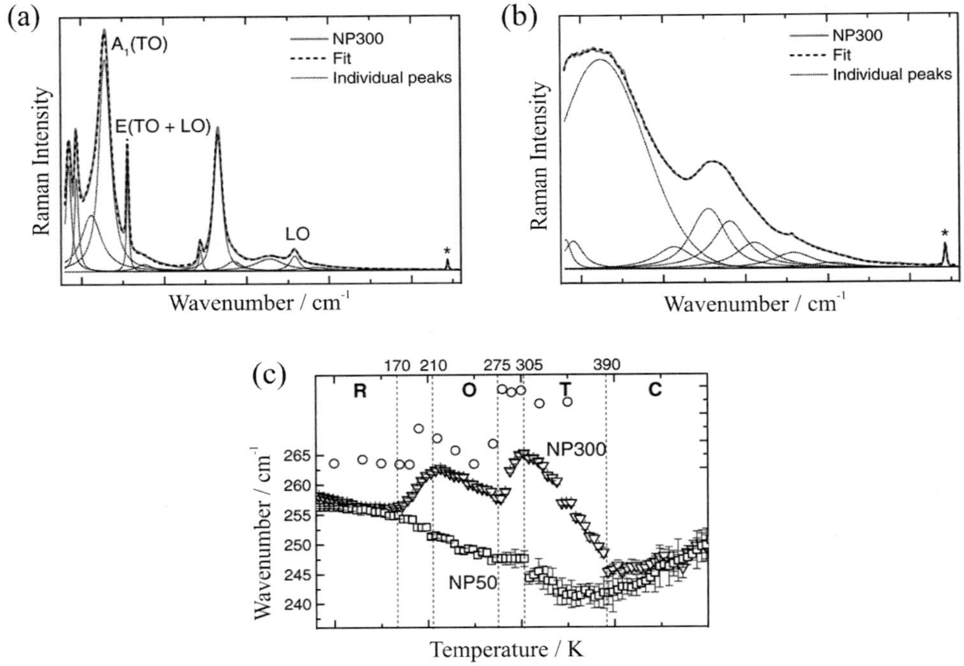

Figure 2.7

Raman spectra of $BaTiO_3$ nanoparticles at (A) 83 K and (B) 503 K; (C) wavenumber shift of the A_1(TO) mode versus temperature for nanoparticles of 300 and 50 nm average sizes. *Adapted from M. Sendova, B.D. Hosterman, R. Raud, T. Hartmann, D. Koury, Temperature-dependent, micro-Raman spectroscopic study of barium titanate nanoparticles, J. Raman Spectrosc. 46 (2015) 25–31 with the permission of John Wiley & Sons.*

material. It is known that Raman mode position will shift with temperature increase/decrease due to the change of the bond length and anharmonic effects. From the peak shift dependence on temperature a calibration curve can be formed and used for material's temperature estimation, but caution should be taken to accurately control the reference material's temperature. It is of great importance to use extremely low excitation powers in order to prevent additional heating due to the excitation power absorption by the sample. Another way to determine the temperature is to measure intensities of a Raman band at the Stokes (I_S) and anti-Stokes (I_{AS}) positions and calculate the temperature from the formula based on the Placzek's approximation

$$\frac{I_{AS}}{I_S} = \frac{(\omega_I + \omega_S)^4}{(\omega_I - \omega_S)^4} \, e^{-\frac{\hbar\omega_S}{k_B T}} \tag{2.14}$$

where ω_I and ω_S are wavenumbers of laser excitation line and a Stokes line, respectively. RS is nowadays readily used to determine the temperature or temperature sensitivity of various nanocrystalline materials.

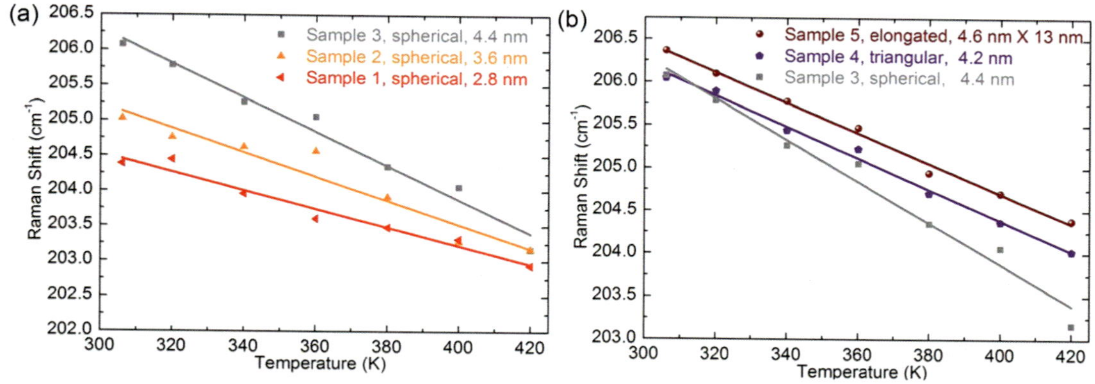

Figure 2.8

A_1(LO) frequency shift as a function of temperature for (A) spherical CdSe NCs of different size, and (B) various shaped CdSe NCs of similar size. *Reprinted from L. Chen, K. Rickey, Q. Zhao, C. Robinson, X. Ruan, Effects of nanocrystal shape and size on the temperature sensitivity in Raman thermometry, Appl. Phys. Lett. 103 (2013) 083107 with the permission of AIP Publishing.*

A study by Chen and coworkers [38] dealt with the effects of different size and shape of CdSe nanocrystals (NCs) on the temperature sensitivity of the A_1(LO) Raman mode shift. This study was performed in the temperature range 300–420 K. In Fig. 2.8 is shown the A_1(LO) mode position dependency on temperature for CdSe NCs of different sizes and shapes.

In the case of CdSe spherical NCs of different size (Fig. 2.8A), it was shown that temperature sensitivity is higher for larger nanocrystals. From the fitting of frequency shift (full lines) based on a model which incorporates thermal expansion and anharmonic effects, it was deduced that in larger nanocrystals anharmonic phonon processes dominate over the decreasing thermal expansion coefficient. In nanocrystals of different shape (triangular and elongated from Fig. 2.8B), the temperature sensitivity decreased because the effect of reduced thermal expansion coefficient overwhelms the anharmonic processes [38]. Since the temperature sensitivity is dependent on crystal size and morphology, this study points to the necessity of finding a balance between nanometric size and morphology for the application of nanomaterials in noncontact thermometry.

Raman thermometry can be also employed for thermal conductivity measurements [39,40]. This is relatively simple method with an advantage of avoiding the influence of thermal contact resistance on the intrinsic thermal conductivity. Applying the Raman shift method, heat is generated by a laser spot and the detected Raman mode shift carries an information on the induced average temperature change. The measured dependence of the peak position on the power dissipated in the material ($\partial\omega/\partial P$) can be later used to estimate thermal conductivity using the premise of radial or plane wave heat propagation away from the laser spot position on the sample [39,40].

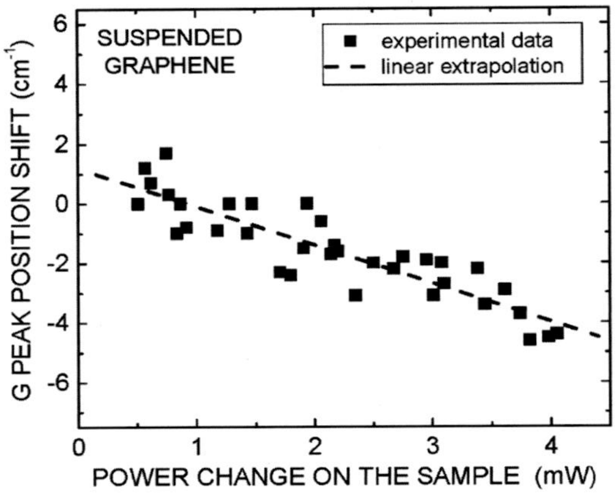

Figure 2.9

G mode position shift as a function of total dissipated power for single-layer graphene. The spectra are collected at RT using 488 nm laser excitation line. *Reprinted with permission from A.A. Balandin, S. Ghosh, W.Z. Bao, I. Calizo, D. Teweldebrhan, F. Miao, C.N. Lau, Superior thermal conductivity of single-layer graphene, Nano Lett. 8 (2008) 902–907. Copyright (2008) American Chemical Society.*

In the case of a graphene flakes placed over a trench made in Si/SiO_2 substrate, [39], and supposing that the G mode position linearly depends on the sample temperature ($\omega = \omega_0 + \chi_G T$), thermal conductivity can be expressed as [39,41].

$$K = \frac{L}{2aW}\chi\left(\frac{\partial\omega}{\partial P}\right)^{-1} \tag{2.15}$$

where L is the lateral dimension of the sample along which the wave propagates, W is the sample width, a, the sample thickness and $\partial\omega$ is a shift of G peak position due to the variation in the heating power ∂P on the sample surface. Fig. 2.9 shows the G mode position shift dependence on the total dissipated power change, from which the slope value, $\partial\omega/\partial P$, can be extracted. The temperature coefficient χ_G can be obtained by explicit Raman measurements of the G mode shift dependency on temperature change in a separate experiment when temperature is changed in a controlled way, i.e., heating of the entire sample in a heating stage.

Using Raman thermometry, thermal conductivity of a single-layer graphene flake was deduced to be ~5300 W/(mK) [39]. This value is much larger compared to other carbon materials which is of significance for this material application. In a similar way, thermal conductivities of individual single-wall and multiwall carbon nanotubes were estimated to be 2400 W/(mK) and 1400 W/(mK) [40].

5. Temperature behavior of acoustic vibrations in nanocrystalline materials studied by low-frequency Raman spectroscopy

Low-frequency Raman acoustic modes have been observed in nanocrystalline materials of various morphologies (spheres, ellipsoids, rods, etc.) and of crystallite size less than tens of nanometers [42–45]. The appearance of these modes in the spectra of nanocrystals results from the geometrical confinement and coupling of longitudinal and transversal acoustic modes within a nanocrystalline particle [46,47]. According to the theory of Lamb, [48], the frequencies of acoustic vibrations are inversely proportional to the particle diameter: $\omega_{ln} = \beta_{ln}/D$, where ω_{ln} is the mode frequency, D is particle diameter and β_{ln} nondimensional eigensolutions of the Navier equation for homogenous elastic sphere. According to the selection rules, only the spheroidal modes with angular quantum numbers $l = 0$ and $l = 2$ are Raman active [46]. The strongest modes in the low-frequency region of the Raman spectra can be used to determine the particle diameter. Intensity of the Raman modes enables the determination of the nanocrystalline size distribution and can be described by the relation [45,49].

$$I(\omega) = \frac{n(\omega) + 1}{\omega} C(\omega) g(\omega) \tag{2.16}$$

where $n(\omega)$ is Bose-Einstein factor, $C(\omega)$ is the mode-radiation coupling factor, $g(\omega)$ is the density of vibrational states. The $l = 0$ and $l = 2$ modes are usually of the highest intensity from which the size distribution can be estimated. Taking the relation between mode frequency and particle diameter into account, in off-resonant conditions, $C(\omega) \sim D \sim 1/\omega$ and the density of states $g(\omega)$ reflects the particle size distribution $N(D) \sim g(\omega_{ln} = \beta_{ln}/D)$. The simple relation can then be used to determine the particle size distribution $N(D)$ [45]:

$$N(D) \sim g(\omega) = \frac{I(\omega)\omega^2}{n(\omega) + 1} \tag{2.17}$$

Therefore, temperature dependent study of low-frequency acoustic modes in nanomaterials can potentially contribute to the pool of noncontact methods for the estimation of nanocrystalline size and/or size distribution change with temperature variation.

Although there is a huge number of papers in the literature dedicated to the temperature-dependent behavior of optical phonons in various nanostructures, little is known about the acoustic phonons dynamics as a function of temperature, despite their relevance in nonradiative relaxation processes or exciton decoherence and dephasing in nanocrystalline solids [50]. Temperature-dependent acoustic mode dynamics can be complex and dependent not only on anharmonicity, but on other parameters like particle size, structural anisotropy, morphology, surface states, or interactions with the charge carriers and therefore deserves more profound investigation.

One of rare studies on temperature dependence of breathing-mode acoustic vibrations in CdSe nanocrystals was presented in the work of Mork et al. [50] The temperature evolution of low-frequency Raman modes of CdSe NCs of several sizes is presented in Fig. 2.10A. All low-frequency Raman modes have shown the expected phonon

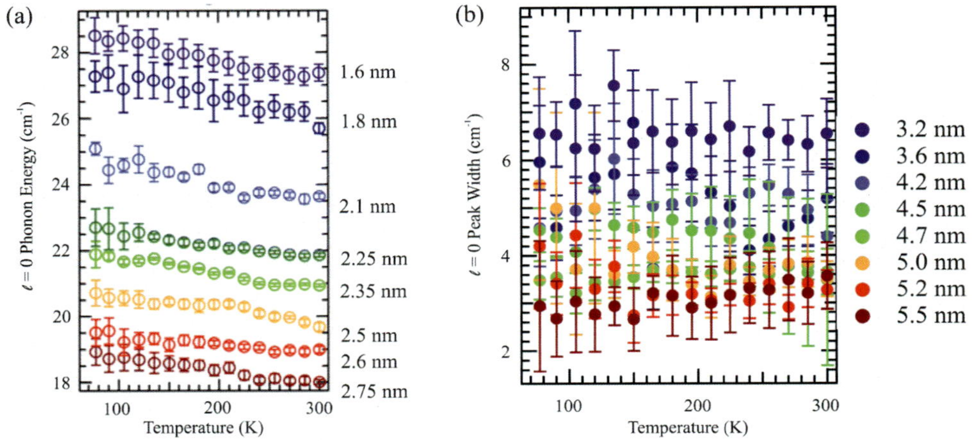

Figure 2.10

Temperature dependence of (A) acoustic phonon energy and (B) linewidth for CdSe NCs of different diameters. *Adapted from A.J. Mork, E.M.Y. Lee, W.A. Tisdale, Temperature dependence of acoustic vibrations of CdSe and CdSe–CdS core-shell nanocrystals measured by low-frequency Raman spectroscopy, Phys. Chem. Chem. Phys. 18 (2016) 28797–28801. Published by the PCCP Owner Societies.*

softening, regardless the particle size, but the magnitude of observed phonon softening could not be explained by Lamb's model only. Since, the acoustic vibrations are dependent not only on size, but also on bulk modulus, density, and transverse and longitudinal sound velocities, their change with temperature has to be taken into account. In the case of CdSe NCs, it was shown that the bulk modulus and density changes in the investigated temperature range were less than 5%, which was insufficient to explain the observed frequency change. It was concluded that organic ligands on the nanoparticle surface, with different elastic modulus, can significantly contribute to the acoustic phonon shift [50]. The linewidth (Fig. 2.10B) has shown no expected change with temperature. Such a temperature-independent behavior of linewidth was ascribed to the strong environmental damping and inhomogenous contribution of different nanocrystalline sizes to the linewidth [50].

A complex temperature-dependent behavior of acoustic phonons was observed in ZnO nanoparticles too [51]. Fig. 2.11 shows the low-frequency Raman spectra of ZnO nanoparticles recorded at elevated temperatures (300–600 K). The prominent mode (around 22/cm) at 300 K, ascribed to spheroidal acoustic mode, shifted toward lower wavenumbers with temperature increase and after 600 K disappeared under the Rayleigh wing. Analyzing the temperature induced shift of the acoustic mode from Fig. 2.11B (open circles), it was deduced that at lower temperatures under 500 K, anharmonic decay processes and bond strength weakening were responsible for mode softening. At higher temperatures (>500 K), fast exponential decrease of peak frequency was attributed to the

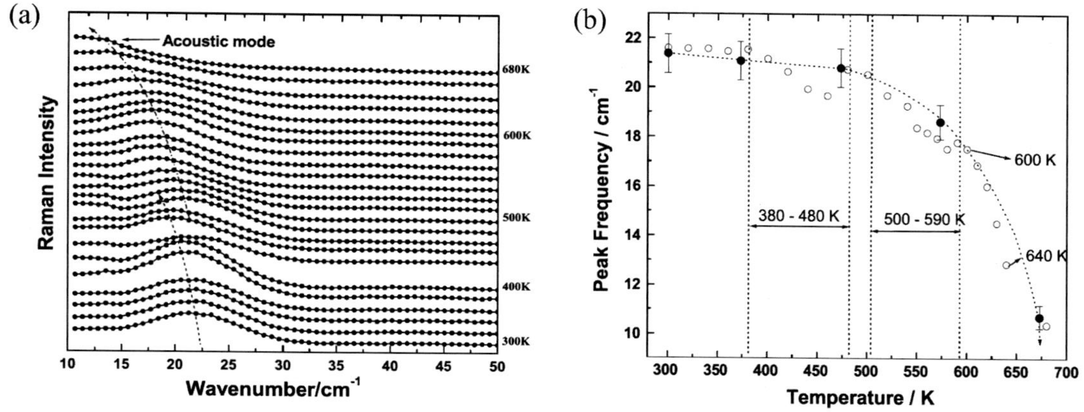

Figure 2.11
Temperature-dependent (A) low-frequency Raman spectra of ZnO nanoparticles and (B) acoustic mode frequency (*open circles*) obtained from the fitting of the acoustic mode [51]. *Black circles* are the calculated values of the peak frequency [51] and dotted line presents expected frequency behavior including anharmonic and size effects. *Reprinted from H.K. Yadav, K. Sreenivas, R.S. Katiyar, V. Gupta, Softening behavior of acoustic phonon mode in ZnO nanoparticles: the effect of impurities and particle size variation with temperature, J. Raman Spectrosc. 42 (2011) 1620–1625 with the permission of John Wiley & Sons.*

increased particle growth and coalescence process. Evident anomalous softening in the temperature ranges 380–480 K and 500–580 K originated from the impurities and reaction byproducts present on the surface of ZnO nanoparticles, the presence of which was proved by thermogravimetric analysis and differential scanning calorimetry measurements [51].

6. Electron-phonon interaction

Investigation of the interaction between electrons and lattice vibrations in nanocrystalline materials is an important topic because it has substantial influence on electronic and thermal transport, optical properties, or superconductivity. RS is convenient method for studying the electron-phonon (e-ph) interaction and enables to quantitatively determine the electron-phonon coupling constant. In metallic and polar or heavily doped semiconductor nanomaterials, the interaction of phonons with the conduction electrons or electron-hole pairs reflects in the Raman spectra as the asymmetric mode shift and broadening. Asymmetric line shapes of the Raman phonon spectra originate from a Fano-type interference effect between the discrete one-phonon states and a continuum of electronic transitions and can be well described with the Fano line shape [52,53].

$$I(\omega) = \sum_i I_{0i} \frac{(\varepsilon_i + q_i)^2}{1 + \varepsilon_i^2} \tag{2.18}$$

where $\varepsilon_i = 2(\omega - \omega_i)/\Gamma_i$, I_{0i}, ω_i and Γ_i are the intensity, bare phonon frequency and linewidth of the ith phonon mode and q_i is the Fano asymmetric parameter for the ith mode. The parameter q controls the asymmetry of the Raman mode with respect to a standard Lorentzian profile, whereas the value $1/q$ serves as a measure of electron-phonon coupling strength. The increase of $1/q$ value indicates stronger coupling and in the limit $|1/q| \to 0$ the Fano profile reduces to the Lorentzian line shape [53].

The electron-phonon coupling strength can also be estimated from the Allen formula [54] which connects the linewidth γ_i of the ith phonon mode with electron-phonon coupling constant λ_i [54,55]

$$\lambda_i = \frac{2g_i\gamma_i}{\pi N_{\varepsilon_f}\omega_{bi}^2} \tag{2.19}$$

where N_{ε_f} is the electronic density of states at the Fermi surface per spin per unit cell, g_i is the mode degeneracy, and ω_{bi} is the bare phonon frequency in the absence of electron-phonon interaction. In the framework of Allen's theory, the electronic density of states N_{ε_f} can be estimated from the relation [55,56].

$$\gamma_i = -\frac{\pi}{2} N_{\varepsilon_f}\omega_{bi}\Delta\omega \tag{2.20}$$

where $\Delta\omega$ presents the difference between the bare phonon frequency and the observed frequency.

A good example of *e*-ph coupling in nanophase materials presents the Raman temperature study on $LiTi_2O_4$ superconducting thin films [57]. In Fig. 2.12A are presented temperature-dependent Raman spectra of three T_{2g} modes (342.2 cm^{-1}, 433.3 cm^{-1}, and 495.2 cm^{-1}) of $LiTi_2O_4$. All T_{2g} modes had asymmetric Fano line shape in the whole temperature range 5–300 K and were well fitted with the Fano function (full lines on Fig. 2.12A). The parameters q, obtained from the fitting of T_{2g} modes from Fig. 2.12A, are displayed in Fig. 2.12B. There is an obvious anomaly in the asymmetry parameters around the temperature of 50 K, where the negative to positive magnetoresistance transition takes place [57]. This anomaly was ascribed to the onset of other competing order like orbital-related state, which can suppress the *e*-ph coupling by modifying the electron density of states [57].

Electron-phonon interactions can be responsible for some anomalous behavior of phonon linewidths and intensities as a function of temperature. As an example, calculations from first principles [58] have proven that the anomalous decrease of E_{2g} (G band) phonon linewidth in graphene is dominated by *e*-ph interactions. In Fig. 2.13 is shown the total

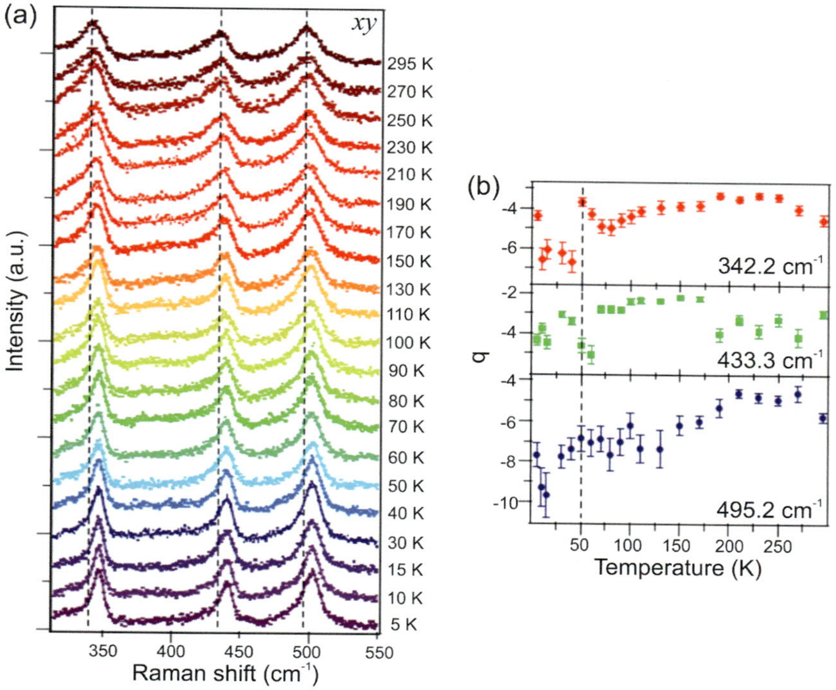

Figure 2.12

Temperature-dependent (A) Raman spectra of three T_{2g} modes of $LiTi_2O_4$ thin films and (B) Fano asymmetry factor q of the T_{2g} modes. *Reprinted figure with permission from D. Chen, Y.-L. Jia, T.-T. Zhang, Z. Fang, K. Jin, P. Richard, H. Ding, Raman study of electron-phonon coupling in thin films of the spinel oxide superconductor* $LiTi_2O_4$*, Phys. Rev. B 96 (2017) 094501. Copyright (2017) by the American Physical Society.*

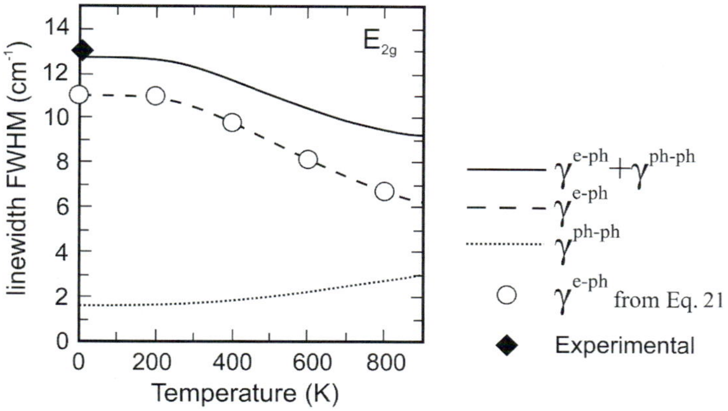

Figure 2.13

Calculated temperature dependence of the total E_{2g} mode linewidth of graphene (*full line*), *e*-ph (*dashed line*), and ph-ph (*dotted line*) contributions, from first principles together with experimental results and the results (*open circles*) based on Eq. (2.21). *Reprinted figure with permission from N. Bonini, M. Lazzeri, N. Marzari, F. Mauri, Phonon anharmonicities in graphite and graphene, Phys. Rev. B 99 (2007) 176802. Copyright (2007) by the American Physical Society.*

linewidth of E_{2g} mode (full line) computed from first principles, whereas dashed and dotted lines present *e*-ph and ph-ph contributions. These calculations are in very good agreement with respect to measurements [58] and have shown that the *e*-ph coupling has dominant role in the temperature dependence of E_{2g} mode linewidth. Furthermore, these calculations are compared with the results (open circles in Fig. 2.13) obtained from the simplified model for temperature dependence of E_{2g} phonon linewidth due to *e*-ph coupling [58,59].

$$\gamma^{e-ph}(T) = \gamma^{e-ph}(0)\left[f\left(-\frac{\hbar\omega_0}{2k_BT}\right) - f\left(\frac{\hbar\omega_0}{2k_BT}\right)\right] \tag{2.21}$$

where $f(x) = [\exp(x)+1]^{-1}$, k_B is the Boltzmann constant and $\hbar\omega_0$ is the E_{2g} phonon energy. As can be seen from Fig. 2.13, the simplified model very well reproduces the calculations for the linewidth behavior of E_{2g} phonon with temperature confirming the dominant role of *e*-ph coupling.

Another temperature-dependent Raman study on nanostructured cuprous oxide (Cu_2O) film (Fig. 2.14), revealed the anomalous temperature behavior of the LO components of

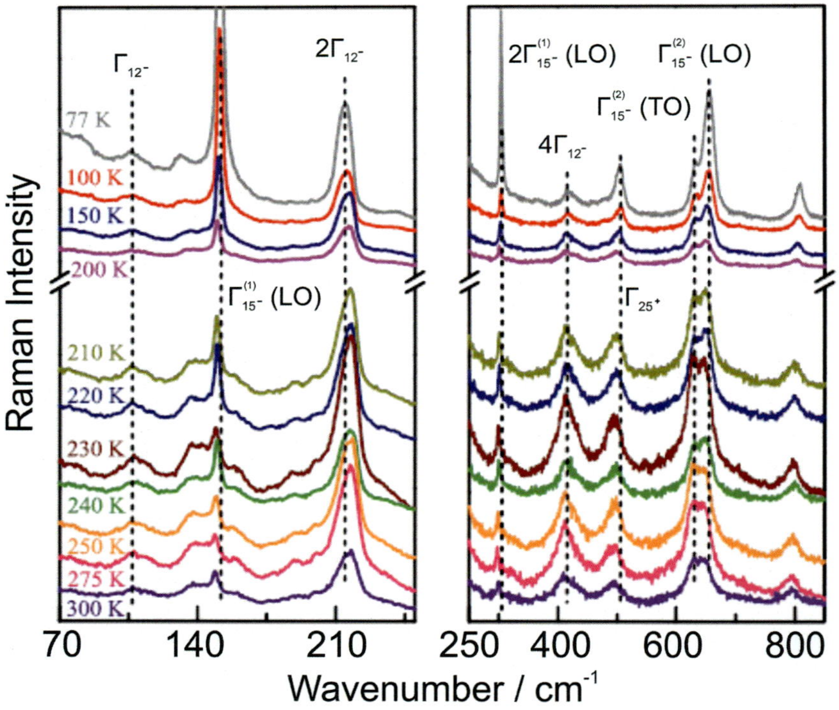

Figure 2.14

Temperature-dependent Raman spectra of Cu_2O film. *Dashed lines* present the guides to the eye of the Raman modes shift with temperature. *Reprinted from W. Yu, M. Han, K. Jiang, Z. Duan, Y. Li, Z. Hu, J. Chu, Enhanced Fröhlich interaction of semiconductor cuprous oxide films determined by temperature-dependent Raman scattering and spectral transmittance, J. Raman Spectrosc. 44 (2013) 142–146 with the permission of John Wiley & Sons.*

the infrared-active polar $\Gamma_{15-}^{(1)}$ and $\Gamma_{15-}^{(2)}$ modes [60]. Namely, an appearance of $\Gamma_{15-}^{(1)}$ and $\Gamma_{15-}^{(2)}$ modes at 151 and 655 cm^{-1} in the Raman spectra from Fig. 2.14 indicated the presence of strong *e*-ph coupling mediated by Fröhlich interaction. At lowest temperature of 77 K, these modes were sharp and of high intensity. With temperature increase, their intensity decreased and after 200 K intensities of both modes were significantly reduced. This study pointed to much stronger Fröhlich interaction at temperatures lower than 200 K. The larger TO-LO splitting of $\Gamma_{15-}^{(2)}$ mode at temperatures below 200 K imply that Fröhlich interaction is very sensitive to temperature [60].

It is worth to mention that Raman studies on nanocrystalline materials [9,61] suggest that the strength of *e*-ph interaction is dependent on nanocrystalline size. Therefore, analyzing the temperature-dependent Raman spectra of nanostructured materials it is important to correctly estimate the contributions from size effects, anharmonicity and *e*-ph coupling on the Raman line profile.

7. Electromagnons in cycloidal multiferroic nanostructures

Raman scattering is an efficient tool for studying the mutual and complex coupling between the magnetic (magnons) and lattice (phonons) excitations. One such example are the single-magnon spin waves with an electric dipole activity, called electromagnons, which can be excited by the electric component of electromagnetic waves and are consequently much stronger coupled to the light than ordinary magnons which interact through magnetic dipole excitations. Up to date, electromagnons were registered in a variety of crystalline antiferromagnets like $CuFe_{1-x}Ga_xO_2$, [62], garnets, [63], multiferroics like RMn_2O_5 (R = Y, Tb, Eu), [64], $Ba_2CoGe_2O_7$, [65], manganites [66].

The presence of electromagnons was first discovered in the low-temperature Raman spectra of cycloidal multiferroic $BiFeO_3$ single crystals [67,68] and recently in $BiFeO_3$ thin films [69]. In Fig. 2.15A is presented the electromagnon Raman spectra of $BiFeO_3$ crystal measured at 7 K in parallel and crossed polarizations in (010) plane.

Two types of electromagnon excitations were registered in the Raman spectra of $BiFeO_3$ crystal, which lie in (cyclon modes, φ) and out (extra-cyclon modes, ψ) of (-12-1) cycloidal plane [67]. The cycloidal plane is formed by the vector of spontaneous polarization P along [111] direction and cycloid propagation defined by the wavevector q along [10-1] direction, as sketched in Fig. 2.15B. These low-energy Raman modes show sudden increase of intensity at temperatures close to the spin reorientation phase transition (140 K) which is additional confirmation of their magnetic nature [67].

In the framework of simple Landau-Ginzburg model for cycloidal multiferroics, Sousa and Moore [70] have shown that the modes which propagate along the cycloidal plane

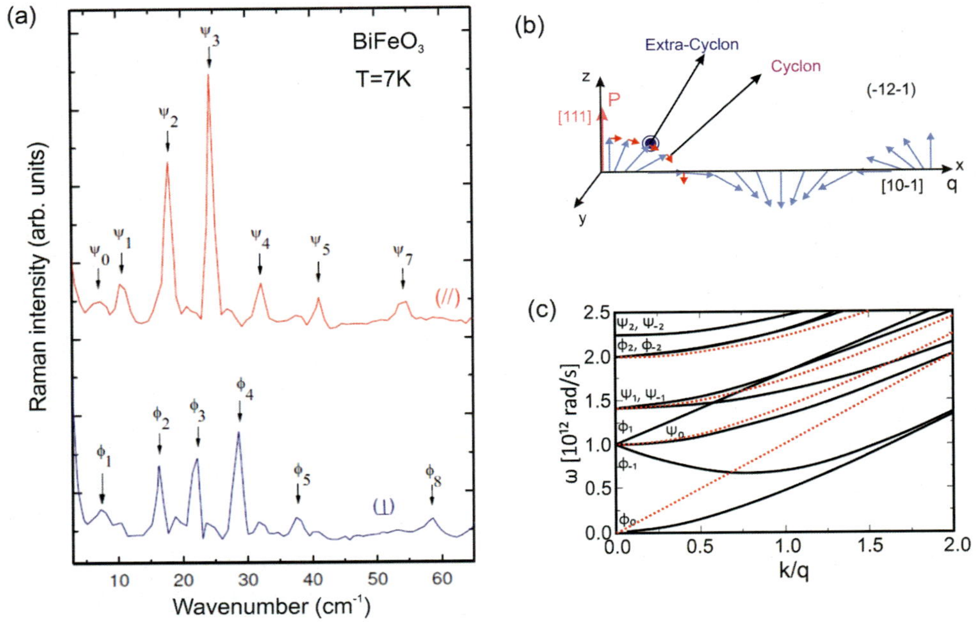

Figure 2.15

(A) Electromagnon Raman spectra of $BiFeO_3$ single crystal obtained using parallel (//) and crossed (⊥) polarizations in the (010) plane; (B) schematic representation of cycloidal magnetic order in $BiFeO_3$ and (C) dispersion curves of cyclon (φ) and extra-cyclon (ψ) modes in the direction perpendicular to (-12-1) cycloidal plane (*full lines*), and along [111] direction (*dashed lines*). *Reprinted figure with permission from M. Cazayous, Y. Gallais, A. Sacuto, R. de Sousa, D. Lebeugle, D. Colson, Possible observation of cycloidal electromagnons in* $BiFeO_3$*, Phys. Rev. Lett. 101 (2008) 037601. Copyright (2008) by the American Physical Society, R. De Sousa, J.E. Moore, Optical coupling to spin waves in the cycloidal multiferroic* $BiFeO_3$*, Phys. Rev. B 77 (2008) 012406. Copyright (2008) by the American Physical Society.*

$(k_y = 0)$, are simple plane waves with different dispersions depending on their coupling to electrical polarization

$$\omega'^2 = c\tilde{k}^2 \ (\varphi \text{ modes}) \tag{2.22}$$

$$\omega'^2 = c\left(\tilde{k}^2 + q^2\right) (\Psi \text{ modes}) \tag{2.23}$$

with $\tilde{k} = (k_x + nq)^2 + k_y^2 + k_z^2$ and n as integer. The propagation along the k_y direction leads to the appearance of small gaps in the propagation frequency due to the pinning of the cycloidal plane by the electrical polarization. The numerical solutions of spin waves propagation [70] in the direction perpendicular to (-12-1) cycloidal plane (full lines) and along [111] direction (dashed lines) are shown in Fig. 2.15C.

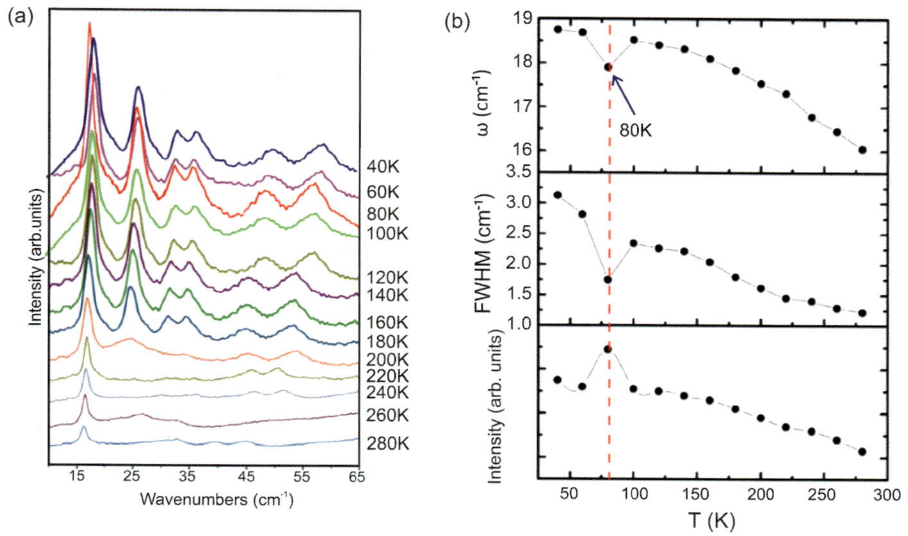

Figure 2.16
Temperature dependence of (A) electromagnon spectra of $BiFeO_3$ thin films in normal configuration (B) wavenumber, linewidth (FWHM), and intensity of electromagnon mode at 18.75 cm^{-1}. *Reprinted from W. Azeem, S. Riaz, A. Bukhtiar, S.S. Hussain, Y. Xu, S. Naseem, Ferromagnetic ordering and electromagnons in microwave synthesized $BiFeO_3$ thin films, J. Magnet. Magnet. Mater. 475 (2019) 60–69, Copyright (2019) with permission from Elsevier.*

Recently, electromagnons have been found in the low-frequency Raman spectra of $BiFeO_3$ thin films [69]. The temperature-dependent electromagnon spectra is shown in Fig. 2.16A. Two intense modes (at 18.75 cm^{-1} and 25.8 cm^{-1}) together with four other modes of lower intensity were observed at 40 K. These two low-energy electromagnon modes are strongly affected by temperature. In Fig. 2.16B is presented the temperature variation of the wavenumber, linewidth and intensity of 18.75 cm^{-1} electromagnon mode. As can be seen, neither frequency, linewidth nor intensity exhibit monotonic behavior. Instead, there is an anomaly at around 80 K. Such a behavior can be correlated with changes in the cycloid spin arrangements at low temperatures like spin reorientation transition in $BiFeO_3$ single crystal [68] or in orthoferrites like $TmFeO_3$ [71].

From these studies, it can be concluded that RS is a suitable optical method not only for detection of electromagnons, but for revealing more about the anharmonicity in magnetic order of cycloidal multiferroic nanostructures.

8. Spin-phonon interaction

RS is a useful tool to elucidate spin-phonon (*s*-ph) coupling mechanism, because the Raman modes in magnetic materials can be sensitive to the magnetic ordering.

Spin-phonon interaction manifests as nontypical temperature dependence of optical phonon frequency, linewidth and integrated intensity, since all these phonon features can be influenced by the exchange coupling between magnetic ions at and below the temperatures of magnetic phase transitions. It is expected that the coupling between the lattice and spin degrees of freedom (spin-phonon coupling) is different for different phonon modes because the magnetic interactions can be complex and the spin-phonon coupling constant may vary even in the case of the same spin-spin interaction [72].

In magnetic materials, the change of a jth phonon mode frequency with temperature can be expressed as:

$$\omega_j(T) - \omega_j(T_0) = \Delta\omega_j(T) = \Delta\omega_{\text{latt}} + \Delta\omega_{\text{anh}} + \Delta\omega_{e-\text{ph}} + \Delta\omega_{s-\text{ph}} \quad (2.24)$$

The first three terms present the change of the phonon frequency due to the lattice expansion/contraction, intrinsic anharmonicity and electron-phonon coupling, whereas the fourth term accounts for the effect of the s-ph contribution. The frequency shift due to s-ph coupling can be expressed as [73,74].

$$\Delta\omega_{s-\text{ph}} = \omega - \omega_0 = -\lambda_{s-\text{ph}}\langle S_i S_j \rangle \quad (2.25)$$

where ω_0 is the eigenfrequency in the absence of s-ph coupling; $S_i S_j$ denotes the spin-spin correlation function for adjacent spins localized at the ith and jth sites and $\lambda_{s-\text{ph}}$ is the s-ph coupling constant which is different for each phonon and may have positive or negative sign [73]. From the mean field theory and considering nearest neighbor interaction it follows that

$$\langle S_i S_j \rangle = S^2 \phi(T) \quad (2.26)$$

where $\phi(T)$ is the normalized short-range order parameter which is estimated from the mean field theory [73] and defined as $\phi(T) = |S_i S_j|/S^2$. According to the mean field theory, $\phi(T)$ decreases with temperature from 1 at $T = 0$ K and falling to 0 at temperature of magnetic phase transition (either T_N or T_C). The $\Delta\omega_{s\text{-ph}}$ can be now expressed as

$$\Delta\omega_{s-\text{ph}} = -\lambda_{s-\text{ph}} S^2 \phi(T) \quad (2.27)$$

On the other hand, according to the mean field approximation the $S_i S_j \propto \left(\frac{M(T)}{M_S}\right)^2$, where $M(T)$ is the temperature-dependent average magnetization per magnetic ion and M_S is the saturation magnetization [75—77]. Therefore, the shift of the Raman mode due to s-ph coupling is $\omega_{s-\text{ph}} \propto \left(\frac{M(T)}{M_S}\right)^2$. In antiferromagnetic (AF) and ferromagnetic (FM) materials spin correlation function $S_i S_{i+1}$ approaches zero for $T > T_N$, T_C and the spin-phonon coupling usually terminates in the paramagnetic phase.

Raman spectroscopy has been utilized to study spin-phonon coupling in various magnetic nanomaterials [11,78–80] through the temperature-dependent variation of the one-phonon or two-phonon Raman spectra. CuO nanowires are a good example to study spin-phonon interaction and the effect of size decrease on the spin-phonon coupling constant [11]. In Fig. 2.17A are presented Raman spectra of in-plane CuO nanowires, with the mean diameter of 120 nm, at different temperatures across the Néel temperature, T_N (~143 K).

Two well-defined modes at around 300 and 348 cm^{-1} of A_g and B_g^1 symmetries were registered at 193 K. With temperature decrease an additional mode at ~231 cm^{-1}

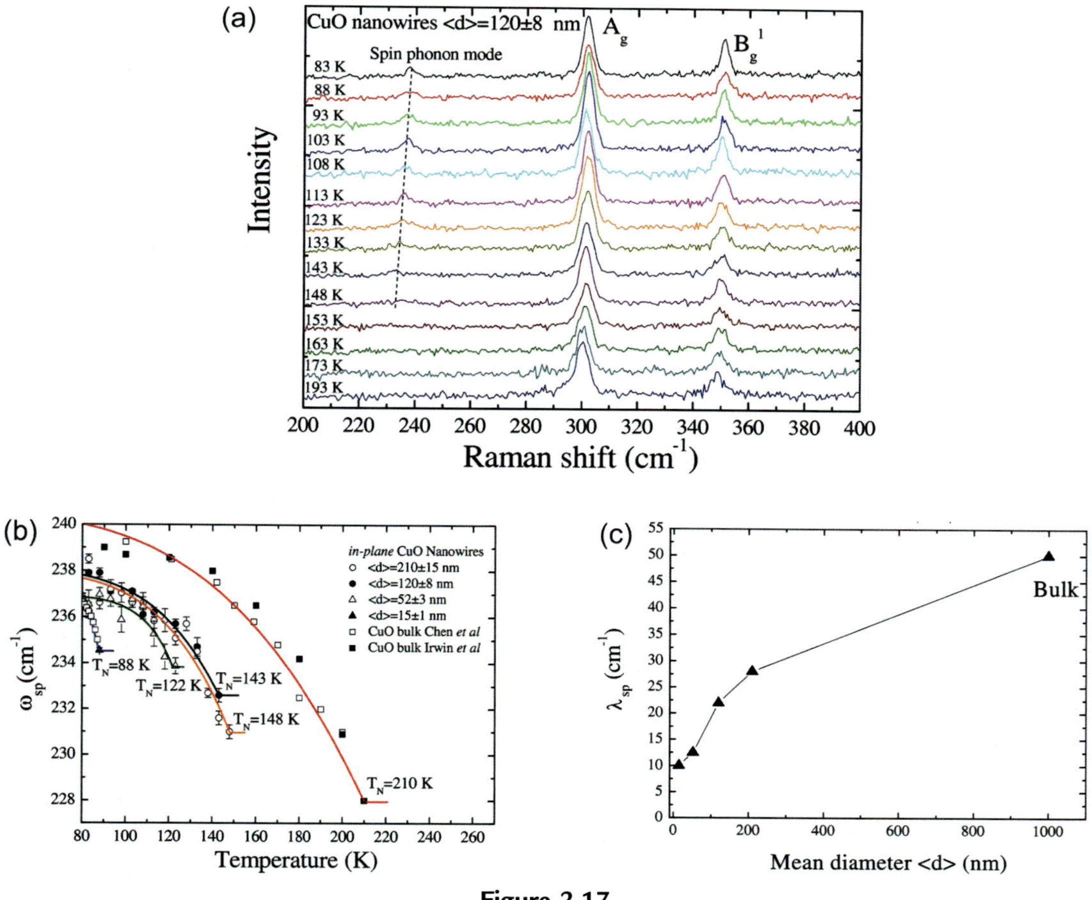

Figure 2.17

(A) Raman spectra of CuO nanowires with mean diameter $<d> = 120 \pm 8$ nm at various temperatures across the T_N; (B) temperature variation of spin-phonon mode frequency and (C) change of spin-phonon coupling constant on nanowires mean diameter. *Reprinted from P.-H. Shih, C.-L. Cheng, S.Y. Wu, Short-range spin-phonon coupling in in-plane CuO nanowires: a low-temperature Raman investigation, Nanoscale Res. Lett. 8 (398) (2013) 1–6 with the permission of SpringerNature.*

appeared. This mode is ascribed to the zone-folded LA phonon from the Z′ point of the Brillouin-zone boundary, which can appear in the Raman spectra below T_N as a consequence of the onset of long-range AF magnetic order and phonon modulation of the exchange constant [72]. This mode exhibits unusual variation both in intensity and frequency with decreasing temperature below T_N. Namely, intensity of this mode abruptly increases, whereas the frequency significantly hardens below T_N. This anomalous behavior results from a strong spin-phonon interaction [72]. The temperature variation of spin-phonon mode frequency for CuO nanowires of different mean diameter and bulk counterpart is shown Fig. 2.17B. The spin-phonon mode temperature variation can be well described by Eq. (2.27) (solid lines) for $S^2 \sim 1/4$[73] using empirical formula for $\phi(T) = 1-(T/T_N)^{\gamma}$ (solid lines) [79]. In such a way T_N and spin-phonon coupling constant $\lambda_{s\text{-ph}}$ can be estimated. As can be seen from Fig. 2.17B and C, with decreasing diameter, T_N and $\lambda_{s\text{-ph}}$ tend to decrease. This implies that the size effect weakens the strong spin-phonon interaction, favorizing the short-range AF ordering [11].

Another example of strong spin-phonon interaction was reported in multiferroic $BiFeO_3$ thin films [80] evidenced through the temperature behavior of the two-phonon Raman modes, $2A_4$, $2E_8$ and $2E_9$. These modes are positioned at around 968, 1110, and 1265 cm^{-1} at RT. Upon heating, as shown in Fig. 2.18A, all Raman modes slightly shift to

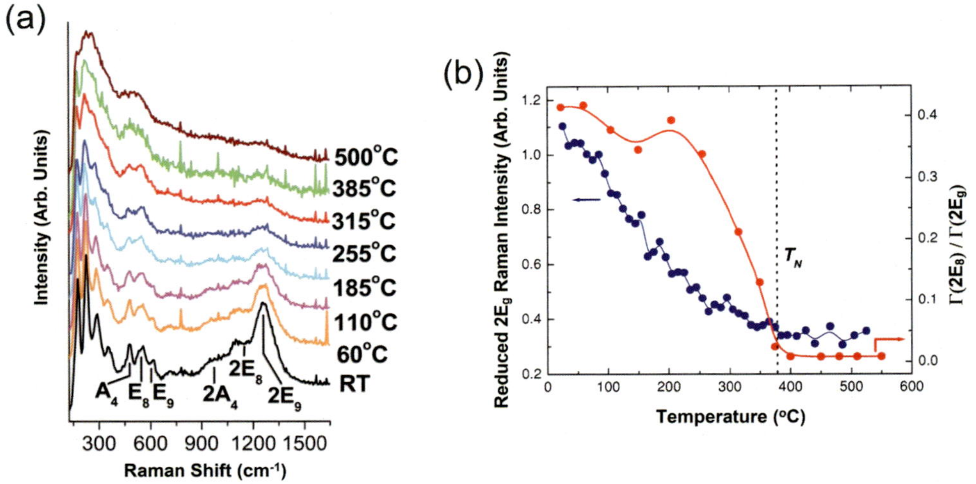

Figure 2.18

(A) Temperature dependence of unpolarized Raman spectra of $BiFeO_3$ film and (B) reduced integrated intensity of 1260 cm^{-1} two-phonon Raman mode (*blue dots*). Temperature dependence of the integrated intensity ratio between 2*E*8 and 2*E*9 two-phonon modes is presented with *red circles*. The *solid lines* are guide to the eye. *Reprinted from M.O. Ramirez, M. Krishnamurthi, S. Denev, A. Kumar, S.-Y. Yang, Y.-H. Chu, E. Saiz, J. Seidel, A.P. Pyatakov, A. Bush, D. Viehland, J. Orenstein, R. Ramesh, V. Gopalan, Two-phonon coupling to the antiferromagnetic phase transition in multiferroic BiFeO₃, Appl. Phys. Lett. 92 (2008) 022511 with the permission of AIP Publishing.*

lower frequency and broaden due to anharmonicity, but the integrated intensity of two-phonon modes drastically decrease, particularly of two-phonon mode $2E_9$. Above the Néel temperature, this mode almost disappeared. Eliminating the contribution from Bose-Einstein population, the reduced intensity of $2E_9$ mode is presented in Fig. 2.18B (blue circles) and compared to the reduced intensity of $2E_8$ mode (red circles). As can be seen, a remarkable decrease of the integrated intensity of $2E_9$ mode is seen when approaching T_N, and above the temperature of magnetic phase transition it is almost constant. The integrated intensity ratio, $\Gamma(2E_8)/\Gamma(2E_9)$, is almost constant up to 200 °C and abruptly decreases in the vicinity of Néel temperature, indicating strong spin-two-phonon coupling in $BiFeO_3$ [80].

9. Summary

This chapter presents a brief review how temperature-dependent RS comes to be of significant importance for investigating various phenomena in nanostructured materials. Raman spectra of diverse nanomaterials in the form of metal-oxide nanopowders, quantum dots, nanowires, nanoribbons, carbon-based, and multiferroic nanostructures were analyzed as examples.

It was shown that the application of RS at elevated temperatures is a very convenient method to study the anharmonic interactions through the evolution of optical phonon modes in nanophase materials. The temperature evolution of the Raman modes can be successfully modeled by PCM which incorporates size, strain, and anharmonic effects and enables to estimate the contribution of each of these effects. It was shown that the size, morphology, and size distribution largely determine which of the anharmonic processes (three or four-phonon anharmonic interaction) will be dominant, pointing out that phonon-phonon interactions can be different in nanomaterials compared to their bulk counterpart. Low-frequency RS provided the information about the acoustic phonons dynamics as a function of temperature in nanocrystalline solids as well. It was demonstrated that the decay dynamics of the acoustic phonons was complex and dependent not only on different size and size distribution, i.e., coalescence process at higher temperatures, but also on organic ligands with different elastic modulus and impurities bound to the nanoparticle surface.

Temperature-dependent Raman scattering has a potential to detect the formation of other structural phases, present even in small concentrations, through the appearance of new peaks characteristic for a certain crystalline structure. As a noncontact method of high spatial resolution, RS can be used in thermometry for the measurements of the local temperature or thermal conductivity of nanomaterials.

RS is a suitable optical method for probing lattice and charge excitations, as well as interplay between them. In the Raman spectra of low-dimensional materials electron-phonon interaction manifests itself by an asymmetric Fano line shape of the phonon modes. The electron-phonon coupling strength can be estimated directly from the fitting of the Raman spectra at different temperatures with the Fano function or from the linewidth of the phonon modes using Allen's formula. Besides, in nanophase materials, the electron-phonon interaction is sensitive to temperature and nanocrystallite size.

Special class of nanomaterials present magnetic and multiferroic nanomaterials. In these materials, it was shown that phonon frequency, linewidth, and integrated intensity may all be influenced by the exchange coupling between magnetic ions and from their temperature dependence it is possible to investigate both spin-ordering and spin-reorientation transitions, revealing more about the complexity of magnetic interactions in nanophase materials.

In summary, from the temperature induced changes of the Raman features (peak position, intensity, or linewidth) it is possible to predict many new thermophysical, electric, and magnetic nanomaterials properties.

Acknowledgment

The authors acknowledge funding by the Institute of Physics Belgrade, through the grant by the Serbian Ministry of Education, Science and Technological Development.

References

[1] C.S.S.R. Kumar, Raman Spectroscopy for Nanomaterials Characterization, Springer-Verlag Berlin Heidelberg, 2012.
[2] H. Richter, Z.P. Wang, L. Ley, The one phonon Raman spectrum in microcrystalline silicon, Solid State Commun. 39 (1981) 625–629.
[3] I.H. Campbell, P.M. Fauchet, The effects of microcrystal size and shape on the one phonon Raman spectra of crystalline semiconductors, Solid State Commun. 58 (1986) 739–741.
[4] L. Donetti, F. Gámiz, N. Rodriguez, F. Jimenez, C. Sampedro, Influence of acoustic phonon confinement on electron mobility in ultrathin silicon on insulator layers, Appl. Phys. Lett. 88 (2006) 122108.
[5] A.C. Gandhi, S.S. Gaikwad, J.-C. Peng, C.-W. Wang, T.S. Chan, S.Y. Wu, Strong electron-phonon coupling in superconducting bismuth nanoparticles, APL Mater. 7 (2019) 031111.
[6] G.S.N. Eliel, M.V.O. Moutinho, A.C. Gadelha, A. Righi, L.C. Campos, H.B. Ribeiro, P.-W. Chiu, K. Watanabe, T. Taniguchi, P. Puech, M. Paillet, T. Michel, P. Venezuela, M.A. Pimenta, Intralayer and interlayer electron–phonon interactions in twisted graphene heterostructures, Nat. Commun. 9 (2018) 1221.
[7] H. Rijckaert, P. Cayado, R. Nast, J.D. Sierra, M. Erbe, P.L. Dominguez, J. Hänisch, K. De Buysser, B. Holzapfel, I. Van Driessche, Superconducting HfO_2-$YBa_2Cu_3O_{7-\delta}$ nanocomposite films deposited using ink-jet printing of colloidal solutions, Coatings 10 (2020) 17.
[8] C.Y. Tsai, H.M. Cheng, H.R. Chen, K.F. Huang, L.N. Tsai, Y.H. Chu, C.H. Lai, W.F. Hsieh, Spin and phonon anomalies in epitaxial self-assembled $CoFe_2O_4$-$BaTiO_3$ multiferroic nanostructures, Appl. Phys. Lett. 104 (2014) 252905.

[9] D.M. Sagar, J.M. Atkin, P.K.B. Palomaki, N.R. Neale, J.L. Blackburn, J.C. Johnson, A.J. Nozik, M.B. Raschke, M.C. Beard, Quantum confined electron-phonon interaction in silicon nanocrystals, Nano Lett. 15 (2015) 1511−1516.
[10] D. Paramanik, S. Varma, Raman scattering characterization and electron phonon coupling strength for MeV implanted InP (111), J. Appl. Phys. 101 (2007) 023528.
[11] P.-H. Shih, C.-L. Cheng, S.Y. Wu, Short-range spin-phonon coupling in in-plane CuO nanowires: a low-temperature Raman investigation, Nanoscale Res. Lett. 8 (2013) 398.
[12] G.S. Doerk, C. Carraro, R. Maboudian, Temperature dependence of Raman spectra for individual silicon nanowires, Phys. Rev. B 80 (2009) 073306.
[13] M. Kazan, S. Pereira, M.R. Correia, P. Masri, Contribution of the decay of optical phonons into acoustic phonons to the thermal conductivity of AlN, Phys. Rev. B 77 (2008) 180302.
[14] J. Kulda, A. Debernardi, M. Cardona, F. de Geuser, E.E. Haller, Self-energy of zone-boundary phonons in germanium: ab initio calculations versus neutron spin-echo measurements, Phys. Rev. B 69 (2004) 045209.
[15] Z. Dohčević-Mitrović, Z.V. Popović, M. Šćepanović, Anharmonicity effects in nanocrystals studied by Raman scattering spectroscopy, Acta Phys. Pol. A 116 (2009) 36−41.
[16] G. Lucazeau, Effect of pressure and temperature on Raman spectra of solids: anharmonicity, J. Raman Spectrosc. 34 (2003) 478−496.
[17] A.A. Maradudin, A.E. Fein, Scattering of neutrons by an anharmonic crystal, Phys. Rev. B 128 (1962) 2589−2608.
[18] R.P. Lowndes, Anharmonic self-energy of a soft mode, Phys. Rev. Lett. 27 (1971) 1134−1136.
[19] M. Balkanski, R.F. Wallis, E. Haro, Anharmonic effects in light scattering due to optical phonons in silicon, Phys. Rev. B 28 (1983) 1928−1934.
[20] G. Morell, W. Pérez, E. Ching-Prado, R.S. Katiyar, Anharmonic interactions in beryllium oxide, Phys. Rev. B 53 (1996) 5388−5395.
[21] P.G. Klemens, Anharmonic decay of optical phonons, Phys. Rev. 148 (1966) 845−848.
[22] T.R. Hart, R.L. Aggarwal, B. Lax, Temperature dependence of Raman scattering in silicon, Phys. Rev. B 1 (1970) 638−642.
[23] Z. Dohčević-Mitrović, M. Grujić-Brojčin, M. Šćepanović, Z.V. Popović, S. Bošković, B. Matović, M. Zinkevich, F. Aldinger, The size and strain effects on the Raman spectra of $Ce_{1-x}Nd_xO_{2-\delta}$ ($0 \leq x \leq 0.25$) nanopowders, Solid State Commun. 137 (2006) 387−390.
[24] Z. Dohčević-Mitrović, M. Grujić-Brojčin, M. Šćepanović, Z.V. Popović, S. Bošković, B. Matović, M. Zinkevich, F. Aldinger, $Ce_{1-x}Y(Nd)_xO_{2-\delta}$ nanopowders: potential materials for intermediate temperature solid oxide fuel cells, J. Phys. Condens. Matter 18 (2006) S2061−S2068.
[25] Z.D. Dohčević-Mitrović, M. Radović, M. Šćepanović, M. Grujić-Brojčin, Z.V. Popović, B. Matović, S. Bošković, Temperature-dependent Raman study of $Ce_{0.75}Nd_{0.25}O_{2-\delta}$ nanocrystals, Appl. Phys. Lett. 91 (2007) 203118.
[26] J. Spanier, R. Robinson, F. Zhang, S.-W. Chan, I. Herman, Size-dependent properties of CeO_{2-y} nanoparticles as studied by Raman scattering, Phys. Rev. B 64 (2001) 245407.
[27] S. Askrabić, Z.D. Dohcević-Mitrović, M. Radović, M. Sćepanović, Z.V. Popović, Phonon-phonon interactions in $Ce_{0.85}Gd_{0.15}O_{2-\delta}$ nanocrystals studied by Raman spectroscopy, J. Raman Spectrosc. 40 (2009) 650−655.
[28] H.K. Yadav, R.S. Katiyar, V. Gupta, Temperature dependent dynamics of ZnO nanoparticles probed by Raman scattering: a big divergence in the functional areas of nanoparticles and bulk material, Appl. Phys. Lett. 100 (2012) 051906.
[29] Y.L. Du, Y. Deng, M.S. Zhang, Variable-temperature Raman scattering study on anatase titanium dioxide nanocrystals, J. Phys. Chem. Solid. 67 (2006) 2405−2408.
[30] K. Gao, Strong anharmonicity and phonon confinement on the lowest-frequency Raman mode of nanocrystalline anatase TiO_2, Phys. Status Solidi 244 (2007) 2597−2604.

[31] M.J. Šćepanović, M. Grujić-Brojčin, Z.D. Dohčević-Mitrović, Z.V. Popović, Temperature dependence of the lowest frequency E_g Raman mode in laser-synthesized anatase TiO_2 nanopowder, Appl. Phys. A 86 (2007) 365–371.
[32] G.R. Hearne, J. Zhao, A.M. Dawe, V. Pischedda, M. Maaza, M.K. Nieuwoudt, P. Kibasomba, O. Nemraoui, J.D. Comins, M.J. Witcomb, Effect of grain size on structural transitions in anatase TiO_2: a Raman spectroscopy study at high pressure, Phys. Rev. B 70 (2004) 134102.
[33] S. Aškrabić, Z. Dohčević-Mitrović, A. Kremenović, N. Lazarević, V. Kahlenberg, Z.V. Popović, Oxygen vacancy-induced microstructural changes of annealed CeO_{2-x} nanocrystals, J. Raman Spectrosc. 43 (2012) 76–81.
[34] J.V. Silveira, L.L. Vieira, J.M. Filho, A.J.C. Sampaio, O.L. Alvesc, A.G. Souza Filho, Temperature-dependent Raman spectroscopy study in MoO_3 nanoribbons, J. Raman Spectrosc. 43 (2012) 1407–1412.
[35] J.M. Todorović, Z.D. Dohčević-Mitrović, D.M. Đokić, D. Mihailović, Z.V. Popović, Investigation of thermostability and phonon-phonon interactions in $Mo_6S_3I_6$ nanowires by Raman scattering spectroscopy, J. Raman Spectrosc. 41 (2010) 978–982.
[36] M. Testa-Anta, M.A. Ramos-Docampo, M. Comesaña-Hermo, B. Rivas-Murias, V. Salgueiriño, Raman spectroscopy to unravel the magnetic properties of iron oxide nanocrystals for bio-related applications, Nanoscale Adv. 1 (2019) 2086–2103.
[37] M. Sendova, B.D. Hosterman, R. Raud, T. Hartmann, D. Koury, Temperature-dependent, micro-Raman spectroscopic study of barium titanate nanoparticles, J. Raman Spectrosc. 46 (2015) 25–31.
[38] L. Chen, K. Rickey, Q. Zhao, C. Robinson, X. Ruan, Effects of nanocrystal shape and size on the temperature sensitivity in Raman thermometry, Appl. Phys. Lett. 103 (2013) 083107.
[39] A.A. Balandin, S. Ghosh, W.Z. Bao, I. Calizo, D. Teweldebrhan, F. Miao, C.N. Lau, Superior thermal conductivity of single-layer graphene, Nano Lett. 8 (2008) 902–907.
[40] Q. Li, C. Liu, X. Wang, S. Fan, Measuring the thermal conductivity of individual carbon nanotubes by the Raman shift method, Nanotechnology 20 (2009) 145702.
[41] S. Ghosh, I. Calizo, D. Teweldebrhan, E.P. Pokatilov, D.L. Nika, A.A. Balandin, W. Bao, F. Miao, C.N. Lau, Extremely high thermal conductivity of graphene: prospects for thermal management applications in nanoelectronic circuits, Appl. Phys. Lett. 92 (2008) 151911.
[42] M. Fujii, T. Nagareda, S. Hayashi, K. Yamamoto, Low-frequency Raman scattering from small silver particles embedded in SiO_2 thin films, Phys. Rev. B 44 (1991) 6243–6248.
[43] L. Saviot, B. Champagnon, E. Duval, A.I. Ekimov, I.A. Kudriavtsev, Size dependence of acoustic and optical vibrational modes of CdSe nanocrystals in glasses, J. Non-Cryst. Solids 197 (1996) 238–246.
[44] A. Diéguez, A. Romano-Rodríguez, Nondestructive assessment of the grain size distribution of SnO nanoparticles by low-frequency Raman spectroscopy, Appl. Phys. Lett. 71 (1997) 1957–1959.
[45] M. Ivanda, K. Furić, S. Musić, M. Ristić, M. Gotić, D. Ristić, A.M. Tonejc, I. Djerdj, M. Mattarelli, M. Montagna, F. Rossi, M. Ferrari, A. Chiasera, Y. Jestin, G.C. Righini, W. Kiefer, R.R. Gonçalves, Low wavenumber Raman scattering of nanoparticles and nanocomposite materials, J. Raman Spectrosc. 38 (2007) 647–659.
[46] E. Duval, Far-infrared and Raman vibrational transitions of a solid sphere: selection rules, Phys. Rev. B 46 (1992) 5795–5797.
[47] M. Montagna, R. Dusi, Raman scattering from small spherical particles, Phys. Rev. B 52 (1995) 10080–10089.
[48] H. Lamb, On the vibrations of an elastic sphere, Proc. Lond. Math. Soc. 13 (1882) 189–212.
[49] R. Shuker, W. Gammon, Raman-scattering selection-rule breaking and the density of states in amorphous materials, Phys. Rev. Lett. 25 (1970) 222–225.
[50] A.J. Mork, E.M.Y. Lee, W.A. Tisdale, Temperature dependence of acoustic vibrations of CdSe and CdSe-CdS core-shell nanocrystals measured by low-frequency Raman spectroscopy, Phys. Chem. Chem. Phys. 18 (2016) 28797–28801.

[51] H.K. Yadav, K. Sreenivas, R.S. Katiyar, V. Gupta, Softening behavior of acoustic phonon mode in ZnO nanoparticles: the effect of impurities and particle size variation with temperature, J. Raman Spectrosc. 42 (2011) 1620–1625.
[52] U. Fano, Effects of configuration interaction on intensities and phase shifts, Phys. Rev. 124 (1961) 1866–1878.
[53] W.-L. Zhang, H. Li, D. Xia, H.W. Liu, Y.-G. Shi, J.L. Luo, J. Hu, P. Richard, H. Ding, Observation of a Raman-active phonon with Fano line shape in the quasi-one-dimensional superconductor $K_2Cr_3As_3$, Phys. Rev. B 92 (2015) 060502.
[54] P.B. Allen, Neutron spectroscopy of superconductors, Phys. Rev. B 6 (1972) 2577–2579.
[55] J. Winter, H. Kuzmany, Landau damping and lifting of vibrational degeneracy in metallic potassium fulleride, Phys. Rev. B 53 (1996) 655–661.
[56] Z.V. Popović, Z.D. Dohčević-Mitrović, N. Paunović, M. Radović, Evidence of charge delocalization in $Ce_{1-x}Fe_x^{2+(3+)}O_{2-y}$ nanocrystals (x=0, 0.06, 0.12), Phys. Rev. B 85 (2012) 014302.
[57] D. Chen, Y.-L. Jia, T.-T. Zhang, Z. Fang, K. Jin, P. Richard, H. Ding, Raman study of electron-phonon coupling in thin films of the spinel oxide superconductor $LiTi_2O_4$, Phys. Rev. B 96 (2017) 094501.
[58] N. Bonini, M. Lazzeri, N. Marzari, F. Mauri, Phonon Anharmonicities in graphite and graphene, Phys. Rev. B 99 (2007) 176802.
[59] M. Lazzeri, S. Piscanec, F. Mauri, A.C. Ferrari, J. Robertson, Phonon linewidths and electron-phonon coupling in graphite and nanotubes, Phys. Rev. Lett. 73 (2006) 155426.
[60] W. Yu, M. Han, K. Jiang, Z. Duan, Y. Li, Z. Hu, J. Chu, Enhanced Fröhlich interaction of semiconductor cuprous oxide films determined by temperature-dependent Raman scattering and spectral transmittance, J. Raman Spectrosc. 44 (2013) 142–146.
[61] H.-M. Cheng, K.-F. Lin, H.-C. Hsu, C.-J. Lin, L.-J. Lin, W.-F. Hsieh, Enhanced resonant Raman scattering and electron-phonon coupling from self-assembled secondary ZnO nanoparticles, J. Phys. Chem. B 109 (2005) 18385–18390.
[62] S. Seki, N. Kida, S. Kumakura, R. Shimano, Y. Tokura, Electromagnons in the spin collinear state of a triangular lattice antiferromagnet, Phys. Rev. Lett. 105 (2010) 097207.
[63] D. Rogers, Y.J. Choi, E.C. Standard, T.D. Kang, K.H. Ahn, A. Dubroka, P. Marsik, C. Wang, C. Bernhard, S. Park, S.-W. Cheong, M. Kotelyanskii, A.A. Sirenko, Adjusted oscillator strength matching for hybrid magnetic and electric excitations in $Dy_3Fe_5O_{12}$ garnet, Phys. Rev. B 83 (2011) 174407.
[64] A.B. Sushkov, R. Valdés Aguilar, S. Park, S.-W. Cheong, H.D. Drew, Electromagnons in multiferroic YMn_2O_5 and $TbMn_2O_5$, Phys. Rev. Lett. 98 (2007) 027202.
[65] I. Kézsmárki, N. Kida, H. Murakawa, S. Bordács, Y. Onose, Y. Tokura, Enhanced directional dichroism of terahertz light in resonance with magnetic excitations of the multiferroic $Ba_2CoGe_2O_7$ oxide compound, Phys. Rev. Lett. 106 (2011) 057403.
[66] A. Pimenov, A.A. Mukhin, V.Y. Ivanov, V.D. Travkin, A.M. Balbashov, A. Loidl, Possible evidence for electromagnons in multiferroic manganites, Nat. Phys. 2 (2006) 97–100.
[67] M. Cazayous, Y. Gallais, A. Sacuto, R. de Sousa, D. Lebeugle, D. Colson, Possible observation of cycloidal electromagnons in $BiFeO_3$, Phys. Rev. Lett. 101 (2008) 037601.
[68] M.K. Singh, R.S. Katiyar, J.F. Scott, New magnetic phase transitions in $BiFeO_3$, J. Phys. Condens. Matter 20 (2008) 252203.
[69] W. Azeem, S. Riaz, A. Bukhtiar, S.S. Hussain, Y. Xu, S. Naseem, Ferromagnetic ordering and electromagnons in microwave synthesized $BiFeO_3$ thin films, J. Magn. Magn Mater. 475 (2019) 60–69.
[70] R. De Sousa, J.E. Moore, Optical coupling to spin waves in the cycloidal multiferroic $BiFeO_3$, Phys. Rev. B 77 (2008) 012406.
[71] S. Venugopalan, M. Dutta, A.K. Ramdas, J.P. Remeika, Magnetic and vibrational excitations in rare-earth orthoferrites: a Raman scattering study, Phys. Rev. B 31 (1985) 1490–1497.
[72] X.K. Chen, J.C. Irwin, J.P. Franck, Evidence for a strong spin-phonon interaction in cupric oxide, Phys. Rev. B 52 (1995) R13130–R13133.

[73] D.J. Lockwood, Spin-phonon interaction and mode softening in NiF_2, Low Temp. Phys. 28 (2002) 505–509.
[74] X.-B. Chen, N.T.M. Hien, K. Han, J.C. Sur, N.H. Sung, B.K. Cho, I.-S. Yang, Raman studies of spin-phonon coupling in hexagonal $BaFe_{12}O_{19}$, J. Appl. Phys. 114 (2013) 013912.
[75] E. Granado, A. García, J.A. Sanjurjo, C. Rettori, I. Torriani, F. Prado, R.D. Sánchez, A. Caneiro, S.B. Oseroff, Magnetic ordering effects in the Raman spectra of $La_{1-x}Mn_{1-x}O_3$, Phys. Rev. B 60 (1999) 11879–11882.
[76] J. Laverdière, S. Jandl, A.A. Mukhin, I.V. Yu, V.G. Ivanov, M.N. Iliev, Spin-phonon coupling in orthorhombic $RMnO_3$ (R=Pr, Nd, Sm, Eu, Gd, Tb, Dy, Ho, Y): a Raman Study, Phys. Rev. B 73 (2006) 214301.
[77] R. Katoch, C.D. Sekhar, V. Adyam, J.F. Scott, R. Gupta, A. Garg, Spin phonon interactions and magnetodielectric effects in multiferroic $BiFeO_3$–$PbTiO_3$, J. Phys. Condens. Matter 28 (2016) 075901.
[78] A. Jaiswal, R. Das, T. Maity, K. Vivekanand, S. Adyanthaya, P. Poddar, Temperature-dependent Raman and dielectric spectroscopy of $BiFeO_3$ nanoparticles: signatures of spin-phonon and magnetoelectric coupling, J. Phys. Chem. C 114 (2010) 12432–12439.
[79] C.-H. Hung, P.-H. Shih, F.-Y. Wu, W.-H. Li, S.Y. Wu, T.S. Chan, H.-S. Sheu, Spin-phonon coupling effects in antiferromagnetic Cr_2O_3 nanoparticles, J. Nanosci. Nanotechnol. 10 (2010) 4596–4601.
[80] M.O. Ramirez, M. Krishnamurthi, S. Denev, A. Kumar, S.-Y. Yang, Y.-H. Chu, E. Saiz, J. Seidel, A.P. Pyatakov, A. Bush, D. Viehland, J. Orenstein, R. Ramesh, V. Gopalan, Two-phonon coupling to the antiferromagnetic phase transition in multiferroic $BiFeO_3$, Appl. Phys. Lett. 92 (2008) 022511.

CHAPTER 3

Brillouin spectroscopy: probing the acoustic vibrations in colloidal nanoparticles

Jeena Varghese, Jacek Gapiński, Mikolaj Pochylski
Faculty of Physics, Adam Mickiewicz University, Poznań, Poland

Chapter Outline

1. Introduction

Over the years, investigation of photon-phonon interactions is witnessing a stimulated interest with major implications in a range of acousto-optic devices [1,2]. Brillouin spectroscopy (BS) has a potential stand in this domain, allowing the direct observation of hypersonic phonons via inelastic light scattering. Stemming from its prediction in 1922,

Design, Fabrication, and Characterization of Multifunctional Nanomaterials. https://doi.org/10.1016/B978-0-12-820558-7.00010-8

BS has a long history in condensed matter research and still undergoing evolution both experimentally and theoretically. It's significance can be overrated by analyzing the multi-disciplinary contributions in recent thrust areas including phononic crystals [3–5], optical fiber sensing [2,6,7], bioimaging [8–11], optomechanical crystals [12,13] or modern micro-spectroscopic designs [14–16]. The basic concept of Brillouin light scattering (BLS) is underlying in spontaneous type scattering in which the incident photons are inelastically scattered from the thermally excited acoustic waves (phonons). BLS can also probe the spin waves (magnons) to study the dynamics of magnon-phonon interactions, with a large horizon of applications to realize in spintronics [17].

Brillouin and Raman spectroscopies have evolved almost in the same years however developed independently as indispensable tools in material characterization [15]. Though the fundamental mechanism remains the same, there exists a considerable difference in the probed frequency range and the detection method. While Raman spectroscopy gives access to optical phonons (frequency shifts in THz range), BLS can have direct information from the acoustic phonons [18]—the quanta of collective oscillations in acoustic field. The induced frequency shifts of BLS come in the GHz range (0.05–5 cm^{-1}); and a typical dispersive grating used in Raman spectroscopy cannot access these vibration modes [8]. This calls for the need of a Fabry-Perot interferometer (FPI), which, thanks to its extraordinary spectral resolution, is critically important in the development of BLS [14,19,20]. However, the early stage developments were limited by the instrumental issues and detector performance. Several experiments in instrumental designs during the years overwhelmed the critical limitations of interferometry. BLS is uniquely in a renaissance age with resolved instrumental issues and new applications are being demonstrated [14]. The predominance of stimulated Brillouin scattering (SBS) in nonlinear optics [21,22] and biomedical research [11,23] are just two examples. Recently, multimodal systems for simultaneous determination of Brillouin-Raman scattering [15,16,24–26] have attained much interest in material science.

As a major highlight, BLS allows an in situ and nondestructive analysis, which is crucial in many experiments, especially with thin films and soft materials. In fact, the photons from a monochromatic laser are inelastically scattered by the acoustic phonons in the medium in a specific direction of the wave vector $\mathbf{q}$ depended on the laser wavelength λ, the scattering angle θ, and the refractive index n of the sample. In typical liquid and solid bulk samples, light from the visible range is scattered on thermally excited phonons with hypersound frequencies, as will be shown in Section 2.1. Contrary to bulk media, in structured, e.g., colloidal materials there exist characteristic resonance frequencies which scale with the dimensions of the system. Thus, the light scattered from a sample containing both bulk and colloidal material should contain spectral components of frequencies dependent purely on the scattering geometry and material constants and components with frequencies defined mostly by the colloidal size, shape, order, and its material properties.

Hypersonic frequencies cannot be easily accessed by vibrating membranes or piezo crystal transducers. But BLS can easily resolve the sound waves propagating through the material using an FPI. Fortunately, there is no need to produce external excitations by a transducer as in ultrasonic pulse echo techniques, since thermal excitations produce them continuously [27]. Pump-probe spectroscopy is another possibility to study the eigen vibrations of nanoparticles. Optical beams from a pump laser impinge on the sample and probe laser measures its femtosecond excitations of the plasmon resonance [27,28]. However, this method works in time domain and from the theoretical point of view, it is easier and more accurate to deal with vibrations in frequency domain using FPI for BLS. Thus, BLS opens a suitable optical pathway to assess elastic properties of soft or highly brittle materials.

In this chapter, we first give an account of the historic development of BLS instrumentation followed by its applications in certain novel systems and a short recall of the theory and setup. Moving forward, the main aim of this chapter is to give an outline to how BLS is used in determining the thermomechanical properties of **colloidal crystals**—ordered array of particles, by probing their acoustic vibrations.

1.1 Historical perspectives

The history of BLS experiments is quite fascinating [29]. In 1922, Léon Brillouin, a French physicist proposed a theoretical prediction for the first time that a coherent light beam can be scattered off by thermally activated elastic waves in a homogeneous medium [30]. The thermal agitation of the particles or the atoms introduces a time and position dependent density fluctuations which propagate as periodic waves. These waves generate variation in material's refractive index, which may act as a diffraction grating for the incident light [9]. The kinematics is consistent to Bragg's law and the energy shift is described by Doppler effect. This phenomenon known as "Brillouin scattering" can be spectroscopically determined by measuring the frequency shift equivalent to the phonon frequency [8]. Later in 1926, Leonid Mandelstam published his independent research on spontaneous scattering problem and is further referred as "Brillouin-Mandelstam scattering" [11].

It took almost a decade for the first experimental confirmation of BLS in liquids and crystals by Gross in 1930 [31,32]. The resultant Brillouin shift reflects only the intermolecular interactions but not much information on material composition or structure. Consequently, the initial research was centered to investigating molecular motions, relaxation process in fluids, anisotropic bonding in organic and inorganic solids and interestingly, phase transitions. Even then, BLS can be connected with material specific spectroscopies (Raman scattering) to elucidate the chemical details [16,33]. However, the early developments were limited due to the unavailability of a proper light source. For any weak-spontaneous scattering technique, enough photon flux should pass through the scattering volume. Conventional light sources like Hg or Zn vapor lamp could measure only a limited species of samples viz. compressed fluids and certain solids [34]. But the 1960s witnessed tremendous progress in spectroscopic applications, shortly after the introduction of lasers [35–37]. Since then, several investigations on viscoelastic properties of amorphous and crystalline solids, fluids and glassy solids with a special focus on polymers came to light [34]. The BLS spectrometer developed by R.Y. Chaio - a combination of He-Ne laser, Fabry-Perot interferometer along with photomultiplier tube-based detection, was a breakthrough in the development of spectrometers. This for the first time allowed a precise determination of hypersonic acoustic waves [37]. One of the first demonstrations for high resolution BLS was in 1942 by Venkateswaran. The earlier versions worked with a pressure cavity in which change of pressure between the two sets of mirrors produced a refractive index change. This system improves the transmitted frequency and thus small frequency shifts could be resolved [8]. Eventually the scanning mode was improved by applying a periodic, voltage ramp through piezoelectric materials, which adjusts the mirror spacing. One cycle produces one spectrum and to improve the signal to noise ratio (S/N), several cycles were added which generates the final BLS spectrum. From the early spectrometers with hours of acquisition period, present spectrometers need minutes or even several seconds. This can be achieved with SBSS and virtually imaged phase array (VIPA) based spectrometers.

The tandem multipass design in BLS was firstly reported by J. R. Sandercock for his studies on ferroelectric (SbSI) materials. This spectrometer introduced two key features: active stabilization and improved spectral contrast. The former helps in continuous monitoring of transmittance and maintaining the mirror parallelism by short correction voltages. Later, the introduction of microprocessors improved the manipulation and control without manual support. Thus the maintenance of a stable transmittance for a given period greatly supported the improvement of BLS instrumentation [38–40]. Similarly, multiple passing of the transmitted light through the interferometer improved the spectral contrast as high as the number of passes [41]. This could easily differentiate the weak resonances from the strong reference signal (Rayleigh line). A new design of tandem type multipass interferometer was reported in BS by Dil [42]. But this was challenged by time control and instability in synchronizing the independent interferometers. Sandercock resolved this issue by introducing a tandem interferometer to the previous platform, the design remains as much popular in BLS instrumentation [8,42]. Recent developments in VIPA based spectrometers enable spontaneous BLS with all spectral components measured at once. This makes faster acquisition and hence quite appreciable among Bio-Brillouin community [9,14,43].

These advancements widened the applicability of BLS studies to several exciting novel system-ferroelectric materials and their phase transitions [44], interface and surface excitations of bulk materials [45], multilayers and thin films [46,47], binary liquid mixtures [48], relaxation dynamics of polymers and amorphous materials [49,50], living cells [51], biomechanics of DNA and proteins [9,11,52], Langmuir-Blodgett films [53], elastic property of spider silk [54], hydration phenomena [55,56,116], NIR-based scattering of opaque materials [8], nanocomposites [58], phononic crystals [59] to the experiments on quasi particles confirming B-E condensation influenced by energy pumping [60].

2. Brillouin spectroscopy

2.1 Brillouin scattering: theory and observables

In a classical view, the inelastic light scattering is because of Bragg diffraction from periodic dielectric constant fluctuations in the material. For spontaneous BLS, this is not induced, rather produced by the thermal excitations (acoustic phonons) in the medium. Not only phonons, other excitations like magnons, plasmons etc. can also contribute to dielectric constant variations in a material further resulting in light scattering in a direction different from incident beam [34]. Here, we describe the spectrum for the simplest case of isotropic bodies.

Considering an isotropic bulk medium in thermodynamic equilibrium, the time-space-number density of atoms, ***N*N(*r*,(), *r*,*t*),** in the thermal motion is given by:

$$N(r,\ t) = N + \Delta N(r,\ t)$$

Where N is the mean number density and $\Delta N(r,\ t)$ is the time-space fluctuation in number density.

Incident light beam interacts with the propagating density fluctuations, which results in Doppler shift of the scattered light [8]. The direction of wave vector ($\mathbf{k_s}$) and frequency (ω_s) of the scattered light depends not only on the incident wave vector ($\mathbf{k_i}$) and angular frequency (ω_i), but also on the frequency (Ω) and wave vector ($\mathbf{q}$) of the thermally activated density fluctuations of the atoms. The light scattering spectrum (i.e., the intensity of scattered light at a certain wave vector and frequency $I(\omega,\ \mathbf{q})$) is obtained from the Fourier transform of auto-correlation of the density fluctuations of wave vector $\mathbf{q}$, $\Delta N(\mathbf{q},t)$.

$$\Delta N(r,\ t) = \int\int_{-\infty}^{\infty} \frac{d^3\,\mathbf{q}}{(2\pi)^2}\frac{d\Omega}{2\pi}\Delta N(\mathbf{q},\ \Omega)\mathrm{e}^{-i2\pi(\mathbf{q}.r-\Omega t)}$$

Depending of the sample under investigation different physical processes are influencing space-time evolution of density fluctuations, which is reflected in the shape of recorded BLS spectrum. In the simple case of Ω. isotropic bulk material, the density fluctuation propagates with material as a plane wave of a given wave vector q and frequency. Regardless of the scattering geometries, all the scattering phenomena must obey the conservation laws:

$$\hbar\omega_s = \hbar\omega_i \pm \hbar\Omega$$
$$\hbar k_s = \hbar k_{\mathbf{i}} \pm \hbar q$$

the +/− sign describes the creation or annihilation of a phonon in the process defined as Stokes (lower frequency side) and anti-Stokes (higher frequency side) scattering, like Raman scattering. In general, for a single scattering geometry, density can fluctuate to produce distinct spectral features: intense elastic line due to Rayleigh scattering and weak inelastic components corresponding to longitudinal and transvers (shear) modes [8]. The inelastic part of the Brillouin spectrum is given by a frequency shift of $\Omega = \pm vq$ relative to the elastic peak, where v is the **velocity of a given vibrational mode (longitudinal or transverse)**, and q is the momentum exchanged in the scatterig process.

As given above, the scattering wave vector $\mathbf{q}$ defined by the difference between incident and scattered beam wave vectors ($\mathbf{q} = \mathbf{k_i} - \mathbf{k_s}$), the magnitude can be given by the law of cosines as [27]:

$$q^2 = k_i^2 + k_s^2 - 2k_i \cdot k_s = 2k_i^2(1 - \cos\theta) = 4k_i^2 \sin^2\frac{\theta}{2}$$

With the condition, $k_s = k_i = 2\pi n/\lambda$

And the magnitude of the $\mathbf{q}$ vector is given by

$$q = 2k_i \sin\frac{\theta}{2} = \frac{4\pi n}{\lambda}\sin\frac{\theta}{2}$$

The phonon frequency (f_s) can be given as

$$f_s = f_i \pm \frac{vq}{2\pi} = f_i \pm \frac{v}{2\pi}\frac{4\pi n}{\lambda}\sin\frac{\theta}{2}$$

Generally acoustic waves are probed with an incident radiation of wavelength in the visible range (hundreds of nm). In scattering experiments, the light waves will interact with the vibrational modes of comparable wavelengths. Thus, BLS probes the vibrational modes by measuring the density fluctuations in a micrometer scale. Typical Brillouin spectra measures the Brillouin intensity with respect to the frequency of acoustic phonons. If the sample is made of colloidal particles of size of the order of phonon wavelength, the density fluctuation will not propagate freely as in the bulk solids or liquids. This localization of vibrating modes is reflected in more complex shape of the BLS spectra [27], will be discussed in Section 3 as."

2.2 BLS instrumentation

Brillouin measurements can be performed with a range of experimental set ups and using different scattering geometries. As any optical scattering technique, BLS also requires minimum three components: a light source, dispersive elements, or optics to resolve frequency shifted beam and a detector that converts the scattered light to a measurable signal. For our application, we restrict this discussion to spontaneous BLS instrumentation. On the other hand, an impulsive stimulated BLS set up can offer higher sensitivity, shorter acquisition times and lower background (from Rayleigh line). This is possible using a pump-probe laser setup in which the sound waves are generated by means of stimulation from a pump laser and the probe laser is scattered by the acoustic waves [61].

Regardless of spectroscopy, an interesting advantage of BLS set up is that the possibility of tailoring the input optics depending on our application. It can be measured as a variable θ spectrometer (as used in measuring anisotropic elasticity of single and polycrystalline materials) or coupled with high pressure or temperature variable setups or 3D imaging (confocal BLS) and so on. The flexibility of its use can be maximized by including multiple optical elements to facilitate the beam paths after all integrated to the same interferometer. Also being similar techniques, both Raman and Brillouin can be simultaneously recorded by varying the detection method, i.e., a monochromator can receive a part of scattered light to detect the Raman resonances and the remaining can be probed by the interferometer [9]. This combined layout can share the light source and microscope objective to better focus the scattered light toward the corresponding detectors. An illustrative diagram for the micro-Brillouin-Raman spectroscopic system and the corresponding measured spectra of a biofilm [62] is shown in Fig. 3.1.

2.2.1 Light source

BLS often provides extremely low signal, which becomes a challenge for the laser light parameters. The light should be almost perfectly monochromatic and no mode hopping or wavelength drift is allowed. Number of laser models fulfilling these conditions is very small because BLS is the only technique capable of detecting such instabilities. Most of them are available in 532 nm wavelength. There are some models in other parts of the visible range (~400 nm, ~600 nm) and in the near IR (~780 nm) [63]. It can also be performed at UV range, however it requires a specific design and optics for the light channel [63,64].

If the application does not require a specific wavelength, an optimum choice is 532 nm because it combines still relatively good scattering efficiency (λ^{-4} factor) with a maximum detector efficiency. Another concern is, the choice of laser wavelength determines the reflective coating on the mirrors so that for a standard model of mirrors the laser color cannot be changed. Recently, new mirrors appeared with reflection coatings suitable for two or three different wavelengths.

Laser power on the sample has to be adjusted to its photostability and to the heat dissipation rate. Typically, for light focused with "normal" lenses ($f = 100-200$ mm) on bulk photostable samples, powers of 100−1000 mW are used. When using microscope objectives, 10 mW is often more than enough. High power lasers after some time of use become unstable at low powers, so that special attenuators have to be used for proper operation.

2.2.2 Scattering geometries

In a BLS experiment, the incident ($\mathbf{k_i}$) and scattered ($\mathbf{k_s}$) wave vectors are determined by the scattering geometry preference. A set of optical components such as mirrors or beam

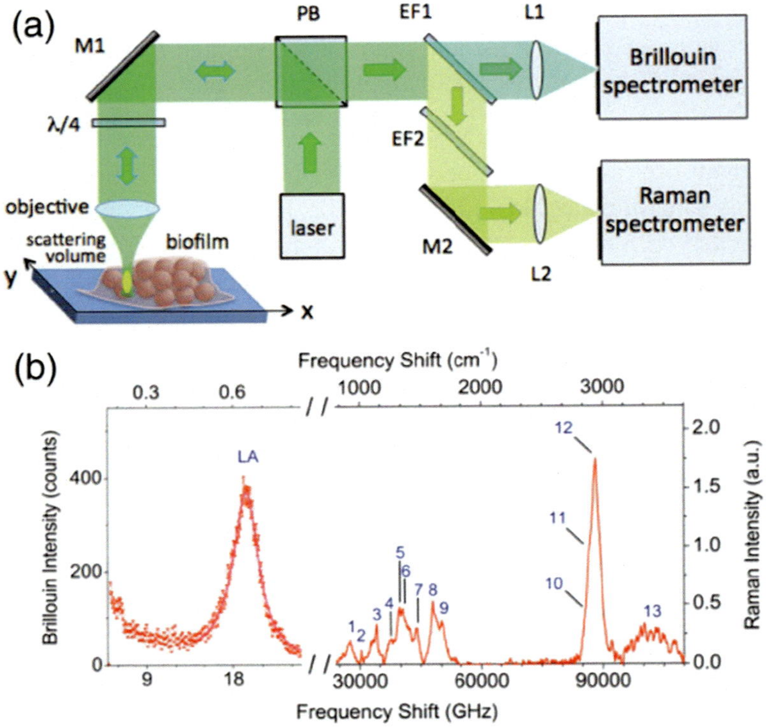

Figure 3.1

(A) A combined setup for the micro-Brillouin-Raman spectroscopic system. (B) a typical BLS (left panel) and Raman (right panel) spectra obtained for a biofilm. The Brillouin peak represents the longitudinal acoustic modes (LA) of the sample. *Adapted with permissions from S. Mattana, M. Alunni Cardinali, S. Caponi, D. Casagrande Pierantoni, L. Corte, L. Roscini, G. Cardinali, D. Fioretto, High-contrast Brillouin and Raman micro-spectroscopy for simultaneous mechanical and chemical investigation of microbial biofilms, Biophys. Chem. 229 (2017) 123–129. https://doi.org/10.1016/j.bpc.2017.06.008. Copyright obtained from Elsevier 2017.*

splitters in between incident and scattered beam path allows to set this scattering geometry. Usually, $\mathbf{k_s}$ is kept constant and the incident beam is positioned at an arbitrary angle, choosing the desired phonon wave vector [8].

The choice of the scattering geometry is dependent on the type of samples to measure [65]. For plane substrates like colloidal films, usually a set of three different scattering geometries is preferred: transmission, reflection, and back scattering geometry [27,47] (Fig. 3.2). The reflection geometry helps to study the phonon propagation normal to the film. For back scattering case, the scattering angle is 180°, and so $\mathbf{k_i}$ and $\mathbf{k_s}$ lies in the same direction. The wave vector will be maximum ($q_{\max} = 2\,k_i$) and its magnitude is given by $q = \frac{4\pi n}{\lambda}$.

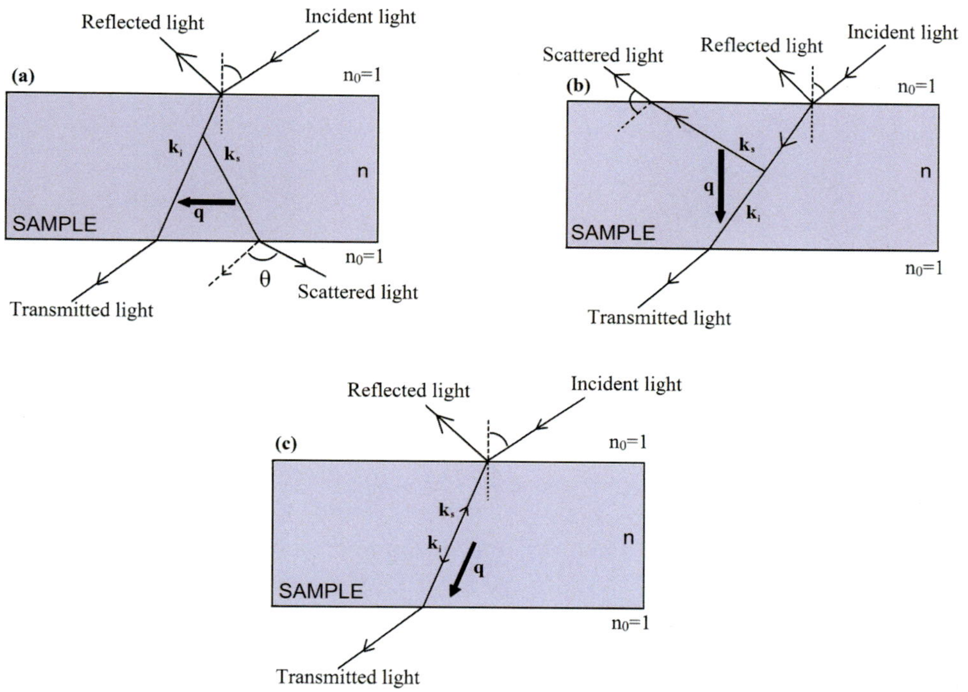

Figure 3.2
The sketch of scattering geometries for colloidal films (A) transmission geometry (B) reflection geometry (C) back scattering geometry.

2.2.3 Experimental setup

A scheme of an experimental setup used to study the colloidal crystals [66] is given in Fig. 3.3. Here, a single mode laser ($\lambda = 532$ nm) is mounted on a goniometer and therefore different scattering angles from 0° to 160° are accessible. The optics includes a focusing lens and a Glan polarizer (vertical polarization) to ensure fully polarized light entering the sample. The sample holder can be equipped with temperature controllers or pressure chambers depending on the type of experiment. The scattered light from the sample is collected by suitable optics and is directed to the pinhole of the detector. The polarization of the scattered light can be varied according to the type of studied vibrations. An analyzer can be used to select either vertical (perpendicular to scattering plane) or horizontal (parallel to the scattering plane) polarization. A multipass tandem FPI (JRS Scientific Instruments) has two interferometers differing slightly in free spectral range (FSR) with mirrors scanned synchronously. The light transmitted through the FPI is detected by a sensitive detector (typically an avalanche photo diode—APD) and analyzed with a multichannel analyzer (up to 1024 channels) triggered by the scan start signal. Thanks to a very good scan linearity and stability, further processing by the computer software is reduced to linear conversion of time data to frequency which leads to formation of complete Brillouin spectra.

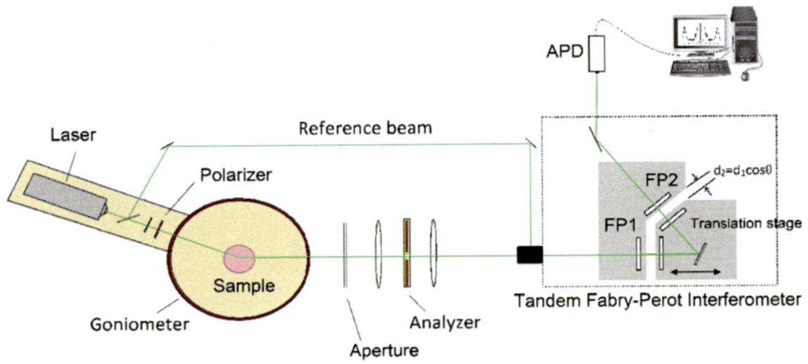

Figure 3.3

Brillouin scattering experimental setup with multi pass tandem Fabry-Perot Interferometer. The laser connected to the goniometer allows to study the dispersion relations by measuring the Brillouin spectra for different **q** values of the scattering vector. *Reproduced with permissions from D. Schneider, P.J. Beltramo, M. Mattarelli, P. Pfleiderer, J. Vermant, D. Crespy, M. Montagna, E.M. Furst, G. Fytas, Elongated polystyrene spheres as resonant building blocks in anisotropic colloidal crystals, Soft Matter 9 (38) (2013) 9129. https://doi.org/10.1039/c3sm50959a. Copyright obtained from RSC publications 2013.*

The alignment stability is achieved by optimizing the position and amplitude of the central line in a feedback loop. In rare cases of VH measurements at right angle the scattered light intensity is low enough to be directly used for this purpose, however typically a part of the laser beam (a reference beam) is directed into the tandem FPI by a mechanical shutter, replacing the very strong Rayleigh line at the moment when the scan goes through the "zero" (unshifted) frequency. To avoid any external disturbances, the whole setup is established on a vibration free optical table [27,67].

3. Acoustic vibrations in colloidal crystals

3.1 Colloidal crystals

For the purpose of this review, colloidal crystals (CCs) will be understood as highly ordered arrays of monodispersed particles [68,69], without any suspending medium required by the standard definition of colloidal suspensions. The presence of the periodic superlattice is quite interesting for applications, as it offers fascinating properties which are remarkably different from standard crystals with atoms, molecules, or ions as repeating units. In recent years, CCs have much interest in coating systems, nanolithography [70], biomedical systems, and materials engineering. Moreover, CCs are attractive due to their ability to control the propagation of classical waves and in realizing components for photonic, phononic and plasmonic devices [59,66,71,72]. The formation of a particle assembly into perfectly ordered coherent architectures usually happens under surrounding effects; for instance gravitational sedimentation, electrophoretic deposition or evaporation [1,73]. From the point of view of optical applications, most interesting are CCs formed of elements that are in micrometer or submicrometer in scale, although there are practically no restrictions concerning the size of the building units [74]. In this size range, the use of microscopic studies is more convenient and the packing and architecture can be conveniently tuned [75,76].

Monodispersed colloidal particles can self-assemble to form face centered cubic (fcc) or hexagonal close packing (hcp) structures with approximately 26 vol% voids [68,69]. And this is how it is used as templates for preparing macroporous materials, often referred as inverse opals. Such structures in white light exposure will produce structural colors by Bragg diffraction, which plays a major role in photonic crystal designs [77,78]. Several bioinspired materials with optical and mechanical functions are also in research now [68].

CCs can be designed in one, two or three dimensions depending on the application. Moreover, the composition can be tuned by binary and polynary crystals or by core-shell and functionalized architectures [1]. A three-dimensional assembly can be simply formed by drop casting the particle suspension on a suitable hydrophilic substrate. Vertical deposition is another commonly used approach by dipping a clean substrate into a colloidal dispersion, which upon evaporation at the contact phases (substrate, dispersion, and air), produces an ordered array [59]. The thickness of such a 3D crystal can be often manipulated by the withdrawal speed of the substrate under specific temperature and humidity controls [1].

Two-dimensional assemblies called colloidal monolayers are potential candidates as templates, filtration membranes, plasmonic devices, lithographic masks [79,80] and a vast number of techniques are available for their fabrication. Two most explored approaches are (1) direct assembly at the solid surface and (2) liquid interface mediated self-assembly. The former includes spin coating, sedimentation, electrostatic deposition, and controlled evaporation. The latter technique is much explored due to the easy fabrication of large area highly ordered close packed assemblies. For example, polystyrene CCs can be formed by introducing the particles to an air-water interface using a syringe pump, which then can be deposited on a treated glass slide using Langmuir-Blodgett technique [69]. An equilibrium among the short range repulsive and long-range attractive forces cause a perfect crystallization.

On the other hand, the assembly of one-dimensional colloidal crystal is demanding, as the isotropic gravitational pull will prevent them from growing in a preferred direction. Even in liquid interface, the surface tension effects may disturb the single chain formation. The modification in the substrate and deposition process can enable its formation, as an example a surface relief grating can be helpful [68,81]. A Langmuir-Blodgett-based assembly of 1D strings [81] is shown in Fig. 3.4.

The characteristics of these structures depend on their component blocks, which needs the complete knowledge regarding their physical properties. For applications, a fundamental knowledge on the thermomechanical properties and the stability of these structures is inevitable. In other words, such information can pave the way to structural engineering

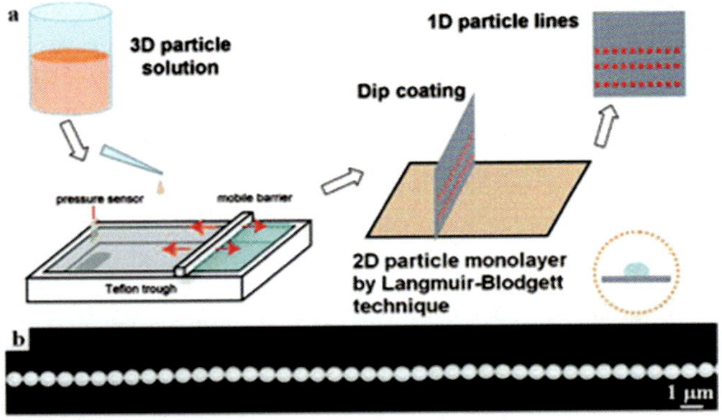

Figure 3.4

1D colloidal crystal fabricated by Langmuir-Blodgett technique (A) Scheme of producing particle strings from the 2D monolayer in Langmuir-Blodgett trough at the contact lines of intersection between the floating colloidal monolayer and the vertical substrate (B) SEM of the particles string formed. *Adapted with permissions from A.R. Tao, J. Huang, P. Yang, Langmuir–Blodgettry of nanocrystals and nanowires, Acc. Chem. Res. 41 (12) (2008) 1662–1673. https://doi.org/10.1021/ar8000525. Copyright obtained from American Chemical Society 2008.*

depending on the specific demand [1]. BLS is a powerful tool for a nondestructive monitoring of the acoustic vibrations in CCs made of submicrometer particles and thereby determining the preferential conditions for tuning their thermomechanical properties [67].

3.2 Vibrational modes of spherical particles

The vibrational modes in free-elastic homogeneous spheres were theoretically identified by H. Lamb in nineteenth century [82]. The modes are classified as spheroidal and torsional and are indicated by three indices (n,l,m) where n refers the radial and (l,m) the angular dependence of the displacement field. A similar kind of vibrational modes is observed in giga spheres like Earth [83]. The torsional modes are tangential and hence involve only shear motions and they do not contribute to the BLS intensity [84]. The spheroidal modes having both shear and stretching motions are BLS active. They can be specified by two indices n, and l $(n = 1,2,3,..)$ and $(l = 0,1,2, \ldots)$ [27]. The BLS active spheroidal modes have frequency close to the phonon frequency in the bulk material i.e., $\omega(n,l) = 2\pi f(n,l) \sim qc_l$, where q is the phonon wave vector magnitude and c_l the longitudinal sound velocity. The intensity of a particular mode (n,l) is depended on the product qR where R is the radius of the particle $(R = d/2)$. Thus a relevant q-dependence is observed in the spectra of spherical particles [84].

Solving the wave equation for the elastic spheres the eigen frequencies are identified as $f(n,l) = A_{n,l}\ c_t/d$ where $A_{n,l}$ is the dimensionless factor dependent on the Poisson's ratio and c_t denotes the transverse sound velocity. The frequencies of the eigen vibrations are

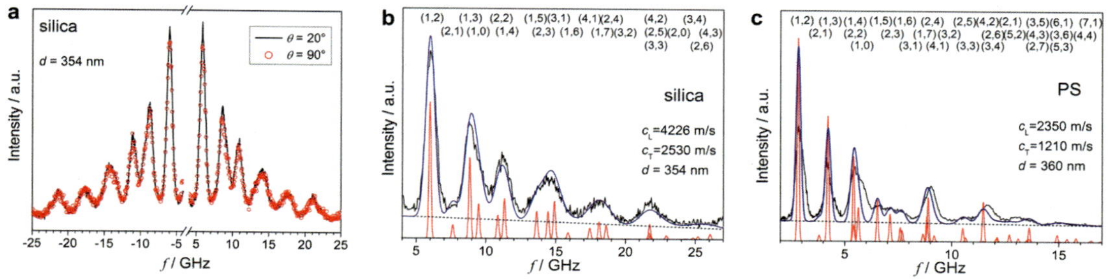

Figure 3.5

(A) BLS spectra of silica nanoparticles (hard colloids) measured at two different scattering geometries; the angle independence on the BLS peak positions for dry samples is clearly demonstrated comparison of experimental (*black lines*) and theoretical (*blue lines*) spectra for (B) silica CCs of 354 nm diameter (C) polystyrene (PS) CCs of 360 nm diameter. The *red lines* show the individual contributions from various eigen modes. *Adapted with permissions from T. Still, M. Mattarelli, D. Kiefer, G. Fytas, M. Montagna, Eigenvibrations of submicrometer colloidal spheres, J. Phys. Chem. Lett. 1 (16) (2010) 2440–2444. https://doi.org/10.1021/jz100774b. Copyright obtained from American Chemical society 2010.*

depended on the particle diameter (*d*), the mass density and elastic modulus. The mechanical properties can be determined by the extraction of the eigen frequencies by suitable theoretical fits of the observed peaks in the BLS spectra [85]. The typical BLS spectra for the silica and polystyrene CC with the eigen modes assigned [84] are given in Fig. 3.5.

3.3 Colloidal crystals as phononic crystals

One among the most exciting developments in modern physics is undoubtedly the manipulation and control of classical wave flow through periodic structures [86,87]. The research on inhibited emission of photons by Yablonovitch and John in 1987 marks the beginning of "*photonic crystals (PhCs)*" which allows the controlled propagation of optical waves through periodic dielectric constant variations [88,89]. The enormous interest in these structures hails from their ability to forbid light with specific frequencies in certain directions, so called "photonic band gaps" [72]. The elastic analogue, "*phononic crystals (PnCs)*" allows propagation of elastic or acoustic waves via periodic modulation of elastic moduli or mass densities [67,86,90]. The origin of the band gap in PnCs is quite similar as they arise due to the destructive interference of diffracted waves in periodic structures so called "Bragg gap." However, electromagnetic and acoustic waves have different nature—the latter are full vector waves having transverse and longitudinal polarizations. For an isotropic medium, the acoustic wave propagation depends on Lame's coefficients and material density instead of just one parameter, the dielectric coefficient in the case of EM waves [67].

In the following, terms sonic, ultrasonic and hypersonic (often found in the literature) refer to frequency ranges of sound waves: acoustic (1–1000 Hz), ultrasound

(10 kHz—100 MHz) and hypersound (>0.1 GHz), respectively. The initial search for phononic materials started with sonic and ultrasonic crystals, which are in macroscopic length scale. These millimeter-sized structures are assembled manually and could be probed by sound transmitting experiments [90—88]. However, hypersonic PnCs have much attention owing to their novel applications including optical modulators and acoustic oscillators [93,94]. Acoustic waves have frequency dependent characteristics; and interestingly hypersonic waves by virtue of their higher frequencies have certain unique features comparing with normal acoustic waves. They can be merely produced by the random thermal vibrations of the atoms in a material (phonons) and are the major heat carriers in dielectrics. Owing to this, such crystals having a potential in controlling the flow of hypersonic waves can be utilized in many important physical phenomena involving heat transport and coupling of phonons with photons and electrons [67].

The design and characterization of hypersonic PnCs is much demanding as the formation of periodic patterns should scale in the submicrometer range. The advances in nanofabrication with an insight to design periodic structures for photonic applications helped in this domain also. As an example, polymer hypersonic PnCs can be fabricated from holographic interference lithography [93,95]. Self-assembly is another promising technique which allows relatively simple and cheap fabrication route for 2D and 3D high quality large area CC. These structures with a well-defined shape and size can offer excellent phononic properties. The presence of binary and ternary crystals as well as modifying the interstitial spaces of the colloidal particles can enrich the class of potential phononic materials [96].

The determination of vibrational modes of the component particles originating from the elastic motion at the nanoscale may depend on the architecture, interfacial properties and geometry of the crystal. Several nondestructive methods can be used, however high frequency resolution and sensitivity is a must. A few eigen modes were actually resolved by Raman scattering [97,98], but with particles of size less than 10 nm. BLS [85,99,100] and pump-probe [101,102] techniques also probe stimulated or spontaneous vibrations confined in the submicrometer particles.

The first observation of hypersonic bandgaps was reported by Cheng et al. [59]. This was performed by using the self-assembly property of particles which allowed the controlled propagation of phonons. A dry colloidal opal film was deposited by the vertical movement of the glass substrate from the particle suspension [103]. The Brillouin spectra of such dry films were however q-independent due to strong multiple scattering (as evident in Fig. 3.6A) occurring from the high elastic form factor and large optical contrast (large difference in refractive index) of polystyrene with surrounding air. This precludes the measurement of the phonon dispersion relation $\omega(\mathbf{q})$. Infiltration using a viscous-nonevaporative (inert for PS) liquid in the subsequent dry samples prevents the multiple

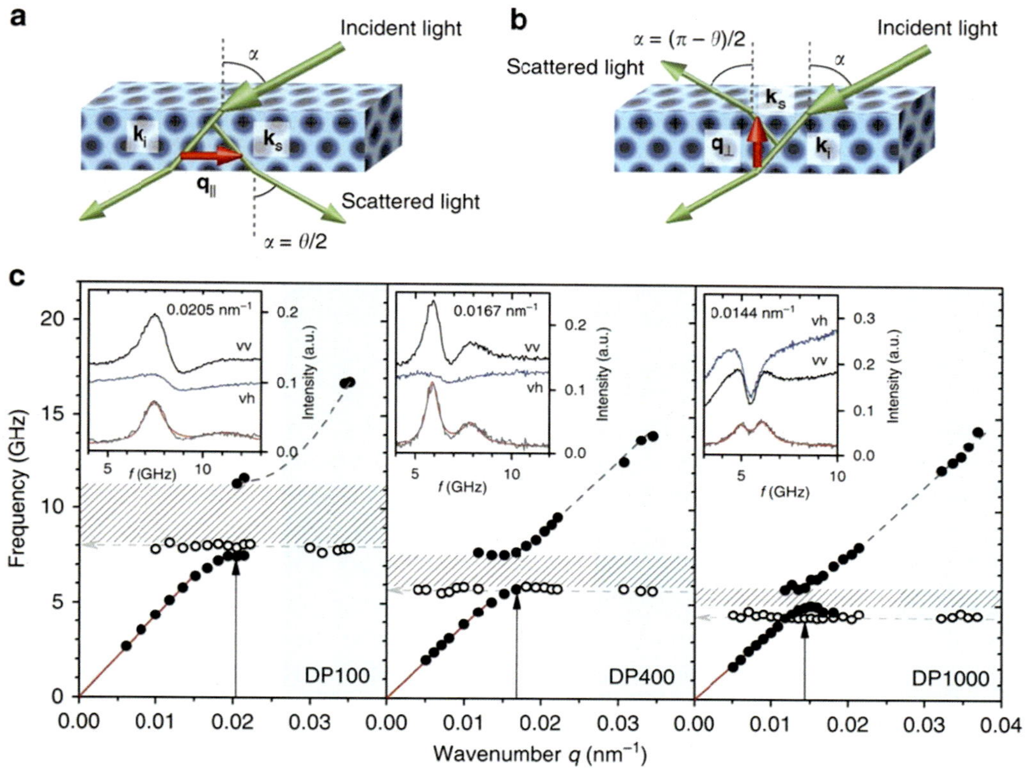

Figure 3.6

(A) Transmission geometry; (B) reflection geometry to probe the phonon propagation via BLS; and (C) the phonon band diagram and BLS spectra (inset) recorded at various polarizations (*vv* and *vh*) for samples with different compositions. The band gap tuning is clearly demonstrated by varying the particle properties. *Adapted with permissions from E. Alonso-Redondo, M. Schmitt, Z. Urbach, C.M. Hui, R. Sainidou, P. Rembert, K. Matyjaszewski, M.R. Bockstaller, G. Fytas, A new class of tunable hypersonic phononic crystals based on polymer-tethered colloids, Nat. Commun. 6 (1) (2015) https://doi.org/10.1038/ncomms9309. Copyright obtained from Springer Nature 2015.*

scattering effects and thus well-defined scattering wave vector magnitude (q) was possible. The splitting of a single Brillouin peak into a doublet across the Brillouin zone (BZ) boundary represented Bragg-gap effect which occurs due to the folding of band into the first BZ. The dispersion relation of the observed longitudinally polarized phonons clearly indicated a Bragg gap at 5 GHz with a width of 0.4 GHz. The phonon propagation in this frequency regime was forbidden. The absence of particle modes after infiltration focused to the hybridization gap between the resonance band ($l = 2$) and continuum acoustic band. This happened because of the weak elastic contrast between the components in the system and therefore the strong dissipation of elastic energy into the liquid medium [104]. The formation of the Bragg gap is depende on the periodic structure length which should be of the order of the phonon wavelength. If it is equal, the scattered acoustic waves interfere constructively and the incident wave is fully reflected, which opens the phononic gap [67]. The width of the Bragg gap was increased by the elastic mismatch between the particles and the fluid. When PS (refractive index, $n = 1.59$) dry opals were infiltrated by PDMS ($n = 1.41$) instead of silicon oil ($n = 1.45$), the increase in Bragg gap is observed. Direct measurement using BLS in these structures can address the phonon propagation and dissipation issues [59].

Several studies on the phononic properties of mesoscopic particles were reported using high resolution BLS, which gives access to resonance modes of particles and dispersion relations of particle assemblies in a liquid host. The hypersonic bandgap in 2D CC were subsequently reported [105]. In addition to Bragg gap, the experimental realization of hybridization gap (HG) was obtained in several systems. The first mention of this is in the BLS experiments of colloidal suspensions [106,107] but clearly demonstrated was in PS and PMMA wet colloidal opals infiltrated by PDMS [5]. This gap originates from the interaction between the bands of quadrupolar eigen mode ($f_{1,2}$) and the effective medium. In order to check the sensitivity of phononic gaps on disorder, hybrid colloidal systems were investigated by BLS. Noncrystalline films of PS spheres—binary mixtures with two different diameters in various compositions were investigated and the BLS spectra showed the superposition nature of the individual components. Interestingly, HG sustains and is inert to the structural disorder, but BG disappears, which is in accordance with the origin of gaps. BG occurs as the effect of crystal structure whereas HG is depended on particle resonances which are unmodified by the sphere arrangements. These materials may have potential applications in vibration isolators and acoustic shields [67].

A novel approach to design hypersonic phononic materials based on polymer tethered colloid (silica -PS brush architecture) [3] was investigated by BLS. This gave evidence to the formation of HG which was robust to the disorder. It explored the property of anisotropic elasticity at the particle-polymer interface. The phononic properties could be suitably tuned by the particle composition, size and the extent of polymerization.

The experimental BLS spectra recorded at transmission geometry and the corresponding dispersion relations for different samples are shown in Fig. 3.6C. Optically polarized (vv) and depolarized (vh) spectra were obtained to study the propagation of phonons having longitudinal and transverse polarizations, respectively. Polarized and depolarized spectra having transverse (vh) and longitudinal (vv) polarizations were obtained to study the phonon propagation. The linear dispersion (red lines) at the low $\mathbf{q}$ range allowed to calculate the c_{eff} that decreases with increase of PS fraction. The spectra show a single band gap (region without black solid circles) with the size of the gap changes in the same way as c_{eff}. The observance of a flat band (open circles) is also evidenced from the band diagram.

3.4 Investigating core-shell architectures

Colloidal composites with polymer materials have increasing importance nowadays. When at the nanoscale, such materials deviate much in properties with its bulk ones due to several effects including interfacial effects, confinement and depletion. The vibrational modes in addition to the elastic motion of individual spheres depend also on the geometrical and architectural characteristics. Still et al. for the first time measured the resonant modes in submicrometer SiO_2-PMMA spheres with constant core radius but varying shell thickness. It is reported as the first measurement of core-shell structures using the nondestructive Brillouin scattering technique. Over pump-probe technique, BLS can probe all the available thermally excited vibrational modes in a single experiment. For Particle eigen modes which are standing elastic excitations, BLS acts as incoherent light scattering method and the q-dependence is lost in analogy to Raman measurements [1].

Fig. 3.7 shows the BLS spectra of bare silica and core-shell particles eigen modes with varying shell thickness. The spectra are fitted with Lorentzian function and nine particle

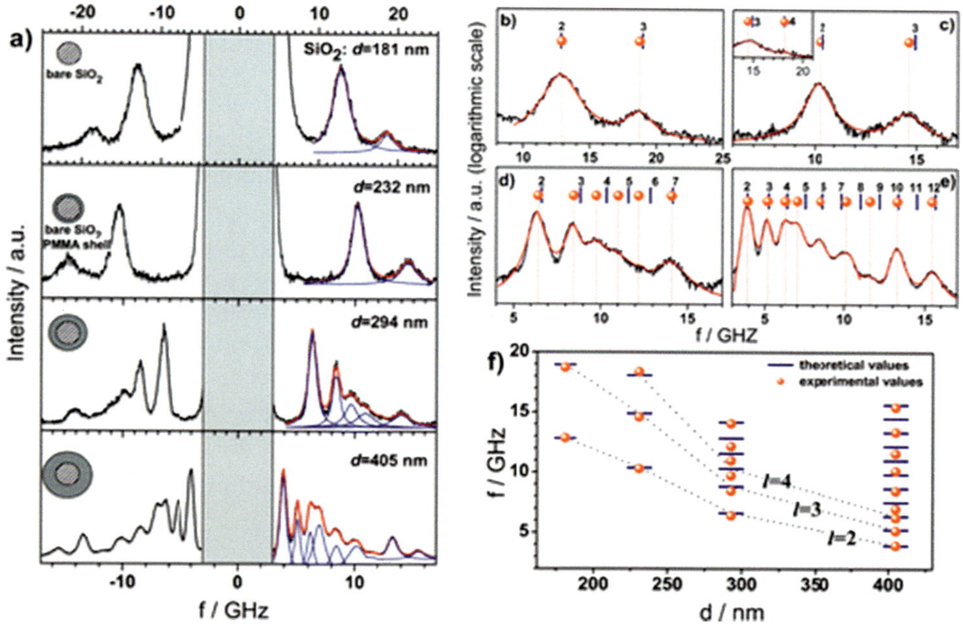

Figure 3.7

(A) BLS spectra of core-shell spheres in comparison to those of bare silica particles. Enlarged view of Stokes sides from (A), as (B) the bare silica (C—E) core-shell structures. (F) Verifying the convergence of experimental and theoretical frequency values. *Adapted with permissions from T. Still, R. Sainidou, M. Retsch, U. Jonas, P. Spahn, G.P. Hellmann, G. Fytas, The "music" of core—shell spheres and hollow capsules: influence of the architecture on the mechanical properties at the nanoscale, Nano Lett. 8 (10) (2008) 3194—3199. https://doi.org/10.1021/nl801500n. Copyright obtained from American Chemical Society 2008.*

vibrational frequencies are identified. The experimental frequency values are represented as orange spheres and the vertical blue lines show the calculated frequency, which agrees with each other. For SiO_2 spheres only two resonant frequencies are present, however the BLS spectra become richer with the increase in PMMA shell thickness. This increase in the resolved modes with particle diameter (d) is theoretically verified as the BLS intensity and is well depended on the product qd. In the conditions of strong multiple scattering, q is approximated as twice the backscattering vector ($q \approx 2k_i$) and the observed modes increase by twice the product of back scattering vector and the diameter of the particle ($2k_i d$) [108].

The elastic characteristics such as longitudinal and transverse sound velocities are not known in nanostructures. An access to these parameters is much important since it can deviate much from the bulk. These parameters are frequency dependent and BLS yields high frequency limiting values which are dependent on the packing type of the crystal, interactions and material glass transition temperature (T_g) [109]. The sound velocities of the silica spheres are calculated from the experimental and theoretical data which are significantly lower than the bulk values indicating the porosity of silica spheres. Thus, the BLS experiment becomes a necessary tool to avoid the wrong assumptions in calculations. Usually the experimental results are verified with theoretical calculations to generate the better fitting of the results. This agreement allowed the calculation of elastic constants and core density in hybrid structures. An exciting increase in the elastic moduli was observed in core-shell structures. The nature of hollow capsules of PMMA after dissolving the silica core were also studied and verified that the elastic constants are not affected by the removal of the core. These studies can be extended to selectively nanostructured colloids [1].

Several core-shell particles were investigated recently via BLS [110,111] tailoring their thermomechanical properties such as glass transition, surface softening, elasticity by varying the shell layers. PS-PMMA (high T_g shell) core-shell structure exhibited increase in the elastic modulus by shell thickness whereas a reverse effect is observed in the PS-PBMA (low T_g shell) particles [110]. Also BLS together with the computational methods determined the elastic properties and vibrational modes of elongated PS spheres (spheroids) [66] in the class of anisotropic CC.

3.5 Temperature dependent BLS

Recently, temperature dependent BLS emerged as a powerful tool for the observation of glassy dynamics in polymer materials under three-dimensional confinement. The eigen modes revealed the glass transition temperature and gave strong evidence for the existence of a mobile surface layer [112]. Dry CC made of polystyrene nanoparticles of different sizes were measured at various temperatures starting from room temperature to above the point of glass transition. The trend in the vibration modes leads to first observation of softening temperature - the point at which the temperature dependent fundamental vibrational frequency $f_{1,2}$ reverses slope just below the glass transition. Interestingly, glass transition temperature is identified by the point where the vibrational modes disappear. The $f_{1,2}$ is dependent on the spherical symmetry and elastic parameters of the particles. Beyond this, another mode of vibration given by interaction-induced mode $f_{1,1}$ was identified which expresses the interparticle interactions in a crystal and is dependent on the thermal properties of particles.

A trend in BLS spectra of $d = 141$ nm sized PS particles is shown in Fig. 3.8 where the fundamental vibrational modes distinctly show the peak broadening and splitting with temperature. This behavior was attributed to the interaction between the particles. As shown in the spectrum obtained at $T = 296$ K, $f_{1,2}$ was fitted with a double Lorentzian function and identified with $2f_1$-f_2 as the exact frequency value. The interaction-induced mode is well represented by $If^2(f)$, which also represented the frequency shifts, broadening of the peak and splitting. As it is more relevant from Fig. 3.8C, frequency decreases with temperature and at a certain point, identified as the softening temperature, it starts increasing. The normalization process for various particles is done by plotting fd versus T. Thus, this method can be an interesting approach to identify the glass transition as well as the elastic modulus of nanoparticles.

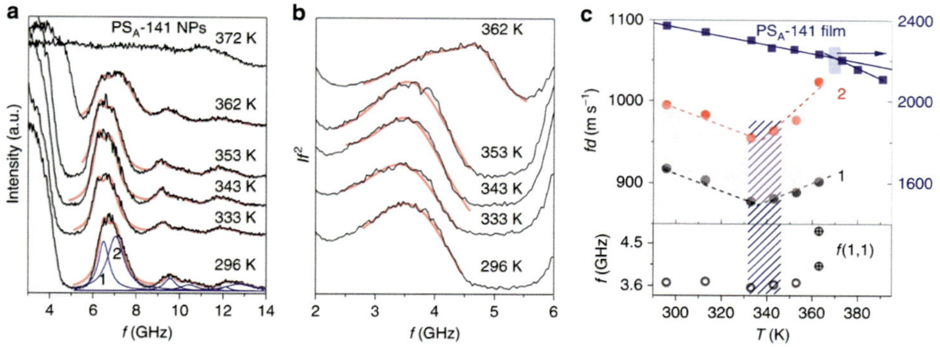

Figure 3.8

(A) BLS spectra of PS nanoparticles of diameter $d = 141$ nm diameter measured in various temperatures (B) Normalized power spectra If^2 versus f depicting interaction-induced mode of vibrations in particles $f_{1,1}$ (C) Temperature dependence of the $f_{1,1}$ (*black open circle*) mode and f_1, f_2 (*black and red solid circles*) of the fundamental vibration mode $f_{1,2}$. The *blue solid squares* stand for the longitudinal sound velocity (c_l) for the PS-141 film annealed at 393 K. The diagonal *blue line* region marks the softening temperature (T_s) and the point at which the eigen modes vanish correspond to the glass transition temperature (T_g). *Adapted with permissions from H. Kim, Y. Cang, E. Kang, B. Graczykowski, M. Secchi, M. Montagna, R.D. Priestley, E.M. Furst, G. Fytas, Direct observation of polymer surface mobility via nanoparticle vibrations, Nat. Commun. 9 (1) (2018). https://doi.org/10.1038/s41467-018-04854-w. Copyright obtained from Nature communications 2018.*

BLS was also used to explore the shape persistence of core particle within a core-shell colloidal sphere [113]. A silica shell of some tens of nm surrounding a polymer core protects the core from changing its shape at a temperature above the glass transition. This was verified by a temperature dependent eigenmode measurement in BLS. The SiO_2 shell acted as a nano armor for PS polymer core above the glass transition and mechanical rigidity of the core is improved compared to bare particles. Thus, BLS provided the single sphere mechanical properties in a colloidal crystal.

The eigen mode spectra of the temperature dependent BLS measurements of core-shell structures are shown in Fig. 3.9. The lowest frequency mode $f_{1,2}$ was highly intense which was quite similar to the bare PS particles spectra. And this was visible even after the melting above its T_g and cooling down to room temperature. This proved the shape persistence of the core particle within the shell which is accompanied by the enhancement of rigidity above its T_g.

3.6 Cold soldering of CCs—pressure dependent BLS

The mechanical reinforcement of polymer CCs by means of hydrostatic gas pressures investigated by in situ BLS was recently reported [70]. The sample was placed in a high-pressure cell and allowed for the BLS measurements at various gas pressures. This treatment was termed "cold soldering" since it was done below the T_g of bulk PS, but the plasticizing effect of the pressurized gas allowed for blending the shell material between touching particles, which resulted in the increase of the particle-particle bonding and thereby the contact area. This was identified by measuring the trend of $f_{1,1}$ vibrational mode in the polystyrene CCs. The procedure was accompanied by temperature treatments to fasten the pressure effects on nanoparticles. The treatments increased the mechanical strength of CCs while maintained their shape and periodicity. A robust nanostructure has an advantage over nano contamination and further environmental hazards [114,115]. Additionally, the malfunctioning of the devices due to fragility of component colloidal structures can be controlled to a great extent [70].

The preferential conditions for the soldering of particle is shown in Fig. 3.10. The region between the softening (T_s) and glass transition temperature (T_g) is the perfect regime for soldering of colloidal particles. These transition temperatures were identified from the temperature dependent BLS measurements at various gas pressures.

3.7 Hypersound tuning and filtering in 2D CCs

The GHz phonon propagation in self-assembled 2D CCs supported by ultrathin Si_3N_4 membranes revealed multiband hypersound filtering [116]. The 2D CCs offered a low cost and large area platform for harnessing GHz phonons and the energy dissipation issues

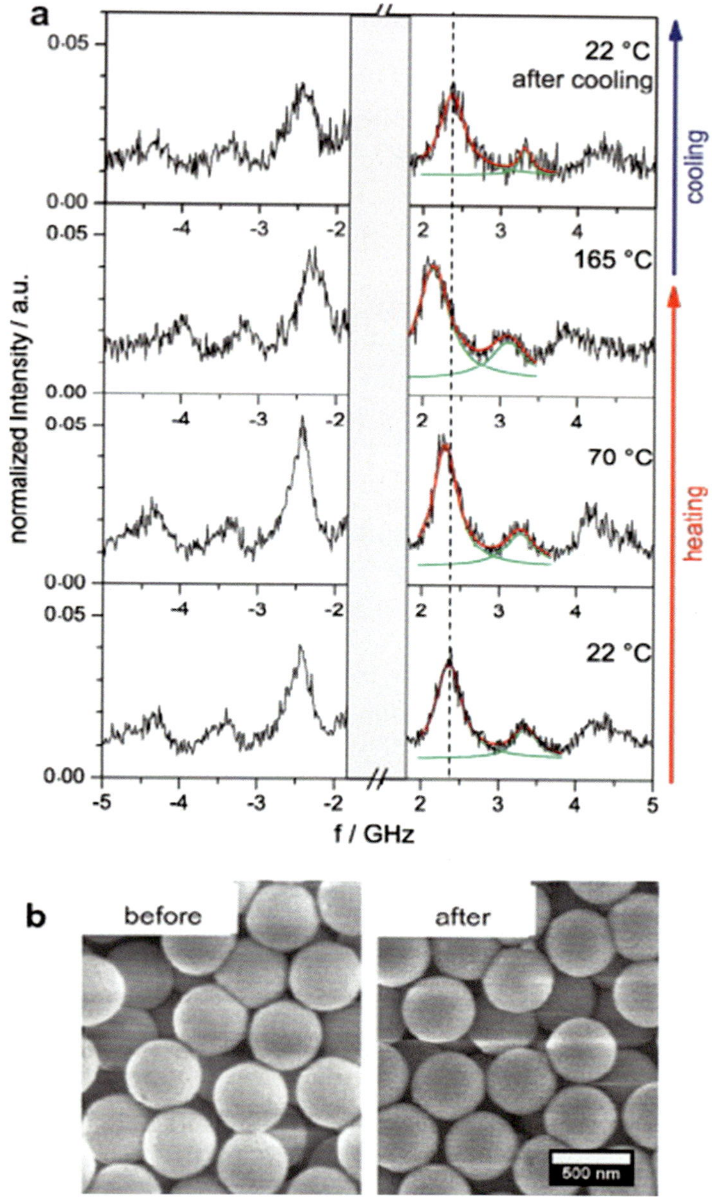

Figure 3.9

(A) Eigenmode spectra of PS-SiO_2 core-shell spheres before and after heating above the glass transition temperature. (B) SEM images of the CC before and after heating and cooling down to room temperature. *Adapted with permissions from T. Still, M. D'Acunzi, D. Vollmer, G. Fytas, Meso-spheres in nano-armor: probing the shape-persistence of molten polymer colloids, J. Colloid Interface Sci. 340 (1) (2009) 42–45. https://doi.org/10.1016/j.jcis.2009.08.022. Copyright obtained from Elsevier 2009.*

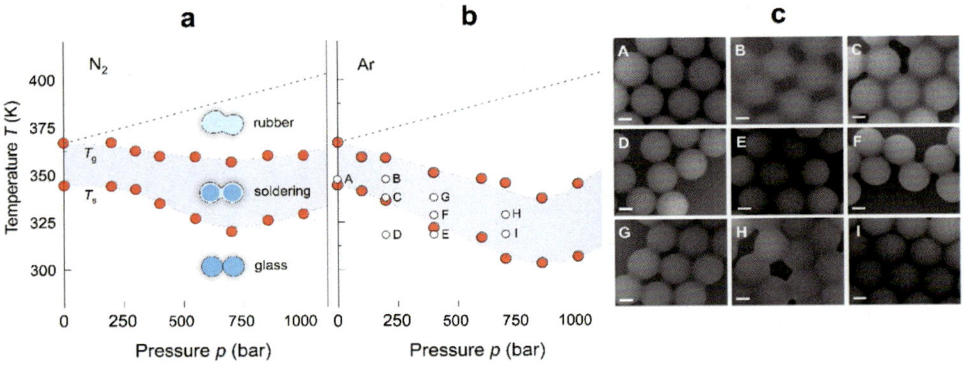

Figure 3.10

Pressure-Temperature phase diagram depicting the preferential conditions (*P*,*T*) of soldering (shaded area) identified for N_2 and Ar gas plasticization in PS colloidal crystals (C) The SEM images of PS CCs after the treatment at (*P*,*T*) corresponding to the points marked in (B). *Adapted from V. Babacic, J. Varghese, E. Coy, E. Kang, E.; Pochylski, M.; Gapinski, J.; Fytas, G.; Graczykowski, B. Mechanical reinforcement of polymer colloidal crystals by supercritical fluids, J. Colloid Interface Sci. 579 (2020) 786–793. https://doi.org/10.1016/j.jcis.2020.06.104.*

were controlled when CCs were supported by an ultrathin membrane. Because of the lower dimensionality of 2D CCs the phonon propagation is restricted to the plane of the membrane, which acts as a wave guide for the propagation of Lamb waves, which can be tailored by the thickness of the membrane. The phonon dispersion relation was measured via micro-Brillouin setup and three different bandgaps were observed. In addition to the Bragg gaps (lattice period), two hybridization gaps originating from particle-membrane contact resonance and particle-particle resonances were identified: the latter being in the super wavelength regime. The occurrence of band gaps allow the propagation of certain frequency range of phonons. This contributed to the hypersound filtering which can be tuned by the particle size governing adhesion, periodicity, and resonances. This was much promising in the further studies on GHz signal processing for future telecommunication systems, optomechanics and nonlinear effects. Thus, BLS provides a better platform for the non-contact measurement of interface mechanics in the submicrometer regime.

4. Conclusion

While approaching the centennial anniversary, Brillouin scattering spectroscopy has developed a prestigious foundation in diverse areas of scientific research. As a nondestructive tool for material analysis, a wide range of exciting systems has already been explored. For instance, BLS can provide a fingerprint of phonon propagation through CCs by probing their acoustic vibrations. This helps in understanding the thermomechanical properties thereby preferentially tuning the materials for specific applications. Temperature and pressure dependent studies provide new insight to widen the spectrum of Brillouin-based studies in fundamental research. Novel approaches and applications have yet to be demonstrated.

Acknowledgments

I am grateful to Dr. Bartlomiej Graczykowski for the guidance and support. Also acknowledging the financial support from the Foundation for Polish Science (POIR.04.04.00-00-5D1B/18) for the project "Photomechanical hybrid membranes nanomechanics (PHNOM)". Special thanks to my colleagues and collaborators in this project.

References

[1] T. Still, R. Sainidou, M. Retsch, U. Jonas, P. Spahn, G.P. Hellmann, G. Fytas, The "music" of Core −Shell spheres and hollow capsules: influence of the architecture on the mechanical properties at the nanoscale, Nano Lett. 8 (10) (2008) 3194−3199, https://doi.org/10.1021/nl801500n.

[2] J.-C. Beugnot, S. Lebrun, G. Pauliat, H. Maillotte, V. Laude, T. Sylvestre, Brillouin light scattering from surface acoustic waves in a subwavelength-diameter optical fibre, Nat. Commun. 5 (1) (2014) 5242, https://doi.org/10.1038/ncomms6242.

[3] E. Alonso-Redondo, M. Schmitt, Z. Urbach, C.M. Hui, R. Sainidou, P. Rembert, K. Matyjaszewski, M.R. Bockstaller, G. Fytas, A new class of tunable hypersonic phononic crystals based on polymer-tethered colloids, Nat. Commun. 6 (1) (2015), https://doi.org/10.1038/ncomms9309.

[4] B. Graczykowski, M. Sledzinska, F. Alzina, J. Gomis-Bresco, J.S. Reparaz, M.R. Wagner, C.M. Sotomayor Torres, Phonon dispersion in hypersonic two-dimensional phononic crystal membranes, Phys. Rev. B 91 (7) (2015) 075414, https://doi.org/10.1103/PhysRevB.91.075414.

[5] T. Still, W. Cheng, M. Retsch, R. Sainidou, J. Wang, U. Jonas, N. Stefanou, G. Fytas, Simultaneous occurrence of structure-directed and particle-resonance-induced phononic gaps in colloidal films, Phys. Rev. Lett. 100 (19) (2008) 194301, https://doi.org/10.1103/PhysRevLett.100.194301.

[6] Y. Muanenda, C.J. Oton, F. Di Pasquale, Application of Raman and Brillouin scattering phenomena in distributed optical fiber sensing, Front. Physiol. 7 (2019) 155, https://doi.org/10.3389/fphy.2019.00155.

[7] W. Zou, X. Long, J. Chen, Brillouin scattering in optical fibers and its application to distributed sensors, in: M. Yasin, H. Arof, S.W. Harun (Eds.), Advances in Optical Fiber Technology: Fundamental Optical Phenomena and Applications, InTech, 2015, https://doi.org/10.5772/59145.

[8] A.B. Singaraju, D. Bahl, L.L. Stevens, Brillouin light scattering: development of a near century-old technique for characterizing the mechanical properties of materials, AAPS PharmSciTech 20 (3) (2019) 109, https://doi.org/10.1208/s12249-019-1311-5.

[9] F. Palombo, D. Fioretto, Brillouin light scattering: applications in biomedical sciences, Chem. Rev. 119 (13) (2019) 7833−7847, https://doi.org/10.1021/acs.chemrev.9b00019.

[10] B.F. Kennedy, P. Wijesinghe, D.D. Sampson, The emergence of optical elastography in biomedicine, Nat. Photonics 11 (4) (2017) 215−221, https://doi.org/10.1038/nphoton.2017.6.

[11] Z. Meng, A.J. Traverso, C.W. Ballmann, M.A. Troyanova-Wood, V.V. Yakovlev, Seeing cells in a new light: a renaissance of Brillouin spectroscopy, Adv. Opt Photon 8 (2) (2016) 300, https://doi.org/10.1364/AOP.8.000300.

[12] J. Chan, A.H. Safavi-Naeini, J.T. Hill, S. Meenehan, O. Painter, Optimized optomechanical crystal cavity with acoustic radiation shield, Appl. Phys. Lett. 101 (8) (2012) 081115, https://doi.org/10.1063/1.4747726.

[13] A.H. Safavi-Naeini, O. Painter, Design of optomechanical cavities and waveguides on a simultaneous bandgap phononic-photonic crystal slab, Opt. Exp. 18 (14) (2010) 14926, https://doi.org/10.1364/OE.18.014926.

[14] Z. Coker, M. Troyanova-Wood, A.J. Traverso, T. Yakupov, Z.N. Utegulov, V.V. Yakovlev, Assessing performance of modern Brillouin spectrometers, Opt. Exp. 26 (3) (2018) 2400, https://doi.org/10.1364/OE.26.002400.
[15] F. Scarponi, S. Mattana, S. Corezzi, S. Caponi, L. Comez, P. Sassi, A. Morresi, M. Paolantoni, L. Urbanelli, C. Emiliani, L. Roscini, L. Corte, G. Cardinali, F. Palombo, J.R. Sandercock, D. Fioretto, High-performance versatile setup for simultaneous Brillouin-Raman microspectroscopy, Phys. Rev. X 7 (3) (2017) 031015, https://doi.org/10.1103/PhysRevX.7.031015.
[16] A.J. Traverso, J.V. Thompson, Z.A. Steelman, Z. Meng, M.O. Scully, V.V. Yakovlev, Dual Raman-Brillouin microscope for chemical and mechanical characterization and imaging, Anal. Chem. 87 (15) (2015) 7519—7523, https://doi.org/10.1021/acs.analchem.5b02104.
[17] D.A. Bozhko, V.I. Vasyuchka, A.V. Chumak, A.A. Serga, Magnon-phonon interactions in magnon spintronics, Low Temp. Phys. 46 (383) (2020), https://doi.org/10.1063/10.0000872.
[18] D. Rouxel, C. Thevenot, V.S. Nguyen, B. Vincent, Brillouin spectroscopy of polymer nanocomposites, in: Spectroscopy of Polymer Nanocomposites, Elsevier, 2016, pp. 362—392, https://doi.org/10.1016/B978-0-323-40183-8.00012-4.
[19] K. Berghaus, J. Zhang, S.H. Yun, G. Scarcelli, High-finesse sub-GHz-resolution spectrometer employing VIPA etalons of different dispersion, Opt. Lett. 40 (19) (2015) 4436, https://doi.org/10.1364/OL.40.004436.
[20] G. Scarcelli, S.H. Yun, Confocal Brillouin microscopy for three-dimensional mechanical imaging, Nat. Photonics 2 (1) (2008) 39—43, https://doi.org/10.1038/nphoton.2007.250.
[21] B.J. Eggleton, C.G. Poulton, P.T. Rakich, M.J. Steel, G. Bahl, Brillouin integrated photonics, Nat. Photonics 13 (10) (2019) 664—677, https://doi.org/10.1038/s41566-019-0498-z.
[22] B. Hafizi, J.P. Palastro, J.R. Peñano, T.G. Jones, L.A. Johnson, M.H. Helle, D. Kaganovich, Y.H. Chen, A.B. Stamm, Stimulated Raman and Brillouin scattering, nonlinear focusing, thermal blooming, and optical breakdown of a laser beam propagating in water, J. Opt. Soc. Am. B 33 (10) (2016) 2062, https://doi.org/10.1364/JOSAB.33.002062.
[23] G. Antonacci, T. Beck, A. Bilenca, J. Czarske, K. Elsayad, J. Guck, K. Kim, B. Krug, F. Palombo, R. Prevedel, G. Scarcelli, Recent progress and current opinions in Brillouin microscopy for life science applications, Biophys. Rev. 12 (3) (2020) 615—624, https://doi.org/10.1007/s12551-020-00701-9.
[24] F. Palombo, M. Madami, N. Stone, D. Fioretto, Mechanical mapping with chemical specificity by confocal Brillouin and Raman microscopy, Analyst 139 (4) (2014) 729—733, https://doi.org/10.1039/C3AN02168H.
[25] S. Caponi, S. Mattana, M. Mattarelli, M. Alunni Cardinali, L. Urbanelli, K. Sagini, C. Emiliani, D. Fioretto, Correlative Brillouin and Raman spectroscopy data acquired on single cells, Data Brief 29 (2020) 105223, https://doi.org/10.1016/j.dib.2020.105223.
[26] K. Brown, A.W. Brown, B.G. Colpitts, in: D. Inaudi, W. Ecke, B. Culshaw, K.J. Peters, E. Udd (Eds.), Combined Raman and Brillouin Scattering Sensor for Simultaneous High-Resolution Measurement of Temperature and Strain, 2006, p. 616716, https://doi.org/10.1117/12.657643. San Diego, CA.
[27] T. Still, High Frequency Acoustics in Colloid-Based Meso- and Nanostructures by Spontaneous Brillouin Light Scattering, Springer Theses; Springer Berlin Heidelberg, Berlin, Heidelberg, 2010, https://doi.org/10.1007/978-3-642-13483-8.
[28] A.V. Akimov, Y. Tanaka, A.B. Pevtsov, S.F. Kaplan, V.G. Golubev, S. Tamura, D.R. Yakovlev, M. Bayer, Hypersonic modulation of light in three-dimensional photonic and phononic band-gap materials, Phys. Rev. Lett. 101 (3) (2008) 033902, https://doi.org/10.1103/PhysRevLett.101.033902.
[29] I.L. Fabelinskiĭ, The prediction and discovery of Rayleigh line fine structure, Phys.-Uspekhi 43 (1) (2000) 89—103, https://doi.org/10.1070/PU2000v043n01ABEH000582.
[30] L. Brillouin, Diffusion de La Lumière et Des Rayons X Par Un Corps Transparent Homogène: influence de l'agitation Thermique, Ann. Phys. 9 (17) (1922) 88—122, https://doi.org/10.1051/anphys/192209170088.
[31] E. Gross, Change of wave-length of light due to elastic heat waves at scattering in liquids, Nature 126 (3171) (1930) 201—202, https://doi.org/10.1038/126201a0.
[32] E. Gross, The splitting of spectral lines at scattering of light by liquids, Nature 126 (3176) (1930) 400, https://doi.org/10.1038/126400a0.

[33] Z. Meng, T. Thakur, C. Chitrakar, M.K. Jaiswal, A.K. Gaharwar, V.V. Yakovlev, Assessment of local heterogeneity in mechanical properties of nanostructured hydrogel networks, ACS Nano 11 (8) (2017) 7690–7696, https://doi.org/10.1021/acsnano.6b08526.
[34] S. Speziale, H. Marquardt, T.S. Duffy, Brillouin scattering and its application in geosciences, Rev. Mineral. Geochem. 78 (1) (2014) 543–603, https://doi.org/10.2138/rmg.2014.78.14.
[35] H.Z. Cummins, R.W. Gammon, Rayleigh and Brillouin scattering IN benzene: depolarization factors, Appl. Phys. Lett. 6 (8) (1965) 171–173, https://doi.org/10.1063/1.1754220.
[36] S.E.A. Hakim, W.J. Comley, Acoustic velocity dispersion in some non-associated organic liquids, Nature 208 (5015) (1965) 1082–1083, https://doi.org/10.1038/2081082a0.
[37] R.Y. Chiao, B.P. Stoicheff, Brillouin scattering in liquids excited by the He–Ne maser, J. Opt. Soc. Am. 54 (10) (1964) 1286, https://doi.org/10.1364/JOSA.54.001286.
[38] M.H. Grimsditch, A.K. Ramdas, Brillouin scattering in diamond, Phys. Rev. B 11 (8) (1975) 3139–3148, https://doi.org/10.1103/PhysRevB.11.3139.
[39] A. Asenbaum, Computer-controlled Fabry-Perot interferometer for Brillouin spectroscopy, Appl. Opt. 18 (4) (1979) 540, https://doi.org/10.1364/AO.18.000540.
[40] S.M. Lindsay, M.W. Anderson, J.R. Sandercock, Construction and alignment of a high performance multipass vernier tandem Fabry–Perot interferometer, Rev. Sci. Instrum. 52 (10) (1981) 1478–1486, https://doi.org/10.1063/1.1136479.
[41] C. Roychoudhuri, M. Hercher, Stable multipass Fabry-Perot interferometer: design and analysis, Appl. Opt. 16 (9) (1977) 2514, https://doi.org/10.1364/AO.16.002514.
[42] J.G. Dil, N.C.J.A. van Hijningen, F. van Dorst, R.M. Aarts, Tandem multipass Fabry-Perot interferometer for Brillouin scattering, Appl. Opt. 20 (8) (1981) 1374, https://doi.org/10.1364/AO.20.001374.
[43] I. Remer, A. Bilenca, High-speed stimulated Brillouin scattering spectroscopy at 780 nm, APL Photon. 1 (6) (2016) 061301, https://doi.org/10.1063/1.4953620.
[44] H. Marquardt, N. Waeselmann, M. Wehber, R.J. Angel, M. Gospodinov, B. Mihailova, High-pressure Brillouin scattering of the single-crystal PbSc_½Ta_½O_{3} relaxor ferroelectric, Phys. Rev. B 87 (18) (2013) 184113, https://doi.org/10.1103/PhysRevB.87.184113.
[45] X. Zhang, J.D. Comins, A.G. Every, P.R. Stoddart, W. Pang, T.E. Derry, Surface Brillouin scattering study of the surface excitations in amorphous silicon layers produced by ion bombardment, Phys. Rev. B 58 (20) (1998) 13677–13685, https://doi.org/10.1103/PhysRevB.58.13677.
[46] J. Milano, M. Grimsditch, Magnetic field effects on the NiO magnon spectra, Phys. Rev. B 81 (9) (2010) 094415, https://doi.org/10.1103/PhysRevB.81.094415.
[47] L. Sui, L. Huang, P. Podsiadlo, N.A. Kotov, J. Kieffer, Brillouin light scattering investigation of the mechanical properties of layer-by-layer assembled cellulose nanocrystal films, Macromolecules 43 (22) (2010) 9541–9548, https://doi.org/10.1021/ma1016488.
[48] F. Aliotta, J. Gapiński, M. Pochylski, R.C. Ponterio, F. Saija, G. Salvato, Excess compressibility in binary liquid mixtures, J. Chem. Phys. 126 (224508) (2007), https://doi.org/10.1063/1.2745292.
[49] S. Cusack, A. Miller, Determination of the elastic constants of collagen by Brillouin light scattering, J. Mol. Biol. 135 (1) (1979) 39–51, https://doi.org/10.1016/0022-2836(79)90339-5.
[50] M. Pochylski, F. Aliotta, R.C. Ponterio, F. Saija, J. Gapiński, Some evidence of scaling behavior in the relaxation dynamics of aqueous polymer solutions, J. Phys. Chem. B 114 (4) (2010) 1614–1620, https://doi.org/10.1021/jp9052456.
[51] G. Scarcelli, W.J. Polacheck, H.T. Nia, K. Patel, A.J. Grodzinsky, R.D. Kamm, S.H. Yun, Noncontact three-dimensional mapping of intracellular hydromechanical properties by Brillouin microscopy, Nat. Methods 12 (12) (2015) 1132–1134, https://doi.org/10.1038/nmeth.3616.
[52] F. Palombo, C.P. Winlove, R.S. Edginton, E. Green, N. Stone, S. Caponi, M. Madami, D. Fioretto, Biomechanics of fibrous proteins of the extracellular matrix studied by Brillouin scattering, J. R. Soc. Interface 11 (101) (2014) 20140739, https://doi.org/10.1098/rsif.2014.0739.
[53] S.V. Adichtchev, N.V. Surovtsev, T.A. Duda, A.G. Milekhin, Brillouin scattering from Langmuir–Blodgett films doped with CdS and CuS nanoclusters, Phys. Status Solidi B 256 (2) (2019) 1800328, https://doi.org/10.1002/pssb.201800328.
[54] K.J. Koski, P. Akhenblit, K. McKiernan, J.L. Yarger, Non-invasive determination of the complete elastic moduli of spider silks, Nat. Mater. 12 (3) (2013) 262–267, https://doi.org/10.1038/nmat3549.

[55] C. Bottari, L. Comez, M. Paolantoni, S. Corezzi, F. D'Amico, A. Gessini, C. Masciovecchio, B. Rossi, Hydration properties and water structure in aqueous solutions of native and modified cyclodextrins by UV Raman and Brillouin scattering, J. Raman Spectrosc. 49 (6) (2018) 1076–1085, https://doi.org/10.1002/jrs.5372.
[56] M. Pochylski, F. Aliotta, Z. Blaszczak, J. Gapiński, Structuring effects and hydration phenomena in poly(ethylene glycol)/water mixtures investigated by Brillouin scattering, J. Phys. Chem. B 110 (41) (2006) 20533–20539, https://doi.org/10.1021/jp0620973.
[57] Mikołaj Pochylski, Jacek Gapiński, Brillouin Scattering Study of Polyethylene Glycol/Water System below Crystallization Temperature, J. Phys. Chem. B 114 (8) (2010) 2644–2649, https://doi.org/10.1021/jp910783j.
[58] G. Maurice, D. Rouxel, B. Vincent, R. Hadji, J.-F. Schmitt, M. Taghite, R. Rahouadj, Investigation of elastic constants of polymer/nanoparticles composites using the Brillouin spectroscopy and the mechanical homogenization modeling, Polym. Eng. Sci. 53 (7) (2013) 1502–1511, https://doi.org/10.1002/pen.23397.
[59] W. Cheng, J. Wang, U. Jonas, G. Fytas, N. Stefanou, Observation and tuning of hypersonic bandgaps in colloidal crystals, Nat. Mater. 5 (10) (2006) 830–836, https://doi.org/10.1038/nmat1727.
[60] S.O. Demokritov, V.E. Demidov, O. Dzyapko, G.A. Melkov, A.A. Serga, B. Hillebrands, A.N. Slavin, Bose–Einstein condensation of quasi-equilibrium magnons at room temperature under pumping, Nature 443 (7110) (2006) 430–433, https://doi.org/10.1038/nature05117.
[61] C.W. Ballmann, Z. Meng, A.J. Traverso, M.O. Scully, V.V. Yakovlev, Impulsive Brillouin microscopy, Optica 4 (1) (2017) 124, https://doi.org/10.1364/OPTICA.4.000124.
[62] S. Mattana, M. Alunni Cardinali, S. Caponi, D. Casagrande Pierantoni, L. Corte, L. Roscini, G. Cardinali, D. Fioretto, High-contrast Brillouin and Raman micro-spectroscopy for simultaneous mechanical and chemical investigation of microbial biofilms, Biophys. Chem. 229 (2017) 123–129, https://doi.org/10.1016/j.bpc.2017.06.008.
[63] S. Gehrsitz, H. Sigg, H. Siegwart, M. Krieger, C. Heine, R. Morf, F.K. Reinhart, W. Martin, H. Rudigier, Tandem triple-pass Fabry–Perot interferometer for applications in the near infrared, Appl. Opt. 36 (22) (1997) 5355, https://doi.org/10.1364/AO.36.005355.
[64] F. Bencivenga, A. Battistoni, D. Fioretto, A. Gessini, J.R. Sandercock, C. Masciovecchio, A high resolution ultraviolet Brillouin scattering set-up, Rev. Sci. Instrum. 83 (10) (2012) 103102, https://doi.org/10.1063/1.4756690.
[65] J.K. Krüger, A. Marx, L. Peetz, R. Roberts, H.-G. Unruh, Simultaneous determination of elastic and optical properties of polymers by high performance Brillouin spectroscopy using different scattering geometries, Colloid Polym. Sci. 264 (5) (1986) 403–414, https://doi.org/10.1007/BF01419544.
[66] D. Schneider, P.J. Beltramo, M. Mattarelli, P. Pfleiderer, J. Vermant, D. Crespy, M. Montagna, E.M. Furst, G. Fytas, Elongated polystyrene spheres as resonant building blocks in anisotropic colloidal crystals, Soft Matter 9 (38) (2013) 9129, https://doi.org/10.1039/c3sm50959a.
[67] T. Still, W. Cheng, M. Retsch, U. Jonas, G. Fytas, Colloidal systems: a promising material class for tailoring sound propagation at high frequencies, J. Phys. Condens. Matter 20 (40) (2008) 404203, https://doi.org/10.1088/0953-8984/20/40/404203.
[68] H. Cong, B. Yu, J. Tang, Z. Li, X. Liu, Current status and future developments in preparation and application of colloidal crystals, Chem. Soc. Rev. 42 (19) (2013) 7774, https://doi.org/10.1039/c3cs60078e.
[69] N. Vogel, M. Retsch, C.-A. Fustin, A. del Campo, U. Jonas, Advances in colloidal assembly: the design of structure and hierarchy in two and three dimensions, Chem. Rev. 115 (13) (2015) 6265–6311, https://doi.org/10.1021/cr400081d.
[70] V. Babacic, J. Varghese, E. Coy, E. Kang, M. Pochylski, J. Gapinski, G. Fytas, B. Graczykowski, Mechanical reinforcement of polymer colloidal crystals by supercritical fluids, J. Colloid Interface Sci. 579 (2020) 786–793, https://doi.org/10.1016/j.jcis.2020.06.104.
[71] E. Ozbay, Plasmonics: merging photonics and electronics at nanoscale dimensions, Science 311 (5758) (2006) 189–193, https://doi.org/10.1126/science.1114849.
[72] J.D. Joannopoulos (Ed.), Photonic Crystals: Molding the Flow of Light, second ed., Princeton University Press, Princeton, 2008.
[73] H.S. Lim, M.H. Kuok, S.C. Ng, Z.K. Wang, Brillouin observation of bulk and confined acoustic waves in silica microspheres, Appl. Phys. Lett. 84 (21) (2004) 4182–4184, https://doi.org/10.1063/1.1756206.

[74] H. Yoshimura, Two-dimensional crystals of apoferritin, Adv. Biophys. 34 (1997) 93–107, https://doi.org/10.1016/S0065-227X(97)89634-7.
[75] H. Cong, W. Cao, Array patterns of binary colloidal crystals, J. Phys. Chem. B 109 (5) (2005) 1695–1698, https://doi.org/10.1021/jp048269i.
[76] A. Imhof, D.J. Pine, Ordered macroporous materials by emulsion templating, Nature 389 (6654) (1997) 948–951, https://doi.org/10.1038/40105.
[77] J.F. Galisteo-López, M. Ibisate, R. Sapienza, L.S. Froufe-Pérez, Á. Blanco, C. López, Self-assembled photonic structures, Adv. Mater. 23 (1) (2011) 30–69, https://doi.org/10.1002/adma.201000356.
[78] G. von Freymann, V. Kitaev, B.V. Lotsch, G.A. Ozin, Bottom-up assembly of photonic crystals, Chem. Soc. Rev. 42 (7) (2013) 2528–2554, https://doi.org/10.1039/C2CS35309A.
[79] M. Rey, T. Yu, R. Guenther, K. Bley, N. Vogel, A dirty story: improving colloidal monolayer formation by understanding the effect of impurities at the air/water interface, Langmuir 35 (1) (2019) 95–103, https://doi.org/10.1021/acs.langmuir.8b02605.
[80] V. Lotito, T. Zambelli, Approaches to self-assembly of colloidal monolayers: a guide for nanotechnologists, Adv. Colloid Interface Sci. 246 (2017) 217–274, https://doi.org/10.1016/j.cis.2017.04.003.
[81] A.R. Tao, J. Huang, P. Yang, Langmuir–Blodgettry of nanocrystals and nanowires, Acc. Chem. Res. 41 (12) (2008) 1662–1673, https://doi.org/10.1021/ar8000525.
[82] H. Lamb, On the vibrations of an elastic sphere, Proc. Lond. Math. Soc. s1–13 (1) (1881) 189–212, https://doi.org/10.1112/plms/s1-13.1.189.
[83] J.-P. Montagner, G. Roult, Normal modes of the Earth, J. Phys. Conf. Ser. 118 (2008) 012004, https://doi.org/10.1088/1742-6596/118/1/012004.
[84] T. Still, M. Mattarelli, D. Kiefer, G. Fytas, M. Montagna, Eigenvibrations of submicrometer colloidal spheres, J. Phys. Chem. Lett. 1 (16) (2010) 2440–2444, https://doi.org/10.1021/jz100774b.
[85] W. Cheng, J.J. Wang, U. Jonas, W. Steffen, G. Fytas, R.S. Penciu, E.N. Economou, The spectrum of vibration modes in soft opals, J. Chem. Phys. 123 (12) (2005) 121104, https://doi.org/10.1063/1.2046607.
[86] M.-H. Lu, L. Feng, Y.-F. Chen, Phononic crystals and acoustic metamaterials, Mater. Today 12 (12) (2009) 34–42, https://doi.org/10.1016/S1369-7021(09)70315-3.
[87] Y.-F. Wang, Y.-Z. Wang, B. Wu, W. Chen, Y.-S. Wang, Tunable and active phononic crystals and metamaterials, Appl. Mech. Rev. 72 (4) (2020) 040801, https://doi.org/10.1115/1.4046222.
[88] E. Yablonovitch, Inhibited spontaneous emission in solid-state physics and electronics, Phys. Rev. Lett. 58 (20) (1987) 2059–2062, https://doi.org/10.1103/PhysRevLett.58.2059.
[89] S. John, Strong localization of photons in certain disordered dielectric superlattices, Phys. Rev. Lett. 58 (23) (1987) 2486–2489, https://doi.org/10.1103/PhysRevLett.58.2486.
[90] J.O. Vasseur, P.A. Deymier, B. Chenni, B. Djafari-Rouhani, L. Dobrzynski, D. Prevost, Experimental and theoretical evidence for the existence of absolute acoustic band gaps in two-dimensional solid phononic crystals, Phys. Rev. Lett. 86 (14) (2001) 3012–3015, https://doi.org/10.1103/PhysRevLett.86.3012.
[91] F.R. Montero de Espinosa, E. Jiménez, M. Torres, Ultrasonic band gap in a periodic two-dimensional composite, Phys. Rev. Lett. 80 (6) (1998) 1208–1211, https://doi.org/10.1103/PhysRevLett.80.1208.
[92] Z. Liu, Locally resonant sonic materials, Science 289 (5485) (2000) 1734–1736, https://doi.org/10.1126/science.289.5485.1734.
[93] T. Gorishnyy, C.K. Ullal, M. Maldovan, G. Fytas, E.L. Thomas, Hypersonic phononic crystals, Phys. Rev. Lett. 94 (11) (2005) 115501, https://doi.org/10.1103/PhysRevLett.94.115501.
[94] M. Maldovan, E.L. Thomas, Simultaneous localization of photons and phonons in two-dimensional periodic structures, Appl. Phys. Lett. 88 (25) (2006) 251907, https://doi.org/10.1063/1.2216885.
[95] J.-H. Jang, C.K. Ullal, T. Gorishnyy, V.V. Tsukruk, E.L. Thomas, Mechanically tunable three-dimensional elastomeric network/air structures via interference lithography, Nano Lett. 6 (4) (2006) 740–743, https://doi.org/10.1021/nl052577q.
[96] J. Wang, S. Ahl, Q. Li, M. Kreiter, T. Neumann, K. Burkert, W. Knoll, U. Jonas, Structural and optical characterization of 3D binary colloidal crystal and inverse opal films prepared by direct Co-deposition, J. Mater. Chem. 18 (9) (2008) 981, https://doi.org/10.1039/b715329e.
[97] E. Duval, A. Boukenter, B. Champagnon, Vibration eigenmodes and size of microcrystallites in glass: observation by very-low-frequency Raman scattering, Phys. Rev. Lett. 56 (19) (1986) 2052–2055, https://doi.org/10.1103/PhysRevLett.56.2052.

[98] A. Courty, A. Mermet, P.A. Albouy, E. Duval, M.P. Pileni, Vibrational coherence of self-organized silver nanocrystals in f.c.c. supra-crystals, Nat. Mater. 4 (5) (2005) 395–398, https://doi.org/10.1038/nmat1366.
[99] M.H. Kuok, H.S. Lim, S.C. Ng, N.N. Liu, Z.K. Wang, Brillouin study of the quantization of acoustic modes in nanospheres, Phys. Rev. Lett. 90 (25) (2003) 255502, https://doi.org/10.1103/PhysRevLett.90.255502.
[100] R.S. Penciu, G. Fytas, E.N. Economou, W. Steffen, S.N. Yannopoulos, Acoustic excitations in suspensions of soft colloids, Phys. Rev. Lett. 85 (21) (2000) 4622–4625, https://doi.org/10.1103/PhysRevLett.85.4622.
[101] D.A. Mazurenko, X. Shan, J.C.P. Stiefelhagen, C.M. Graf, A. van Blaaderen, J.I. Dijkhuis, Coherent vibrations of submicron spherical gold shells in a photonic crystal, Phys. Rev. B 75 (16) (2007) 161102, https://doi.org/10.1103/PhysRevB.75.161102.
[102] J.H. Hodak, A. Henglein, G.V. Hartland, Size dependent properties of Au particles: coherent excitation and dephasing of acoustic vibrational modes, J. Chem. Phys. 111 (18) (1999) 8613–8621, https://doi.org/10.1063/1.480202.
[103] C.-A. Fustin, G. Glasser, H.W. Spiess, U. Jonas, Parameters influencing the templated growth of colloidal crystals on chemically patterned surfaces, Langmuir 20 (21) (2004) 9114–9123, https://doi.org/10.1021/la0489413.
[104] I.E. Psarobas, A. Modinos, R. Sainidou, N. Stefanou, Acoustic properties of colloidal crystals, Phys. Rev. B 65 (6) (2002) 064307, https://doi.org/10.1103/PhysRevB.65.064307.
[105] T. Gorishnyy, J.-H. Jang, C. Koh, E.L. Thomas, Direct observation of a hypersonic band gap in two-dimensional single crystalline phononic structures, Appl. Phys. Lett. 91 (12) (2007) 121915, https://doi.org/10.1063/1.2786605.
[106] X. Jing, P. Sheng, M. Zhou, Theory of acoustic excitations in colloidal suspensions, Phys. Rev. Lett. 66 (9) (1991) 1240–1243, https://doi.org/10.1103/PhysRevLett.66.1240.
[107] J. Liu, L. Ye, D.A. Weitz, P. Sheng, Novel acoustic excitations in suspensions of hard-sphere colloids, Phys. Rev. Lett. 65 (20) (1990) 2602–2605, https://doi.org/10.1103/PhysRevLett.65.2602.
[108] M. Montagna, Brillouin and Raman scattering from the acoustic vibrations of spherical particles with a size comparable to the wavelength of the light, Phys. Rev. B 77 (4) (2008) 045418, https://doi.org/10.1103/PhysRevB.77.045418.
[109] W. Cheng, R. Sainidou, P. Burgardt, N. Stefanou, A. Kiyanova, M. Efremov, G. Fytas, P.F. Nealey, Elastic properties and glass transition of supported polymer thin films, Macromolecules 40 (20) (2007) 7283–7290, https://doi.org/10.1021/ma071227i.
[110] E. Kang, B. Graczykowski, U. Jonas, D. Christie, L.A.G. Gray, D. Cangialosi, R.D. Priestley, G. Fytas, Shell architecture strongly influences the glass transition, surface mobility, and elasticity of polymer core-shell nanoparticles, Macromolecules 52 (14) (2019) 5399–5406, https://doi.org/10.1021/acs.macromol.9b00766.
[111] E. Kang, H. Kim, L.A.G. Gray, D. Christie, U. Jonas, B. Graczykowski, E.M. Furst, R.D. Priestley, G. Fytas, Ultrathin shell layers dramatically influence polymer nanoparticle surface mobility, Macromolecules 51 (21) (2018) 8522–8529, https://doi.org/10.1021/acs.macromol.8b01804.
[112] H. Kim, Y. Cang, E. Kang, B. Graczykowski, M. Secchi, M. Montagna, R.D. Priestley, E.M. Furst, G. Fytas, Direct observation of polymer surface mobility via nanoparticle vibrations, Nat. Commun. 9 (1) (2018), https://doi.org/10.1038/s41467-018-04854-w.
[113] T. Still, M. D'Acunzi, D. Vollmer, G. Fytas, Mesospheres in nano-armor: probing the shape-persistence of molten polymer colloids, J. Colloid Interface Sci. 340 (1) (2009) 42–45, https://doi.org/10.1016/j.jcis.2009.08.022.
[114] Y. Chae, Y.-J. An, Effects of micro- and nanoplastics on aquatic ecosystems: current research trends and Perspectives, Mar. Pollut. Bull. 124 (2) (2017) 624–632, https://doi.org/10.1016/j.marpolbul.2017.01.070.
[115] A. Haegerbaeumer, M.-T. Mueller, H. Fueser, W. Traunspurger, Impacts of micro- and nano-sized plastic particles on benthic invertebrates: a literature review and gap analysis, Front. Environ. Sci. 7 (2019) 17, https://doi.org/10.3389/fenvs.2019.00017.
[116] B. Graczykowski, N. Vogel, K. Bley, H.-J. Butt, G. Fytas, Multiband hypersound filtering in two-dimensional colloidal crystals: adhesion, resonances, and periodicity, Nano Lett. (2020), https://doi.org/10.1021/acs.nanolett.9b05101.

CHAPTER 4

In-situ microstructural measurements: coupling mechanical, dielectrical, thermal analysis with Raman spectroscopy for nanocomposites characterization

Isabelle Royaud[1], Marc Ponçot[1], David Chapron[2], Patrice Bourson[2]
[1]Université de Lorraine, CNRS, IJL, Nancy, France; [2]Université de Lorraine, CentraleSupélec, LMOPS, Metz, France

Chapter Outline

Design, Fabrication, and Characterization of Multifunctional Nanomaterials. https://doi.org/10.1016/B978-0-12-820558-7.00017-0

1. Introduction

From a technological point of view, vibrational spectroscopy techniques have evolved considerably in recent years. They are faster, have better spectral and spatial resolutions, and are more sensitive. Moreover, when combined with statistical analysis methods, they allow atline/online or real-time measurements with times compatible with industrial processes. These technological evolutions (more powerful sources, more sensitive detectors, and use of optical fibers) also lead to the miniaturization of instruments which can now be used in industrial or hostile environments. But this requires an upstream study or learning of Raman to respond and model this measurement to identify the spectroscopic parameter(s) related to the physicochemical properties and then to build a model.

In this chapter, we will present several examples of results of coupling experimental techniques and in situ measurements from various thesis works.

We will show through these selected examples that filler addition and its particle size have an influence on the Raman spectroscopy measurement; in particular, we will show that this response can be related to the size or volume of defects present in the material and also that a complementary information can be drawn from in situ coupling of experiments to interpret the various results obtained (thermal or dielectric).

2. What is the advantage of in-situ or real-time measurements?

In the laboratory, real-time measurements allow

- a better knowledge of the materials,
- a better understanding of kinetics,
- monitoring and surveillance of physicochemical properties during reactions,
- coupling with other techniques and obtaining correlated information, and
- a better understanding of chemical and physical properties.

For industry, this type of measurement allows

- direct control of production,
- to reduce imperfections and nonconforming products.
- to provide the customer with a product that meets specifications,
- to improve the process (reliability, efficiency, and safety),
- to better understand the different stages of the process, and
- to identify the key steps of the process in order to optimize them.

3. Technological innovations and online measures

Over the last few decades, numerous technological innovations have contributed to making vibrational spectroscopy more and more efficient and suitable for in situ measurements. For example, these measurements have gone from a characteristic acquisition time of a few thousand seconds to a few seconds in 10 years. This is thanks to increasingly powerful sources, in particular powerful and compact lasers, but also to a set of elements in the measurement chain: sensitive detectors, high-performance optical filters and modulators, adapted optical fibers, electronic systems for fast data acquisition and processing, and finally thanks to probes adapted to the measurement or industrial environment.

The implementation of an optical spectroscopy measurement system on a process requires learning the technique by correlations with proven reference techniques and then by modeling.

Vibrational spectroscopy techniques such as Raman spectroscopy are particularly well suited to perform in situ measurements in real time. In particular, they are sensitive, fast and nondestructive. Also, depending on the type of optical probe used, it is possible to make measurements in the heart of the process, using immersion probes for example, or noninvasive measurements, such as through an optical window. Finally, the measurement can be offset, i.e., an optical fiber can be used between the probe on the process and the spectrometer located far from the industrial process.

Depending on the needs related to the process and the measurement, equipment with lower performance than a laboratory spectrometer can be used. The main constraint is to have a measurement time compatible with the characteristic time of the industrial process. In this context, it is often necessary to combine it with statistical tools (chemometrics) to best model the measurement or the property to be measured.

4. What is the probed volume?

The question arises as to what to measure with an in situ Raman spectroscopy measurement.

Raman spectroscopy is first of all a laser, in particular a laser focused on the sample. Therefore, it is possible to know the size of the laser spot and consequently the volume probed by the Raman spectrometer. the focusing of a Gaussian beam (the laser) is called the waist and allows us to know the size of the laser spot but also the depth of field and the volume probed.

It depends on the wavelength of the laser, the size of the beam, the focal length used in the case of a lens but also the numerical aperture in the case of a microscope objective. In Fig. 4.1, we compare the size of the observed area as a function of, e.g., the numerical aperture used for a microscope objective (red circles) or the focal length of a lens (orange circles) in comparison with e.g., a SEM image of spherolites in isotactic polypropylene.

But this also results in a different volume and depth analyzed, a microscope objective will scan only a small depth in the sample contrary to a lens that can analyze the complete thickness of the sample (Fig. 4.2).

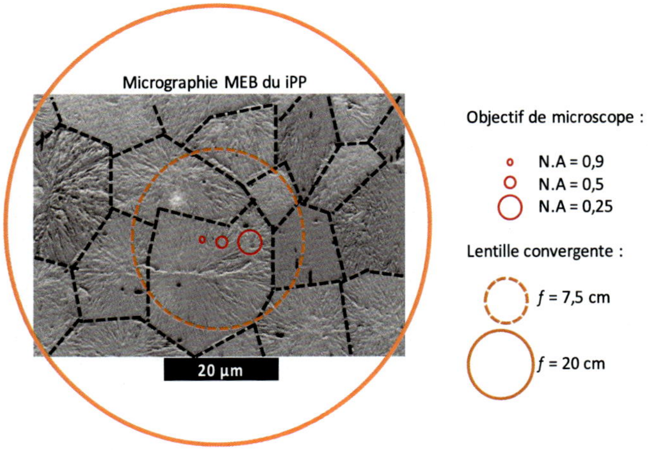

Figure 4.1
Surface of measurements by microscope objectives and lenses of different focal lengths with respect to the surface of a polymer and its spherulitic structure. (Ph.D. S. Chaudemanche 2013 2013LORR0263 [1].)

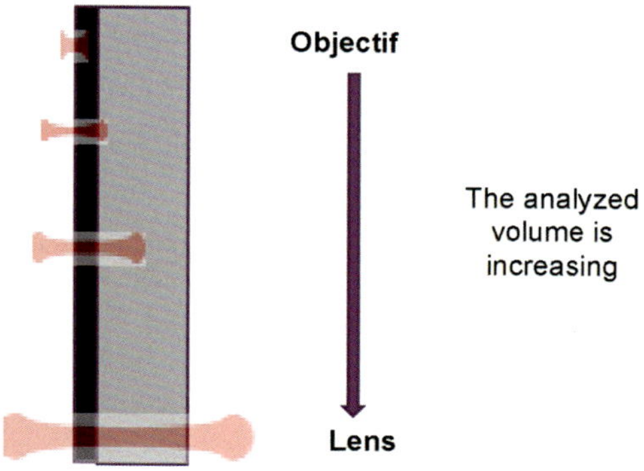

Figure 4.2
Schematic representation of the depth of field in a material comparison between microscope objectives and lenses. (Ph.D. S. Chaudemanche 2013 2013LORR0263 [1].)

It is a great advantage of optical spectroscopy, and especially with the Raman spectroscopy, to have the possibility, depending on the choice of the optical configuration of the experimental set-up, to analyze either the surface or the volume and even the entire volume of the sample.

5. DSC/Raman coupling system to describe the thermal microstructural behavior of thermoplastic polymers

Contributors: M. Vietmann, S. Chaudemanche, K. Ben Hafsia, M. Donnay, S. Saidi, A. Filliung.

To acquire the vibrational signatures of polymers by Raman spectroscopy in situ during thermal treatments, the acquisition parameters of the spectrometer have been specifically set in order to allow at least to record one spectrum every Celsius degree during heating or cooling (Fig. 4.3).

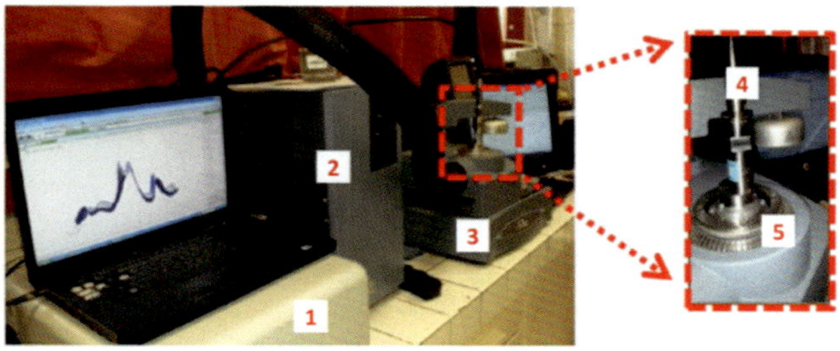

Figure 4.3
Experimental device for thermal characterization used in the laboratory [2,3]. 1: Raman spectrometer, 2: Cooling cell (Refrigerated Cooling System 90), 3: Differential Scanning Calorimeter Q200, 4: Raman probe, 5: Furnace.

5.1 Semicrystalline polymorphism identification

PVDF is a semicrystalline thermoplastic polymer that crystallizes in several phases α, β, γ, and so forth. Different methods have been used to characterize the polymer materials and to follow for example their crystallization kinetics. These techniques can be complementary, in particular to identify the different phases of a polymer, in this case PVDF [3].

Differential scanning calorimetry (DSC) is a technique based on measuring the difference in energy required to keep the sample and a reference at the same temperature when subjected to the same temperature variations. But in the case of a mutiphase polymer such as PVDF, as shown on the DSC curve (Figs. 4.4 and 4.5), this technique only gives an average value and it is not possible to identify which phase is the phase at the origin of the energy variations and therefore to allow a complete interpretation of this thermogram.

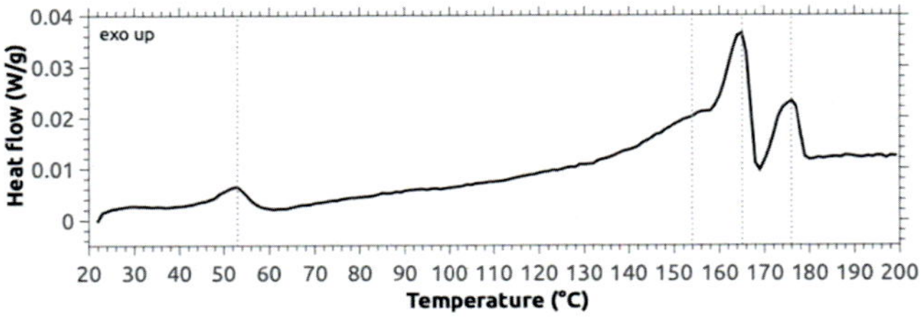

Figure 4.4
DSC thermogram (5°C/min) of a PVDF/PMMA sample [3].

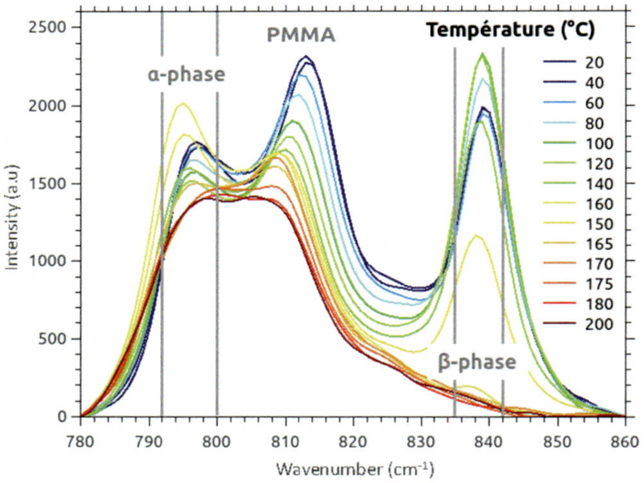

Figure 4.5
part of the Raman spectrum of a PVDF/PMMA sample with the assignment of the peaks to the phases α, β as a function of temperature.

As shown for example in Marie Vietmann Ph.D. thesis [3,4], it is possible with Raman spectroscopy to identify in the spectrum, specific signatures (peaks) of the phases present in the material.

We have therefore, as shown in Fig. 4.3, coupled the two experiments (DSC and Raman spectrometry) and thus simultaneously monitored the thermal and vibrational behaviors of the material, with a heating rate 5°C/min and acquisition parameters 10s spectra/every 12s like this corresponds to a spectrum every °C [3,4].

As shown in Fig. 4.6, the temperature evolution of the Raman peaks specific to each of the phases can be followed in a distinct way, thus revealing their specific thermal behavior. This allows a clear interpretation of the DSC thermogram.

First of all, the large exothermic peak that appears on the DSC thermogram from T = 45°C, synonymous with crystallization, could be better defined, but above all the crystalline phase that recrystallizes could be identified. Indeed, phase β describes a marked growth on the Raman spectra, while phase α shows very little evolution. The Raman spectrum analysis thus makes it possible to affirm that it is the crystalline phase β and it alone that recrystallizes when the sample is heated, before 120°C. The recrystallization of the β-phase is consistent with the thermal history of the sample and its crystallization kinetics.

The analysis of Raman spectra also made it possible to delineate the phenomena of crystallization and fusion, which is essential for enthalpy and crystallinity calculations. Thus, the crystallization of the β phase stops at a temperature of T = 117°C, and starts to melt, until T = 156°C where it has totally disappeared. The originality of the polymorphism of PVDF is that shortly after the beginning of the melting of phase β, the phase α has in turn described a recrystallization step of about 117–156°C. The observation of the evolution of

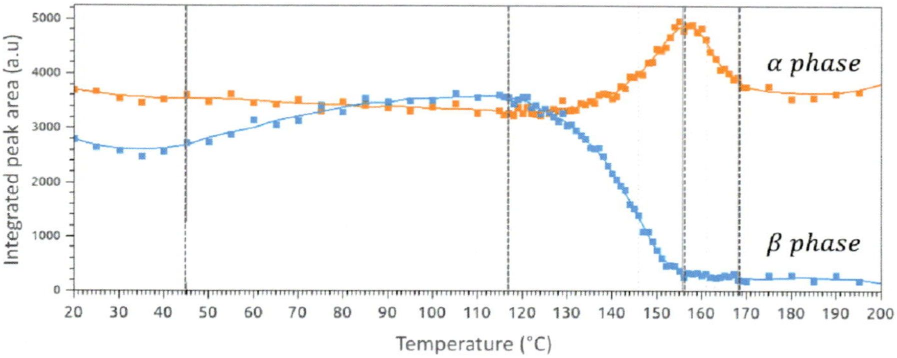

Figure 4.6
Evolution of the Raman peaks of the phases α, β as a function of the temperature of a PVDF sample [3].

the Raman spectra thus confirms the hypothesis of the quasi-instantaneous recrystallization of phase α from the melting of phase β [3].

The peak at 176°C is not observable in Raman and corresponds to the gamma phase revealed by infrared spectroscopy [4]. The band at 176°C is not observable in Raman and corresponds to the gamma phase revealed by infrared spectroscopy [5].

This example shows the complementarity between the techniques and the contribution of in situ Raman spectroscopy in the interpretation of DSC experiments and thermograms.

5.2 Microstructural transitions and crystallinity ratio criteria

At the time of thermal treatments, semicrystalline thermoplastics such as polyethylene terephthalate demonstrates important microstructural changes that could drastically modify its properties in use. For instance, in case of nonisothermal conditions, the PET ability to crystallize is highly dependent on the heating or cooling rates. Simultaneous DSC/Raman analysis could then help Raman spectra to reveal the microstructure of PET whatever the external thermal conditions and history. Since Raman spectroscopy is quite transportable, the rapidity of acquisition could provide determinant information of the material structural state both on an industrial production line and when used as an object.

Regarding the architecture of PET macromolecules, they can arrange themselves following two different conformational states: "*Trans*" and "*Gauche*," either of the ethylene glycol (EG) segment or of the ester function. It was reported in the literature that PET is able to crystallize only under a global "*Trans*" conformation state of both segments. In the opposite, the amorphous morphology is made of a major fraction of "*Gauche*" conformations but can contain also "*Trans*" ones [6–8]. The Raman vibrational bands assignments were widely reported by numerous authors in the literature [9–14].

The 1725 cm^{-1} vibrational band is one of the most intense in the PET vibrational spectra and corresponds to the stretching of the C=O bond of the terephthalic acid. This band is narrower in the semicrystalline material spectrum than in the one of the amorphous as it can be observed in Fig. 4.7A. [15,16]. Shifts of bands position to higher or lower values are due to mechanical and/or thermal overloads leading to the compression (higher wavenumbers) or the stretching (lower wavenumbers) of the molecular bonds. This evolution of the C=O vibrational band indicates a compressive state in two steps due to the temperature increase and the changes in the geometrical structure of the monomeric unit from a disordered amorphous phase to a regular crystalline state (Fig. 4.7B). Its full width at half maximum (FWHM) can be correlated to the polymer density (Fig. 4.7C) [9,17].

The 1118 cm^{-1} band characterizes the C–H rocking in the benzene ring and the C–O stretching in the EG unit [8,18,19]. The 1096 cm^{-1} band characterizes the combination of C(O)–O/C–C/C–C (EG) symmetric stretching bonds and describes

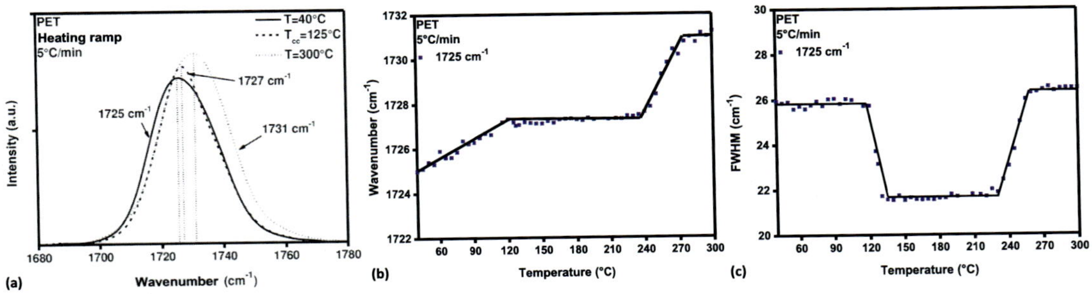

Figure 4.7

(A) Evolution of the 1725 cm^{-1} vibrational band attributed to the elongation of the C=O double bond during heating from the initial quasi-amorphous state (Ben Hafsia, 2016 [16]). (B) Evolution of the 1725 cm^{-1} vibrational band position. (C) FWHM evolution of the 1725 cm^{-1} vibrational band (Ben Hafsia et al., 2016 [15]).

the evolution of the conformation from "*Gauche*" to "*Trans*" [13,20]. Fig. 4.8A shows the evolution of the 1096 cm^{-1} vibrational band from the glassy amorphous state at 40°C to the semicrystalline state at 190°C just before melting. The 1096 cm^{-1} vibrational band intensity increases with increasing temperature which enlightens the initially quasi-amorphous PET crystallization.

In case of amorphous PET, the 1096 cm^{-1} vibrational band still exists and exhibits an initial intensity value $\left(I_{\overline{\upsilon}(1096\ \mathrm{cm}^{-1})a}\right)$, which statistically keeps constant. A Raman criterion which is taking into account this *trans* conformations persistent proportion at the amorphous state is presented in Eq. (4.1) and can be considered as an accurate expression allowing the quantitative determination of the crystallinity ratio of PET:

$$r_c^{\mathrm{Raman}} = \frac{I_{\overline{\upsilon}(1096\ \mathrm{cm}^{-1})} - I_{\overline{\upsilon}(1096\ \mathrm{cm}^{-1})a}}{I_{\overline{\upsilon}(1118\ \mathrm{cm}^{-1})} + I_{\overline{\upsilon}(1096\ \mathrm{cm}^{-1})}} \tag{4.1}$$

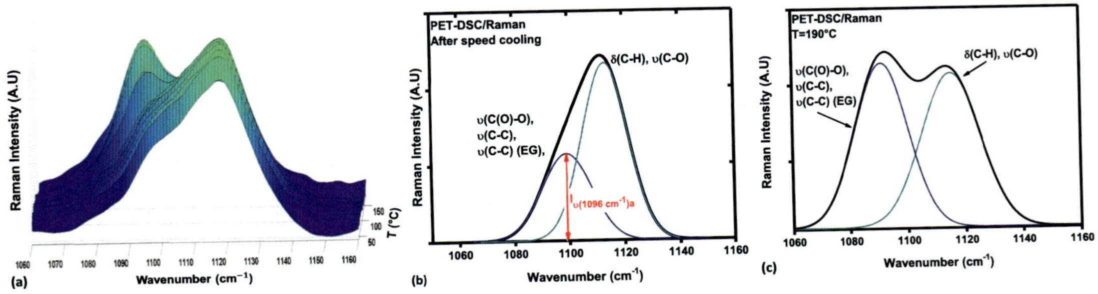

Figure 4.8

(A) Evolution as a function of temperature from 40°C to 190°C at 5°C/min of the Raman spectra of the initially quasi-amorphous PET, (B) quasi-amorphous PET Raman spectra at 40°C, (C) semicrystalline PET Raman spectra at 190°C (Ben Hafsia et al., 2016 [15]).

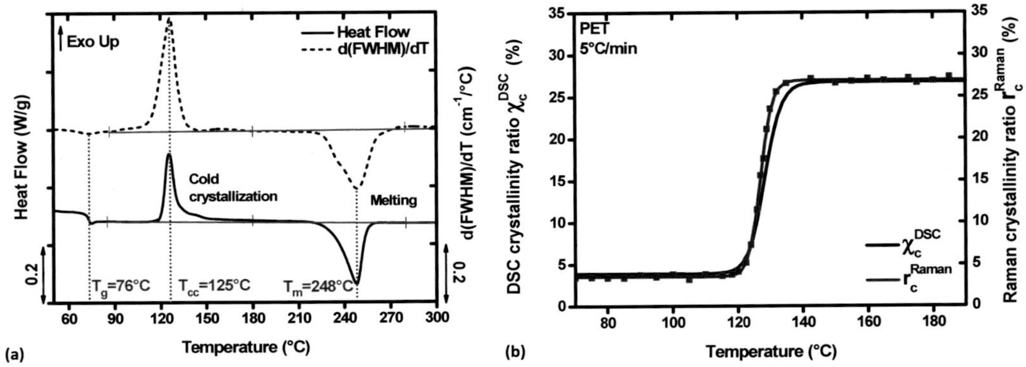

Figure 4.9

(A) Comparison between the DSC thermogram and the differential evolution of FWHM of the 1725 cm^{-1} vibrational band d(FWHM)/dT (derivative of Fig. 4.7C curve) plotted as a function of temperature at a constant heating rate of 5°C/min. (B) Evolutions of the crystallinity ratio by Raman spectroscopy (Eq. 4.1) and DSC [16].

where $I_{\overline{\upsilon}(1096\ \mathrm{cm}^{-1})}$, $I_{\overline{\upsilon}(1096\ \mathrm{cm}^{-1})a}$ and $I_{\overline{\upsilon}(1118\ \mathrm{cm}^{-1})}$ are the intensities of the 1096 cm^{-1}, 1096 cm^{-1} in the 100% amorphous state and 1118 cm^{-1} vibrational bands, respectively.

Regarding the evolution of the 1725 cm^{-1} FWHM of Fig. 4.7C, its derivative as a function of temperature is plotted in Fig. 4.9A. The DSC thermogram obtained in situ is also given. Perfectly similar shapes are observed. Thus, it is shown that the d(FWHM)/dT allows the determination of the characteristic temperatures of the material, glass transition, cold crystallization, and melting which corresponds exactly to the characteristic temperatures determined by DSC (onset, maximum, and endpoint).

The evolutions of r_c^{Raman} and degree of crystallinity determined simultaneously by DSC as a function of the temperature are plotted in Fig. 4.9B. A good correlation is observed in terms of shapes and values. Only a slight shift to higher temperatures can be noticed in case of DSC results and can be explained by the fact that Raman spectroscopy acquisitions are performed in volume independently of the thermal conductivity of the PET which is not the case of the DSC technique. Hence, Raman results would be more accurate than those obtained by DSC.

6. Monitoring of mechanical properties of composites: fillers influence

Contributors: J. Martin, M. Ponçot, S. Chaudemanche, K. Ben Hafsia, S. Saidi, D. Hermida Merino.

6.1 Raman/traction/wide-angle X-ray scattering (WAXS) coupling system: macromolecular chains orientation and crystallinity

The coupling system of VidéoTraction, wide-angle X-ray scattering (WAXS), and Raman spectroscopy especially designed to establish the narrow relationship between polymer

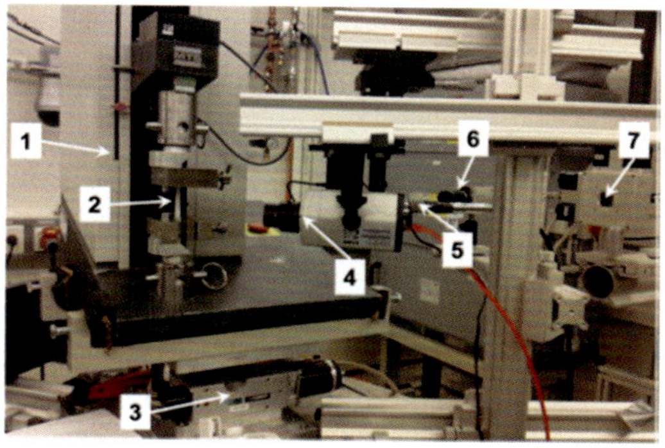

Figure 4.10
(1) Tensile test machine, (2) Dumbbell-shape specimen, (3) Z-motorized stage, (4) Video camera, (5) Raman head of probe, (6) VidéoTraction system camera, and (7) X-rays monochromator [21].

blends microstructure evolution and mechanical behaviors and properties was used at the DESY laboratory of PETRA III synchrotron (Figs. 4.10 and 4.11). The work focused on neat and filled iPP matrix (μ-talc and $CaCO_3$).

The crystalline macromolecular chains orientation is estimated from Raman spectra by performing acquisitions in particular polarization geometry of the incident and scattered radiations where both the incident and scattered radiations are polarized along the stretching direction. Following these conditions, the ratio $I_{973}\ \mathrm{cm}^{-1}/I_{998}\ \mathrm{cm}^{-1}$ of the integrated intensity of the two particular Raman scattering bands at 973 cm^{-1} (symmetric stretching mode of the C—C skeletal backbones in crystalline phase) and 998 cm^{-1} (rocking mode of CH_3 lateral alkyl groups in crystalline phase), enable to estimate orientation of crystalline phase chains in the tensile direction [6,21—27].

The modifications of the mechanical properties due to the gradual complexity of formulations (addition of EPR nodules and μ-talc particles) were explained in terms of influence of the micromechanisms of deformation during a uniaxial stretching at true strain rates. Good correlations are obtained between orientation factors determined by WAXS and Raman spectroscopy. Singularities were highlighted: WAXS is more efficient at lower strains to observe the prior reorientation of the crystalline chains, however at the time of plasticity the signal is interfered by the development of a crystalline mesophase; Raman spectroscopy is not sensitive to the reorientation phenomenon at low strain values due to the initial laser polarization in the stretching direction; however, it enables to follow the orientation until the end of the mechanical test. The true intrinsic mechanical behaviors are then determined and give finally the real mechanical behavior of blends evicting the inside voids. Similar strain hardenings were obtained whatever the blend when the volume strain correction is applied. However, macromolecular orientation levels are

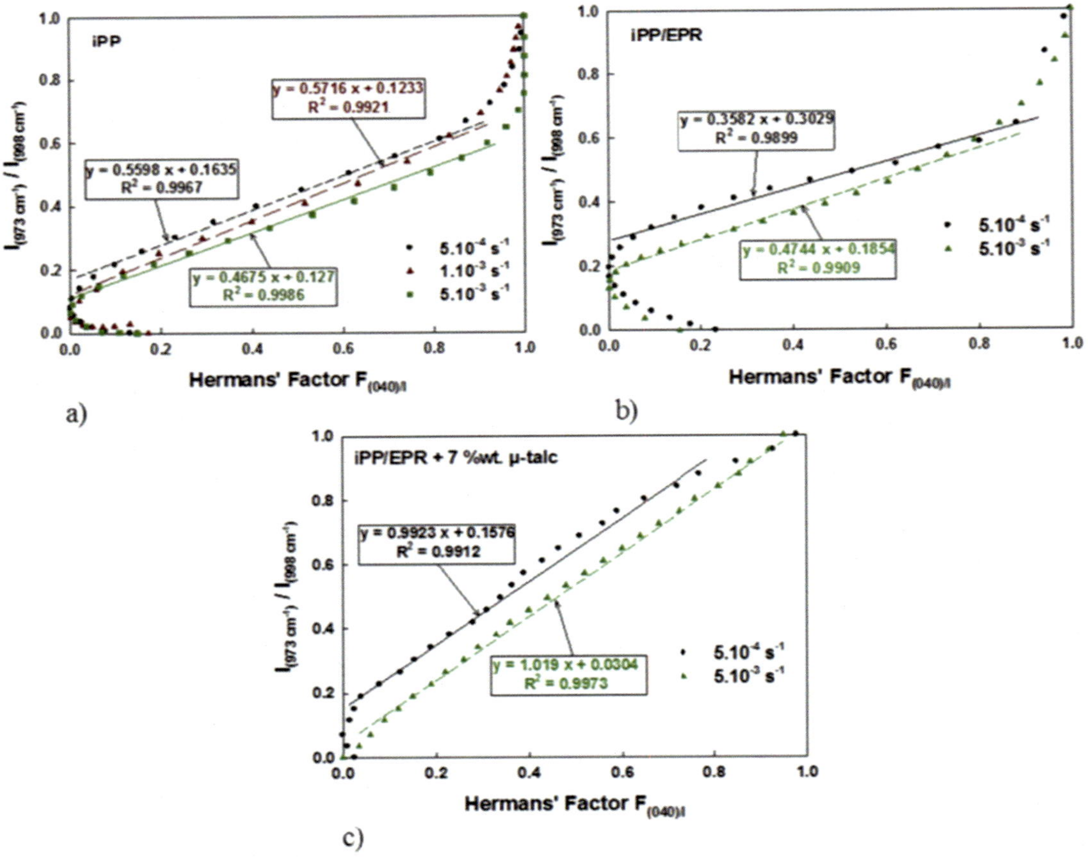

Figure 4.11

Correlations between the evolutions of the normalized orientation factors determined by Raman spectroscopy and 2DWAXS at the ambient temperature as a function of the true strain rates for (A) iPP, (B) iPP/EPR, and (C) iPP/EPR +7 wt.% µ-talc [21,22,24].

totally different. It was explained by the dimensions of the analyzed volume of the used experimental techniques which are too large to locally measure the orientation gathered right in the microfibrils, amount of matter between two adjacent voids. The more elevated is the volume strain; the lower is the measured macromolecular orientation whereas similar stress-hardening behaviors are obtained.

A direct consequence of this deformation micromechanism, which is the orientation of the macromolecular chains of the amorphous and semicrystalline phases in the direction of the mechanical stress, is a change in the crystallinity of the material. Indeed, it is possible to observe either a decrease (crystallite rupture by bending then "unwinding") or an increase ("stress-induced crystallization") or both simultaneously and successively. As explained previously for PET, the width at half-height of the vibrational band at 1727 cm^{-1} correlates with the density, and thus the crystallinity ratio of the material. Melveger determined an experimental criterion relating half-height width to density [6]. From the theoretical densities of the amorphous and crystalline phases, the half-height width can be directly related to the crystallinity rate [28]. The criterion, noted r^{1727}, is therefore analogous to a crystallinity ratio in the same way as r_c^{Raman} and is calculated from the following equations:

$$\rho = \frac{305 - \Delta\nu_{1/2}}{209} \quad (4.2)$$

where ρ is the semicrystalline polymer density and $\Delta\nu_{1/2}$ the width at half-height of the band at 1727 cm^{-1}.

$$r^{1727} = \frac{\rho - \rho_a}{\rho_c - \rho_a} \quad (4.3)$$

where ρ_a is 100% amorphous polymer density (1.335 g/cm^3 [29]) ρ_a is 100% crystalline polymer density (1.479 g/cm^3 [30]) for PET.

In case of films, whatever the sampling direction (LD or TD for the machine/longitudinal and the transversal directions, respectively), Donnay showed in Fig. 4.12A that crystallinity indexes decrease during uniaxial stretching applying the Raman criteria of Eq. (4.3) [30].

Everall et al. proposed that the 1727 cm^{-1} band is the result of the contribution of several vibrational bands. There would be three or four, localized at 1721, 1726, and 1730 cm^{-1} or at 1716, 1724, 1732, and 1740 cm^{-1}, respectively. The conformational states related to these different bands are not always well-attributed in the literature, but it would be

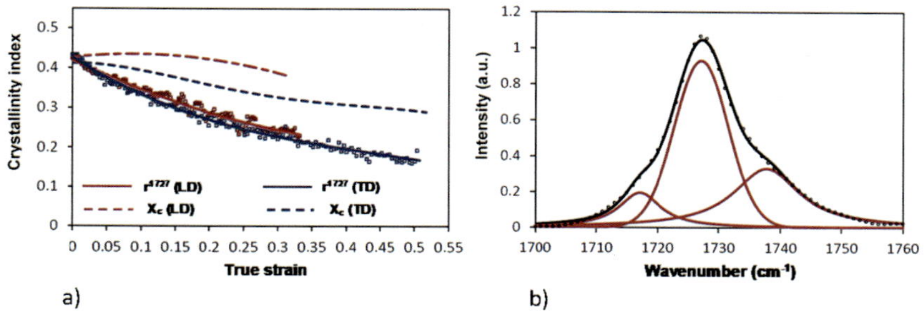

Figure 4.12

(A) r^{1727} Raman criteria and WAXS crystallinity index (X_c) evolutions as a function of the axial true strain for samples stretched toward the machine and the transversal directions (LD and TD, respectively). The true strain rate is 5.10^{-4} s^{-1}. (B) Desummation in three bands of the 1727 cm^{-1} vibrational band at room temperature [30].

possible to separate the contributions of different conformational states. For the spectra collected in the thesis of Donnay in 2017 [30], a three-band desummation was proposed and the result is shown in Fig. 4.12B. They are centered at 1717, 1727 and 1738 cm^{-1}. These positions are similar to the three-band desummation proposed by Adar et al. in 1985 and taken up by Lesko et al. in 2000 [31,32].

Their evolutions are still very similar regardless of the LD (Fig. 4.13A) or TD (Fig. 4.13B) films' orientation involved during uniaxial stretching. In both cases, there is an increase of the 1727 cm^{-1} band area and a decrease of the 1717 and 1738 cm^{-1} bands areas. These observations are consistent with a decrease in crystallinity and an elongation of the macromolecular chains toward the mechanical direction solicitation. The decrease in the area of the band at 1717 cm^{-1}, assumed to correspond to states of rotation out of the plane of the aromatic ring, is explained by the orientation of the chains requiring more efficient stacking. The increase in *trans* conformations in the ethylene segment following orientation contributes to this more efficient stacking. Similarly, ester groups in the aromatic ring plane are required, especially for the formation of mesophases that occurs during PET deformation. The decrease in crystallinity is correctly followed by the band area at 1738 cm^{-1}, especially in the LD sample. For the TD sample, the decrease appears to be smaller by Raman than by WAXS.

Nevertheless, the conclusions drawn here are encouraging and the results are consistent with the assumptions made about the assignment of the three desummated bands. Furthermore, using the technique of statistical analysis of spectra by chemometrics, Everall et al. were able to differentiate four vibrational bands grouped under the main 1727 cm^{-1} band [33]. This point is still in progress and needs more works to enlighten all the potential information that Raman spectroscopy would allow.

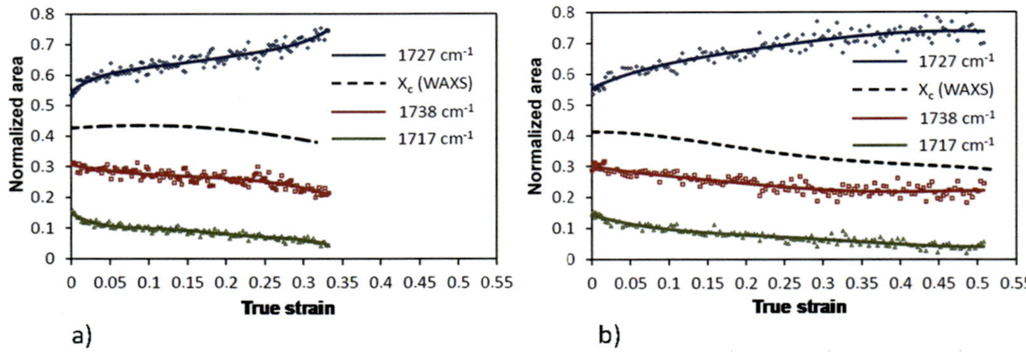

Figure 4.13
Evolutions of the normalized areas of the vibrational bands at 1717, 1727 and 1738 cm^{-1} during the tensile deformation of stretched PET film in MD (A) and TD (B) directions, with the parallel crystallinity index by WAXS. The strain rate is 5.10^{-4} s^{-1} [30].

6.2 Raman/light scattering/traction coupling system: volume damage

As developed by Chaudemanche et al. [34], the experimental device enabling the quantitative characterization of the incoherent light transport in materials consists of a light source focused on the surface of a turbid specimen. The laser radiation is produced with an incoherent light source which is first collimated into an optic fiber. Then, a laser line tunable filter selects the desired wavelength with very narrow bandwidth (as low as 0.3 nm), allowing then to probe the light scattering at the same wavelength as used in the Raman apparatus. A CCD camera collects the halo of the backscattered light emerging from the polymer surface. Specimens are placed in a minitensile machine (Kammrath and Weiss), which ensures symmetric displacement of both grips. The treatment of backscattered image consists in calculating the decrease in intensity (integrated over 360°) as a function of the radius (distance from the laser impact). A model of the light transport in scattering medium allows to identify a single parameter which quantifies the turbidity evolution of the specimen. This parameter is the transport length l_{TR} which corresponds to the mean free path of photons inside the matter in the diffusion approximation of scattering. The experimental device just described is presented in Fig. 4.14.

Scattering is the ability of matter to disperse light in all directions in space. This diffusion, which is achieved without loss of energy, depends on the size of the diffusing elements. When the size of the particles is very small compared to the wavelength, it is called Rayleigh scattering also called molecular scattering, the distribution of the scattering is isotropic. But for larger scatters, the spatial distribution of the scattered light is then no longer isotropic. Mie scattering is generally considered to occur for scatters with diameters equal to and larger than one 10th of the wavelength.

A correlation between the measurement of the incoherent light transport and the evolution of the Raman intensity during uniaxial stretching is presented in Fig. 4.15. The plateau from which the transport length is constant is enlightened for each iPP blend.

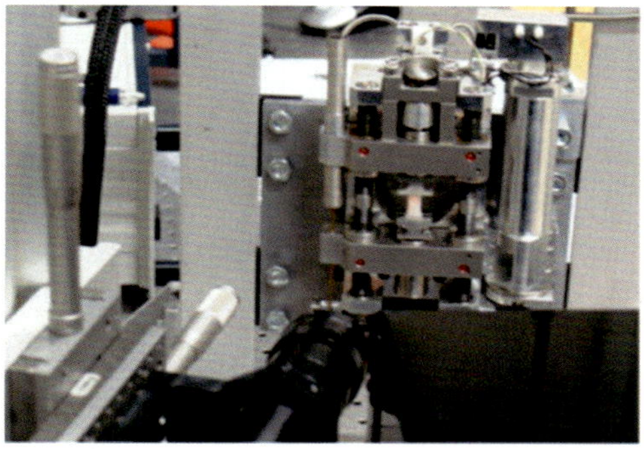

Figure 4.14
Photography of the light scattering experiment coupled to a tensile test [34].

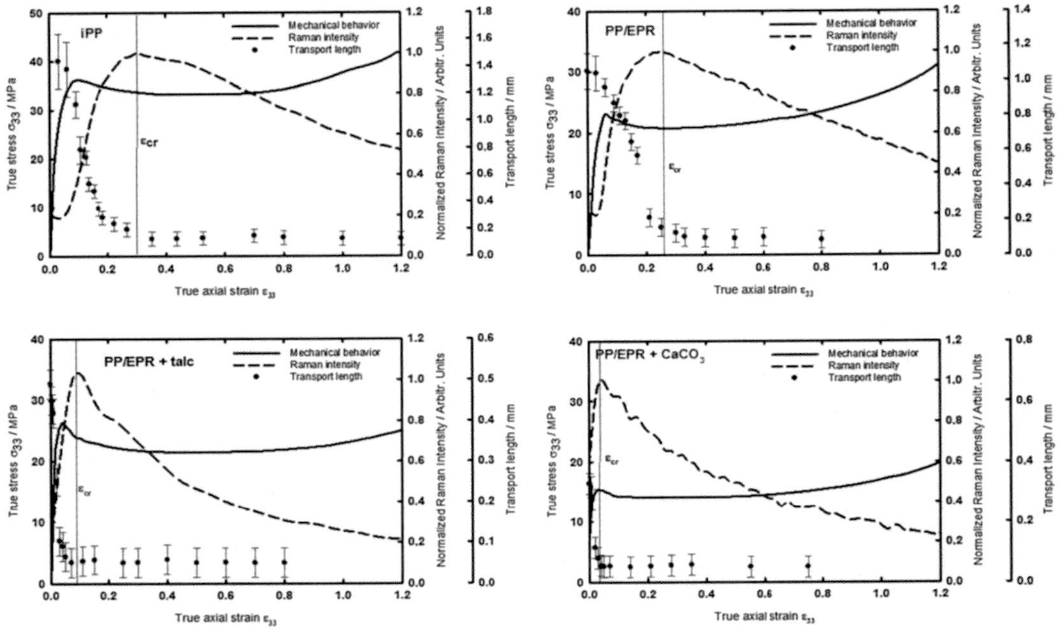

Figure 4.15
Comparison of the light and Raman scattering results in correlation with the true mechanical behavior of each material at the true strain rate of 5.10^{-3} s^{-1} [34].

Martin [22] and Ponçot [23,35] proposed a Raman criterion of volume damage for polymer film. This criterion is based on the principle that the Raman intensity is proportional to the quantity of material probed [36]. In this study, it was observed that the Raman intensity was dependent on the light scattering [35]. Indeed, before ε_{cr} (critical strain defined as the strain value at the time of the maximal normalized and integrated Raman intensity), the Raman backscattered intensity increases as the light transport length in the matter decreases. After ε_{cr}, the transport length is constant; thus, the volume probed by Raman spectroscopy is constant. This allows to use the Raman criterion from ε_{cr}. This criterion is noted:

$$\varepsilon_v^{\text{Raman}} = 1 - \frac{I_{\varepsilon_{33}}}{I_{\max}} \quad \text{with } \varepsilon_{33} \geq \varepsilon_{cr} \tag{4.4}$$

Fig. 4.16 shows the results for volume damage obtained by the VidéoTraction system and by Raman spectroscopy.

For filled materials, the Raman measurement starts at low strain values. There are some differences between the volume damage obtained by Raman and by VidéoTraction, but they are quite similar. For iPP and unfilled PP/EPR, large differences were observed between volume damage values obtained by Raman spectroscopy and VidéoTraction

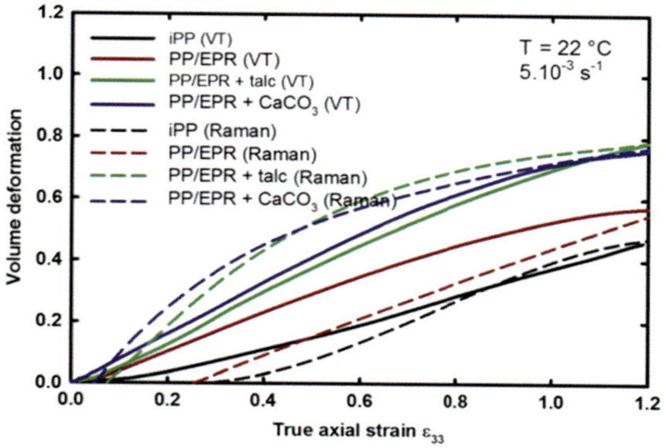

Figure 4.16

Comparison of the volume damage determined by Raman and VidéoTraction for each material at a true strain rate of 5.10^{-3} s^{-1} [22].

system. However, during the plastic stress-hardening regime ($\varepsilon_{33} \geq 0.8$) there is a convergence of values obtained by the two techniques. One explanation is that the Raman measurements do not take into consideration the cavities, which are smaller than the wavelength. But in the stress-hardening area, cavities have grown to larger sizes, so they are included in the Raman measurement analyzed volume. The differences between volume strain measurements can also find their sources in the VidéoTraction measurement method. The transverse isotropy hypothesis and the fact that the videometric measurement is made at the surface can lead to variations.

6.3 Raman/WAXS/SAXS coupling system: correlation between atomic and micrometric scales

In Sarah Saidi Ph.D. thesis work, it was possible to make in situ measurements between high-energy X-ray diffraction measurements on a synchrotron line (Dubble) simultaneously with Raman spectrometry measurements in a DSC cell, and this on a polymer (PVDF). The experimental setup is given in Fig. 4.17 [37].

The experimental conditions are Time-resolved simultaneous small-angle X-ray scattering (SAXS) and WAXS experiments (wavelength = 0.1 nm), were performed in the European synchrotron radiation facility (ESRF), at DUBBLE (BM26B, Grenoble, France). Acquisition of simultaneous SAXS and WAXS patterns with an exposure time of 15 s were collected with Pilatus 1M detector (981 × 1043 pixels of 172 × 172 μm placed at a distance of 3.5 m from the sample cell) and a Pilatus 300K detector (1472 × 195 pixels of 172 × 172 μm placed at a distance of 0.31 m), respectively. Raman spectroscopy in reflection mode has been coupled to time-resolved simultaneous SAXS and WAXS, and was performed using an RXN1 spectrometer from Kaiser Optical Systems. Raman spectra

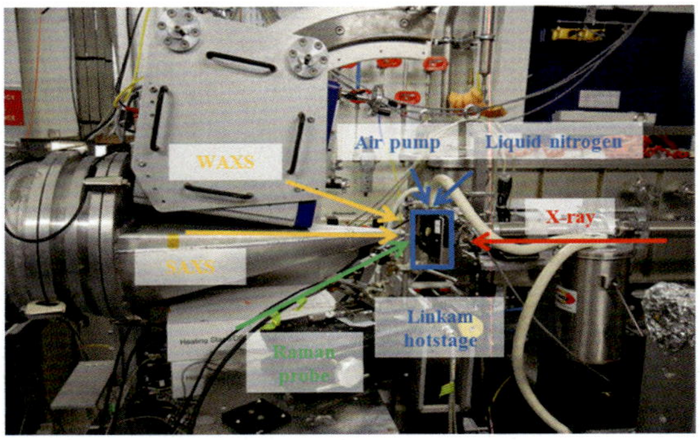

Figure 4.17
Experimental setup of this coupling of experiences (SAXS, WAXS, Raman, and DSC).

were obtained with an acquisition time of 10 s for each spectrum in a window of 15 s to be similar to the simultaneous SAXS/WAXS acquisition, which leads to the recording of one spectrum every 1.25°C. The acquisition time of the Raman spectra followed the corresponding X-ray exposure time (15 s) and therefore, an acquisition period of 15 s each spectrum with an acquisition time of 10 s per spectrum [37].

Fig. 4.18 shows the result of the three simultaneous measurements as a function of temperature. Thus, both the nanostructure and the vibration behavior of the sample as a function of temperature can be measured on the same sample in the same experimental conditions, in this case during the second heating.

Fig. 4.19 presents the monitoring of three microstructural parameters of the sample as a function of temperature, like crystallinity, long period, and amount of alpha phase, this allows us to correlate and monitor measurements at different scales from angstrom to micrometer and as shown in Fig. 4.20 give additional information at different scales on the microstructure of the sample. For example, in this figure we can follow, as a

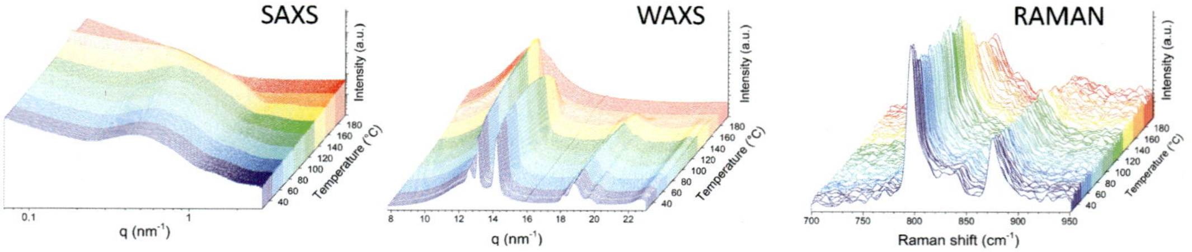

Figure 4.18
Simultaneous measurements (SAXS, WAXS, Raman) as a function of temperature during the second heating in DSC experiment.

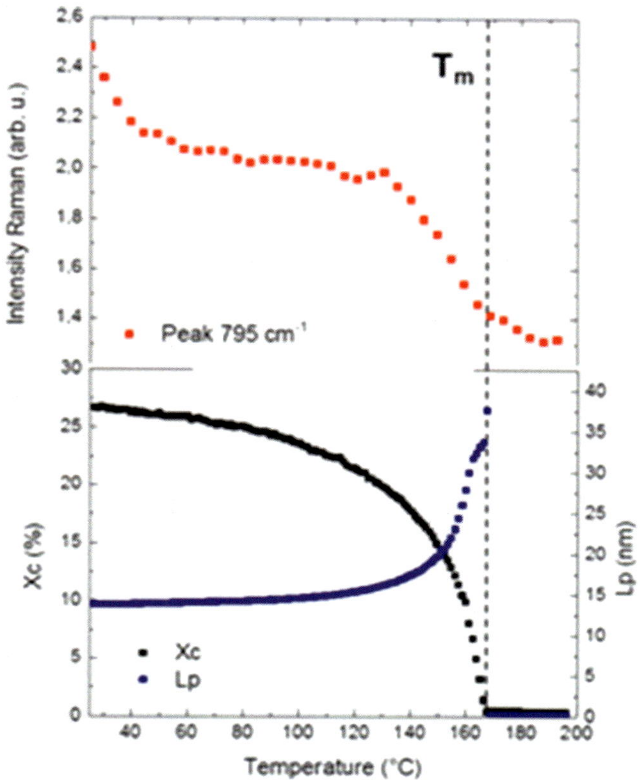

Figure 4.19
Simultaneous measurements (SAXS, WAXS, Raman) as a function of temperature during the second heating in DSC experiment.

function of a temperature cycle of the DSC, and simultaneously, the crystallization given in the WAXS, the long period (*Lp*) by the SAXS and the tracking (here the red dots) of a Raman line linked to phase A. We can thus follow both the vibrational behavior and have information on the nanostructure of the material. And very important point under the same conditions and on the same sample without having to move it to a laboratory.

Fig. 4.20 shows the different information that can be accessed with its simultaneous techniques, it also shows the different information on the structure of the material ranging from macromolecules to micrometric structure. It clearly shows the complementarity and contribution of the techniques and the importance of coupling them to the same measurement. It also shows the range of information that can be deduced from a Raman spectrum provided that a thorough learning and interpretation of the spectrum is carried out.

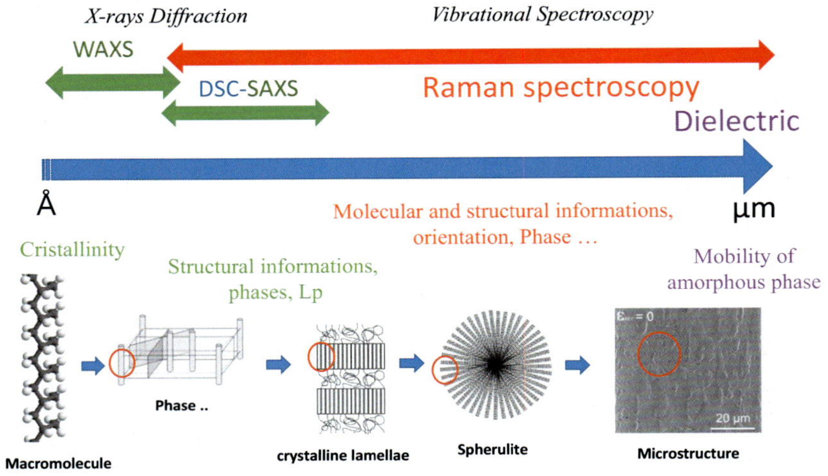

Figure 4.20
Complementarity of techniques for the study of the microstructure of a polymer ranging from the macromolecule to its macrostructure and corresponding to additional information necessary to understand the properties of the material.

7. Feasibility of in-situ coupling with dielectric dynamic analysis

Contributors: A. Letoffe, M. Donnay, S. Ogier, K. Ben Hafsia, A. Serghei, L. David, A. Kallel, A. El Amrani, Z. Ayadi.

7.1 Introduction on dielectric dynamic analysis

Dynamic dielectric analysis is a spectroscopic technique which allows to meet three objectives:

- Follow the evolution of the microstructure, of the crosslinking (see Section 7.4) of the orientation in polymers and nanocomposites through macromolecular mobility after all types of morphology changes in use (see Section 7.2).
- Characterize relaxation phenomena, such as the relaxation associated with the glass transition, and compare them with those detected by dynamic mechanical analysis in positions, amplitudes, distributions of relaxation times (see Section 7.3).
- Measure dielectric parameters (complex, real, and imaginary permittivity), loss factor (tan δ), of polymer dielectric materials and nanocomposites with a polymer matrix (specific study of interfacial phenomena according to nanostructuring, see Section 7.3) as a function of temperature and frequency.

Polymers and nanocomposites undergo an evolution of the molecular mobility of the macromolecular chains after changes in morphology following aging, for example. Dynamic dielectric analysis makes it possible to follow these changes through the relaxation phenomena of the macromolecular chains (cooperative relaxation associated with the glass transition for example). Specifically, for nanocomposites with a polymer matrix, dielectric analysis serves as a probe to study the interfacial phenomena between matrix and reinforcements according to the morphology, nanostructuration and aging of the material. These relaxation phenomena will be compared with those detected by dynamic mechanical analysis (see Section 7.3), which is considered as a more classical type of relaxation spectroscopy.

The reduction in the size of fillers in the nanocomposites leads to a large increase in the surface/volume ratio of the reinforcement and therefore to the development of an important interface. The creation of such an interface with lower volume fractions makes it possible to significantly improve certain properties (mechanical, electrical, thermal, and so forth). In addition, it is well recognized that the performance of these nanomaterial materials with a polymer matrix depends both on the characteristics of the nanoreinforcement (nature, geometry, orientation, and volume fraction) and on the characteristics of the matrix. However, in addition to the intrinsic properties of each of these basic constituents, the reinforcement-matrix interface plays a decisive role in the design and monitoring of such materials, both in terms of mechanical behavior as well as electrical and physicochemical behavior. Furthermore, mechanical characterization alone can provide qualitative information on the quality of the matrix-reinforcement adhesion since the charge transfer from the matrix to the reinforcement takes place through the interface. Spectroscopic techniques such as dielectric, are complementary to mechanical measurements; they prove to be essential since they probe the interface directly and therefore make it possible to better elucidate the damage phenomena in nanocomposites. Thus, many works have been developed in this direction in recent years, particularly in terms of functionalization. However, rarely do the results presented highlight correlations between measured quantities obtained by different characterization techniques (see perspectives of Section 3.3). This technique makes it possible to follow, under an alternating electric field of variable frequency depending on the temperature, the displacement of conducting species (electrons, ions, impurities, water, etc.) and/or the reorientation of electric dipoles linked to macromolecules. It is thus possible to probe the evolution in relaxation phenomena and in the molecular mobility of polymeric chains during aging and/or during any other microstructural changes (such as crosslinking as seen in Section 3.4).

7.2 Feasibility of in-situ coupling dynamic dielectric analysis and Videotraction for an amorphous and semicrystalline thermoplastic (PET)

7.2.1 State of the art on in-situ coupling dynamic dielectric and mechanical spectroscopy

In addition to identifying the molecular relaxations taking place in the material following the temperature variation, dielectric analysis allows us to understand the evolution of these same relaxations as a function of the deformation of the material. The molecular mobility of thermoplastic polymer materials is of paramount importance in establishing their physical and mechanical properties given their wide fields of application. Very little work has been done to study in real time the influence of stretching on the evolution of different molecular relaxations. Dargent et al. [38] and Hakme et al. [39] studied the effect of the molecular orientation of PET and PEN respectively on molecular relaxations. The PET being initially amorphous, was subjected to a tensile test at a temperature T = 95°C (above Tg). After stretching the material is quenched with cold air to freeze its microstructural state at different rates of deformation. The PEN was stretched at temperatures lower (100°C) and higher (160°C) than its glass transition temperature Tg. In both cases, the measurements by dielectric spectroscopy were carried out in post mortem conditions. Fig. 4.21 compares the dielectric spectrum of amorphous PEN, stretched and thermally crystallized. The stretched PEN at 100°C is characterized by a β relaxation of greater amplitude than in the case of an amorphous PEN while the PEN stretched at 160°C shows a secondary relaxation of lower amplitude. On the other hand, the primary relaxation α shows a shift toward the highest temperatures with a decrease in amplitude when the material is stretched or crystallized at a temperature below Tg. Hakme et al. [39] have shown that the glass transition temperature increases when the material is stretched at $T > Tg$ and decreases when the material is stretched at $T < Tg$. This shows that plastic deformation and crystallization have an opposite effect on the molecular mobility of the material. Plastic deformation in the glassy state can also have an effect on β-subglass relaxation. The deformation can increase the amplitude of relaxation when the material is deformed at $T < Tg$ due to an ordered structure which increases both the dynamics and the number of relaxed entities.

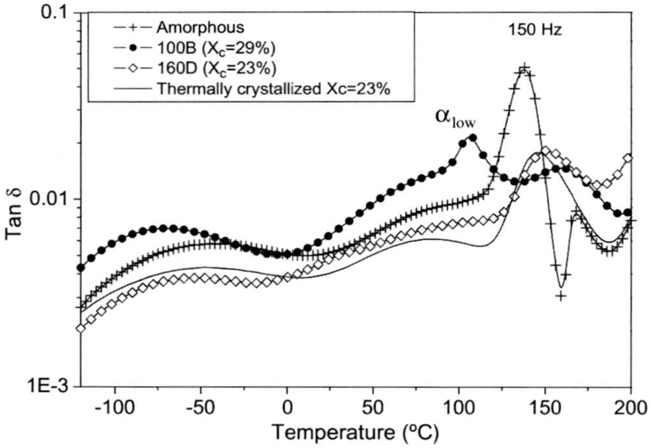

Figure 4.21

Evolution of the loss factor as a function of the temperature at 150 Hz for an amorphous PEN, (100B) stretched at 100°C at $\lambda = 3.2$, (160D) stretched at 160°C at $\lambda = 5.2$, and thermally crystallized at 160°C for 6 h [39].

7.2.2 Experimental setup of coupling tensile test with dynamic dielectric analysis

Dynamic dielectric spectroscopy is a technique for measuring the dielectric properties of a material as a function of frequency and/or temperature. This characterization tool is ideal for the analysis of molecular dynamics at short and long distance and the transport of conducting species in polymers. Indeed, this technique covers a wide frequency range from 10^{-6}–10^{12} Hz [40,41]. The principle of dielectric spectroscopy is based on the application of a sinusoidal voltage on a sample located between two metal electrodes and on the analysis of the amplitude and the phase shift of the response of the material in sinusoidal current [42] (Fig. 4.22).

The response of the material to the applied electric field is given by the complex permittivity.

(Eq. 4.1): $\varepsilon^*(\omega) = \varepsilon'(\omega) - i\varepsilon''(\omega)$ (Eq. 4.1) where $\varepsilon'(\omega)$ and $\varepsilon''(\omega)$ are respectively the relative permittivity and the dielectric loss of the material [44]. These values allow us to calculate the dissipation factor tan $\delta = \varepsilon'(\omega)/\varepsilon''(\omega)$ as a function of frequency. The dielectric measurements made during this work were carried out in real time during a tensile test. To do this, a portable traction machine was designed in the Polymer Materials Engineering (IMP) laboratory at Lyon1 University by Pr. Laurent David and Dr. Anatoli Serghei. It is equipped with two electrodes which will be positioned, without contact, on either side of the nonmetallized PET sample in order to follow in real time the molecular mobility of the material during its deformation. The quantities which have been measured (ε, $\varepsilon''$and tan δ) have been obtained at different deformation rates. The electrodes are rectangular, 2 millimeters thick ($18 \times 11 \times 2$ mm^3). The tests were carried out under the following experimental conditions: in multifrequency: from 10 to 10^6 Hz, at constant temperature: room temperature, in situ with a tensile test at 1 mm/min crosshead speed.

The test pieces used have a thickness e of approximately 0.5 mm. Their two ends were perforated in order to be able to fix it to the jaws of the traction machine. The experimental setup and the geometry of the test pieces are presented in Fig. 4.23.

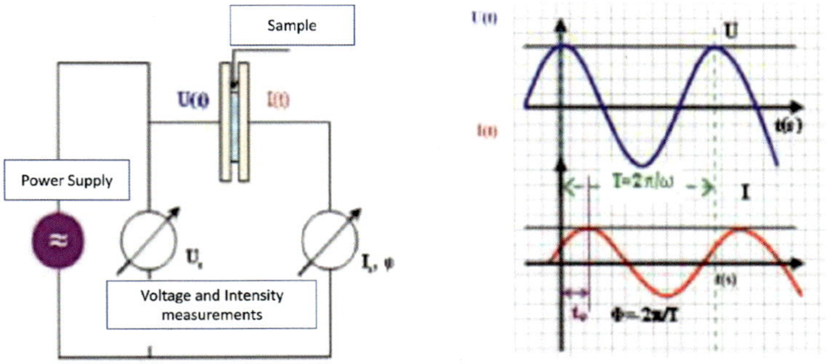

Figure 4.22
Principle of dielectric dynamic measurements [43].

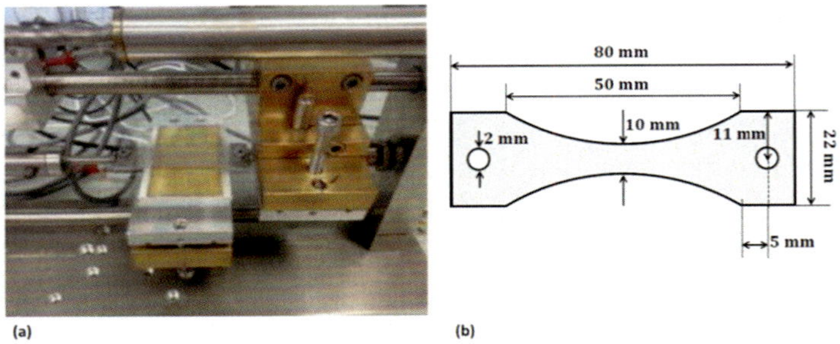

Figure 4.23
(A) Experimental set-up of coupling plane tensile test with dielectric dynamic analysis; and (B) geometry of the test specimens used.

7.2.3 Study of the molecular mobility of quasi-amorphous and semicrystalline PET by in situ coupling of dynamic dielectric spectroscopy and a tensile test

The almost amorphous PET samples have a crystallinity ratio of 3% and the semicrystalline PET samples have a crystallinity ratio of 29%.

Figs. 4.24 and 4.25 show the changes in dielectric constants for a frequency as a function of the deformation and for a deformation rate as a function of the frequency respectively.

Figs. 4.24 and 4.25 show that the total signal of dielectric constants recorded for amorphous PET, as a function of deformation (Fig. 4.24) or as a function of frequency (Fig. 4.25), is higher than that recorded for semi-PET crystalline. In addition, the intensity of the β relaxation observed in Fig. 4.25B and C is higher for the quasi-amorphous PET than for the semicrystalline PET. In conclusion in this part, the results from a coupling tensile test/dynamic dielectric analysis were presented in order to correlate the phenomenon of molecular mobility with the mechanical behavior of the material.

Given the frequency range studied and the highly oriented structure obtained following the stress on our study material and given that this structure is characterized by the arrangement of the PET monomer according to certain types of conformation, only the orientation of the dipoles has been studied. By applying a sinusoidal electric voltage to an undeformed material, an electric field appears at the electrodes thus inducing the orientation of the dipoles distributed on the macromolecular chain according to the direction of the applied electric field (Fig. 4.26).

The changes in dielectric properties, quasi-amorphous PET, and semicrystalline PET have been described in terms of variation of the relative permittivity ε', the dielectric loss ε'' and the dissipation factor tan δ as a function of the true deformation ε_{33} for different frequency values and as a function of frequency for different true deformation rates characteristics of the true behavior of the materials under study.

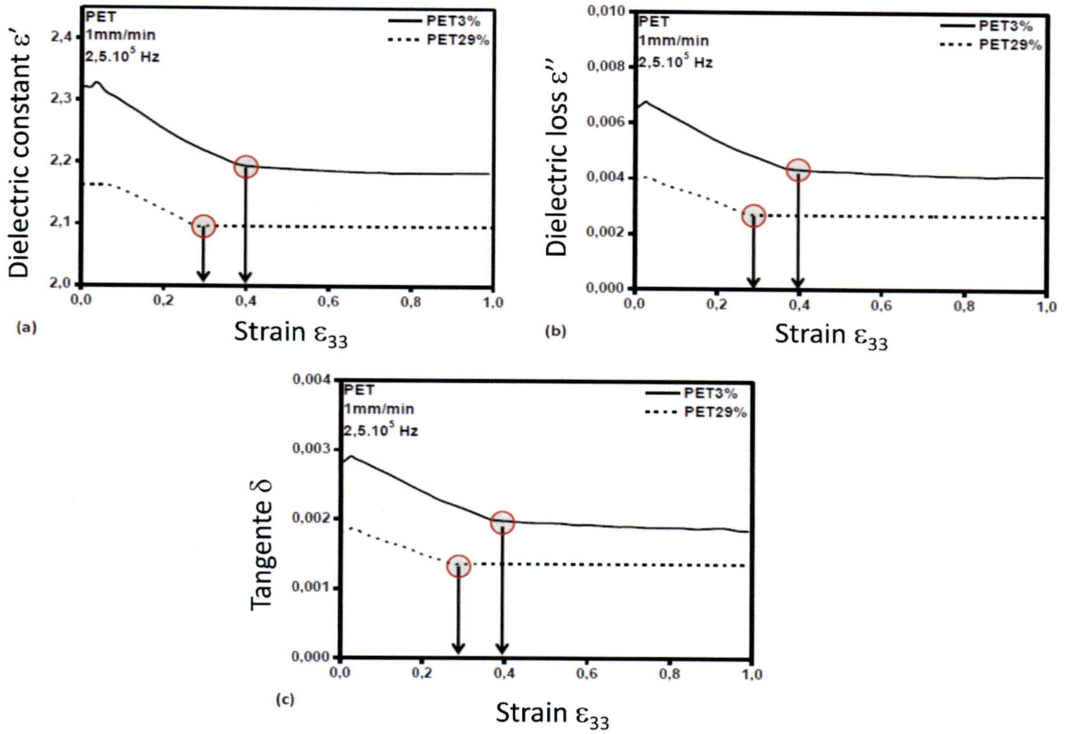

Figure 4.24

Changes in (A) ε', (B) ε'' and (C) tan δ as a function of the deformation for a frequency of $2.5.10^5$ Hz for quasi-amorphous PET and semicrystalline PET deformed during a tensile test at 25°C and a speed of 1 mm/min.

In order to make the link between the evolution of the dielectric properties and the microstructural changes taking place during the deformation of PET, Fig. 4.27 shows the evolution of the tensile curve for quasi-amorphous PET and semicrystalline PET and the dielectric loss ε'' recorded at the same frequency value $f = 5.10^5$ Hz (fairly high value in order to avoid noise signals at low frequencies).

- Step 1: The viscoelastic domain is characterized by a motion of disentanglement of the macromolecular chains and their displacement relative to each other without giving rise to a particular orientation. At very low values of deformation (purely elastic domain), the movement of the macromolecular chains of the amorphous phase is characterized by a modification of the strong and weak inter and intramolecular bonds [46]. In this zone, the material always presents a disordered isotropic structure which would explain the constant evolution of the dielectric loss.
- Step 2: By going beyond the purely elastic domain as the deformation increases until the plasticity threshold (viscoelastic domain), the movement of the chains is

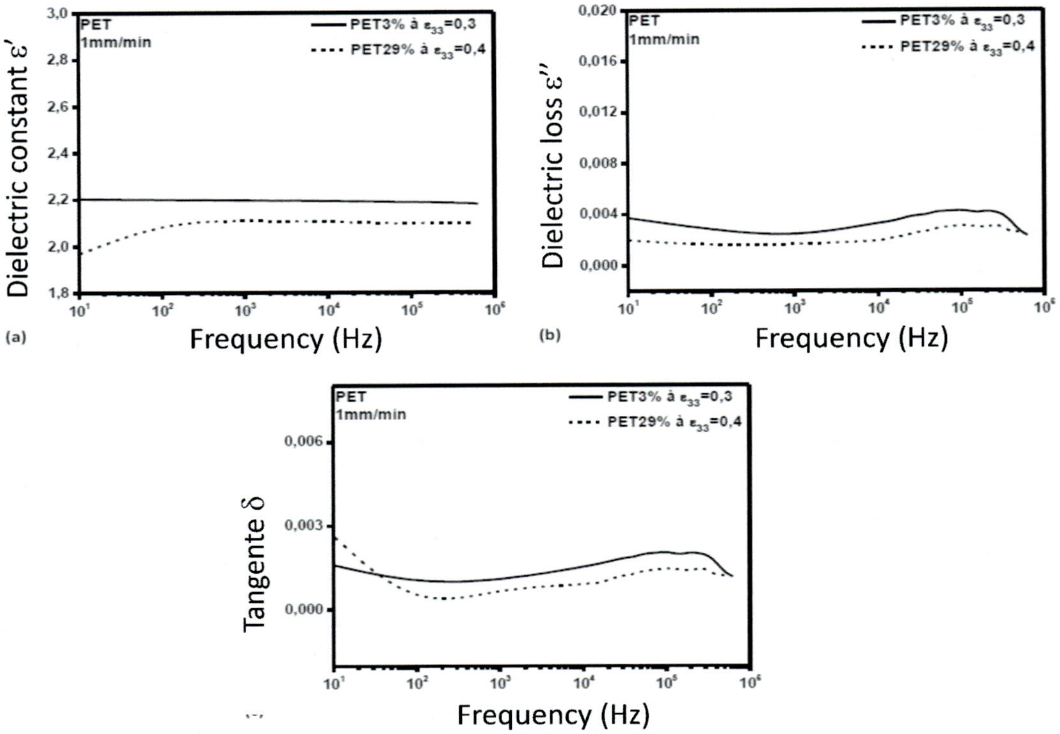

Figure 4.25

Evolution of (A) ε', (B) ε'' and (C) tan δ as a function of frequency for quasi amorphous PET at $\varepsilon_{33} = 0.4$ and semicrystalline PET at $\varepsilon_{33} = 0.3$ deformed during a tensile test at 25°C and a speed of 1 mm/min.

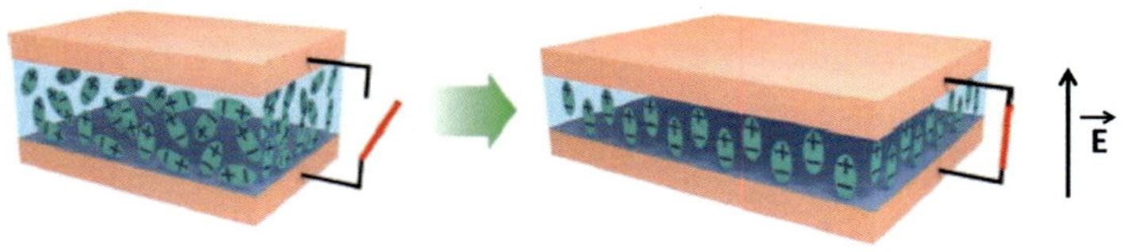

Figure 4.26

Principle of the reorientation of dipoles in a electric field for a dielectric polymer [45].

characterized by conformational rotations of the amorphous macromolecular chains generating an entropy elasticity of the material [47]. The cohesion between the chains is therefore relatively low and the deformability is high: the interlamellar amorphous chains undergo viscous flows [48]. Since PET is characterized by its polarity due to the presence of electronegative oxygen atoms providing dipoles in the ester groups (C=O and C–O), the rotating conformation indeed characterizes a relaxation phenomenon due

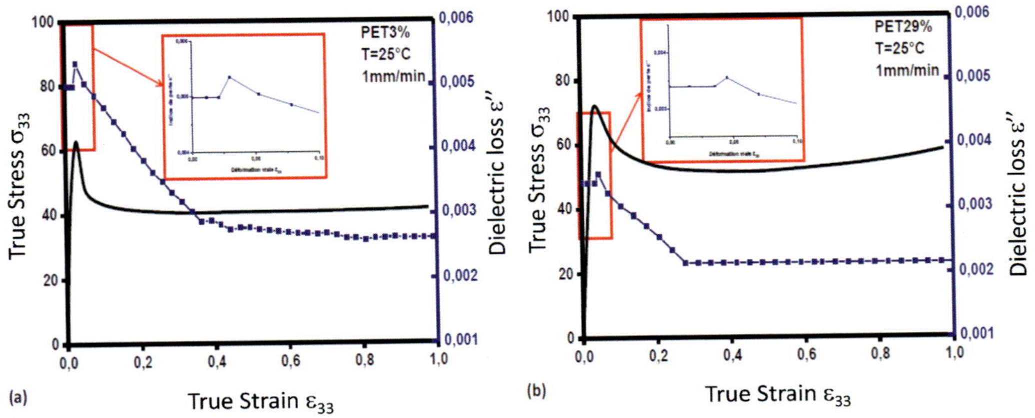

Figure 4.27

Diagram juxtaposing the mechanical behavior curve and the evolution of the dielectric loss ε'' recorded at the same frequency value $f = 5.10^5$ Hz for (A) quasi-amorphous PET and (B) semicrystalline PET deformed in plane tensile test at 1 mm/min ($5.10^{-4}s^{-1}$) at room temperature [2].

to the rotation of the dipoles of the material in the direction of the applied electric field. These relaxations generate a release of energy by the material which implies the increase of the value of the dielectric loss up to the level of the plasticity threshold ($\varepsilon_{33} = \varepsilon_{\text{threshold}}$).

- Step 3: As the Hermans orientation factor begins to decrease appreciably from the plasticity threshold, macromolecular chains begin to orient gradually in the direction of the stress applied. The C—O dipoles initially oriented in the direction of the electric field begin to change direction and will orient in the direction of stress which will explain the gradual decrease in the dielectric loss. During its deformation from the initial amorphous or semicrystalline isotropic state, PET develops a certain ordered structure called smectic phase (mesophase) characterized by an interreticular distance less than the length of the axis $\vec{c}$ of the triclinic crystal structure. Although different from the triclinic structure, this mesophase is characterized by an arrangement of the monomers similar to that of semicrystalline PET. The structure of the PET monomer responsible for the crystallinity and characterized by a tTt conformation: trans (t) conformation of the CO—O—CH_2—C group, trans (T) conformation of the EG group and a trans orientation of the ester group.
- Step 4: At large values of deformation, ε'' reaches a minimum and continues to evolve steadily until the end of the test. In this step, the material develops a fibrillar structure characterized by a periodic mesophase and a maximum orientation of the macromolecular chains in the direction of traction. The dipoles of the C—O bonds are therefore oriented perpendicular to the direction of the applied electric field while the dipoles of the carbonyl groups (C = O) will cancel out since they are arranged in trans conformation. In order to discuss the effect of the deformation and the molecular arrangement during a

tensile test on the relaxations likely to take place in the material, the evolution of the dielectric parameters according to the frequency applied for each rate deformation characteristic of the various zones detailed previously has been traced. It should be noted that the intensity of the β relaxation increases slightly between $\varepsilon_{33} = 0$ and $\varepsilon_{33} = \varepsilon_{\text{threshold}}$ corresponding to the deformation interval of the viscoelastic domain. Indeed, this domain is characterized by a disordered molecular structure which slightly increases the local movements of the amorphous phase [48]. As the deformation increases, the macromolecular chains tend to arrange themselves in the direction of the stress (and perpendicular to the applied electric field) thus forming an oriented amorphous phase ranging from the nematic structure to the periodic smectic phase. This structure is characterized by the fall in the intensity of the secondary relaxation β due to the decrease in the fraction of the disordered amorphous phase during the deformation whatever its initial morphology. Indeed [48] have shown that in the case of the deformation of PEN, whose structure is similar to PET, below its glass transition temperature, the β relaxation, of the same origin as that of PET, follows two trends: an increase in intensity due to its initial disordered mobile amorphous structure having no arrangement at very low values of the deformation then a progressive decrease in intensity of this same relaxation caused by the development of an ordered rigid amorphous phase (mesophase).

7.3 Comparisons of dielectric dynamic analysis with mechanical dynamic analysis for postmortem characterization of deformed states and characterization of filler/matrix interfaces in PP nanocomposites grafted MAH and crosslinked with polyether diamine

The authors particularly thank Professor Ali Kallel, Ms. Taktak Sirine and Ms. Agrebi Fatma for their invaluable help in carrying out dielectric dynamic analysis in University of Sfax (Tunisia).

7.3.1 Dynamic dielectric analysis characterization of crosslinked PP nanocomposites filled with graphite

Due to the absence of polarity of the macromolecular chain of polypropylene, dielectric spectroscopy was not used for the characterization of the polypropylene matrices, but this technique is particularly interesting for the analysis of nanocomposites based on PP in particular for the charge/matrix interfaces. Fig. 4.28 shows the evolution of the relative permittivity and the dielectric loss as a function of the frequency for a cross-linked polypropylene, which is used as a basis for the development of nanocomposites. The crosslinking method, described in different work [49,50] is based on the chemical reaction between pluri-amine groups molecules and anhydride maleic groups graft on the PP macromolecular chain.

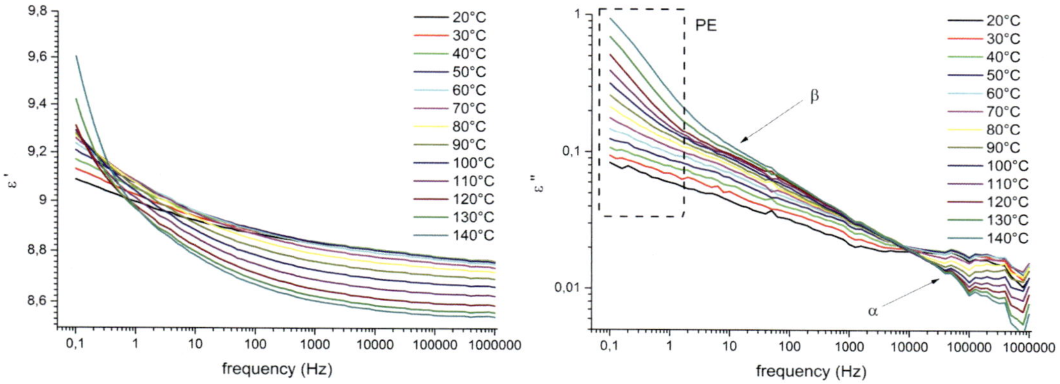

Figure 4.28

Dielectric isothermal spectra of Crosslinked PP matrix CA 100 THF 100 1: 1 [49].

Three relaxation mechanisms can be observed on the dielectric spectrum, the first one between 10^3 Hz and 1 MHz corresponds to the relaxation α of polypropylene, associated with the crystalline phase of the material [50]. The second relaxation which occurs around 10^2 Hz can be associated with the glass transition of polypropylene (β relaxation). It can also be associated with the polar group grafted with polypropylene (carboxylic acid, amide and imide) due to crosslinking [51,52]. These two relaxations are also observed by DMA [49]. The third relaxation conventionally found in polypropylene is the γ relaxation associated with the movements of methyl groups in the polypropylene chain. This is not visible here, because it is at a lower temperature. Finally, electrode polarization relaxation of the initial matrix is observed at low frequencies as a wide dispersion of ε' values as already observed in Ref. [53]. Fig. 4.29 shows the dielectric spectrum of the

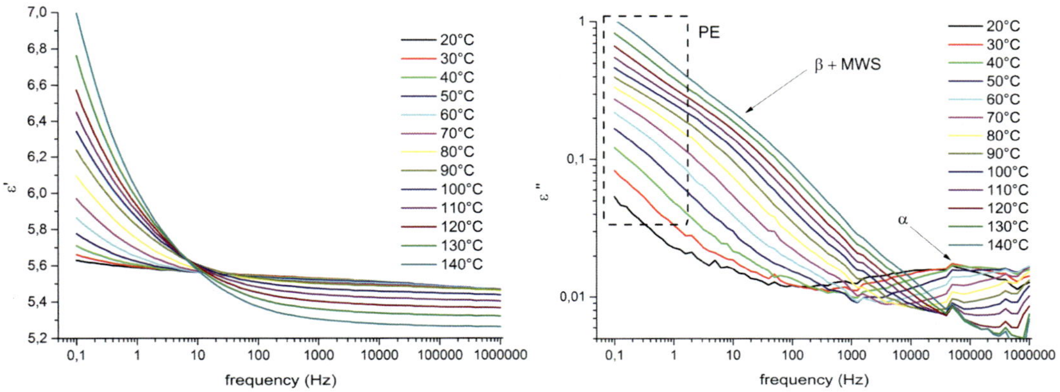

Figure 4.29

Evolution of ε' and ε'' as a function of frequency for the different temperatures for the nanocomposite filled with 1% w graphite [49].

nanocomposite filled with 1% w of graphite (so called KNG 180 which is the graphite used in this study and was purchased from Knano). The nanocomposites were produced by reactive extrusion, the nanofillers is introduced during the crosslinking step of the production with different massif concentration [49]. The different relaxations observed for the virgin matrix are presented here independently of the addition of the KNG 180 filler. The main change observed is a new relaxation present at high temperatures. This relaxation is attributed to the Maxwell-Wagner-Sillars (MWS) relaxation. This relaxation is associated with the conducting species accumulation at the interphase between the nanocharge and the matrix, which have very different permittivity, and strongly depends on the conductivity of the material, the nature of the filler, its surface rugosity, its volume or weight fraction, but also on the compatibility with the matrix [54,55]. Fig. 4.30 shows the comparison of the values of ε' and ε'' at 100°C for the different nanocomposites, as a function of frequency. An overall increase in the values of ε' and ε'' is observed with the carbon content, except for the two materials filled with 3% and 5% w of graphite which have extremely close values. This development is consistent with the addition of nanofiller. A significant change in the shape of the curve at high frequencies is also present, which indicates a change in the MWS relaxation.

The deconvolution of the curve ε'' for the different nanocomposites, carried out from the function of Havriliak-Negami (HN) makes it possible to measure the intensity of the interfacial polarization $\Delta\varepsilon_{MWS} = \varepsilon_s - \varepsilon_\infty$ associated to the MWS relaxation [56], see Eqs. (4.5) and (4.6).

$$\varepsilon'' = \sum_i \left[\frac{\Delta\varepsilon_i \sin \beta\phi}{\left[1 + 2(\omega\tau_{HNi})^{\alpha_i} \cos\left(\alpha_i \frac{\pi}{2}\right) + (\omega\tau_{HNi})^{2\alpha_i}\right]^{\frac{\beta_i}{2}}} \right] + j\frac{\sigma_{dc}}{\varepsilon_0\, \omega} \tag{4.5}$$

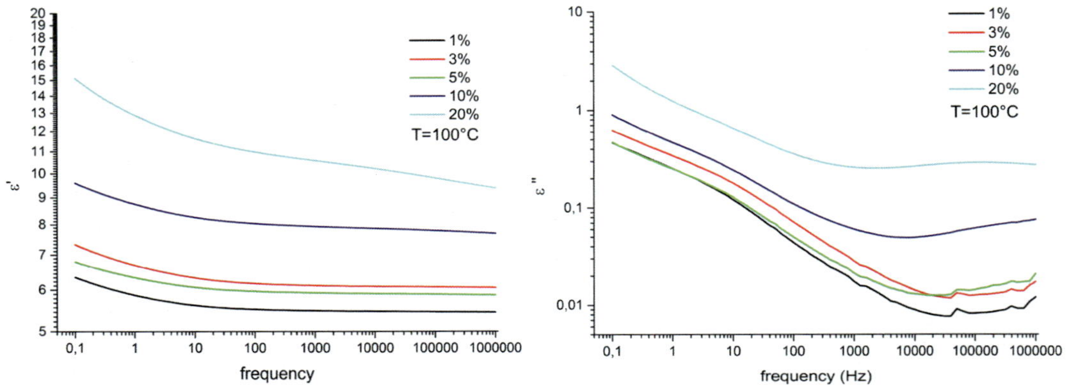

Figure 4.30
Comparison of the dielectric spectra obtained for different nanocomposites filled with graphite at 100°C [49].

$$\text{Avec,}\ \phi = \arctan\frac{(\omega\tau)^{\alpha}\sin\left(\alpha\frac{\pi}{2}\right)}{1+(\omega\tau)^{\alpha}\cos\left(\alpha\frac{\pi}{2}\right)} \quad et \quad \tau_{\max,\varepsilon} = \tau_{HN}\left[\frac{\sin\left(\frac{\alpha\beta\pi}{2+2\beta}\right)}{\sin\left(\frac{\alpha\pi}{2+2\beta}\right)}\right] \tag{4.6}$$

Table 4.1 summarizes the values of $\Delta\varepsilon_{MWS}$ for the different nanocomposites filled with graphite at different temperatures.

One notes an evolution of the values of $\Delta\varepsilon_{MWS}$ with the temperature for the same matrix. This evolution is explained by the greater mobility of charge carriers at the filler/matrix interphase at high temperatures. This result agrees with the literature, an increase in the temperature implying an increase in the intensity of the polarization and a shift of the relaxation toward the high frequencies. This is due to the greater ease of charge carriers at the interphase to be polarized at high temperatures [53]. One most interesting result is the evolution of the values of the intensity of polarization with the concentration in nanocharge of KNG 180. This evolution makes it possible to study the evolution of the interaction filler/filler and filler/matrix. A decrease in the values of the intensity of the polarization of the interphase between the nanocomposite filled with 1% w and 3% w is observed. This evolution with the increase in reinforcement is explained by the interactions, whether attractive or repulsive, of the KNG fillers between them and with the matrix. The decrease observed is explained here by the greater compatibility between the filler and the matrix and the increase in the interphase surface which implies a decrease in the mobility of the dipoles and their capacity to relax. On the other hand, the intensity of the polarization presents an increase at the highest concentrations (>5% w) in KNG 180. The increase in the carbon concentration, and the beginning of their agglomerations that can be observed with other experimental techniques implies a decrease in the filler to filler distances and the increase in their interactions, to the detriment of the filler/matrix interaction. In addition, the formation of aggregates limits the contact surface between the polymer matrix and the carbon nanofiller. In general, a decrease in the intensity values of the polarization can be associated with an increase in filler/matrix interactions, and therefore with an improvement in the dispersion of the nanofiller and in its compatibility with the polymer. On the contrary,

Table 4.1: Evolution of the intensity of the MWS relaxation for the different nanocomposites filled with graphite (KNG 180).

Temperature (°C)	KNG 180 1%w	KNG 180 3%w	KNG 180 5%w	KNG 180 10%w
100°C	0.24	0.18	0.26	0.27
110°C	0.27	0.19	0.29	0.32
120°C	0.28	0.26	0.32	0.35
130°C	0.35	0.34	0.39	0.42
140°C	0.40	0.39	0.43	0.54

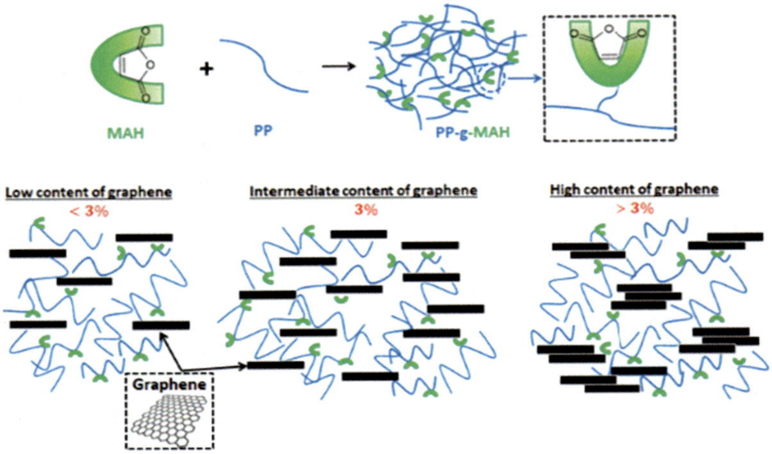

Figure 4.31
Diagram of the filler/matrix compatibility as a function of the weight percentage of graphite present.

an increase in $\Delta\varepsilon_{MWS}$ implies a degradation of the dispersion and an increase in the filler/matrix repulsion. It is therefore possible to characterize the state of dispersion of the nanofiller by following the evolution of the interfacial relaxation intensity. The final properties of the nanocomposite, in particular mechanical properties, strongly depend on the quality of the dispersion, but also on the compatibility filler/matrix, it is therefore particularly important to characterize this compatibility [50–57].

As summarized in Fig. 4.31, for low graphite concentrations, <3% w, the carbon nanofiller has a good dispersion within the matrix and good compatibility with the matrix. The matrix/filler interaction is maximum for the material loaded at 3% w, indicating that the reinforcing effect will be maximum at this concentration. Finally, the increase in $\Delta\varepsilon_{MWS}$ values for the highest concentration indicates a degradation of the filler dispersion, which agrees with the previous results. In conclusion, it can be noticed that the different matrices have a good graphite dispersion up to 5% w, with maintenance of the form factor of the nanofillers, and little or no agglomerates, but a distance between graphite fillers too large to allow electrical percolation due to low exfoliation. As the previous results suggested, the addition of 10% and 20% w of carbon nanofiller does not correct this problem, the fillers then have a degradation of their form factors, and the formation of large agglomerates within the matrix.

7.3.2 Dynamic dielectric characterization of crosslinked nanocomposites PP filled with graphite functionalized by plasma treatment

In this part, the same elaboration protocol is used to produce a new type of nanocomposites filled with a plasma-treated KNG 180 graphite [58] with the objective to

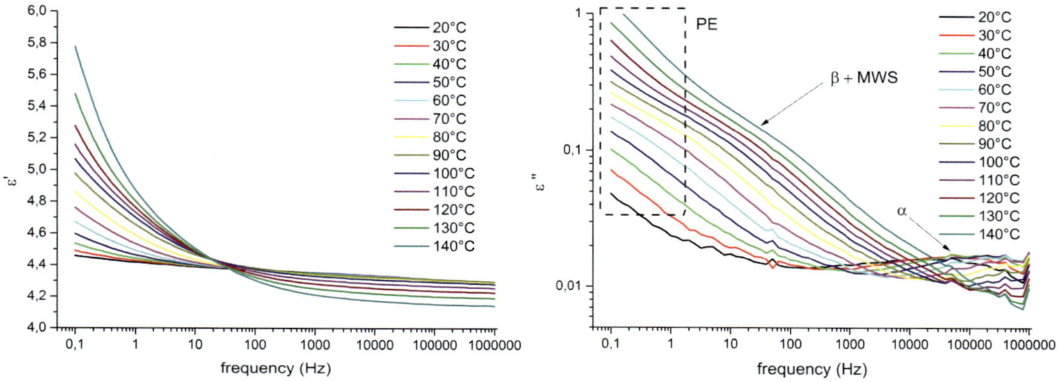

Figure 4.32
Dielectric spectra showing ε' and ε'' as a function of frequency for the nanocomposite filled with 1% w of graphite KNG 180 treated by plasma [49].

improve charge-matrix compatibility. This compatibility has been characterized by dynamic dielectric spectroscopy. The nanocomposites are then filled with plasma-treated KNG 180 graphite to improve charge-matrix compatibility and have been characterized by dynamic dielectric spectroscopy. Fig. 4.32 shows the dielectric spectra of the material filled with 1% m of KNG 180 graphite functionalized by plasma treatment.

Different relaxations are observed for PP, as well as electrode polarization and MWS relaxation, it is therefore possible to characterize the quality of the dispersion and the compatibility of the functionalized fillers with the polymer matrix. Fig. 4.33 shows the comparison of the dielectric spectra of the various nanocomposites filled with KNG 180 between 1% and 5% w treated or not.

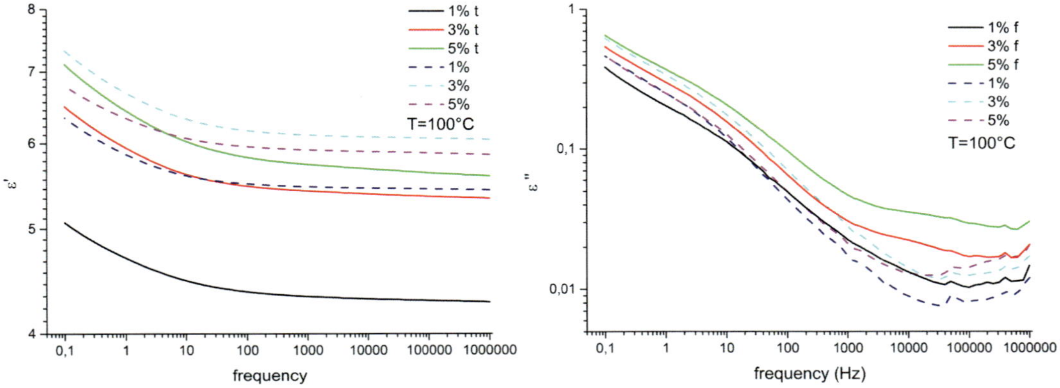

Figure 4.33
Comparison of the dielectric spectra ε' and ε'' as a function of frequency for the different nanocomposites filled with 1% and 5% w graphite treated (f) or not at 100°C [49].

Table 4.2: Evolution of the intensity of the MWS relaxation for the various nanocomposites filled with graphite KNG 180 treated by plasma (values measured at equivalent concentration for not treated graphite KNG 180 are given for comparison, the treated fillers are called *f* for functionalized).

Temperature (°C)	KNG 180 *f* 1%w	KNG 180 1%w	KNG 180 *f* 3%w	KNG 180 1%w
100°C	0.19	0.24	0.10	0.18
110°C	0.20	0.27	0.12	0.19
120°C	0.22	0.28	0.13	0.26
130°C	0.29	0.35	0.15	0.34
140°C	0.35	0.40	0.19	0.39
Temperature (°C)	**KNG 180 *f* 5%w**	**KNG 180 5%w**		
100°C	0.46	0.26		
110°C	0.49	0.29		
120°C	0.51	0.32		
130°C	0.39	0.39		
140°C	0.43	0.43		

The evolution of the values of ε' and ε'' here also shows an increase with the graphite concentration, despite values of ε' slightly lower than equivalent concentration for the filler treated by plasma (except for 1%w). The deconvolution of the different curves made it possible to calculate the values of $\Delta\varepsilon_{MWS}$, these values are given in Table 4.2.

As for the range of nanocomposite loaded with initial KNG 180 graphite, an increase in the values of $\Delta\varepsilon_{MWS}$ with temperature for the same matrix is observed. This development can be explained by the greater mobility of charge carriers at high temperatures [59,60].

On the contrary, a decrease in the values of $\Delta\varepsilon_{MWS}$ is observed at equivalent temperature between the nanocomposite filled with 1% w and the 3% w treated KNG 180 graphite. This development indicates an increase in matrix/filler interactions and is consistent with the increase in the surface interaction between filler and matrix. A significant increase is then observed between the nanocomposite filled at 3% w and at 5% w of filler treated. This change may indicate a significant deterioration in filler/matrix interactions, but also a significant increase in filler/filler interactions. Since results from TGA, from DRX but also observations by SEM and TEM have not shown a degradation of the dispersion of the nanofillers between these two concentrations, it is more coherent to associate this evolution with a greater interaction between fillers. A consequence of plasma treatment is to decrease the thickness of the nanofillers, the latter can be distributed more evenly within the matrix reducing the distance between fillers and increasing their interaction. Despite this improvement, it was not possible to achieve a dispersion allowing percolation. At a filler rate and equivalent temperature, the nanocomposites loaded with 1% and 3% m of graphite treated with plasma has values of $\Delta\varepsilon_{MWS}$ lower than the values estimated for

Table 4.3: Evolution of the resistivity of functionalized KNG 180 filled PP nanocomposites.

PP nanocomposites	KNG 180 *f* 1% m	KNG 180 *f* 3% m	KNG 180 *f* 5% m
Resistivity (10^{13} Ω cm)	2,8	2,9	2,9

the nanocomposites filled with not treated graphite KNG 180. Therefore, these different materials all show a good dispersion of the fillers, this difference in values can logically be associated with better compatibility of the oxidized fillers with the polymer matrix. Hence, these results confirm the improvement in the matrix/filler interactions due to the functionalization of the graphene sheets by the plasma treatment. Table 4.3 gives the resistivity measurement results for the different matrices.

Although resistivity values are lower than those measured for nanocomposites produced with unmodified nanofillers, the measured values remain high and indicate an insulating behavior for the nanocomposites under study. Hence, despite a better exfoliation obtained by plasma treatment as it was shown by the study of the filler dispersion within the matrices, percolation has not been achieved in this type of nanocomposites.

7.3.3 Outlook: evaluation of the potential of the correlation between dielectric and mechanical properties in order to characterize the reinforcement-matrix interfaces of nanocomposites

In the last decade, numerous composite materials have been developed in lightning structure applications in order to cope with the specifications of transport industries. One of the key conditions to obtain composites with good properties is a good matrix/fillers interface quality. It plays an important role in the mechanical behavior by transmitting the load from the matrix to the reinforcement and therefore ensure the nondecohesion of the material during the loading. Nanocomposites have been developed in order to enhance the interface by having a better surface to volume ratio between the matrix and reinforcements. In order to characterize the quality of the interface, mechanical characterization with static and dynamic testing gives multiple information on the behavior of the interface and its capacity to pass the loading from the matrix to the nanoreinforcements. On the other hand, dielectric characterization is also widely used. It can directly control the quality of the interface and its behavior. In this part some preliminary results on the correlation between the mechanical and dielectric results of grafted PP reinforced with functionalized graphite are shown [47,48]. The three diagrams shown below (Figs. 4.34–4.36) illustrate how to identify the key parameters that influence the quality of the interface by linking the mechanical and the dielectric behavior.

It is well accepted that the performance of nanocomposite materials with a polymer matrix depends both on the characteristics of the nanoreinforcement (nature, geometry, orientation, and volume fraction) and on the characteristics of the matrix. But, in addition to the intrinsic properties of each of the basic constituents, the reinforcement-matrix

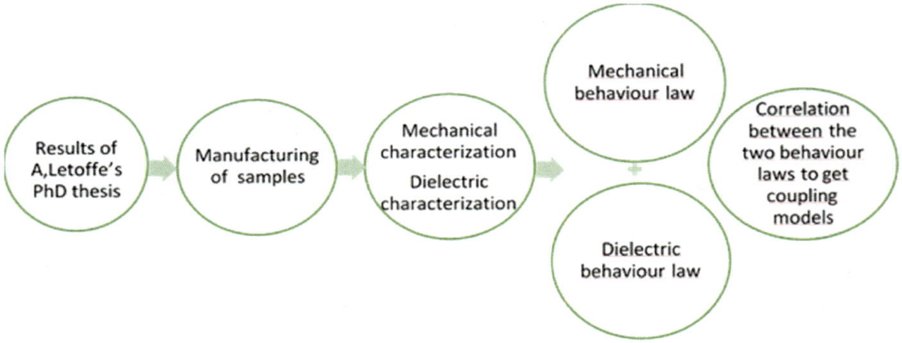

Figure 4.34
Synoptic diagram illustrating the perspective approach of this part.

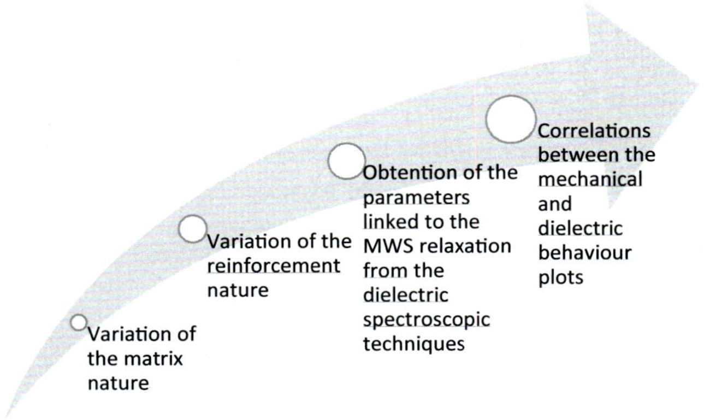

Figure 4.35
Synoptic diagram explaining the experimental approach of this perspective part.

interface plays a decisive role in the design and monitoring of such materials, both in terms of mechanical behavior as well as electrical and physicochemical behavior. Furthermore, mechanical characterization alone can provide qualitative information on the quality of the matrix-reinforcement adhesion since the load transfer from the matrix to the reinforcement takes place through the interface.

Dielectric spectroscopic techniques are complementary to mechanical measurements; they prove to be essential since they probe the interface directly and therefore make it possible to better elucidate the damage phenomena of nanocomposites. Thus, many works have been developed in this direction in recent years. However, rarely do the results presented highlight correlations between measured quantities obtained by different characterization techniques. In this context, as Fig. 4.35 shows, this section describes the contribution of the skills of LaMaCoP (University of Sfax-Tunisia) in dynamic dielectric measurements

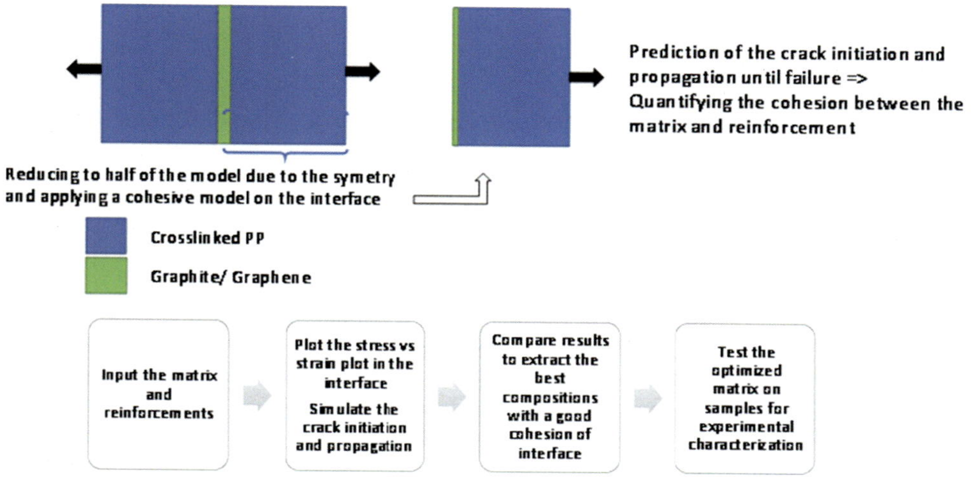

Figure 4.36
Synoptic diagram explaining the modeling approach of this perspective part.

[59] and those of the IJL (University of Lorraine-Nancy) with regard to thermomechanical properties and microstructures relationships of polymer composites via the development of innovative experimental couplings (Raman, WAXS-SAXS-TomoX, 3D stereocorrelation, VideoTraction , ESPI) allowing the study as close as possible to the actual conditions of use of their behavior [60]. The main objective is then to better define the properties of interfaces in heterogeneous systems in general and above all to establish international standards in terms of correlation between mechanical measurements (under static and dynamic stresses), dielectric measurements, and measurements, both optical and X-ray. These characterization techniques are complementary and they make it possible to determine the optimal conditions for the implementation of nanocomposites having the best performance. The experiments will initially be carried out on crosslinked polypropylene and reinforced with functionalized nanoparticles (graphite, graphene) for structural reduction in the automotive sector. Subsequently a study of the damage to composites [61] combining experimental, numerical and theoretical approaches will be made to understand and simulate the damage mechanisms in these materials. This study essentially contributes to:

- Better understand the quality of the interface as a function of the choice of the matrix and the nanoreinforcements (nature, rate, morphology, functionalization of the nanofillers, etc.) via the quantities influencing the dielectric mechanical properties.
- Carry out comparative studies according to the implementation making it possible to determine the adequate nanocomposite used for applications with lightning of structure and recyclable.

- Acquire and determine the parameters influencing the damage mechanisms until the rupture of these nanocomposite materials such as: (1) understanding the damage mechanisms under static loading and fatigue; (2) highlighting the interaction between the cracks and in particular the effect of the immediate vicinity of the defect on its growth as well as the relationship between the density of the defects and the elastic constants; and (3) the evaluation of the effect of the damage on the elastic constants of these nanocomposites.
- Establish international standards in terms of coupling between the two techniques used.

7.4 Feasibility of coupling dynamic dielectric analysis and Raman spectroscopy for monitoring in-situ crosslinking of a polymer based on acrylic resin crosslinkable under UV (apply to encapsulation of photovoltaic cells)

Kinetic monitoring of photopolymerization in situ has been shown to be possible with Raman spectroscopy and dielectric analysis techniques taken separately from Refs. [62–64]. Indeed, kinetic monitoring by dielectric measurements is possible as soon as the study material has residual ionic species and Raman spectroscopy makes it possible to follow the evolution of chemical species during photopolymerization. Previous studies have highlighted possible correlations between FTIR and dielectric spectroscopic measurements [65,66] and between Raman and dielectric spectroscopic measurements [67] for kinetic monitoring of polymerization. However, coupling in situ the two Raman/dielectric techniques to follow the photopolymerization is innovative because it represents a simultaneous approach from both a physical and chemical point of view. Indeed, the dynamic dielectric analysis allows a physical kinetic follow-up following the change in mobility of the residual ionic charges within the matrix which is induced by the crosslinking of the network on a macroscopic scale, while Raman spectroscopy allows the following of the chemical kinetics of the species formed during crosslinking on a microscopic scale. Thus, coupling the two analysis methods makes it possible to obtain additional information on the photopolymerization/crosslinking of the system based on UV crosslinkable resins used for the encapsulation of photovoltaic cells. It is with this in mind that an experimental device has been set up capable of allowing the coupling of these two techniques. The experimental device includes a dielectric measuring cell which contains the sample to be studied. This cell consists of a polytetrafluoroethylene (PTFE) plate cut into a U shape and placed between two brass electrodes used to conduct the current through the sample. The dimensions of the analysis cell are $26.0 \times 6.5 \times H$ in mm (length $\times$ width $\times$ H = variable height). The two plates are connected to the frequency generator (Novocontrol). The liquid formulation is placed in the hollow of the PTFE cell and then covered with quartz slide transparent to UV, to isolate it from external oxygen. This roughly reproduces the manufacturing conditions of the photovoltaic module according to the liquid encapsulation process (Fig. 4.37).

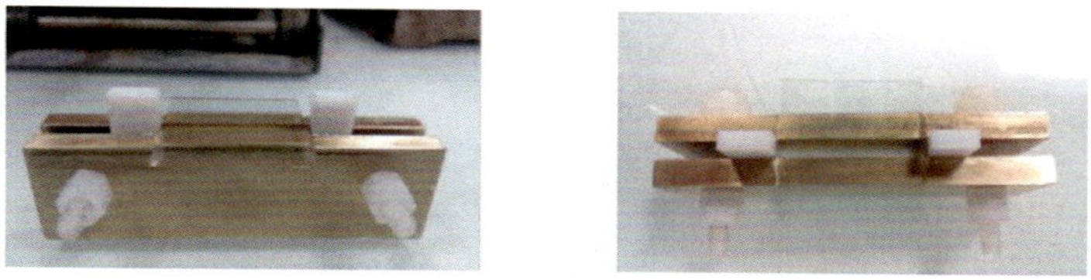

Figure 4.37
Measuring cell for kinetic monitoring of photopolymerization by Raman spectroscopy coupled in situ with dynamic dielectric analysis.

The UV emission source is arranged perpendicularly above the analysis cell. The Raman probe is attached to the collimator at an angle of about 45degrees to the sample. On the other side of the collimator, a pipe carrying nitrogen is placed above the sample (see Fig. 4.38) to limit the dioxygen inhibition effect during the polymerization.

During the photopolymerization, the Raman spectra measured correspond to the average of two acquisitions of 3 s each. The measurements are spaced every 8 s. As for dielectric measurements, a preliminary analysis is necessary to determine the frequency of measurement of the conductivity. As a reminder, the conductivity is directly related to the mobility of free conducting charges (ionic impurities for example in the case of this example). The frequency analysis of the conductivity of the resin at room temperature is illustrated in Fig. 4.39.

The dielectric response is of the resistive type between 1 and 1000 Hz. Beyond 3.10^3 Hz, limit frequency linked to the time necessary for a free charge to pass, the response is mainly capacitive. The choice was made to follow the kinetics of the photopolymerization by dielectric analysis of the conductivity at 10 Hz (σ_{DC} on Fig. 4.39). An example of kinetic monitoring of the resin formulation with a thickness of 12 mm is presented in Fig. 4.40.

The times required to reach 90% conversion are summarized in Fig. 4.41 according to the kinetic monitoring methods and according to the sample thickness.

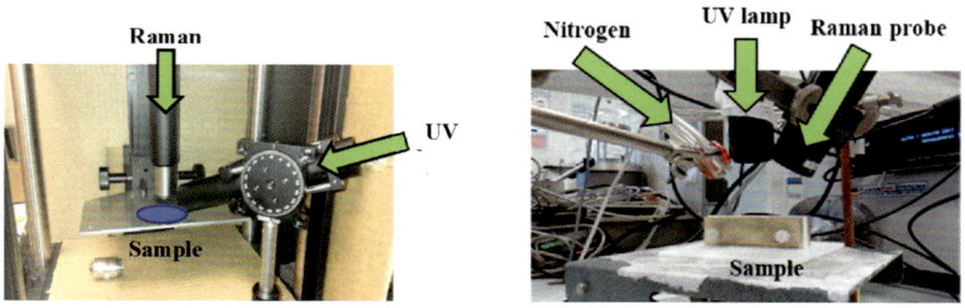

Figure 4.38
Left, experimental setup for kinetic monitoring of photopolymerization by in situ coupling of Raman spectroscopy and dielectric analysis. Right, the system being measured under UV irradiation.

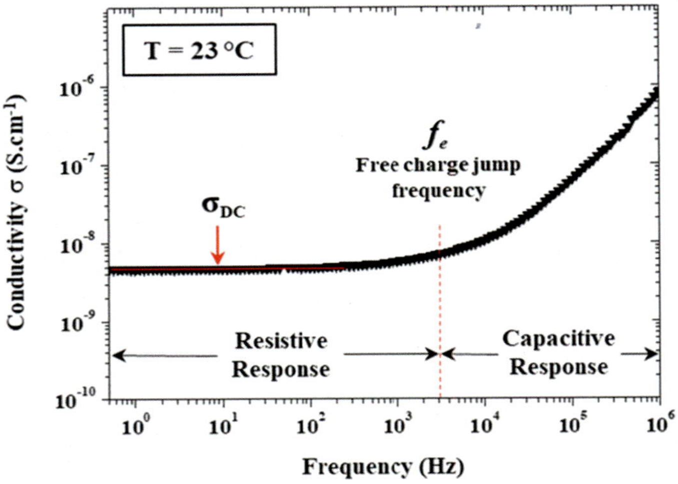

Figure 4.39
Frequency analysis of the conductivity of the resin formulation at room temperature.

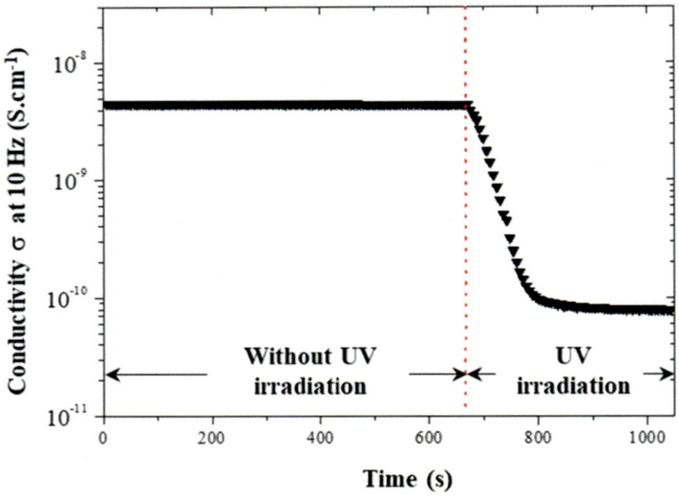

Figure 4.40
Evolution of the conductivity at 10 Hz as a function of time during the photopolymerization of a 12 mm thick resin sample.

For a better understanding of the kinetic differences between the conversions obtained by dielectric spectroscopy and those measured by Raman spectroscopy, the latter are plotted one according to the other in Fig. 4.42.

Thus, the conversions evaluated by dielectric analyses are much faster than those observed by Raman spectroscopy. The very clear difference observed between the conversion by

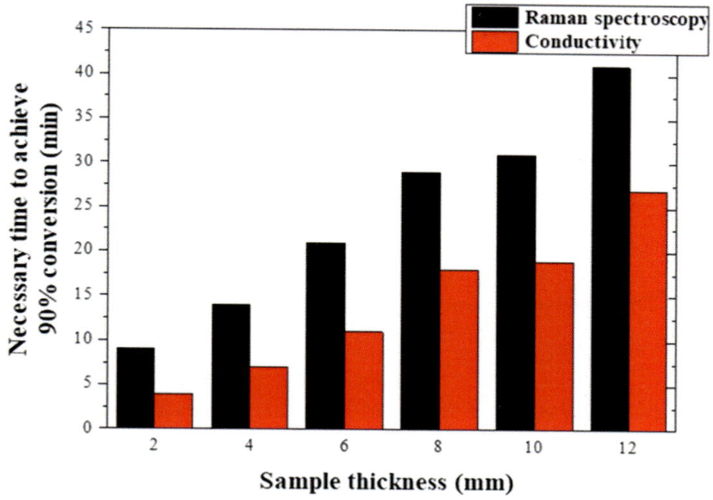

Figure 4.41

Time required to reach 90% variation in the parameters used as a kinetic polymerization monitoring index. The intensity of the spectral band at 1638 cm^{-1} of the C = C bond is followed by Raman spectroscopy and the conductivity at 10 Hz by dynamic dielectric analysis.

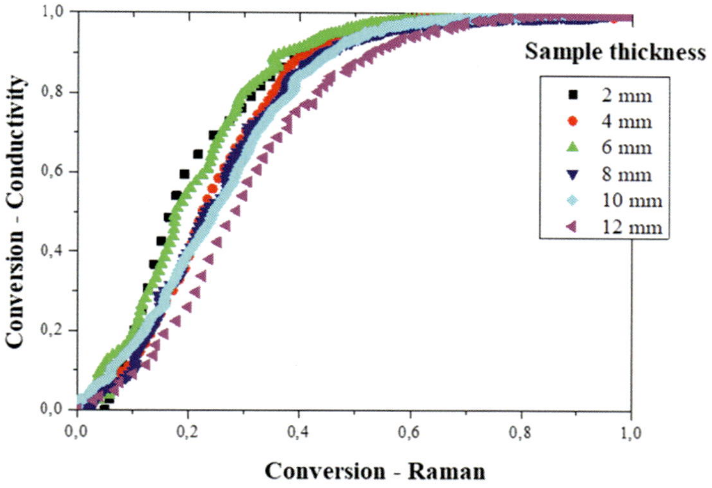

Figure 4.42

Conversion measured by the variation of the conductivity at 10 Hz for the formulated resin as a function of the conversion measured by Raman spectroscopy.

dielectric analysis and that by Raman could be attributed to two factors: the difference in power of UV lamp irradiation, and the local effect of the inhibition of oxygen (limited by the nitrogen) which has more impact on Raman spectroscopic measurements than on the

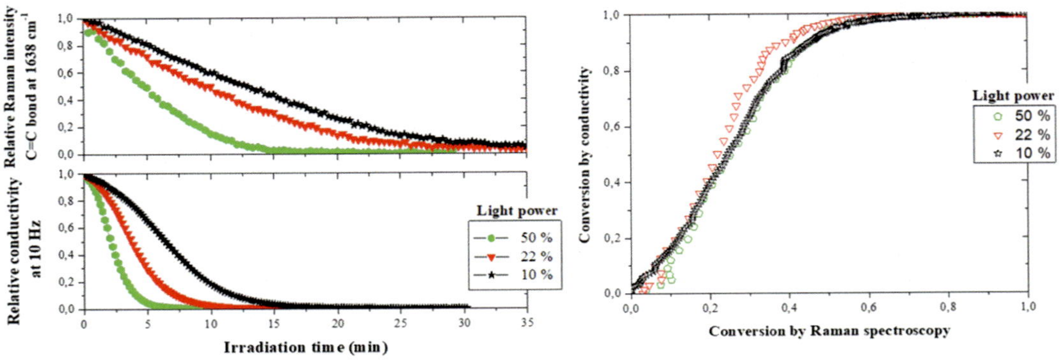

Figure 4.43

On the left, the conversion kinetics are followed by Raman spectroscopy (top) and by dielectric analysis (bottom) for different powers of UV lamp. On the right, the conversion kinetics of the two methods are compared for the three powers of UV irradiation.

dielectric analysis of conductivity (because the latter gives a global measurement). However, whatever the sample thickness, the correlation between the kinetics measured by the two methods gives the same tendency: at 60% chemical conversion, the conductivity, reflecting the mobility of the ions, reaches its minimum. In addition, the measurement of the analysis volume of the Raman spectrometer shows that the analysis is carried out at a depth of about 20 mm. Hence, the two types of measurements evaluate the evolving sample in a global manner. It was then question to see later if the difference came from the irradiation power of the lamp. The second series of experiments highlights the influence of the UV irradiation power on the polymerization kinetics of the resin (Fig. 4.43).

The more the power of the UV lamp increases, the higher the speed of polymerization is. In addition, conversions measured by dielectric analysis are also much faster than those measured by Raman spectroscopy. However, whatever the UV lamp power, the conversion by monitoring the conductivity reaches a maximum of 60% of the chemical conversion observed by Raman spectroscopy. This nonlinearity between the two conversions can be explained by a phenomenon of percolation during the gelation of the system [68]. As the system polymerizes and crosslinks, which occurs simultaneously due to the presence of bi- or polyunsaturated monomers in the formulations, the migration of charge-bearing ions is greatly slowed down or even stopped. The conductivity thus reaches its minimum plateau even though the chemical polymerization reaction, observed by Raman spectroscopy, continues. According to this description, the percolation threshold determined here experimentally, is reached for 60% of chemical conversion rate (double bonds). It thus appears that the radical crosslinking which occurs in parallel with the polymerization causes the gelling of the system toward 60% of chemical conversion. In conclusion, this part deals with the feasibility of monitoring the conversion kinetics of a family of liquid resin encapsulant. Control of the conversion kinetics under the conditions of irradiation closest to the process could be carried out by monitoring the glass transition temperature of the acrylic formulation. The dioxygen inhibitory effect on the conversion kinetics was then characterized by confocal Raman spectroscopy. Dioxygen has shown to have an inhibitory effect up to 250 μm in sample depth. In summary, the conversion could be measured in real time thanks to the in situ and simultaneous coupling of Raman spectroscopy with dynamic dielectric analysis.

8. Conclusions

In this chapter, we have shown by a few examples how Raman spectrometry can be coupled in situ to various techniques and how it can provide additional and complementary information to these techniques. Table 4.4 summarizes the contribution of

Table 4.4: Contribution of the additional information provided by in situ Raman spectrometry to the techniques used for the characterization of polymers.

Technics	Main informations	Contribution and informations from coupling with Raman spectrometry
Chromatography	Selective composition, 1 equipment for one element, measurements of traces or very low concentrations	Global sample information
Rheology	Viscosity, gel effect	Phase information, polymer conversion, modeling of polymerizations, confirmation gel effect ...
DSC	X_c, T_f, T_g ...	Proportion cristalline/amorphe phase, X_c, phase-specific response, amorphous phase information
X-rays	Structural and nanostructural informations, phases, information atomic scale ...	Amorphous phase information, additional or complement phase informations, molecular vibrations and their environments
Dielectric measurements	Relaxational phenomena, molecular mobility, glass transition, local motions	Changes of molecular mobility linking to crosslinking, aging, crystallization
Tensile test	Strain, stess and local deformations	Orientations, stresses, local deformations, crystal parameters (X_c, ...) information about amorphous phase

Raman spectrometry, particularly in situ, coupled to the various classical techniques which are used to characterize polymeric materials.

8.1 Chromatography measurements

This is a very sensitive, selective technique that allows traces or very low concentrations to be measured but requires a machine for one element to be measured. It is relatively long and expensive. After learning Raman spectrometry by chemometric techniques with chromatographic measurements as a reference method [69], one can use Raman spectroscopy to have more specific and selective information, and thus allow to measure from a single Raman spectrum, each element of a compound with a very good accuracy.

8.2 Rheology measurements

With these measurements, one has access essentially to viscosity and gel effects as a function of physical parameters (temperature, speed …), one measures a measure of a physical property. Raman spectrometry, correlated with this viscosity measurement, allows to highlight information on the phases, the polymer conversion which will allow to model the polymerization and to confirm for example the gel effects by monitoring the intensity of the specific Raman lines [70].

8.3 DSC measurements

It is a global measurement that gives the different functional temperatures of the polymer and its crystallinity. Raman spectrometry provides, as a function of temperature and in correlation with DSC measurements, specific information on the evolution of the phases, the crystalline phase but also the amorphous phase [4].

8.4 X-rays measurements

X-ray diffraction measurements give information on the structure, from the phases of the polymer to atomic scales. The contribution of Raman spectrometry is to have additional information on the amorphous phase and phase tracking, but also on the molecular vibrations and their environment [24].

8.5 Dynamic dielectric measurements

Dielectric dynamic measurements give essentially information on the mobility of macromolecules through the evolution of relaxational phenomena, essentially on the amorphous phase confined or not by crystalline phase. Coupling in situ with Raman spectrometry provides information in terms of changes in molecular mobility linked to crosslinking, water uptake, interface between matrix and filler in composite, aging and crystallization [71]. Moreover in situ coupling of dielectric dynamic spectroscopy and tensile test measurements has been proved to be possible as shown in Ref. [72].

8.6 Tensile test

Tensile test analysis coupled to in situ Raman spectroscopy can be obtained and allow to obtain information about macromolecular chain orientation and microstructure up to break. Many information may be deduced from these in situ measuremnets on amorphous phase and cristallinity [27].

As we have just shown, Raman spectrometry, used in real time or as in situ measurement, is a very rich technique that allows us to provide a great number of information on polymers. However, it requires a thorough study of the Raman response as a function of the parameters to be measured, especially a learning process and therefore correlations with reference techniques. This is often done by using chemometric tools which offer very interesting and innovative new possibilities and open new perspectives [73].

References

[1] S. Chaudemanche, Étude de l'endommagement des polymères, au cours de la déformation plastique, selon différents modes de sollicitation, University of Lorraine, 2013. LORR026.
[2] K.B. Hafsia, Identification des micro-mécanismes de déformation du PET amorphe et semi-cristallin in situ au cours d'un essai mécanique, University of Lorraine, 2016.
[3] M. Veitmann, Influence de la mise en œuvre et du vieillissement sur la cristallinité de films PVDF, University of Lorraine, 2015.
[4] M. Veitmann, D. Chapron, S. Bizet, S. Devismes, J. Guilment, I. Royaud, M. Poncot, P. Bourson, Thermal behavior of PVDF/PMMA blends by differential scanning calorimetry and vibrational spectroscopies (Raman and Fourier-Transform Infrared), Polym. Test. 48 (2015) 120–124, https://doi.org/10.1016/j.polymertesting.2015.10.004.
[5] C. Leonard, J.L. Halary, L. Monnerie, Crystallization of poly(vinylidene fluoride)-poly(methylmethacrylate) blends : analysis of the molecular parameter controlling the nature of the poly(vinylidene fluoride) crystalline phase, Macromolecules 21 (1988) 2988–2994.
[6] A.J. Melveger, Laser-raman study of crystallinity changes in poly(ethylene terephthalate), J. Polym. Sci. Part 2 Polym. Phys. 10 (2) (1972) 317–322.
[7] T. Lippert, F. Zimmermann, A. Wokaun, Surface analysis of excimer-laser-treated polyethylene-terephthalate by surface-enhanced Raman scattering and x-ray photoelectron spectroscopy, Appl. Spectrosc. 47 (11) (1993) 1931–1942.
[8] D. Kawakami, Deformation-induced phase transition and superstructure formation in poly(ethylene terephthalate), Macromolecules 38 (1) (2005) 91–103.
[9] G. McGraw, American chemical society meeting, Polym. Preprint. 11 (1970) 1122.
[10] B.J. Bulkin, M. Lewin, F.J. DeBlase, Conformational change, chain orientation, and crystallinity in poly (ethylene terephthalate) yarns: Raman spectroscopic study, Macromolecules 18 (12) (1985) 2587–2594.
[11] J. Štokr, Conformational structure of poly(ethylene terephthalate). Infra-red, Raman and n.m.r. spectra, Polymer 23 (5) (1982) 714–721.
[12] G. Ellis, FT Raman study of orientation and crystallization processes in poly (ethylene terephthalate), Spectrochim. Acta Mol. Biomol. Spectrosc. 51 (12) (1995) 2139–2145.

[13] T. Manley, D. Williams, Structure of terephthalate polymers infra-red spectra and molecular structure of poly(ethylene terephthalate), Polymer 10 (1969) 339–384.
[14] M.W. Ward, Infra-red and Raman spectra of poly (m-methylene terephthalate) polymers, Polymer 18 (4) (1977) 327–335.
[15] K. Ben Hafsia, M. Ponçot, D. Chapron, I. Royaud, P. Bourson, A novel approach to study the isothermal and non-isothermal crystallization kinetics of poly(ethylene terephthalate) by Raman spectroscopy, J. Polym. Res. 23 (5) (2016) 1–14.
[16] K. Ben Hafsia, Identification des micro-mecanismes de déformation du PET amorphe et semi-cristallin in situ au cours d'un essai mecanique (Ph.D. thesis), Lorraine University, 2016.
[17] B.H. Stuart, Polymer crystallinity studied using Raman spectroscopy, Vib. Spectrosc. 10 (2) (1996) 79–87.
[18] R. Paquin, M.H. Limage, P. Colomban, Micro-Raman study of PET single fibres under high hydrostatic pressure: phase/conformation transition and amorphization, J. Raman Spectrosc. 38 (9) (2007) 1097–1105.
[19] S.K. Liang, Infrared spectra of high polymers: part IX. Polyethylene terephthalate, J. Mol. Spectrosc. 3 (1) (1959) 554–574.
[20] P. Colomban, J. Corset, Foreword to the special issue on Raman (micro) spectrometry and materials science, J. Raman Spectrosc. 30 (10) (1999) 863–866.
[21] S. Chaudemanche, Caractérisation in situ del'endommagement volumique par spectroscopie raman et rayons x de différents polypropylènes deformés en traction uniaxiale (Ph.D. thesis), Lorraine University, France, 2013.
[22] M. Ponçot, Comportements tehrmomécaniques de polymères chargés selon différents chemins de déformation et traitements thermiques (Ph.D. thesis), Lorraine University, France, 2009.
[23] J. Martin, Etude par spectroscopie Raman du polypropylène isotactique au cours de sa déformation uniaxiale (Ph.D. thesis), Lorraine University, France, 2009.
[24] M. Ponçot, J. Martin, S. Chaudemanche, O. Ferry, T. Schenk, J.P. Tinnes, D. Chapron, I. Royaud, A. Dahoun, P. Bourson, Complementarities of high energy WAXS and Raman spectroscopy measurements to study the crystalline phase orientation in polypropylene blends during tensile test, Polymer 80 (2015) 27–37, https://doi.org/10.1016/j.polymer.2015.10.040.
[25] J. Martin, M. Ponçot, P. Bourson, A. Dahoun, J.M. Hiver, Study of the crystalline phase orientation in uniaxially stretched polypropylene by Raman spectroscopy: validation and use of a time-resolved measurement method", Polym. Eng. Sci. 51 (2011) 1607–1616.
[26] M. Ponçot, J. Martin, J.M. Hiver, D. Verchère, A. Dahoun, Study of the dimensional instabilities of laminated polypropylene films during heating treatments, J. Appl. Polym. Sci. 125 (5) (2012) 3385–3395.
[27] J. Martin, M. Ponçot, J.M. Hiver, P. Bourson, A. Dahoun, Real time Raman spectroscopy measurements to study the uniaxial tension of isotactic polypropylene, J. Raman Spectrosc. 44 (5) (2013) 776–784.
[28] R.P. Daubeny, C.W. Bunn, The crystal structure of polyethylene terephthalate, Proc. R. Soc. A 226 (1954) 531–542.
[29] Y.Y. Tomashpolskii, G.S. Markova, An electron diffraction study of the crystalline structure of polyethylene terephthalate by means of the fourier synthesis, Polym. Sci. 6 (1964) 316–324.
[30] M. Donnay, Etude des mécanismes de déformation de membranes polymères poreuses pour applications biomédicales (Ph.D. thesis), Lorraine University, France, 2017.
[31] F. Adar, H. Noether, Raman microprobe spectra of spin-oriented and drawn filaments of poly(ethylene terephthalate), Polymer 26 (1985) 1935.
[32] C.C.C. Lesko, J.F. Rabolt, R.M. Ikeda, B. Chase, A. Kennedy, Experimental determination of the fiber orientation parameters and the Raman tensor of the 1614 cm^{-1} band of poly(ethylene terephthalate), J. Mol. Struct. 521 (2000) 127–136.
[33] N. Everall, P. Tayler, J. Chalmers, D. MacKerron, R. Ferwerda, J.V. der Maas, Study of density and orientation in poly(ethylene terephthalate) using fourier transform Raman spectroscopy and multivariate data analysis, Polymer 35 (1994) 3184–3192.

[34] S. Chaudemanche, M. Ponçot, S. André, A. Dahoun, P. Bourson, Evolution of the Raman spectroscopy backscattered intensity used to analyze the micromechanisms of deformation of various polypropylene blends in situ during a uniaxial tensile test, J. Raman Spectrosc. 45 (2014).

[35] M. Ponçot, F. Addiego, A. Dahoun, True intrinsic mechanical behaviour of semi-crystalline and amorphous polymers: influences of volume deformation and cavities shape, Int. J. Plast. 40 (2013) 126–139.

[36] P. Colomban, J.C. Badot (Eds.), Proton Conductors: Solids, Membranes and Gels -Materials and Devices, Cambridge University Press, Cambridge, 1992.

[37] S. Saidi, D. Hermida-Merino, A. Bytchkov, G. Mant, G. Portale, F. Bargain, S. Bizet, J. Guilment, M. Ponçot, I. Royaud, D. Chapron, P. Bourson, Probing polymer phase transitions by in-situ thermal analysis coupled to small and wide-angle X-ray scattering combined with Raman spectroscopy, J. Synchrotron Radiat. (2020).

[38] E. Dargent, E. Bureau, L. Delbreilh, A. Zumailan, J.M. Saiter, Effect of macromolecular orientation on the structural relaxation mechanisms of poly(ethylene terephthalate), Polymer 46 (9) (2005) 3090–3095.

[39] C. Hakme, I. Stevenson, L. David, G. Seytre, G. Boiteux, Effect of orientation and crystallization on dielectric and mechanical relaxations in uniaxially stretched poly(ethylene naphthalene 2,6 dicarboxylate)(PEN) films, J. Non-Cryst. Solids 352 (42–49) (2006) 4746–4752.

[40] J.P. Runt, J.J. Fitzgerald, Dielectric Spectroscopy of Polymeric Materials: Fundamentals and Applications, American Chemical Society, 1997.

[41] F. Kremer, Dielectric spectroscopy—yesterday, today and tomorrow, J. Non-Cryst. Solids 305 (1) (2002) 1–9.

[42] G. Williams, D.K. Thomas, Phenomenological and molecular theories of dielectric and electrical relaxation of materials, Novocontrol. Appl. Note Dielec. 3 (1998) 1–29.

[43] N.G. McCrum, B.E. Read, G. Williams, Anelastic and Dielectric Effects in Polymeric Solids, 1967.

[44] A. Kahouli, Étude des propriétés physico-chimiques et (di)-électriques du parylène C en couche mince, PhD University of Grenoble, 2011.

[45] B. Kim, Y.D. Park, K. Min, J.H. Lee, S.S. Hwang, S.M. Hong, B.H. Kim, S.O. Kim, C.M. Koo, Electric actuation of nanostructured thermoplastic elastomer gels with ultralarge electrostriction coefficients, Adv. Funct. Mater. 21 (17) (2011) 3242–3249.

[46] L. Lin, A. Argon, Structure and plastic deformation of polyethylene, J. Mater. Sci. 29 (2) (1994) 294–323.

[47] (a) C. G'Sell, A. Dahoun, Evolution of microstructure in semi-crystalline polymers under large plastic deformation, Mater. Sci. Eng. 175 (1–2) (1994) 183–199;
(b) A. Létoffé, S.M. García-Rodríguez, S. Hoppe, N. Canilho, O. Godard, A. Pasc, I. Royaud, M. Ponçot, Switching from brittle to ductile isotactic polypropylene-g-maleic anhydride by crosslinking with capped-end polyether diamine, Polymer 164 (2019) 67–78.

[48] (a) S. Castagnet, J.-L. Gacougnolle, P. Dang, Correlation between macroscopical viscoelastic behaviour and micromechanisms in strained α polyvinylidene fluoride (PVDF), Mater. Sci. Eng. 276 (1) (2000) 152–159;
(b) A. Létoffé, S. Hoppe, R. Lainé, N. Canilho, A. Pasc, D. Rouxel, R.J. Jiménez Riobóo, S. Hupont, I. Royaud, M. Ponçot, Resilience improvement of an isotactic polypropylene-g-maleic anhydride by crosslinking using polyether triamine agents, Polymer 179 (2019) 121655.

[49] A. Létoffé, (Ph.D. Thesis), University of Lorraine, 2020.

[50] A. Ridhore, J.P. Jog, A dynamic mechanical and dielectric relaxation study of PP-g-MAH/clay nanocomposites, Open Macromol. J. 6 (2012) 53–58.

[51] A. Motori, G. Montanari, A. Saccani, F. Patuelli, Electrical conductivity and polarization processes in nanocomposites based on isotactic polypropylene and modified synthetic clay, J. Polym. Sci., Part B: Polym. Phys. 45 (2007) 705–713.

[52] M. Bohning, H. Goering, A. Fritz, K. Brzezinka, G. Turky, A. Schonhals, B. Schartel, Dielectric study of molecular mobility in poly (propylene-graft-maleic anhydride)/clay nanocomposites, Macromolecules 38 (2005) 2764–2774.

[53] F. Agrebi, N. Ghorbel, A. Ladhar, S. Bresson, A. Kallel, Enhanced dielectric properties induced by loading cellulosic nanowhiskers in natural rubber: modeling and analysis of electrode polarization, Mater. Chem. Phys. 200 (2017) 155–163.
[54] X. Jin, S. Zhang, J. Runt, Observation of a fast dielectric relaxation in, semicrystalline poly(ethylene oxide), Polymer 43 (2002) 6247e6254.
[55] S. Havriliak, S.J. Havriliak, Comparaison of the Havriliak-Negami and stretched exponential functions, Polymer 37 (1996) 4107e4110.
[56] A. Triki, M. Guicha, M.B. Hassen, M. Arous, Z. Fakhfakh, Studies od dielectric relaxation in natural fibres reinforced unsaturated polyester, J. Mater. Sci. 46 (2011) 3698–3707.
[57] A. Ladhar, M. Arous, H. Kaddami, Z. Ayadi, A. Kallel, Correlation between the dielectric and the mechanical behavior of cellulose nanocomposites extracted from the rachis of the date palm tree, in: Paper Presented at the IOP Conference Series: Materials Science and Engineering, 2017.
[58] S. Cuynet, M. Poncot, S. Fontana, A. Letoffe, G. Henrion, C. Herold, I. Royaud, T. Belmonte, Method of Exfoliation and/or Functionalization of Lamellar Objects and Associated Device, Patent Deposited, 12th of March 2020, N° FR2002479.
[59] Z. Ghallabi, M. Samet, M. Arous, A. Kallel, G. Boiteux, I. Royaud, G. Seytre, Giant permittivity and low dielectric loss in three phases $BaTiO_3$/carbon nanotube/polyvinylidene fluoride composites, J. Adv. Phy. 3 (2014) 87–91.
[60] M. Donnay, M. Ponçot, J.-P. Tinnes, T. Schenk, O. Ferry, I. Royaud, In situ study of the tensile deformation micro-mechanisms of semi-crystalline poly(ethylene terephthalate) films using synchrotron radiation x-ray scattering, Polymer 117 (2017) 268–281.
[61] M. Loukil, Z. Ayadi, J. Varna, ESPI analysis of crack face displacements in damaged laminates, Compos. Sci. Technol. 94 (2014) 80–88.
[62] M.E. Nichols, C.M. Seubert, W.H. Weber, J.L. Gerlock, A simple Raman technique to measure the degree of cure in UV curable coatings, Prog. Org. Coating. 43 (2001) 226–232.
[63] W.S. Shin, X.F. Li, B. Schwartz, S.L. Wunder, G.R. Baran, Determination of the degree of cure of dental resins using Raman and FT-Raman spectroscopy, Dent. Mater. 9 (1993) 317–324.
[64] J. Steinhaus, B. Hausnerova, T. Haenel, M. Großgarten, B. Möginger, Curing kinetics of visible light curing dental resin composites investigated by dielectric analysis (DEA), Dent. Mater. 30 (2014) 372–380.
[65] K. Zahouily, C. Decker, E. Kaisersberger, M. Gruener, Real-time UV cure monitoring, Eur. Coating J. (2003) 14–34.
[66] K. Xu, S. Zhou, L. Wu, On-line and in-real-time monitoring of UV photopolymerization of nanocomposites with microdielectrometry, Prog. Org. Coating 65 (2009) 237–245.
[67] R. Hardis, J.L.P. Jessop, F.E. Peters, M.R. Kessler, Cure kinetics characterization and monitoring of an epoxy resin using DSC, Raman spectroscopy, and DEA, Compos. Part Appl. Sci. Manuf. 49 (2013) 100–108.
[68] R.A.L. Jones, Soft Condensed Matter, OUP Oxford, 2002.
[69] A. Filliung, D. Chapron, P. Bourson, G. Finqueneisel, A. Riondel, J. Guilment, Raman spectroscopy and chemometrics for quantitative analysis of complex flows in an industrial transesterification process, J. Raman Spectrosc. 45 (10) (2014) 941–946.
[70] M.C. Chevrel, S. Hoppe, D. Meimaroglou, D. Chapron, P. Bourson, J. Wilson, P. Ferlin, L. Falk, A. Durand, Application of Raman spectroscopy to characterization of residence time distribution and online monitoring of a pilot-scale tubular reactor for acrylic acid solution copolymerization, Wiley-VCH Verlag, Macromol. React. Eng. 10 (4) (2016). Special Issue: Transfer of Polymer Production Processes.
[71] F. Kremer, A. Schönhals, Broadband Dielectric Spectroscopy, Springer Verlag, 2003, ISBN 978-3-642-56120-7.

[72] M. Samet, V. Levchenko, G. Boiteux, G. Seytre, A. Kallel, A. Serghei, Electrode polarization vs. Maxwell-Wagner-Sillars interfacial polarization in dielectric spectra of materials: characteristic frequencies and scaling laws, J. Chem. Phys. 142 (2015) 194703.
[73] N. Brun, M. Ponçot, P. Bourson, Raman correlation spectroscopy: a method studying physical properties of polystyrene by the means of multivariate analysis, Chemometr. Intell. Lab. Syst. 128 (2015) 77–82.

CHAPTER 5

Positron annihilation spectroscopy for defect characterization in nanomaterials

Ann Rose Abraham[1], **P.M.G. Nambissan**[2]

[1]*Department of Physics, Sacred Heart College (Autonomous), Kochi, Kerala, India;*
[2]*Applied Nuclear Physics Division, Saha Institute of Nuclear Physics, Kolkata, India*

Chapter Outline

Design, Fabrication, and Characterization of Multifunctional Nanomaterials. https://doi.org/10.1016/B978-0-12-820558-7.00016-9

1. Introduction

Positron annihilation spectroscopy (PAS) [15] has proved to be a unique and versatile experimental technique for the study of defects in materials. It deals with the process of electron-positron annihilation and its exploitation for defect analysis in materials. The preferential sensitivity of positrons toward microdefects like vacancies, their clusters and buried layers, surfaces and interfaces, which are not assessable by other techniques, makes it an attractive tool for many materials science problems [11].

In crystalline solids, atomic-level defects such as vacant lattice sites and impurities play important roles in modifying their structure and properties. Especially such imperfections often determine the physical characteristics of the materials such as mechanical properties, electrical conductivity, diffusivity or light emission and these can be appropriately molded for various novel applications if a qualitative and quantitative accounting of the defects and defect interactions can be priory made available [10]. There are many theoretical approaches associated with point defects that are based on density functional methods or Monte Carlo simulations [7]. PAS offers a systematic and efficient experimental study of structural defects in nanomaterials [7].

The impact of PAS in the field of condensed matter and materials science research is enormous [8]. It has many advantages over other conventional experimental methods. It is a highly sensitive and reliable method for the investigation of defects even in very low concentrations in materials. Conventional tools such as X-ray diffraction (XRD) technique can give measurable changes only when the defect concentration is high enough and transmission electron microscopy (TEM) till in recent times was limited by its finite resolving power. In this context, the preferential sensitivity of positrons toward defects, which are not accessible by other methods, makes the spectroscopy using them as investigative probes an attractive tool for materials research. The exceptional niche of PAS [9] over other common techniques for vacancy defect analysis such as optical microscopy (OM), TEM, neutron scattering (nS), atomic force microscopy (AFM), scanning tunneling microscopy (STM), and X-ray scattering is illustrated in Fig. 5.1. Another key advantage of PAS over other techniques is that it is a nondestructive method and hence the sample can be reused. PAS provides information on the atomic scale. It is now well established as a technique for detection and characterization of defects and their clusters and for monitoring their evolution under controlled changes of experimental parameters like temperature, chemical composition, etc.

2. Fundamentals of positron annihilation spectroscopy

PAS is a characterization method that helps in the estimation of the local electron density and determination of the atomic structure at the site chosen by the electrostatic interaction

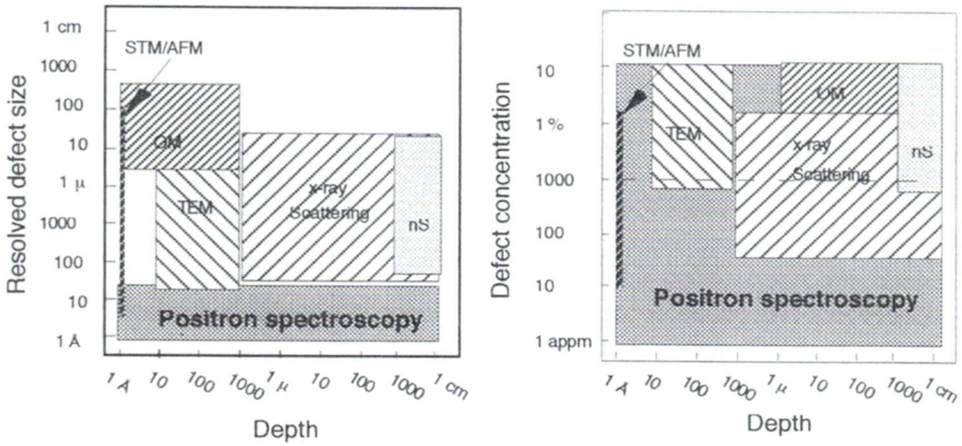

Figure 5.1
A diagrammatic representation of the high sensitivity of PAS over other techniques in determining low concentrations of defects. *Reprinted from R.H. Howell, T.E. Cowan, J.H. Hartley, P.A. Sterne, Positron beam lifetime spectroscopy at Lawrence Livermore National Laboratory, in: AIP Conference Proceedings, vol. 392, American Institute of Physics, 1997, no. 1, pp. 451–454, with the permission of AIP Publishing.*

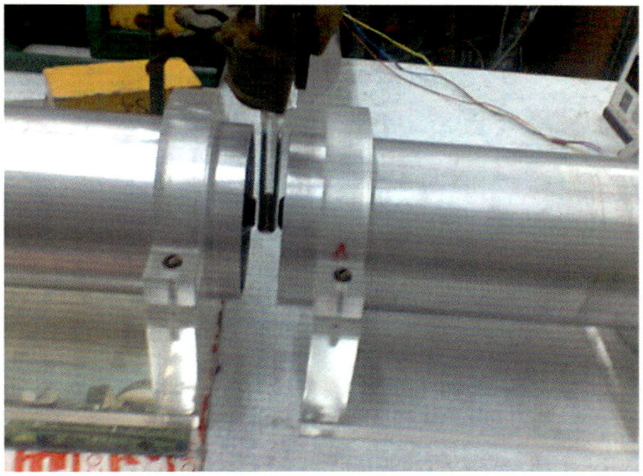

Figure 5.2
Experimental arrangement with the ^{22}Na source immersed in nanomaterial sample in the glass tube and aligned in line with the scintillation detectors.

of the positron with the electrons. Positrons can be used to study atoms, molecules, and imperfections like defects in solids and processes involving structural or phase transitions involving changes in the surrounding medium.

2.1 Introduction to positron

The Schrodinger equation, the fundamental equation of wave mechanics, did not satisfy the requirement of invariance postulated by the theory of special relativity. This problem was solved by P. A. M. Dirac who introduced the now-famous electron equation known after his name, and which is one of the most important developments in modern physics. Dirac pointed out that the equation which represents the motion of an electron should be invariant under the Lorentz transformation. The solution of the Dirac equation led to the fine structure of spectral lines and to the spin of the electron. But, to solve the Dirac equation, it was necessary to assume that electron can exist in two sets of quantum states, one of positive energy and other of negative energy. The possible values of energy of a free electron are either greater than $+ mc^2$ or lower than $-mc^2$ and no possible energies could exist between these limits. A state of negative energy has no real physical meaning either. According to quantum theory, an electron can make a discontinuous transition from one energy state to another and so there is no way of ruling out a jump into a negative energy state from a positive energy state.

The issue was resolved in 1933 when C. D. Anderson, during the course of a study of cosmic rays by means of a cloud chamber, observed particles with the same mass as that of electrons, and each with an electric charge equal in magnitude but opposite in sign to that of the electron. This particle was given the name *positron* and identified with the Dirac hole.

In terms of Dirac's theory, the production of a positron (e^+) is interpreted as follows. A photon having energy greater than $2m_oc^2$ can move up an electron (e^-) from a negative energy state to a positive energy state. This disappearance of an electron from a state of negative energy leaves a hole, which means the appearance of a positron. The appearance of an electron in positive energy state means the appearance of an ordinary electron. This explains the creation of a pair of particles (e^+ and e^- pairs).

2.2 Principles of electron-positron interaction and annihilation

Being thus the *anti*particle of the electron, the positron will eventually meet with the end of its existence by its interaction with the electron and the process is termed as annihilation. When the annihilation of a positron with an electron takes place, their rest masses are converted into energy, which are emitted in the form of gamma ray photons. Gamma rays produced due to the annihilation of a positron and electron yield very accurate information regarding the electronic structure and properties of the material medium as discussed in the forthcoming sections.

The most commonly used experimental methods for observing positron annihilation in matter are based on measuring the positron lifetime, the angular correlation of the annihilation gamma rays and the Doppler shift induced broadening of the energy spectrum of the annihilation radiation [18]. The positron lifetime measures the electron density, whereas the angular correlation and Doppler broadening of the annihilation radiations reflects the electron momentum distribution. It is thus possible to investigate experimentally the local heterogeneous structures embedded in the bulk of the material such as vacancies, atomic clusters, superlattices and device structures, quantum dots as well as free volumes and voids in solids and even in biological materials.

2.3 Positron lifetime measurements

As stated earlier, positron is the antiparticle of the electron, which meant a unit positive charge and the same mass as that of the electron. For the purpose of experiments in the laboratory, positrons are obtained from radioactive nuclide sources like ^{22}Na, ^{64}Cu, etc. They emit positrons with sufficient energy so as to penetrate any material. Positrons getting emitted in the process of beta decay can slow down if captured in the bulk of the material under study to thermal velocities within a time period of about 1−10 ps (picoseconds). Such positrons with only thermal energies ($\sim k_BT$) annihilate with the electrons of the material either from the bulk or after getting trapped by neutral or negatively charged point defects or extensive defect clusters. It is the electromagnetic interaction that makes the annihilation of e^+ and e^- pairs possible. The electron-positron pair annihilation process proceeds by the emission of gamma rays, which are captured by efficient radiation detectors for further measurements and analysis.

^{22}Na is the most widely used β^+ decaying isotope and it also emits a gamma ray of energy 1.276 MeV in the transition

$$^{22}\mathrm{Na} \rightarrow {}^{22}\mathrm{Ne} + e^{+} + \Upsilon\ (1.276\ \mathrm{MeV}) \tag{5.1}$$

The positron and the nuclear photon (1.276 MeV) are emitted almost simultaneously and hence the latter is treated as the birth signal for the former as the intermediate time gap is negligible compared to the lifetime of positrons in any material [15]. The positron annihilates with an electron in the material surroundings, resulting into the conversion of the rest masses of both into two gamma rays in a time ∂t and moving in opposite directions each with energy equal to the rest mass equivalent of either the electron or the positron [13]. The detection of any one of them will yield the death signal for the positron. Thus, the detection of the nuclear photon (1.276 MeV) serves as a birth signal to denote that the positron has landed in the medium, while the detection of one of the 0.511 MeV photons indicates that the positron has got annihilated. This is very useful in and as the measurement of the lifetime ∂t of the positron. The conservation of the linear momentum and energy demands that the two annihilation photons should move in exactly opposite directions.

$$e^{+} + e^{-} \rightarrow \gamma_1 + \gamma_2 \tag{5.2}$$

The distribution of the ∂t values for a number of events, measured in a PAS lifetime experiment, yields information regarding the total electron density $n_e(\mathbf{r})$ in the region of positron-electron annihilation. The rate of positron annihilation λ, which is the reciprocal of the positron lifetime, is given by the overlap integral of the electron and positron densities [15] as,

$$\lambda = \pi r_0^2 c \iiint \rho^-(r)\, \rho^+(r) d^3r \tag{5.3}$$

where r_0 is the classical electron radius, and c is the velocity of light.

The positron lifetime in a crystalline solid with homogeneous electron density n_e can be found out from the Dirac equation

$$\tau_b = \left(\pi r_0^2 c \xi(r) n_e\right)^{-1} \tag{5.4}$$

where $\xi(\mathbf{r})$ is the electron density enhancement factor, which accounts for the local increase in the density of electrons due to Coulomb attraction by the positron with the electrons around. In metals, it is given as

$$\xi(r) = 1 + 1.23\, r_s + 0.8295\, r_s^{3/2} - 1.26\, r_s^2 + 0.3826 r_s^{5/2} + 0.167\, r_s^3 \tag{5.5}$$

where r_s is the radius of a fictitious sphere containing one conduction electron. For example, $\xi(r) = 2$ for Al [15].

2.4 Doppler broadening measurements

When a positron annihilates with an electron, it produces two gamma rays, which are of energies approximately 511 keV each and which go in opposite directions. In fact, both the energies of the gamma rays and the directions of their emission are determined by the conservation of linear momentum. For this reason, the gamma ray in the direction of the positive projection of the electron momentum will have a higher energy and the gamma ray in the opposite direction correspondingly will have a lower energy. Thus the two gamma rays produced on electron-positron annihilation do not have exactly equal energies but vary over a wide range 511 ± ΔE keV. This effect that the momentum of the electrons has on the spectrum of energies of the gamma rays is called Doppler broadening. It may be mentioned once again in this context that the momenta of the positrons become negligible due to the rapid thermalization they undergo after entering the material. This phenomenon results into a broadening of the annihilation gamma ray energy spectrum in excess of the resolution broadening of the detector. It reflects information about the electron momentum distribution in the annihilation environment [4] as the magnitude of Doppler shift in energy (ΔE) is related to the electron momentum p as

$$\Delta E = 1/2pc \tag{5.6}$$

The core electrons having higher momentum contribute more to the broadening of the spectrum and the events of annihilation with the low momentum valence or free electrons usually populate the peak region of the spectrum in a realistic experiment. Measuring ΔE distribution will therefore yield a true representation of the electron momentum distribution within the material. A high pure germanium detector with good energy resolution (i.e., full width at half maximum ~ 1.0–1.4 keV at 511 keV) is used to observe the Doppler broadening of the annihilation gamma ray spectrum.

A lineshape parameter S (shape) is conventionally used to help monitor the evolution of defects in materials under changes in experimental conditions. The S-parameter is calculated as the ratio of counts under a fixed number of channels around the peak (central area) to the total counts (total area) under the spectrum. To make it more precise, it is the ratio of the area falling under the energy segment 511±~0.6 keV to the area under the total spectrum (511±~8 keV), giving thus the fraction of low momentum electrons annihilated by the positrons. It is highly sensitive to positron annihilation within the defects since there is a considerably reduced probability for annihilation with the high momentum core electrons of the surrounding atoms or ions.

2.5 Angular correlation measurements

The kinetic energy and linear momentum of the positron reduce to negligible values due to the rapid thermalization after entering into the material and the momentum of the electron needs to be conserved. Therefore, the two annihilation gamma rays have to move with angular deviation $\theta = p/mc$, instead of in exactly opposite directions. Measuring the angular correlation, i.e., the distribution of the number of gamma rays at each angle for a fixed duration of time is the method of sampling the electron momentum distribution within a material. This technique had been popular for the determination of Fermi momentum and mapping the Fermi surfaces of materials.

2.6 Positron trapping

Positrons preferably reside in or rather diffuse through the interatomic spaces due to the Coulomb repulsion by the positive ion cores. At the defect sites, the potential sensed by the positron is lowered due to the reduction in repulsion and, as a result, a localized positron state at the defect has a lower energy than the delocalized (free) positron. The transition from the delocalized state to the localized one is called positron trapping [13]. Additional exponential components occur in measured positron lifetime spectra as a consequence of positron trapping.

In a crystalline solid, the positron starts feeling the repulsion from the atomic nucleus. However, if the lattice site is vacant, there is no repulsion and the positron gets trapped at the vacancy. Thus, the ultimate path of the positron can be affected by the presence of defects.

The lifetimes of free positrons, i.e., those not trapped in any kind of defects or surfaces, in most metals are of the order of 100–250 ps, while in ionic solids, it lies in the 100–400 ps range. The presence of defects, at which the positron is trapped, leads to increase in the positron lifetime relative to that for the free positron. This is because, at the vacancy sites, the electron density is lower than the average electron density of the material. In a homogeneous defect-free media, all positrons annihilate with the same rate λ_b, which is a characteristic of the given material. Thus a stream of positrons entering into a material containing defects will give rise to a situation governed by the two rate equations [15].

$$\frac{dn_b}{dt} = -\lambda_b n_b - \kappa_d n_b \tag{5.7a}$$

and

$$\frac{dn_d}{dt} = -\lambda_d n_d - \kappa_d n_b \tag{5.7b}$$

Here λ_b stands for the rate of annihilation in the bulk and κ_d is the rate of trapping of positrons from the bulk into the defects. n_b and n_d respectively stand for the number of positrons in the bulk and the defect states at a given instant t. The solution of the above rate equations leads to

$$N(t) = \sum_{i=1}^{n} I_i \exp(-t/\tau_i) \tag{5.8}$$

where the annihilation rates λ_s and the positron lifetimes τ_s are the respective reciprocals of each other. In the nonrelativistic approximation, the cross-section of two-photon annihilation of a free positron and a free electron increases with decreasing relative velocity v of the colliding particles as

$$\sigma_{2r} = \sigma_D = \pi\, r_0^2 \frac{c}{v} \tag{5.9}$$

where r_0 is the classical electron radius and c is the velocity of light. As v -> 0, the cross-section σ_D becomes infinite. However, the positron annihilation rate tends to a finite limit given by

$$\lambda_D = \sigma_D \upsilon n_e = \pi r_0^2 c n_e \tag{5.10}$$

where n_e is the density of electrons at the site of annihilation of the positron.

2.7 Positronium formation

In many insulators and low electron density solids like polymers and plastics, the positron can form a bound state known as positronium (Ps) with a host electron. Ps is a hydrogen-like atomic state [15]. Positronium forms within the final section of a positron track due to the combination of the positron with one of the surrounding electrons.

$$e^{+} + e^{-} \rightarrow Ps \tag{5.11}$$

The Ps atom has binding energy of 6.1 eV and radius of 0.106 nm. Ps atom has two ground states, the singlet (parapositronium, pPs) state and the triplet (orthopositronium, oPs) state. The self-annihilation of pPs has a lifetime of 125 ps, while oPs has a much longer lifetime of 142 ns, although this can be shortened to a few nanoseconds in condensed matter by pick-off process involving an electron with opposite spin [15].

3. Experimental methods of positron annihilation spectroscopy

Positron annihilation spectroscopic experiments essentially involve the detection of gamma rays using detectors, the processing of the signals in nuclear electronic circuits and the final storage of the data in a multichannel analyzer. The quantities most needed are the number of photons arriving at the detector per unit time and their energies. When a gamma ray falls on a material medium, it is rapidly captured by the medium through any one or two of the interaction process, viz, photoelectric effect, Compton scattering and pair production. The energy thus absorbed, in turn, causes excitation and ionization of the molecules of the material. These are the main processes through which the photon energy is captured within the detector materials. The process of conversion of the captured energy into measurable electrical signals will depend on the type of the detector used, as discussed below.

3.1 Scintillation detector

A scintillation detector works on the principle of electronic excitation and fast de-excitation of a species of atoms and, in the process, emitting the excess energy in the form of photons. For this to happen, the absorbed gamma ray energy is first distributed among tens of thousands of atoms as the electronic excitation energies are of the order of a few tens of electron volt (eV). Hence, when the excited electrons fall back to the ground state, they emit photons whose frequencies and wavelengths match with the UV-visible region of the electromagnetic spectrum. When these photons get transmitted to the photocathode of a photomultiplier tube (PMT, described below), photoelectrons are produced. Thus the absorbed gamma ray energy is converted into electric signals. Materials like anthracene, naphthalene, certain alkyl halides and organic phosphors have the desirable properties of very fast (~nanoseconds) decay time as well as transparency to visible or ultraviolet radiation. More recently, barium fluoride (BaF_2) has become the fastest scintillator (or phosphor) in use as around 20% of the electrons de-excite within an ultra fast time of 0.6 ns, giving out scintillations of wavelength ~ 220 nm (UV region). The remaining 80%, which is a slow component (decay time ~ 620 ns), constitute the light output (of wavelength ~ 310 nm; again in the UV region) helps in the excellent energy resolution in comparison with those could be obtained with other common scintillators in use.

3.2 Photomultiplier tube (PMT)

The PMT converts the scintillations from the phosphor into electrical pulses by photoelectric emission at the cathode of the tube. Light from the phosphor strikes the cathode, which is usually made of photosensitive material such as a combination of two or three alkaline metals, and ejects electrons (photoelectric effect). The tube has several electrodes called dynodes to which progressively higher positive potentials are applied. The photoelectrons are accelerated in the electrostatic field between the cathode and the first dynode, which is at a positive potential relative to that of the cathode, and strike the dynode. The accelerated electrons impart enough energy to the electrons in the dynode to eject more in number of them. There may be as many as about 5 to 10 secondary electrons for each incoming electron and the process is repeated at the succeeding dynodes. The number of secondary electrons thus gets multiplied as the electrons are accelerated from dynode to dynode. The output current, or pulse, at the anode is normally due to more than a million times the number of electrons originally emitted from the cathode. Each photon incident on the phosphor is thus able to produce a pulse, and the pulses are fed to an electronic system where they are processed for further analysis and storage.

3.3 Multichannel analyzer

The basic function of the multichannel analyzer (MCA) is the conversion of the analog pulse into proportionate (to its amplitude) number of digital signals in an analog to digital converter (ADC) and store in its memory. ADC thus assigns a digital value to the input pulse in accordance with its amplitude. The memory is arranged as a stack of locations, from first address (channel number 1) to the maximum location (channel number 2^n; n normally is 10–14). Once a pulse has been processed by the ADC, the content of that memory location is increased by one count.

MCAs are also optionally provided with another linear gate that is controlled by the output from a single channel analyzer (SCA). The gate is opened and input pulses are passed onto the MCA circuitry only if the pulses meet the amplitude criteria set by the SCA. Thus, input pulses that are either smaller or larger than the region of interest set by the SCA limits, referred to as the lower-level discriminator (LLD) and upper-level discriminator (ULD), are rejected (Knoll, [19]).

The signals generated from the PMTs are passed through constant-fraction discriminators (CFDs) to minimize the spread (known as "walk problem") in timing. This happens due to the signals of different amplitudes from the PMT as a result of difference in the absorbed quantities of the incident gamma ray energies. There are basically three ways of gamma ray energy absorption in matter, viz., photoelectric absorption, Compton effect and pair production (Knoll, [19]). The signals from the two PMTs through the two CFDs are then led to the start and stop inputs of a time-to-amplitude converter (TAC). The amplitude of the TAC output signal is proportional to the time delay δt between the two signals. The side channels employing properly set amplifiers and SCAs ensure that only the gamma rays of the selected energy (1.276 MeV on the start side and 0.511 MeV on the stop side) are allowed to pass through and generate the output from a fast coincidence unit. This pulse is fed to the TAC as its strobe input, the arrival of which is made mandatory for the TAC to send the output to the MCA for analysis and storage. Thus only correlated events are finally stored in the multichannel analyzer. The collected positron lifetime spectra are then analyzed using the universally accepted computer code PALSfit [12].

4. Experimental procedure for positron annihilation measurements

Positron annihilation measurements are normally carried out using the ^{22}Na radioactive isotope, which acts as the positron source [15]. The experiments are normally performed by placing the source in sandwich geometry with two identical specimens of the sample (Fig. 5.2). Before going to its details, certain preliminary information are discussed in the following sections.

4.1 Positron source preparation

Nickel (Ni) foil of about 2 μm thickness is annealed in vacuum in a high temperature furnace at 1000–1200°C to remove all the defects in it. A solution of sodium bicarbonate or sodium chloride containing radioactive ^{22}Na ions, dissolved in dilute hydrochloric acid, is then deposited on to the annealed foil using a syringe or micropipette. As many drops are deposited as required to obtain a radioactivity of about 10–15 μCi, which is monitored by a radiation dosimeter. The nickel foil is then carefully folded to conceal the active source region within its fold. This makes a powerful ^{22}Na source, which can emit positrons into the sample to be studied using PAS. All these are done with necessary precautions like use of a thermo luminescent dosimeter (TLD) badge. (A TLD measures ionizing radiation exposure by permanent atomic excitation until it is made to emit visible light when the radiation-sensitive material in it is heated.)

In the case of samples in powder form, the Ni foil containing the radioactive ^{22}Na source is kept immersed in the volume of the powder taken in a glass tube. The powdered sample covers the source from all sides in sufficient thickness to ensure that the annihilations of all the emitted positrons take place within the sample and no positrons reach the glass wall. The glass tube is continuously evacuated using a vacuum pump during the experiment to remove and prevent the entry of air, moisture and other gases. This ensures that positronium atoms, if formed within the intergranular region, are not quenched by the gas molecules and the measured positronium lifetime and its intensity do not yield erroneous results of intergranular volume dimensions and concentration [1,2].

4.2 Recoupling of scintillator crystals with photomultiplier tubes

The time resolution of the positron lifetime spectrometer is an important parameter that needs to be maintained as best and unchanged when positron lifetime measurements are done, especially on a series of samples with one or two physical parameters varying. Proper coupling of the scintillator crystals with XP2020Q PMTs (having quartz end-windows that transmit the ultraviolet photons) is done to ensure good performance. The scintillator crystals are covered with white Teflon tape at sides to reflect back the scintillation photons trying to escape through the surfaces. The emitted ultraviolet radiation passes through the quartz window of the PMT to its photocathode and ejects photoelectrons, which are subsequently multiplied at the array of dynodes to ultimately generate the signal at the anode.

Recoupling of the scintillator with the PMT is advisable when more than two Gaussian peaks are observed in the resolution spectrum (described elsewhere). Recoupling is done with utmost care to improve the detector resolution. The ability of a detector to resolve the small differences in the pulse amplitudes is referred to as its resolution. The better the resolution, the better a detector or spectrometer in totality is able to distinguish between two events of closely lying amplitudes.

4.3 Optimization of PMT voltages

Before starting the positron annihilation experiments, optimization of the operating voltages of the photomultiplier tubes (PMTs) is to be done by making use of the gamma rays from a ^{60}Co source. ^{60}Co source emits two prompt gamma rays of energy 1.173 and 1.332 MeV. It is used to monitor the full width at half maximum (FWHM) of the time-correlated coincidence spectrum recorded by a spectrometer. During the optimization process, the voltage of one of the PMTs, say PMT A, is varied while the voltage of PMT B is kept constant and the FWHM for each setting is noted until the minimum value of FWHM is obtained. The CFD thresholds and the lower and upper-level thresholds of the SCAs are carefully readjusted to accept the photopeaks of the two gamma rays from ^{60}Co source on both sides after each change of the voltage and before the data is acquired. This corresponded to the optimum voltage for the PMT A, whose voltage is varied.

The optimum operational voltage for PMT A is found to be $V_A = +2250$ V (Fig. 5.3). Similarly the operating voltage of PMT B is also optimized and found to be $V_B = +2050$ V (Fig. 5.4). Both PMTs are set at their optimized voltages and the

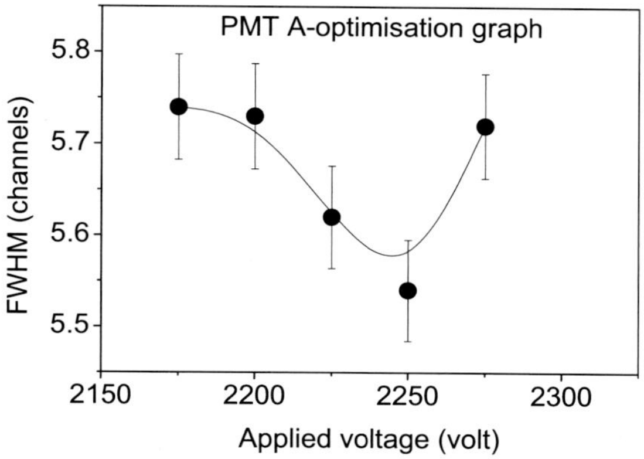

Figure 5.3
FWHM versus applied voltage for PMT A.

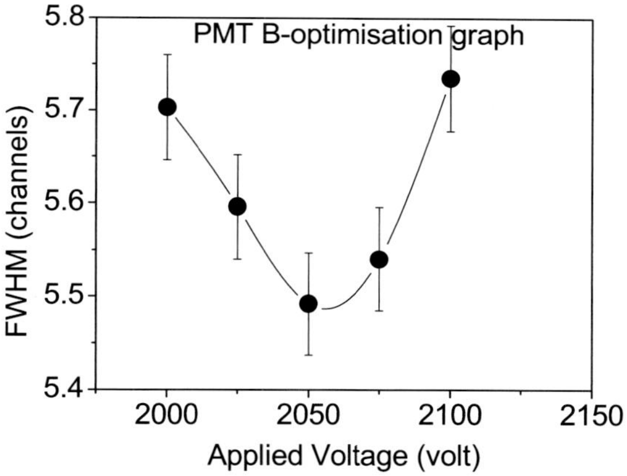

Figure 5.4
FWHM versus applied voltage for PMT B.

minimum FWHM value is obtained to be 5.719 channels. The channel constant (time/channel) used in the experiments discussed here was 24.92 ps (Fig. 5.5), which gave the FWHM as 142 ps.

For the Doppler broadening measurements of the samples, the energy resolutions of the two HPGe detectors were monitored by measuring the FWHM of the spectrum of 1.332 MeV gamma rays from a ^{60}Co source (it also emits another gamma ray of energy 1.173 MeV). The resolution at the energy of interest (i.e., the annihilation gamma ray energy of $511 \pm \Delta E$ keV) is then obtained from the relation FWHM $\propto$ (energy)$^{1/2}$ and is found to be around 1.3 keV.

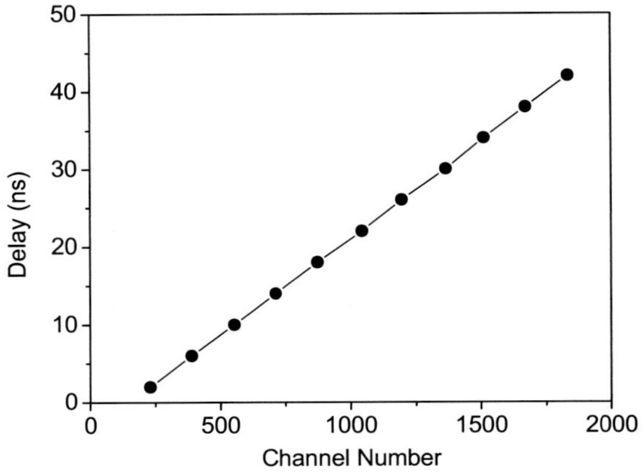

Figure 5.5
Delay versus channel number.

4.4 Source correction

A fraction of positrons, usually up to 10%–20%, gets annihilated in the source material itself and the supporting foil and also at the sample source interfaces while trying to enter into the sample material. The following lifetimes will be present in all the spectra, which need to be corrected for obtaining the positron lifetimes and intensities due to the sample material.

1. A fraction of positrons is annihilated by the electrons of the source material, i.e., $NaHCO_3$ or NaCl in dilute HCl, in which case the positron lifetime is normally about 350–400 ps.
2. Positron annihilation due to the electrons of the Ni foil covering the source which gives a lifetime of about 108 ps.
3. When the positrons travel from the source to the sample, a few of them get back-scattered due to the change in medium and gives a long lifetime of about ~1–2 ns.

An empirical formula for calculating the intensity of positrons annihilating in the foil is as follows.

$$I_{foil} = 0.324Z^{0.98}d^{3.45Z^{-0.41}} \tag{5.12}$$

where Z is the atomic number of the sample material and d is the foil thickness in mg/cm^2.

The intensities of the positron lifetimes from the source and the source-sample interfaces are normally estimated by first taking the positron lifetime spectrum of a pure,

Table 5.1: Positron lifetimes and their respective intensities due to the source.

Origin of the components	Ni foil	Source material	Backscattering from the source-sample interfaces
Lifetimes (ns)	0.1102	0.3983	1.5755
Intensities (%)	65.089	34.0036	0.9074

well-annealed and preferably single crystalline sample, the positron lifetime in which is previously known. For example, single crystalline Al with purity 99.999% and annealed in vacuum in a furnace of temperature ~600°C for 2–4 h and slow-cooled, can serve as an ideal reference sample in this context. Positron annihilation lifetime spectrum of the Al reference sample was taken with the source intended to be used in the actual experiments and was analyzed by using the PALSfit program [12]. The data of this spectrum are analyzed by fixing the positron lifetime in the sample (170 ps for Al) and allowing the other components to converge around the different contributions from the source. The fraction of positron annihilation due to the Ni foil is calculated using Eq. (5.12) and its value is obtained as 16.75%. The total fraction of positron annihilation due to the source is 25.74% which is the sum of the positron annihilation due to the Ni foil, elemental electrons of the source material and back scattering. The various lifetime components and their intensities due to the source are given in Table 5.1. These lifetimes and intensities are subtracted as "source correction" from any spectra acquired using the particular source before the residual spectra is analyzed for the positron lifetimes and intensities due to the sample.

4.5 Positron annihilation spectroscopic experiments

For the characterization of nanomaterials, PAS is carried out by implementing two different measurement schemes, i.e., positron lifetime and Doppler broadening [14]. Both give information on the electronic structural aspects of the material [11].

4.5.1 Positron lifetime measurements

In order to carry out positron lifetime measurements, BaF_2+XP2020Q PMTs have been used as the gamma ray detectors. The various electronic circuits include amplifiers, single channel analyzers, coincidence unit, constant-fraction discriminators and time-to-amplitude converters. The TAC in the gamma-gamma ray coincidence setup produces the pulse with proportionate amplitude and the correlated events are recorded in the multichannel analyzer. The data is finally acquired in a multichannel analyzer (MCA). The side channels employing properly set amplifiers and SCAs ensure that only genuine positron lifetime events are ultimately recorded and stored. The "positron lifetime spectra" are then

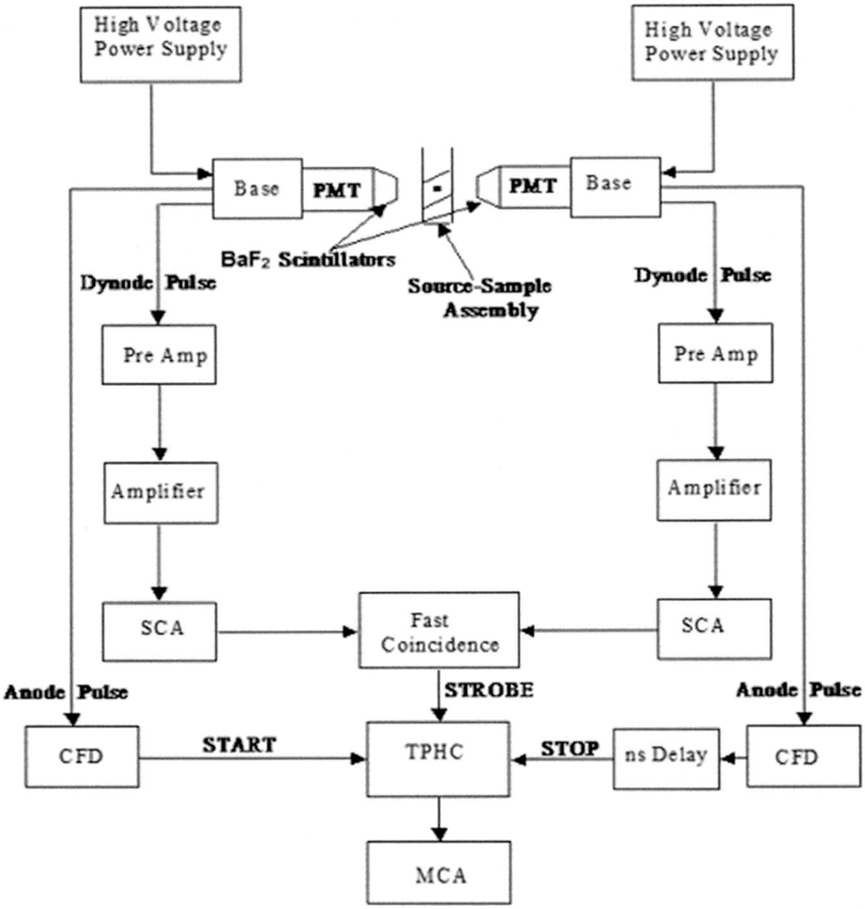

Figure 5.6
Experimental set up for positron lifetime experiments.

acquired using the computation software Genie-Gamma Acquisition and Data Analysis and analyzed using the PALSfit software [12]. The experimental set-up used in measuring the positron lifetime is shown in Fig. 5.6.

4.5.2 Doppler broadening measurements

For Doppler broadening measurements, a high pure Germanium (HPGe) detector is used, which is operated at liquid nitrogen temperature to ensure the minimization of the leakage current due to thermally generated charge carriers [17]. The measurements are carried out using HPGe detector and other electronic devices such as amplifier and MCA. The coincidence Doppler broadening spectroscopic (CDBS) arrangement which uses two HPGe detectors helps to suppress the nuclear background (Fig. 5.7) [3].

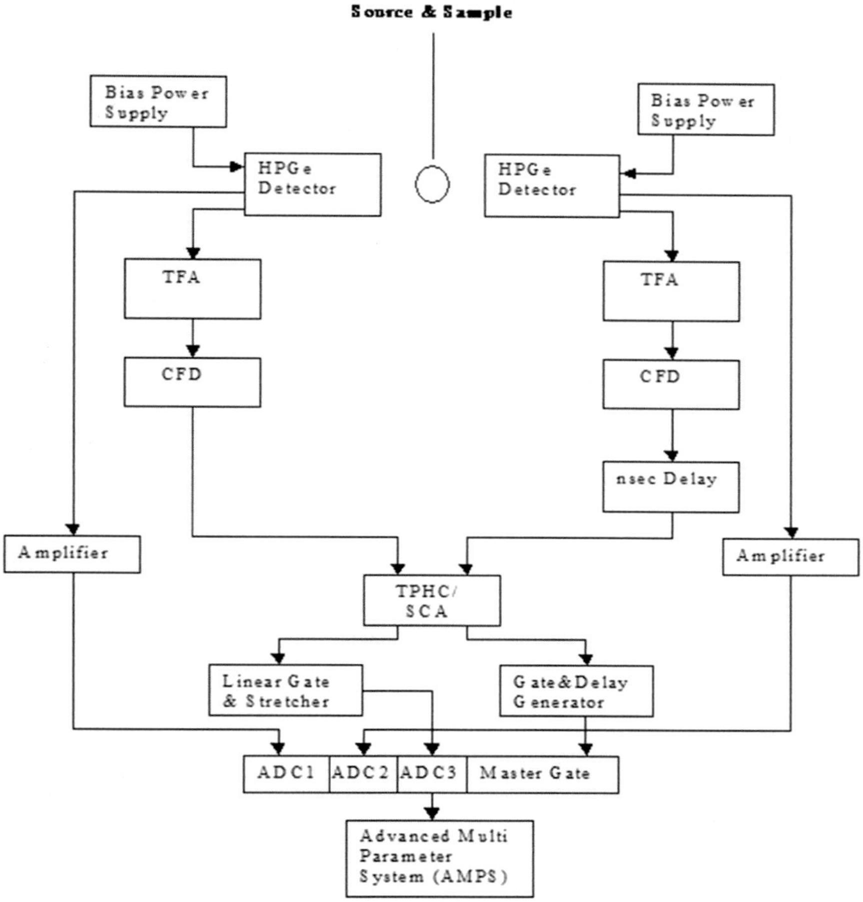

Figure 5.7
The experimental arrangement for coincidence Doppler broadening measurements.

5. PAS - results in nanomaterials

5.1 Positron lifetime measurements in nanomaterials

Defects are inherent in nanomaterials because of their non-stoichiometric composition. PAS is helpful in finding out the type or nature of the defects present in the sample, their sizes and concentrations [6]. In nanocrystalline samples, positrons get annihilated at the crystallite/particle surfaces and interfaces also [5,16]. This happens when the particle sizes are less than the thermal diffusion length of positrons in the nanomaterial (~50—100 nm). A typical positron lifetime spectrum of $Mg_{1-x}Ca_xFe_2O_4$ nanocrystalline system is shown in Fig. 5.8.

The lifetime data are analyzed using the PALSfit program and the best possible reduced chi-square fit with acceptable standard deviation is obtained for three lifetime components.

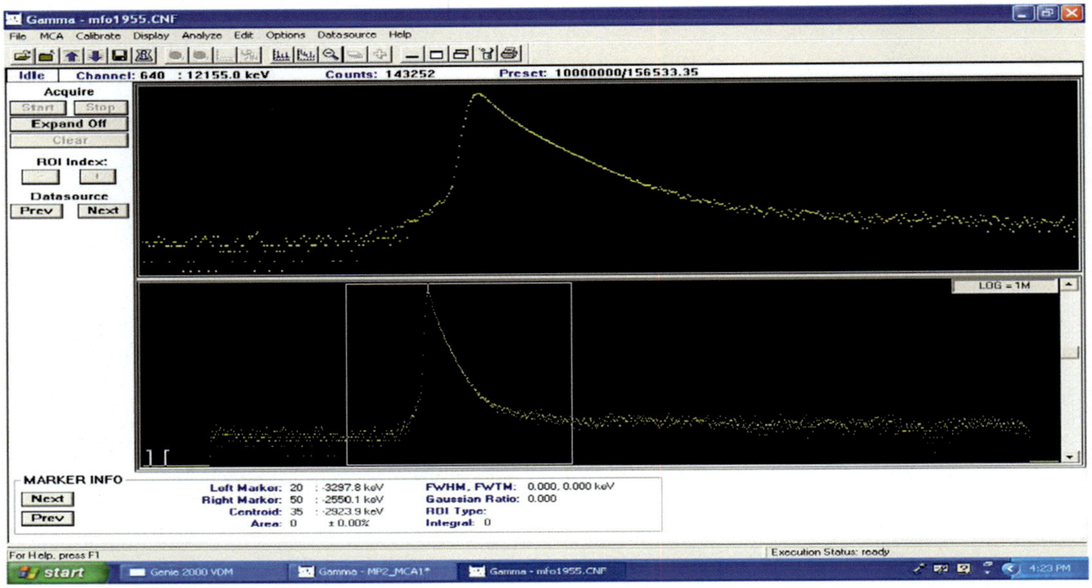

Figure 5.8
A typical positron lifetime spectrum.

The three lifetimes are named as τ_1, τ_2 and τ_3 in the increasing order of their magnitudes and the corresponding intensities are named as I_1, I_2 and I_3 irrespective of their magnitudes. From the Doppler broadened spectra, we also calculated the S-parameter, as explained earlier.

As mentioned above, the positron lifetimes τ_1, τ_2 and τ_3 are obtained for the best possible reduced chi-square fit using PALSfit program. The intensities I_1, I_2 and I_3 represent the percentage of positrons having lifetimes of τ_1, τ_2 and τ_3 respectively. Intensity 'I_i' can be interpreted as the number of positrons having a lifetime of 'τ_i' in a sample when the total number of positrons entering the sample is considered as $I_1+I_2+I_3 = 100\%$.

The shortest positron lifetime τ_1 generally represents the lifetime of positrons in the bulk material or defect-free material and reduced by the trapping of positrons at the defects. In other words, it is supposed to be the admixture of the lifetimes of those positrons annihilating within the bulk region of the sample with a lifetime τ_b characteristic of the electron density of the sample and the Bloch state residence time of the trapped positrons [15,18]. Since the samples used in these experiments are nanocrystalline in nature, with average crystallite sizes less than ~50 nm, the nontrapped positrons will diffuse out to the surfaces of the crystallites before annihilation. Hence $\boldsymbol{\tau_1}$ in this case represents only the Bloch state residence time of trapped positrons. It is the maximum time positron can spend within the bulk of the material before annihilation. τ_1 must be less than τ_b in such cases.

$$\frac{1}{\tau_1} = \frac{1}{\tau_b} + \kappa_d \tag{5.13}$$

$$\frac{1}{\tau_b} = \frac{I_1}{\tau_1} + \frac{I_2}{\tau_2} + \frac{I_3}{\tau_3} \tag{5.14}$$

Annihilation on the crystallite or grain surfaces, on the other hand, will result in very large lifetimes for positrons, of the order of 450–500 ps. In this case, such lifetimes have got admixed with the lifetime of positrons trapped in the vacancy type defects within the crystallites to give the resultant second component τ_2. The admixing is a result of the limited instrumental resolution as well as the inability of the PALSfit program to resolve more components from the positron lifetime data. Each positron lifetime component is therefore an average of closely lying lifetimes and the specific components need to be separated out using well-formulated theoretical models.

In the intercrystallite region, where electron density is low, positron can form positronium (Ps). The longest lifetime τ_3 in the analysis is far separated from the other lifetimes and is the pick-off annihilation lifetime of oPs atoms formed within the extended intercrystallite regions (Three-fourth of the Ps atoms will be oPs, i.e., spin triplet, since the magnetic quantum numbers can be −1, 0 or +1. The remaining one-fourth will be pPs, i.e., spin singlet, and have only one allowed state). The presence of a small insignificant fraction ($I_3/3$) of pPs is ignored in the discussion. The results of positron annihilation experiments on calcium doped magnesium ferrite samples are reported [1,2]. The variation of positron lifetimes and the relative intensities with the concentration (x) of Ca^{2+} ions in the $Mg_{1-x}Ca_xFe_2O_4$ samples are given in Fig. 5.9A and B. The detailed interpretations of the values and variations of the different positron lifetimes and the intensities are presented in the reference cited [1,2].

We also had defined the mean positron lifetime τ_m as

$$\tau_m = (\tau_1 I_1 + \tau_2 I_2 + \tau_3 I_3)/(I_1 + I_2 + I_3) \tag{5.15}$$

and its variation gives a cumulative picture of the defect-interaction processes taking place in the samples due to the substitution within the samples. The variation of the mean positron lifetime τ_m, calculated for the calcium doped $Mg_{1-x}Ca_xFe_2O_4$ samples, is presented in Fig. 5.10.

5.2 Results of Doppler broadening measurements

A typical Doppler broadened spectrum of $Mg_{1-x}Ca_xFe_2O_4$ nanomaterials is shown in Fig. 5.11.

The spectrum obtained from Doppler broadening is not analyzed directly because there are no proper mathematical functions to define the detector resolution and an exact method to

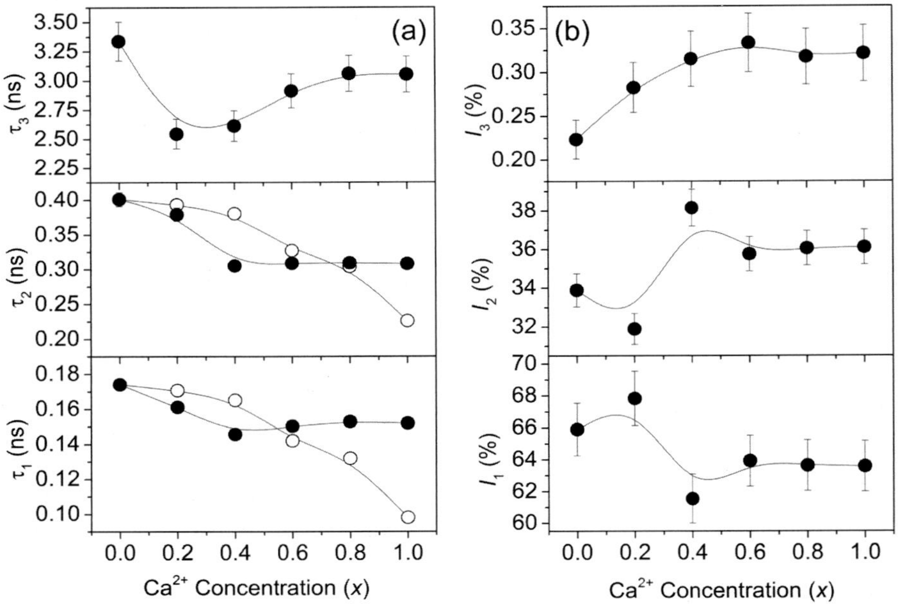

Figure 5.9

(A) The positron lifetimes and (B) the relative intensities versus the concentration (x) of Ca^{2+} ions in the $Mg_{1-x}Ca_xFe_2O_4$ samples. *Reproduced from A.R. Abraham, B. Raneesh, D. Sanyal, S. Thomas, N. Kalarikkal, P.M.G. Nambissan, Defect-focused analysis of calcium-substitution-induced structural transformation of magnesium ferrite nanocrystals, New J. Chem. 44 (4) (2020) 1556–70 with permission from the Center National de la Recherche Scientifique (CNRS) and The Royal Society of Chemistry.*

deconvolute it from the spectrum obtained. Various lineshape parameters are therefore used in order to derive qualitative information regarding the physical changes taking place when the sample is subjected to external experimental treatments. The S-parameter is calculated as the ratio of the central area to the total area where area is the total number of counts in all the channels devoid of the background. Thus, the Doppler broadened spectrum is analyzed by calculating the S-parameter, which is defined as the ratio of the central (511 ± 0.6 keV) area to the total (511 ± 8 keV) area, where the word area represents the integral counts minus the background.

In the work reported on defect-focused analysis of calcium-substitution-induced structural transformation of magnesium ferrite nanocrystals [1,2], the variations of the two lineshape parameters in the $Mg_{1-x}Ca_xFe_2O_4$ samples are illustrated in Fig. 5.12. The W-parameter is one which represents the fraction of high momentum core electrons annihilated by the positrons and are calculated from the area under two symmetrically placed regions of the "wings" or falling regions of the spectrum on either side and dividing by the same total area used in the calculation of the S-parameter. The identical and opposite trends of

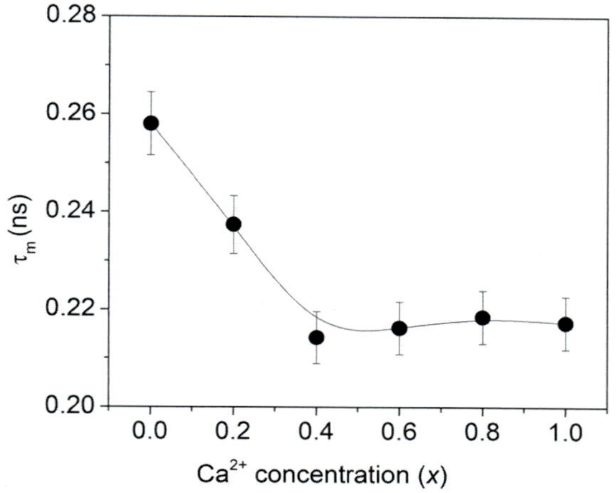

Figure 5.10

The variation of the mean positron lifetime versus the concentration (x) of Ca^{2+} ions in the $Mg_{1-x}Ca_xFe_2O_4$ samples. *Reproduced from A.R. Abraham, B. Raneesh, D. Sanyal, S. Thomas, N. Kalarikkal, P.M.G. Nambissan, Defect-focused analysis of calcium-substitution-induced structural transformation of magnesium ferrite nanocrystals, New J. Chem. 44 (4) (2020) 1556–70 with permission from the Center National de la Recherche Scientifique (CNRS) and The Royal Society of Chemistry.*

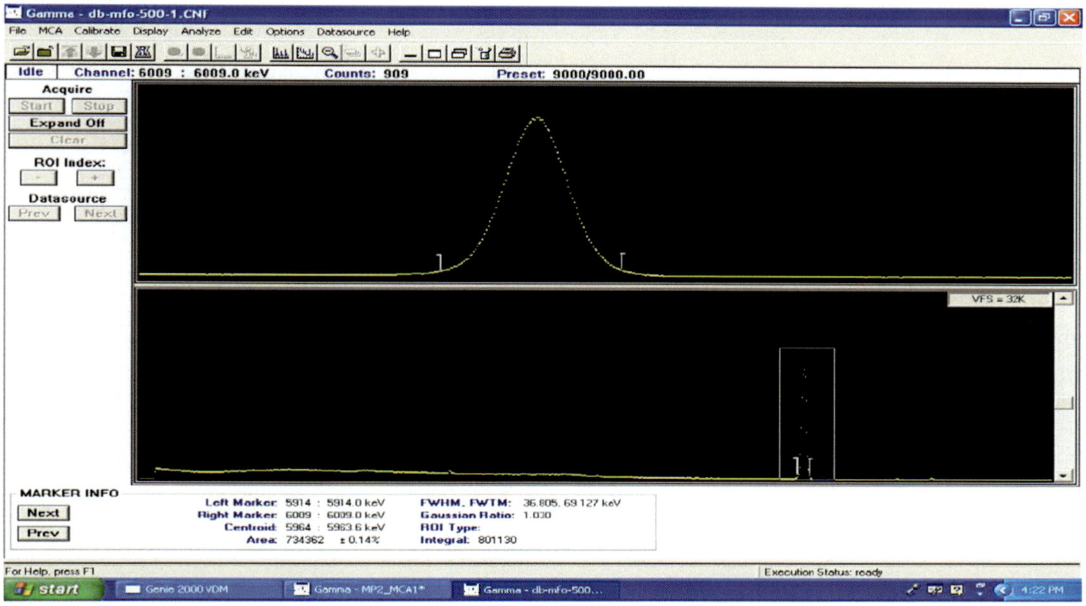

Figure 5.11

A typical Doppler broadened spectrum.

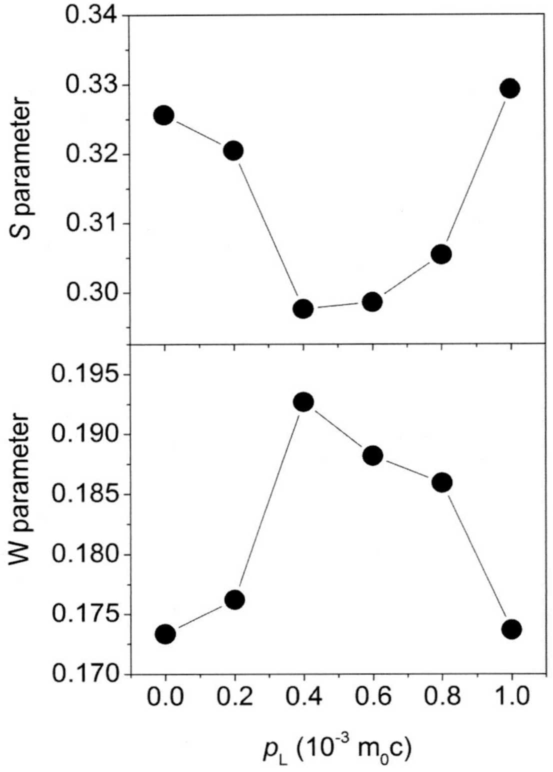

Figure 5.12

Variation of S and W parameters with the Ca^{2+} concentration in the $Mg_{1-x}Ca_xFe_2O_4$ samples. *Reproduced from A.R. Abraham, B. Raneesh, D. Sanyal, S. Thomas, N. Kalarikkal, P.M.G. Nambissan, Defect-focused analysis of calcium-substitution-induced structural transformation of magnesium ferrite nanocrystals, New J. Chem. 44 (4) (2020) 1556–70 with permission from the Center National de la Recherche Scientifique (CNRS) and The Royal Society of Chemistry.*

variation of the lineshape parameters with the Ca^{2+} concentration, particularly at $x = 0.4$ imply a redistribution of the electron momentum due to the occurrence of structural changes evident in the nanocrystallites.

6. Summary and conclusions

Defect spectroscopy [20] has become very important for investigation of nanomaterials for identifying their exotic features and properties, as defects play a significant role in modifying the properties of materials. The presence of defects is one aspect that needs very special investigation, though other conventional experimental techniques might ignore or overlook their significance. PAS has become very useful in this context and by monitoring the variations of the positron lifetimes, their intensities, and the Doppler broadened lineshape parameters [23], we could obtain valuable information regarding the structural transformation [20,22] that is simultaneously taking place in materials [20,21]. A distinct advantage of PAS is that it is possible to carry out these studies even at low vacancy concentrations, which are below the limit of sensitivity of conventional electron microscopy techniques. The present chapter contains a brief introduction to the principles of PAS and its potential is exemplified with the help of some results obtained from our studies on calcium-modified magnesium ferrite nanosystems.

References

[1] A.R. Abraham, B. Raneesh, D. Sanyal, S. Thomas, N. Kalarikkal, P.M.G. Nambissan, Defect-focused analysis of calcium-substitution-induced structural transformation of magnesium ferrite nanocrystals, New J. Chem. 44 (4) (2020) 1556–1570.

[2] A.R. Abraham, B. Raneesh, P.M.G. Nambissan, D. Sanyal, S. Thomas, N. Kalarikkal, Defects characterisation and studies of structural properties of sol–gel synthesised $MgFe_2O_4$ nanocrystals through positron annihilation and supportive spectroscopic methods, Phil. Mag. 100 (1) (2020) 32–61.

[3] P. Asoka-Kumar, M. Alatalo, V.J. Ghosh, A.C. Kruseman, B. Nielsen, K.G. Lynn, Increased elemental specificity of positron annihilation spectra, Phys. Rev. Lett. 77 (10) (1996) 2097–2100.

[4] W. Brandt, Positron dynamics in solids, Appl. Phys. 5 (1) (1974) 1–23.

[5] S. Chakrabarti, S. Chaudhuri, P.M.G. Nambissan, Positron annihilation lifetime changes across the structural phase transition in nanocrystalline Fe_2O_3, Phys. Rev. B 71 (6) (2005) 064105.

[6] J. Cyriac, R.M. Thankachan, B. Raneesh, P.M.G. Nambissan, D. Sanyal, N. Kalarikkal, Positron annihilation spectroscopic studies of Mn substitution-induced cubic to tetragonal transformation in $ZnFe_{2-x}Mn_xO_4$ (x=0.0–2.0) spinel nanocrystallites, Phil. Mag. 95 (35) (2015) 4000–4022.

[7] S. Gottschalk, H. Hahn, A.G. Balogh, P. Werner, H. Kungl, M.J. Hoffmann, A positron lifetime study of lanthanum and niobium doped Pb $(Zr_{0.6}Ti_{0.4})O_3$, J. Appl. Phys. 96 (12) (2004) 7464–7470.

[8] V.I. Grafutin, E.P. Prokop'ev, Positron annihilation spectroscopy in materials structure studies, Rus. Acad. Sci. Phys. Uspekhi 45 (1) (2002) 59.

[9] R.H. Howell, T.E. Cowan, J.H. Hartley, P.A. Sterne, Positron beam lifetime spectroscopy at Lawrence Livermore National Laboratory, in: AIP Conference Proceedings, vol. 392, American Institute of Physics, 1997, pp. 451–454, no. 1.

[10] R. Krishnan, D.D. Upadhyaya, Defect structure studies using positron annihilation spectroscopy, Pramana 24 (1–2) (1985) 351.

[11] P.M.G. Nambissan, Defect characterization in nanomaterials through positron annihilation spectroscopy, in: S. Sinha, N.K. Navani, J.N. Govil (Eds.), Nanotechnology: Synthesis and Characterization, vol. 2, Studium Press LLC, Houston, 2013, pp. 455–491.

[12] J.V. Olsen, P. Kirkegaard, N.J. Pedersen, M. Eldrup, PALSfit: a new program for the evaluation of positron lifetime spectra, Phys. Status Solidi C 4 (10) (2007) 4004–4006.

[13] I. Procházka, Positron annihilation spectroscopy, Mater. Struct. 8 (2) (2001) 55.

[14] O. Shpotyuk, I. Adam, Z. Bujňáková, P. Baláž, Y. Shpotyuk, Probing sub-atomistic free-volume imperfections in dry-milled nanoarsenicals with PAL spectroscopy, Nanoscale Res. Lett. 11 (1) (2016) 1–7.

[15] R.W. Siegel, Positron annihilation spectroscopy, Annu. Rev. Mater. Sci. 10 (1) (1980) 393–425.

[16] R.M. Thankachan, J. Cyriac, B. Raneesh, N. Kalarikkal, D. Sanyal, P.M.G. Nambissan, Cr^{3+} substitution induced structural reconfigurations in the nanocrystalline spinel compound $ZnFe_2O_4$ as revealed from X-ray diffraction, positron annihilation and Mössbauer spectroscopic studies, RSC Adv. 5 (80) (2015) 64966–64975.

[17] A. Wagner, M. Butterling, M.O. Liedke, K. Potzger, R. Krause-Rehberg, Positron annihilation lifetime and Doppler broadening spectroscopy at the ELBE facility, in: AIP Conference Proceedings, vol. 1970, AIP Publishing LLC, 2018, p. 040003 (1).
[18] R.N. West, Positron studies of condensed matter, Adv. Phys. 22 (1973) 263–383.
[19] G.F. Knoll, F. Glein. Radiation Detection and Measurement, Wiley, New York, 1989.
[20] A.R. Abraham, B. Raneesh, S. Joseph, P.M. Arif, P.M.G. Nambissan, D. Das, D. Rouxel, O. Samuel Oluwafemi, S. Thomas, N. Kalarikkal, Magnetic performance and defect characterization studies of core–shell architectured $MgFe_2O_4$@$BaTiO_3$ multiferroic nanostructures, Phys. Chem. Chem. Phys. 21 (17) (2019) 8709–8720.
[21] A. Das, A.C. Mandal, S. Roy, P.M.G. Nambissan, Positron annihilation studies of defects and fine size effects in nanocrystalline nickel oxide, J. Exp. Nanosci. 10 (8) (2015) 622–639.
[22] P.M.G. Nambissan, Doping effects in wide band gap semiconductor nanoparticles: lattice variations, size changes, widening band gaps but no structural transformations!, Nanomater. Phys. Chem. & Biol. Appl. (2018) 37.
[23] P.M.G. Nambissan, C. Upadhyay, H.C. Verma, Positron lifetime spectroscopic studies of nanocrystalline $ZnFe_2O_4$, J. Appl. Phys. 93 (10) (2003) 6320–6326.

CHAPTER 6

The use of organ-on-a-chip methods for testing of nanomaterials

Ippokratis Pountos[1,2], Rumeysa Tutar[3], Nazzar Tellisi[1,2], Mohammad Ali Darabi[4,5], Anwarul Hasan[6], Nureddin Ashammakhi[4,5,7,8]

[1]Academic Department of Trauma and Orthopedics, University of Leeds, Leeds, United Kingdom; [2]Chapel Allerton Hospital, Leeds Teaching Hospitals, Leeds, United Kingdom; [3]Department of Chemistry, Faculty of Engineering, Istanbul University-Cerrahpasa Avcılar, Istanbul, Turkey; [4]Center for Minimally Invasive Therapeutics (C-MIT), University of California, Los Angeles, Los Angeles, CA, United States; [5]Department of Bioengineering, Henry Samueli School of Engineering, University of California, Los Angeles, Los Angeles, CA, United States; [6]Department of Mechanical and Industrial Engineering, College of Engineering, Qatar University, Doha, Qatar; [7]Department of Radiological Sciences, David Geffen School of Medicine, University of California, Los Angeles, Los Angeles, CA, United States; [8]Department of Biomedical Engineering, College of Engineering, Michigan State University, MI, United States

Chapter Outline

1. Introduction

Nanomedicine is a rapidly evolving field, which combines advances made in medicine, genetics, basic science, proteomics, and technology. It is a tool for the investigation, manipulation, and ultimately the control of atoms, molecules, and objects with size

Design, Fabrication, and Characterization of Multifunctional Nanomaterials. https://doi.org/10.1016/B978-0-12-820558-7.00008-X

ranging from 1 to 100 nm [1]. These materials can be synthetic or natural materials, and due to their size they have the ability to translocate efficiently across cell membrane of the cells [1]. Taking into account that most of the inner cellular functions occur on a nanoscale level, nanomedicine has the ability to alter these cellular functions.

Many researchers have already explored medical treatments and devices based on nanotechnology, often referred to as nanotherapeutics, to increase treatment efficacy and sensitivity and adding new therapeutic modalities to our armamentarium. The most common application of nanomaterials has been the design of carriers that deliver therapeutic payload to diseased cells. Such nanocarriers include liposomes, dendrimers, organic polymer nanoparticles, micelles, inorganic mesoporous silica nanoparticles, and many others [2]. These synthetic and natural polymer carriers have been designed to encapsulate therapeutic agents and carry them to specific sites in the body [3]. Cancer treatment is among the most commonly studied area, where various nanoparticles have been engineered aiming to extend the systemic circulatory half-life of chemotherapeutic drugs, allow diffusion from blood vessels in the tumor, with enhanced permeability, selectively attach to tumor cells, identify the location and boundaries of the tumor, and finally release chemotherapeutic substances [3—5]. In addition, gold and silver nanoparticles are two materials that have been used for the treatment of cancer and infection [6,7]. Gold nanoparticles have been used in tumor chemotherapy, radiotherapy, photodynamic, and gene therapy [7]. In a recent report, silver nanoparticles have found to have a dual effect against bacteria [6]. They induced a strong antimicrobial effect but also inhibited quorum sensing, which is the process of chemical communication used by bacteria to regulate virulence [6]. Current evidence has suggested that the use of nanotechnology can improve the bioavailability of medicines by protecting them from degradation and deactivation [3]. They can incorporate a controlled release mechanism and alter the pharmacokinetics of the drugs inducing a greater response [3,8]. With regard to medicines, the use of nanomaterials can improve drug solubility and overall safety [3,8]. There are still, however, several challenges to address. Some authors reported rapid clearance from circulation and limited capacity to cross-biological barriers such as the blood-brain barrier, as critical issues [9,10].

In addition to the use of nanomaterials as carriers of chemotherapeutic substances, other applications have also emerged. Metal-containing functionalized nanoparticles can be used for the imaging of cancer, extend therapeutic indices, and improve early diagnosis [11]. Alternatively, photonic nanomaterials that can either emit or absorb or respond to light like nanoparticles, semiconductor quantum dots, plasmonic metal, organic carbon and others can be utilized in diagnosis of a range of pathologies [12]. Microfabricated devices like for example the neural prosthetic devices have allowed recording from neural tissues and the control of their functions [13]. More sophisticated devices often referred to as nanobionics, involve devices that can communicate information from inside the body or targeted cells and allow the analysis and possible control of the function of the cells [14].

Despite the impressive developments in the field of functional nanomaterials, commercialization of such advantages has been slow. Very few systems have found clinical applications and are used in the clinical practice. Among those that reached the clinical pipeline and granted approval for clinical use by Food and Drug Administration (FDA) include Doxil (pegylated liposomal doxorubicin) and Abraxane (albumin-bound paclitaxel) [15,16]. Collectively, 20 years of research has resulted in 50 nanomaterials achieving FDA approval [17]. This slow bench-to-bedside translation rates seems to be multifactorial. One of the most critical challenges however, is the absence of robust pre-clinical platform for tissue culture. These platforms should be biomimetic recapitulating the in vivo environment, hence, able to predict the effect of these nanoparticles within the human body. Animal in vivo testing was once the mainstay prior to clinical applications. However, errant pharmacokinetics and diversity of outcome between experimental animal studies and humans had led to skepticism and abandonment of some of these experiments prior to clinical trials [18]. Organ-on-a-chip (OoC) technology is a recently developed method that could bridge the gap and aims to enable a solid preclinical testing. The aim of this chapter is to present our current understanding on the use of OoC method for testing nanomaterials.

2. Organ-on-a-chip

OoC system is a novel biomimetic platform that utilizes three-dimensional (3D) tissue constructs and microfluidic networks to mimic tissue microenvironment. Viscous forces rather than inertia control the manipulation of fluid within these channels [19]. These systems have predetermined geometries and structure, which is specific for the target tissue [20]. Cells within these in vitro platforms experience physiologic stresses, concentration gradients, mechanical forces and fluid flow that are normally generated in native in vivo microenvironment [19,20]. OoC systems have their origins in the technologies of chemical separations and semiconductors; they are employed to assess biologically relevant phenomena at the microscale, utilizing only minute volumes of samples [21]. Techniques such as 3D printing and soft lithography have allowed the design and prototyping of such devices [22], and those of 3D bioprinting have allowed the printing of cells and organoids into OoC systems [23].

At present OoC platforms exist for the majority of the human tissues. Lung-on-a-chip systems were among the first to emerge [20,24]. This system mimicked the alveolar-capillary composition and used two chambers, one to culture alveolar epithelia and the other endothelial cells [24]. A flexible, microporous membrane was used to separate the two chambers. Since then, the field has expanded and at present systems capable of portraying organ-level functions on a chip also exist for liver, kidney, heart, blood vessels, blood-brain-barrier, breast, and other organs [25–27] (Fig. 6.1). In addition, more complex systems including interconnected modules are also available and aim to analyze different inter-organ interactions [26,28]. An example include the work of Skardal et al., who developed a multi-OoC system composed of lung, heart, and units [28]. Authors analyzed the interactions between different organs following drug administration [28]. They proposed that it is crucial to integrate multiple tissue in the in vitro studies as in addition to drug efficacy the potential side effects can also be studied [28].

OoC platforms can be an excellent tool for preclinical testing. It is a miniaturized system that mimics the most vital aspects of in vivo and in vitro world and enables scalability.

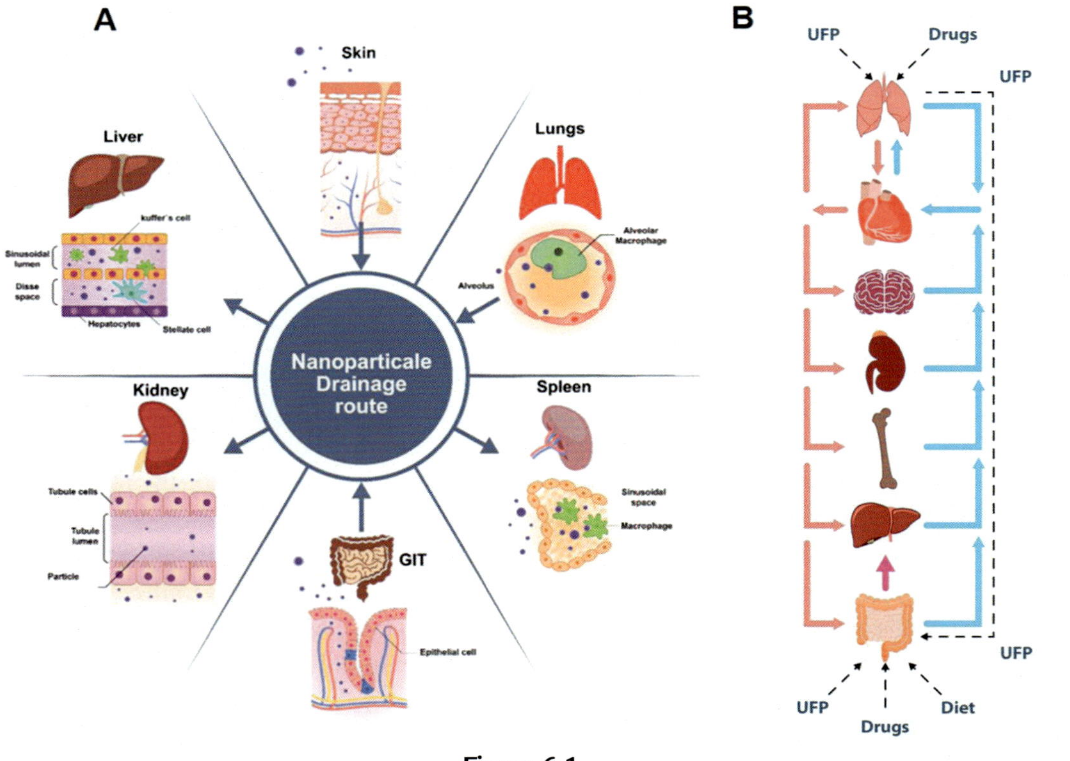

Figure 6.1

The multi-organ OoC concept. (A, B) Different organs can be integrated to capture the effect of nanomaterials on a specific organs and their systemic effects. *Reproduced from N. Ashammakhi, M.A. Darabi, B. Çelebi-Saltik, R. Tutar, M.C. Hartel, J. Lee, S.M. Hussein, M.J. Goudie, M.B. Cornelius, M.R. Dokmeci, A. Khademhosseini, Microphysiological systems: next generation systems for assessing toxicity and therapeutic effects of nanomaterials, Small Methods 4 (2020) 1900589, with permission from Wiley-VCH.*

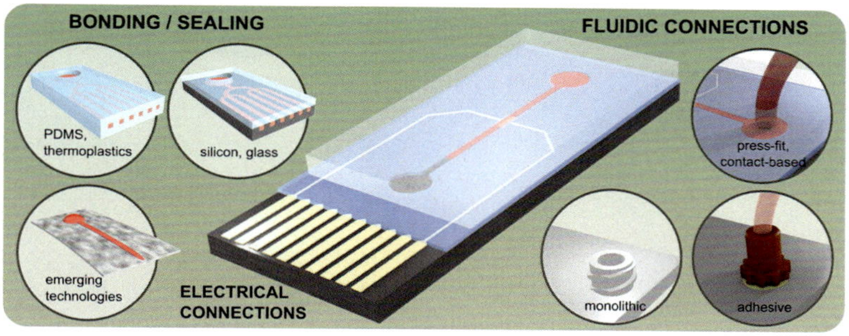

Figure 6.2
The anatomy of OoC platform. Microfluidic devices need to be sealed, connected though electrical connection and carry fluidic connections. *Reproduced from Y. Temiz, R.D. Lovchik, G.V. Kaigala, E. Delamarche, Lab-on-a-chip devices: how to close and plug the lab? Microelectron. Eng. 132 (2015) 156–175, with permission from Microelectronic Engineering.*

Continuous flow and identical culture conditions ensures the same quality over time which eliminates the batch-to-batch variability [21] (Fig. 6.2). The volume of fluids and reagents inside the channels is minimal which makes this technology cost and time effective [21]. Finally, its use requires minimal training and/or input from health care professionals and avoids the need for animal testing. The aforementioned advantages make OoC platforms a tool that could allow us to draw solid preclinical data in a reproducible way.

3. Organ-on-a-chip platforms for testing nanomaterials

OoC platforms are versatile and can be used to test the effect of nanomaterials in accordance to the to target tissues. This includes the analysis of nanomaterials used as drug delivery systems, analysis of compounds, and evaluation of the nanoparticles produced. The following platforms are currently available:

3.1 Cancer-on-a-chip platform

Cancer treatment could be considered the main focus of nanomedicine. The majority of studies involve nanomaterials as drug delivery vehicles. The outcome is assessed following supplementation of static cancer cells line culture with these compounds. For example, in a lung and breast cancer treatment models the use of human albumin nanoparticles for paclitaxel targeted delivery was evaluated using human breast cancer (cell line MCF-7) and human lung cancer (cell line A549) [29]. Although the results of these studies were favorable, the importance of 3D biomimetic fluidic systems to imitate cancer microenvironment was deemed essential.

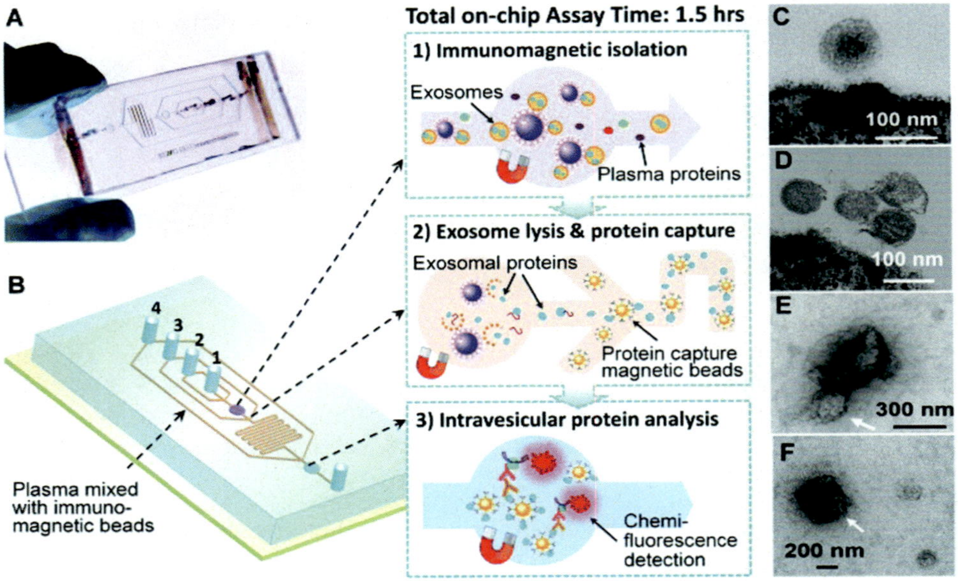

Figure 6.3

Microfluidic system for exosome analysis. (A) Image of prototype, (B) (C—F) Images of exosomes from patients with nonsmall-cell lung cancer (C) and ovarian cancer (D) isolated using microfluidic system. Nonsmall-cell lung cancer and ovarian cancer exosomes conjugated with the antibodies are shown in (E) and (F) respectively. *Reproduced from M. He, J. Crow, M. Roth, Y. Zeng, A.K. Godwin, Integrated immunoisolation and protein analysis of circulating exosomes using microfluidic technology, Lab Chip 14 (19) (2014) 3773—3780, with permission from the Royal Society of Chemistry.*

In cancer detection, microfluidics could be a powerful tool in the identification of cancer, evaluation of drug sensitivity, progress monitoring and prognosis [23] (Fig. 6.3). Siemer et al. investigated geometrically patterned microfluidic platforms with surfaces coated with antibodies which can isolate cancer cell labeled with nanomaterials and achieve high purification capacity [30]. Gold nanoparticles carrying up to 95 aptamers were used in conjunction to a laminar flow flat channel microfluidic device to recognize and capture leukemia cancer cells [31]. With this microfluidic device, an increase in the capture of human acute leukemia cells was increased from 49% using aptamer alone to 92% using gold nanoparticle-aptamer from a cell mixture [31].

In cancer treatment, microfluidics can bridge the gap between static cultures and in vivo testing. Albanese et al., reported a tumor-on-a-chip platform (using tumor spheroids), that was capable of achieving real-time analysis of nanoparticle accumulation [32]. It was shown that the diameter of the nanoparticles limits their penetration into the tissue and that receptor targeting can help to improve nanoparticle retention. Ozkan et al., developed a vascularized breast tumor and tumorigenic liver microenvironment-on-a-chip [33].

Investigators developed a dynamic method, which allowed the measurement of the transport of drug and nanoparticles, and the treatment efficacy and the levels of hepatotoxicity. They analyzed the permeability of each tissue in terms of nanoparticle accumulation, suggesting that this platform can be used to assess the effect of drug and nanoparticle properties and the tissue microenvironment, on transport, efficacy, selectivity and toxicity [33]. Pernia et al., carried out a feasibility study to evaluate magnetic thermoresponsive polymers for controlled release of chemotherapeutic drugs using a microfluidic platform [34]. The study was performed without using cells in PBS but authors suggested that with further development it would be feasible to study drug delivery to cells using these thermoresponsive carriers [34].

3.2 Liver-on-a-chip platforms

Liver is the main organ in our bodies responsible for the metabolism of nutrients and proteins, and for the detoxification and metabolism of drugs. Liver-on-a-chip systems have been designed as physiological models and have been studied as preliminary platforms for in vitro drug metabolism and enzyme induction [35]. Drug hepatotoxicity and lack of efficacy are two major concerns during drug discovery and testing. Static in vitro models utilizing hepatocytes result in poor performance because many original functions of hepatocytes are lost ex vivo. Liver-on-a-chip platforms proved to be a more realistic option, resulting in strong similarities in cellular behaviors to those seen in vivo [36,37]. As an example, Powers et al., developed a liver culture device and demonstrated the formation of aggregates of liver cells representing the liver acinar structure, which maintained their viability for up to two weeks [38]. A study with similar concept was undertaken by Ma et al., where an integrated liver-on-a-chip platform was fabricated to analyze drug metabolism based toxicity. Hepatic cell carcinoma cell line (HepG2) was chosen as target cell type and human liver microsomes (HLM) were used as a bioreactor [39]. In another study, Esch et al., evaluated the use of body-on-a-chip system in investigating the effect of oral uptake of carboxylated polystyrene nanoparticles [40]. They tried to simulate liver tissue and gastrointestinal system in an OoC platform to identify the damage caused by the nanoparticles taken by the liver. This system mimicked the liquid portions of the organs and showed that these nanoparticles have the potential of causing gastrointestinal and liver tissue injury following consumption. The results obtained through in vitro studies showed that 50 nm carboxylated polystyrene nanoparticles lead to liver damage at the cellular level.

Multi-organ-chip systems have been also designed to overcome failure of single OoC to capture secondary and systemic toxicity [41]. For example, a microfluidic system, in which human liver and skin tissue were studied together, was developed by Wagner et al., and used for long-term culturing and substance testing [20].

3.3 Heart-on-a-chip platforms

Cardiovascular system has attracted significant interest both in terms of studying novel treatment strategies and side effects resulting from systemic treatment of other organs. Poor screening is a major cause of failure and a key aspect of carrying nanomaterial-based innovations forward. In particular, cardiotoxicity resulting from cardiopharmaceuticals is one of the critical issues affecting the cardiac muscle [42]. Several authors have developed physiologically relevant heart-on-a-chip platforms utilizing cardiomyocytes, which could mimic the physiology of the heart. These platforms make human cardiac tissue functions more amenable for screening of the unexpectedly high toxicity for clinical trials of drug materials [43]. For instance, Annabi et al., developed a system that was coated using cross-linkable methacrylated tropoelastin (MeTro) and gelatin methacryloyl (GelMA) hydrogels to increase the binding of viable primary cardiomyocyte (CM) cells in the microfluidic device for increasing the cardiac tissue functions [44].

In a heart-on-a-chip platform designed for in vitro drug testing by Grosberg et al., data collection and analysis were performed in real time for both physiological and pharmacological applications [45]. Ahn et al., microfabricated an artificial heart-on-chip platform to evaluate the toxicity of nanomaterials used for the treatment of cardiovascular disorders [46]. Extracellular matrix environment mimicked the myocardium by having polydopamine (PDA)/polycaprolactone (PCL) nanofibers. These fiber-based scaffolds made the chip system most useful to understand engineered nanomaterial effects. The cardiotoxicity of titanium dioxide (TiO_2) and Ag nanoparticles was determined using cardiac cells that were seeded on fiber scaffolds [46].

3.4 Lung-on-a-chip platforms

Lungs are responsible for gas exchange between the body and the environment. At the same time, they offer a potentially appealing route for drug administration. The lungs have a massive absorbing surface of about 100 m^2 and aerosol nanomaterials have the potential of crossing the alveoli-capillary barrier [47]. In fact there is a significant number of studies analyzing the effect of nanomaterials on lung cell functions, however, the vast majority involve static cultures of lung cells [48,49]. Burges et al., fabricated the microplatform to mimic lung environment and understand chronic respiratory problems. This system was divided into two channels, representing blood capillary and gas pathway. This biohybrid platform was known as one of the first artificial lung-on-a-chip systems [50]. Huh et al., developed a mimetic lung architecture for constructing alveolar-capillary barrier [24]. They evaluated the toxic and inflammatory effects of the silica nanoparticles on lung tissue with the organ-on-chip system. Tissue-tissue interactions were investigated for functional alveolar breathing to understand nanoparticle absorption and cytotoxicity [24].

Another application of lung-on-a-chip system was reported by Zhang et al., who fabricated a chip model to evaluate toxicity effects of the nanoparticles used in lung pulmonary diseases [51]. The zinc oxide (ZnO) and TiO_2 nanoparticles, which are commonly used in medical care were chosen to understand their pulmonary toxicity. This platform was prepared using three parallel channels, two different cell types and 3D Matrigel membrane [51]. Future studies should involve the development of an alveolar-capillary platform, which combines multiple cell structures without any complicated culturing environment, and may be modeled for the development of nanomaterials [52].

3.5 Kidney-on-a-chip platforms

Human kidney is an important organ being responsible for water metabolism, blood filtration and the excretion of body wastes. Kidney filtration barriers are very important for separating the vasculature from the urinary spaces. This natural filtration system also selects the blood chemistry components such as ions and molecules [53]. In nanomaterial research, it is important that the treatment does not cause advert toxicity to the kidneys [54]. Animal testing is the mainstay of evaluation of renal toxicity in preclinical testing [54]. Various kidney-on-a-chip platforms have been developed [55] to open new avenues for studying kidney disease and for developing treatment of the renal disorders [56]. It is important to understand the mechanism that causes nephrotoxicity in order to achieve better results with drug development and testing studies [41]. In one study, in vivo biological distributions of siRNA loaded dextran nanogels used as nanocarriers in the kidney filtration barrier were investigated in detail by Naeye et al. using kidney filtration barriers [57]. Following intravenous injection, siRNA loaded dextran nanogels were found to accumulate in the kidneys [57]. A significant delay of the transition of siRNA from kidney to bladder occurred, as opposed to the injection of free siRNA [57]. It was suggested that a dissociation of siRNA from its carrier may be induced by components of the glomerular filtration barrier [57].

3.6 Placenta-on-a-chip platforms

The uptake, transfer and accumulation of nanoparticles in the placental barrier in fetal blood causes both fetal and placental inflammation. When this occurs, fetal growth syndrome (FGR) and fetal infections are observed [58]. Therefore, the effect of environmental exposure to nanoparticles on pregnant women has been analyzed though a placenta-on-a-chip system. In one study, Yin et al. utilized a placental barrier-on-a-chip to analyze the effect of TiO_2 nanoparticles, which are known to induce various effects such as oxidative stress, cell apoptosis, barrier permeability, and maternal immune cell reaction [59]. Investigators suggested that these effects were greatly influenced even with low concentration of these nanoparticles highlighting the potential detrimental effects involved in pregnant women. In another study, Schuller et al. developed placenta-on-a-chip system to mimic the in vivo placental microenvironment and studied the effect of the silicon dioxide (SiO_2), TiO_2, and ZnO nanomaterials on placental cells [60]. Using this platform, nanoparticle toxicity affecting placental cytotrophoblast (Bewo) cells and placental barrier was evaluated [60].

3.7 Other organ-on-a-chip platforms

Other OoC platforms that are important for testing of nanomaterials are the brain [61] and blood vessels [62]. Brain-on-a-chip system has been studied with combining different regions of the brain together. For example, Dauth et al., developed a multicompartment brain-on-a-chip platform based on the use of different neurons created from different brain regions in vitro. With this multilayered brain-on-a-chip system, it was investigated how the drug, phencyclidine, which is known to induce schizophrenia-like disorder symptoms, affects different parts of the brain [61]. The application of microdevice platforms in neuroscience research was reported by Taylor et al., who developed a microdevice, which was composed of different compartments to understand neurological problems such as Alzheimer's disease [63]. OoC systems have been used to investigate the axonal degeneration of neurons resulting from toxicity caused by nanoparticles [52]. Mahto et al., prepared a platform to mimic somal, and axonal parts of neurons [64]. This platform was tested by using surface-modified quantum dots and neuron-like PC12 cells. The cells were cultured inside the chips to investigate the cytotoxicity of the surface-modified quantum dots. It is worth noting that quantum dots have been shown to be toxic to neural cells based on surface stability of the modifying ligands [64].

Blood vessels have also been studied in vitro using organ-on-chip platforms to understand the diffusion and toxicity of the nanoparticles. Wu et al., prepared a microfluidic device which was divided two parts, one to simulate tissue and the other the blood vessels in order to understand diffusion of the quantum dots between capillary and tissue. The main part of the chip system was defined for the capillary and nanoparticles, while the other part was defined for the cell culture, HepG2 cells [62]. In another application, Kim et al. fabricated a chip platform that mimics blood vessel and its environment to study the effect of mesoporous silica nanoparticles on platelets. Different nanoparticle doses were investigated. The viability of the platelets was found unaffected by exposure to mesoporous silica [65].

4. Challenges, future directions, and conclusions

Currently, the application of microfluidic OoC systems in nanomedicine is in its early stages. The use of nanoparticles is slow in reaching the clinical practice due to the prolonged testing and significant financial risks taken by the industry and research institutions. The use of microfluidic OoC platform in nanomaterial testing could be used as development instrument, where the need of such tool is evident [71,72]. The potential of OoC could be expanded to multi-organ testing in a single set of experiments. For example, Marchmeyer et al., have showed that multiple organs, such as human skin, kidney, intestine, and liver could be included in the same experiments [66]. This will allow us to understand the systemic effect of nanomaterials including the interactions between different human organs. However, robust sensors to capture and record different physiological parameters need to be developed. Technologies that could record multiple responses ranging from cell viability, biochemical changes and genetic expressions are needed. This analysis at microscale level at the moment is difficult and requires assay technology that is systemized and automated. Another major challenge is our lack of understanding of several cellular processes and functions. It is well accepted that we lack important knowledge in regard to human physiology. Hence, decoding these cellular and molecular processes will allow us to correlate the data received from the organ-on-a-chip platforms with those seen in vivo in humans. Cell sourcing remains an unsolved crucial issue together with the culture conditions and media. For example, some commercially available cell lines contain a mixture of animal and human cells due to the inherent difficulties isolating such cells [67,68]. Differentiation of cells remains nonphysiologic and is forced by the action of growth factors and medicines [69]. The results of such experiments could often be misleading. In addition to the need for proof-of-concept studies and data demonstrating correlation between in vitro and in vivo observations, extensive work should also be carried out on the nanomaterials per se. It has been realized that due to the size, charge and hydrophobicity, some nanoparticles have undesired effects on humans and especially on the coagulation system [70]. These safety issues can be missed when analysis is performed using single OoC assay.

The use of OoC platforms in nanomedicine, provides new and exciting opportunities to expand this field of research and bring novel clinical applications based on nanomaterials. Protocol and assay validation with proof of reproducibility should be a primary focus. The creation of independent validation bodies and databases are required to safeguard reliability, safety and transparency. Strong links between researchers, regulatory, and industrial partners remain critical for further advancement of OoC platforms. Given the rapid expansion of the field of microfluidics experienced in the past few years, it is likely that OoC will revolutionize the process of preclinical testing. This will allow fast evaluation of the effectiveness and safety of a range of nanomaterials, reducing the time required to reach clinical trials for their potential translational use.

References

[1] S. Su, P.M.1 Kang, Systemic review of biodegradable nanomaterials in nanomedicine, Nanomaterials 10 (4) (2020) pii: E656.
[2] J.L. Markman, A. Rekechenetskiy, E. Holler, J.Y. Ljubimova, Nanomedicine therapeutic approaches to overcome cancer drug resistance, Adv. Drug Deliv. Rev. 65 (13–14) (2013) 1866–1879.
[3] Z. Cheng, Z.A. Al, J.Z. Hui, V.R. Muzykantov, A. Tsourkas, Multifunctional nanoparticles: cost versus benefits of adding targeting and imaging capabilities, Science 338 (2012) 903–910.
[4] H. Liu, Y. Yang, Y. Liu, J. Pan, J. Wang, F. Man, W.2 Zhang, G. Liu, Melanin-like nanomaterials for advanced biomedical applications: a versatile platform with extraordinary promise, Adv. Sci. 7 (7) (2020) 1903129.
[5] G. Jin, X. Zhao, F. Xu, Therapeutic nanomaterials for cancer therapy and tissue regeneration, Drug Discov. Today 22 (9) (2017) 1285–1287.
[6] S. Ilk, G. Tan, E. Emül, N. Sağlam, Investigation the potential use of silver nanoparticles synthesized by propolis extract as N-acyl-homoserine lactone-mediated quorum sensing systems inhibitor, Turk. J. Med. Sci. 50 (4) (2020) 1147–1156, https://doi.org/10.3906/sag-2004-148.
[7] M. Fan, Y. Han, S. Gao, H. Yan, L. Cao, Z. Li, X.J. Liang, J. Zhang, Ultrasmall gold nanoparticles in cancer diagnosis and therapy, Theranostics 10 (11) (2020) 4944–4957.
[8] D. Peer, J.M. Karp, S. Hong, O.C. Farokhzad, R. Margalit, R. Langer, Nanocarriers as an emerging platform for cancer therapy, Nat. Nanotechnol. 2 (2007) 751–760.
[9] V. Agrahari, V. Agrahari, A.K. Mitra, Next generation drug delivery: circulatory cells-mediated nanotherapeutic approaches, Expet Opin. Drug Deliv. 14 (2017) 285–289.
[10] A.C. Anselmo, S. Mitragotri, Cell-mediated delivery of nanoparticles: taking advantage of circulatory cells to target nanoparticles, J. Contr. Release 190 (2014) 531–541.
[11] R. Tutar, A. Motealleh, A. Khademhosseini, N.S. Kehr, Functional nanomaterials on 2D surfaces and in 3D nanocomposite hydrogels for biomedical applications, Adv. Funct. Mater. (2019) 1904344.
[12] H. Kim, S. Beack, S. Han, M. Shin, T. Lee, Y. Park, K.S. Kim, A.K. Yetisen, S.H. Yun, W. Kwon, S.K. Hahn, Multifunctional photonic nanomaterials for diagnostic, therapeutic, and theranostic applications, Adv. Mater. 30 (10) (2018).
[13] P. Fattahi, G. Yang, G. Kim, P.M.R. Abidian, A review of organic and inorganic biomaterials for neural interfaces, Adv. Mater. 26 (12) (2014) 1846–1885.
[14] F.R. Maia, R.L. Reis, J.M.4 Oliveira, Nanoparticles and microfluidic devices in cancer research, Adv. Exp. Med. Biol. 1230 (2020) 161–171.
[15] P.H. Sugarbaker, O.A. Stuart, Pharmacokinetics of the intraperitoneal nanoparticle pegylated liposomal doxorubicin in patients with peritoneal metastases, Eur. J. Surg. Oncol. (19) (2019) 30380–30384, pii: S0748-7983.
[16] L. Jena, E. McErlean, H.2 McCarthy, Delivery across the blood-brain barrier: nanomedicine for glioblastoma multiforme, Drug Deliv. Transl. Res. 10 (2) (2020) 304–318.
[17] D. Bobo, K.J. Robinson, J. Islam, K.J. Thurecht, S.R. Corrie, Nanoparticle-based medicines: a review of FDA-approved materials and clinical trials to date, Pharmaceut. Res. 33 (2016) 2373–2387.
[18] J.H. Lin, Species similarities and differences in pharmacokinetics, Drug Metab. Dispos. 23 (1995) 1008–1021.
[19] J. Wang, Y. Song, Microfluidic synthesis of nanohybrids, Small 13 (18) (2017).
[20] I. Wagner, E.M. Materne, S. Brincker, U. Sussbier, C. Fradrich, M. Busek, F. Sonntag, D.A. Sakharov, E.V. Trushkin, A.G. Tonevitsky, R. Lauster, U. Marx, A dynamic multi-organ-chip for long-term cultivation and substance testing proven by 3D human liver and skin tissue co-culture, Lab Chip 13 (2013) 3538–3547.
[21] P.M. Valencia, O.C. Farokhzad, R. Karnik, R. Langer, Microfluidic technologies for accelerating the clinical translation of nanoparticles, Nat. Nanotechnol. 7 (10) (2012) 623–629.
[22] G.M. Whitesides, The origins and the future of microfluidics, Nature 442 (2006) 368–373.

[23] K.L. Fetah, B.J. DiPardo, E.M. Kongadzem, J.S. Tomlinson, A. Elzagheid, M. Elmusrati, A. Khademhosseini, N. Ashammakhi, Cancer modeling-on-a-chip with future artificial intelligence integration, Small 15 (50) (2019) e1901985.
[24] D. Huh, B.D. Matthews, A. Mammoto, M. Montoya-Zavala, H.Y. Hsin, D.E. Ingber, Reconstituting organ-level lung functions on a chip, Science 328 (5986) (2010) 1662–1668.
[25] H.J. Kim, D. Huh, G. Hamilton, D.E. Ingber, Human gut-on-a-chip inhabited by microbial flora that experiences intestinal peristalsis-like motions and flow, Lab Chip 12 (2012) 2165–2174.
[26] C. Moraes, J.M. Labuz, B.M. Leung, M. Inoue, T.H. Chun, S. Takayama, On being the right size: scaling effects in designing a human-on-a-chip, Integr. Biol. 5 (2013) 1149–1161.
[27] D. Huh, H. Fujioka, Y.C. Tung, N. Futai, R. Paine 3rd, J.B. Grotberg, S. Takayama, Acoustically detectable cellular-level lung injury induced by fluid mechanical stresses in microfluidic airway systems, Proc. Natl. Acad. Sci. U. S. A. 104 (2007) 18886–18891.
[28] A. Skardal, S.V. Murphy, M. Devarasetty, I. Mead, H.W. Kang, Y.J. Seol, Y. Shrike Zhang, S.R. Shin, L. Zhao, J. Aleman, A.R. Hall, T.D. Shupe, A. Kleensang, M.R. Dokmeci, S. Jin Lee, J.D. Jackson, J.J. Yoo, T. Hartung, A. Khademhosseini, S. Soker, C.E. Bishop, A. Atala, Multi-tissue interactions in an integrated three-tissue organ-on-a-chip platform, Sci. Rep. 7 (1) (2017) 8837.
[29] D. Ding, X. Tang, X. Cao, J. Wu, A. Yuan, Q. Qiao, J. Pan, Y. Hu, Novel self-assembly endows human serum albumin nanoparticles with an enhanced antitumor efficacy, AAPS PharmSciTech. 15 (1) (2013) 213–222, https://doi.org/10.1208/s12249-013-0041-3.
[30] S. Siemer, D. Wünsch, A. Khamis, Q. Lu, A. Scherberich, M. Filippi, M.P. Krafft, J. Hagemann, C. Weiss, G.B. Ding, R.H. Stauber, A. Gribko, Nano meets micro-translational nanotechnology in medicine: nano-based applications for early tumor detection and therapy, Nanomaterials 10 (2) (2020) pii: E383.
[31] W. Sheng, T. Chen, W. Tan, Z. Hugh Fan, Multivalent DNA nanospheres for enhanced capture of cancer cells in microfluidic devices, ACS Nano 7 (8) (2013) 7067–7076.
[32] A. Albanese, A.K. Lam, E.A. Sykes, J.V. Rocheleau, W.C. Chan, Tumour-on-a-chip provides an optical window into nanoparticle tissue transport, Nat. Commun. 4 (2013) 2718.
[33] A. Ozkan, N. Ghousifam, P.J. Hoopes, T.E. Yankeelov, M.N. Rylander, In vitro vascularized liver and tumor tissue microenvironments on a chip for dynamic determination of nanoparticle transport and toxicity, Biotechnol. Bioeng. 116 (5) (2019) 1201–1219.
[34] M. Pernia Leal, A. Torti, A. Riedinger, R. La Fleur, D. Petti, R. Cingolani, R. Bertacco, T. Pellegrino, Controlled release of doxorubicin loaded within magnetic thermo-responsive nanocarriers under magnetic and thermal actuation in a microfluidic channel, ACS Nano 6 (2012) 10535–10545.
[35] A. Sivaraman, J.K. Leach, S. Townsend, T. Iida, B.J. Hogan, D.B. Stolz, R. Fry, L.D. Samson, S.R. Tannenbaum, L.G. Griffith, A microscale in vitro physiological model of the liver: predictive screens for drug metabolism and enzyme induction, Curr. Drug Metabol. 6 (2005) 569–591.
[36] N. Ashammakhi, E. Elkhammas, A. Hasan, Translating advances in organ-on-a-chip technology for supporting organs, J. Biomed. Mater. Res. B Appl. Biomater. 107 (6) (2018) 2006–2018, https://doi.org/10.1002/jbm.b.34292.
[37] C.E. Suurmond, S. Lasli, F.W. Dolder, A. Ung, H. Kim, P. Bandaru, J.L. Kang, H.-J. Cho, S. Ahadian, N. Ashammakhi, R. Mehmet, D. Junmin Lee, A. Khademhosseini, Liver-on-a-Chip: in vitro human liver model of nonalcoholic steatohepatitis by coculturing hepatocytes, endothelial cells, and Kupffer cells, Adv. Healthcare Mater. 8 (24) (2019) 1970094.
[38] M.J. Powers, K. Domansky, M.R. Kaazempur-Mofrad, A. Kalezi, A. Capitano, A. Upadhyaya, P. Kurzawski, K.E. Wack, D.B. Stolz, R. Kamm, L.G. Griffith, A microfabricated array bioreactor for perfused 3D liver culture, Biotechnol. Bioeng. 78 (3) (2002) 257–269.
[39] B. Ma, G. Zhang, J. Qin, B. Lin, Characterization of drug metabolites and cytotoxicity assay simultaneously using an integrated microfluidic device, Lab Chip 9 (2009) 232–238.

[40] M.B. Esch, G.J. Mahler, T. Stokol, M.L. Shuler, Body-on-a-chip simulation with gastrointestinal tract and liver tissues suggests that ingested nanoparticles have the potential to cause liver injury, Lab Chip 14 (16) (2014) 3081–3092.

[41] N. Ashammakhi, M.A. Darabi, B. Çelebi-Saltik, R. Tutar, M.C. Hartel, J. Lee, S.M. Hussein, M.J. Goudie, M.B. Cornelius, M.R. Dokmeci, A. Khademhosseini, Microphysiological systems: next generation systems for assessing toxicity and therapeutic effects of nanomaterials, Small Methods 4 (2020) 1900589.

[42] J.-J. Zhu, Y.-Q. Xua, J.-H. He, H.-P. Yua, C.-J. Huang, J.-M. Gao, Q.-X. Dong, Y.-X. Xuan, C.-Q. Lia, Human cardiotoxic drugs delivered by soaking and microinjection induce cardiovascular toxicity in zebrafish, J. Appl. Toxicol. 34 (2) (2013) 139–148.

[43] E. Cimetta, A. Godier-Furnemont, G. Vunjak-Novakovic, Bioengineering heart tissue for in vitro testing, Curr. Opin. Biotechnol. 24 (2013) 926–932.

[44] N. Annabi, S. Selimovic, A. Cox, J. Ribas, M. Afshar Bakooshli, D. Heintze, A.S. Weiss, D. Cropek, A. Khademhosseini, Hydrogel-coated microfluidic channels for cardiomyocyte culture, Lab Chip 13 (2013) 3569–3577.

[45] A. Grosberg, P.W. Alford, M.L. McCain, K.K. Parker, Ensembles of engineered cardiac tissues for physiological and pharmacological study: heart on a chip, Lab Chip 11 (2011) 4165–4173.

[46] S. Ahn, H.A.M. Ardoña, J.U. Lind, F. Eweje, S.L. Kim, G.M. Gonzalez, Q. Liu, J.F. Zimmerman, G. Pyrgiotakis, Z. Zhang, J. Beltran-Huarac, P. Carpinone, B.M. Moudgil, P. Demokritou, K.K. Parker, Mussel-inspired 3D fiber scaffolds for heart-on-a-chip toxicity studies of engineered nanomaterials, Anal. Bioanal. Chem. 410 (2018) 6141–6154.

[47] P. Zarogoulidis, N.K. Karamanos, K. Porpodis, K. Domvri, H. Huang, W. Hohenforst-Schmidt, E.P. Goldberg, K. Zarogoulidis, p Zarogouldis, et al., Vectors for inhaled gene therapy in lung cancer. Application for nano oncology and safety of bio nanotechnology, Int. J. Mol. Sci. 13 (2012) 10828–10862, 17290–17291.

[48] R.J. Vandebriel, W.H. De Jong, A review of mammalian toxicity of ZnO nanoparticles, Nanotechnol. Sci. Appl. 5 (2012) 61–71.

[49] R. Govender, A. Phulukdaree, R.M. Gengan, K. Anand, A.A. Chuturgoon, Silver nanoparticles of Albizia adianthifolia: the induction of apoptosis in human lung carcinoma cell line, J. Nanobiotechnol. 11 (2013) 5.

[50] K. Burgess, H.H. Hu, W. Wagner, W. Federspiel, Towards microfabricated biohybrid artificial lung modules for chronic respiratory support, Biomed. Microdevices 11 (2009) 117–127.

[51] M. Zhang, C. Xu, L. Jiang, J. Qin, A 3D human lung-on-a-chip model for nanotoxicity testing, Toxicol. Res. 7 (2018) 1048.

[52] S.K. Mahto, V. Charwat, P. Ertl, B. Rothen-Rutishauser, S.W. Rhee, J. Sznitman, Microfluidic platforms for advanced risk assessments of nanomaterials, Nanotoxicologhy (2014) 1–15.

[53] D.C. Eaton, J. Pooler, Vander's Renal Physiology, eighth ed., McGraw-Hill Medical Publishing, 2013.

[54] R. Su, Y. Li, D. Zink, L.H. Loo, Supervised prediction of drug-induced nephrotoxicity based on interleukin-6 and -8 expression levels, BMC Bioinf. 15 (Suppl. 16) (2014) S16.

[55] N. Ashammakhi, K. Wesseling-Perry, A. Hasan, E. Elkhammas, Y.S. Zhang, Kidney-on-a-chip: untapped opportunities, Kidney Int. 94 (6) (2018) 1073–1086, https://doi.org/10.1016/j.kint.2018.06.034.

[56] N. Kamaly, J.C. He, D.A. Ausiello, O.C. Farokhzad, Nanomedicines for renal disease: current status and future applications, Nature 12 (2016) 738–753.

[57] B. Naeye, H. Deschout, V. Caveliers, B. Descamps, K. Braeckmans, C. Vanhove, J. Demeester, T. Lahoutte, S.C. De Smedt, K. Raemdonck, In vivo disassembly of IV administered siRNA matrix nanoparticles at the renal filtration barrier, Biomaterials 34 (2013) 2350–2358.

[58] M.S. Poulsen, T. Mose, L.L. Maroun, L. Mathiesen, L.E. Knudsen, E. Rytting, Kinetics of silica nanoparticles in the human placenta, Nanotoxicology 9 (2015) 79–86.
[59] F. Yin, Y. Zhu, M. Zhang, H. Yu, W. Chen, J. Qin, A 3D human placenta-on-a-chip model to probe nanoparticle exposure at the placental barrier, Toxicol. Vitro 54 (2019) 105–113.
[60] P. Schuller, M. Rothbauer, S.R.A. Kratz, G. Höll, P. Taus, M. Schinnerl, J. Genser, N. Bastus, O.H. Moriones, V. Puntes, B. Huppertz, M. Siwetz, H. Wanzenböck, P. Ertl, A lab-on-a-chip system with an embedded porous membrane-based impedance biosensor array for nanoparticle risk assessment on placental Bewo trophoblast cells, Sensor. Actuator. B Chem. 312 (2020) 127946.
[61] S. Dauth, B.M. Maoz, S.P. Sheehy, M.A. Hemphill, T. Murty, M.K. Macedonia, A.M. Greer, B. Budnik, K.K. Parker, Neurons derived from different brain regions are inherently different in vitro: a novel multiregional brain-on-a-chip, J. Neurophysiol. 117 (3) (2017) 1320–1341.
[62] J. Wu, Q. Chen, W. Liu, Y. Zhang, J.-M. Lin, Cytotoxicity of quantum dots assay on a microfluidic 3D-culture device based on modelling diffusion process between blood vessels and tissues, Lab Chip 12 (2012) 3474–3480.
[63] A.M. Taylor, S.W. Rhee, C.H. Tu, D.H. Cribbs, C.W. Cotman, N.L. Jeon, Microfluidic multicompartment device for neuroscience research, Langmuir 19 (5) (2003) 1551–1556.
[64] S.K. Mahto, T.H. Yoon, S.W. Rhee, Cytotoxic effects of surface modified quantum dots on neuron-like PC12 cells cultured inside microfluidic devices, Biochip J. 4 (2010c) 82–88.
[65] D. Kim, S. Finkenstaedt-Quinn, K.R. Hurley, J.T. Buchman, C.L. Haynes, On-chip evaluation of platelet adhesion and aggregation upon exposure to mesoporous silica nanoparticles, Analyst 139 (2014) 906–913.
[66] I. Maschmeyer, A.K. Lorenz, K. Schimek, T. Hasenberg, A.P. Ramme, J. Hübner, M. Lindner, C. Drewell, S. Bauer, A. Thomas, N.S. Sambo, F. Sonntag, R. Lauster, U. Marx, A four-organ-chip for interconnected long-term co-culture of human intestine, liver, skin and kidney equivalents, Lab Chip 15 (12) (2015) 2688–2699.
[67] M.M. Laronda, J.E. Burdette, J.J. Kim, T.K. Woodruff, Recreating the female reproductive tract in vitro using iPSC technology in a linked micro- fluidics environment, Stem Cell Res. Ther. 4 (2013). S13–S.
[68] S.L. Eddie, J.J. Kim, T.K. Woodruff, J.E. Burdette, Microphysiological mod- eling of the reproductive tract: a fertile endeavor, Exp. Biol. Med. 239 (2014) 1192–1202.
[69] I. Pountos, D. Corscadden, P. Emery, P.V. Giannoudis, Mesenchymal stem cell tissue engineering: techniques for isolation, expansion and application, Injury 38 (Suppl. 4) (2007) S23–S33.
[70] A.N. Ilinskaya, M.A. Dobrovolskaia, Nanoparticles and the blood coagulation system. Part II: safety concerns, Nanomedicine 8 (6) (2013) 969–981.
[71] Y. Temiz, R.D. Lovchik, G.V. Kaigala, E. Delamarche, Lab-on-a-chip devices: how to close and plug the lab? Microelectron. Eng. 132 (2015) 156–175.
[72] M. He, J. Crow, M. Roth, Y. Zeng, A.K. Godwin, Integrated immunoisolation and protein analysis of circulating exosomes using microfluidic technology, Lab Chip 14 (19) (2014) 3773–3780.

CHAPTER 7

Electroanalytical techniques: a tool for nanomaterial characterization

Sijo Francis[1], Ebey P. Koshy[1], Beena Mathew[2]

[1]Department of Chemistry, St. Joseph's College, Moolamattom, Kerala, India; [2]School of Chemical Sciences, Mahatma Gandhi University, Kottayam, Kerala, India

Chapter Outline

Design, Fabrication, and Characterization of Multifunctional Nanomaterials. https://doi.org/10.1016/B978-0-12-820558-7.00004-2

1. Introduction

Electroanalytical chemistry is the branch of chemical analysis that employs electrochemical methods to obtain information related to amounts and characteristics of chemical species. In these techniques, an electrode probe is used for the qualitative and the quantitative analysis of compounds under investigation. In quantitative electrochemical measurements, potential of the cell is being measured, which in turn is related to their respective chemical properties. The method is highly selective as they are able to control the potential of an electrode, which makes it possible to determine the electrochemical spectrum of electroactive species in the solution.

2. Electrochemical techniques

In classical electrochemical analysis (amperometric and potentiometric) the information on the electrode response of the analyte when the electrolyte, solvent, or surface is modified is studied [1]. For potentiometric and potentiostatic types of analysis require at least two electrodes, which together constitute an electrochemical cell. One of the two electrodes responds to target analyte and is termed as indicator (working) electrode. The second one is termed as a reference electrode, which is of constant potential and is independent of the properties of the solution. Potentiometry is a static technique in which the information about the sample composition is obtained from the measurement of the potential established across as membrane materials. In potentiometric titrations, the variation in cell potential is monitored as a function of the volume of reagent added, whereas in direct potentiometry, the cell potential is determined and related to the activity of concentration of individual species.

Electrochemical sensors use an electrochemical cell consisting minimum of two electrodes: a working electrode and a reference electrode. Current flowing in a reduction/oxidation processes at the working electrode surface in connection with a counterelectrode is measured. It includes cyclic voltammetry, stripping voltammetry, differential pulse voltammetry, linear sweep voltammetry, and so forth. Measurement based on current-potential response, current for a specific period, concentration of analyte on stripping etc are named as voltammetry, amperometry, and stripping techniques. Nature of the analyte, type, morphology and size of electrode used are the main influential factors determine the response of the analysis.

2.1 Coulometry

Coulometry is an electroanalytical method which involves the measurement of the quantity of electricity (coulombs) needed to convert the analyte quantitatively to a different oxidation state. Coulometric analysis is an application of Faraday's law of electrolysis.

That is the extent of chemical reaction at an electrode is directly proportional to the quantity of electricity passing through the electrode. The sample mass, molecular mass, no. of electrons in the electrode and the number of electrons passed during the experiment are all related to Faraday's law

$$W = MQ/96487\,n$$

where, M is the molecular mass of substance liberated or consumed, n is the number of electrons, W is the weight of the substance produced or consumed during electrolysis involving, Q coulomb of charge, and F is Faraday constant.

2.2 Voltammetry

In voltammetry information about an analyte is obtained by measuring the current as the potential is varient. The potential Q varied arbitrarily with step by step or continuously and the actual current value is measured as the dependent variables. The potential is varied to cause oxidation or reduction of electroactive species at the electrode. The resultant current is proportional to the concentration of the electrochemical species. The commonly employed voltammetric techniques are cyclic voltammetry, pulse voltammetry, and stripping analysis. In polarography, polarizable microelectrodes like dropping mercury electrode, is used as cathode and whose response is obtained combined mass transport phenomena like diffusion, migration, and convection.

2.3 Cyclic voltammetry

Cyclic voltammetry is the most widely used electroanalytical technique for acquiring qualitative information about electrochemical reactivity. In the most versatile electroanalytical techniques cyclic voltammetry, out of the three electrode cell systems, a novel working electrodes is used [2]. A platinum wire is the counterelectrode and saturated calomel electrode is used as the reference electrodes in the electrochemical studies. Cyclic voltametry is useful for the study of interfacial processes and soluble reaction intermediates [3]. In cyclic voltammetry, the current response of small stationary microelectrode is recorded as a function of triangular potential waveform. Pulse voltammetric techniques are aimed at lowering the deduction limit of voltammetric measurements.

2.4 Stripping voltammetry

Stripping analysis is an extremely sensitive electrochemical technique for measuring trace samples like metals. It is a two-step technique. First step is the deposition step that involves the electrolytic deposition of a small portion of the metal ions in solution onto the working electrode to preconcentrate the samples. This is followed by stripping; which involves the dissipation of the deposit. The analyte deposited is either by an oxidation/reduction potential and current while it is stripped in the reverse scan is measured. Thus two stripping processes are in use, anodic and cathodic stripping depending on the nature of potential used.

Stripping voltammetry using carbon nanotubes (CNTs) electrode surface modifier is an efficient method for heavy metals detection. The heavy metal contamination in the human body can be monitored by analyzing the biological matrices like blood, sweat, or urine. There exists a direct relationship between the heavy metal content level and human metabolic effects [1]. Glassy carbon electrodes coated with polyaniline-multiwalled CNT composite is used for the qualitative detection of detection of Pb^{2+} using the electroanalytical technique square wave anodic stripping voltammetry [4]. Stripping voltammetry is an efficient method for detection and estimation of heavy metals and CNT-coated electrodes are used for the purpose. Bare CNTs and functionalized CNTs are employed for the electrode preparation purpose. Functionalization by carboxyl, amino, and thiol groups are common [5].

2.5 Differential pulse voltammetry

Differential pulse voltammetry (DPV)—at low glutamate levels, a sharp sensitive electrode response is observed and at high concentration level, saturation is seen as according to Michaelis—Menten kinetics when the analysis is performed in the concentration range of 20–300 μM [6].

2.6 Electrochemiluminescence

Fluorescent carbon quantum dots are a rising star due inexpensive preparation procedure and ardent applications. Pristine carbon dots has emission at 365 nm arising from $\pi \rightarrow \pi^*$ electron transition and can be utilized in light emitting purposes [7]. Carbon nanotubes provide an excellent method for the instant detection of environment pollutants. Photoluminescent ionic complex sensors are used for the sensing of organic dyes. Usually uncharged dyes in the family of polymethine can be easily tailorable due to their molecular structure characteristics [8].

3. Carbon nanomaterials

Carbon nanomaterials contain sp [2] graphitic carbon and are seen in dimensionalities ranging from 0D to 3D. They are generally classified into one-dimensional, two-dimensional, three-dimensional, and hybrid carbonaceous materials. Major nanoallotropes of carbon used in electroanalytical and electrocatalytical purpose are carbon nanohorns (0D), CNTs (1D), reduced graphene oxide (2D), and graphite (3D). Carbon nanomaterials like graphene and CNTs have exceptionally good electrochemical properties including high electrical conductivity due to high electron mobility and exploited in electrochemical sensing purposes [9].

3.1 CNTs

Electrochemical biosensors having various interfaces like nanoparticles, CNTs, and metal oxides are found in literature. The advent of nanotechnology manipulates the properties of materials at the atomic level. The peculiar electrochemical properties of CNTs are unique among the members in the carbon family because of their unique structure. The C—C backbone with a rich π-electron conjugation is the main features of CNTs. Multiplex-functionalized SWNT is used in molecular sensing devices. Doping on SWCNTs largely improve the electroanalytical response of CNTs. Groups like nitro and amino sometimes add or withdraws electrons from CNTs and modifies electrochemical responses [10]. Among the four platinum functionalized nanocarbon materials, Pt—NP@CNTs material show excellent electroanalytical response [11]. Conducting polymer nanomaterials, and many other conducting nanowires are also found in many industrial applications.

3.1.1 Heavy metals

Highly sensitive voltammetric detection of heavy metals on CNTs is known by their effective increase of surface area of the electrode film. The peculiar properties of MWCNTs include excellent electrical conductivity, greater adsorptive capability, and large surface area. The electrochemical response of MWCNT modified electrode is greater than that of SWCNT-modified electrode and is attributed to two reasons. Primarily due to the high affinity of heavy metals toward the engineered MWCNTs and the functional groups embedded on the MWCNT also has a contribution toward it. Secondly, as a result of surface area increase, the total number of active centers in the adsorbent increase and this in turn affect the rate of reaction on MWCNT matrix [12]. Simultaneous and separate detection of Hg^{2+}, Cd^{2+}, Pb^{2+}, and Cr^{2+} at the ppb level using CV are known [13]. Linear ion-exchange rates of were reported in the CNT-coated composite electrodes [14].

The glassy carbon electrodes coated with multi-walled carbon nanotube-poly(pyrocatechol violet) composite along with bismuth film is an efficient electrochemical sensor for the simultaneous detection of $Cd^{2+}$and Pb^{2+}ions using stripping voltammery. The specific and selective interaction between $Cd^{2+}$and Pb^{2+}ions and the modified electrode enables the simultaneous determination of these heavy metal ions in traces from real water samples collected from tap or well. Conductivity and chelation capacity of the polymeric film has many intrinsic advantages like low cost preparation, good stability, easy regeneration, and needs no refreshing of bismuth film. It offers a means for the massive production of devices for the sensing of heavy metal pollutants that has an adverse effect on the environmental and makes a crisis to humanity [15]. Electrochemical stripping performance of a nanoparticle-based heavy metal sensor on carbon nanostructured material electrodes, such as carbon nanotubes are used for the multidetection of Cd^{2+}, Pb^{2+}, Cu^{2+} and Hg^{2+} ions from seawater [16].

Nitrogen-containing CNTs are used for the electrochemical determination of heavy metal ions. The electrocatalytic activities of the N-doped carbonaceous materials is used as a probe to examine Cd, Pb and Cu metal ions by means of square wave stripping voltammetry. It is a sensitive and selective method for the determination of Pb and Cd metals without any preconcentration as in the case of stripping voltammetry. The multiple nitrogen functional groups situated on the surface of carbon nanotubes enabled a better sensitivity for the both individual and simultaneous determination of Pb and Cd metal [17].

CNT-based heavy metal ion sensor electrodes do not create less toxicity and are much more eco-friendly. They improve selectivity and specificity toward an analyte, when more ions are present in the test samples. Selectivity can be further improved by chelating reagents like ionophores and Schiff bases. Heavy metals content in the waste water effluents, river water, oil field water, lake water, electroplating effluents, wheat flour etc. can be sensed using the CNT-based working electrodes.

3.1.1.1 CNT-polymer adsorbents

Carbon nanotubes modified by polyamidoamine dendrimer is used to remove Cu^{2+} and Pb^{2+} ions from aqueous solutions. The nanocomposite performed as a superadsorbent following Langmuir type isotherm and pursue pseudosecond order kinetic model for the removal of organic and inorganic pollutants in liquid phase due to high specific areas, and high electrostatic bonding forces. Heavy metals from wastewaters can be removed off by this procedure and adsorption performance can be improved by functionalization on CNTs. Hyperbranched polymer with tree-like structures are called dendrimers. Terminal amine groups in the dendrimers enable easy complexation with heavy metals [18].

3.1.1.2 MWCNT-biopolymers

Chitosan is a biopolymer and its low cost, and renewable origin makes then suitable for electroanalysis by stripping strategy of heavy metals. Cross-linked chitosan-carbon nanotube film modified glassy carbon electrode can be used for the study of copper, cadmium and lead by an adsorption mechanism in waste soil in situ. MWCNT is easily acid functionalized and the carboxylic, amino, and hydroxyl groups in Chitosan enable easy complexation, chelation, electrostatic interaction, and ion-exchange processes by mainly adsorption mechanism [19].

3.1.1.3 CNT-based potentiometric techniques: ion selective electrodes

It is customary to call potentiometric sensors as a zero-current technique. Ion selective electrodes are used for sensing and quantitative estimation of potential pollutants.

Neutral species like hydrocarbons, organophosphate pesticides, and cations like lanthanide ions, and anions like halide ions can be sensed using ion selective electrodes [20]. CNT-coated electrodes are a promising electrode, because of the electrical and mechanical capabilities to sense and monitor toxic metals from different fields of ecological relevance. Potentiometric detection of heavy metals involves the measurement of potential of a reference electrode when dipped in a sample solution. According to Nernst equation the potential change depends on the analyte concentration. Some sensitive ionophores like 1,3-bis(2- methoxybenzene)triazene can be used for the detection of heavy metals especially Hg^{2+}. MWCNTs with diameter 8–15 nm is also applied on carbon paste structure and the electrical conductivity and sensitivity of MWCNT enables carbon paste electrode composition to selectively bind Hg(II) ions and quantitative estimation in the ppb level. Carbon paste electrode is an ion selective electrode with high operation speed with affordable rate and simple preparation. The sensitivity of carbon paste electrodes depends mainly on the chemical nature and properties of additives used to modify the electrode. MWCNT exhibit fascinating optoelectronic distinguishable from the other carbonaceous nanomaterials [21]. Wearable potentiometric sensors are used for calibration and the validation in medical field, especially in on-body measurements [22].

3.1.2 Biomolecules

CNTs can accumulate the biomolecules such as DNA which aids in enhancement of the selectivity and sensitivity of the probe used. The electroanalytical particulates of CNTs, mainly SWCNT include selectivity, sensitivity, and fast electrochemically reversible responses at normal temperature and pressure. CNTs are exclusively used as electrochemical transducers for coating of electrode. Metal decorated CNTs are also employed in sensing and catalytic purposes [23].

By introducing photolithography, carbon nanotubes are vertically aligned and a nano electrode array performed and is used as electrochemical glutamate biosensor. The performance level in cyclic voltammetry is in the range of 0.1–20 μM and 0.182 mA/ mM cm^2 and is thus adapted clinically as biomarkers. These types of in vivo sensing microelectrodes provide high speed mass transport, improved current density, low ohmic resistance and improved signal sensitivity. These biosensors have high capacity toward the oxidation of H_2O_2 and nicotinamide adenine dinucleotide (NADH) and produce electrochemical signals with significant amplification. In general, the electrochemical response of enzymatic biosensor strongly depends on pH, temperature, and the amount of coenzyme used. Sigmoidal electrochemical curve obtained repeatedly in a CV experiment usually indicate the steady-state mass transfer in the scan rate regime.

A modified electrode fabricated using nanosilver-multiwalled CNT composites are used for the sensing of hydrogen peroxide an oxidizing reagent using cyclic voltammetry. CNTs can modify the electrical and mechanical properties of the nanocomposites.

Medical samples of hydrogen peroxide which is used for disinfection were satisfactorily analyzed using this CNT-modified electrode [24]. Green synthesized silver nanoparticles can be used for electrochemical sensing of Hg from real water sample from lake water. Phytochemical encapsulated silver nanoparticles form aggregate with Hg(II) ions and a color change from brown to black due to new Hg-complex formation is observed [25].

3.1.3 Fungicides

Electroanalytical sensing of benomyl can be done by nickel ferrite/multiwalled carbon nanotubes modified glassy carbon electrode. Benomyl, is an agricultural fungicide have an adverse impact on human health. Real sample analysis using mushroom, apple, and spinach were conducted using techniques CV and DPV sowed reliable and reproducible results [26].

Zn(II) ion imprinted polymers are layered on MWCNT are fabricated and sensing study were carried out by cyclic voltammetry and DPV. Specific and selective binding of the heavy metal Zn is happened presence of other metal ions [27]. An electrochemical sensor for the heavy metal lead ions can also fabricated on the backbone of using multiwalled carbon nanotube and can be employed successfully to discriminate lead ions from food samples [28]. Multiwalled carbon nanotube supports Ni(II) ion binding from electroplating and steel industries [29].

3.1.4 Organic molecules

Organic molecules like catechol and catechin in fruits at the nanoscale were measured by means of cyclic voltammetry employing laccase immobilization on carbon nanomaterials like nitrogen-doped multiwalled carbon nanotubes and graphene oxide [30]. Detection of Ciprofloxacin is possible by polyaniline film incorporated with multi-walled carbon nanotubes and β—cyclodextrin [31]. Carbon nanomaterials functioned as electrochemical sensor in the voltammetric determination of methylparabens in the nanomolar level in cosmetic samples.

3.1.5 Natural receptors

Natural receptors have molecular specificity and are intracellular or extracellular receptors. Among these, G-protein-coupled receptors are function as sensors for dopamine, hormones, trimethylamine, etc. The electrical biosensors comprise of two components—a nonbiological part and a biological component. Biological recognition can be possible electrical properties of the transistors and by discrimination ability of biomolecules. The electrical properties are improved by a process called chemical immobilization, which include modification by amino acid groups or by some surface coating. Certain natural receptors operate by surface plasmon resonance and found utility in daily life in the areas of food, beverage, clinical and agricultural, and environmental monitoring. G-protein-coupled receptors are also a special category of natural receptors [32].

3.2 Nanomaterials and nanostructures

Electrochemical sensors and biosensors have exceptional attributes such as easy setup, portability, and sensitivity. Electrochemical sensors based on analytical principles like luminescence, amperometric behavior, and impedance responses are known. Functional electrodes nanomaterials and nanostructures can sense small molecules like glucose. Bimetallic PtCu nanochains can be utilized to detect glucose very specifically in the presence of other biomolecules like ascorbic acid, uric acid, and dopamine. Hydrogen peroxide is of practical importance in bioanalysis, and food safety. Electrochemical sensor formed by deposition of Pd—Pt and Pd—Au NPs on spectrographic graphite is used for H_2O_2 reduction. Copper sulfide NPs-decorated rGO used for H_2O_2 sensor in human serum and urine samples and find utility in medical diagnosis. GO—CNTs—Pt nanocomposites have electrocatalytic capacities. Nitrogen and boron when doped together in graphene sheet has exceptional selectively to detect H_2O_2 when glucose and AA are also present [5].

3.3 Carbon nanofibers

Carbon nanofibers are one-dimensional nanomaterials and one among the class of carbon materials with large surface-to-volume ratio and good electrical conductivity. Carbon nanofibers are graphene layered cylindrical nanostructures with stacked cups, or plates. Fullerene—carbon nanofibers in paraffin oil have excellent voltammetric and amperometric responses in surface water to detect at trace levels of diclofenac, an antiinflammatory drug [33]. An oil-free carbon nanofiber can be used for the sensing of glucose levels when nickel is electrodeposited on it. Using cyclic voltammetry and amperometry measurements oxidation of glucose is evaluated with beneficial fast, low limit, and high selectivity [34].

3.4 Carbon ionic liquids

Ionic liquids are a class of materials with high electrical conductivity, have great applications in green chemistry. It replaces nonconductive organic binder. Among the known ionic liquid electrodes carbon ionic liquid electrode have excellent properties than graphite and pyridinium-based ionic liquid electrodes. Multiwalled and single-walled CNTs provide a conducting phase for the construction efficient composite electrodes [35].

3.5 Conducting polymer nanomaterials

Conducting polymer nanomaterials contains conjugated π-electron systems. They exhibit good optical and electrical properties. Polypyrrole, polyaniline, and polythiophene are some examples.

3.6 Graphene

Graphene is a type of 2D nanomaterials with π-conjugated carbon layers having one atom thickness. The oxidation of graphite oxide and functionalization on the surface of the graphene layers is beneficial. The multiple functional groups provide an enhanced layer separation and improved hydrophilicity. Graphene oxide is produced by the oxidation of graphite oxide and hydrophilicity allows many pleasing properties. Graphene has hexagonal crystal structure, which renders extraordinary thermal, electrical, and mechanical characteristics. Two-dimensional sensing material can interact with the analyte in different ways. The active sites on the sensor must be recognized by the complementary binding sites of the guest analytes by noncovalent interactions. In this case, site modifications can be made to some extend prior to synthesis. Electronic properties like electrostatic potential difference, permittivity, and conductivity parameters of the system may facilitate effective interaction between the analyte and sensing material in electroanalytical sensing [36]. Facile and simultaneous electrochemical estimation of bisphenol A, 8-hydroxy-2′deoxyguanosine, and hydroquinone in urine samples is possible by employing differential pulse voltametry using screen-printed carbon electrodes coated with graphene and MWCNTs in a specified concentration [37].

The carbohydrate-functionalized carbon nanotubes could be employed for the quick identification of microorganisms in water, soil, and human samples. Nanopores of multilayered graphene—Al_2O_3 functioned as DNA sensors. Electrodes functionalized using multilayered graphene is able to detect urea molecules by an ion channel mimetic mechanism [38]. Graphene-based macroscopic structure with flexible ribbons have good fiber solar cell performance and by tailoring its find applications as sensors, and electrodes [39]. The supercapacitors shows the CNT-grown graphene sheets provide excellent electrochemical performance while investigated using cyclic voltammetry through reliable conductive channels [40].

3.7 Carbon-based quantum particles

Quantum confinement is observable in carbon-based quantum particles like spherical carbon quantum dots and graphene quantum dots. Carbon-based quantum particles have size typically 1—2 nm and graphene quantum dots has a size less than 10 nm. Carbon quantum particles are a member of zero-dimensional nanomaterials and they exhibit biomedical and sensing applications within the time scale of certain minutes. Their photoluminescent capabilities vary with their size and HOMO-LUMO band gap. There is only morphological distinction between carbon and graphene quantum dots are possible. Personalized medicine has ardent applications and benefits of these quantum dots.

Electroanalytical properties of carbon-based quantum dots are adequate by means of their large surface area, electronic conductivity, and catalytic capabilities.

Electrochemical performance, mainly cyclic voltammograms largely dependent on pH and better results, are obtained in acidic conditions. Electrochemical sensing of hydrogen peroxide and glucose is possible based on carbon-based quantum dot-octahedral cuprous oxide matrix. The chronic disorder in human blood caused by high glucose content and subsequent kidney failures and heart problems can be timely monitored by fluorescence-based CQD sensors. Heavy metal contamination of water can be detected by fluorescence quenching by CQDs through surface bonding and an alteration in their fluorescence intensity can be seen.

Carbon-based electrodes have wider anodic and cathodic potential and are cost-effective. Carbon-based quantum dots is an essential components in novel high performing electrochemical sensors via cyclic voltammetric and differential pulse voltammetric analysis. Metal-ion sensing, dopamine sensing, and hydrogen peroxide in biological systems are possible with high demand of practical applications. Electrochemical biosensors can effectively detect Aah50, the toxic fraction of the *Androctonus australis hector*. The electronic transport between the SWNT and the copper ions substrate and active sites of enzyme can be improved by nanocrystalline CQDs [41].

4. Conclusions

Electrochemical analytical methods are cost-effective, and simple techniques for the determination of electroactive species in solutions. It has advantages over chromatography, and spectroscopy because of poor cost, high accuracy, and reliability. Electrochemical sensors open new avenues in the current research. Electrochemical biosensors enable on-site sensing applications which include health care, food quality control, and environmental monitoring in fields without much investment [42]. Industrial interests of electroanalytical devices include in conductive devices, energy storage, hydrogen storage, optical devices, and environmental monitoring systems.

References

[1] T. Wang, W. Yue, Carbon nanotubes heavy metal detection with stripping voltammetry: a review paper, Electroanalysis 29 (10) (2017) 2178–2189.
[2] N. Karthigayan, et al., Synthesis and characterization of NiFe2O4, CoFe2O4 and CuFe2O4 thin films for anode material in Li-ion batteries, Nanomater. Nanotechnol. 7 (2017), 1847980417711084.
[3] J. Hui, et al., Advanced electrochemical analysis for energy storage interfaces, Anal. Chem. 91 (1) (2018) 60–83.
[4] Z. Wang, et al., Glassy carbon electrode coated with polyaniline-functionalized carbon nanotubes for detection of trace lead in acetate solution, Thin Solid Films 519 (15) (2011) 5280–5284.
[5] C. Zhu, et al., Electrochemical sensors and biosensors based on nanomaterials and nanostructures, Anal. Chem. 87 (1) (2015) 230–249.
[6] A. Gholizadeh, et al., Fabrication of sensitive glutamate biosensor based on vertically aligned CNT nanoelectrode array and investigating the effect of CNTs density on the electrode performance, Anal. Chem. 84 (14) (2012) 5932–5938.

[7] Y. Qin, et al., Oxygen containing functional groups dominate the electrochemiluminescence of pristine carbon dots, J. Phys. Chem. C 121 (49) (2017) 27546–27554.
[8] P. Lutsyk, et al., A sensing mechanism for the detection of carbon nanotubes using selective photoluminescent probes based on ionic complexes with organic dyes, Light Sci. Appl. 5 (2) (2016) e16028.
[9] A.C. Power, et al., Carbon nanomaterials and their application to electrochemical sensors: a review, Nanotechnol. Rev. 7 (1) (2018) 19–41.
[10] M.E. Itkis, et al., Networks of semiconducting SWNTs: contribution of midgap electronic states to the electrical transport, Acc. Chem. Res. 48 (8) (2015) 2270–2279.
[11] J. Muñoz, et al., Synthesis of 0D to 3D hybrid-carbon nanomaterials carrying platinum (0) nanoparticles: towards the electrocatalytic determination of methylparabens at ultra-trace levels, Sensor. Actuator. B Chem. 305 (2020) 127467.
[12] Y. Lu, et al., A review of the identification and detection of heavy metal ions in the environment by voltammetry, Talanta 178 (2018) 324–338.
[13] S. Deshmukh, et al., Red mud-reduced graphene oxide nanocomposites for the electrochemical sensing of arsenic, ACS Appl. Nano Mater. 3 (5) (2020) 4084–4090.
[14] L. Fu, et al., Facile and simultaneous stripping determination of zinc, cadmium and lead on disposable multiwalled carbon nanotubes modified screen-printed electrode, Electroanalysis 25 (2) (2013) 567–572.
[15] M.A. Chamjangali, et al., A voltammetric sensor based on the glassy carbon electrode modified with multi-walled carbon nanotube/poly (pyrocatechol violet)/bismuth film for determination of cadmium and lead as environmental pollutants, Sensor. Actuator. B Chem. 216 (2015) 384–393.
[16] G. Aragay, J. Pons, A. Merkoçi, Enhanced electrochemical detection of heavy metals at heated graphite nanoparticle-based screen-printed electrodes, J. Mater. Chem. 21 (12) (2011) 4326–4331.
[17] A. Joshi, T.C. Nagaiah, Nitrogen-doped carbon nanotubes for sensitive and selective determination of heavy metals, RSC Adv. 5 (127) (2015) 105119–105127.
[18] B. Hayati, et al., Super high removal capacities of heavy metals (Pb2+ and Cu2+) using CNT dendrimer, J. Hazard Mater. 336 (2017) 146–157.
[19] K.-H. Wu, et al., Electrochemical detection of heavy metal pollutant using crosslinked chitosan/carbon nanotubes thin film electrodes, Mater. Exp. 7 (1) (2017) 15–24.
[20] D.W. Kimmel, et al., Electrochemical sensors and biosensors, Anal. Chem. 84 (2) (2012) 685–707.
[21] M.H. Mashhadizadeh, S. Ramezani, M.K. Rofouei, Development of a novel MWCNTs–triazene-modified carbon paste electrode for potentiometric assessment of Hg (II) in the aquatic environments, Mater. Sci. Eng. C 47 (2015) 273–280.
[22] M. Cuartero, M. Parrilla, G.A. Crespo, Wearable potentiometric sensors for medical applications, Sensors 19 (2) (2019) 363.
[23] L. Zhang, et al., Disintegrative activation of Pd nanoparticles on carbon nanotubes for catalytic phenol hydrogenation, Catal. Sci. Technol. 6 (4) (2016) 1003–1006.
[24] P. Yang, et al., Nano-silver/multi-walled carbon nanotube composite films for hydrogen peroxide electroanalysis, Microchim. Acta 162 (1–2) (2008) 51–56.
[25] M. Sebastian, A. Aravind, B. Mathew, Green silver-nanoparticle-based dual sensor for toxic Hg (II) ions, Nanotechnology 29 (35) (2018) 355502.
[26] Q. Wang, et al., One-Step fabrication of a multifunctional magnetic nickel ferrite/multi-walled carbon nanotubes nanohybrid-modified electrode for the determination of benomyl in food, J. Agric. Food Chem. 63 (19) (2015) 4746–4753.
[27] M. Sebastian, B. Mathew, Carbon nanotube-based ion imprinted polymer as electrochemical sensor and sorbent for Zn (II) ion from paint industry wastewater, Int. J. Polym. Anal. Char. 23 (1) (2018) 18–28.
[28] M. Sebastian, B. Mathew, Ion imprinting approach for the fabrication of an electrochemical sensor and sorbent for lead ions in real samples using modified multiwalled carbon nanotubes, J. Mater. Sci. 53 (5) (2018) 3557–3572.
[29] A. Aravind, M. Sebastian, B. Mathew, Unmodified silver nanoparticles based multisensor for Ni (II) ions in real samples, Int. J. Environ. Anal. Chem. 99 (4) (2019) 380–395.

[30] S.A. Aguila, et al., A biosensor based on Coriolopsis gallica laccase immobilized on nitrogen-doped multiwalled carbon nanotubes and graphene oxide for polyphenol detection, Sci. Technol. Adv. Mater. 16 (5) (2015) 055004.
[31] J.M.P.J. Garrido, et al., β—Cyclodextrin carbon nanotube-enhanced sensor for ciprofloxacin detection, J. Environ. Sci. Health A 52 (4) (2017) 313—319.
[32] O.S. Kwon, et al., Conducting nanomaterial sensor using natural receptors, Chem. Rev. 119 (1) (2018) 36—93.
[33] S. Motoc, et al., Enhanced Electrochemical response of diclofenac at a fullerene—carbon nanofiber paste electrode, Sensors 19 (6) (2019) 1332.
[34] M. Adabi, M. Adabi, Electrodeposition of nickel on electrospun carbon nanofiber mat electrode for electrochemical sensing of glucose, J. Dispersion Sci. Technol. (2019) 1—8.
[35] A. Safavi, et al., Comparative study of carbon ionic liquid electrodes based on different carbon allotropes as conductive phase, Fullerenes Nanotub. Carbon Nanostruct. 21 (6) (2013) 472—484.
[36] Z. Meng, et al., Electrically-transduced chemical sensors based on two-dimensional nanomaterials, Chem. Rev. 119 (1) (2019) 478—598.
[37] J.-S. Chen, et al., Multiwalled carbon nanotubes/reduced graphene oxide nanocomposite electrode for electroanalytical determination of bisphenol A, 8-hydroxy-2'-deoxyguanosine and hydroquinone in urine, Int. J. Environ. Anal. Chem. 100 (7) (2020) 774—788.
[38] H. Radecka, et al., Electrochemical Sensors and Biosensors Based on Self-Assembled Monolayers: Application of Nanoparticles for Analytical Signals amplification, in: Functional Nanoparticles for Bioanalysis, Nanomedicine, and Bioelectronic Devices, vol. 1, American Chemical Society, 2012, pp. 293—312.
[39] J. Sun, et al., Macroscopic, flexible, high-performance graphene ribbons, ACS Nano 7 (11) (2013) 10225—10232.
[40] R. Kumar, et al., Self-assembled hierarchical formation of conjugated 3D cobalt oxide nanobead—CNT—graphene nanostructure using microwaves for high-performance supercapacitor electrode, ACS Appl. Mater. Interfaces 7 (27) (2015) 15042—15051.
[41] K. Nekoueian, et al., Carbon-based quantum particles: an electroanalytical and biomedical perspective, Chem. Soc. Rev. 48 (15) (2019) 4281—4316.
[42] N. Wongkaew, et al., Functional nanomaterials and nanostructures enhancing electrochemical biosensors and lab-on-a-chip performances: recent progress, applications, and future perspective, Chem. Rev. 119 (1) (2018) 120—194.

CHAPTER 8

Magnetron sputtering for development of nanostructured materials

Ajit Behera[1], Shampa Aich[2], T. Theivasanthi[3]
[1]Department of Metallurgical & Materials Engineering, National Institute of Technology, Rourkela, Odisha, India; [2]Department of Metallurgical & Materials Engineering, Indian Institute of Technology, Kharagpur, West Bengal, India; [3]International Research Center, Kalasalingam Academy of Research and Education, Krishnankoil, Tamil Nadu, India

Chapter Outline

Design, Fabrication, and Characterization of Multifunctional Nanomaterials. https://doi.org/10.1016/B978-0-12-820558-7.00002-9

Abbreviations

AC	alternating current
Ag	silver
Al	aluminum
Al_2O_3	Aluminum oxide
Ar	Argon
Au	gold
CD	compact disk
Co	cobalt
Co-Cr	cobalt-chromium
Co-Pt	cobalt-platinum
Cr–Ni	chromium-nickel
Cu–Ni	copper-nickel
CVD	chemical vapor deposition
DC	direct current
DVD	digital versatile disc
ECR	electron cyclotron resonance
eV	electron Volt
Fe	iron
Fe_2O_3	iron(III) oxide
Ge	germanium
GHz	giga Hertz
MBE	molecular-beam epitaxy
MEMS	microelectromechanical system
MHz	megahertz
Mo	molybdenum
N_2	nitrogen
NbN	Niobium nitride
Ni–Cr	nickel-chromium
Ni–Ti	nickel-titanium
O_2	oxygen
RF	radio frequency
Si	silicon
Si_3N_4	Silicon nitride
SiO_2	silica
Sn	tin

TiN	Titanium nitride
TiO_2	titania
W	tungsten
WC	Tungsten carbide
WO_3	tungsten trioxide
Zn	zinc
ZnO	zinc oxide

1. Introduction

Demand on the applications of nanostructured materials in various industries is due to the unbeatable nanotechnology that provides unique characteristics retrieves from the dimensional range 1−100 nm. Nanomaterials constitute the nanostructured materials by the help of nanoparticles, nanolayers, nanofiber, nanocomposite, fullerenes, nanotube, nanowires, nanopillars, and quantum dots [1,2].

The applications of the nanostructure materials cover a wide area in chemical industries, biomedical industries, aviation industries, electronic industries, automotive industries, and household appliances (manufacturing) industries. The focus point is on how to fabricate nanostructured materials economically with respect to the application, mostly either in physical vapor deposition method or chemical vapor deposition method or hybridized method. A number of attempts has been carried out depending on the top-down approach or bottom-up approach. The top-down approach includes mechanical attrition, mechanochemical milling, lithography, laser ablation, sputtering, flame pyrolysis, electrospinning, and electrospraying [3,4].

The bottom-up approach includes thermo-chemical vapor deposition, plasma-assisted deposition, molecular beam epitaxy, sol-gel technique, solvo-thermal synthesis, chemical reduction and self-assembly techniques [5,6]. Among all the techniques, magnetron sputtering is the widespread process fully scalable for mass production and widely adopted in industries and research sectors.

2. What is magnetron sputtering?

Sputtering technique is one of a physical vapor deposition processes to develop the thin film, was pioneered in 1852 and is gained commercial success in the 1960s in microelectromechanical system industry [7]. In this process the atomic-scale materials ejected from the target materials is deposited on the exposed surface of substrate. The magnetron is generally employed as the sputtering source now a day.

Magnetron utilizes strong electric and magnetic fields and plays a dominant role to guide the plasma particle toward the target material. The powerful magnet is responsible to confine the glow discharge plasma to the region closest to the target. In magnetic field, the electron follows a helical path around the magnetic line of force, experiences more ionizing collisions with the gaseous neutrals near the target. Here the plasma is created and the positively charged ions from the plasma are accelerated by an electrical field and strike the negative electrode (target) with a sufficient force to eject atoms. The atoms ejected from the target have a wide energy distribution up to tens of eV (100,000 K) [8].

The path of the atom from target-to-substrate depends on vacuum condition, degree of ionization, power level, gas pressure, substrate bias etc. Only a small fraction of the ejected particles are ionized in the path, i.e., up to 1%. During sputtering, 95% of incident energy goes into the target and only 5% of incident energy is carried off by target atoms (typical energy of 5−100 eV) [9].

These fractions of highly energized atom balastically move in a straight line to deposit on the substrate. Also secondary electrons are generated from the target, accelerated away from the target and have a main role to retain the ionization in plasma. The ballistic impact of atom to the substrate can be controlled by changing the background gas pressure. In nonreactive sputtering process, generally inert Ar is taken as sputtering gas. In magnetron sputtering, if the target material is conductive, then it is connected to direct current (DC) power supply and if the target material is insulator then it is connected to radio frequency (RF) power supply. Commercially, the magnetron sputtering sources are available in a variety of geometric forms such as tubular, circular, and rectangular. The deposition rate in magnetron sputtering can be improved by maintaining the higher density of ions. It improves ionization process that leads to the increasing of the sputter yield ultimately. The sputter yield can be calculated by the ratio of the number of atoms knocked out from the target to the number of the incident atom to the target [10].

A typical magnetron sputtering process is given in Fig. 8.1. The sputter yield influenced by the incident ion energy, the total size (masses) of the ions, the binding energy of atom at the target surface, sputtering gas, sputter power, substrate temperature, substrate bias, and the angle of incident ions. Fig. 8.2A and B show the sputtering deposition process and sputtering mechanism respectively.

3. Market size of magnetron sputtering

The global magnetron sputtering system market is anticipated to increase as per the market demand with an appreciable rate during the forecast period from the year 2020 to 2027 [11]. Global sputtering targets and evaporation materials market growth rate is expected to increase in the coming years due to the increasing demands of several industries especially from electronics, automobile part, energy, optical, biomedical and other related industries. The sputtering targets and evaporation materials market revenue analysis for the year 2017 has been estimated to be $3036 million. Again, the predicted value from the year 2018 to 2023 grows at a CAGR of 2.2% to reach up to $3.4 billion [12].

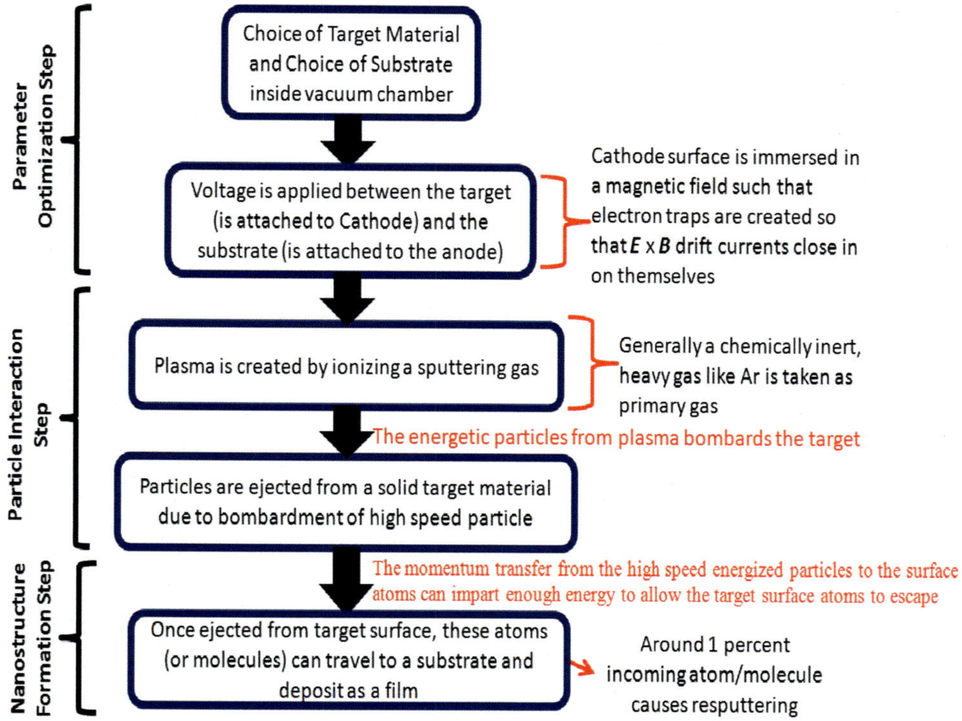

Figure 8.1
Steps involved with sputtering deposition.

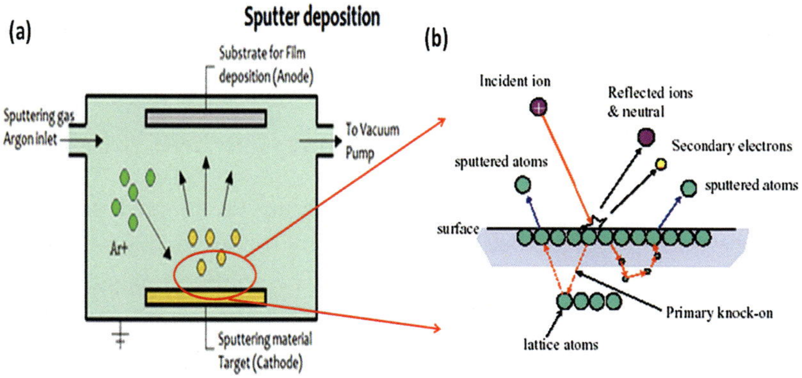

Figure 8.2
(A) Sputtering chamber at the time of deposition, (B) mechanism of deposition (sputtering/resputtering atoms).

Currently, in this year projected horizon, the market was raising at a constant rate and with various key players. The current key players in magnetron sputtering system market report are: Showa Shinku, Shincron, Torr International Inc., Von Ardenne, Buhler Leybold Optics, ULVAC, Hongda Vacuum, Angstrom Engineering, SKY Technology and Angstrom Sciences Inc. Sputtering targets and evaporation material coatings has great demand (with wide range of applications) in semiconductor components, automobile mirrors, displays, industrial tools, sensors, data storage devices, batteries, photovoltaic cells, optical instruments, decorative products, and for enhancing the properties of metal surface [13].

4. Advantages of magnetron sputtering

Various advantages are given below due to which magnetron sputtering have high demand when comparing to other deposition processes [14,15]:

- Compatible to develop the nanostructured materials using variety of materials such as metals, alloys, oxides and composite materials.
- Gives a dense film as comparison to other deposition processes.
- Extremely high adhesion of films with the substrate.
- High deposition rates.
- High-purity of the film formation.
- Possible to sputter on the heat-sensitive substrate.
- Magnetron sputtering process can operate at lower pressures.
- Can be modified the substrate surface by the reaction (reactive plasma deposition).
- Gives excellent uniformity in large-area substrates like architectural glass.
- The top-down approach performance of sputtering can be adopted in the advanced epitaxial growth process.
- The target of the sputtering process provides a stable, long-lived ionization source.
- The space adjustment between the source and the substrate is possible.

5. Magnetrons sputtering techniques in nanostructure fabrication

The use of external magnetic field is mainly differentiating the magnetron sputtering from glow discharge. The magnetic field can be developed by the application of permanent magnet or electromagnet or combination of the both. The confined magnetically plasma is responsible to sputter the atom from cathode target. There are several techniques of magnetron sputtering deposition utilized for the fabrication of nanostructured materials which are discussed below:

5.1 DC magnetron sputtering processes

Direct current magnetron sputtering is generally used for the deposition of metal, alloy and compound nanostructure. Here the magnetron is applied to deposit conducting materials by DC sputtering technique. The cost involved in this technique is lower than the RF sputtering, due to simple design and manufacturing. DC power consumption is depending on the size of the target. DC magnetron sputtering is well adopted for the fabrication of a microelectromechanical system (MEMS) due to the factors: a conductive tracks in direct contact with the Si-wafer; easy compositional control; uniform deposition throughout the cross-section [16].

5.2 RF magnetron sputtering

DC magnetron sputtering is unable to sputter the low conducting materials and insulator materials, because of nonpassing of the current through the target materials and the problem of charge accumulation. This limitation can be efficiently replaced by using RF magnetron sputtering. To overcome this stage, the alternating current with high frequency is required. In RF magnetron sputtering, the bias of anode-cathode is varied at a high rate. During sputtering at high frequency fluctuating field, the produced ion and electron have dissimilar mobilities that cover the different distances in each half cycle. The matching of impedance is a requirement in RF magnetron sputtering to ensure the maximum power is gained into the plasma.

RF magnetron sputtering has the advantages to deposit not only insulator but also conducting materials, polymer, ionic and covalent compounds to sputter away at rates depending on their specific sputtering yield. The sputter yield of ionic and covalent compound materials is much lower than the metallic materials. Dielectric materials can be deposited with sputtering yields only 10% of the metals. The power consumed for ceramic or insulator material deposition by RF is much less when compared to DC magnetron sputtering. Therefore, the deposition rate for insulator is very much lower than DC magnetron sputtering. When comparing to DC magnetron sputtering equipment, RF magnetron sputtering equipment is very complex and more expensive [17].

5.3 Ion-beam magnetron sputtering

Ion-beam sputtering process has a provision to control the energy as well as the flux of ions independently. Here the target is external to the ion source. The ion source is generated from the hot filament ionization gauge without use of any magnetic field.

Then the ionized matter is confined by using magnetic field present in the magnetron and accelerated by electric field flows from a grid toward a target. Once the ion left the source, they are neutralized by electrons from a second external filament. These neutralized atoms bombard on the target surface, which can eject the surface atoms either from the insulating or conducting targets. There is a pressure difference generated between the chamber of ion generation and the chamber of substrate. The inlet of gas is present at the source chamber. So, there is less of a chance of contamination. This type of magnetron sputtering device required more maintenance [18].

5.4 Reactive magnetron sputtering

In reactive magnetron sputtering process, the incoming atoms from the target undergo a chemical reaction due to the presence of reactive gas inside the sputtering chamber. Generally various reactive gases can be introduced such as O_2 and N_2 to develop the thin films such as oxide, nitride as well. The reactions occur between the path from target surface and the substrate surface just before deposition. The chemical composition of the film depends on the pressure of the primary inert gas and reactive gas [19].

5.5 Inductively coupled plasma-magnetron sputtering

This is an advanced version of DC/RF magnetron sputtering coupled with nonresonant induction coil. Inductively coupled plasma is used to increase the ionized discharge in the magnetron sputtering. Inductively coupled plasma coil is placed in between the cathode target and the anode substrate, parallel to the substrate. A typical inductive call operate at 13.56 MHz using a 50 Ω RF power supply. The Inductively coupled discharges are generally controlled in the pressure range 1–50 mTorr and applied power 200–1000 W results around 1016–1018 m^{-3} electron density. The ionized discharge increases linearly when the power and pressure increase. The ion flux and the energy of the ions are influenced by the applied rf power and the substrate bias independently. This is the main advantage over a conventional DC magnetron sputtering discharge where the ion flux and the ion bombarding energy cannot be change independently without changing the magnetic configuration [20].

5.6 Microwave amplified magnetron sputtering

Another way to increase the ionization in magnetron sputtering is to introduce a supplementary electron cyclotron resonance (ECR) in general DC magnetron sputtering. This process can generate high plasma densities (1017–1018 m^{-3}) generally using lower working pressures (0.1–10 mTorr) with low plasma potential. The microwave generator is present at the place between the cathode and the anode for which it is possible to increase, an electron density of the order 1 m^{-3} × 1017 m^{-3}. The magnetic coils are present around the chamber periphery to achieve ECR conditions. The ECR discharges are generally controlled at a microwave frequency around 2.45 GHz with a strong magnetic field. The deposition rate in microwave amplified magnetron sputtering can be increased with the microwave power and also increasing discharge gas pressure [21].

5.7 High power impulse magnetron sputtering

The maximum power in conventional DC magnetron sputtering is limited due to the thermal load on the target given by the incoming high energized positive ions. To overcome this limitation, the power is introduced in pulses with high power densities of the order of kW/cm^2. By reducing the duty cycle (on-time divided by the cycle-time), a corresponding increase in power during the on-time can be achieved. By this process high plasma densities can be produce with the order of 1019 m^{-3} [22].

6. Magnetron sputtering for fabrication of NiTi smart materials

Among the advanced material systems, the smart-structural system has high demand in structural and functional applications in various industries from aerospace industry to medical industry. By the expected applications for MEMS, various techniques were tried for the fabrication of Ni–Ti thin films. From a practical point of view, magnetron sputtering results a perfect shape memory effect in thin film as that of bulk materials [23].

In smart material sputtering process, utilization of inert gas like Ar is essential for the formation of the thin film. Properties such as transformation temperatures, shape memory behavior and pseudoelasticity of the sputtered NiTi films are sensitive to target composition, target power, gas pressure, target-to-substrate distance, deposition temperature, substrate bias, substrate rotation, good configuration of target, and the deposition environment. DC and RF magnetron sputtering process is used for NiTi shape memory thin film production, in which separately, Ni is connected with the DC and Ti is connected with the RF power to maintain the sputter yield [24]. The nanostructured film composition is controlled from separate targets by confined power ratio. It is very important to limit the impurities (typically oxygen, hydrogen, and carbon) to prevent the loss of shape memory effect and to prevent brittleness and deterioration.

7. Variation of parameters in magnetron sputtering deposition

7.1 Ar gas pressure

By increasing the Ar pressure, the resultant nanostructure gives a low dense thin film with fewer structural defects. At very low working pressures, there are fewer collisions and the adatoms arrive at the substrate surface with high energies and results atomic peening. By atomic peening, the surface atoms are pushed into the interior of the film, increasing the density with creating the compressive stresses. By gradual increase in gas pressure, the momentum of the adatom atoms decreases by higher collision, energy decreases, peening effect decreases and the compressive stresses decreases. Further increase in pressure, decreases density of the film, increases the tensile intrinsic stress of film as well. High gas pressure reduces the energy of adatom and leads to low surface mobilities. The film grows in a columnar fashion and results in low dense film [25].

7.2 Target-to-substrate distance

The target-to-substrate distance suggests the energies of adatoms coming from the target and influences the deposition efficiency, film density, structural integrity, and stress. The small target-to-substrate distances with low gas pressure produces a film of intrinsic compressive stresses which become tensile as the Ar pressure increases. This distance should be maintained in accordance with the gas pressure. Increasing of target-to-substrate distance increases the gas pressure. On the other hand, decreasing the distance lowers the gas pressure [26].

7.3 Substrate temperature

Higher substrate temperature leads to lower the cooling rate and higher diffusivities of adatoms which results in partial crystallization in the films. At low substrate temperature, the film consists of a continuous network of uniformly wide and homogeneously distributed grain boundary like structures. This can be attributed to the lack of surface migration of adatoms to form aggregates which lead to the porous boundaries. The increase in the grain size occurs at high substrate temperature. At higher substrate temperature, the surface diffusion of adatoms increase results in denser and compact film. It is also observed that few grains are more elevated than the neighboring grains along with a reduction in density of grain boundaries [27].

7.4 Substrate bias

Most of the case sputtering deposition efficiency depends on the substrate bias. Ion bombardment during the film growth process is usually achieved in the sputtering process by means of applying small negative bias voltage at the substrate end in order to improve the quality of sputter deposited films. It is expected that the substrate-bias voltage might have a considerable effect on the film crystallinity, microstructure as well as the resputtering rate from the film. It is customary to introduce bias to the substrate during film deposition. Throughout the process, the low energy ion bombardment of the growing film may preferentially remove loosely bonded adatoms from the surface and increase their diffusion on the substrate [28].

7.5 Substrate rotation

Substrate rotation is required for uniform distribution of adatoms. The rpm depends on the velocity of an atomic particle launching toward the substrate. Substrate rotation has a significant effect on structural and mechanical properties. In recent years, the possibility to grow nearly arbitrary shaped nanostructures by oblique incidence deposition under controlled substrate rotation in a process commonly known as glancing angle deposition has gained significant attention due to the manifold of possible applications such as photonic crystals, polarizing filters, and humidity sensors or pressure sensors [29].

7.6 Degree of ionization

The ionization rate increases by increase in applied power as well as increase in pulse length. Additional equipment may be coupled in between the cathode and anode like inductive coil or microwave coil to improve the degree of ionization. The comparison suggests an even higher ion to neutral ratio for the high-induced-plasma discharge. The calculation of the ionized flux fraction has also been made by weight gain differences on a floating and a positively biased substrate [30].

7.7 Deposition rate

In the high impulse discharge magnetron sputtering, the deposition rate is generally found to be lower than in a conventional DC magnetron sputtering discharge at the same average power. The reduction in deposition rate is due to the presence of sputtered material close to the target. In this case, many of the metallic ions will be attracted back to the target surface by the cathode potential [31].

In the presence of a confining magnetic field, as in magnetron sputtering, not all the applied voltage drops over the sheath region that forms in front of the target, but a fraction of the applied voltage will penetrate the bulk plasma and create a potential gradient inside the plasma, referred to as the plasma presheath. If a sputtered atom is ionized inside this region, it must have enough kinetic energy to escape this potential in order to reach the substrate. However, if the ion is attracted back to the target it will act as a sputtering particle. A reduction in the deposition rate is then expected to occur especially for metals with a low self-sputtering yield. The deposition rate can be optimized by varying the magnetic confinement [32].

7.8 Stress

Stress in the film plays a dominant role in film/substrate adhesion. The main factors responsible for the stress evolution in nanostructured materials is due to: the thermal expansion mismatch between developed nanostructure and the substrate causes the intrinsic tensile stress; the small density difference between amorphous and crystalline such as in NiTi. A wide range of residual stress levels (either tensile or compressive) was found in the sputtered films. Large residual stress could lead to either film cracking or delamination under tension and buckling under compression [33].

8. Nanocomposite coatings by magnetron sputtering

By sputtering deposition, there are two way to fabricate the nanocomposite: one is codeposition, and second one is multilayer deposition followed by annealing. In codeposition process, several targets are exposed to plasma to deposit simultaneously on the substrate. In multilayered deposition process, only one target is exposed to plasma and all other targets are covered by the shutter. After one by one nano-scale deposition of the target, next target starts to deposit. In this way alternate nanolayer deposition happens. Once deposition finished, annealing started in vacuum to homogenize the thin film throughout the whole thickness. Vacuum is necessary to avoid oxidation. A little diffusion already happened in the thin film, due to three reasons: setting the substrate temperature during magnetron sputtering deposition; increasing of bias to the substrate; during sputtering/resputtering interaction [34].

Khan et al. have prepared Ag-Permalloy (Ni76Fe19Ag5 and Ni72Fe18Ag10) thin films using DC reactive magnetron sputtering method in 5n (99.999%) pure Ar gas atmosphere. Also, they have prepared the Ni72Fe18Ag10 thin film in Ar + N_2 gas atmosphere (with aggregate flow rate 60 sccm and total pressure 0.55 Pa). The films have been deposited on the flexible polymer substrate, i.e., polyethylene terephthalic ester (PETE) using mosaic alloy target at room temperature. To prepare flexible soft magnetic thin films, polymer substrate is more suitable than glass substrates (brittle and nondeformable). By comparing these films preparation methods, it is observed that the reactive gas (N_2) makes changes in the chemical composition of the thin films. Further, addition of N_2 reduces the average film deposition rate. In reactive nitrogen atmosphere, average film deposition rate is 1.8 nm/min and in Ar atmosphere 2.05 nm/min [35].

SEM images and EDS analysis of the as-deposited (done with 115 W sputtering power) thin films on PETE substrate are shown in Fig. 8.3. Thin films deposited in Ar gas only are shown in Fig. 8.3A and B. Their average grain size is 25.0 and 23.2 nm respectively. The surface of the films is smoother. Fig. 8.3C shows the film deposited in Ar + N_2 gas

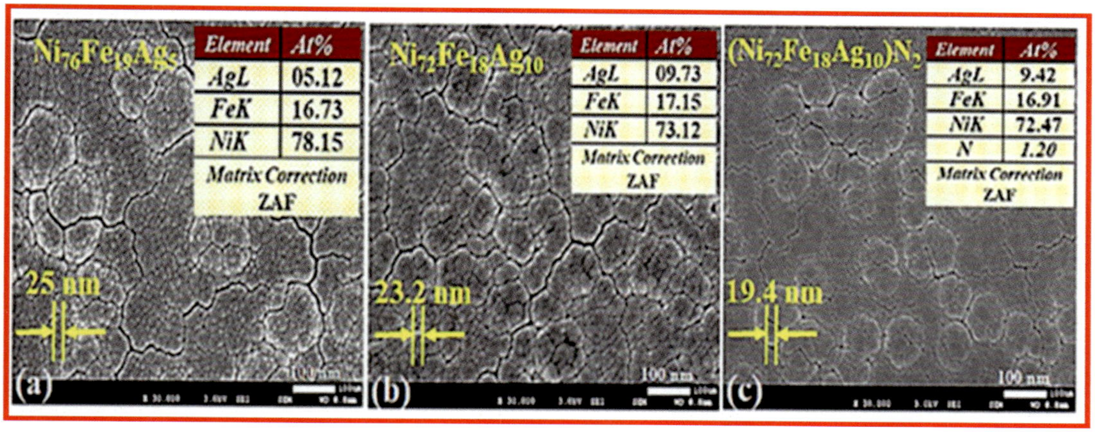

Figure 8.3
SEM images and EDS analysis of nanostructured silver doped permalloy thin films deposited at 115 W. (A) $Ni_{76}Fe_{19}Ag_5$; (B) $Ni_{72}Fe_{18}Ag_{10}$ and (C) $(Ni_{72}Fe_{18}Ag_{10})N_2$. Ar gas has been used for deposition of sample (A) and (B). For sample (C), Ar + N_2 gas has been used [35].

mixture. It shows that the grains have fine surface and their size is 19.4 nm. It is observed from the EDS analysis of Fig. 8.3C that nitrogen has mixed. The reactive gas (N_2) has modified the chemical composition of the film [35].

9. *Applications*

9.1 *Electronic industries*

Magnetron sputtering uses to produce the metallizing for microelectronic circuits and chip carriers, electrical resistance films (Ni—Cr for strain gauges). Magnetic films are another important area for magnetron sputtering. Magnetic thin films are used predominantly for read-write storage of data for computer equipment. Storage magnetic films (Co, Fe, Co—Pt, Co—Cr, Cu—Ni) for magnetic storage devices, floppy discs, tapes and thin-film magnetic heads, opto-storage devices (compact discs and video discs), corrosion-resistant films (Cr—Ni), bonding for interlayers, hermetic seals, brazing for lead and gas sensors [36,37].

Ion-beam sputtering has found application in the manufacture of thin-film heads for disk drives. Magnetron sputtering is also can be seen in computer hard disk, gate dielectric, thin-film transistors, low-emissivity coatings on glass, interlayer dielectric, on-chip interconnects and interlevel bias, passive thin film components, printed circuit boards, sensors, surface acoustic wave devices, double-pane window assemblies, super mirror as well [38,39].

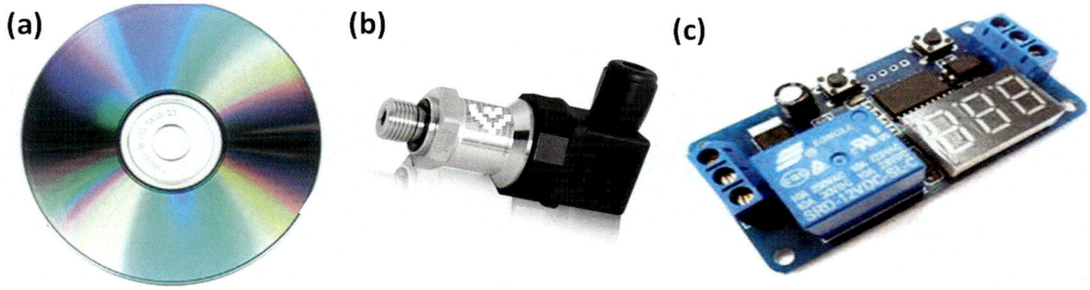

Figure 8.4
(A) Sputter deposited CD for magnetic storage application, (B) thin-film coated sensor, (C) integrated circuits.

Sputtered NiTi film acts as an antivibration damping structure. It is responsible for minimizing the vibrations during the operations of the hard disk drive [40]. Sputtering is one of the main processes of manufacturing optical waveguides. It is applied for making of efficient photovoltaic solar cells [41]. Sputtering is also used as the process to deposit the metal (e.g., Al) layer during the fabrication of CDs and DVDs (Fig. 8.4A). Sputter deposited sensor and integrated circuit are given in Fig. 8.4B and C.

9.2 Automobile and aerospace industries

In automobile industries, the magnetron sputtering deposition can be seen in automobile wheels, rims, mirror, headlights and taillights, auto trim components, drive train bearings and components which proved to be a better surface resistance application. Auto car wheel, plastic/metal cutlery are used chrome magnetron sputtering of vacuum coating machine [42]. In aerospace industries, the magnetron sputtering deposition is useful for heads-up cockpit displays, jet turbine engines, mirrors for optical and X-ray telescopes, and night vision equipment [43]. Some of the magnetron sputtered automotive and aero-industries application have shown in Fig. 8.5.

9.3 Energy harvesting industries

In energy harvesting industries, sputtering deposition is utilized in gas turbine blade coatings, solar panels as well [44]. Though other processes are used for the production of the silicon, sputtering is widely used to produce a back reflector layer (made predominantly with Al) and front contacts (made with transparent conducting oxides) [45]. In addition to the traditional Al and transparent conducting oxides coatings, these technologies employ additional sputtered layers such as Mo and intrinsic ZnO [46].

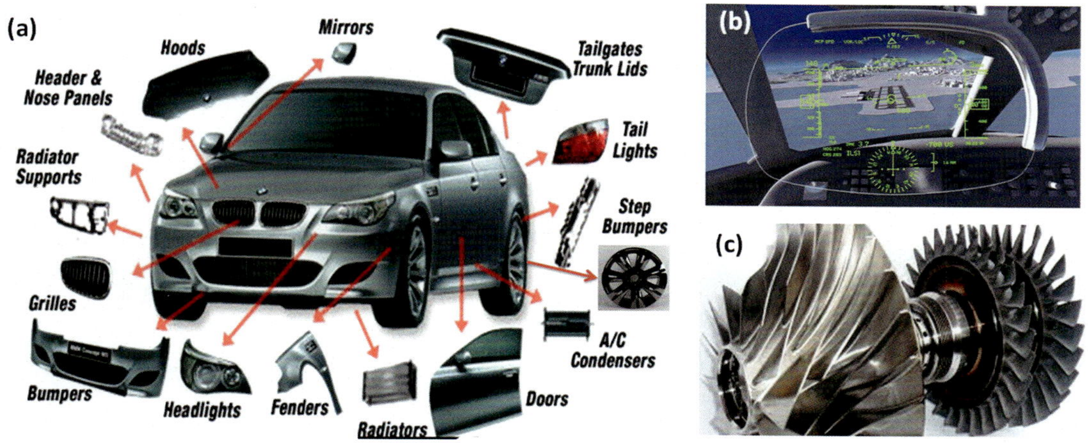

Figure 8.5
Sputter deposited (A) interior and exterior trim components in automobile, (B) heads-up cockpit displays of aerocraft, and (C) surface area of aero-turbine engine.

Intrinsic ZnO can be produced using midfrequency AC sputtering at much higher rates than the traditional RF technology used with planar cathodes. The magnetron sputtering is also used to fabricate: (1) antireflective coatings (to allow to solar energy to pass through protective coverings such as glass), (2) reflective coatings for solar collectors, and (3) absorbing coatings to absorb the maximum solar radiation [47]. Fig. 8.6 shows the solar panel prepared by utilizing the magnetron sputtering deposition technique.

9.4 Biomedical industries

Generally, biomedical equipments are biocompatible, wear resistance, chemically inert, and very hard in nature. Magnetron sputtering is used to deposit diamond-like carbon coatings deposited on other materials in order to provide many of the beneficial properties

Figure 8.6
Magnetron sputter deposited solar panel.

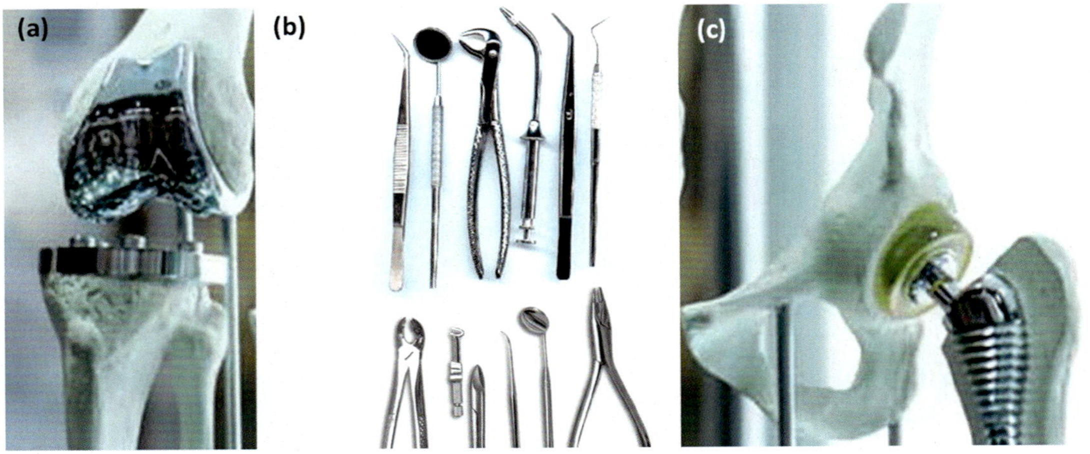

Figure 8.7
Hard surface deposition by RF magnetron sputtered (A) knee implant material, (B) surgical instruments, and (C) hip implant.

which minimizes friction [48]. Generally, Ti or Ti-based alloys and other biocompatible materials are used to sputter on the medical instruments [49]. Nitinol-smart material coated on the knee implant, hip implant, and surgical instrument using magnetron sputtering are shown in Fig. 8.7.

9.5 Textile industries

Magnetron sputtering technology can deposit metal or nonmetal films on the surface of textiles (substrate) such as polyester, cotton, linen, silk, wool, polyamide, polylactic acid, and polypropylene. This kind of film deposition is done by selecting appropriate sputtering process, different target materials and ambient gases [50]. The main structures of the textiles substrate include woven fabrics, knitted fabrics, and nonwoven fabrics. Here the sputtering target materials are metals (Cu, Ti, Ag, Al, W, Ni, Sn, Pt, etc.) or nonmetals (Si, graphite, etc.) as well as metal oxides (TiO_2, Fe_2O_3, WO_3, ZnO etc.) and nonmetal oxide (SiO_2, graphene oxide, etc.) [51,52].

Also, ceramic materials and single or multi-layer composite nanofilms formed with polymers (such as polyimide and polytetrafluoroethylene) can be deposited using this method. It have single or compound functions such as electromagnetic shielding, UV protection, antistatic, antibacterial, conductive, or waterproof properties [53]. A metal coated fibers and a heat resistance kitchen gloves prepared by magnetron sputtering deposition has been shown in Fig. 8.8.

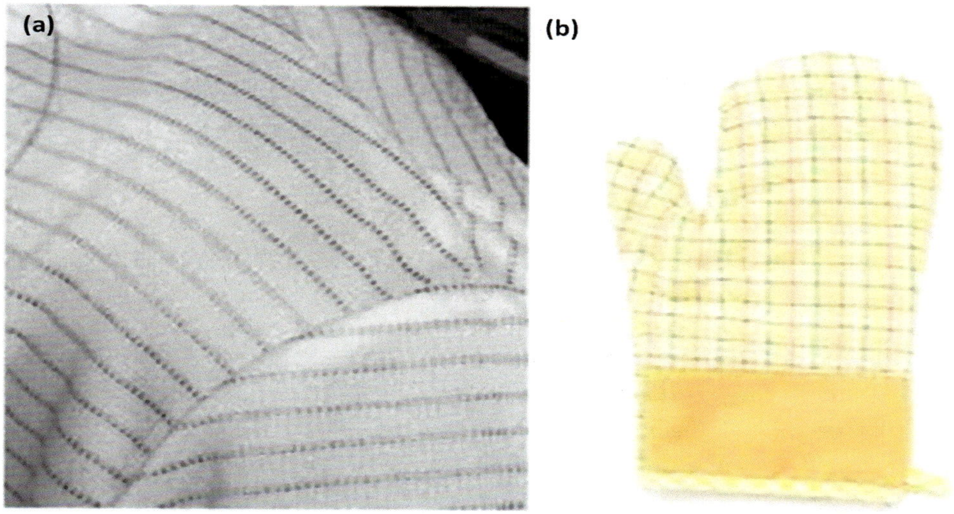

Figure 8.8
(A) Metal coated fibers by magnetron sputtering, (B) silicon sputtered heat resistant kitchen gloves.

9.6 Other industries

Sputtering technology spread a large area in glass coating either developing the transparent or opaque film. A large industry has developed around tool bit coating using sputtered nitrides, such as TiN, creating the familiar gold colored hard coat [54]. DC sputtering can be used for bonding films and other metallurgical films. Sputtered films of Cr, Ti, Ni, Cu, and Au have been applied successfully for the subsequent soldering and brazing treatments onto glass or ceramic substrates [55].

Superconducting films such as NbN, Si_3N_4, and mixed oxide compositions can be conveniently prepared by RF sputtering [56]. Wear-resistant coatings can be deposited by RF magnetron sputtering from compound targets or by reactive sputtering. Material such as WC, Si_3N_4, $A1_2O_3$, and TiN can be sputtered to give very adherent hard coatings of thicknesses up to several micrometers [57]. Comparing to the CVD and MBE, magnetron sputtering is also an alternative method to grow GeSn alloys for optoelectronic applications [58].

A high-quality SiGeSn/GeSn multiple quantum well structure have also been grown on Ge substrate by sputtering epitaxy. In this structure, 6% Sn content of the GeSn layer mainly contributes for the functions of GeSn photo-detector [59]. Some sputtering applications used in optics industries are antireflective/antiglare coatings, cable communications, laser lenses, optical filters for achromatic lenses, spectroscopy; wear resistance coatings: anticorrosion coatings, antiseize coatings, dies and molds, sewing needles, tool and drill bit hardening; in decorative: appliance trim, building glass, building hardware, clothing, jewelry, packaging, plumbing fixtures, and toys [60,61].

10. Limitations of magnetron sputtering

Sputtering is relatively simple, versatile process and most popular deposition technique. However, it has some limitations which are discussed below [62–64]:

- It contains the expensive targets, which limits the scope of their applications.
- A particular designed target is necessary to develop according to the sputtering machine to carry out the sputtering process.
- After few depositions, an erosion track is found on the target. For further deposition, it gives poor efficiency of sputtering yield.
- The target utilization can be increased by flattening the magnetic field lines parallel to the target surface.
- Most of the energy incident on the target becomes heat, which must be removed.
- In reactive sputter deposition, the gas composition must be carefully controlled to prevent poisoning the sputtering target.
- Compressive stresses may be generated in films during sputtering caused by the energetic species bombarding the substrate. In extreme cases, such stresses may cause the cracking and the delamination of the film from the substrate.
- Nonlinearity, complex thermomechanical behavior, ineffectiveness for precise and complex motion control, and force tracking.
- Sputtering is performed using a vacuum chamber in which the target(s) and the substrate are placed; thereafter the chamber is evacuated to vacuum.

Despite these excellent properties, full integration of smart alloy thin films into MEMS is limited. Their stoichiometry sensitivity has restricted the manufacturability and reliability of these materials in MEMS. The large exothermic heat of transformation limits the cycle lifetime. The poor machinability limits the many of its applications.

11. Influencing factors in magnetron sputtering

It is analyzed and observed from the various literatures that the magnetron sputtering depends on several factors such as angle/energy of the incident ions, total mass of the ions, binding energy of atom at the target, sputtering gas/power and substrate temperature/bias. Some of the factors are influencing the sputtering yield/output. They are enumerated below.

- ➢ Maintaining the higher density of ions improves the ionization process and sputter yield/deposition rate in magnetron sputtering.
- ➢ The ionized discharge can be increased by increasing the power and pressure.
- ➢ Increasing the microwave power and gas pressure increases the deposition rate in microwave amplified magnetron sputtering.
- ➢ The pressure of primary inert gas and reactive gas (O_2 and N_2) determines the chemical composition of oxide/nitride thin films produced in reactive magnetron sputtering process.
- ➢ Increased target-to-substrate distance leads to increase gas pressure.
- ➢ Increasing Ar gas pressure decreases the density and structural defects of the resultant thin films.
- ➢ Reducing the impurities (oxygen, hydrogen, and carbon) during smart material film formation improves shape memory effect and prevents brittleness and deterioration.
- ➢ Low conducting materials and insulator materials cannot be sputtered in DC magnetron sputtering due to nonpassing of the current in target materials and charge accumulation. Utilization of the RF magnetron sputtering and high frequency AC solve this limitation.
- ➢ Energy of ions and flux of ions can be controlled in ion-beam sputtering process.
- ➢ The RF power and substrate bias have influences on the ion flux/energy of the ions.
- ➢ High energized positive ions limit the maximum power in DC magnetron sputtering by thermal load on the target.
- ➢ High substrate temperature supports to yield denser and compact film.
- ➢ Applying of small negative bias voltage at the substrate end improves the quality of sputter deposited films.
- ➢ Low distance between sputtered material and the target attracts the metallic ions back to the target. It leads to the reduction in deposition rate (can be tuned by magnetic confinement).

12. Conclusion

This chapter has given a brief description on magnetron sputtering and its market demand over the manufacturing of thin films and nanostructured coatings. A distinct number of advantages give a unique position of sputtering deposition among all other deposition processes. It has been demonstrated that there are various magnetron sputtering techniques such as DC magnetron sputtering processes, RF magnetron sputtering, ion-beam magnetron sputtering, reactive sputtering, inductively coupled plasma-magnetron sputtering, microwave amplified magnetron sputtering, high power impulse magnetron sputtering, and high-target-utilization sputtering. NiTi smart materials fabrication using magnetron sputtering has been discussed. Various process parameters are associated practical problems are depicted briefly. Formation of nanocomposite using two different way of magnetron sputtering has been discussed. A wide range of application from micro/nanoelectronics, smart structure to surface modifications of textile has been discussed with some of its process limitations.

Acknowledgment

The authors express thanks to their colleagues for the assistances in this work. Also, they acknowledges the assistances of National Institute of Technology, Rourkela-769008, (India), Indian Institute of Technology, Kharagpur-721302, (India) and International Research Center, Kalasalingam Academy of Research and Education (Deemed University), Krishnankoil-626 126, (India).

References

[1] A. Rani, R. Reddy, U. Sharma, P. Mukherjee, P. Mishra, A. Kuila, L.C. Sim, P. Saravanan, A review on the progress of nanostructure materials for energy harnessing and environmental remediation, J. Nanostruct. Chem. 8 (2018) 255–291, https://doi.org/10.1007/s40097-018-0278-1.

[2] P. Samyn, A. Barhoum, T. Öhlund, A. Dufresne, Review: nanoparticles and nanostructured materials in papermaking, J. Mater. Sci. 53 (2018) 146–184, https://doi.org/10.1007/s10853-017-1525-4.

[3] Y. Wang, Y. Xia, Bottom-up and top-down approaches to the synthesis of monodispersed spherical colloids of low melting-point metals, Nano Lett. 4 (10) (2004) 2047–2050, https://doi.org/10.1021/nl048689j.

[4] G. Guisbiers, S. Mejía-Rosales, Francis leonard deepak, nanomaterial properties: size and shape dependencies, J. Nanomater. 2012 (2012), https://doi.org/10.1155/2012/180976. Article ID 180976, 2 pages.

[5] L. Lin, H. Xu, H. Gao, X. Zhu, V. Hessel, Plasma-assisted nitrogen fixation in nanomaterials: fabrication, characterization, and application, J. Phys. Appl. Phys. 53 (2020) 133001, https://doi.org/10.1088/1361-6463/ab5f1f.

[6] S.C. Warren, M.R. Perkins, A.M. Adams, M. Kamperman, A.A. Burns, H. Arora, E. Herz, T. Suteewong, H. Sai, Z. Li, J. Werner, J. Song, U. Werner-Zwanziger, J.W. Zwanziger, M. Grätzel, F.J. DiSalvo, U. Wiesner, A silica sol-gel design strategy for nanostructured metallic materials, Nat. Mater. 11 (2012) 460–467, https://doi.org/10.1038/nmat3274.

[7] G. Soto, H. Tiznado, W. de la Cruz, A. Reyes, Synthesis of ReN3Thin films by magnetron sputtering, J. Mater 2014 (2014), https://doi.org/10.1155/2014/745736. Article ID 745736, 9 pages.

[8] Y. Kudriavtsev, A. Villegas, A. Godines, R. Asomoza, Calculation of the surface binding energy for ion sputtered particles, Appl. Surf. Sci. 239 (3–4) (2005) 273–278, https://doi.org/10.1016/j.apsusc.2004.06.014.

[9] J. Musil, J. Vlcek, P. Baroch, Chapter 3-Magnetron discharges for thin films plasma processing, materials surface processing by directed energy techniques, Eur. Mater. Res. Soc. Ser. (2006) 67–110, https://doi.org/10.1016/B978-008044496-3/50004-6.

[10] K. Wasa, 2 - Sputtering Phenomena, Handbook of Sputtering Technology, second ed., 2012, pp. 41–75, https://doi.org/10.1016/B978-1-4377-3483-6.00002-4.

[11] See. https://medium.com/@rakeshgupta070791/magnetron-sputtering-system-market-size-share-growth-trends-analysis-and-forecast-2019-to-2027-b7cef7f09700. 16/04/2020.

[12] See. www.marketwatch.com/press-release/magnetron-sputtering-system-market-size-share-2020-global-growth-new-updates-trends-industry-expansion-market-demand-development-status-opportunities-challenges-and-forecast-by-2026-2020-04-16. 16/04/2020.

[13] See. www.marketgrowthinsight.com/7708/magnetron-sputtering-system-market. 16/04/2020.

[14] A. Behera, Synthesis of Nanostructure Multilayered Ni/Ti Thin Film by Magnetron Sputtering and Effect of Annealing, PhD thesis. Submitted in, Indian Institute of Technology, Kharagpur, 2016.
[15] A. Behera, R. Suman, S. Aich, S.S. Mohapatra, Sputter-deposited Ni/Ti double-bilayer thin film and the effect of intermetallics during annealing, Surf. Interface Anal. 49 (7) (2017) 620–629, https://doi.org/10.1002/sia.6201.
[16] I. Safi, Recent aspects concerning DC reactive magnetron sputtering of thin films: a review, Surf. Coating. Technol. 127 (2–3) (2000) 203–218, https://doi.org/10.1016/S0257-8972(00)00566-1.
[17] P.F. Carcia, R.S. McLean, M.H. Reilly, G. Nunes Jr., Transparent ZnO thin-film transistor fabricated by rf magnetron sputtering, Appl. Phys. Lett. 82 (2003) 1117, https://doi.org/10.1063/1.1553997.
[18] G. Abadias, Y.Y. Tse, P. Guérin, Interdependence between stress, preferred orientation, and surface morphology of nanocrystalline TiN thin films deposited by dual ion beam sputtering, J. Appl. Phys. 99 (2006) 113519, https://doi.org/10.1063/1.2197287.
[19] J. Musil, P. Baroch, J. Vlček, K.H. Nam, J.G. Han, Reactive magnetron sputtering of thin films: present status and trends, Thin Solid Films 475 (1–2) (2005) 208–218, https://doi.org/10.1016/j.tsf.2004.07.041.
[20] K. Pedersen, J. Bøttiger, M. Sridharan, M. Sillassen, P. Eklund, Texture and microstructure of Cr_2O_3 and $(Cr,Al)_2O_3$ thin films deposited by reactive inductively coupled plasma magnetron sputtering, Thin Solid Films 518 (15) (2010) 4294–4298, https://doi.org/10.1016/j.tsf.2010.01.008.
[21] L. Zhang, X.D. Su, Y. Chen, Q.F. Li, V.G. Harris, Radio-frequency magnetron sputter-deposited barium hexaferrite films on Pt-coated Si substrates suitable for microwave applications, Scripta Mater. 63 (5) (2010) 492–495, https://doi.org/10.1016/j.scriptamat.2010.05.013.
[22] J.T. Gudmundsson, High power impulse magnetron sputtering discharge, J. Vac. Sci. Technol. 30 (2012) 030801, https://doi.org/10.1116/1.3691832.
[23] Y. Fu, H. Du, Magnetron Sputtered TiNiCu Shape Memory Alloy Thin Film for MEMS Applications, in: F.E.H. Tay (Ed.), Materials & Process Integration for MEMS, Microsystems, vol. 9, Springer, Boston, MA, 2002, https://doi.org/10.1007/978-1-4757-5791-0_4.
[24] A.L.L. Daniel, A.P. Ehiasarian, Study of the effect of RF-power and process pressure on the morphology of copper and titanium sputtered by ICIS, Surf. Coating. Technol. 327 (2017) 200–206, https://doi.org/10.1016/j.surfcoat.2016.10.018.
[25] N. Madaoui, L. Bait, K. Kheyar, N. Saoula, Effect of argon-oxygen mixing gas during magnetron sputtering on TiO_2 coatings, Adv. Mater. Sci. Eng. 2017 (2017), https://doi.org/10.1155/2017/4926543. Article ID 4926543, 6 pages.
[26] R. Wuhrer, W.Y. Yeung, Effect of target–substrate working distance on magnetron sputter deposition of nanostructured titanium aluminium nitride coatings, Scripta Mater. 49 (3) (2003) 199–205, https://doi.org/10.1016/S1359-6462(03)00264-1.
[27] V. Bukauskas, S. Kaciulis, A. Mezzi, A. Mironas, G. Niaura, M. Rudzikas, I. Šimkienė, A. Šetkus, Effect of substrate temperature on the arrangement of ultra-thin TiO_2 films grown by a dc-magnetron sputtering deposition, Thin Solid Films 585 (2015) 5–12, https://doi.org/10.1016/j.tsf.2015.04.007.
[28] A.S. Reddy, G.V. Rao, S. Uthanna, P.S. Reddy, Influence of substrate bias voltage on the properties of magnetron sputtered Cu_2O films, Phys. B Condens. Matter 370 (1–4) (2005) 29–34, https://doi.org/10.1016/j.physb.2005.08.041.
[29] N. Panich, Y. Sun, Effect of substrate rotation on structure, hardness and adhesion of magnetron sputtered TiB_2 coating on high speed steel, Thin Solid Films 500 (1–2) (2006) 190–196, https://doi.org/10.1016/j.tsf.2005.11.055.
[30] J.T. Gudmundsson, Ionized physical vapor deposition (IPVD): magnetron sputtering discharges, J. Phys. Conf. 100 (2008) 082002, https://doi.org/10.1088/1742-6596/100/8/082002.
[31] S.D. Ekpe, L.W. Bezuidenhout, S.K. Dew, Deposition rate model of magnetron sputtered particles, Thin Solid Films 474 (1–2) (2005) 330–336, https://doi.org/10.1016/j.tsf.2004.09.007.
[32] R. Alvarez, J.M. Garcia, Martin, M.C.L. Santos, V. Rico, F.J. Ferrer, C. Jose, A.R. Gonzalez-Elipe, A. Palmero, On the deposition rates of magnetron sputtered thin films at oblique angles, Plasma Process. Polym. 11 (6) (2014) 571–576, https://doi.org/10.1002/ppap.201300201.

[33] R. Daniel, K.J. Martinschitz, J. Keckes, C. Mitterer, The origin of stresses in magnetron-sputtered thin films with zone T structures, Acta Mater. 58 (7) (2010) 2621–2633, https://doi.org/10.1016/j.actamat.2009.12.048.
[34] J. Musil, J. Vlček, Magnetron sputtering of hard nanocomposite coatings and their properties, Surf. Coating. Technol. 142–144 (2001) 557–566, https://doi.org/10.1016/S0257-8972(01)01139-2.
[35] W. Khan, Q. Wang, X. Jin, Effect of target composition and sputtering deposition parameters on the functional properties of nitrogenized Ag-permalloy flexible thin films deposited on polymer substrates, Materials 11 (2018) 439, https://doi.org/10.3390/ma11030439.
[36] See. https://en.wikipedia.org/wiki/Magnetic_storage. 18/04.2020.
[37] See. https://www.google.com/url?sa=t&rct=j&q=&esrc=s&source=web&cd=8&cad=rja&uact=8&ved=2ahUKEwiEhYK_2vfoAhUoxzgGHUesCmoQFjAHegQIBhAB&url=https%3A%2F%2Fwww.emerald.com%2Finsight%2Fcontent%2Fdoi%2F10.1108%2FACMM-09-2019-2175%2Ffull%2Fpdf%3Ftitle%3Dappraisal-on-corrosion-performances-of-crni-tialncrni-and-crnialsub2subosub3subtiosub2sub-coatings-on-h13-hot-work-mold&usg=AOvVaw1JSPMhHgvKnzg8doCPO3uQ. 18/04.2020.
[38] R.W. Johnson, T.L. Phillips, R.C. Jaeger, S.F. Hahn, D.C. Burdeaux, Multichip thin-film technology on silicon, IEEE Trans. Compon. Hybrids Manufac. Technol. 12 (2) (June 1989), https://doi.org/10.1109/33.31423.
[39] See. https://products.inficon.com/en-us/nav-products/product/detail/sputteringsensor/. 18/04.2020.
[40] B. Yan, K. Wang, Z. Hu, C. Wu, X. Zhang, Shunt damping vibration control technology: a review, Appl. Sci. 7 (2017) 494, https://doi.org/10.3390/app7050494.
[41] J.P. Vilcot, B. Ayachi, T. Aviles, P. Miska, Full sputtering deposition of thin film solar cells: a way of achieving high efficiency sustainable tandem cells? J. Electron. Mater. 46 (2017) 6523–6527.
[42] See. http://www.iam.kit.edu/awp/english/270.php. 19/04.2020.
[43] Y. Taga, Recent progress of thin film technology in the automobile industry, Surf. Interface Anal. 22 (1–12) (1994) 149–155, https://doi.org/10.1002/sia.740220134.
[44] V.V. Savinkin, P. Vizureanu, A.V. Sandu, T.Y. Ratushnaya, A.A. Ivanischev, A. Surleva, Improvement of the turbine blade surface phase structure recovered by plasma spraying, Coatings 10 (1) (2020) 62, https://doi.org/10.3390/coatings10010062.
[45] J. Müller, B. Rech, J. Springer, M. Vanecek, TCO and light trapping in silicon thin film solar cells, Sol. Energy 77 (6) (2004) 917–930, https://doi.org/10.1016/j.solener.2004.03.015.
[46] B. Szyszka, W. Dewald, S.K. Gurram, A. Pflug, C. Schulz, M. Siemers, V. Sittinger, S. Ulrich, Recent developments in the field of transparent conductive oxide films for spectral selective coatings, electronics and photovoltaics, Curr. Appl. Phys. 12 (2012) S2–S11, https://doi.org/10.1016/j.cap.2012.07.022.
[47] C. Atkinson, C.L. Sansom, H.J. Almond, C.P. Shaw, Coatings for concentrating solar systems – a review, Renew. Sustain. Energy Rev. 45 (2015) 113–122, https://doi.org/10.1016/j.rser.2015.01.015.
[48] M. Fedel, 4- Blood Compatibility of Diamond-like Carbon (DLC) Coatings, Diamond-Based Materials for Biomedical Applications, Woodhead Publishing Series in Biomaterials, 2013, pp. 71–102, https://doi.org/10.1533/9780857093516.1.71.
[49] N. Eliaz, Corrosion of metallic biomaterials: a review, Materials 12 (3) (February 2019) 407, https://doi.org/10.3390/ma12030407.
[50] B.L. Gorberg, A.A. Ivanov, O.V. Mamontov, V.A. Stegnin, V.A. Titov, Modification of textile materials by the deposition of nanocoatings by magnetron ion-plasma sputtering, Russ. J. Gen. Chem. 83 (2013) 157–163, https://doi.org/10.1134/S1070363213010350.
[51] S. Shahidi, J. Wiener, M. Ghoranneviss, Plasma-Enhanced Vapor Deposition Process for the Modification of Textile Materials, 2016, https://doi.org/10.5772/62832.
[52] S. Shahidi, B. Moazzenchi, M. Ghoranneviss, A review-application of physical vapor deposition (PVD) and related methods in the textile industry, Eur. Phys. J. Appl. Phys. 71 (2015) 31302, https://doi.org/10.1051/epjap/2015140439.
[53] X.Q. Tan, J.Y. Liu, J.R. Niu, J.Y. Liu, J.Y. Tian, Recent progress in magnetron sputtering technology used on fabrics, Materials 11 (10) (2018) 1953, https://doi.org/10.3390/ma11101953.

[54] G. Mah, C.W. Nordin, J.F. Fuller, Structure and properties of sputtered titanium carbide and titanium nitride coatings, J. Vac. Sci. Technol. 11 (371) (1974), https://doi.org/10.1116/1.1318623.
[55] C. Xin, W. Liu, N. Li, J. Yan, S. Shi, Metallization of Al_2O_3 ceramic by magnetron sputtering Ti/Mo bilayer thin films for robust brazing to Kovar alloy, Ceram. Int. 42 (8) (2016) 9599–9604, https://doi.org/10.1016/j.ceramint.2016.03.044.
[56] Z. Han, X. Hu, J. Tian, G. Li, G. Mingyuan, Magnetron sputtered NbN thin films and mechanical properties, Surf. Coating. Technol. 179 (2–3) (2004) 188–192, https://doi.org/10.1016/S0257-8972(03)00848-X.
[57] A. Baptista, F. Silva, J. Porteiro, J. Míguez, G. Pinto, Sputtering physical vapour deposition (PVD) coatings: a critical review on process improvement and market trend demands, Coatings 8 (11) (2018) 402, https://doi.org/10.3390/coatings8110402.
[58] J. Yang, H. Hu, Y. Miao, L. Dong, B. Wang, W. Wang, H. Su, R. Xuan, H. Zhang, High-quality GeSn layer with Sn composition up to 7% grown by low-temperature magnetron sputtering for optoelectronic application, Materials 12 (17) (2019) 2662, https://doi.org/10.3390/ma12172662.
[59] J. Zheng, S. Wang, Z. Liu, H. Cong, C. Xue, C. Li, Y. Zuo, B. Cheng, Q. Wang, GeSn p-i-n photodetectors with GeSn layer grown by magnetron sputtering epitaxy, Appl. Phys. Lett. 108 (2016) 033503, https://doi.org/10.1063/1.4940194.
[60] R. Das, S. Ray, Transparent conducting zinc oxide as anti-reflection coating deposited by radio frequency magnetron sputtering, Indian J. Phys. 86 (23–29) (2012).
[61] See. www.osapublishing.org/abstract.cfm?uri=OIC-2001-ThD1. 20/04/2020.
[62] A. Behera, S. Aich, Characterization and properties of magnetron sputtered nanoscale NiTi thin film and the effect of annealing temperature, Surf. Interface Anal. 47 (2015) 805–814, https://doi.org/10.1002/sia.5777.
[63] A. Behera, S. Aich, A. Behera, A. Sahu, Processing and characterization of magnetron sputtered Ni/Ti thin film and their annealing behaviour to induce shape memory effect, 4th International Conference on Materials Processing and Characterization, Materials Today: Proceedings, Mater. Today Proc. 2 (2015) 1183–1192, https://doi.org/10.1016/j.matpr.2015.07.030.
[64] D.K. Maurya, A. Sardarinejad, K. Alameh, Recent developments in R.F. Magnetron sputtered thin films for pH sensing applications-an overview, Coatings 4 (2014) 756–771, https://doi.org/10.3390/coatings4040756.

PART II

Design and fabrication of nanomaterials

SECTION A

Development of magnetic nanoparticles

CHAPTER 9

Synthesis and characterization of magnetite nanomaterials blended sheet with single-walled carbon nanotubes

Indradeep Kumar
Vels Institute of Science, Technology & Advanced Studies (VISTAS), Chennai, Tamil Nadu, India

Chapter Outline

Design, Fabrication, and Characterization of Multifunctional Nanomaterials. https://doi.org/10.1016/B978-0-12-820558-7.00003-0

Nomenclature

1D	One-Dimensional
2D	Two-Dimensional
Å	Angstrom (1exp.-10)
cm	centimeter
EDAX	Energy Dispersive X-ray Analysis
FCM	$Fe_aC_bMg_{1-(a+b)}$
FESEM	Field Emission Scanning Electron Microscope
GPa	Giga Pascal (unit)
HCl	Hydro Choleric Acid
HRTEM	High-Resolution Transmission Electron Microscopy
nm	nanometer (unit)
°C	Degree Celsius
sccm	standard cubic centimeters per minute (unit)
Scm-1	Siemens per centimeter
TA	Thermal Analysis
TPa	Tera Pascal (unit)
W/mk	Watts per meter Kelvin
XRD	X-ray powder Diffraction

1. Introduction

Carbon is an astonishing element, because of its existence in numerous forms of allotropes. Carbon in the form of nanotubes (CNTs) plays, a dominant role in industrial, scientific, and nanotechnological research. In a few decades, carbon nanotubes have acquired central attraction in the nanoscience domain [1]. There is no doubt that now, among all the fields of research, nanotechnology's synthesis, characterization, and applications of CNTs is one of the most active fields, which leads to a recommence the interest in synthesizing many different forms of carbon nanomaterials like helices, fibers, and graphene. Morphology diversity of carbon allows flexibility during alteration of the properties. Nanomaterial having nonlinear or helical morphology. The reported nanomaterial was first considered the curiosity's effort, which was focused on their prevention rather than the synthesis. After a long time in the 1990s, it again stimulated after the discovery of carbon nanotubes, which led to the attention in the carbon fiber and tubes, mostly with abnormal morphology like helices, spring, and so forth, and it was expected that nanomaterials, which having helical morphology, should have unique and similar chemical and physical properties as their macro components. The formation of carbonaceous material in helical form from carbon precursors in the presence of catalyst by using a bottom-up approach is likely to move by a similar procedure for synthesizing straight tubes. In the nanoscale domain, the nanoparticle's properties cannot be observed in bulk or counterpart of atoms. Therefore it is a great deal to prepare nanomaterials that provide desired characteristics. Because of synthesizing simplicity in the laboratory and its elegant characteristics, iron-oxide nanoparticles (IONs) gains a unique place in between other oxides of metal of nanosized.

Iron oxide nanostructure-based material is exploited for various research purposes because of its magnetic, optical, electrical, and catalytic behaviors. Despite that, in any situations, it is required to combine IONs with other nanostructured materials to forms enhanced properties nanocomposites. Concerning these single-walled carbon nanotubes (SWCNTs), a valuable option had great potential for synthesizing the nanocomposites. This hybridized nanocomposites can be used for developing a new class of electrical and mechanical components, or in other words, in electromechanical applications like transformers, inductor, bobbins, and other electromechanical components, to increase the life span of the components and to prevents from the hazard because of malfunctioning, like overheating causes fire, and so forth. This chapter describes Fe_3O_4(IONs)-SWCNTs nanocomposite for electromechanical applications and starts with the introductory information and preparation of IONs and SWCNTs separately. Various iron oxides are available in nature; among them, mostly hematite, maghemite, and magnetite are found in nature. By reducing or oxidizing the annealing process, iron oxide can be converted into other forms.

2. Materials and methodology

Research objectives were achieved through three hypothesis-driven tasks. Task one focuses on synthesizing highly dispersed and stable nanoparticles like iron oxide nanopowder and a blend of SWCNTs. The optimal stabilizers will be determined by SWCNTs based on effectiveness, environmentally friendliness, and cost [3]. Then, IONs will be synthesized with the aid of selected low-cost and "green" stabilizers [2]. Task two was characterized and tested the particle size, surface area, and long-term stability of the stabilized nanoparticles, and was measured with state-of-the-art microscopic and spectroscopic techniques. Task three was validated with the bench-scale experimental results through pilot-scale testing and test the feasibility of formed material.

3. Fabrication of single-walled carbon nanotubes

SWCNTs are used in many applications, either accumulated as a thin film or dispersed in powder form. However, SWCNTs commercial purpose is still challenging, so their execution can be accomplished by combining the conventional formation procedures. Due to various researchers working on SWCNTs, so fabricating process is getting advanced continuously. Wang et al. investigated composite CNTs having multifunctionalities, such as higher electrical conductivity, greater Young's modulus, and high strength. They found these properties along the length. Higher volume fraction, perfect molecule alignment, co-existence, and less waviness of the CNTs and a novel fabricating procedure was used for commercial production purposes.

3.1 Experimental procedure

3.1.1 Materials

The gases used in the synthesizing of SWCNTs are acetylene, nitrogen, and hydrogen, in which acetylene having a purity of 99.9%, is used as a carbon source; nitrogen, having a purity of 99%, is used as a carrier gas, while hydrogen of 99% purity is used as a reducing agent. For the purification of SWCNTs, commercial-grade hydrochloric acid is used. For the solvent, double distilled water is used. All the used chemicals are of analytical grade and hence used without filtering it.

3.2 Synthesis of combustive catalyst

3.2.1 Synthesis of FNM catalyst

The solution combustion synthesis technique is used for synthesizing the combustive catalyst $Fe_aNi_bMg_{1-(a+b)}O$ (FNM). Magnesium Nitrate, Nickel Nitrate, and Ferrous Nitrates are used in stoichiometric form FNM. Various molar composition forms such as $Fe_{0.25}Ni_{0.10}Mg_{0.65}O$ (FNM-a), $Fe_{0.20}Ni_{0.15}Mg_{0.65}O$ (FNM-b), and $Fe_{0.15}Ni_{0.20}Mg_{0.65}O$ (FNM-c), is prepared by keeping the constant molar ratio of Magnesium Nitrate.

3.3 Synthesis and purification of single-walled carbon nanotubes

3.3.1 Schematic diagram of chemical vapor deposition apparatus

The technique of combustion catalyst synthesizes the SWCNTs by chemical vapor deposition (CVD). CVD apparatus contains a tube horizontally with heating filament and a program controller for controlling the required temperature for reaction. Mass flow meter and various gas sources are connected with the horizontal tube, which is made up of alumina, as shown in Fig. 9.1. A schematic diagram of the chemical vapor test apparatus for synthesizing SWCNTs is shown in Figs. 9.5A and 9.5B. The tube has input and output terminals. For maintaining the temperature, the filament is winded all around the tube. Using a programmable controller placed at the center of the tube and the temperature is maintained. For regulating the mass flow rate of various gasses according to the requirement, a digital mass flow meter is attached to the gas cylinder.

3.3.2 Synthesis of single-wall carbon nanotubes

The combustive catalyst CVD technique was carried out at normal atmospheric pressure in the tube. The catalytic reaction is for SWCNTs preparation is carried out with various composition of catalyst as formed earlier. Around 100 mg of combustive catalyst are

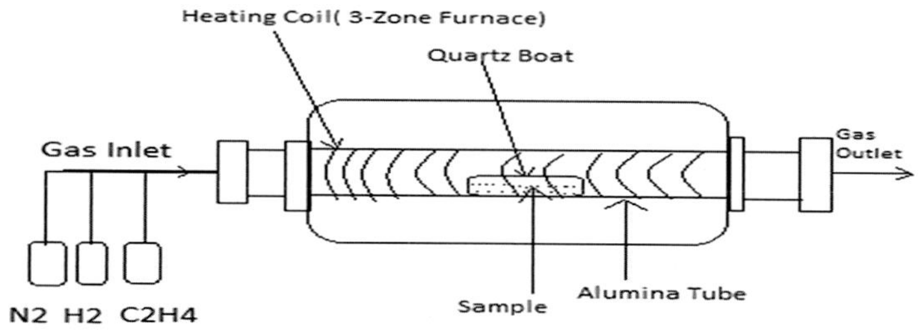

Figure 9.1
Chemical vapor deposition apparatus schematic diagram.

kept on a quartz boat, further kept in the horizontal tube center. Under the nitrogen atmosphere, the catalyst is heated with a flow rate of 100sccm for decreasing the catalytic surface until inside the tube temperature reaches 650°C. The further reaction is carried out by changing the flow rate from 10 to 100 sccm at various temperatures ranging between 650°C to 850°C for 10 min. After completing the reaction, the furnace is cool to room temperature in the nitrogen's atmosphere.

The mass of the SWCNTs deposited by acetylene decomposition is calculated when the below equation uses the catalyst:

$$\text{SWCNTs Accumulated } (\%) = \frac{Mtot - Mcat}{Mcat} \times 100\%$$

Mtot is the total mass of acetylene after completing the reaction, and *Mcat* is the mass of catalyst before starting the reaction.

3.3.3 Purification of single-walled carbon nanotubes

SWCNT, which is synthesized, is further treated for purification with hydrochloric acid to remove the metal particles present in the FNM catalyst. This process is carried out by stimulating the SWCNT, which is synthesized, with a concentrated aqueous hydrochloric acid solution (35%–40% by weight) at room temperature for 45–50 min. After that, is solution is drenched and washed by the use of distilled water till the product gets stable, and then SWCNT is dried out around 80–85°C for 5–6 h.

4. Preparation methods of iron oxide nanoparticles

The shape and size of nanostructures, size distributions, and chemistry of the surface has a significant impact on behaviors and features of IONs. In practical aspects, preparation methods have a crucial role in determining the impurities present, the degree of a structural defect in the particles, and the distribution of such defects by manipulating the IONs' behaviors.

IONs can be prepared in many different ways, but here we discuss just four most commonly used, extensively exploited for synthesizing IONs method, i.e., sol-gel, coprecipitation, hydrothermal, and microemulsion methods [29—31].

4.1 Hydrothermal method

This method is one of the most common and useful preparation methods for IONs and other inorganic nanocrystals, especially for metal oxides and metals. In this method, the reaction is carried out in aqueous media in a reactor where the temperature is more than 200°C and pressure more than 1500 kPa for IONs preparation. At this pressure and temperature, the metal salt is undergone for hydrolyzing and dehydration by water use. As a result, supersaturation is achieved due to significantly less solubility of the acquired metal oxides in an aqueous medium at a much higher temperature and pressure. The higher temperature increases the dehydration rates, which causes the high diffusivity of reactants in such conditions. Because of the very low solubility of metal oxides, higher supersaturation can be achieved, and hence, excellent crystal can be prepared finally. Since very high pressure and temperature are used, the nanocrystals' magnetic property and quality improve significantly due to the synergistic effect. The parameters such as pressure, temperature, type of concentration, precursor, and reaction time can be altered to obtain the nanocrystals' required shape and size. This method is environmentally and ecofriendly since there is no requirement for organic solvent or posttreatments. Because of the features mentioned above, hydrothermal methods have been used widely to synthesize nanoparticles, metal oxide as powder, and single crystals. The effect of precursor concentration, pressure, temperature, and reaction time on morphology of particle and size by this method were investigated by Hao et al. [4]. The particles' size distribution and size increase with an increase in the precursor concentration, while the reaction time affects the average particle size more than feed concentration. The effect of reaction time and temperature can be comprehended by bearing in mind that synthesizing particles occurs in two steps: first, nucleation, and second, crystal growth. At very high pressure and temperature, the crystal growth process is slower than the nucleation process, and hence, smaller particles size obtained while larger particles can be formed by longer reaction time where the determining factor is crystal growth [4—7].

4.2 Coprecipitation

One of the simplest, cheapest, and most environmentally friendly methods for preparing IONs is simultaneous precipitation of Fe^{3+} and Fe^{2+} ions in aqueous essential nature media. In a medium with pH between 9 and 14, completely Fe_3O_4 can be precipitated while the molar ratio of Fe^{2+}:Fe^{3+} is 1:2 in a nonoxidizing environment. Since Fe_3O_4 is not highly stable; hence, it can be converted into maghemite by oxidation by oxygen present in the solution phase; hence, nonoxidizing or oxygen-free media is necessary to form magnetite. The coprecipitation method is very frequently used for the synthesis of Fe_3O_4 nanoparticles in the aqueous phase. Generally, in this method, ammonia solution or sodium hydroxide is used to precipitate Fe^{3+} and Fe^{2+}. The surface of IONs produced having many OH groups, and these IONs easily can be dispersed in media having aqueous. Refeit and Olowe suggested a mechanism that they considered $Fe(OH)_2$ to act as intermediate for Fe_3O_4 preparation. According to the mechanism, oxidation of $Fe(OH)_2$ by oxygen to FeOOH, the precipitation of Fe^{2+} by alkaline, and the combination of FeOOH and $Fe(OH)_2$ to prepare Fe_3O_4. Hence, they suggested that only using Fe^{2+} as a precursor can the nanoparticle be prepared by coprecipitation technique in air. Hui et al. formed a large-scale Fe_3O_4 nanoparticle, which is hydrophilic, showed particle size controlling possibilities from 20 to 40 nm range by changing the precursor's parameters ionic strength and temperature during the investigations. They used $FeSO_4.4H_2O$ as a precursor to producing magnetite [10–12].

4.3 Sol-gel

Sol-gel methods refer to the hydrolysis and condensation of an alkoxide as precursors and prepare a sol, resulting from nanoparticles' dispersion. Generally, ION formation, iron alkoxide, and iron salts such as nitrates, chlorides, and acetates undergo different-different types of hydrolysis and condensation reactions. Further condensation and inorganic polymerization result in a 3D network of metal oxide named wet gel. As the reaction is carried out at room temperature, further heat treatment is applied to convert in a state of crystals. Generally, water is used as a solvent, but hydrolyzing precursor acid or base can be used. Necessary catalysis results in a colloidal gel, while acid catalysis forms a polymeric gel. Properties of the obtained products highly depend on the hydrolysis and condensation rate. At a more controlled and slower rate of hydrolysis, a smaller size can be formed. Temperature, solution of composition, and pH of the solution also determine the particle size. Hence it is essential to control the hydrolysis and condensation rate of gel precursor and other reduction or oxidation reactions during the gelling stage and heat treatment to convert crystal structure since there has a significant effect on porosity and structural properties in final products. Cui et al. prepared monodispersed Fe_3O_4, α-Fe_2O_3, and γ-Fe_2O_3 nanoparticles using a low-temperature sol-gel technique the same starting material and same precursor. The formation method includes the $FeCl_2$ reacting in boiling ethanol with propylene oxide to form a gel-type solution, further dried [8,9].

4.4 Nanoemulsion method

Nanoemulsion can be defined as the isotropic dispersion of relatively two or more immiscible liquids or blends thermodynamically stable and stabilized by anionic, cationic, or/and nonionic, which is widely used for obtaining IONs. Depending upon the relative concentrations, molecules of surfactants self-assembled in various structures in the mixture. A much different technique can be used to synthesized nanoparticles using the nanoemulsion method. For example, Fe_3O_4 IONs can be mixed with SWCNTs (30 PPM and 60 PPM) (shown in Fig. 9.2) in blend form to produce nanoparticles in the form of precipitate using the nanoemulsion method shown in Fig. 9.3.

The above shown dispersed phase gives a confined environment for preparing nanoparticles. The surfactant-covered water pools provide a micro- or nanoenvironment for synthesizing nanoparticles and, at the same time, limited the nanoparticle growth. The size of the droplets of nanoemulsion is directly related to the ratio of water to surfactant. The particle size may also change due to various parameters such as the surfactant's flexibility and reactants' concentration. Nanoparticles can be prepared by combining a precipitating substance to nanoemulsion or microemulsion. This nanoemulsion or microemulsion has primary reactants that are dissolved in an aqueous solution. These precipitating substances may be either a gas or liquid, like hydrogen or hydrazine [13,14].

5. Synthesis of Fe_3O_4-SWCNT-IONs sheet

The shape and size of nanoparticles, distribution of size, and surface chemistry have a more significant impact on their characteristics. The synthesizing method, in particular aspects, plays a key role. The synthesizing method also determines the degree of defect in structure and foreign particles or impurities present in it and their distribution. Although Fe_3O_4 IONs sheet can be formed in many ways, we use the nanoemulsion method [7], as shown in Fig. 9.4.

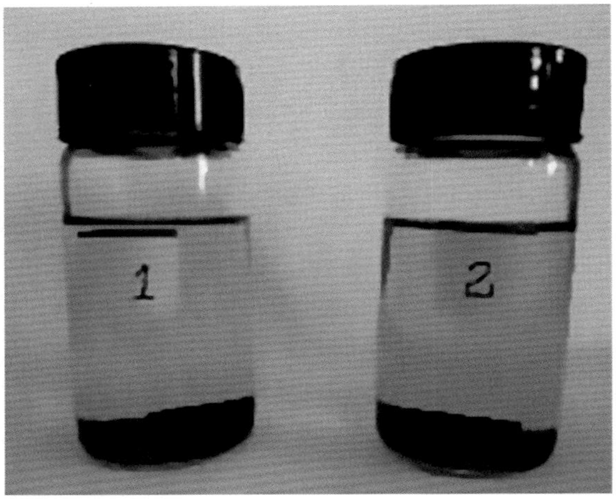

Figure 9.2
SWCNT 30 PPM and 60 PPM blend in ethanol.

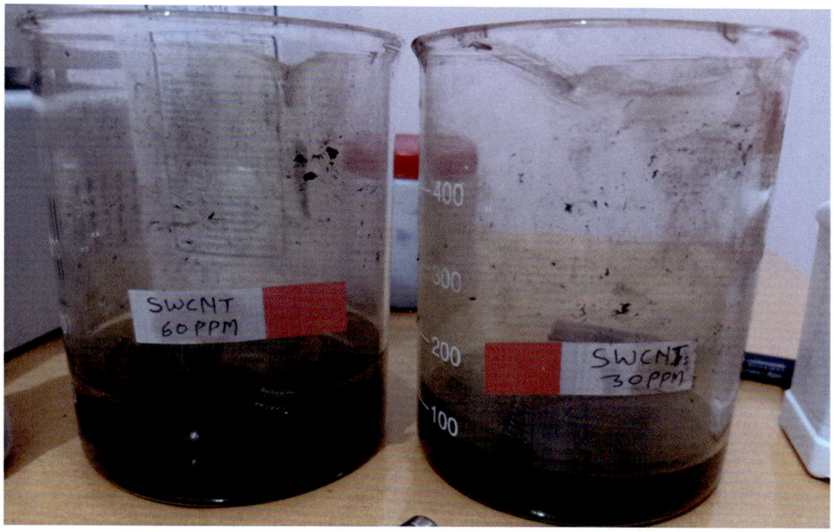

Figure 9.3
Nanoparticle in the form of precipitate using nanoemulsion.

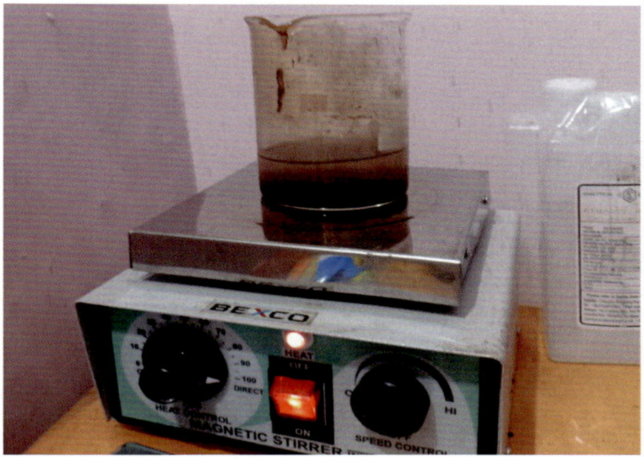

Figure 9.4
Formation of Fe_3O_4 IONS- SWCNTs blend.

5.1 Nanoemulsion method

Nanoemulsion can be defined as the isotropic dispersion of relatively two or more immiscible liquids or blends thermodynamically stable and stabilized by anionic, cationic, or/and nonionic, which is widely used for obtaining IONs. Depending upon the relative concentrations, molecules of surfactants self-assembled in various structures in the

Figure 9.5A
Synthesized Fe_3O_4-SWCNT-IONs sheet.

mixture. A much different technique can be used to synthesized nanoparticles using the nanoemulsion method. For example, Fe_3O_4 IONs can be mixed with SWCNTs (30 PPM and 60 PPM) in blend form to produce nanoparticles in the form of a sheet using the nanoemulsion method shown in Figs. 9.5A and 9.5B.

5.2 Gas-phase technique

This technique of synthesizing magnetite nanomaterials blended with SWCNT precipitates the gaseous phase's reliable product obtained by various reactions like disproportionation hydrolysis and pyrolysis, reduction, or oxidation. In the CVD technique, a precursor in the gaseous stream is supplied continuously by a gas delivery hose in the combustion chamber kept in a vacuum and has a temperature of more than 900°C. The CVD reactions are carried out in a combustion chamber where the products combine and form nanoparticles. Agglomeration and development of the nanoparticle sheet are mitigated by rapid expansion at the combustion chamber's outlet of the two-phase gas stream. The prepared nanoparticle sheet is further heat treated in different-different gas streams, which provides structural and compositional modifications and particle purifications; conversion is required shape and size, morphology, and crystallization. By this method, we get the higher quality of magnetite nanomaterials blended sheet with SWCNTs, but the yield is meager, and hence for commercial applications, it is very challenging. The variables like the concentration of the magnetite nanoparticles, concentration of SWCNTs, heating times, and gas-phase impurity must be precisely controlled to get the pure nanosheet.

5.3 Liquid-phase technique

In this technique, the coprecipitation method is used to use ethanol as a solvent or aqueous solution. A stoichiometric mixture of Fe_30_4 IONs and SWCNTs will react in alcoholic

Figure 9.5B
Synthesized $Fe_3O_{4\text{-}}$SWCNT-IONs sheet.

media or aqueous media to yield a homogeneous mixture of Fe304 IONs-SWCNTs a sheet structure. The size and phase of the Fe_30_4 IONs-SWCNTs particles depending on the concentration of the variables and the pH value of the solution. By adjusting the pH value and concentration or strength of ions, the particle size can be controlled. Nanoparticles formed in liquid phase coprecipitation have a higher ratio of surface area to volume. Hence it reduces their surface energy by aggregating in the solution. By adding dispersing agents, the suspension of nanoparticles can be stabilized. This technique provides ease of surface treatment as well as better yields and less expensive.

6. Results and discussion

As we know, the nanomaterials' shape and size, surface properties, and size distribution have a significant impact on the nanosheets' behavior and characteristics, formation method having a pivotal role in practical purposes. Formation methods also determine the nanosheet's impurity, and the degree of defects in structure and their distribution over the surface can be achieved. This research aimed to synthesize low-cost or cost-effective Iron-based highly dispersive nanosheets that can be used in electromechanical applications. The aim was achieved by (1) synthesizing a new type of stabilized iron-based nanoparticle using a green stabilizer at a lower cost; (2) test the stabilized nanoparticles' effectiveness for making small electrical components; and (3) test the formed sheet's electrical components feasibility. This research was tested the following three key research hypotheses: (1) high strength-to-weight ratio (>5); (2) almost negligible rusting ability;

the stabilized nanoparticles will offer much greater resistivity and longevity over currently used nonstabilized iron particles for rusting; (3) Damage-tolerant systems consist of nanoscale approaches to increase system robustness through improved interlaminar interfaces, repair mechanisms, and health monitoring; and (4) thermal protection and control: thermal management system gives the lightweight approach to protect systems from getting damaged due to uncontrolled thermal cycling [24–28].

6.1 Characterization techniques

The phase purity, microstructure, and morphology of both the samples are characterized by transmission electron microscopy, scanning electron microscope, and electron diffraction. Magnetic properties of the synthesized IONs-SWCNTs are investigated at room temperature by Vibrating Sample Magnetometer by applying the magnetic field. Scanning electron microscopy (SEM) image is taken using a Carl Zeiss Auriga SEM-FEB in field emission scanning electron microscope. All SEM samples are sputter-coated with a thin layer of gold to avoid charge over it [15–23].

6.1.1 Transmission electron microscopy, electron diffraction, and scanning electron microscope

Transmission electron microscopy (TEM) image of Fe_3O_4-SWCNTs (30 PPM) in Fig. 9.6A and B. and Fe_3O_4-SWCNTs (60 PPM) in Fig. 9.6D and E. Fig. 9.6C shows the electron diffraction pattern for Fe_3O_4-SWCNTs. TEM images show two-dimensional SWCNTs plates with size about 30–50 nm and 3–5 nm in thickness. A larger plate of Fe_3O_4 IONs can be seen clearly in gray color particle in Fig. 9.6A. After reducing the SWCNTs, a particle having dark in color became covered by Fe_3O_4-IONs with light gray

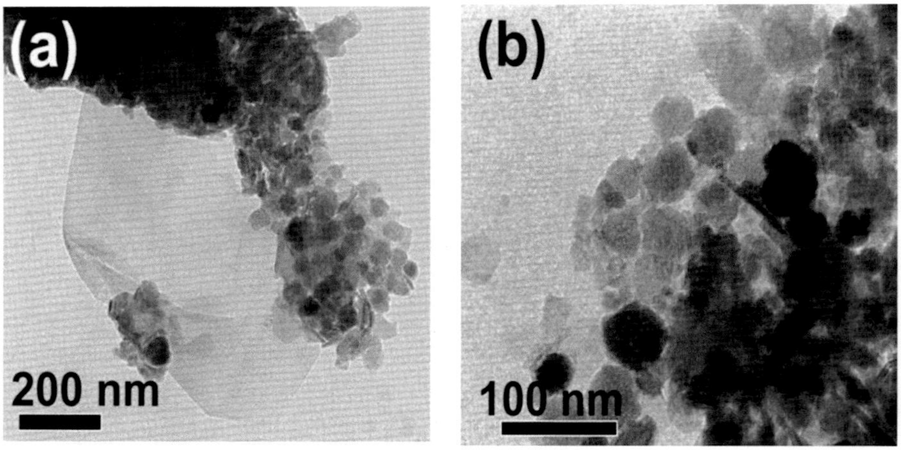

Figure 9.6A and B
TEM images of Fe_3O_4-SWCNTs (30 PPM).

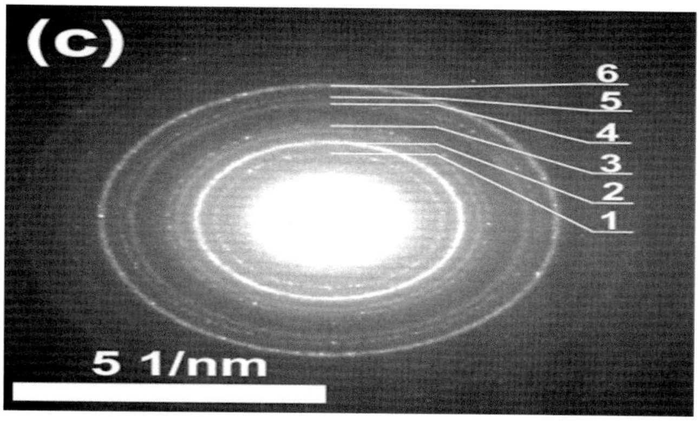

Figure 9.6C
Electron Diffraction pattern for Fe_3O_4-SWCNTs.

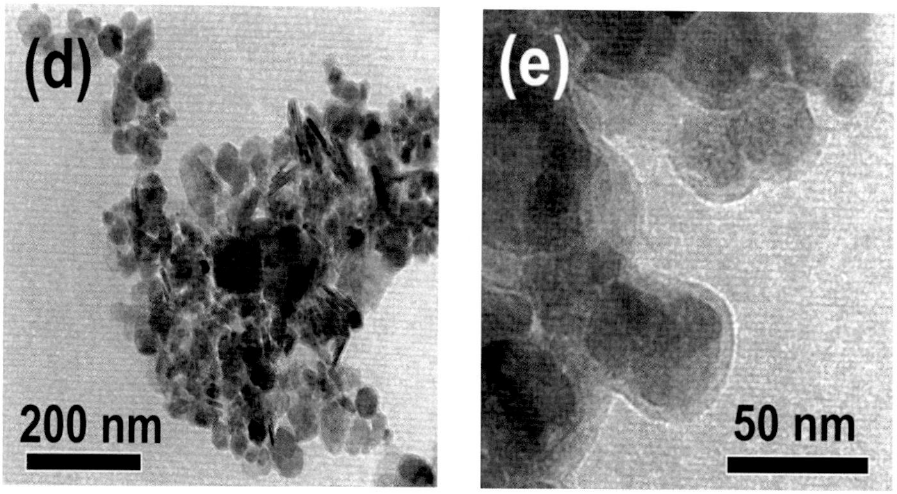

Figure 9.6D and E
TEM images of Fe_3O_4-SWCNTs (60 PPM).

color. Hence the composite of Fe_3O_4-IONS and SWCNTs forms under annealing at 900°C. The XRD technique investigated the crystal phase and structure of the Fe_3O_4IONs-SWCNTs. Fig. 9.7 shows the XRD patterns of Fe_3O_4IONs alone and Fe_3O_4IONs-SWCNTs nanocomposites. Both show similar diffraction peaks, which are indexed to the (015), (108), (120), (125), (034), (126), (024), (235), and (300). Fig. 9.8 shows the tubular morphology HRTEM images of (A–B) Fe_3O_4 IONs and (C–D) Fe_3O_4 IONs-SWCNTs (Insets show SAED patterns). Fig. 9.9 shows FESEM images and EDX spectra of (A–B) Fe_3O_4IONs and (C–D) Fe_3O_4 IONs-SWCNTs.

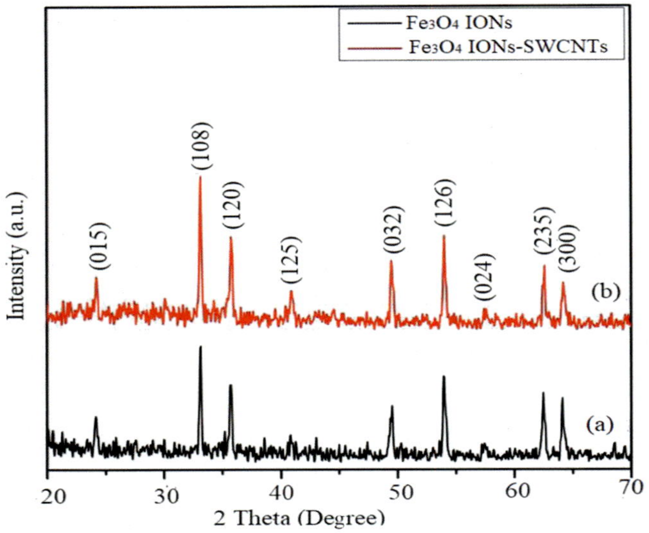

Figure 9.7

XRD patterns of Fe_3O_4 IONs and Fe_3O_4 IONs-SWCNTs.

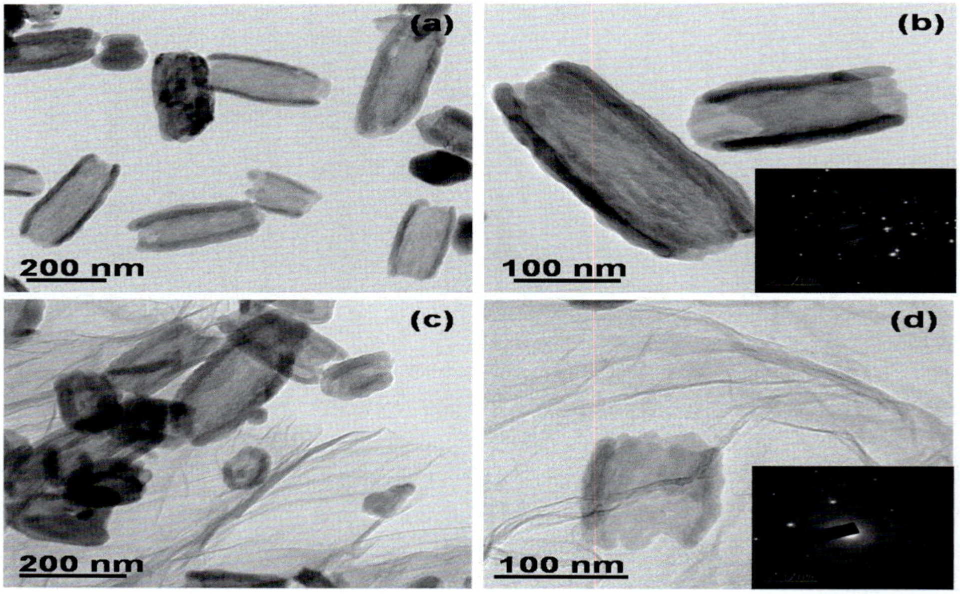

Figure 9.8

HRTEM images of (A–B) Fe_3O_4 IONs and (C–D) Fe_3O_4 IONs-SWCNTs (insets show SAED patterns).

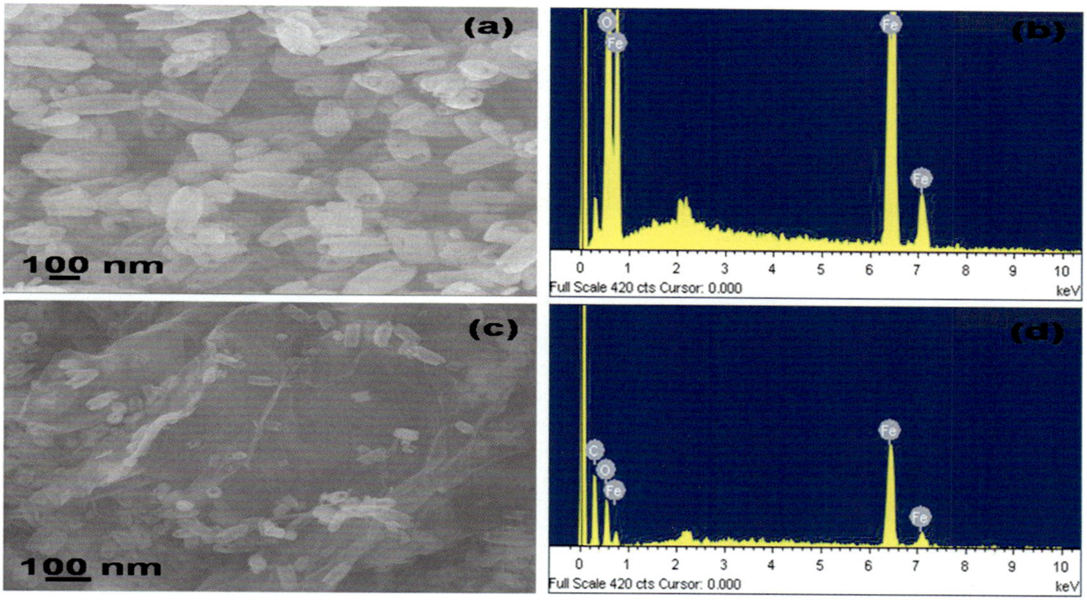

Figure 9.9

FESEM images and EDX spectra of (A—B) Fe_3O_4 IONs and (C—D) Fe_3O_4 IONs-SWCNTs.

7. Conclusion

This research directly addresses the priority need for research "concerning nanotechnology applications in electromechanical components." A new class nanosheet was synthesized by combining nanopowder of IONs and SWCNTs, physically stable, thermally robust, and environmentally friendly nanosheets and characterized. A cost-effective, higher strength-to-weight technology based on the new materials was developed for electrical components like transformers, inductors, bobbins, and other electromechanical components to increase the components' life span and prevent it hazard because of malfunctioning, like overheating causes fire, etc.

Using the nanoemulsion method in which thermodynamically stable two immiscible substances stabilized by nonionic, anionic, or cationic surfactants, the nanoclay of mixture can be obtained through nanopowder of iron oxide and SWCNTs. The formed nanosheets from the nanoclay having improved electromechanical properties like high strength-to-weight ratio, almost negligible rusting ability, damage-tolerance, and thermal protection and control was achieved.

Over the past few decades, carbon nanotubes' production has advanced its structure, quality, and quantity. By considering this, carbon nanomaterials of various structures like carbon nanorod, carbon nano coil, graphene, and so forth were formed in huge quantities.

During the experiment, it was found that in the CVD method, controlling the size of the catalyst and carbon source, helical SWCNTs can be formed in bulk quantity for industrial purposes. Hence by careful manipulation of the factors mentioned above, we can produce our desired shape and size of SWCNTs as when we required.

References

[1] A. Ali, H. Zafar, M. Zia, I.U. Haq, A.R. Phull, J.S. Ali, A. Hussain, Synthesis, characterization, applications and challenges of iron oxide nanoparticles, Nanotechnol. Sci. Appl. 9 (2016) 49–67.

[2] W. Wei, W. Zhaohui, Y. Taekyung, J. Changzhong, K. Woo-Sik, Recent progress on magnetic iron oxide nanoparticles: synthesis, surface functional strategies and biomedical applications, Sci. Technol. Adv. Mater 16 (2015) 023501.

[3] S.G. Leonardi, A. Mirzaei, A. Bonavita, S. Santangelo, P. Frontera, F. Pantò, P.L. Antonucci, G. Neri, A comparison of the ethanol sensing properties of -iron oxide nanostructures prepared via the sol-gel and electrospinning techniques, Nanotechnology 27 (2016) 075502.

[4] Y. Hao, A.S. Teja, Continuous hydrothermal crystallization of _—Fe_2O_3 and Co_3O_4 nanoparticles, J. Mater. Res. 18 (2011) 415–422.

[5] M. Shandilya, R. Rai, J. Singh, Review: hydrothermal technology for smart materials, Adv. Appl. Ceram. 115 (2016) 354–376.

[6] C. Hui, C. Shen, T. Yang, L. Bao, J. Tian, H. Ding, C. Li, H.J. Gao, Large-scale Fe_3O_4 nanoparticles soluble in water synthesized by a facile method, J. Phys. Chem. C 112 (2008) 11336–11339.

[7] D. Ramimoghadam, S. Bagheri, S.B.A. Hamid, Progress in electrochemical synthesis of magnetic iron oxide nanoparticles, J. Magn. Magn. Mater. 368 (2014) 207–229.

[8] H. Cui, Y. Liu, W. Ren, Structure switch between α-Fe_2O_3, γ-Fe_2O_3 and Fe_3O_4 during the large scale and low temperature sol–gel synthesis of nearly monodispersed iron oxide nanoparticles, Adv. Powder Technol. 24 (2013) 93–97.

[9] U.K. Sahu, M.K. Sahu, S.S. Mahapatra, R.K. Patel, Removal of As(III) from aqueous solution using Fe_3O_4 nanoparticles: process modeling and optimization using statistical design, Water Air Soil Pollut. 228 (2017) 45.

[10] S. Liang, J. Li, F. Wang, J. Qin, X. Lai, X. Jiang, Highly sensitive acetone gas sensor based on ultrafine Fe_2O_3nanoparticles, Sens. Actuators B Chem. 238 (2017) 923–927.

[11] P. Oulego, M.A. Villa-García, A. Laca, M. Diaz, The effect of the synthetic route on the structural, textural, morphological and catalytic properties of iron(III) oxides and oxyhydroxides, Dalton Trans 45 (2016) 9446–9459.

[12] I.M. Mirza, A.K. Sarfraz, S.K. Hasanain, Effect of surfactant on magnetic and optical properties of α-Fe_2O_3 nanoparticles, Acta Phys. Pol. A 126 (2014) 1280–1287.

[13] T. Lu, J. Wang, J. Yin, A. Wang, X. Wang, T. Zhang, Surfactant effects on the microstructures of Fe_3O_4 nanoparticles synthesized by microemulsion method, Colloids Surfaces A Physicochem. Eng. Asp. 436 (2013) 675–683.

[14] X.-M. Lin, A.C.S. Samia, Synthesis, assembly and physical properties of magnetic nanoparticles, J. Magn. Magn. Mater. 305 (2006) 100–109.

[15] N.R. Stradiotto, H. Yamanaka, M.V.B. Zanoni, Electrochemical sensors: a powerful tool in analytical chemistry, J. Braz. Chem. Soc. 14 (2003) 159–173.

[16] G. Ulrich, V. Winfried, Z. Jens, Recent developments in electrochemical sensor application and technology—a review, Meas. Sci. Technol. 20 (2009) 042002.

[17] S. Jiang, E. Hua, M. Liang, B. Liu, G. Xie, A novel immunosensor for detecting toxoplasma gondii-specific IgM based on goldmag nanoparticles and graphene sheets, Colloids Surfaces B Biointerfaces 101 (2013) 481–486.

[18] P.B. Luppa, L.J. Sokoll, D.W. Chan, Immunosensors—principles and applications to clinical chemistry, Clin. Chim. Acta 314 (2001) 1–26.
[19] S. Zhang, B. Du, H. Li, X. Xin, H. Ma, D. Wu, L. Yan, Q. Wei, Metal ions-based immunosensor for simultaneous determination of estradiol and diethylstilbestrol, Biosens. Bioelectron. 52 (2014) 225–231.
[20] X. Chen, H. Yan, Z. Shi, Y. Feng, J. Li, Q. Lin, X. Wang, W. Sun, A novel biosensor based on electro-co-deposition of sodium alginate-Fe_3O_4-graphene composite on the carbon ionic liquid electrode for the direct electrochemistry and electrocatalysis of myoglobin, Polym. Bull. 74 (2017) 75–90.
[21] H. Derikvand, A. Azadbakht, An impedimetric sensor comprising magnetic nanoparticles–graphene oxide and carbon nanotube for the electrocatalytic oxidation of salicylic acid, J. Inorg. Organomet. Polym. Mater. 27 (2017) 901–911.
[22] X. Shi, J. Lu, H. Yin, X. Qiao, Z. Xu, A biomimetic sensor with signal enhancement of ferriferrous oxide-reduced graphene oxide nanocomposites for ultratrace levels quantification of methamidophos or omethoate in vegetables, Food Anal. Methods 10 (2017) 910–920.
[23] B. Sun, X. Gou, R. Bai, A.A.A. Abdelmoaty, Y. Ma, X. Zheng, F. Hu, Direct electrochemistry and electrocatalysis of lobetyolin via magnetic functionalized reduced graphene oxide film fabricated electrochemical sensor, Mater. Sci. Eng. C 74 (2017) 515–524.
[24] A. Halder, M. Zhang, Q. Chi, Electrocatalytic applications of graphene–metal oxide nanohybrid materials, in: L.E. Norena, J.-A. Wang (Eds.), Advanced Catalytic Materials—Photocatalysis and Other Current Trends, InTech, Rijeka, Croatia, 2016, p. 14.
[25] Z. Yu, H. Li, J. Lu, X. Zhang, N. Liu, X. Zhang, Hydrothermal synthesis of Fe_2O_3/graphene nanocomposite for selective determination of ascorbic acid in the presence of uric acid, Electrochim. Acta 158 (2015) 264–270.
[26] S. Yang, G. Li, G. Wang, D. Deng, L. Qu, A novel electrochemical sensor based on Fe_2O_3 nanoparticles/N-doped graphene for electrocatalytic oxidation of L-cysteine, J. Solid State Electrochem. 19 (2015) 3613–3620.
[27] Y. Wu, C. Hu, M. Huang, N. Song, W. Hu, Highly enhanced electrochemical responses of rutin by nanostructured Fe_2O_3/RGO composites, Ionics 21 (2015) 1427–1434.
[28] S. Radhakrishnan, K. Krishnamoorthy, C. Sekar, J. Wilson, S.J. Kim, A promising electrochemical sensing platform based on ternary composite of polyaniline-Fe_2O_3-reduced graphene oxide for sensitive hydroquinone determination, Chem. Eng. J. 259 (2015) 594–602.
[29] C. Wu, Q. Cheng, L. Li, J. Chen, K. Wu, Synergetic signal amplification of graphene-Fe_2O_3 hybrid and hexadecyltrimethylammonium bromide as an ultrasensitive detection platform for bisphenol A, Electrochim. Acta 115 (2014) 434–439.
[30] M. Du, T. Yang, X. Guo, L. Zhong, K. Jiao, Electrochemical synthesis of Fe_2O_3 on graphene matrix for indicator-free impedimetric aptasensing, Talanta 105 (2013) 229–234.
[31] B. Sun, Y. Gou, Y. Ma, X. Zheng, R. Bai, A.A.A. Abdelmoaty, F. Hu, Investigate electrochemical immunosensor of cortisol based on gold nanoparticles/magnetic functionalized reduced graphene oxide, Biosens. Bioelectron. 88 (2017) 55–62.

CHAPTER 10

Magnetic nanocomposite: synthesis, characterization, and applications in heavy metal removal

Jitendra Kumar Sahoo[1], **Harekrushna Sahoo**[2]
[1]*Department of Basic Science and Humanities, GIET University, Gunupur, Odisha, India;*
[2]*Department of Chemistry, National Institute of Technology, Rourkela, Odisha, India*

Chapter Outline

1. Introduction

Use of recycled wastewater for drinking purposes is a major challenging task. Due to the increase of population, urbanization, overuse of pesticides in agriculture, industrial effluents are highly responsible for wastewater generation. There are different types of water contaminants such as organic contaminants like benzene hexachloride, acrylamide, phenol, carbon tetrachloride, and different organic dyes like congo red, methyl blue, methyl orange, malachite green, rhodamine b, and inorganic pollutants like cadmium, chromium, arsenic, fluoride, lead, and mercury [1–5]. The most toxic inorganic pollutants like fluorides and heavy metal ions are important nutrients in certain permissible limit, but above the permissible limit that threats to human health and environment. The maximum

Design, Fabrication, and Characterization of Multifunctional Nanomaterials. https://doi.org/10.1016/B978-0-12-820558-7.00021-2

concentration and permissible limit of the inorganic pollutants in drinking water have been set by World Health Organization (WHO) are shown in Table 10.1. Different metals are existing in different oxidation state and the toxicity is varied in oxidation state. Inorganic pollutants such as arsenic ($As^{+3}$and As^{+5}), chromium ($Cr^{+3}$and Cr^{+6}), lead (Pb^{+2}), mercury (Hg^{+2}), and cadmium (Cd^{+2}) are difficult to biodegradable, poisonous at low concentration and due to their small size easily accumulate in the environment and living organism. Heavy metals enter into our body via drinking water, air, and food. Some metals like selenium, copper, and zinc are important in trace quantities to maintain the vital metabolic processes in humans; however if the concentration increases it shows reverse effect on living organism like respiratory disorders, chromosomal fractures, mutagenic, lungs damage, nausea, kidney damage, cancer, and so forth [6–9]. From environmental safety point of view, before discharging the polluted water in to natural water stream is required to remove all the inorganic pollutants. Several removal process have been used like precipitation, osmosis, activated carbon filter, ion exchange, membrane separation, adsorption, photocatalysis, nanofiltration, and solvent extraction [10,11]. Among all removal techniques, researchers focused on adsorption because of its simple operation, low residual product generation, and cost effectiveness [12–14].

Any materials with very small scale, i.e., approximately 1–100 nm, are called nanomaterials. They may be one dimensional, two dimensional, and three dimensional materials [5,15]. Nanomaterials containing nanorods, nanowires, nanocomposites, nanotubes, and nanoparticles possess unique properties like small dimension structure and high aspect ratio due to which they are used in composite material, nanocatalysis, biomedical application, adsorption, and filter media [16–18]. Nanocomposites are the composite systems in which at least one of the constituent phase in the nanometer range (1 nm = 10^{-9}). In the past 11 years, nanocomposite materials have proven as suitable alternative over microcomposites and monolithic adsorbents [19].

Table 10.1: The maximum contamination levels of toxic inorganic pollutants.

Heavy metals	WHO permissible limits (mg/L)
Arsenic	0.010
Mercury	0.002
Chromium	0.100
Nickel	0.100
Cobalt	0.002
Thallium	0.002
Barium	2.000
Beryllium	0.004
Antimony	0.006
Fluoride	1.500
Asbestos	0.010
Selenium	0.050

In past decades, a number of researchers are interested on theoretical and experimental approaches of synthesis, characterization, and application of different inorganic structures like ceramics, metal oxides, and nanocomposite materials that resulted in multidisciplinary research fields. Among various inorganic nanoparticles, metal oxide has very much attention from engineering and science point of view. Due to the higher density and small size, metal oxide nanomaterials demonstrate unique chemical and physical prosperities [20]. Metal oxide has different properties such as fuel cells, magnetic, photoelectrochemical, mechanical, thermal, optical, electrochemical, optoelectronic, and catalytic properties [21]. Metal oxide has a large application in various fields like wastewater treatment like degradation and adsorption, catalysis, and sensors [22]. Some of the metal oxides such as copper oxide (CuO), magnesium oxide (MnO_2), cerium oxide (CeO_2), zirconium oxide (ZrO_2), iron oxide (Fe_2O_3, Fe_3O_4), and titanium oxide (TiO_2) have been widely used in wastewater treatment and many other purpose. Among all metal oxides iron oxide nanoparticles have greater interest because of its high surface are compare to volume ratio, excellent magnetic character, using external magnetic field easily separated from aqueous solution, comparatively low cost, small size, low toxicity, and environmental friendly nature [23,24]. The iron oxides have different phases like FeO, $Fe(OH)_3$, Fe_3O_4, Fe_4O_5, Fe_4O_3, FeOOH, polymorphs of Fe_2O_3 (α-Fe_2O_3, and γ-Fe_2O_3), and so on. Among various phase, three main phase like hematite (α-Fe_2O_3), maghemite (γ-Fe_2O_3), and magnetite (Fe_3O_4) are greater interest due to their applications like magnetic properties, electrical, drinking water treatment, ferrofluid technology, optical properties, gas sensing, and magneto caloric refrigerant [25–27]. The Fe_3O_4 have lot of impact on water remediation like easy and fast production, high adsorption capacity, easy separation, rapid uptake, etc. [28]. The separation of Fe_3O_4 in aqueous solution is easy by external magnetic field after adsorption of toxic substance [29,30]. Therefore, most of the researcher uses Fe_3O_4 for removal of inorganic and organic contaminants from wastewater. Today some chief, nontoxic, and environmental friendly materials support iron oxide to removing inorganic pollutants to achieve high adsorption efficiency because these materials can form stable chelates with heavy metal ions [31,32].

This chapter summerizes the synthesis of iron oxide and modified or functionalized iron oxide (Functionalize materials like activated carbon, biochar, polymer, amine functionalize) are used as adsorbents for the removal of inorganic toxic environmental pollutants (mainly chromium, arsenic, lead, mercury, flouride, cadmium, nickel, and cobalt) from aqueous solution. The functionalized iron oxide materials have novel nanocomposite materials for efficient removal of toxic ions and improved the maximum adsorption capacity. The adsorption affinity, mechanisms, and factors sorption capacity are mainly focused here.

2. Preparation of iron oxide-functionalized magnetic nanocomposites

2.1 Preparation of iron oxide magnetic nanoparticles

Different methods like biological, chemical and physical are available for preparation of iron oxide. Most of the researchers have greater interest on chemical method because bulk nanoparticles is formed with required shapes, sizes, desired morphologies, and with new important scientific properties [33]. There are different experimental methods available for the preparation of iron oxide nanoparticles such as microemulsion, thermal decomposition, precipitation, hydrothermal electrochemical process, sol-gel synthesis, sonochemical synthesis, and laser pyrolysis [34–37]. Among all synthetically methods, most of the researchers focused on coprecipitation method for iron oxide because of less time to prepare, large scale production, simple operation and good yield. Coprecipitation method of iron oxide briefly described as follows.

Iron oxide (Fe_3O_4) was synthesized using previously reported chemical coprecipitation method, which was shown in Fig. 10.1 with some modification [38]. Briefly, 4 g of anhydrous $FeCl_2$ and 12 g of anhydrous $FeCl_3$ were dissolved in 50 mL of 0.1 M HCl solution. Further, the mixture was added slowly to 500 mL of 1.5 M NH_3 solution until reaches pH 10, afterward, stirred for 2 h at the temperature 40°C. A black precipitate of

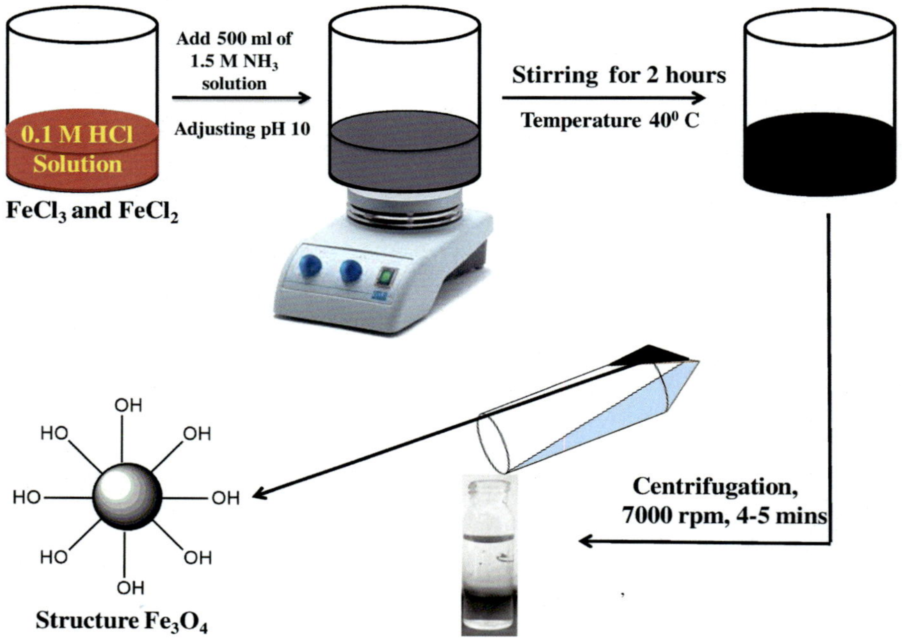

Figure 10.1
Schematic representation for the chemical synthetic procedure of iron oxide via co-precipitation method.

Fe_3O_4 magnetic nanoparticle was formed after stirring for 2 h, which was collected by centrifugation and washed three times with distilled water and two times with ethanol, and then dried at 70°C.

2.2 Characterization of magnetic nanoparticles (Fe_3O_4)

Compeán-Jasso et al. [39], have reported characterization of iron oxide nanoparticles. In FT-IR spectrum of Fe_3O_4 nanoparticles shows 441 cm^{-1}attribute to the shifting of the υ_2 band of the Fe—O bond of bulk iron oxide phase, two strong peaks at 583 cm^{-1} and 628 cm^{-1}attribute to the presence of Fe—O and two peaks at 3431 cm^{-1} and 1622 cm^{-1} attribute to the stretching and bending frequency of water molecules. The X-ray diffraction (XRD) patterns of Fe_3O_4, they got six intense peaks at $2\theta = 30.3$ degrees (220), 35.7 degrees (311), 43.5 degrees (400), 53.9 degrees (422), 57.5 degrees (511), and 63.0° (440) corresponds to the structure of Fe_3O_4 according to the JCPDS no. 019—0629. Since the XRD patterns of magnetite and maghemite have very similar, they performed Mössbauer spectroscopy technique to distinguish the oxidation state of iron as magnetite contain Fe^{+2} oxidation state and maghemite contain Fe^{+2} and Fe^{+3} oxidation states. The obtained Mössbauer spectrum data were fitted into two sextets patterns corresponding to the Fe^{+3} tetrahedral and octahedral site of iron (Fe^{+2} and Fe^{+3}) (hyperfine field were 433.6 KOe and 471.3 KOe finally), which corroborates the presence of magnetite. The sextets with quadruple splitting, hyperfine field, and isomer shift of one are 0.019 mm/s, 433.6 KOe, and 0.317 mm/s respectively, and the other 0.001 mm/s, 471.3 T, and 0.228 mm/s relative to the iron metal, suggesting the existence of two iron with different oxidation states considered as ferrimagnetic magnetite. Sahoo et al. [11], characterize TEM analysis and found the average particles size of iron oxide was 12.15 nm.

3. Removal of inorganic pollutants from water using iron oxide nanoparticles

In the present section we will describe current literature results related to the application of iron oxide nanomaterials for the removal of heavy metal ions mainly chromium, arsenic, lead, mercury, flouride, cadmium, nickel, and cobalt. Generally, the interaction of heavy metal ions and adsorbent containing active functional group or higher surface area plays an essential role in maximum adsorption capacity. In this section, the adsorption heavy metal ions on iron oxide and functionalize iron oxide are focused. Some important parameters like adsorption affinity, adsorption mechanism, factors (such as temperature, time, and pH), and effect of adsorption capacity are mainly focused.

3.1 Removal of heavy metal ions using iron oxide nanoparticles

Giraldo et al. [40] synthesized Fe_3O_4 magnetic nanoparticles by coprecipitation method. The nanoparticles size was 8 nm. The authors were tested the nanoparticles toward metal ions, i.e., Pb(II), Cu(II), Zn(II), and Mn(II) adsorption. The experimental result was shows that the adsorption capacity of Fe_3O_4 magnetic nanoparticles is maximum for Pb(II) and minimum for Mn(II), likely due to a various electrostatic attraction between cations of heavy metal and negatively charged adsorbents site, mainly related to hydrated ionic radii of the investigated heavy metals. They vary some parameters like pH, time, and temperature, which influence the adsorption of metal ions and the experimental results shows that the adsorption was strongly influenced by pH and temperature.

Zhang et al. [41] Study the adsorption of Cr(VI) and Cu(II) in mixed solution on magnetite (Fe_3O_4) nanoparticles via batch experiments. In batch adsorption experiments, they influence various parameters such as, adsorbent dosage, temperature, initial metal ion concentration, pH, and coexisting ions. The optimum pH value for Cr(VI) and Cu(II) adsorption is 4. The mechanism of adsorption was investigated through the study of surface properties of magnetite, the presence of ions, and the influence of pH and zeta potential. The result shows that pH has an influence on adsorption for Cr(VI) and Cu(II). The removal efficiency increases with an increase in adsorbent dosage. The maximum adsorption capacity of Cr(VI) and Cu(II) reached 8.67 and 18.61 mg/g. Both the metal ions followed Langmuir isotherm and pseudosecond-order model. Due to the good magnetic properties, it could easily be separated from aqueous solution using magnetic field.

Al-Saad et al. [42] Investigate the adsorption behavior of the iron oxide nanoparticles and its adsorption study was applicable to purify the water from Al(III), As(III), Cd(II), Co(II), Cu(II), and Ni(II). The adsorption of various metal ions was studied under various conditions like pH, contact time, dosage, temperature, and metal ion concentration. They reported the maximum percentage removal of both As(III) and Cu(II) reached more than 95%, while the other metal had percentage between 35% and 65%.

Mayo et al. [43] Studied the nanoscale iron oxides for the As(III) and As(V) adsorption focusing on magnetite (Fe_3O_4) nanoparticles. They reported the particle size of Fe_3O_4 could influence on the adsorption and desorption behavior of As(III) and As(V). They estimate that as the particle size decline from 300 to 12 nm the adsorption capacities incline to about 200 times for As(III) and As(V). Accordingly, this increase is higher than expected surface area. They also reported the arsenic removal is different in both bulk and nanoscale iron oxide.

Nassar et al. [44] Prepared cost-effective iron oxide nanoadsorbents that provide high adsorption capacity, simple regeneration, and separation and higher adsorption rate.

They suggest iron oxide nano adsorbents had been employed for the adsorption of Pb(II) ions from contaminated water by batch adsorption techniques. The maximum equilibrium reached within 30 min and the amount of Pb(II) removal efficiency increases with increase in temperature. The optimum pH value 5.5. Furthermore, the addition of coexisting cations such as Ca(II), Ni(II), Co(II), and Cd(II) had no remarkable effect on Pb(II) adsorption. The adsorption equilibrium followed Langmuir isotherm as well as Freundlich isotherm models. The thermodynamic results shows, the Pb(II) adsorption on Fe_3O_4 magnetic nanoparticles indicate spontaneous, endothermic, and physical in nature. The reusable capacity of nanoadsorbents proven that Fe_3O_4 magnetic nanoparticles could be employed repeatedly without impacting its removal capacity. Lingamdinne et al. modified tangerine peel extract on iron oxide nanoparticles to enhance the removal efficiency. It was observed that 99% of Pb(II) adsorption achieved with 0.6 g/L dosage, 10 mg/L initial Pb(II) concentration at room temperature.

Chowdhury and Yanful et al. [1] Experimented for adsorption of arsenic and chromium using mixed magnetite and meghemite nanaoparticles from aqueous solution. At optimum pH conditions, 96%–99% uptake of arsenic was monitor under controlled pH condition. Accordingly, at pH two and initial concentration of 1.5 mg/L for As(III) and As(V) species, the maximum adsorption capacity was studied at the values 3.69 and 3.71 mg/g, whereas the adsorption capacity of Cr(VI) was observed 2.4 mg/g at same pH but the initial concentration was 1 mg/L. Their consequence also explains the limitation of both arsenic species and chromium uptake by the nanoscale magnetite-maghemite mixture in the presence of a competing anion such as phosphate.

Sahu et al. [45] Prepared an efficient, cost-effective, and eco-friendly iron oxide using Exttran (biodegradable surfactant) by microemulsion method to enhance the adsorption efficiency of As(III) from aqueous solution. The equilibrium isotherm follows Langmuir isotherm. The maximum adsorption capacity of As(III) was found to be 7.18 mg/g at pH 7.7 in room temperature.

3.2 Removal of heavy metal ions using activated carbon-modified iron oxide nanocomposites

Kakavandi et al. [46] Synthesized powder activated carbon mobilized with iron oxide nanoparticles (Fe_3O_4@C) as an efficient adsorbent toward Pb(II) from aqueous solution. The maximum adsorption capacity was 71.4 mg/g at 50°C. The Fe_3O_4@C was a reusable after washing with HCl solution. Parlaylcl and Pehlivan [47] prepared magnetic-plum stone (*Prunus nigra*) activated carbon (m-PSAC) with phosphoric acid in N_2 atmosphere. The m-PSAC was novel for the adsorption of Cu(II) and Pb(II) ions from aqueous solution. The adsorption data showed that removal of Cu(II) and Pb(II) followed Langmuir isotherm model and pseudosecond-order kinetic model. From Langmuir isotherm, the adsorption capacity of Cu(II) and Pb(II) were 48.3 and 80.6 mg/g, respectively. The adsorption capacity of Pb(II) on m-PSAC was quite higher than Fe_3O_4@C. Mahmouda et al. [48] in 2019 prepared activated carbon (AC) and oxidized activated carbon (OAC) modified on Magnetite Fe_3O_4 as an efficient adsorption of Cu(II) from waste water. They found that the removal efficiency of Cu(II) was 96.1% after 30 min. The adsorption process followed pseudosecond-order kinetics and the equilibrium data follow Langmuir isotherm.

Jain and the group members [49] are prepared iron oxide/activated carbon (Fe_3O_4/AC) by coprecipitation methods. The Fe_3O_4/AC used for the removal of Cr(VI), Cu(II), and Cd(II). The adsorption capacity of Cr(VI), Cu(II), and Cd(II) was 4.4, 2.7, 2.9 mg/g, which was quite lower. Bharath et al. [50] prepared activated carbon derived from mesoporous peanut shell (PSAC) at low temperature (500°C) pyrolysis methods. After that, Fe_3O_4/PSAC nanocomposite was synthesized via one-pot hydrothermal synthesis method at 180°C. The Fe_3O_4/PSAC nanocomposite exhibit highly mesoporous structure with high surface area (680 m^2/g). The Fe_3O_4/PSAC nanocomposite was novel materials for Cr(VI) removal with adsorption capacity 24.5 mg/g at 1.2.

Imam and Zango [51] prepared activated carbon (derived from coconut shell) modified on iron oxide (AC- Fe_3O_4) for Ni(II) adsorption from aqueous solution. The adsorption capacity of Ni(II) on AC- Fe_3O_4 was 8.2 mg/g. In 2020 Ngan et al. [52] Experiments to adsorb Ni(II) from aqueous using lotus pods and iron (III) chloride (Fe_3O_4 −NPs/AC). The advantages combined activated carbon of lotus seed and Fe_3O_4 nanoparticles shows excellent adsorption efficiency for Ni(II) ions with the maximum adsorption capacity 50.7 mg/g which was larger than the AC- Fe_3O_4 at optimal conditions (time 60 min, pH 6, dosage 4.0 g/L and room temperature). The adsorbent having good reusable efficiency because after five cycle the removal efficiency 86.2%.

Sahu et al. [53] adsorb As(III) and As(V) from aqueous solution using Fe_3O_4/CSAC adsorbent. The Fe_3O_4/CSAC was prepared by the thermal method using cigarette soot activated carbon (CSAC) as a template. The surface area of Fe_3O_4/CSAC was 575.6 m^2/g and pore size 6.8 nm and the particle size was less than 10 nm. The adsorption efficiency was depend on pH, 91% of As(III) and 93% of As (V) removed by the adsorbent at pH seven and pH 3. The adsorption equilibrium followed Langmuir isotherm with adsorption capacity 80.9 mg/g for As(III) and 107.9 mg/g for As(V), respectively. Some coexisting anions had quite impact on As(III) and As(V). The Fe_3O_4/CSAC adsorbent reusable after washing with 0.5 M NaOH solution, after four cycles without any major decrease in removal efficiency. In the mechanism part, hydroxyl groups present in the Fe_3O_4/CSAC adsorbent played a key role for As(III) and As(V) adsorption.

3.3 Removal of heavy metal ions using biochar-modified iron oxide nanocomposites

Biochar is a carbon-based material prepared by pyrolysis method in the absence of oxygen. Biochar have excellent adsorption capacity of heavy metals but they have some limitations like difficult to separate from solution and reusable after adsorptions. To overcome these limitations, current researcher modified biochar on iron oxide to enhance the applications properties. Son et al. [54] developed an engineered magnetic-biochar by pyrolyzing waste marine macroalgae as a feedstock, and we doped iron oxide particles to impart magnetism. They compare the physicochemical properties and heavy metal adsorption with pinewood sawdust biochar and reported marine algae-based magnetic-biochar exhibit greater potential. The magnetic-biochar partially reduced its adsorption efficiency of heavy metal as compare to bare biochar because the iron oxide blocks the pores of biochar surface but due to the magnetic nature of magnetic-biochar, it is easily separated from aqueous solution. The optimum concentration of the iron loading solution for the magnetic-biochar was to be 0.025–0.05 mol/L. The magnetic-biochar has excellent removal techniques toward Cu(II) than Zn(II) and Cd(II). The removal corresponds to the presence of different adsorption sites like carboxy (-COOH) and hydroxyl (-OH).

Another researcher Shafiee et al. [55] and the group members are prepared oak power modified magnetic (OP/Fe_3O_4) composite for three heavy metal adsorption like Pb(II), Co(II), and Ni(II) ions from aqueous solution. They got from XRD data OP/Fe_3O_4 and OP were in crystalline form. They perform different kinetic and isotherm models and got pseudosecond-order kinetics and Freundlich isotherm are best fit. This supports the adsorption follows heterogeneous surface mechanism. From thermodynamic data indicates that the process of the metal ion adsorption is exothermic and spontaneous.

Lalhmunsiama et al. [56]. prepared dried biomass of *Chlorella vulgaris* (CV) modified on magnetic materials and successively employed in the adsorption of Cd(II) and Pb(II) from aqueous solution. Higher uptake capacity observed for both the heavy metals at wide range of initial adsorbate concentrations and pH. If you increases the electrolyte ($NaNO_3$) concentrations, the uptake of Cd(II) affected but Pb(II) not affected. This implied that Pb(II) ions were aggregated on the solid surface forming inner sphere complexes whereas Cd(II) was adsorb with weaker forces. Furthermore the adsorption mechanism of Cd(II) shows weak electrostatic attraction and Pb(II) ions shows chemically bound with the amino group of magnetic-CV. In 2020 Li et al. [57] and the comembers are fabricate different biochars like rice husk and coconut husk. Wheat stalk with magnetic Fe_3O_4 for the adsorption of Pb(II). The maximum adsorption capacity of Pb(II) removal was higher that other reported adsorbents like magnetic-CV and OP/Fe_3O_4.

3.4 Removal of heavy metal ions using polymer-modified iron oxide nanocomposites

The surface modification of polymer on iron oxide nanoparticles has been great impact on higher adsorption capacities toward heavy metal ions. Polymeric modification provides a number of advantages on iron oxide such as high mechanical strength, better chemical stability, and biocompatibility. Many scientists are reported various polymers fabricated iron oxide to improve the adsorption properties. For example, a magnetic nanoparticles fabricated with poly(maleic anhydride)-grafted-poly(vinyl alcohol) were reported by Liu et al. [58]. The author illustrate the efficiency of magnetic poly(maleic anhydride)-grafted-poly(vinyl alcohol) nanocomposite for the purify of water containing Ni(II), Ag(I), Cd(II), Pb(II) and Co(II) ions. The adsorption properties is purely pH sensitive and the mechanism of adsorption elucidate as pH- driven changes in electrostatic interaction and molecular morphology. After five consecutive adsorption/desorption cycles, the uptake efficiency 80%. Similarly, Badruddoza et al. [59] present another polymer like carboxymethyl-β-cyclodextrin (CM-β-CD) modified Fe_3O_4 nanoparticles for selective removal of Ni(II), Cd(II), Pb(II) ions from water. The equilibrium of adsorption reached in 45 min and the adsorption capacity of Ni(II), Cd(II), and Pb(II) were 13.2, 27.7, and 64.5 mg/g, at room temperature. The Pb(II) has preferentially high adsorption capacity as compare to Ni(II) and Cd(II), which could explain by hard soft acid base (HSAB) theory. Another polymer iminodiacetic acid grafted poly(glycidylmethacrylate-maleic anhydride) modified on magnetic materials (MNCPs) were presented by Hasanzadeh et al. [60]. The experiment results that the MNCPs showed high tendency toward the adsorption of Pb(II) and Cd(II).

Martinez-Cabanas et al. chitosan functionalized iron oxide nanoparticles for the effective adsorption of As(V). A green approach has been used for the preparation of iron oxide nanoparticles using extract of eucalyptus derived from plant. The batch adsorption experiments shows the As(V) removal with small equilibrium time and good adsorption capacity at neutral pH [61]. In another experiments, magnetic nanoparticles modified with hydroxyapatite (MNHAP) and tested for the adsorption of Cd(II) and Zn(II) [62].

3.5 Removal of heavy metal ions using amine-functionalized iron oxide nanocomposites

Amine functionalizes magnetic nanocomposites form complexation with heavy metal ions which increases the adsorption capacity. Huang and Chen illustrate the use of novel magnetic iron oxide adsorbent functionalize with polyacrylic acid (PAA) and diethylenetriamine (DETA) for the removal of Cu(II) and Cr(VI) [63]. The adsorption capacity of Cu(II) was higher than Cr(VI). Another researcher fabricated acrylic acid (AA) and crotonic acid (CA) grafted iron oxide nanoparticles modified with

3-aminopropyltriethoxysilane (APTES) for the removal of various heavy metals in given order Pb(II)>Cu(II)>Zn(II)>Cd(II) [64]. The order of the adsorption attribute to various complex capacity of the carbonyl interaction with the metal ions.

Lin et al. [65] using simple technique to synthesize amine-functionalized magnetic iron oxide nanoparticles (MIONPs-NH_2), which ascribe good adsorption properties for cationic-type heavy metals like Cu(II), Zn(II), Cd(II), Pb(II), and Ni(II). The mechanistic art, the amino groups of MIONPs-NH_2 form complexation with heavy metals. Sahoo et al. [66] prepared EDTA functionalized iron oxide nanocomposite (FAE) and APTES used as across linker in between them. The FAE nanocomposite found to be good adsorbent for Pb(II), Cd(II), Ni(II), Co(II), and Cu(II) with uptake capacity 11.31, 13.88, 7.64, 4.86, and 78.67 mg/g, respectively. They explain the small size of Cu(II) increase the uptake capacity. Similarly, another researcher functionalized chelating ligand, i.e., triethylenetetramine with mesoporous superparamagnetic Fe_3O_4 nanoparticles [67]. The functionalized materials had been ascribe to the adsorption Cu(II) from waste water. Tan et al. [68] adsorbed Pb(II) ions from aqueous solution using amino-functionalized Fe_3O_4 nanoparticles.

Shen et al. [69] demonstrated the mechanism of Cr(VI) on the surface of magnetic polymer grafted with tetraethylene pentaamine, i.e., TEPA-NMPs. The batch adsorption results shows the removal of Cr(VI) on the surface of synthesized materials due electrostatic interaction and also the reduction of Cr(VI) into Cr(III) through charge transfer on the adsorbent surface. In addition Shen et al. [70] prepared a number of core-shell Fe_3O_4 nanoparticles fabricated with different multi amino groups i.e., NH_2-NMPS for the removal of Cu(II) and Cr(VI) ions in sole metal ion system and coexisting metal ion system. The removal process was found to be highly dependent on pH of the medium for both Cu(II) and Cr(VI) ions. Liu et al. [71] synthesized cost-effective and environmental friendly humic acid (HA) modified Fe_3O_4 nanoparticles (Fe_3O_4/HA) for the removal of Pb(II), Cd(II), Cu(II) and Hg(II). The authors compare the stability and adsorption capacity of Fe_3O_4/HA as compared to Fe_3O_4 nanoparticles.

4. Conclusion

This chapter concludes with the development of different AC, polymer, amine-functionalized, and biochar modified iron oxide nanoadsorbent toward the removal of heavy metal ions (mainly As, Cd, Cr, Hg, Pb, Ni, Co, and Cu) from aqueous solution. The adsorbents have been specific layered functionality structure, high surface area, and more numbers of binding sites such as carboxyl, hydroxyl, and amine functional groups improve the removal efficiency. In a mechanistic way, the electrostatic interaction, low pH values were important factors for the removal of anionic pollutants like arsenic and chromium, while a high pH value favor the removal of cationic pollutants like (Cd, Hg, Pb, Ni, Co, and Cu). The negatively charged metal atom ions have adsorbed on adsorbent surface through specific and/or nonspecific adsorption. Whereas, positively charged metal atom ions have electrostatic interaction, ion exchange, and complex formation. Furthermore, the impact of various parameters such as adsorbent dosage, initial time, temperature, initial concentration of metal ion and ionic strength of solution on removal process have also been summarized. Most of the adsorbents followed Langmuir isotherm and pseudosecond-order kinetic model. The reusable of the adsorbent could be appreciated with desorption agents mainly as HCl or NaOH or various alcohol solution. The removal efficiency was found to be decreased with the increasing number of reusable cycles.

References

[1] S.R. Chowdhury, E.K. Yanful, Arsenic and chromium removal by mixed magnetite e maghemite nanoparticles and the effect of phosphate on removal, J. Environ. Manag. 91 (2010) 2238–2247.

[2] M. Bhaumik, A. Maity, V.V. Srinivasu, M.S. Onyango, Enhanced removal of Cr (VI) from aqueous solution using polypyrrole/Fe_3O_4 magnetic nanocomposite, J. Hazard Mater. 190 (2011) 381–390.

[3] A. Ajmal, I. Majeed, N. Malik, Principles and mechanisms of photocatalytic dye degradation on TiO 2 based photocatalysts, RSC Adv. (2014) 37003–37026.

[4] J. Xu, Z. Cao, Y. Zhang, Z. Yuan, Z. Lou, X. Xu, X. Wang, A review of functionalized carbon nanotubes and graphene for heavy metal adsorption from water: preparation , application, and mechanism, Chemosphere 195 (2018) 351–364.

[5] X. Lu, C. Wang, Y. Wei, One-dimensional composite nanomaterials: synthesis by electrospinning and their applications, Small 5 (2009) 2349–2370.

[6] J.M. Chem, L. Fan, C. Luo, M. Sun, H. Qiu, Synthesis of graphene oxide decorated with magnetic cyclodextrin for fast chromium removal, J. Mater. Chem. 22 (2012) 24577–24583.

[7] I.E. Mejias, J.D. Mangadlao, H.N. Nguyen, R.C. Advincula, D.F. Rodrigues, Graphene oxide functionalized with ethylenediamine triacetic acid for heavy metal adsorption and anti-microbial applications, Carbon 77 (2014) 289–301.

[8] W. Peng, H. Li, Y. Liu, S. Song, A review on heavy metal ions adsorption from water by graphene oxide and its composites, J. Mol. Liq. 230 (2017) 496–504.

[9] M. Ahmaruzzaman, Industrial wastes as low-cost potential adsorbents for the treatment of wastewater laden with heavy metals, Adv. Colloid Interface Sci. 166 (2011) 36–59.

[10] J.f. Peng, Y.h. Song, P. Yuan, X.Y. Cui, G.l. Qiu, The remediation of heavy metals contaminated sediment, J. Hazard Mater. 161 (2009) 633–640.

[11] J.K. Sahoo, A. Kumar, L. Rout, J. Rath, P. Dash, H. Sahoo, An investigation of heavy metal adsorption by hexa-dentate ligand-modified magnetic nanocomposites, Separ. Sci. Technol. 53 (2017) 863–876.

[12] D. Corte, A. Toro-labbe, Improving as (III) adsorption on graphene based surfaces: impact of chemical doping, Phys. Chem. Chem. Phys. 17 (2015) 12056–12064.

[13] O. Duman, S. Tunc, T.G. Polat, B.K. Bozo, Synthesis of magnetic oxidized multiwalled carbon application in cationic Methylene Blue dye adsorption, Carbohydr. Polym. 14 (2016) 79–88.

[14] V.A. Online, J. Zhang, C. Shi, J. Huang, B. Yan, Q. Liu, H. Zeng, Poly(acrylic acid) functionalized magnetic graphene oxide nanocomposite for removal of methylene blue, RSC Adv. 5 (2015) 32272–32282.

[15] J. Zhu, S. Wei, L. Zhang, Y. Mao, J. Ryu, N. Haldolaarachchige, D.P. Young, Z. Guo, Electrical and dielectric properties of polyaniline-Al_2O_3 nanocomposites derived from various Al_2O_3 nanostructures, J. Mater. Chem. 21 (2011) 3952–3959.

[16] J.A. Barreto, W.O. Malley, M. Kubeil, B. Graham, H. Stephan, L. Spiccia, Nanomaterials: applications in cancer imaging and therapy, Adv. Healthc. Mater. 23 (2011) 18–40.
[17] M.S. Mauter, M. Elimelech, Environmental applications of carbon-based nanomaterials, Environ. Sci. Technol. 42 (2008) 5843–5859.
[18] L.R. Khot, S. Sankaran, J. Mari, R. Ehsani, E.W. Schuster, Applications of nanomaterials in agricultural production and crop protection: a review, Crop Protect. 35 (2012) 64–70.
[19] P. Henrique, C. Camargo, K.G. Satyanarayana, F. Wypych, Nanocomposites: synthesis, structure, properties and new application opportunities, Mater. Res. 12 (2009) 1–39.
[20] M. Fernandez-Garcia, A. Martinez-Arias, J.C. Hanson, J.A. Rodriguez, Nanostructured oxides in chemistry: characterization and properties, Chem. Rev. 104 (2004) 4063–4104.
[21] L. Vayssieres, On the design of advanced metal oxide nanomaterials, Int. J. Nanotechnol. 1 (2004) 1–40.
[22] G. Oskam, Metal oxide nanoparticles: synthesis, characterization and application, J. Sol. Gel Sci. Technol. 37 (2006) 161–164.
[23] K.R. Reddy, W. Park, B.C. Sin, J. Noh, Y. Lee, Synthesis of electrically conductive and superparamagnetic monodispersed iron oxide-conjugated polymer composite nanoparticles by in situ chemical oxidative polymerization, J. Colloid Interface Sci. 335 (2009) 34–39.
[24] J. Zhu, K.Y.S. Ng, D. Deng, Micro single crystals of hematite with nearly 100% exposed 104 Facets: preferred etching and lithium storage, Cryst. Growth Des. 14 (2014) 2811–2817.
[25] H. Xu, X. Wang, L. Zhang, Selective preparation of nanorods and micro-octahedrons of Fe_2O_3 and their catalytic performances for thermal decomposition of ammonium perchlorate, Powder Technol. 5 (2008) 176–180.
[26] M. Mohapatra, S. Anand, Synthesis and applications of nano-structured iron oxides/hydroxides - a review, Int. J. Eng. Sci. Technol. 2 (2010) 127–146.
[27] C. Gregor, M. Hermanek, D. Jancik, J. Pechousek, J. Filip, J. Hrbac, R. Zboril, The effect of surface area and crystal structure on the catalytic efficiency of iron(III) oxide nanoparticles in hydrogen peroxide decomposition, Eur. J. Inorg. Chem. 2010 (2010) 2343–2351.
[28] R. Akhbarizadeh, M.R. Shayestefar, E. Darezereshki, Competitive removal of metals from wastewater by maghemite nanoparticles: a comparison between simulated wastewater and AMD, Mine Water Environ. 33 (2014) 89–96.
[29] L. Tan, J. Xu, X. Xue, Z. Lou, J. Zhu, A. Baig, Multifunctional nanocomposite Fe_3O_4 @SiO_2-mPD/SP for selective removal of Pb(II) and Cr(VI) from aqueous solutions, RSC Adv. 4 (2014) 45920–45929.
[30] A.F. Ngomsik, A. Bee, D. Talbot, G. Cote, Magnetic solid-liquid extraction of Eu(III), La(III), Ni(II) and Co(II) with maghemite nanoparticles, Separ. Purif. Technol. 86 (2012) 1–8.
[31] M. Rahim, M.R.H.M. Haris, Application of biopolymer composites in arsenic removal from aqueous medium: a review, J. Radiat. Res. Appl. Sci. 8 (2015) 255–263.
[32] J. Sun, X. Li, Q. Zhao, J. Ke, D. Zhan, Novel V_2O_5/$BiVO_4$/TiO_2 nanocomposites with high visible-light induced photocatalytic activity for the degradation of toluene, J. Phys. Chem. C 118 (2014) 10113–10121.
[33] M. Mahmoudi, A. Simchi, M. Imani, Recent advances in surface engineering of superparamagnetic iron oxide nanoparticles for biomedical applications, J. Iran. Chem. Soc. 7 (2010) S1–S27.
[34] C. Pascal, J.L. Pascal, F. Favier, M.L.E. Moubtassim, C. Payen, Electrochemical synthesis for the control of γ-Fe_2O_3 nanoparticle size. morphology, microstructure, and magnetic behavior, Chem. Mater. 11 (1999) 141–147.
[35] K.V.P.M. Shafi, A. Ulman, Y. Xingzhong, N.-L. Yang, C. Estournès, H. White, M. Rafailovich, Sonochemical synthesis of functionalized amorphous iron oxide nanoparticles, Langmuir 17 (2001) 5093–5097.
[36] Y. Lu, Y. Yin, B.T. Mayers, Y. Xia, Modifying the surface properties of superparamagnetic iron oxide nanoparticles through A Sol–Gel approach, Nano Lett. 2 (2002) 183–186.

[37] C.M. Bautista, O. Bomati-Miguel, D.P.M. Morales, C.J. Serna, S. Veintemillas-Verdaguer, Surface characterisation of dextran-coated iron oxide nanoparticles prepared by laser pyrolysis and coprecipitation, J. Magn. Magn Mater. 293 (2005) 20–27.

[38] B. Tang, L. Yuan, T. Shi, L. Yu, Y. Zhu, Preparation of nano-sized magnetic particles from spent pickling liquors by ultrasonic-assisted chemical co-precipitation, J. Hazard Mater. 163 (2009) 1173–1178.

[39] M.E. Compeán-Jasso, F. Ruiz, J.R. Martínez, A. Herrera-Gómez, Magnetic properties of magnetite nanoparticles synthesized by forced hydrolysis, Mater. Lett. 2 (2008) 4248–4250.

[40] L. Giraldo, A. Erto, J.C. Moreno-Piraján, Magnetite nanoparticles for removal of heavy metals from aqueous solutions: synthesis and characterization, Adsorption 19 (2013) 465–474.

[41] J. Zhang, S. Lin, M. Han, Q. Su, L. Xia, Z. Hui, Adsorption properties of magnetic magnetite nanoparticle for coexistent Cr(VI) and Cu(II) in mixed solution, Water 12 (2020) 446.

[42] K.A. Al-Saad, M.A. Amr, D.T. Hadi, R.S. Arar, M.M. AL-Sulaiti, T.A. Abdulmalik, N.M. Alsahamary, J.C. Kwak, Iron oxide nanoparticles: applicability for heavy metal removal from contaminated water, Arab. J. Nuc. Sci. & Appl. 45 (2012) 335–346.

[43] J.T. Mayo, C. Yavuz, S. Yean, L. Cong, H. Shipley, W. Yu, The effect of nanocrystalline magnetite size on arsenic removal, Sci. Technol. Adv. Mater. 8 (2007) 71–75.

[44] N.N. Nassar, Rapid removal and recovery of Pb(II) from wastewater by magnetic nanoadsorbents, J. Hazard Mater. 184 (2010) 538–546.

[45] U.K. Sahu, M.K. Sahu, S.S. Mahapatra, R.K. Patel, Removal of as(III) from aqueous solution using Fe_3O_4 nanoparticles: process modeling and optimization using statistical design, Water, Air, Soil Pollut. 228 (2017) 45.

[46] B. Kakavandi, R.R. Kalantary, A.J. Jafari, S. Nasseri, A. Ameri, A. Esrafili, A. Azari, Pb(II) adsorption onto a magnetic composite of activated carbon and superparamagnetic Fe_3O_4 nanoparticles: experimental and modeling study, Soil Air Water 43 (2015) 1157–1166.

[47] Ş. Parlayıcı, E. Pehlivan, Removal of metals by Fe_3O_4 loaded activated carbon prepared from plum stone (*Prunus nigra*): kinetics and modelling study, Powder Technol. 317 (2017) 23–30.

[48] A.S. Mahmouda, N.A. Youssefa, A.O.A.E. Nagab, M.M. Selimc, Removal of copper ions from wastewater using magnetite loaded on active carbon (AC) and oxidized active carbon (OAC) support, J. Sci. Res. Sci. 36 (2019) 226–247.

[49] M. Jain, M. Yadav, T. Kohout, M. Lahtinen, V.K. Garg, M. Sillanpaa, Development of iron oxide/activated carbon nanoparticle composite for the removal of Cr(VI), Cu(II) and Cd(II) ions from aqueous solution, Water Resour. Ind. 20 (2018) 54–74.

[50] G. Bharath, K. Rambabu, F. Banat, A. Hai, A.F. Arangadi, N. Ponpandian, Enhanced electrochemical performances of peanut shell derived activated carbon and its Fe_3O_4 nanocomposites for capacitive deionization of Cr(VI) ions, Sci. Total Envirno. 691 (2019) 713–726.

[51] S.S. Imam, Z.U. Zango, Magnetic nanoparticle (Fe_3O_4) impregnated onto coconut shell activated carbon for the removal of Ni(II) from aqueous solution, IJRCE 8 (2018) 9–15.

[52] T.T.K. Ngan, L.T.T. Nhi, L.H. Sinh, T.D. Lam, N.Q. Vinh, P.T. Minh, L.V. Thuan, Facile synthesis of Fe_3O_4 nanoparticles loaded on activated carbon developed from Lotus seed pods for removal of Ni(II) ions, J. Nano Res. 61 (2020) 1–17.

[53] U.K. Sahu, S. Sahu, S.S. Mahapatra, R.K. Patel, Cigarette soot activated carbon modified with Fe_3O_4 nanoparticles as an effective adsorbent for As(III) and As(V): material preparation, characterization and adsorption mechanism study, J. Mol. Liq. 243 (2017) 395–405.

[54] E.-B. Son, K.-M. Poo, J.-S. Chang, K.-J. Chae, Heavy metal removal from aqueous solutions using engineered magnetic biochars derived from waste marine macro-algal biomass, Sci. Total Environ. 615 (2018) 161–168.

[55] M. Shafiee, R. Foroutan, K. Fouladia, M. Ahmadlouydarab, B. Ramavandi, S. Sahebi, Application of oak powder/Fe_3O_4 magnetic composite in toxic metals removal from aqueous solutions, Adv. Powder Technol. 30 (2018) 544–554.

[56] Lalhmunsiama, P.L. Gupta, J. Hyunhoon, D. Tiwari, K. S.-H, L. Seung-Moka, Insight into the mechanism of Cd(II) and Pb(II) removal by sustainable magnetic biosorbent precursor to *Chlorella vulgaris*, J. Taiwan Inst. Chem. E 71 (2017) 206–213.
[57] Y. Li, X. Zhang, P. Zhang, X. Liu, L. Han, Facile fabrication of magnetic bio-derived chars by co-mixing with Fe_3O_4 nanoparticles for effective Pb^{2+} adsorption: properties and mechanism, J. Clean. Prod. 262 (2020) 121350.
[58] X. Liu, J. Guan, G. Lai, Q. Xu, X. Bai, Z. Wang, S. Cui, Stimuli-responsive adsorption behavior toward heavy metal ions based on comb polymer functionalized magnetic nanoparticles, J. Clean. Prod. 253 (2020) 119915.
[59] A.Z.M. Badruddoza, Z.B.Z. Shawon, T.W.J. Daniel, K. Hidajat, M.S. Uddin, Fe_3O_4/cyclodextrin polymer nanocomposites for selective heavy metals removal from industrial wastewater, Carbohydr. Polym. 91 (2013) 322–332.
[60] R. Hasanzadeh, P.N. Moghadam, N. Bahri-Laleh, M. Sillanpaa, Effective removal of toxic metal ions from aqueous solutions: 2-bifunctional magnetic nanocomposite base on novel reactive PGMAMAn copolymer@Fe3O4 nanoparticles, J. Colloid Interface Sci. 490 (2017) 727–746.
[61] M. Martínez-Cabanas, M. López-García, J.L. Barriada, R. Herrero, M.E.S.d. Vicente, Green synthesis of iron oxide nanoparticles: development of magnetic hybrid materials for efficient as (V) removal, Chem. Eng. J. 301 (2016) 83–91.
[62] Y. Feng, J.L. Gong, G.M. Zeng, Q.Y. Niu, H.Y. Zhang, C.G. Niua, J.H. Deng, M. Yan, Adsorption of Cd (II) and Zn (II) from aqueous solutions using magnetic hydroxyapatite nanoparticles as adsorbents, Chem. Eng. J. (2010) 487–494.
[63] H. Huang, D.H. Chen, Rapid removal of heavy metal cations and anions from aqueous solutions by an amino-functionalized magnetic nano-adsorbent, J. Hazard Mater. 163 (2009) 174–179.
[64] F. Ge, M.M. Li, H. Ye, B.X. Zhao, Effective removal of heavy metal ions Cd^{2+}, Zn^{2+}, Pb^{2+}, Cu^{2+} from aqueous solution by polymer-modified magnetic nanoparticles, J. Hazard Mater. 211–212 (2012) 366–372.
[65] S. Lin, L. Liu, Y. Yang, K. Lin, Study on preferential adsorption of cationic-style heavy metals using amine-functionalized magnetic iron oxide nanoparticles (MIONPs-NH_2) as efficient adsorbents, Appl. Surf. Sci. 407 (2017) 29–35.
[66] J.K. Sahoo, A. Kumar, L. Rout, J. Rath, P. Dash, H. Sahoo, An investigation of heavy metal adsorption by hexa-dentate ligand-modified magnetic nanocomposites, Separ. Sci. Technol. 53 (2018) 863–876.
[67] J. Gao, Y. He, X. Zhao, X. Ran, Y. Wuc, Y. Su, J. Dai, Single step synthesis of amine-functionalized mesoporous magnetite nanoparticles and their application for copper ions removal from aqueous solution, J. Colloid Interface Sci. 481 (2016) 220–228.
[68] Y. Tan, M. Chen, Y. Hao, High efficient removal of Pb (II) by amino-functionalized Fe_3O_4 magnetic nano-particles, Chem. Eng. J. 191 (2012) 104–111.
[69] H. Shen, J. Chen, H. Dai, L. Wang, M.H.Q. Xia, New insights into the sorption and detoxification of chromium (VI) by tetraethylenepentamine functionalized nanosized magnetic polymer adsorbents: mechanism and pH effect, Ind. Eng. Chem. 52 (2013) 12723–12732.
[70] H. Shen, S. Pan, Y. Zhang, X. Huang, H. Gong, A new insight on the adsorption mechanism of amino-functionalized nano-Fe_3O_4 magnetic polymers in Cu (II), Cr(VI) co-existing water system, Chem. Eng. J. 183 (2012) 180–191.
[71] J.F. Liu, Z. Shanzhao, G. Binjiang, Coating of Fe_3O_4 magnetic nanoparticles with humic acid for high efficient removal of heavy metals in water, Environ. Sci. Technol. 42 (2008) 6949–6954.

CHAPTER 11

Iron-based functional nanomaterials: synthesis, characterization, and adsorption studies about arsenic removal

M. Tripathy[1], S.K. Sahoo[1], M. Mishra[2], Garudadhwaj Hota[1]
[1]Department of Chemistry, National Institute of Technology, Rourkela, Odisha, India; [2]Department of Life Science, National Institute of Technology, Rourkela, Odisha, India

Chapter Outline

1. Introduction

Contamination of water due to the gradual increase in population and industrialization is the most stimulating concern for every environmentalist because it is continuously growing environmental threats to the entire living world. The leading threats of water pollution are mainly due to the presence of different toxic metal ions such as arsenic, lead, chromium, mercury, and cadmium in the water system. Among all, arsenic causes extremely carcinogenic and toxic effects on human health [1]. The increase in arsenic poisoning in both surface and groundwater is mainly due to the effluents coming from

Design, Fabrication, and Characterization of Multifunctional Nanomaterials. https://doi.org/10.1016/B978-0-12-820558-7.00013-3

various anthropogenic sources (i.e., agricultural chemicals, coal fly ash, sewage sludge, petroleum refining, ceramic manufacturing industries, and mining waste, etc.). Apart from this, few natural sources like volcanic emissions, erosion of rocks/soils, and weathering also subsidize arsenic concentration in water. Naturally, arsenic obtains more than in 200 various mineral forms, out of these around 60% are arsenates, 20% are sulfides, and sulfosalts and the remaining 20% are arsenides, arsenite, oxides, silicates, and elemental arsenic. Also, the leading arsenic holding arsenopyrite mineral is obtained by the interference of granitic magma and orogenesis [2]. In drinking water, arsenic mainly obtained as inorganic (arsenite As(III) and arsenate As(V)) and organic forms (methyl and dimethyl arsenic compounds). Considering the high noxiousness of arsenic, WHO has retained the maximum acceptable limit of arsenic in drinking water is about 0.01 mg/L. Due to, the gradual increase in arsenic poisoning in most of the developing countries, it acquires the attention of most of the researchers to develop advanced techniques for its elimination. For the treatment of such arsenic-contaminated water various methods such as oxidation, coprecipitation, ion-exchange, adsorption, and membrane filtration are the most common [3].

Among all, adsorption is considered the best one because it does not implicate expensive instrumentation and long method, easy operation, and safe to handle, effective even at low adsorbate concentration. Thus it can be used for both lab tests and commercial purposes [4]. However, adsorption has also certain shortcomings, such as lack of appropriate adsorbents with greater adsorption efficiency, less lifespan of synthesized adsorbents, and lack of commercially applicable adsorbents system. Furthermore, the same material is not applicable for the treatment of other water contaminants. For the treatment of polluted water, various adsorbents such as synthetic activated carbon, ferrous material, surfactants, industrial byproducts and wastes, mineral products, and iron-based soil amendment were used previously [1,5,6]. Considering the above issues, over the past two decades researchers have been trying to develop different nanoadsorbent materials for remediation of arsenic. Advanced nanomaterials have attracted today's researchers in solving various existing problems including environmental issues. The basic difference between buck-materials and nanomaterials is that the former have intraparticle diffusion properties that reduce the adsorption rate and existing ability of the adsorbent, while the latter have low diffusion resistance that improved its applicability toward adsorption application. Also, nanomaterials possess some unique properties like ultrasmall size, high aspect ratio, catalytic potential, a greater number of active sites, easy separation, and high reactivity that facilitate better sorption capacity [7,8].

For the synthesis of nanoadsorbents, different metal-based compounds such as iron, aluminum, magnesium, titanium, and cupper have been used for remediation of arsenic from water. Also, polymer-based nanomaterials play a crucial role in adsorption applications. Although nanoadsorbents contribute to improved performance as compared to other adsorbents for the sorption of arsenic, it also possesses few key challenges for batter applications, that includes nonavailability of economically affordable nanomaterials and the other is its toxicity and biocompatibility. The actual applicability of the materials mainly dependents on their types, availability, synthesis procedure, nontoxicity and low cost. Considering the above challenges researchers are trying to develop iron-based nanoadsorbents due to its easy availability, nontoxicity, low cost, and biocompatibility nature for arsenic sorption application [9–11]. However, iron-based nanomaterials possess certain limitation which includes the irreversible accumulation of individual nanoparticles, which may hinder their effectiveness toward sorption. Therefore, iron oxides nanomaterials have been functionalized with different functionality to improve their dispersibility and performance. Other distinctive properties of well-dispersed functional nanomaterials are that they can be directly implemented for in situ elimination of arsenic from contaminated water [2,12,13].

This chapter gives a comprehensive idea about the synthesis of iron-based nanomaterials mainly functional iron-based nanostructures and the basic characterization techniques required to determine the crystal phase, functionality, stability, morphological characteristics, surface area, and magnetic properties. Furthermore, the application of such materials toward remediation of arsenic and also the basic adsorption mechanism occurs between iron-based adsorbents and arsenic species.

2. Iron-based nanomaterials

The specialty of iron-based nanomaterials toward remediation of the toxic heavy metal ion is due to their outstanding characteristics like simplicity, resourcefulness, easy availability, and nontoxicity. The basic common phases of iron oxides and their hybrid architectures discussed in this chapter are the magnetic phase-magnetite (Fe_3O_4) and maghemite (γ-Fe_2O_3) and nonmagnetic, hematite (α-Fe_2O_3), and ferric hydroxide (FeOOH) [14]. These phases with their hybrid system can be engineered by a wide range of techniques such as coprecipitation, hydrothermal/solvothermal, thermal decomposition, electrospinning, chemical vapor deposition, and sol-gel. The structural and morphological features of the nanomaterials can be modified into different morphologies like nanoparticles [15], flowerlike structures [16,17], nanotubes [18], nanofibers [19], nanorods [20], and nanochains [21].

2.1 Synthesis

2.1.1 Synthesis of nonmagnetic and magnetic iron-based nanomaterials and their composites

2.1.1.1 Nonmagnetic iron-based nanomaterials

Hematite (α-Fe_2O_3) is the nonmagnetic phase of iron-oxide. The use of the nonmagnetic phase α-Fe_2O_3 for various applications like sensors, biomedical, catalysis, and environmental remediation is mainly because of its high stability, nontoxicity, easy synthesis procedure, and resourcefulness. The synthesis of various hierarchical architecture of α-Fe_2O_3 with different shapes and sizes is of great attention because of its unique properties compared to bulk materials [22]. Here we emphasize to interpret few recently reported nonmagnetic α-Fe_2O_3 nanostructures and their hybrid architectures for remediation of arsenic from water.

Cao et al. synthesized flowerlike α-Fe_2O_3 nanostructures for the removal of heavy metal ion (i.e., As(V) and Cr(VI)). Initially, $FeCl_3 \cdot 6H_2O$ (5 mmol) and urea (7.5 mmol) were mixed together in anhydrous ethanol (100 mL of 99%). In a Teflon-lined autoclave, 40 mL of prepared solution was taken and sealed in a programmable microwave oven. The oven was heated to 150°C in 2 min by microwave irradiation and then preserved the system at that temperature for another 30 min under microwave heating. Then the oven was cooled to room temperature. Afterward, the system was cooled to room temperature and the α-Fe_2O_3 precipitate was collected by subsequent centrifugation and washing and then dried at 80°C for 3 h [17]. The SEM and TEM micrograph of the prepared α-Fe_2O_3 nanostructures is presented in Fig. 11.1.

Tang et al. synthesized ultrafine α-Fe_2O_3 nanoparticles by the solvent thermal method and apply them for removal of arsenic from both lab-prepared and natural samples under neutral pH conditions. The synthesis process involves three different steps. Firstly 1.13 g of anhydrous $FeCl_3$ was dissolved into 80 mL ethanol in a beaker to prepare 0.1 M $FeCl_3$ solution, subsequently 0.8 g NaOH was added to it and the reaction was continuing up to 1 h at room temperature conditions under constant magnetic stirring. Then the obtained yellowish-brown suspension was transferred into a 100 mL Teflon-lined autoclave and placed in a preheated oven at 150°C for 2 h. Finally, the obtained red precipitate was washed three times using distilled water and once using ethanol and dried at 80°C for 12 h to get ultrafine α-Fe_2O_3 nanoparticles [23]. The crystal phase, surface morphology, and phase purity of the prepared α-Fe_2O_3 nanoparticles were determined by X-ray diffraction (XRD), transmission electron microscope (TEM), and high resolution transmission electron microscope (HRTEM) technique, as shown in Fig. 11.2.

Also, Alijani et al. synthesized an effective adsorbent (i.e., zero-valent iron-doped MWCNT) for remediation of aqueous arsenic by using natural α-Fe_2O_3 as a precursor via in situ CVD technique [22]. Huo et al. synthesized modified hydrous ferric oxide nanoparticles for effective removal of As(V) from water [8].

From the aforementioned studies we observed that nonmagnetic iron-based nanomaterials play a vital role in the removal of arsenic from water, but these materials have certain limitations like separation problem. To overcome such limitations, researchers are trying to develop magnetic iron-based nanomaterials for environmental remediation.

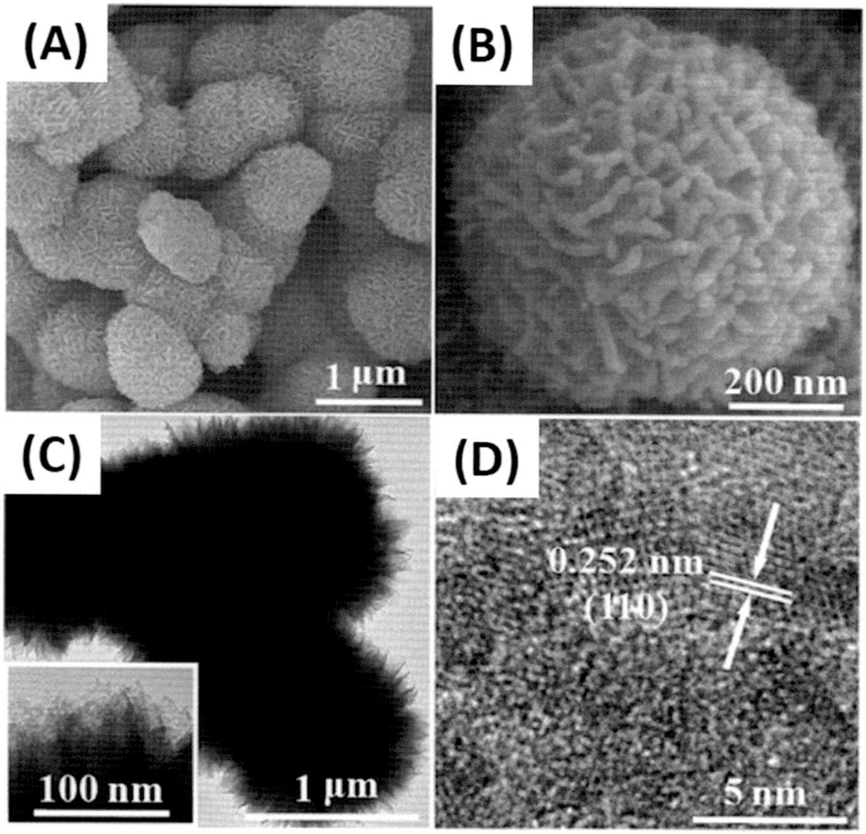

Figure 11.1
(A and B) Low and high-magnification SEM image, (C) TEM image, and (D) HRTEM image of flowerlike α-Fe_2O_3 nanostructures.

2.1.1.2 Magnetic iron-based nanomaterials

Separation and regaining of sorbent materials after the adsorption of lethal pollutants from water is a foremost contest. Magnetic forms of iron-oxide materials or their hybrid nanostructures offer a feasible result to collect and remove noxious species. This benefit is due to the easy separation of the sorbent materials by using a simple external magnet. There have been many reports on the use of magnetic nanoparticles for cleanup of heavy metals including arsenic [24], chromium [24–26], cobalt [27], lead [26], and nickel [27], etc. from aqueous solutions. The experimental conditions for the adsorption of these ions onto the magnetic nanoparticles differ. However, here the highlights covered are the morphology of the nanoadsorbent, how it was synthesized, and the adsorption efficiency toward remediation of arsenic (III/V) from aqueous medium.

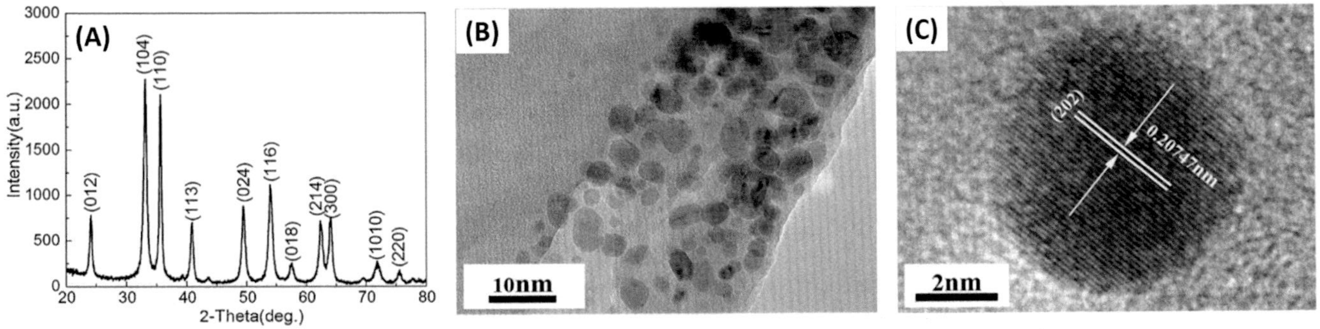

Figure 11.2
(A) XRD pattern, (B) TEM image, and (C) HRTEM image of α-Fe_2O_3.

As magnetite nanoparticles have a strong affinity toward adsorption of As(III) and As(V) and simply regenerated by using external magnetic fields, so these nanoparticles are ideal adsorbents for adsorption of arsenic species. The arsenic adsorption studies along with the effect of different aquatic contaminants like sulfate, silica, calcium, bicarbonate, dissolved organic matter, phosphate, magnesium, and iron was conducted by Shipley et al. [28]. Adsorption experiments took place in a sealed vessel containing 100 mg/L As(V) with 0.5 g/L magnetite nanoparticles in Houston tap water and 39 mg/L total arsenic-contaminated groundwater from Brownsville, TX with 0.5 g/L magnetite nanoparticles. In the case of spiked Houston tap water, 83 mg/L of As(V) was removed within 1 h. But for the arsenic-contaminated groundwater, the concentration of arsenic decreased to 10 mg/L within 10 min and 5 mg/L within 1 h. The experimental data obtained were well fitted with the Langmuir isotherm model. Also, the results confirm that the prepared adsorbents have the capacity to remove arsenic from water in the presence of other competitive species (i.e., phosphates and carbonates). The use of characteristics wastewater and drinking water samples revels the commercial applicability of the adsorbent, but according to the author, it needs more study for industrial applications.

Leus et al. synthesized the embedding of γ-Fe_2O_3 engineered nanoparticles in the covalent triazine framework (CTF-1) for the removal of arsenic and mercury from water. At first, they synthesized CTF-1 by adopting a reported method [29]. Briefly, 10 g $ZnCl_2$ and 2 g 1,4-dicyanobenzene were mixed together, subsequently sealed and heated to 400°C for 4 h and kept at this temperature for 40 h under inert conditions. The obtained black product was stirred in water and 1M HCl at room temperature for 24 h to remove $ZnCl_2$. Then the obtained product was filtered and dried at 120°C under vacuum for further analysis. The γ-Fe_2O3 nanoparticles were prepared in situ by adding $FeCl_3 \cdot 6H_2O$ (406.9 mg) and $FeCl_2 \cdot 4H_2O$ (152.6 mg) into a 100 mL aqueous suspension having 0.5 g CTF-1 and stirred for 2 h under inert condition, followed by the addition of 15 mL NH_3 solution. Then the obtained γ-Fe_2O_3@CTF-1 materials were filtered and washed properly with water and acetone and dried under vacuum at 90°C for 12 h. Before evaluating the adsorption efficiency toward remediation of arsenic and mercury of the prepared materials they characterize the adsorbents by several characterization techniques. Also, special attention is given to the kinetics, presence of other compounds, use of domestic wastewater, and regeneration of the adsorbent [30].

Also, Lin et al. synthesized γ-Fe_2O_3 nanoparticles by a simple coprecipitation method. Briefly, $FeCl_3.6H_2O$ and $FeCl_2.4H_2O$ (2.2:1 M ratio) were dissolved in deionized water. The solution was then mixed with NH_3 solution (25%) at a ratio of 4:1 (v/v), which was determined experimentally, and the resulting mixture was stirred for 20 min at 1000 rpm. Then the obtained γ-Fe_2O_3 nanoparticles were washed properly with distilled water and collected by using an external magnet. Then the prepared adsorbents were applied for remediation of As(III) and As(V). Afterward, the adsorption mechanism was explained by

FTIR and XPS analyses. Also, adsorption kinetics and isotherms at different temperatures were measured. The effects of pH and competing ions on the adsorption were also examined. Finally, the regeneration and recycling of the magnetic γ-Fe_2O_3 nanoparticles were investigated. An understanding of the adsorption mechanism and demonstration of the adsorption, regeneration, and reuse of the magnetic γ-Fe_2O_3 nanoparticles, as has been the key aim [31].

Mou et al. synthesized magnetic iron-oxide chestnut-like hierarchical nanostructures for the removal of arsenic from water. At first, the precursor Fe_2O_3 chestnut-like amorphous core/ γ-phase shell hierarchical nanoarchitectures (CAHNs) were synthesized by the solvothermal process. In a brief synthetic procedure, at first 0.24 mmol $SnCl_2 \cdot 2H_2O$ and 0.5 mL $Fe(CO)_5$ were added to 80 mL of N, N-dimethylformamide (DMF) and the solution was transferred into a 100 mL Teflon-lined stainless steel autoclave and then sealed for 8 h at 200°C. After cooling the prepared brown-colored product was centrifuged and washed repeatedly and dried in vacuum at 40°C for 8 h. To obtain γ-Fe_2O_3 and Fe_3O_4 CHNs, the obtained product were annealed at 300 and 500°C under the protection of N_2 for 4 h, respectively [32]. The typical SEM images of the prepared nanomaterials are presented in Fig. 11.3.

2.1.2 Synthesis of iron-oxide nanoparticles from natural sources

In the last decade, great attention has been given to adsorbents based on natural waste and readily-available materials, trying to develop low-cost strategies for the synthesis of eco-friendly adsorbents for water treatment applications, also to utilize natural west product obtained from the ecosystem. So here we have also trying to represent few recently reported techniques for synthesis of iron-oxide nanoadsorbents from natural sources by green synthesis techniques and have used for arsenic remediation application.

Zeng et al. synthesized magnetic particle adsorbent (MPA), using iron-containing sludge by the solvothermal process by referring to a previously reported method for the removal of As(V) [33]. Briefly, 2.025 g iron sludge and 10.8 g anhydrous sodium acetate were added to 60 mL of ethylene glycol with proper stirring. Then the solution was transferred to a 100 mL reaction pot and stirred for 30 min. Afterward, the reaction pot was placed in a drying oven and calcined at 180°C for 10 h. After the solvothermal reaction, the black sediment was collected and washed with deionized water and anhydrous ethanol and dried in an electric blast oven at 60°C for 6 h. The physicochemical properties of prepared MPA were studied by VSM, XPS, XRD, SEM, and BET analysis. Then the adsorption capacity of the prepared adsorbents toward the removal of As(V) was studied by laboratory experiments [34].

Majumder et al. have developed a simple green technique for the synthesis of iron-oxide nanoparticles (IONPs) for the removal of arsenic from water and sludge. Here the waste banana peel extract and anhydrous $FeCl_3$ were used as the initial precursor materials for the synthesis of IONPs. The key advantages of banana peel include they are reached in different

Figure 11.3
SEM images of the Fe_2O_3 CAHNs (A, B), and the products obtained by calcining the Fe_2O_3 CAHNs at 300°C (C, D) and 500°C (E, F) under the protection of N_2 for 4 h, respectively. B, D, and F show the SEM images of the corresponding cracked structures.

phenolic compounds and antioxidants and they have a higher reducing ability as compared to other fruit peel. At first, the collected waste banana peel from kitchen waste were washed thoroughly and dried in an oven at100°C for 3–4 h for complete removal of moisture. Then the obtained black products were ground properly and boiled in distilled water for 30 min for the preparation of brown-colored, polyphenol-rich banana peel extract. Then 1:1(v/v) ratio of peel extract and $FeCl_3$ were mixed together and subjected to continuous heating at 80°C and stirring (500 rpm) until the color changes from yellowish to black indicated the formation of IONPs and subsequently washed and oven-dried gives the required NPs [35].

Cabanas et al. developed an appropriate organic/inorganic magnetic hybrid adsorbent (i.e., iron-oxide encapsulated chitosan beads) by green synthetic technique and apply them for efficient removal of As(V) [11]. The details procedure was shown in Fig. 11.4.

Vieira et al. have synthesized a simple inexpensive adsorbent material by using brown marine algae Sargassum muticum coated with iron-oxy(hydroxides) for adsorptive removal of As(III) and As(V) [36].

From the above all synthesis strategy and their application toward adsorption, it is clear that iron-based nanomaterials play a significant role for adsorptive removal of arsenic (III/V) from contaminated water. But pristine iron-oxide nanoparticles show few significant shortcomings like high agglomeration rate, low specific surface area, and low stability as compared to incorporated or functionalized nanoarchitectures. To overcome such limitations, researchers try to develop an advanced synthetic strategy to develop nanoadsorbents (like the hollow, porous and functionalized materials) having high surface area and improved stability.

2.1.3 Synthesis of hollow and porous iron-oxide nanoarchitectures

Nowadays the foremost attention of every material scientist is to develop a low cost, high surface area, high aspect ratio (i.e., surface to volume ratio) and materials having more active sites within the nano-dimensions (1–100 nm) for energy, environment, and biomedical applications. So considering the above concerns here we emphasize to interpret the details insight into the formation and arsenic remediation application of hollow and porous iron-oxide nanoarchitectures from aqueous medium.

Wang et al. have reported a simple template free synthetic technique for the synthesis of uniformly distributed urchinlike α-FeOOH hollow spheres for superior water treatment application. In a brief synthetic procedure, 0.111 g of $FeSO_4 \cdot 7H_2O$ was dissolved in 40 mL of the glycerol-water mixture with different amounts of glycerol such as 5 mL (sample A), 3 mL (sample B), and 2 mL (sample C). Then the solution was stirred for about 10 min to obtained a transparent solution and then loaded into a 50 mL Teflon-lined stainless steel autoclave followed by heating at 120°C for 24 h in an electric oven. Then the autoclave was cooled down automatically to room temperature and the obtained

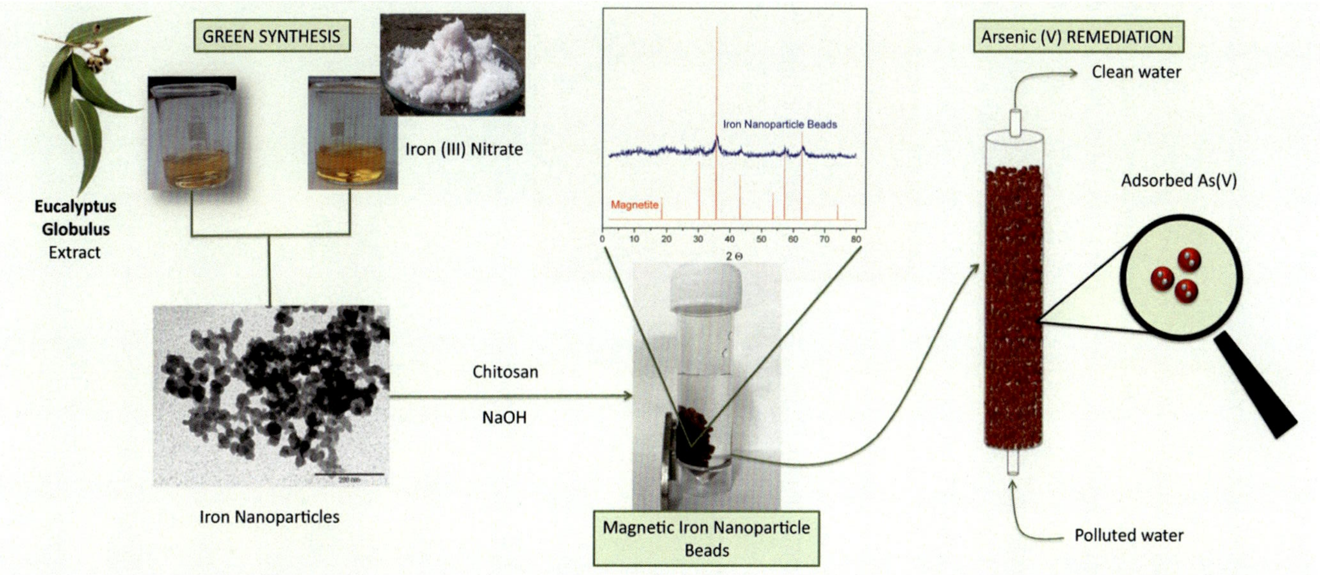

Figure 11.4

Procedure for synthesis of iron-oxide encapsulated chitosan beads.

precipitate was washed with deionized water and ethanol several times and dried overnight at 60°C [37]. The details formation strategy of the α-FeOOH hollow spheres is presented in Fig. 11.5.

Jia et al. proposed a simple template-free solution method for synthesis of hollow magnetic porous Fe_3O_4/α-FeOOH microspheres having more number of surface hydroxyl groups and carbonate like species and apply them for removal of As(V), Cr(VI), Cd(II) and Hg(II) ion from aqueous solution. In a typical synthesis process, under room temperature conditions, a particular amount of $FeSO_4.7H_2O$ (0.1 MOL/L) was dissolved in a mixture of distilled water and ethylene glycol(EG). Then 0.4 mol/L concentration of urea was added under stirring conditions and the prepared solution was transferred into a conical flask with a stopper and heated at 100 C for 12 h. After completion of the reaction, a brown precipitate was obtained, which on washed several times and finally dried in an oven at 80°C for 6 h [38]. The FESEM and TEM micrograph of the prepared materials is presented in Fig. 11.6.

Yu et al. also synthesized a superior bulk adsorbent of porous Fe_2O_3 nanocubes-impregnated graphene aerogel (PGA/PeFe_2O_3) for enhanced removal of high concentration arsenic from wastewater. As PGA/PeFe_2O_3 is bulky in nature, so it could be directly used for the column process for large scale applications [39].

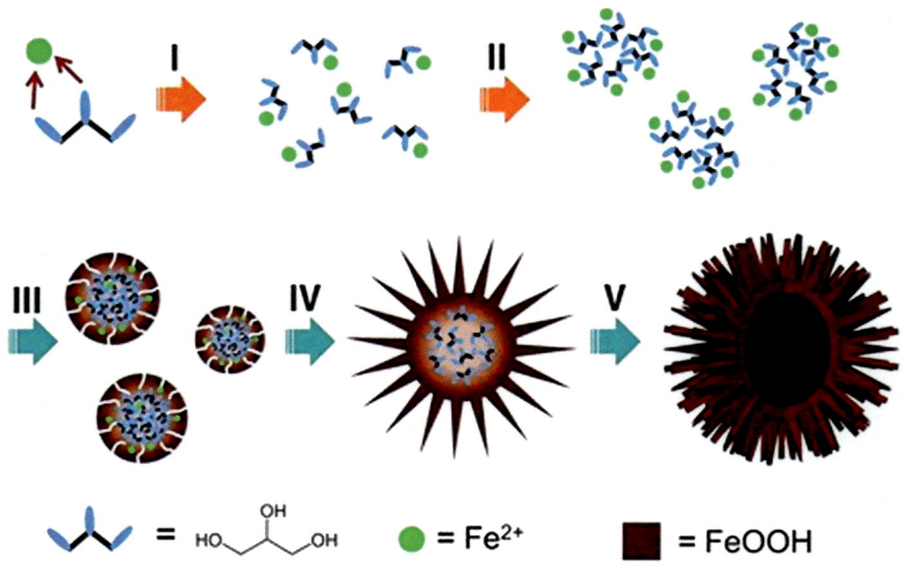

Figure 11.5
Schematic illustration of the morphological evolution process of the as-obtained urchin-like hierarchical α-FeOOH spheres: (I) formation of Fe(II)-glycerol complex; (II) formation of aggregate and quasi-emulsion; (III) hydrolysis and oxidation of Fe(II) to form initial FeOOH shell around the aggregates; (IV) further growth of nanorods on the shell; (V) formation of urchin-like hollow structures.

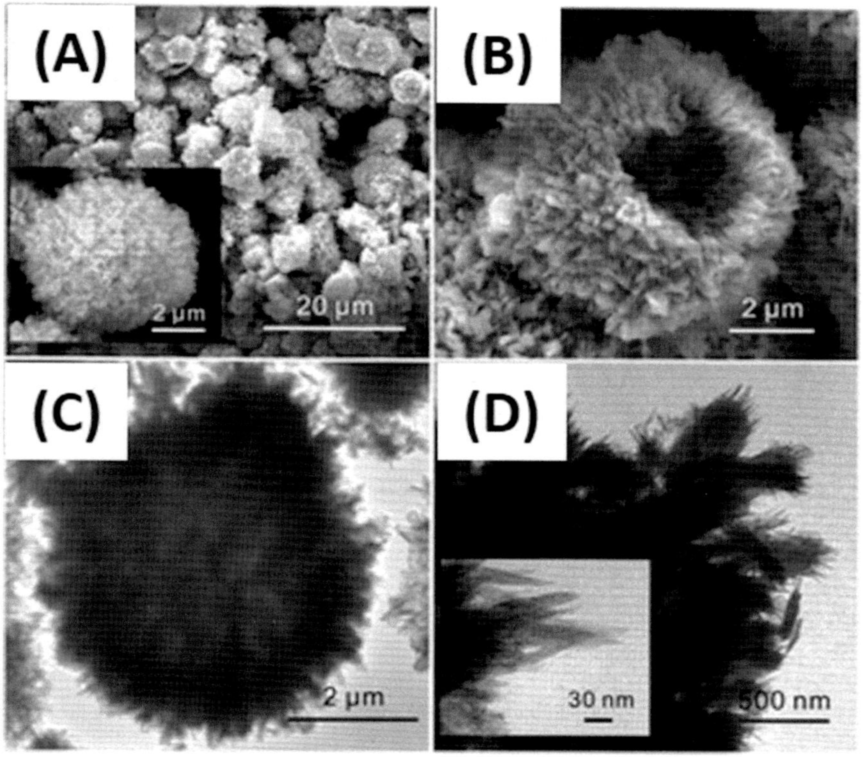

Figure 11.6
(A, B) FESEM and (C, D) TEM images of the Fe_3O_4/α-FeOOH hollow microspheres.

These hollow and porous materials also possess certain shortcomings, so researchers trying to develop a more advanced adsorbent system by engineered the surface of iron-oxide nanomaterials by convenient functional group or functionalized the surface of other supporting materials by iron-oxide nanoarchitectures, to improve the stability and other physicochemical properties. In the next part of the chapter, we explain the synthesis of functionalized iron-based nanomaterials.

2.1.4 Synthesis of surface-functionalized iron-based nanomaterials

In this part, we try to highlight a few recent approaches, how the adsorption efficiency toward arsenic of the iron-oxide NPs increases when its surface is engineered by various chemical species. This type of approach enhances the adsorption efficiency with a low agglomeration rate and high stability to the adsorbent. In order to increase efficiency and to avoid interference from other metals ions, iron-oxide NPs have been functionalized to improve their adsorption ability, creating a more effective process.

Nata et al. synthesized surface-functionalized bacterial cellulose (BC) nanofibrils with aminated magnetite nanoparticles (MH) by the simple one-pot solvothermal process.

Briefly, an appropriate amount of 1,6-hexanediamine, $FeCl_3.6H_2O$, and BC pellicle in ethylene glycol were magnetically stirred for 24 h, subsequently transferred to an autoclave and solvothermally treated at 200°C for 6 h. The presence of amine-functionalized magnetite nanoparticles on the surface of cellulose nanofibrils increases the thermal and mechanical properties also increase the amine content on the surface of BC. Then the synthesized adsorbents were applied for removal of arsenic from water [40]. The details synthesis strategy is given in Fig. 11.7.

Morillo et al. synthesized a novel forager sponge-loaded superparamagnetic iron-oxide nanoparticles for efficient removal of As(III) and As(V). To avoid oxidation of Fe^{2+} to Fe^{3+}, the synthesis needs a continuous bubbling of nitrogen. At first, $FeCl_36H_2O$ was dissolved with deoxygenated 0.2 M HCl to prepare the stock solution of Fe^{3+} in a chloride medium. Also, 0.7 M NH_4OH stock solution was deoxygenated under N_2 atmosphere and heated to 70°C. Then both the stock solution were mixed together. Afterward, $FeCl_2$ was added to the above solution in a ratio of 1:2 of Fe^{2+}/Fe^{3+} and continue the stirring under N_2 atmosphere for the aging of nanoparticles. Then the resulting suspension was centrifuged and washed several times. A subsequent redispersion in a deoxygenated aqueous solution of tetramethylammonium hydroxide (TMAOH) 0.01 M (pH = 12) let to obtain SPION in a stable suspension for 6–8 months under deoxygenated atmosphere. The key purpose of this work is to examine the role of superparamagnetic iron-oxide nanoparticles (SPION) toward the adsorptive removal of As(III) and As(V) ions from water under acidic conditions [41].

In the same year Morillo et al. also developed a simple and convenient method for synthesis of 3-mercaptopropanoic acid-coated superparamagnetic iron-oxide nanoparticles (3MPA-SPION) for removal of As(V) from aqueous solution [42]. The proposed mechanism of 3MPA-SPION interaction is illustrated below (Fig. 11.8).

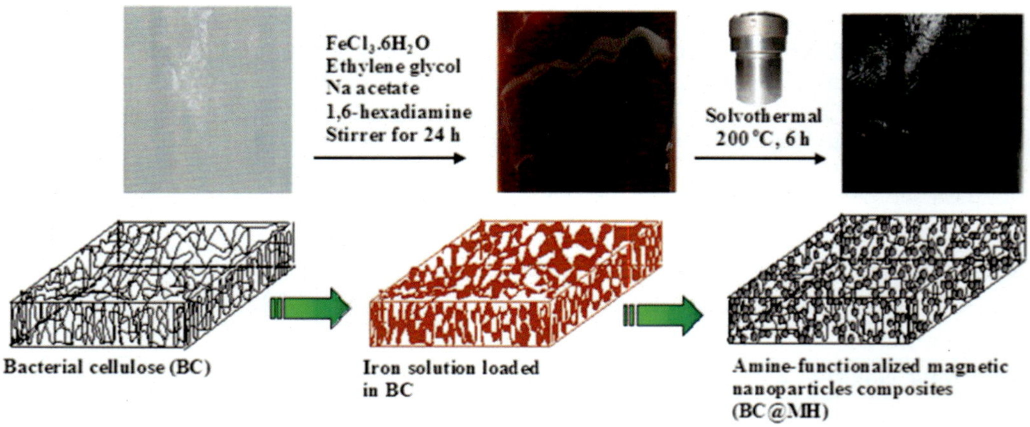

Figure 11.7
Procedure for synthesis of BC@MS.

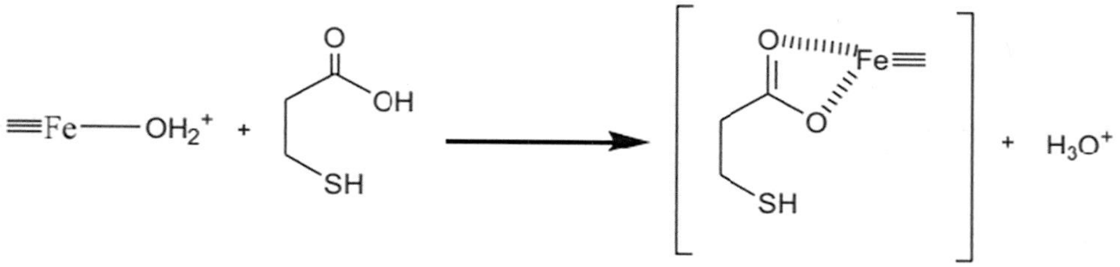

Figure 11.8
Proposed mechanism of 3MPA-SPION interaction where Fe ≡ represents the SPION.

In the case of functionalized materials, the presence of good living groups like OH_2^+ on the SPION surface enhances the adsorption behavior toward the removal of As(V) from the water sample.

Bringas et al. reported an excellent functionalized magnetic material which possesses high separation ability after adsorption. The synthesis process involves three steps. The first step involves the synthesis of superparamagnetic iron-oxide nanoparticles (SPINPs) by the coprecipitation method by using $FeCl_3.6H_2O$ and $FeCl_2.4H_2O$ as the precursor materials. The second step involves the coating of SPIONs by a mesoporous silica layer by using tetraethyl orthosilicate (TEOS) in the presence of the cationic surfactant cetyltrimethylammonium bromide (CTAB). Finally, the third step involves postgrafting of N-[3-(trimethoxysilyl) propyl] ethylenediamine (TMPED) followed by the protonation and Fe^{3+} organic coordination of the amino groups. The key focus of this work is on the evaluation of the adsorption equilibrium of As(V) polluted groundwater on the functionalized SiO_2/Fe_3O_4 magnetic nanoparticles [5].

Feng et al. synthesized the ascorbic acid-coated superparamagnetic magnetite NPs of diameter less than 10 nm for the removal of arsenic from water. At first, a clear solution was prepared by adding 1 mmol of $FeCl_3 \cdot 6H_2O$ to 40 mL of distilled water by continuous stirring. Then 3 mmol of ascorbic acid and 5 mL of $N_2H_4 \cdot H_2O$ (50% v/v) were added to the above solution. The prepared solution was then transferred to a Teflon-line stainless still autoclave and heated at 180 C for 8 h. The obtained precipitate was then centrifuged and washed properly with distilled water and ethanol and dried under vacuum at 60°C for 6 h. After successful synthesis, the prepared material was applied toward the removal of As(III) and As(V) from the water medium. The efficiency of these NPs were calculated by fitting adsorption data to Langmuir isotherm, which gave 16.56 and 46.06 mg/g capacity for As(V) and As(III), respectively [43].

Jin et al. also synthesized CTAB functionalized magnetite NPs by a simple coprecipitation method for removal of As(V). The modified iron-oxide material was easily separated from solutions within 5.0 min with a simple magnetic process because of the magnetic properties of the materials with a saturation magnetization of 67.2 emu/g. The adsorption of As(V) on the adsorbent surface attains an equilibrium within 2 min, with a maximum adsorption capacity of 23.07 mg/g. The modified adsorbent was regenerated in alkali solutions and more than 85% As(V) was removed in the fifth regeneration/reuse cycle [44].

Prem Singh et al. fabricated a cost-effective thiol-tethered bifunctional iron composite (TH-Fe) adsorbent system as a potential commercial applicant for arsenic (III/V) sorption application. In a brief synthesis procedure, thiomer was dissolved in 0.1 M $FeCl_3$ solution and the pH of the solution was maintained at ~3.0. The thiomer can act as a reducing agent and reduces Fe^{3+} to Fe^{2+}. Now the solution of 1.5 M NaOH was added to this solution under vigorous stirring at 60°C for 2 h. The black precipitate thus obtained was washed with water for further application [45]. The details synthesis scheme is presented in Fig. 11.9.

3. Characterization of iron-based nanoadsorbents

After the successful synthesis of adsorbent materials, it must be characterized properly to determine the crystal structure, phase purity, morphology, magnetic properties, and specific surface area. In this section, we discuss a few essential instrumental characterization techniques for the characterization of synthesized nanoadsorbents. Table 11.1 summarizes essential characterization techniques and what information we get after characterization.

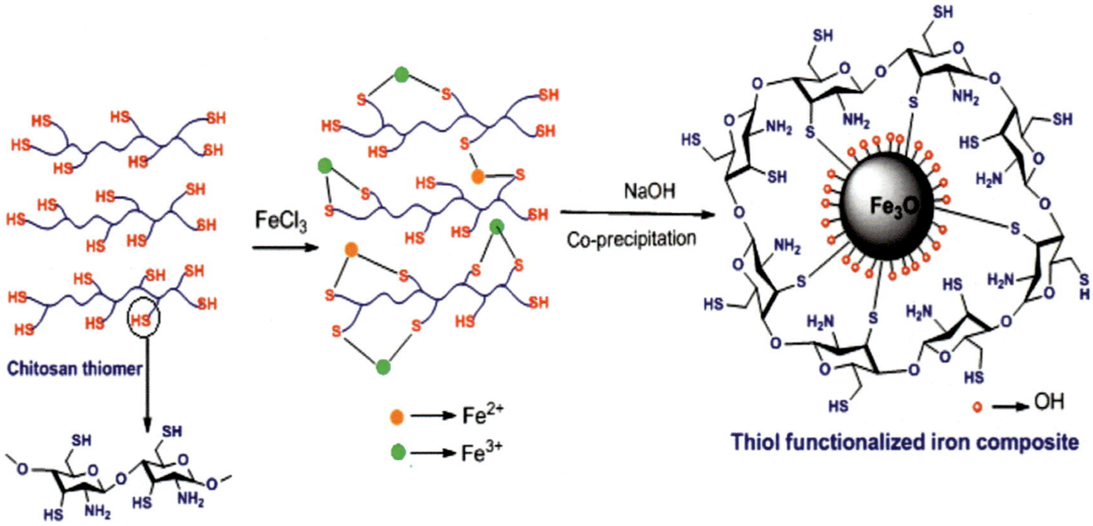

Figure 11.9
Schematic representation for the synthesis of TH-Fe.

Table 11.1: Basic instrumental techniques for characterization of nanoadsorbents.

Characterization techniques	Physicochemical properties
X-ray diffraction (XRD)	Mostly used to characterize crystalline and polycrystalline/semicrystalline materials for shape, size, and structural determination.
Fourier transform infrared spectroscopy (FTIR)	Bioconjugate, surface properties such as structure and conformation.
Field emission scanning electron microscope (FE-SEM)	Distribution, shape, aggregation, dispersion.
Transmission electron microscope (TEM)	Shape heterogeneity, size and size navigation, dispersion, accumulation, determination of crystal phase from HRTEM.
Raman scattering	Hydrodynamic size (complimentary data obtained from IR)
Vibrating sample magnetometer (VSM)	Magnetic properties.
Brunauer, Emmett, and Teller (BET) analysis	Specific surface area, pore size, pore volume.
Zeta-potential	The stability concerning charge on the surface.
X-ray photoelectron spectroscopy (XPS)	Electronic structure in terms of oxidation state and binding configuration.

By taking the following example we can demonstrate, how different characterization techniques are essential to determine the physicochemical properties of prepared nanoadsorbents.

Feng et al. synthesized superparamagnetic high surface area ascorbic acid-coated Fe_3O_4 nanoparticles by a simple environmentally friendly template-free hydrothermal route and apply them toward the removal of arsenic from water [43]. Here they characterize the sample by different characterization techniques like XRD, Raman, FTIR, TEM, VSM, Zeta potential, and BET surface area measurement.

The XRD pattern of the ascorbic acid-coated Fe_3O_4 nanoparticles is shown in Fig. 11.10A. The diffraction peaks at $2\theta = 30.1°$, $35.3°$, $43.0°$, $57.2°$, and $62.6°$ match well with the magnetite phase (JDPCS: 19–0629). By using the strongest peak (i.e., $2\theta = 35.3°$ (3 1 1)) with the help of Scherer's formula the average diameter of the particle is found to be 5 nm.

To further support the XRD result the sample is further characterized by Raman technique. Raman technique is an alternative tool for differentiating the various structural phases of iron-oxide. Fig. 11.10B depicts the Raman spectrum of the sample at room temperature. The peak at 668 cm^{-1} typical indicates the characteristic of Fe_3O_4, while for γ-Fe_2O_3, three strong peaks at 695, 500, and 352 cm^{-1} are generally found. Hence, both the XRD and Raman data confirmed the Fe_3O_4 phase, not γ-Fe_2O_3.

The FTIR spectra of ascorbic acid and ascorbic acid-coated Fe_3O_4 nanoparticles are presented in Fig. 11.10C. The Fe_3O_4 nanoparticle sample as well as ascorbic acid feature bands in the range of 3160–3430 cm^{-1}, assignable to the O–H vibration, the bands at

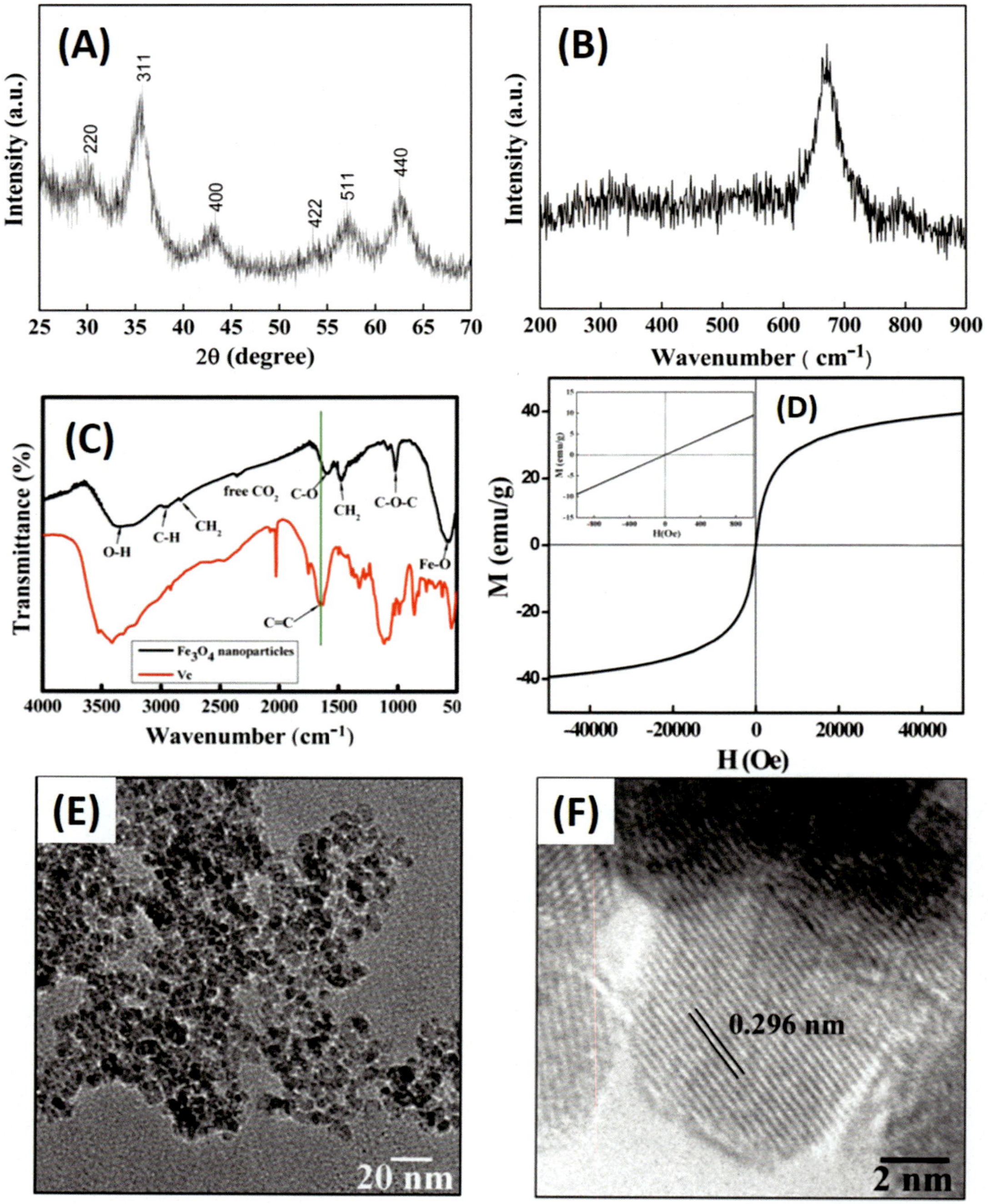

Figure 11.10
(A) XRD, (B) Raman, (C) FTIR, (D) VSM, (E) TEM, and (F) HRTEM images of ascorbic acid-coated Fe_3O_4 nanoparticles.

2943 and 2976 cm^{-1} to C—H stretching vibration, the bands at 2835 and 1476 cm^{-1} to CH_2 stretching modes, and the one at 1602 cm^{-1} to C=O stretching vibrations. For the Fe_3O_4 nanoparticle sample, the band at 580 cm^{-1} is a typical characteristic for Fe—O. However, the presence of two strong bands at 1024 and 1091 cm^{-1}, assignable to C—O—C symmetric stretching mode, and the absence of the band at 1651 cm^{-1}, assignable to C=C stretching vibration mode, point to the fact that ascorbic acid is probably oxidized to dehydroascorbic acid under basic conditions.

Fig. 11.10D shows the magnetic hysteresis loop of ascorbic acid-coated Fe_3O_4 nanoparticles which signifies the superparamagnetic characteristics of the materials (due to the S-shaped loop) with saturation magnetization value of 40 emu/g.

Fig. 11.10E and F indicate the TEM and HRTEM images of the prepared nanoparticles respectively, which indicates the sample consists of uniformly distributed nanoparticles with 5 nm diameter. Also from the HRTEM image, it is clear that the product is highly crystalline despite the small particle size. The typical lattice fringe spacing, determined to be 0.296 nm, corresponds to the spacing of the (2 2 0) planes of Fe_3O_4.

The sample is also characterized by BET surface area measurement and the surface area of the ascorbic acid-coated Fe_3O_4 nanoparticle powder is 178.48 m^2/g, which is much higher than that of Fe_3O_4 sample obtained without ascorbic acid (15.63 m^2/g). From the Zeta potential analysis, they also predict that the prepared adsorbent shows better adsorption efficiency with a pH range of 2—7. After successful synthesis and characterization, the prepared adsorbents can be used for the remediation of As(III) and As(V) from aqueous medium.

4. Adsorptive removal of arsenic from water by using iron-based nanoadsorbents

Basically, the adsorption process proceeds through a single step or through a combination of multiple-step like film or external diffusion, pore diffusion, surface diffusion and adsorption on the pore surface [46]. It was also reported that adsorption of arsenic on the adsorbent surface proceeds through three steps as mentioned below.

- surface migration (electrostatic attraction).
- deprotonation (or dissociation) of complexed aqueous As(III) or As(V).
- surface complexation.

The common pathways of adsorption are illustrated in Fig. 11.11. Initially, adsorbate (As(III)/As(V)) is diffused on the external surface of the nanoadsorbents due to diffusion potential characterized by the concentrations of adsorbate and available external surface area on the adsorbent. After diffusion on the external surface of the adsorbent, adsorbate is diffused on the available pores of the adsorbent. All the available exposed active sites are

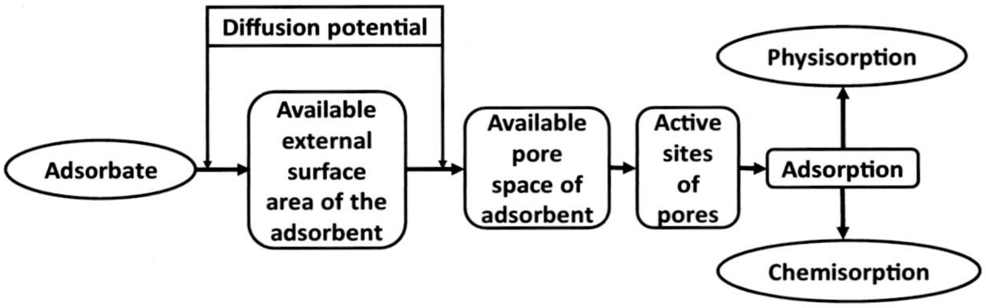

Figure 11.11
Common pathway of the adsorption process.

occupied during the adsorption process governed by either physisorption or chemisorption. Also, few works of literature reported that electrostatic and surface complexation govern the adsorption behavior of arsenic on the adsorbent surface. Badruzzaman et al. supported Fick's second law of diffusion (i.e., adsorption rates are inversely proportional to the square of the particle radius); according to that adsorption kinetics is faster for smaller particles. But latter Vignola et al. demonstrate that this law is not applicable to all types of adsorbents. For example, in the case of small particles of zeolite (having a high surface area), the rate of adsorption is lesser than that of a larger crystal. This may be due to the diffusion of adsorbate onto adsorption sites more quickly on larger crystals rather than the smaller ones. Also, few literatures illustrate the formation of the inner-sphere surface complex during adsorption of both the arsenic species [2,47].

As we mentioned earlier Tang et al. [23] synthesized ultrafine α-Fe_2O_3 nanoparticles by a simple solvent thermal method and apply them for adsorptive removal of arsenic from both lab-prepared and natural water samples. They follow the Batch adsorption process for the removal of arsenic under normal temperature conditions. Initially required amount of adsorbents were added to the arsenic-contaminated water and stirred magnetically to disperse α-Fe_2O_3 nanoparticles to confirm a good interaction with arsenic ions. Then the sample was centrifuged for 30 min and one drop of concentrated HCl was added to the clear solution to avoid the potential oxidation of As(III) to As(V). After that, the residual arsenic solution on the treated sample was determined by an atomic fluorescence spectrophotometer. The adsorption experiment results demonstrated that they were effective, especially at low equilibrium arsenic concentrations, in removing both As(III) and As(V) from lab-prepared and natural water samples. Near the neutral pH, the adsorption capacities of the α-Fe_2O_3 nanoparticles on As(III) and As(V) from lab-prepared samples were found to be no less than 95 mg/g and 47 mg/g, respectively. Here the electrostatic attraction between the positively charged surface (due to protonated surface hydroxyl group) of the α-Fe_2O_3 and the negative charge of arsenic species is mostly responsible for the adsorption.

Also, they represent the synthesis of the TH–Fe composite system for the removal of arsenic from water in the synthesis section [45]. They also follow the batch adsorption experiment and the residual arsenic concentration in the solution was determined by cyclic voltammetry. The effect of pH, coexisting anions (SO_4^{2-}, NO_3^-, Cl^-, and PO_4^{3-}), the effect of adsorbent dosage, adsorption isotherm, adsorption kinetics, and regeneration studies were also carried out to recognize the applicability of composite in real conditions. To predict the possible sorption mechanism (Fig. 11.12), the treated adsorbent samples were characterized by FTIR, SEM-EDX, TEM, Raman, and XRD techniques.

The synthesized bifunctional TH−Fe composite exhibited almost complete (>99%) and quick removal of both arsenic species, via coordinate covalent bonds (Fig. 11.12), within the recommended 10 ppb range of the WHO. Sorption characteristics via isotherm models conclude the Freundlich isotherm and confirm that the removal of As(III) and As(V) on TH−Fe composite is chemisorption via coordinate covalent bonds. The maximum sorption capacities are found to be 88 mg/g for As(V) and 91 mg/g for As(III) at pH 7.0. Moreover, the bifunctionalized composite reserves its performance efficiency in reusability without any noticeable decrease for nine cycles (residual concentration <3.5 ppb). Thus, the TH−Fe composite with unique arsenic adsorption−desorption properties has the potential to be a promising economical adsorbent for complete removal of arsenic from water without the pre-/posttreatment and pH adjustment. Moreover, the high adsorption capacities and superior selectivity make the TH−Fe composite a capable candidate for the development of a point-of-use filter for commercial application.

Also, there are so many reported iron-based nanomaterials developed by the different research groups as summarizes in Table 11.2.

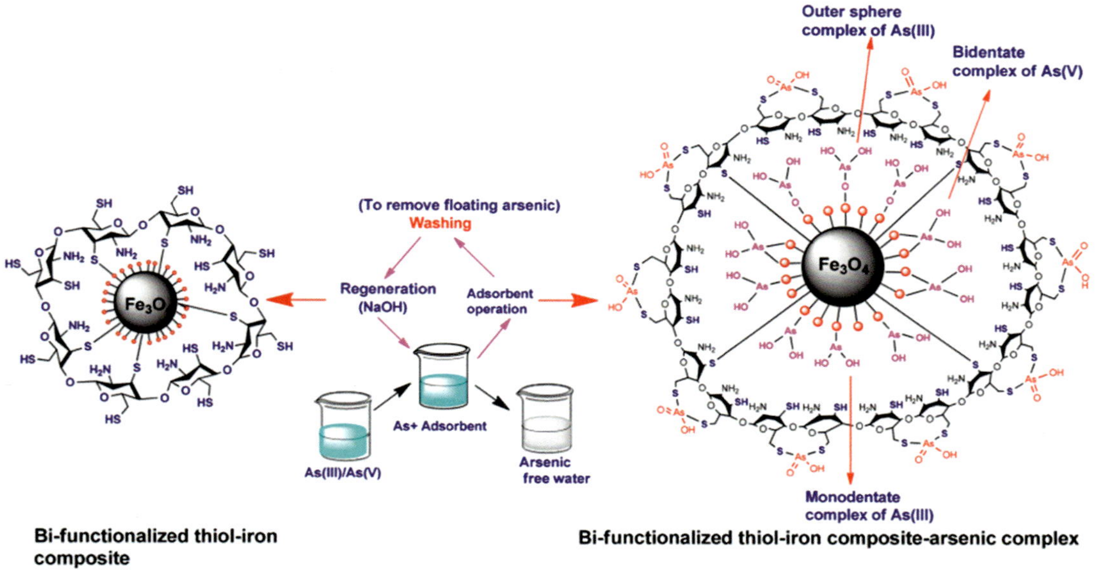

Figure 11.12
Adsorption mechanism of arsenic on Th−Fe composite.

Table 11.2: Removal of arsenic species by different iron-oxide based nanomaterials.

Adsorbent	Sample	Synthetic method	Optimum pH	Isotherm	Kinetics	Maximum As(III) adsorption capacity	Maximum As(V) adsorption capacity	References
α-Fe_2O_3	Lab prepared and natural water	Solvent thermal	7	Langmuir	Pseudo 2nd order	95	47	[23]
Fe_3O_4 loaded PCL	Lab prepared water		3–9	Langmuir	Pseudo 2nd order	32	28	[48]
Fe_3O_4-silica	Lab prepared water	Coprecipitation	8	Langmuir	Pseudo 2nd order	14.7	121	[49]
Magnetite–RGO composite	Lab prepared water	Coprecipitation	7	Freundlich	Pseudo 2nd order			[50]
Fe_3O_4–GO	Lab prepared water	Chemical precipitation	4 and 7	Freundlich	Pseudo 2nd order	85	38	[51]
HIO-alginate beads	Lab prepared water	Chemical precipitation	Acidic	Langmuir	Pseudo 2nd order	47.8	45.1	[52]
Yeast cross-linked Fe_3O_4	Lab prepared water	Nanoprecipitation method	Acidic	Langmuir	Pseudo 2nd order		28.70	[53]
Mixed Fe_3O_4-γ-Fe_2O_3	Natural groundwater and lab prepared water	Commercially available	2	Langmuir	Pseudo 2nd order	3.69	3.71	[54]

5. Basic mechanism of arsenic adsorption on iron-oxide surface

Generally, iron-oxide possesses more surface hydroxyl groups in the water system. Thus it either binds hydrogen (H^+) ion or releases H^+ ion, depending on the pH of the solution. Due to the presence of OH^{2+}, OH^-, and O^- functional groups on the hydrated iron-oxide surface, it shows outstanding adsorption performance. Both the arsenic species (III/V) represent metal (As) and ligand (O) characters, so they can bind with iron oxides via surface complexation or ligand exchange process. The adsorption of arsenic may occur through the ligand exchange of OH^- and OH^{2+} functional groups and complete with the inner-sphere complex formation [55]. For example, partial dissociation of H_3AsO_4 release H^+ ion which form H_2O with OH^- and leave space for anion binding. Basic adsorption reactions of As(III) and As(V) are shown below.

$$Fe(OH)_3(s) + H_3AsO_4 \Longleftrightarrow FeAsO_4.2H_2O + H_2O$$

$$\equiv FeOH^{\bullet} + H_3AsO_3 \Longleftrightarrow \equiv FeH_2AsO_3 + H_2O$$

$$\equiv FeOH^{\bullet} + AsO_4^{3-} + 3H^+ \Longleftrightarrow \equiv FeH_2AsO_4 + H_2O$$

$$\equiv FeOH^{\bullet} + AsO_4^{3-} + 2H^+ \Longleftrightarrow \equiv FeHAsO^{4-} + H_2O$$

Here, the physisorption or chemisorption process occurs between adsorbate and adsorbent for surface complexation. The arsenic species may interact with iron-oxide by intramolecular or extramolecular interactions or surface adsorption. The formation of a complex between iron-oxide surface and arsenic species have been confirmed by using Fourier transform infrared spectroscopy (FTIR), and Extended X-ray absorption fine structure (EXAFS) techniques [56]. These techniques suggest that monodentate or bidentate complexes can be formed between arsenic and iron-oxide as shown in Fig. 11.13. Also, it is observed that, if surface coverage is low monodentate complex is formed, but for higher surface coverage bidentate complex is formed. The most thermodynamically stable form of the complex between arsenic and iron-oxide surfaces are bidentate binuclear complexes. The bidentate mononuclear complex should be assigned to the As—O—O—As structure. However, both monodentate and bidentate mononuclear complexes are thermodynamically unstable. Therefore, the most accepted mechanism of As(III) and As(V) adsorption onto iron-oxide surface is ligand exchange in the bidentate binuclear, inner-sphere complexation mode. But, As(III) sorption on iron-oxide surface can also form the outer-sphere complexes.

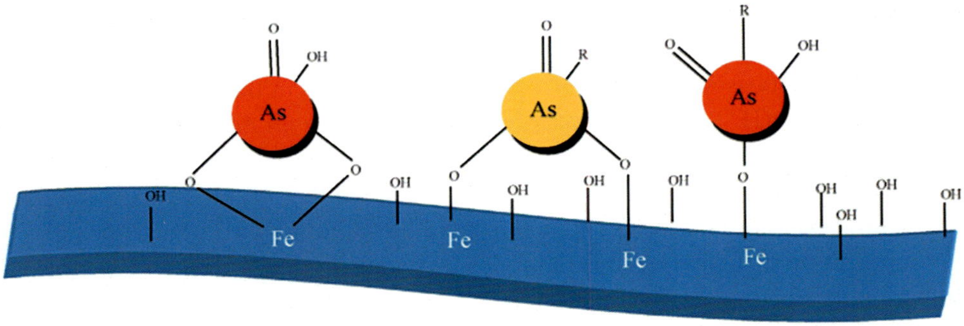

Figure 11.13
Adsorption mechanism due to complex formation between the iron-oxide surface and arsenic species.

The modified iron-oxide surface contains different functionality that provides various active sites for adsorption of arsenic. Thus the process of adsorption follows variable degrees. This makes mechanism a complicated phenomenon, as inner and outer both the complexes may be formed. Therefore, the adsorption of arsenic on the iron-oxide surface still possesses a large conflict after a lot of work. Also, no unanimity is found regarding the structure of the complexes.

6. Conclusion

This chapter provides a comprehensive knowledge about the synthesis, characterization, and adsorptive applications of various iron-based nanoarchitectures toward the adsorptive removal of As(III) and As(V) species from water stream. From the overall literature review it is clear that by adopting the various synthetic strategy, different iron-based and their functional materials can be synthesized. Various nanostructures like nanospheres, nanocube, nanoflowers, functional iron-oxide, and their hybrid architectures can provide greater specific surface area having a greater number of surface active sites. Therefore, these materials can be promisingly used for the effective adsorption of various arsenic species. The formation, functionality, and morphological analysis can be investigated by various instrumental techniques as mentioned in the text. From the mechanistic approach, it is concluded that the adsorption of arsenic species onto the adsorbent surface is mainly governed by a physicochemical approach (i.e., electrostatic attraction and surface complexation). Hence this chapter encourages the readers and researchers to develop various iron-based nanostructures that have significantly outstanding adsorption performance toward arsenic remediation and water treatment applications.

Conflict of interest

The authors declare no competing financial interest.

Acknowledgment

The authors acknowledge BRNS (Project No. 37(3)/14/27/2017-BRNS), Govt. of India for providing financial support, and also NIT Rourkela, India (Odisha) for providing the research facility to carry out this work.

References

[1] V.D. Martinez, E.A. Vucic, D.D. Becker-santos, L. Gil, W.L. Lam, Arsenic exposure and the induction of human cancers, J. Toxicol. 2011 (2011), https://doi.org/10.1155/2011/431287.
[2] S. Lata, S.R. Samadder, Removal of arsenic from water using nano adsorbents and challenges: a review, J. Environ. Manag. 166 (2016) 387–406, https://doi.org/10.1016/j.jenvman.2015.10.039.
[3] U. Hellriegel, T.L. Luu, J. Hoinkis, V.T. Luong, E.E. Ca, Iron-based subsurface arsenic removal technologies by aeration: a review of the current state and future prospects, Water Res. (2018) 133, https://doi.org/10.1016/j.watres.2018.01.007.
[4] F.S. Cannon, Preloading hydrous ferric oxide into granular activated carbon for arsenic removal, Environ. Sci. Technol. 42 (2008) 3369–3374.
[5] E. Bringas, J. Saiz, I. Ortiz, Removal of As(V) from groundwater using functionalized magnetic adsorbent materials: effects of competing ions, Separ. Purif. Technol. 156 (2015) 699–707, https://doi.org/10.1016/j.seppur.2015.10.068.
[6] K. Gupta, T. Basu, U.C. Ghosh, Sorption characteristics of arsenic (V) for removal from water using agglomerated nanostructure iron (III) - zirconium (IV) bimetal mixed oxide, J. Chem. & Eng. Data (2009) 2222–2228.
[7] Z. Liu, J. Chen, Y. Wu, Y. Li, J. Zhao, P. Na, Synthesis of magnetic orderly mesoporous -Fe_2O_3 nanocluster derived from MIL-100 (Fe) for rapid and efficient arsenic (III,V) removal, J. Hazard Mater. 343 (2018) 304–314, https://doi.org/10.1016/j.jhazmat.2017.09.047.
[8] L. Huo, X. Zeng, S. Su, L. Bai, Y. Wang, Enhanced removal of As(V) from aqueous solution using modified hydrous ferric oxide nanoparticles, Nat. Publ. Gr. (2017) 1–12, https://doi.org/10.1038/srep40765.
[9] S. Chatterjee, S. De, Adsorptive removal of arsenic from groundwater using chemically treated iron ore slime incorporated mixed matrix hollow fiber membrane, Separ. Purif. Technol. 179 (2017) 357–368, https://doi.org/10.1016/j.seppur.2017.02.019.
[10] B. Lan, Y. Wang, X. Wang, X. Zhou, Y. Kang, L. Li, Aqueous arsenic (AS) and antimony (Sb) removal by potassium ferrate, Chem. Eng. J. 292 (2016) 389–397, https://doi.org/10.1016/j.cej.2016.02.019.
[11] M. Martínez-cabanas, M. López-garcía, J.L. Barriada, R. Herrero, M.E.S. De Vicente, Green synthesis of iron oxide nanoparticles. Development of magnetic hybrid materials for efficient As(V) removal, Chem. Eng. J. 301 (2016) 83–91, https://doi.org/10.1016/j.cej.2016.04.149.
[12] N. Kalel, A. Rahim, M. Yusoff, S. Krishna, Z. Abdul, S. Salmiati, High concentration arsenic removal from aqueous solution using nano-iron ion enrich material (NIIEM) super adsorbent, Chem. Eng. J. 317 (2017) 343–355, https://doi.org/10.1016/j.cej.2017.02.039.
[13] U.K. Sahu, M.K. Sahu, Removal of As(III) from aqueous solution using Fe_3O_4 nanoparticles: process modeling and optimization using statistical design, water, air, Soil Pollut. (2017), https://doi.org/10.1007/s11270-016-3224-1.
[14] P. Zito, H.J. Shipley, RSC advances inorganic nano-adsorbents for the removal of heavy metals and arsenic: a review, RSC Adv. 5 (2015) 29885–29907, https://doi.org/10.1039/C5RA02714D.
[15] Z. Cheng, Z. Gao, W. Ma, Q. Sun, B. Wang, X. Wang, Preparation of magnetic Fe_3O_4 particles modified sawdust as the adsorbent to remove strontium ions, Chem. Eng. J. 209 (2012) 451–457, https://doi.org/10.1016/j.cej.2012.07.078.
[16] M.Y. Nassar, I.S. Ahmed, T.Y. Mohamed, M. Khatab, Nanostructures for textile dye Removal, RSC Adv. (2016) 20001–20013, https://doi.org/10.1039/c5ra26112k.

[17] C. Cao, J. Qu, W. Yan, J. Zhu, Z. Wu, W. Song, Low-cost synthesis of flowerlike α -Fe_2O_3 nanostructures for heavy metal ion removal: adsorption property and mechanism, Langmuir (2012), https://doi.org/10.1021/la300097y.
[18] A. Roy, J. Bhattacharya, Removal of Cu (II), Zn (II) and Pb (II) from water using microwave-assisted synthesized maghemite nanotubes, Chem. Eng. J. 211–212 (2012) 493–500, https://doi.org/10.1016/j.cej.2012.09.097.
[19] Q. Liu, L. Zhong, Q. Zhao, C. Frear, Y. Zheng, Synthesis of Fe_3O_4/polyacrylonitrile composite electrospun nano fiber mat for effective adsorption of tetracycline, ACS App. Mater. & Inter. (2015), https://doi.org/10.1021/acsami.5b04598.
[20] H. Karami, Heavy metal removal from water by magnetite nanorods, Chem. Eng. J. 219 (2013) 209–216, https://doi.org/10.1016/j.cej.2013.01.022.
[21] H. Liang, B. Xu, Z. Wang, Superior capability for heavy metal ion removal, Mater. Chem. Phys. 141 (2013) 727–734, https://doi.org/10.1016/j.matchemphys.2013.05.070.
[22] H. Alijani, Z. Shariatinia, Effective aqueous arsenic removal using zero valent iron doped MWCNT synthesized by in situ CVD method using natural a -Fe_2O_3 as a precursor, Chemosphere 171 (2017) 502–511, https://doi.org/10.1016/j.chemosphere.2016.12.106.
[23] W. Tang, Q. Li, S. Gao, J. Ku, Arsenic (III, V) removal from aqueous solution by ultrafine α-Fe_2O_3 nanoparticles synthesized from solvent thermal method, J. Hazard. Mater. 192 (2011) 131–138, https://doi.org/10.1016/j.jhazmat.2011.04.111.
[24] A. Zayed, Z. Bin, Z. Shawon, T. Rahman, K. Wei, K. Hidajat, M. Shahab, Ionically modified magnetic nanomaterials for arsenic and chromium removal from water, Chem. Eng. J. 225 (2013) 607–615, https://doi.org/10.1016/j.cej.2013.03.114.
[25] Y. Lei, F. Chen, Y. Luo, Three-dimensional magnetic graphene oxide foam/Fe_3O_4 nanocomposite as an efficient absorbent for Cr (VI) removal, J. Materi. Sci. (2014) 4236–4245, https://doi.org/10.1007/s10853-014-8118-2.
[26] L. Tan, J. Xu, X. Xue, Z. Lou, J. Zhu, S.A. Baig, X. Xu, Multifunctional nanocomposite Fe3O4@SiO2–mPD/SP for selective removal of Pb(II) and Cr(VI) from aqueous solutions, RSC Adv. 4 (2014) 45920–45929, https://doi.org/10.1039/c4ra08040h.
[27] A. Ngomsik, A. Bee, D. Talbot, G. Cote, Magnetic solid – liquid extraction of Eu (III), La (III), Ni (II) and Co (II) with maghemite nanoparticles, Separ. Purif. Technol. 86 (2012) 1–8, https://doi.org/10.1016/j.seppur.2011.10.013.
[28] H.J. Shipley, S. Yean, A.T. Kan, M.B. Tomson, Research article a sorption kinetics model for arsenic adsorption to magnetite nanoparticles, Environ. Sci. & Pollut. Res. (2010) 1053–1062, https://doi.org/10.1007/s11356-009-0259-5.
[29] P. Kuhn, M. Antonietti, A. Thomas, Porous, covalent triazine-based frameworks prepared by ionothermal synthesis, Angewandte Chem. Int. Ed. (2008) 3450–3453, https://doi.org/10.1002/anie.200705710.
[30] K. Leus, K. Folens, N. Ricci, H. Perez, M. Filippousi, M. Meledina, M.M. Dîrtu, S. Turner, G. Van Tendeloo, Removal of arsenic and mercury species from water by covalent triazine framework encapsulated γ -Fe_2O_3 nanoparticles 353 (2020) 312–319, https://doi.org/10.1016/j.jhazmat.2018.04.027.
[31] S. Lin, D. Lu, Z. Liu, Removal of arsenic contaminants with magnetic c -Fe_2O_3 nanoparticles, Chem. Eng. J. 211–212 (2012) 46–52, https://doi.org/10.1016/j.cej.2012.09.018.
[32] F. Mou, J. Guan, L. Xu, W. Shi, Magnetic iron oxide chestnutlike hierarchical nanostructures: preparation and their excellent arsenic removal capabilities, ACS Appl. Material. & Interf. (2012), https://doi.org/10.1021/am300814q.
[33] S. Zhu, S. Fang, M. Huo, Y. Yu, Y. Chen, X. Yang, Z. Geng, A novel conversion of the groundwater treatment sludge to magnetic particles for the adsorption of methylene blue, J. Hazard Mater. 292 (2015) 173–179, https://doi.org/10.1016/j.jhazmat.2015.03.028.
[34] H. Zeng, C. Yin, T. Qiao, Y. Yu, J. Zhang, D. Li, As(V) removal from water using a novel magnetic particle adsorbent prepared with iron-containing water treatment residuals, ACS Sustain. Chem. Eng. 6 (2018) 14734–14742, https://doi.org/10.1021/acssuschemeng.8b03270.

[35] A. Majumder, L. Ramrakhiani, D. Mukherjee, U. Mishra, A. Halder, Green synthesis of iron oxide nanoparticles for arsenic remediation in water and sludge utilization, Clean Technol. Environ. Policy (2019), https://doi.org/10.1007/s10098-019-01669-1.
[36] R.C. Vieira, A.M.A. Pintor, R.A.R. Boaventura, M.S. Botelho, Arsenic Removal from Water Using Iron-Coated Seaweeds, vol. 192, 2017, pp. 224–233, https://doi.org/10.1016/j.jenvman.2017.01.054.
[37] B. Wang, H. Wu, L. Yu, R. Xu, T. Lim, X. Wen, D. Lou, Template-free formation of uniform urchin-like α -FeOOH hollow spheres with superior capability for water treatment, Adv. Mater. (2012) 1111–1116, https://doi.org/10.1002/adma.201104599.
[38] Y. Jia, X. Yu, T. Luo, M. Zhang, J. Liu, X. Huang, Two-step self-assembly of iron oxide into three-dimensional hollow magnetic porous microspheres and their toxic ion adsorption mechanism, Dalton Trans. (2013) 1921–1928, https://doi.org/10.1039/c2dt32522e.
[39] X. Yu, Y. Wei, C. Liu, J. Ma, H. Liu, S. Wei, W. Deng, J. Xiang, S. Luo, Chemosphere ultrafast and deep removal of arsenic in high-concentration wastewater: a superior bulk adsorbent of porous Fe_2O_3 nanocubes-impregnated graphene aerogel, Chemosphere 222 (2019) 258–266, https://doi.org/10.1016/j.chemosphere.2019.01.130.
[40] I.F. Nata, M. Sureshkumar, C. Lee, One-pot preparation of amine-rich magnetite/bacterial cellulose nanocomposite and its application for arsenate removal, RSC Adv. (2011) 625–631, https://doi.org/10.1039/c1ra00153a.
[41] D. Morillo, G. Pérez, M. Valiente, Efficient arsenic (V) and arsenic (III) removal from acidic solutions with Novel Forager Sponge-loaded superparamagnetic iron oxide nanoparticles, J. Colloid Interface Sci. 453 (2015) 132–141, https://doi.org/10.1016/j.jcis.2015.04.048.
[42] D. Morillo, A. Uheida, G. Pérez, M. Muhammed, M. Valiente, Arsenate removal with 3-mercaptopropanoic acid-coated superparamagnetic iron oxide nanoparticles, J. Colloid Interface Sci. 438 (2015) 227–234, https://doi.org/10.1016/j.jcis.2014.10.005.
[43] L. Feng, M. Cao, X. Ma, Y. Zhu, C. Hu, Superparamagnetic high-surface-area Fe_3O_4 nanoparticles as adsorbents for arsenic removal, J. Hazard Mater. 217–218 (2012) 439–446, https://doi.org/10.1016/j.jhazmat.2012.03.073.
[44] Y. Jin, F. Liu, M. Tong, Y. Hou, Removal of arsenate by cetyltrimethylammonium bromide modified magnetic nanoparticles, J. Hazard Mater. 227–228 (2012) 461–468, https://doi.org/10.1016/j.jhazmat.2012.05.004.
[45] P. Singh, S. Sharma, K. Chauhan, R.K. Singhal, Fabrication of economical thiol-tethered bifunctional iron composite as potential commercial applicant for arsenic sorption application, Ind. Eng. Chem. Res. 57 (2018) 12959–12972, https://doi.org/10.1021/acs.iecr.8b03273.
[46] C.H.S. Gulipalli, B. Prasad, K.L. Wasewar, Batch study, equilibirum and kinetics of adsorption of selenium using rice husk ash (RHA), J. Eng. Sci. & Technol. 6 (2011) 586–605.
[47] M. Pena, X. Meng, C. Jing, Adsorption mechanism of arsenic on nanocrystalline titanium dioxide, Environ. Sci. & Technol. (2006) 1257–1262, https://doi.org/10.1021/es052040e.
[48] D. Setyono, S. Valiyaveettil, Multi-metal oxide incorporated microcapsules for effi cient As(III) and As(V) removal from water, RSC Adv. (2014) 53365–53373, https://doi.org/10.1039/c4ra09030f.
[49] J. Saiz, E. Bringas, I. Ortiz, Functionalized magnetic nanoparticles as new adsorption materials for arsenic removal from polluted waters, J. Chem. Technol. & Biotechnol. (2014) 2013, https://doi.org/10.1002/jctb.4331.
[50] V. Chandra, J. Park, Y. Chun, J.W. Lee, I. Hwang, K.S. Kim, Removal 4 (2010) 3979–3986.
[51] Y. Yoon, W. Kyu, T. Hwang, D. Ho, W. Seok, J. Kang, Comparative evaluation of magnetite – graphene oxide and magnetite-reduced graphene oxide composite for As(III) and As(V) removal, J. Hazard. Mater. 304 (2016) 196–204.
[52] A. Sigdel, J. Park, H. Kwak, P. Park, Arsenic removal from aqueous solutions by adsorption onto hydrous iron oxide-impregnated alginate beads, J. Ind. Eng. Chem. 35 (2016) 277–286, https://doi.org/10.1016/j.jiec.2016.01.005.

[53] S.R. Kumar, V. Jayavignesh, R. Selvakumar, K. Swaminathan, N. Ponpandian, Facile synthesis of yeast cross-linked Fe_3O_4 nanoadsorbents for efficient removal of aquatic environment contaminated with As(V), J. Colloid Interface Sci. 484 (2016) 183−195, https://doi.org/10.1016/j.jcis.2016.08.081.

[54] S.R. Chowdhury, E.K. Yanful, Arsenic and chromium removal by mixed magnetite e maghemite nanoparticles and the effect of phosphate on removal, J. Environ. Manag. 91 (2010) 2238−2247, https://doi.org/10.1016/j.jenvman.2010.06.003.

[55] S.I. Siddiqui, S.A. Chaudhry, Iron oxide and its modified forms as an adsorbent for arsenic removal: a comprehensive recent advancement, Process Saf. Environ. Protect. 111 (2017) 592−626, https://doi.org/10.1016/j.psep.2017.08.009.

[56] S.D.J. Essilfie-dughan, M.J. Hendry, Arsenate adsorption onto hematite nanoparticles under alkaline conditions: effects of aging, J. Nanopart. Res. (2014), https://doi.org/10.1007/s11051-014-2490-3.

SECTION B

Development of perovskite nanomaterials

CHAPTER 12

Development of perovskite nanomaterials for energy applications

Arunima Reghunadhan[1,2], **A.R. Ajitha**[2,3]

[1]Department of Chemistry, Milad-E-Sherif Memorial College, Kayamkulam, Alappuzha, Kerala, India; [2]International and Interuniversity Centre for Nanoscience and Nanotechnology, Mahatma Gandhi University, Kottayam, Kerala, India; [3]Department of Chemistry, Newmann College, Thodupuzha, Idukki, Kerala, India

Chapter Outline

1. Introduction

As energy demand and consumption is increasing day to day, usage of nonrenewable fossil energy sources will cause serious harmful effect to the environment. Hence fabrication and

Design, Fabrication, and Characterization of Multifunctional Nanomaterials. https://doi.org/10.1016/B978-0-12-820558-7.00020-0

utilization of renewable energy and advanced energy storage devices are particularly urgent. Here comes the necessity of the development of novel, low cost, and eco-friendly energy storage systems. The three main types of electrical energy and conversion devices are batteries, electrochemical supercapacitors, and dielectric capacitors. In comparison with batteries and electrochemical supercapacitors, dielectric capacitors have advantages of high power density, wide temperature range, high charge, and discharge rate, long cycle time, and are a type of energy storage device with great potential [1]. Current research focused on the fabrication of more sustainable energy technologies and materials to reduce the usage of fossil fuels. In which perovskite composites materials have an inevitable importance [2,3].

The materials, structurally similar to calcium titanium oxides discovered in 1839 by Count Lev Aleksevich Von Perovski, a Russian mineralogist, are considered as the origin of a new class called perovskites. From their discovery they have been used as different components in photovoltaics. Even after years, they remain in their place in the energy storage devices. Additions and modifications have been employed to improve the properties of perovskite materials. Halide materials, transition metal-based materials, polymer-based materials, nanomaterial incorporated, etc. are some of the available materials under the class of perovskites. The perovskite materials have been employed in catalysis, sensing, energy devices, conductive materials, etc. They have also been utilized in photochromic, electrochromic, image storage, switching, filtering, and surface acoustic wave signal processing devices.

2. Structure of perovskites

Perovskites are inorganic caged materials having an orthorhombic structure and general formula ABX_3 (A and B are cations and X will be suitable anions). The schematic diagram of the perovskite structure is given in Fig. 12.1. On analyzing the structure, the A type of cations belongs to s, d or f block elements whereas B will be possessing a smaller size compared to A. The B type belong to transition metals having smaller radius. The A type cation will occupy the corners of a cubic unit cell. B will occupy the center and the anions (X) will be at the face centers of the crystal lattice. The stabilization of the structure is due to the arrangement of sixfold coordinated B atoms and 12-fold coordinated A atoms.

3. Properties of perovskite nanomaterials

They have superior properties such as high-absorption coefficient, long-range ambipolar charge transport, low exciton-binding energy, high dielectric constant, ferroelectric properties, etc. These properties make them suitable for energy-related applications such as in photovoltaics. They have catalytic, conductive, semiconductor, metallic and superconducting characteristics (Scheme 12.1).

Some perovskites were found to have magnetic and paramagnetic properties. There are ferroelectric perovskites such as $BaTiO_3$, $PbTiO_3$, dielectric such as $(Ba,Sr)TiO_3$,

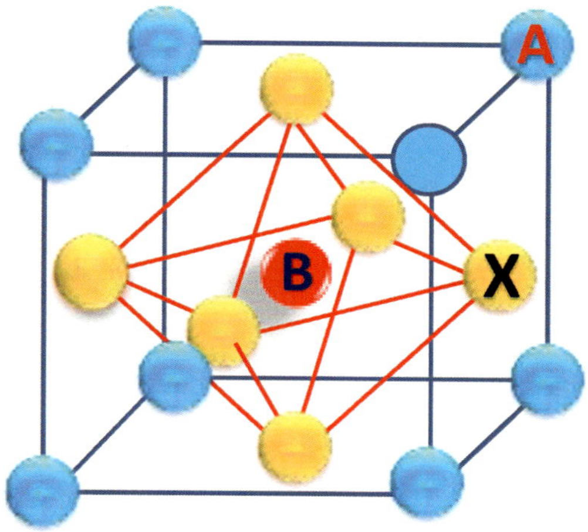

Figure 12.1
Crystal structure of perovskite.

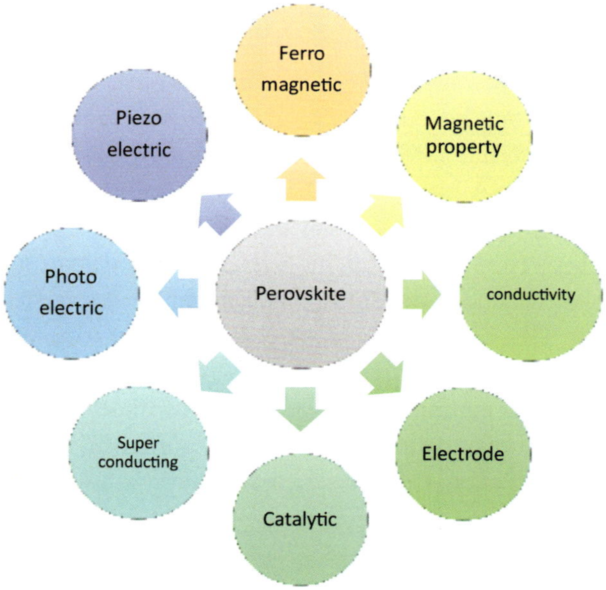

Scheme 12.1
Different properties processed by perovskite nanomaterials.

piezoelectric such as $Pb(Zr,Ti)O_3$, electrostrictive such as $Pb(Mg,Nb)O_3$, magnetoresistant $(La,Ca)MnO_3$ perovskites, and multiferroic, $BiFeO_3$ [4,5]. Examples for some of the perovskite showing insulating, metallic, magnetic and superconducting properties are given in Fig. 12.2.

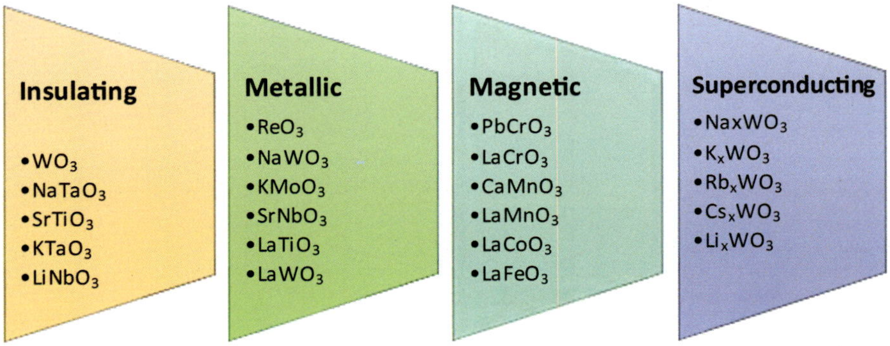

Figure 12.2
Properties of perovskite materials and examples.

4. Types of perovskite materials

The basic molecule in the perovskite class is calcium titanate, $CaTiO_3$. Perovskite materials can be classified in a number of ways. In one classification, the formula structure is considered and according to that we have halide perovskites and inorganic oxide perovskites. In this type of classification, the difference is in the X atom in the ABX_3 crystal lattice. X is either halide or oxide. In addition to typical materials, there are so many perovskites structures that have technological importance. Depending on the morphology and composition, the perovskite materials can be divided in to nanocrystals, nanoparticles, composites, etc. A broad classification is possible when we consider the type of crystal structure. According to that they can be classified into stoichiometric, nonstoichiometric and layered. The nonstoichiometric perovskites are identified by the defect present in them. Either the cations or the anion can be absent from their lattice point creating a vacancy. The third type is being utilized in the advanced applications and are widely used. They include halide type perovskites, organic-inorganic hybrids and super conducting cuprates and some additional special structures materials. A detailed investigation is beyond the scope of this chapter. It is reported by Mitchell and coworkers in very detailed manner [6] (Scheme 12.2).

5. Methods of synthesis of perovskite materials

Perovskite materials can be synthesized in a number of ways. They are:

(a) Solid-state reactions
In this the starting material as well as the final product will be in the solid state. Here the carbonates, nitrates, oxides, etc. will be directly combined with A and B type of cations. The mixture is then calcinated at high temperatures for prolonged time and final materials are obtained after grinding and sieving [7]

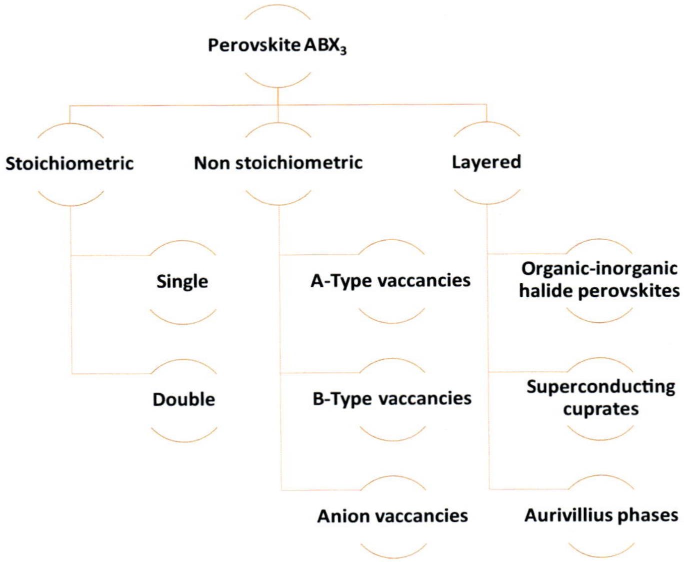

Scheme 12.2
Classification of perovskite materials.

(b) Gas phase reactions
This technique is mainly employed for depositing perovskite films. The electron vapor deposition method, magnetic sputtering, molecular beam epitaxy, thermal evaporation, etc. are associated with gas phase synthesis [8—10].

(c) Wet chemical methods
Wet chemical methods are the common method utilized in the case of nanomaterials. This method involves the sol-gel and precipitation methods. The precipitation method incorporates various precipitating agents like citrates, oxalate, hydroxide, etc. different methods of synthesis are represented in Fig. 12.3. The solution based synthesis can also be done using the following methods.
   (i) Precipitation
   (ii) Thermal treatment
   (iii) Microwave assisted synthesis
   (iv) Plasma assisted synthesis

6. Characterization techniques used for perovskites

The important techniques used for the perovskite materials are it crystal structure analysis, thermal measurements, conductive measurements, power efficiency and morphology. The materials synthesized for specific applications such as catalysis, solar cells, sensors, etc. requires additional characterization techniques. The X-ray diffraction methods are

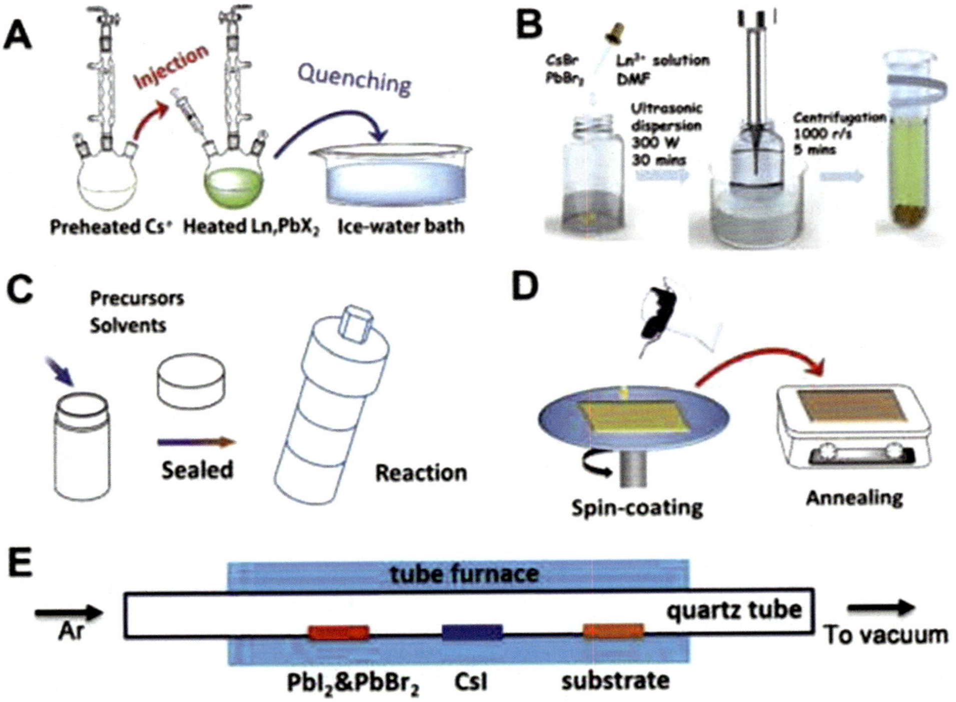

Figure 12.3

Schematic representation of the different synthetic methods employed for perovskites [11–13]. *Reproduced with permission from Ref. Z. Zeng, Y. Xu, Z. Zhang, Z. Gao, M. Luo, Z. Yin, et al., Rare-earth-containing perovskite nanomaterials: design, synthesis, properties and applications, Chem. Soc. Rev. (2020). https://doi.org/10.1039/c9cs00330d.*

employed for having idea about the particle size, the crystal parameters like a, b and c, the lattice volume, etc. [15–17]. Morphology of the perovskite materials are often understood using scanning electron microscope, transmission electron microscope and atomic force microscope (Fig. 12.4).

7. Developments in the field of perovskite-based energy materials

1839: During an expedition, Gustav Rose discovered the mineral based on $CaTiO_3$ in the Ural Mountains. The mineral was named "perovskite" after the Russian mineralogist Lev Aleksevich von Perovski.

1892: H.L. Wells, G.F. Campbell, P.T. Walden and A.P. Wheeler prepared compounds based on cesium, lead and halides from aqueous solutions.

1947: Philips (Eindhoven, the Netherlands) introduced barium titanate for production of condensers.

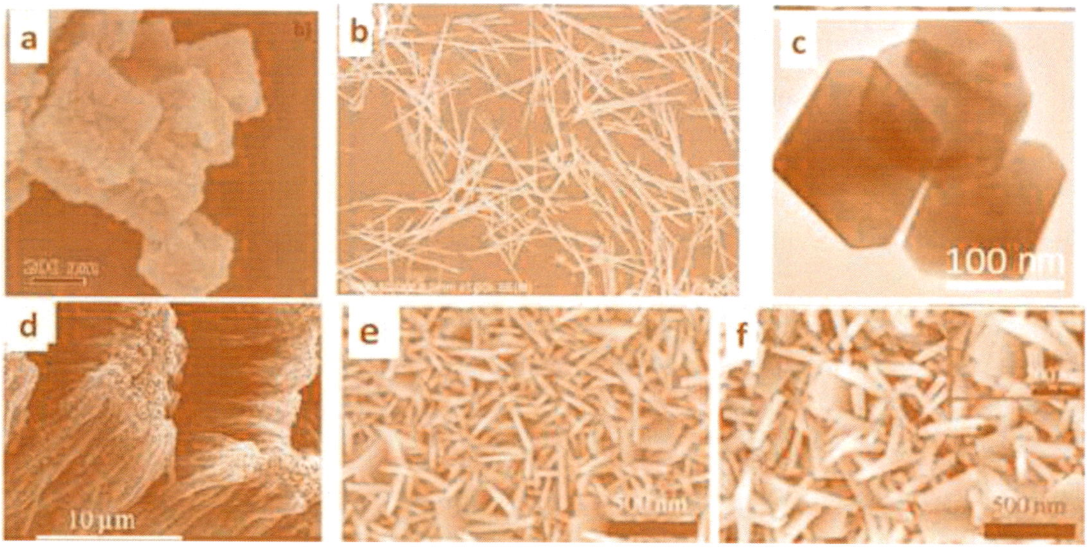

Figure 12.4
SEM and TEM images of perovskites having different morphologies such as (A) cubes (B) wires (C) crystals (D) tubes (E) sheets and (F) particles.

1955: Western Electric (New York, N.Y.) reported the use of ferroelectric crystalline oxides with perovskite structure for fabrication of electromechanical transducers.

1957: C.K. Møller from the Chemical Laboratory at the Royal Veterinary and Agricultural College (Copenhagen, Denmark) evaluated the microstructure of the compounds produced by H.L. Wells and his collaborators and found that they had a perovskite structure.

1957: Siemens (Munich, Germany) developed barium titanate-based resistors.

1959: Clevite (Cleveland, Ohio) introduced perovskite materials in the fabrication of piezoelectric resonators for electromechanical filters.

1962: A.E. Ringwood proposed that the Earth's lower mantle is made primarily of $MgSiO_3$ perovskite.

1964: Compagni e Generale d'Electricité (Paris, France) developed perovskite-based solid electrolytes for fuel cells.

1971: Corning Glass Works (Corning, N.Y.) reported the use of perovskite oxides in frits for glass-ceramic articles.

1971: Exxon Research Engineering (Linden, N.J.) developed perovskite-based cathode catalysts for electrochemical cells used to convert alcohols into ketones.

1975: Hitachi (Tokyo, Japan) manufactured the first gas sensors based on oxide perovskites.

1978: D. Weber/University of Stuttgart (Stuttgart, Germany) developed the first organic-inorganic halide perovskites.

1979: NGK Insulators (Nagoya, Japan) introduced a honeycomb structural body based on barium titanate for use as a heating element.

1981: GTE Laboratories (Waltham, Mass.) introduced lasers based on perovskite crystals.

1988: Ferranti Plc (Oldham, U.K.) developed a superconducting composition with a perovskite structure.

1988: Sharp (Osaka, Japan) developed a thermoelectric material with a perovskite structure composed of a rare earth element and a transition metal.

1994: D.B. Mitzi et al./IBM (Yorktown Heights, N.Y.) developed luminescent organic-inorganic halide perovskites for light-emitting devices.

1996: Boeing North America (Seal Beach, Calif.) introduced cesium-germanium halide salts with perovskite structure as nonlinear optical crystals for optoelectronics.

1999: Murase Chikao et al./National Institute of Advanced Industrial Science & Technology (Tokyo, Japan) created an optical absorption layer for a solar cell using a rare earth oxide having a perovskite crystal structure.

2000 onwards: New processes for fabrication of solar cells based on perovskite materials were developed leading to an increase of activities in this field, the creation of the first devices, and the first.

8. Perovskite nanomaterials

The wide acceptability of perovskite materials in the variety of applications suggested the reduction of size of the materials to the Nano regime. The conversion to nano size will increase the efficiency and performance of the materials where they have been utilized. As considering about the normal nanomaterials, the nano perovskites also have certain advantages. The nano perovskites are composed of nanowires, nanocrystals, nano particles, nanofilms, quantum dots etc. [18] (Figs. 12.5 and 12.6).

The reduction to nano size will increase the photoelectric and magnetic properties. A deep discussion about all types, their synthetic aspect and property analysis is beyond the scope of this chapter.

8.1 Perovskite nanomaterials as energy storage devices

When dealing with the energy-related devices, the solar cells are the field where perovskites are widely used. The most common types of them are halides and oxides. Halides possess limited adjustability and poor stability to environment [13,19,20].

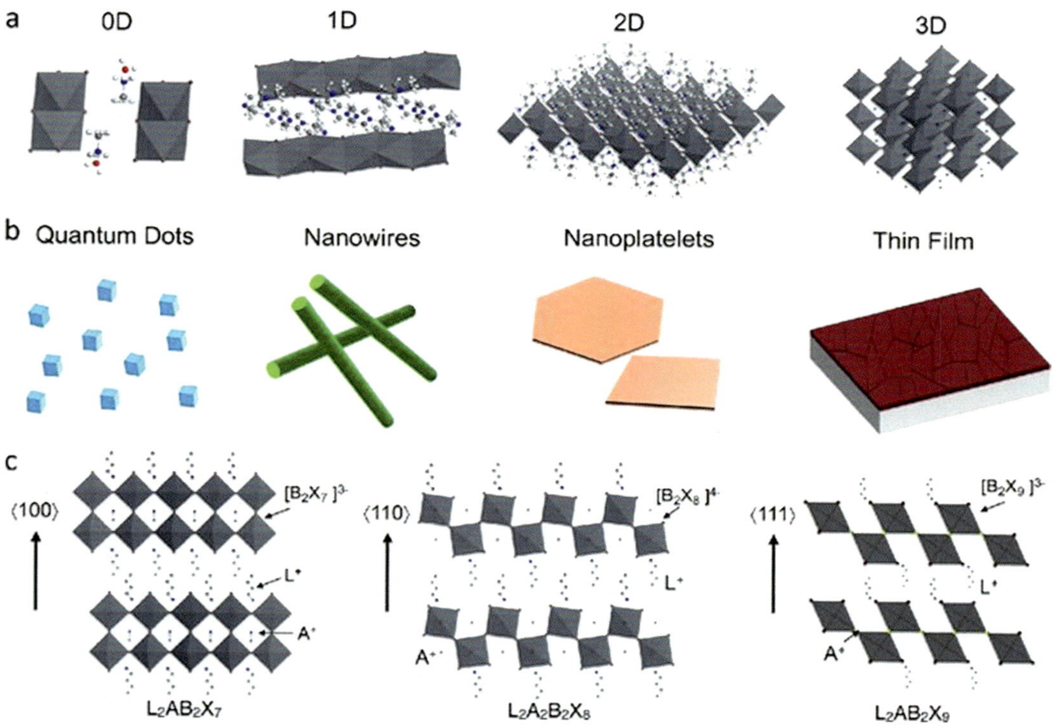

Figure 12.5

Different types of perovskite nanostructures [14]. *Republished with permission from Ref. E. Shi, Y. Gao, B.P. Finkenauer, A. Akriti, A.H. Coffey, L. Dou, Two-dimensional halide perovskite nanomaterials and heterostructures, Chem. Soc. Rev. (2018). https://doi.org/10.1039/c7cs00886d.*

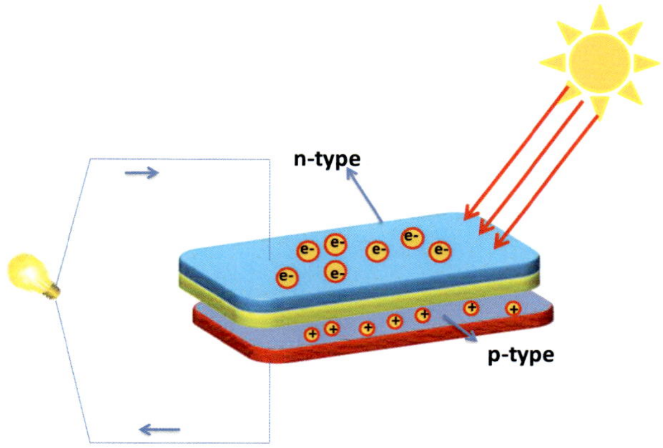

Figure 12.6

Mechanism of a photovoltaic cell.

8.1.1 Photovoltaics based on perovskites

Photovoltaics, which are commonly referred to as solar cells, are the widely used alternatives for the conventional electricity producing methods. They are devices or materials that can convert the energy from sunlight to electrical energy or electricity. The first solar cell was constructed from semiconductor selenium materials coated with a thin layer of gold. From this simple semiconductor device, the development journey of solar cells reached in silicon based materials after a long term research in the semiconductor materials. Doped silicon materials were found to be photosensitive and this property later lead to the development of silicon based solar cells.

The basic principle of photovoltaics is the photovoltaic effect. The effect was first discovered by Edmond Becquerel in 1839. It can be explained as:

The sunlight is comprised of photons. Each photon carries a particular amount of energy, that means the photons are quantized and the amount of energy depends on the wavelength. The photovoltaics are made up of semiconductor devices which contain junctions. When sunlight falls on the surface of such a material, a photon having a particular energy is falling. Then these photons can be reflected from the surface, absorbed by the surface or can transmit through the material. When photons are absorbed by the materials, it results in the photovoltaic effect. For each absorbed photon, an electron gets displaced from the material's atom. The semiconductor materials are designed in such a way to ensure the easy flow of electrons toward the exposed surface. When the free electrons move toward the front surface, a difference of electrical charge density will be created between the front and the rear surface of the material. This difference results in generation of potential difference in the solar cell. A simple schematic representation is given in Fig. 12.1.

8.1.2 Structure of perovskite-based solar cells

The Perovskite solar cell has progressed from dye-sensitized solar cells (DSSCs). The earlier stages of perovskite solar cells were based on solid state electrolyte, but the difficulty in dissolution and decomposition of the materials were the major drawbacks. When photo excited, electrons and holes are generated at the conduction bands (CBs) and valence bands (VBs) of a perovskite respectively. The electrons will move to the CB of TiO_2, while the holes will move to the VB of the hole transport material (HTM) layer (Fig. 12.7).

Perovskite photovoltaics have a wide bandgap. This creates an opportunity in pairing them up with a low band gap photovoltaic technology, which will result in improved efficiency and will matter in a highly competitive market where system costs depend on efficiencies. In addition, perovskite solar cells offer additional attributes like flexibility, semitransparency, thin-film, lightweight, and low processing costs.

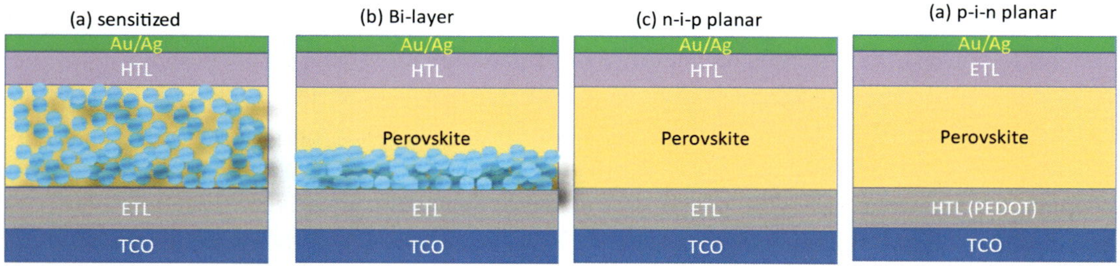

Figure 12.7
Layered structure of different types of perovskite-based solar cell.

8.1.3 Specification required for photovoltaic materials

When selecting a material for solar cell fabrication or any other photovoltaic application. The first and foremost requirement is the energy conversion efficiency. It is defined as the ability of a material to convert solar energy to electrical energy or the percentage of power converted (from absorbed light to electrical energy) and collected. The other factors are thermodynamic efficiency, quantum efficiency, maximum power point etc.

8.1.4 Limitations of ordinary solar cells

- One of the major disadvantages of the silicon based solar cells is that the materials for the construction is not available in the pure form. Silicon in the elemental form is rare to find and the processing of its common compound SiO_2 requires a great effort to get purified
- The cost for the material and fabrication of ordinary solar cells/photovoltaics is high
- Another disadvantage is the weight and rigidity of the materials
- Efficiency decreases when temperature increases
- Toxic materials are used
- Conventional solar cells are their power conversion efficiency, which has been stuck at 25% for 15 years
- Lifetime is short

8.1.5 Advantages of perovskite nanomaterials

- High Optical Absorption
- Tunable Band Gap
- Low Recombination
- High Open-Circuit Voltage (Voc)
- Rapidly Increasing Efficiency

8.1.6 Perovskite nanomaterials in photovoltaics

The halide perovskite materials are favorite candidates in solar cells or generally photovoltaics. The defects of shortcomings of the perovskite materials can be eliminated by nano structure inclusion or conversion to nanoscale. The quantum dots, one-dimensional and two-dimensional nanomaterials, nanohybrids of perovskites, etc. are currently used in the solar cells. In the layered perovskite solar cells, the organometallic hybrids are catching much attention because of the large coefficient of absorption, ease of processing, low recombination time, low exciton-binding energies, etc. But they have low moisture stability and moisture get easily trapped inside the organometallic materials. In order to overcome this defect, the material should be modified with polymers or nanomaterials should be fabricated. The nanomaterial modified and nano included perovskites are often referred to as low dimensional perovskites. In the structure of low dimensional perovskites, the ID materials consist of extended BX_6^{4-} octahedral along one axis and in 2D, it is extended to two dimensions. In the case of zero-dimensional perovskites, we observe disconnected octahedral clusters [21].

Core-shell particles can be employed to enhance the properties of the perovskite solar cells. Yang et al. reported Au—Pt—Au core-shell particles with optical an dielectrical properties incorporated to solar cells. They were mixed with the TiO_2 mesoporous layer. They showed a power conversion efficiency of 8.1%. The enhanced efficiency can be attributed to the reorganization of the electrons and holes on the surface of the core-shell nanoparticles [22]. Au@SiO_2 nanoparticles incorporated meso-superstructured organometallic halide perovskite-based solar cells could deliver a device efficiency of up to 11.4%. With the incorporation of nanoparticles, there was and enhancement of charge carriers [23—25]. Light absorption capacity of the PSC can be further enhanced by the plasmonic nanoparticles. Novel morphologies are always applied to examine the efficiency. In comparison with the nanoparticles, sheets and wires, popcorn-shaped nanoalloys of Au and Ag showed better absorption characteristics when added to solar cell structures. Maximum power conversion with the unit efficiency (PCE) is rising from 8.9% to 10.3%, i.e., 15.7% rise, with the aid of plasmonic popcorn-shaped nanoparticles [26]. The absorption characteristics and the voltage-current profile for the popcorn nanoparticle modified PSC is depicted in Fig. 12.8.

Nanostructured perovskites are fabricated by co-precipitation route from ethyl amine ($C_2H_5NH_2$) and hydrochloric acid as the precursors with aqueous solution of $Pb(CH_3COO)_2.3H_2O$. The obtained materials had the formula $C_2H_5NH_3^+PbCl_3^-$. The materials had a band gap of 1.57eV and the morphology confirmed the grain size in the nano dimension [30].This could be a promising materials in the future solar cell applications. Seed-mediated method was developed by GeO_2 nanoparticles (NPs) for growing crystal perovskite films. By tuning the size of the GeO_2 nanoparticles, the quality

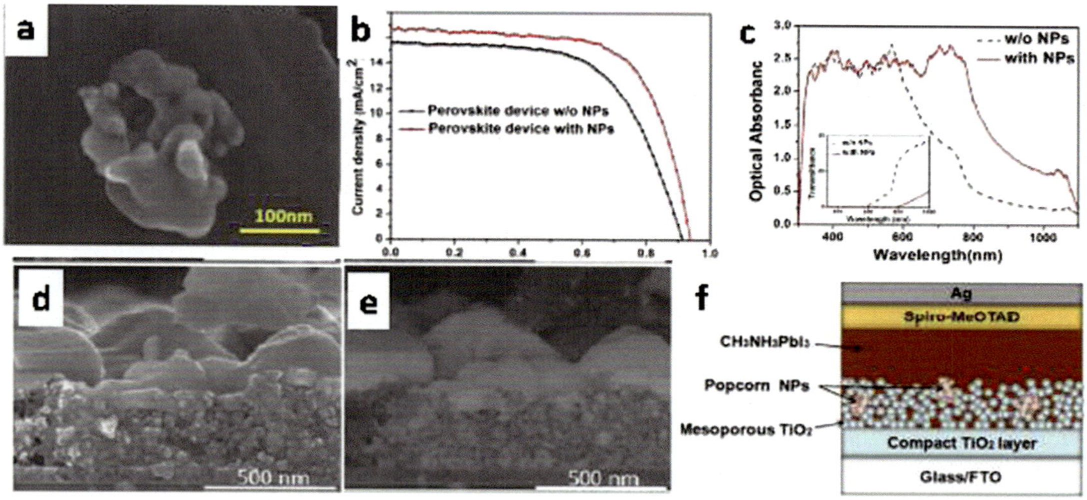

Figure 12.8

(A) SEM image of the popcorn-shaped nanoparticle (B) current-voltage characteristics of solar cells (C) light absorption characteristics (D) and (E) the mesoporous modified perovskites on TiO_2 (F) the layer structure of the solar cell [26]. *Adapted with permission from Ref. Z. Lu, X. Pan, Y. Ma, Y. Li, L. Zheng, D. Zhang, et al., Plasmonic-enhanced perovskite solar cells using alloy popcorn nanoparticles, RSC Adv. 5 (2015) 11175–11179. https://doi.org/10.1039/c4ra16385k.*

of perovskite material could be enhanced [31]. Regulation of the size of perovskite crystal grain, passivation of the grain boundary and bridging of grain are the keys to achieving high efficiency in solar cells with perovskite. This can be done with a small amount of single-walled carbon nanotubes attached to an active layer of perovskite [32].

There are a large number of reports in which the perovskite solar cells are modified with graphene sheets, nanoparticles and other nano sized materials as different components [33–41].

8.1.7 Perovskite/polymer nanocomposite based photovoltaics

The metal–halide perovskites are the normal materials used in the energy applications. Handling of imperfect grain boundaries in polycrystalline perovskite films is vital to take full advantage of the optoelectronic properties and stability of the film and the resultant devices. The grain boundaries act as the trigger for the degradation of the materials. The imperfect grain boundaries are more vulnerable to heat and moisture deprivations which proliferate inwards into the grain interiors and bring on the physical and electrical decoupling of individual grains hence it reduces the device performance. So the modification of the perovskite structure is essential to improve its efficiency. Many small molecules with Lewis bases were tested to decrease the defects and some of them found useful. But, the high degree of instability and high diffusion coefficients of small molecules probably create difficulties in including them in the realistic devices operating under harsh environments. Here comes the thought of using macromolecules or polymers. The long chain polymers can effectively mask up the defects and can improve the properties of the perovskite crystals forming a polymer/perovskite composite material.

The perovskite materials were employed for solar cells/photovoltaics 2 decades ago. Since then they are the most favorite candidates in the industry. The perovskite materials when selected for solar cell applications, organic-inorganic type materials are considered. In these materials, the A group/atoms will be from the organic side and will be placed on the vertices of the fcc lattice, the cations like Pb^{2+}, Sn^{2+}, Ba^{2+}, etc. and the anions like halide ions will be on the center and the apex of the octahedron. These materials will have high-energy absorption and high dielectric constant.

Polymers, the macromolecules are everywhere in our lives. When comes into technology, they are the inevitable part. In the case of photovoltaics and solar cells, polymers can be utilized in a number of ways. A large number of reports are available in solar cells based on polymers. The key role of polymers in photovoltaics is the construction of the hole transport layers (HTL). New materials with high efficiency and low are always in search. Polymers are also incorporated in the coating of the solar cells to improve the absorption. The low band gap polymers and block-copolymers are utilized in the active layer to interfacial materials and cross-linkers. The works can be divided into three main categories (a) polymeric HTMs, (b) polymeric electron transport materials (ETMs) and (c) polymeric templating agents. Yang et al. reported the usage of macromolecules as cross linker in perovskite crystals to enhance the properties and to minimize the defects in the grain boundaries [42]. When perovskite materials are synthesized from the solution process, there is always the chance of generation of defected grain boundaries. Su et al. resolved this problem by adding a polymer, polyethylene glycol in the perovskite layer nanotitania layer and they obtained a power efficiency increase of 13.2% [43]. The microscopic analysis made clear that the polymer could reduce the surface roughness of the film from 117.07 to 73.04 nm. This indicate that the polyethylene glycol (PEG) can hold up the growth and aggregation of perovskite crystals. Thus a continuous film can be obtained which will increase the absorption of solar energy.

The key role of polymers in photovoltaics is the construction of the HTL. The polymers as hole transport layer materials have received much attention than the other applications. The hole transport layer can effectively transfer across the bulk, but to avoid the recombination of charges at the interface, the HTMs should have the electron blocking properties also. The HTM should have a HOMO (highest occupied molecular orbital) layer to accept electrons and high conductivity and hole mobility. The low band gap polymers are also employed as the hole transport layer. Low band gap thiophene-based polymers, such as PCPDTBT, PCDTBT, or PCBTDPP were studied as the HTM. Polyfluorenes, thiophene-based polymers, polytriarylamines, polyanilenes, etc. are also used as hole transport material. Poly(3-hexylthiophene-2,5-diyl) (P3HT) is another HTM that has been extensively studied.

Capacitors have great importance in the case of energy technology systems because dielectric capacitors have an intrinsic high power density. Capacitors exhibits fast energy uptake and delivery. Generally, materials with good dielectric constant, low dielectric loss and high breakdown voltage have been preferred for energy storage applications. Nowadays polymer-based capacitors have more importance, because in a particular electronic device capacitors occupying more than 25% of the volume and weight of the device. Polymer-based capacitors can find a solution for this due to its lightweight and flexibility properties. Ceramics based polymer composites can make a change in the electronic fields. Nano incorporated polymer nanocomposites have more ability to store and harvest electrical energy. Current research more focused on the nano electronic devices, hence polymer nanocomposites have inevitable importance. It was very clear from many studies that polymers have high dielectric breakdown strength along with low energy density, while ceramic fillers show high values of dielectric permittivity and low breakdown strength. Hence their combination will offer a system with enhanced dielectric properties. Ceramic nanoparticles are promising fillers to improve the dielectric properties and functionality of the nanocomposites, due to their high permittivity and temperature dependent polarization. There are several studies reported based on the energy storage applications of perovskite-based polymer microcomposites and nanocomposites.

Manika and *Psarras* prepared a nanocomposite of Epoxy/$BaTiO_3$ for energy storage application. They prepared epoxy polymer composites with 1, 3, 7, 10, 12 and 15 phr $BaTiO_3$. In their study they reported that the coefficient of energy efficiency (n_{eff}) increased with filler content at ambient conditions and reaching the highest value of 58.2% for the 7 phr $BaTiO_3$ nanocomposite. They found that the prepared composite nano dielectrics can act as compact capacitive energy storing systems [44] (Table 12.1).

Table 12.1: coefficient of energy efficiency (*n*eff(%)) at for three different applied voltage levels 50, 100 and 200V for the time instant t = 10s at 30°C for all specimens.

	n_{eff} [%]		
$BaTiO_3$ content [phr]	50 V	100 V	200 V
0	17.11	28.02	13.23
1	12.31	43.80	7.06
3	58.16	55.83	54.56
7	49.33	55.72	5820
10	41.86	39.84	50.42
12	41.85	43.58	48.62
15	38.33	39.68	50.52

Mayeen et al. synthesized and analyzed PVDF-TrFE based nanocomposites of both functionalized and nonfunctionalized BaTiO3 (BTO). They studied the dielectric, piezoelectric, ferroelectric, and magnetoelectric properties and found that the properties were enhanced with filler content in the polymer matrix. Better improvement was observed for functionalized samples. They also reported that the prepared composite can be considered as a good candidate for the fabrication of smart energy storage, magneto electric as well as energy harvesting device applications [45].

Polymer-based ceramic materials could be act as a good candidate for modern technology. As already discussed, even though polymers have high breakdown field strength, its low dielectric constant limits its dielectric application. But according to recent reports incorporation of ceramic materials with high dielectric constant into polymers can offer new materials to dielectric application especially for energy storage application. Among various polymers certain polymers such as PVDF, polyimide etc. have excellent thermal stability and flexibility and are capable in high temperature environments.

Even though polymer-based composites with high aspect ratio fillers exhibit higher dielectric constant, the agglomeration and incompatibility between fillers and polymer matrix are some challenges facing in this field. Some studies reported that by using different morphology and structures of ceramic materials such as core-shell structures, multiphase composite structures, etc. will offer a composite material with improved dielectric and energy storage properties. Wan et al. prepared $BaTiO_3$ (BT) nanofibers incorporated polymer composite film with polyimide (PI) matrix. And systematically investigated the dielectric and energy storage properties of all composites investigated. They discovered that the developed composite films under a reducing (H_2) atmosphere have excellent dielectric properties, compared with that under Air and O_2. The dielectric constant PI composite films with 20 wt.% BT-fibers reached up to 17.6 which was about four times greater than that of pure PI ($\varepsilon r = 4.1$). Composite film with 15 wt.% BT exhibited a maximum energy storage density of $Ue = 6.12$ J/cm^3 and it is about 150% of pure PI. Their reports open up an effective method to tune the dielectric and energy storage properties of ferroelectric/polymer composites.

Thus according to reported works it can be say that dielectric polymer nanocomposites with high dielectric constant ceramics have great importance in energy storage devices to fulfill the requirements of modern electric power systems. The increased interest for their potential applications is mainly due to their intrinsic charge–discharge capability to store and release the electrical energy through dielectric polarization and depolarization. Recent research is mainly focused on the polymer composite with functionalized ceramics (Fig. 12.9).

Wang et al. selected PVDF for preparation of polymer composites due to their relative high breakdown strength in comparison with most common polymers. In their study

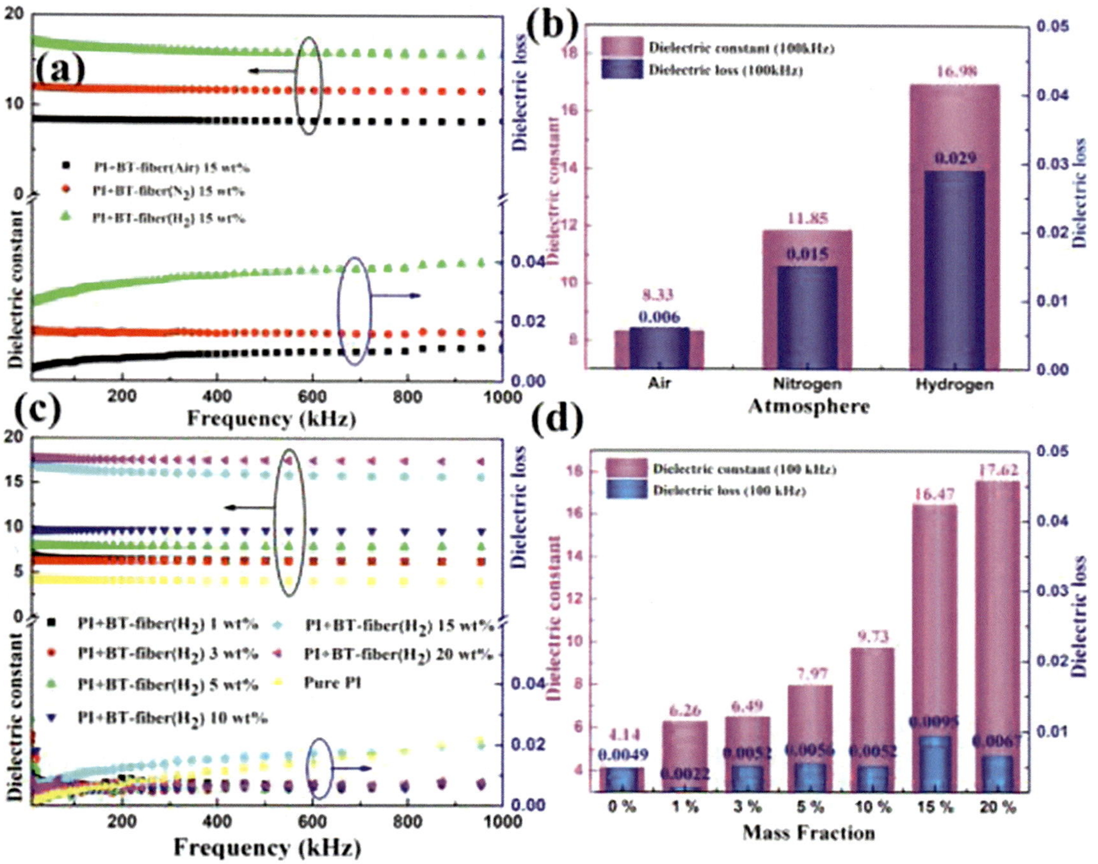

Figure 12.9
(A) Frequency dependence of the dielectric constant and loss of composite films with BT-fibers (15 wt.%) sintered under Air, N_2 and H_2 atmospheres, (B) corresponding dielectric constant and loss at 100 kHz. (C) Frequency dependence of the dielectric constant and loss of PI composite films on the mass fractions of the BT-fibers (H_2), (D) the corresponding dielectric constant and loss of composite films at 100 kHz [27]. *Republished with permission from Ref. B. Wan, H. Li, Y. Xiao, S. Yue, Y. Liu, Q. Zhang, Enhanced dielectric and energy storage properties of* $BaTiO_3$ *nanofiber/polyimide composites by controlling surface defects of* $BaTiO_3$ *nanofibers, Appl. Surf. Sci. 501 (2020) 144243. https://doi.org/10.1016/j.apsusc.2019.144243.*

bio-inspired fluoro-polydopamine functionalized $BaTiO_3$ nanowires ($BaTiO_3$ NWs) were incorporated into a fluoropolymer to realize flexible polymer nanocomposites with high-energy storage capability. They reported that rationally designed fluoro-polydopamine shell layers may increase the movement of charge carriers in the interface between polymer and nanowires and prevent the agglomeration of fillers in the polymer matrix, resulting in lower dielectric loss and smaller leakage current densities in comparison with the nanocomposite on the basis of raw nanowires. They reported that the functionalization of $BaTiO_3$ NWs by bio-inspired fluoro-polydopamine has guaranteed both the increase of dielectric constant and the maintenance of breakdown strength, which is crucial for the substantial increase of energy storage capability [1,46].

An article was reported very recently by Wang et al. describing energy storage properties of barium titanate-based PVDF polymer composites. In this paper the energy storage theory, mechanisms and a detailed study of BT/PVDF-based nanocomposite were included. Some new ideas for the development of BT/PVDFbased nanocomposites with better energy density also discussed in this paper [47].

8.2 Piezoelectric nanoperovskites

As mentioned in the properties of perovskites, they possess piezoelectric characteristics. The nanopiezo electric perovskites are more efficient than their microcounter parts. Piezoelectric materials can generate electricity from mechanical energy or they are the devices which convert mechanical energy into electrical energy. The electricity thus generated is called piezo electricity [48,49]. Piezoelectric perovskite nano materials are employed as sensors, actuators, power generators, etc. One of the normal techniques employed for developing piezoelectric nanomaterials having perovskite structure involves lead ions incorporation and lead incorporated structures shoes improved piezoelectric properties [50]. The most common material in this category is the lead zirconate titanate (PZT) But due to the harmful nature of lead toward the environment, the researchers developed lead-free nanomaterials [51–54].

Formamidinium lead halide nanoparticles ($FAPbBr_3$ NPs) having perovskite were developed and they were incorporated in the polymer PVDF to fabricate energy harvesting nanogenerators. They displayed an output voltage of 30 V and current density of 6.2 $\mu A/cm^2$. The ac generated from this system can be utilized to fabricate light-emitting diode systems [55]. The same polymeric system with lead-free approach was put forward by Lee et al. and they reported the incorporation of barium titanate into the PVDF-TrFE membrane could lead to highly efficient piezoelectric materials that can be utilized in smart wearable electronics and autonomous nanosystems. They also possess biomedical energy harvesting efficiency. Their output energy is very much comparable to the PZT materials [55]. A well aligned one-dimensional nanostructure-based flexible energy harvesters can be fabricated from barium titanate nanowires. These arrays could successfully convert maximum open-circuit voltage of about 15 V, a maximum short-circuit current of nearly 400 nA, and an effective power of 0.27 μW as a result of repeated bending deformations [56].

One-dimensional nanomaterials are extensively employed as piezo electric materials. Characterization of zero-dimensional and one-dimensional perovskites reveals that most of them have the capacity for storing energy and also give response to mechanical stress. They are suitable for the nanogenerator concept. The Pb (Zr, Ti) O_3 and $BaTiO_3$ based perovskite nano generators can be fabricated using wet chemical synthesis and aligned nanowires can be obtained. Nanowires of Nb can be added to PDMS polymeric system, so that a very stable and high piezoelectric output (an open-circuit voltage of 3.2 V, a closed circuit current of 72 nA, and a power density of 0.6 mW/cm^3) is obtained considering the volume fraction of the $NaNbO_3$. Same system presents two morphologies which affect its performance. Nano cubes and nanowires were fabricated and were selected for nanocomposite preparation. The power generation and morphological comparison suggest that Since one-dimensional nanowires are superior to zero-dimensional nanocubes in percolation, the NG based on nanowires could effectively deliver generated piezoelectricity to the outside electrodes compared to a nanocube-based one [28] (Fig. 12.10).

An interesting concept of arbitrary bending energy harvesting nanogenerators are synthesized using $PbTiO_3$ nanotube arrays. The materials can be used to synthesize a series of other perovskites. The fiber like materials were synthesized from TiO_2 nanotubes as positive templates using hydrothermal and anodic oxidation methods. The advantages pointed out were the usage of flexible core could lead to electrodes which can provide sufficient connection, synthesis of mass quantity, easy alteration of dimensions like length,

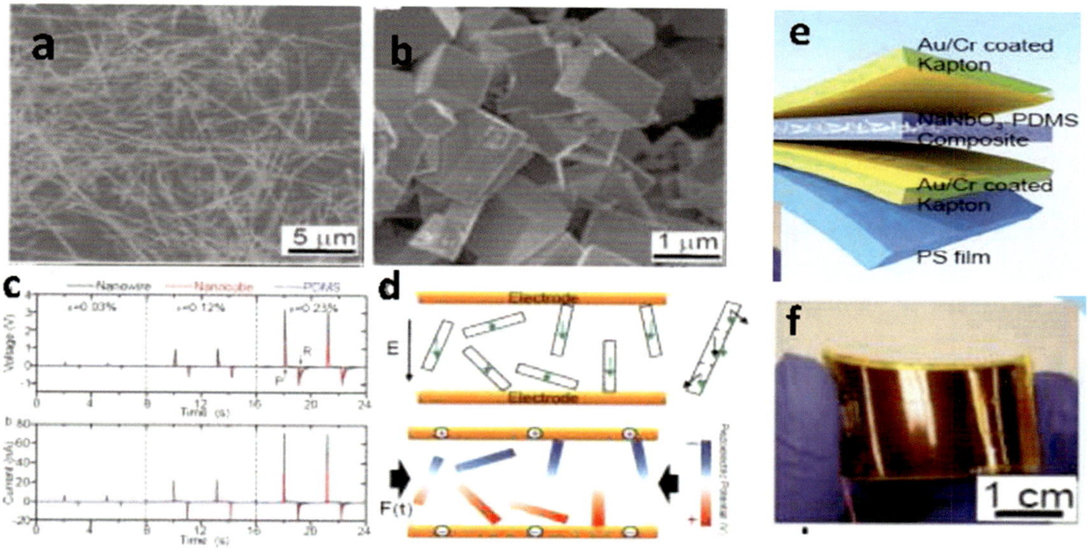

Figure 12.10
(A) nanowire and (B) nanocube of $NaNbO_3$. (C) represents the comparison of power generation of nanogenerators using nanowires and nanocubes (D) mechanism of nanogenerator (E) the layered structure of piezoelectric device (F) fabricated device. *Republished with permission from Ref. J.H. Jung, M. Lee, J.I. Hong, Y. Ding, C.Y. Chen, L.J. Chou, et al., Lead-free NaNbO$_3$ nanowires for a high output piezoelectric nanogenerator, ACS Nano 5 (2011) 10041–10046. https://doi.org/10.1021/nn2039033.*

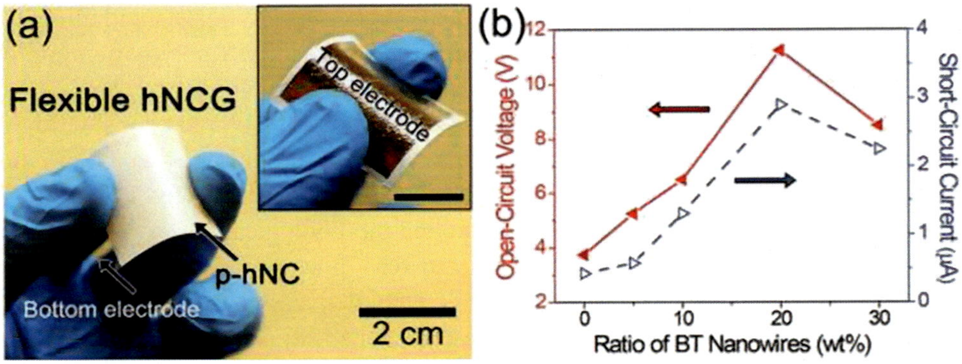

Figure 12.11

(A) The flexible barium titanate/PVDF nanogenerators and (B) the generated output from the piezoelectric [29]. *Republished with permission from Ref. C.K. Jeong, C. Baek, A.I. Kingon, K.I. Park, S.H. Kim, Lead-free perovskite nanowire-employed piezopolymer for highly efficient flexible nanocomposite energy harvester, Small 14 (2018) 1–8. https://doi.org/10.1002/smll.201704022.*

diameter and shape of base units and possibility of synthesis of Ba, Bi, Na, K, Sr, Nb, or La derivatives with perovskite structure. The fibers had a core-shell structure and this enhanced the response and the output voltage was 620 mV and current density of 1.0 nA/cm^2 [57].

Sol-gel synthesized lead zirconate titanate nanowires using nanochannel alumina template having diameter of 45 nm shows a significant piezo response. The size, shape and aspect ratio influences the response [58]. Barium titanate–PVDF nanocomposite-nanogenerator was invented by Jeong et al. The nanocomposite material can be directly attached to a hand to scavenge energy using a human motion and they were flexible and could be used as wearables and they displayed a harvest output of up to 14 V and current density of 4 μA, which is very high when compared to the current levels of piezo ceramic film-based flexible energy harvesters. Finite element analysis method simulations study that the outstanding performance of hNCG devices attributes to not only the piezoelectric synergy of well-controlled BT NWs and within P(VDF-TrFE) matrix, but also the effective stress transferability of piezo polymer [29] (Fig. 12.11).

8.3 Fuel cells based on perovskites

Fuel cells are devices having high efficiency. They are the devices which converts fuels into electrical energy with the help of an oxidant through set of chemical reactions. They are in a simple sense the developed form of the heat engines. There are different types of fuel cells such as solid-oxide fuel cell, polymer electrolyte fuel cell, phosphoric acid fuel cell, alkaline fuel cell, high temperature fuel cells, energy storage fuel cells, etc. In addition to their efficiency, low emission is another advantage. The most common fuel cell is the H_2 $_-O_2$ fuel cell. The components are very simple and involve a fuel, electrode and electrolytes [59].

The perovskite materials are connected with the solid-oxide fuel cells. SOFCs showed many features compared to alternative cell types, such as high-energy storage performance, cheap materials, low fuel impurity susceptibility, low environmental impacts, environmental stability and excellent fuel versatility. The main components are a cathode, an anode and an electrolyte and they convert the chemical energy directly in to electricity. The cathode often reduces the oxygen by electrons and at anode reduction of fuel takes place. Perovskites can be used as all the components in a fuel cell. Some of the perovskite materials used in fuel cells are $La(Sr)MnO_3$, $La(Sr)CoO_3$, $LaNi(Fe)O_3$, $SrTi(Fe)O_3$ and $CaTi(Fe)O_3$ as cathodes $Sr(Ti,Nb)O_3$ is used as an anode material. Materials such as doped $BaCe(Zr)O_3$ and $La(Sr)Ga(Mg)O_3$ are examples of electrolytes [60].The cathode materials selected for the solid-oxide fuel cell(SOFC) should have properties like chemical and physical compatibility with the electrolyte, good electronic conductivity, catalytic activity, porosity and stable microstructure. Lanthanum manganite having the formula La MnO_3 is the common material selected for the cathodic material in most of the SOFCs [61]. Nano fibrous lanthanum cobaltate, $LaCoO_3$ is employed as cathode where the nanofibers are fabricated using electro deposition of the metal hydroxide on carbon nanotube template. The anode-supported SOFCs with nanofibrous perovskite cathodes on zirconia and ceria scaffolds showed high and stable performance of 0.95 and 1.27 W/cm^2 at 800°C. It is assumed that the special nanostructure of the cathode contributes to the improvement of electrochemical properties by providing a large number of active reaction sites and by enabling mass transport through the porous nanofibrous framework [62]. Dramatically high performance can be achieved by fabricating the cathodes using infiltration method. $Sm_{0.6}Sr_{0.4}CoO_{3-\delta}$ (SSC) perovskite nanoparticles having diameter of about 20–80 nm were incorporated into the cathodes of La_{1-x} $SrxMnO_{3-\delta}$ yttria stabilized zirconia. The SSC particles are in diameter, and intimately adhere to the pore walls of the preformed LSM-YSZ cathodes. The enhancement is attributed to electro catalytic activity of SSC [63]. $Ba_{0.5}Sr_{0.5}Fe_{0.8}Cu_{0.2}O_{3-\delta}$ nanofibers produced by sol-gel electrospinning process in which a mixture of polyvinylpyrrolidone and acetic acid was used as the spinning assist and barium, strontium, iron and copper nitrates were being used as precursors for nanofibers production. In addition to an electrical conductivity of 69.54 Scm^{-1} at 600°C, the high porosity and surface area of the $Ba_{0.5}Sr_{0.5}Fe_{0.8}Cu_{0.2}O_{3-\delta}$ cathodes show the ability of the nanofibers to act as efficient cathode materials for solid-oxide fuel cells at moderate temperatures [63]. $La_{1-x}Sr_xCo_{1-y}Fe_yO_{3-\delta}$ were prepared by a chemical precipitation method can be applied in low temperature solid-oxide fuel cells as cathode materials [64].Perovskite structures can be decorated with nanomaterials and they serve as potential materials for energy storage and conversion. The layered structure of perovskites can be modified with nanomaterials by insitu growth technique. One such material is layered

Table 12.2: List of perovskite materials that are employed as electrode materials in SOFC.

Perovskite	Use	Operational temperature in °C	Maximum power density mW/cm^2
$Ba_{0.5}Sr_{0.5}Co_{0.8}Fe_{0.2}O_{3-\delta}$	Cathode	600	1010
		500	402
$Ba_{0.5}Sr_{0.5}Co_{0.2}Fe_{0.8}O_{3-\delta}$	Cathode	800	266
$La_{0.7}Sr_{0.3}Co_{0.5}Fe_{0.5}O_3$	Cathode		
$La_{0.6}Sr_{0.4}Fe_{0.8}Co_{0.2}O_3$	Cathode	800	
$LaBaCuFeO_{5+x}$	Cathode	700	327
		550	105
$LaBaCuCoO_{5+x}$	Cathode	700	432
		550	171
$Y_{0.8}Ca_{0.2}BaCoFeO_{5+\delta}$	Cathode	650	426
$La_{0.75}Sr_{0.25}Cr_{0.5}Mn_{0.5}O_3$	Anode		
$Sr_2MgMnMoO_{6-\delta}$	Anode	800	438
$La_{0.8}Sr_{0.2}Cr_{0.97}V_{0.03}O_3$	Anode	800	

$La_{0.8}Sr_{1.2}Fe_{0.9}Co_{0.1}O_{4\pm\delta}$ and is applied as a redox stable and active electrode for SOFC. Incorporation of adequate amount of cobalt into this structure influence the cathodic and anodic performance. The resultant material can serve both as anode and cathode [65]. In a similar study, a novel perovskite oxide with A-site deficiency $Sm_{0.70}Sr_{0.20}Fe_{0.80}Ti_{0.15}Ru_{0.05}O_{3-\delta}$ was designed for using in symmetric SOFCs. The perovskite oxides were exfoliated with Ruthenium nanoparticles to improve the properties [66]. One of the key problems for practical perovskite catalysts assisted by engineered metallic nanoparticles is the combined attainment of a sufficient nanoparticle population with consistent distribution as well as long-lasting high efficiency. One of the simple manufacturing method to avoid this is the crystal reconstruction of double perovskite under a reducing atmosphere and will spontaneously contribute to the development of ordered layered oxygen deficiency and the segregation of massively and finely dispersed nanoparticles [67]. A large number of nanoparticles are incorporated in perovskite structures to fabricate the hybrid nano perovskite materials. The heavy metals such as rare earths have been successfully incorporated in halide perovskites and oxide perovskites. The lanthanide and actinide cations are explored to a wide extent in the solid-oxide fuel cells. The table below is showing some of the perovskites that are utilized in the solid-oxide fuel cells as cathodes and anodes. The optimum temperature of working is also mentioned (Table 12.2).

9. Conclusion

Structurally interesting and commercially important perovskite materials when reduced to nanoscale can work wonders in terms of their applicability and properties. The perovskites and their nanoversions are chiefly employed as energy harvesting materials and in devices having energy storage capacity. They have been employed in solar cells as HTL, in solid-oxide fuel cells as cathodes and anodes, and as nanogenerators. The very important applications of perovskites are mentioned in this chapter and the variety of perovskites make the applications countless. A large number of metallic and organic-inorganic combinations lead to hybrid materials. The widespread acceptability of perovskites is due to their conductivity, electric (Piezo, ferro), magnetic, and catalytic activities.

References

[1] Y. Wang, M. Yao, R. Ma, Q. Yuan, D. Yang, B. Cui, et al., Design strategy of barium titanate/polyvinylidene fluoride-based nanocomposite films for high energy storage, J. Mater. Chem. 8 (2020) 884–917, https://doi.org/10.1039/C9TA11527G.
[2] M.I. Ahmed, A. Habib, S.S. Javaid, Perovskite solar cells: potentials, challenges, and opportunities, Int. J. Photoenergy (2015), https://doi.org/10.1155/2015/592308.
[3] M.I.H. Ansari, A. Qurashi, M.K. Nazeeruddin, Frontiers, opportunities, and challenges in perovskite solar cells: a critical review, J. Photochem. Photobiol. C Photochem. Rev. (2018), https://doi.org/10.1016/j.jphotochemrev.2017.11.002.
[4] N.G. Park, Perovskite solar cells: an emerging photovoltaic technology, Mater. Today (2015), https://doi.org/10.1016/j.mattod.2014.07.007.
[5] Q. Fu, X. Tang, B. Huang, T. Hu, L. Tan, L. Chen, et al., Recent progress on the long-term stability of perovskite solar cells, Adv. Sci. (2018), https://doi.org/10.1002/advs.201700387.
[6] R.H. Mitchell, M.D. Welch, A.R. Chakhmouradian, Nomenclature of the perovskite supergroup: a hierarchical system of classification based on crystal structure and composition, Mineral. Mag. (2017), https://doi.org/10.1180/minmag.2016.080.156.
[7] T. Baikie, Y. Fang, J.M. Kadro, M. Schreyer, F. Wei, S.G. Mhaisalkar, et al., Synthesis and crystal chemistry of the hybrid perovskite $(CH_3NH_3)PbI_3$ for solid-state sensitised solar cell applications, J. Mater. Chem. (2013), https://doi.org/10.1039/c3ta10518k.
[8] K. Liang, D.B. Mitzi, M.T. Prikas, Synthesis and characterization of organic-inorganic perovskite thin films prepared using a versatile two-step dipping technique, Chem. Mater. (1998), https://doi.org/10.1021/cm970568f.
[9] J. Liu, Y. Xue, Z. Wang, Z.Q. Xu, C. Zheng, B. Weber, et al., Two-dimensional $CH_3NH_3PbI_3$ perovskite: synthesis and optoelectronic application, ACS Nano (2016), https://doi.org/10.1021/acsnano.5b07791.
[10] Gas-phase Synthesis of Nanoparticles, 2017, https://doi.org/10.1002/9783527698417.
[11] Q. Hu, Z. Li, Z. Tan, H. Song, C. Ge, G. Niu, et al., Rare earth ion-doped $CsPbBr_3$ nanocrystals, Adv. Opt. Mater. (2018), https://doi.org/10.1002/adom.201700864.
[12] L. Huang, Q. Gao, L.D. Sun, H. Dong, S. Shi, T. Cai, et al., Composition-graded cesium lead halide perovskite nanowires with tunable dual-color lasing performance, Adv. Mater. (2018), https://doi.org/10.1002/adma.201800596.
[13] Z. Zeng, Y. Xu, Z. Zhang, Z. Gao, M. Luo, Z. Yin, et al., Rare-earth-containing perovskite nanomaterials: design, synthesis, properties and applications, Chem. Soc. Rev. (2020), https://doi.org/10.1039/c9cs00330d.
[14] E. Shi, Y. Gao, B.P. Finkenauer, A. Akriti, A.H. Coffey, L. Dou, Two-dimensional halide perovskite nanomaterials and heterostructures, Chem. Soc. Rev. (2018), https://doi.org/10.1039/c7cs00886d.
[15] I. Spanopoulos, W. Ke, C.C. Stoumpos, E.C. Schueller, O.Y. Kontsevoi, R. Seshadri, et al., Unraveling the chemical nature of the 3D "hollow" hybrid halide perovskites, J. Am. Chem. Soc. (2018), https://doi.org/10.1021/jacs.8b01034.

[16] A. Rachel, S.G. Ebbinghaus, M. Güngerich, P.J. Klar, J. Hanss, A. Weidenkaff, et al., Tantalum and niobium perovskite oxynitrides: synthesis and analysis of the thermal behaviour, Thermochim. Acta (2005), https://doi.org/10.1016/j.tca.2005.08.010.

[17] M. Bradha, T. Vijayaraghavan, S.P. Suriyaraj, R. Selvakumar, A.M. Ashok, Synthesis of photocatalytic $La_{(1-x)}A_xTiO_{3.5-\delta}$ (A=Ba, Sr, Ca) nano perovskites and their application for photocatalytic oxidation of Congo red dye in aqueous solution, J. Rare Earths (2015), https://doi.org/10.1016/S1002-0721(14)60397-5.

[18] W. Zhang, G.E. Eperon, H.J. Snaith, Metal halide perovskites for energy applications, Nat. Energy (2016), https://doi.org/10.1038/nenergy.2016.48.

[19] X. Wang, T. Hisatomi, Z. Wang, J. Song, J. Qu, T. Takata, et al., Core–shell-structured $LaTaON_2$ transformed from $LaKNaTaO_5$ plates for enhanced photocatalytic H_2 evolution, Angew. Chem. (2019), https://doi.org/10.1002/ange.201906081.

[20] J.S. Kang, J.Y. Kim, J. Yoon, J. Kim, J. Yang, D.Y. Chung, et al., Room-temperature vapor deposition of cobalt nitride nanofilms for mesoscopic and perovskite solar cells, Adv. Energy Mater. (2018), https://doi.org/10.1002/aenm.201703114.

[21] C. McDonald, C. Ni, P. Maguire, P. Connor, J.T.S. Irvine, D. Mariotti, et al., Nanostructured perovskite solar cells, Nanomaterials 9 (2019), https://doi.org/10.3390/nano9101481.

[22] B. Wang, X. Zhu, S. Li, M. Chen, N. Liu, H. Yang, et al., Enhancing the photovoltaic performance of perovskite solar cells using plasmonic Au@Pt@Au core-shell nanoparticles, Nanomaterials 9 (2019), https://doi.org/10.3390/nano9091263.

[23] W. Zhang, M. Saliba, S.D. Stranks, Y. Sun, X. Shi, U. Wiesner, et al., Enhancement of perovskite-based solar cells employing core-shell metal nanoparticles, Nano Lett. (2013), https://doi.org/10.1021/nl4024287.

[24] M. Saliba, W. Zhang, V.M. Burlakov, S.D. Stranks, Y. Sun, J.M. Ball, et al., Plasmonic-induced photon recycling in metal halide perovskite solar cells, Adv. Funct. Mater. (2015), https://doi.org/10.1002/adfm.201500669.

[25] M.D. Brown, T. Suteewong, R.S.S. Kumar, V. D'Innocenzo, A. Petrozza, M.M. Lee, et al., Plasmonic dye-sensitized solar cells using core-shell metal-insulator nanoparticles, Nano Lett. (2011), https://doi.org/10.1021/nl1031106.

[26] Z. Lu, X. Pan, Y. Ma, Y. Li, L. Zheng, D. Zhang, et al., Plasmonic-enhanced perovskite solar cells using alloy popcorn nanoparticles, RSC Adv. 5 (2015) 11175–11179, https://doi.org/10.1039/c4ra16385k.

[27] B. Wan, H. Li, Y. Xiao, S. Yue, Y. Liu, Q. Zhang, Enhanced dielectric and energy storage properties of $BaTiO_3$ nanofiber/polyimide composites by controlling surface defects of $BaTiO_3$ nanofibers, Appl. Surf. Sci. 501 (2020) 144243, https://doi.org/10.1016/j.apsusc.2019.144243.

[28] J.H. Jung, M. Lee, J.I. Hong, Y. Ding, C.Y. Chen, L.J. Chou, et al., Lead-free $NaNbO_3$ nanowires for a high output piezoelectric nanogenerator, ACS Nano 5 (2011) 10041–10046, https://doi.org/10.1021/nn2039033.

[29] C.K. Jeong, C. Baek, A.I. Kingon, K.I. Park, S.H. Kim, Lead-free perovskite nanowire-employed piezopolymer for highly efficient flexible nanocomposite energy harvester, Small 14 (2018) 1–8, https://doi.org/10.1002/smll.201704022.

[30] A. Dhar, A. Dey, P. Maiti, P.K. Paul, S. Roy, S. Paul, et al., Fabrication and characterization of next generation nano-structured organo-lead halide-based perovskite solar cell, Ionics (2018), https://doi.org/10.1007/s11581-017-2256-x.

[31] Y.H. Lou, M. Li, Z.K. Wang, Seed-mediated superior organometal halide films by GeO_2 nano-particles for high performance perovskite solar cells, Appl. Phys. Lett. (2016), https://doi.org/10.1063/1.4941416.

[32] H.-S. Lin, S. Okawa, Y. Ma, S. Yotsumoto, C. Lee, S. Tan, et al., Polyaromatic nanotweezers on semiconducting carbon nanotubes for the growth and interfacing of lead halide perovskite crystal grains in solar cells, Chem. Mater. 32 (2020) 5125–5133, https://doi.org/10.1021/acs.chemmater.0c01011.

[33] J.H. Im, J. Luo, M. Franckevičius, N. Pellet, P. Gao, T. Moehl, et al., Nanowire perovskite solar cell, Nano Lett. (2015), https://doi.org/10.1021/acs.nanolett.5b00046.

[34] M. Batmunkh, C.J. Shearer, M.J. Biggs, J.G. Shapter, Nanocarbons for mesoscopic perovskite solar cells, J. Mater. Chem. (2015), https://doi.org/10.1039/c5ta00873e.

[35] H. Liu, Z. Huang, S. Wei, L. Zheng, L. Xiao, Q. Gong, Nano-structured electron transporting materials for perovskite solar cells, Nanoscale (2016), https://doi.org/10.1039/c5nr05207f.
[36] S. Paek, P. Schouwink, E.N. Athanasopoulou, K.T. Cho, G. Grancini, Y. Lee, et al., From nano- to micrometer scale: the role of antisolvent treatment on high performance perovskite solar cells, Chem. Mater. 29 (2017) 3490–3498, https://doi.org/10.1021/acs.chemmater.6b05353.
[37] H. Lu, J. Sun, H. Zhang, S. Lu, W.C.H. Choy, Room-temperature solution-processed and metal oxide-free nano-composite for the flexible transparent bottom electrode of perovskite solar cells, Nanoscale 8 (2016) 5946–5953, https://doi.org/10.1039/C6NR00011H.
[38] M.M. Lee, J. Teuscher, T. Miyasaka, T.N. Murakami, H.J. Snaith, Efficient hybrid solar cells based on meso-superstructured organometal halide perovskites, Science 338 (2012) 643 LP–647, https://doi.org/10.1126/science.1228604.
[39] G.-H. Kim, H. Jang, Y.J. Yoon, J. Jeong, S.Y. Park, B. Walker, et al., Fluorine functionalized graphene nano platelets for highly stable inverted perovskite solar cells, Nano Lett. 17 (2017) 6385–6390, https://doi.org/10.1021/acs.nanolett.7b03225.
[40] N. Phung, A. Abate, The impact of nano- and microstructure on the stability of perovskite solar cells, Small 14 (2018) 1802573, https://doi.org/10.1002/smll.201802573.
[41] Z. Li, C. Liu, G. Ren, W. Han, L. Shen, W. Guo, Cations functionalized carbon nano-dots enabling interfacial passivation and crystallization control for inverted perovskite solar cells, Sol. RRL 4 (2020) 1900369, https://doi.org/10.1002/solr.201900369.
[42] T.H. Han, J.W. Lee, C. Choi, S. Tan, C. Lee, Y. Zhao, et al., Perovskite-polymer composite cross-linker approach for highly-stable and efficient perovskite solar cells, Nat. Commun. (2019), https://doi.org/10.1038/s41467-019-08455-z.
[43] C.Y. Chang, C.Y. Chu, Y.C. Huang, C.W. Huang, S.Y. Chang, C.A. Chen, et al., Tuning perovskite morphology by polymer additive for high efficiency solar cell, ACS Appl. Mater. Interfaces (2015), https://doi.org/10.1021/acsami.5b00052.
[44] G.C. Manika, G.C. Psarras, Barium titanate/epoxy resin composite nanodielectrics as compact capacitive energy storing systems, Express Polym. Lett. 13 (2019) 749–758, https://doi.org/10.3144/expresspolymlett.2019.63.
[45] A. Mayeen, M.S. Kala, M.S. Jayalakshmy, S. Thomas, D. Rouxel, J. Philip, et al., Dopamine functionalization of $BaTiO_3$: an effective strategy for the enhancement of electrical, magnetoelectric and thermal properties of $BaTiO_3$-PVDF-TrFE nanocomposites, Dalton Trans. (2018), https://doi.org/10.1039/c7dt03389c.
[46] G. Wang, X. Huang, P. Jiang, Bio-inspired fluoro-polydopamine meets barium titanate nanowires: a perfect combination to enhance energy storage capability of polymer nanocomposites, ACS Appl. Mater. Interfaces 9 (2017) 7547–7555, https://doi.org/10.1021/acsami.6b14454.
[47] H. Li, F. Liu, B. Fan, D. Ai, Z. Peng, Q. Wang, Nanostructured ferroelectric-polymer composites for capacitive energy storage, Small Methods 2 (2018) 1700399, https://doi.org/10.1002/smtd.201700399.
[48] A.S. Bhalla, R. Guo, R. Roy, The perovskite structure - a review of its role in ceramic science and technology, Mater. Res. Innovat. (2000), https://doi.org/10.1007/s100190000062.
[49] M.M. Vijatović, J.D. Bobić, B.D. Stojanović, History and challenges of barium titanate: part II, Sci. Sinter. (2008), https://doi.org/10.2298/SOS0803235V.
[50] U. Kenji, Glory of piezoelectric perovskites, Sci. Technol. Adv. Mater. (2015).
[51] P.K. Panda, Review: environmental friendly lead-free piezoelectric materials, J. Mater. Sci. (2009), https://doi.org/10.1007/s10853-009-3643-0.
[52] T. Zheng, J. Wu, D. Xiao, J. Zhu, Recent development in lead-free perovskite piezoelectric bulk materials, Prog. Mater. Sci. (2018), https://doi.org/10.1016/j.pmatsci.2018.06.002.
[53] Y. Saito, H. Takao, T. Tani, T. Nonoyama, K. Takatori, T. Homma, et al., Lead-free piezoceramics, Nature (2004), https://doi.org/10.1038/nature03028.
[54] S.O. Leontsev, R.E. Eitel, Progress in engineering high strain lead-free piezoelectric ceramics, Sci. Technol. Adv. Mater. (2010), https://doi.org/10.1088/1468-6996/11/4/044302.

[55] R. Ding, X. Zhang, G. Chen, H. Wang, R. Kishor, J. Xiao, et al., High-performance piezoelectric nanogenerators composed of formamidinium lead halide perovskite nanoparticles and poly(vinylidene fluoride), Nanomater. Energy 37 (2017) 126−135, https://doi.org/10.1016/j.nanoen.2017.05.010.

[56] D.Y. Hyeon, K.I. Park, Vertically aligned piezoelectric perovskite nanowire array on flexible conducting substrate for energy harvesting applications, Adv. Mater. Technol. (2019), https://doi.org/10.1002/admt.201900228.

[57] Y.B. Lee, J.K. Han, S. Noothongkaew, S.K. Kim, W. Song, S. Myung, et al., Toward arbitrary-direction energy harvesting through flexible piezoelectric nanogenerators using perovskite $PbTiO_3$ nanotube arrays, Adv. Mater. 29 (2017), https://doi.org/10.1002/adma.201604500.

[58] X.Y. Zhang, X. Zhao, C.W. Lai, J. Wang, X.G. Tang, J.Y. Dai, Synthesis and piezoresponse of highly ordered $Pb(Zr_{0.53}Ti_{0.47})O_3$ nanowire arrays, Appl. Phys. Lett. 85 (2004) 4190−4192, https://doi.org/10.1063/1.1814427.

[59] L. Carrette, K.A. Friedrich, U. Stimming, Fuel cells: principles, types, fuels, and applications, ChemPhysChem 1 (2000) 162−193, https://doi.org/10.1002/1439-7641(20001215)1:4<162::AID-CPHC162>3.0.CO;2-Z.

[60] M. Gazda, P. Jasinski, B. Kusz, B. Bochentyn, K. Gdula-Kasica, T. Lendze, et al., Perovskites in solid oxide fuel cells, Solid State Phenom. 183 (2012) 65−70, https://doi.org/10.4028/www.scientific.net/SSP.183.65.

[61] D.P. Tarragó, B. Moreno, E. Chinarro, V.C. de Sousa, Perovskites used in fuel cells, Perovskite Mater. Synth. Charact. Prop. Appl. (2016), https://doi.org/10.5772/61465.

[62] S.U. Rehman, R.-H. Song, T.-H. Lim, S.-J. Park, J.-E. Hong, J.-W. Lee, et al., High-performance nanofibrous $LaCoO_3$ perovskite cathode for solid oxide fuel cells fabricated via chemically assisted electrodeposition, J. Mater. Chem. 6 (2018) 6987−6996, https://doi.org/10.1039/C7TA10701C.

[63] C. Lu, T.Z. Sholklapper, C.P. Jacobson, S.J. Visco, L.C. De Jonghe, LSM-YSZ cathodes with reaction-infiltrated nanoparticles, J. Electrochem. Soc. 153 (2006) A1115, https://doi.org/10.1149/1.2192733.

[64] D. Radhika, A.S. Nesaraj, Chemical precipitation and characterization of multicomponent perovskite oxide nanoparticles − possible cathode materials for low temperature solid oxide fuel cell, Int. J. Nano Dimens. (IJND) 5 (2014) 1−10, https://doi.org/10.7508/ijnd.2014.01.001.

[65] J. Zhou, T.H. Shin, C. Ni, G. Chen, K. Wu, Y. Cheng, et al., In situ growth of nanoparticles in layered perovskite $La_{0.8}Sr_{1.2}Fe_{0.9}Co_{0.1}O_{4-\delta}$ as an active and stable electrode for symmetrical solid oxide fuel cells, Chem. Mater. (2016), https://doi.org/10.1021/acs.chemmater.6b00071.

[66] W. Fan, Z. Sun, Y. Bai, K. Wu, J. Zhou, Y. Cheng, In situ growth of nanoparticles in a-site deficient ferrite perovskite as an advanced electrode for symmetrical solid oxide fuel cells, J. Power Sources (2020), https://doi.org/10.1016/j.jpowsour.2020.228000.

[67] Y.F. Sun, Y.Q. Zhang, J. Chen, J.H. Li, Y.T. Zhu, Y.M. Zeng, et al., New opportunity for in situ exsolution of metallic nanoparticles on perovskite parent, Nano Lett. (2016), https://doi.org/10.1021/acs.nanolett.6b02757.

CHAPTER 13

Development of PVDF-based polymer nanocomposites for energy applications

Sreelakshmi Rajeevan[1,2], Thomasukutty Jose[1], Runcy Wilson[3], Soney C. George[1]

[1]Centre for Nanoscience and Technology, Amal Jyothi College of Engineering, Kottayam, Kerala, India; [2]APJ Abdul Kalam Technological University, CET Campus, Thiruvananthapuram, Kerala, India; [3]Department of Chemistry, St. Cyril's College, Kilivayal, Kerala, India

Chapter Outline

1. Introduction

Owing to the ever-increasing energy crisis arising from the destruction of traditional fossil fuels (e.g., coal, gasoline, gas, etc.), the production of sustainable and renewable energy from natural energy reservoirs are the most critical concern of the 21st century. Because of the unreliable existence of such renewable energy supplies, efficient energy storage systems are in urgent need for storing and generating electricity. Researchers and engineers are profoundly fascinated by the development of innovative multifunctional materials and technologies to transform certain sources of energy such as wind, chemical and mechanical resources into electrical energy [1]. Polymer nanocomposites possess unusual physicochemical properties that cannot be achieved when working solely on individual components. Polymer nanocomposites have pulled in noteworthy research interests due to their promising flexibility and mechanical properties; extended the domain of applications from environmental redress, energy storage, electromagnetic (EM) absorption, sensing and actuation, transport and safety, defense system, information

Design, Fabrication, and Characterization of Multifunctional Nanomaterials. https://doi.org/10.1016/B978-0-12-820558-7.00015-7

industry, and so on. Broad research works have been completed to grow high dielectric polymer or polymer nanocomposites (PNCs) for both energy and environmental problems [2]. The incorporation of nanofillers into polymers will contribute to distinct properties of each component's synergistic effects in the PNCs. The aspect ratio (height to width) of the nanofiller, the complexity of the nanofiller/polymer matrix interface, and the percolation level are essential facets of the configuration and actions of the PNCs [3]. The nanofillers are capable of supplying the PNCs with both structural and functional strengthening. For instance, the introduction of nanofillers with intrinsic high modulus has witnessed substantial changes in mechanical strength in the PNCs. The homogenous dispersion of the nanofillers and the strong interfacial associations between the polymer matrix and the nanofillers are fundamentally liable for the altogether improved mechanical properties. Covalent bonding between the filler and matrix is known to be the most powerful way of rising the interfacial shear to enhance stress transfer. In addition, owing to the excellent thermal and electrical conductivities of the nanofillers, nanocomposites were developed with expanded applications for sensors, nanogenerators, electromagnetic interference (EMI) shielding materials; and electrodes, separators and binders for both Li-ion batteries and supercapacitors. Among the different types of polymer nanocomposites, polyvinylidene fluoride (PVDF)-based nanocomposites have a significant role in the field of energy storage and energy-saving applications [4]. They have been extensively studied because of their excellent piezoelectric, pyroelectric, and ferroelectric properties. The chapter gives an overview of the synthesis and development of PVDF based nanocomposites for energy storage and energy-saving applications.

1.1 Polyvinylidene fluoride (PVDF)

PVDF is a semicrystalline thermoplastic polymer which consists of repeated molecular monomer units of ($-CH_2-CF_2-$) in a long chain as shown in Fig. 13.1. The linear hydrofluorocarbon polymer has been known since the 1960s and it contains about 59.4 percent of fluorine and 3 percent of hydrogen. The significant level of characteristic crystallinity (close to 60%) gives stiffness and tough, mechanical strength, creep resistance and other properties [5,6]. PVDF is generally synthesized from 1,1-difluoroethylene (VD_2 or VF_2) by the free-radical polymerization, a monomer commonly synthesized from acetylene or vinylidene chloride. Although it is synthesized commercially by emulsion or suspension polymerization technique. The degree of polymerization of PVDF ranges from 1000 to 2500 VDF units PVDF's physical and electrical characteristics depend on the molecular weights, molecular weight distributions, chain arrangements, crystalline form, and chaining defects. The interest for PVDF has been driven by various components, for example, mechanical quality, lightweight, flexibility, high dielectric constant, high impact resistance, wettability, and electrochemical stability. PVDF possesses the trademark obstruction towards harsh chemical environments as in fluoropolymers and also has excellent resistance towards thermal, ultraviolet, and weather changes [7].

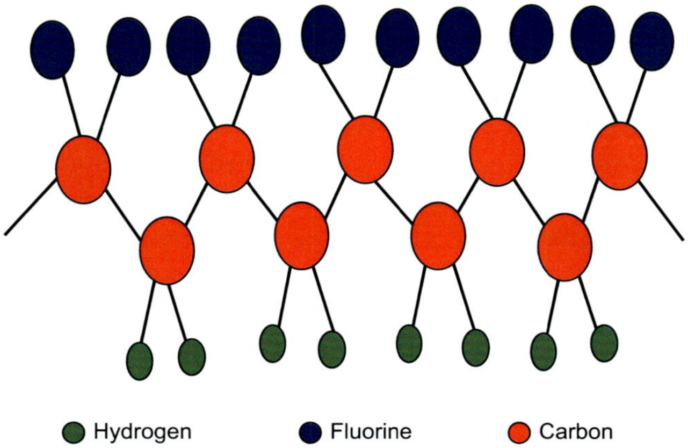

Figure 13.1
Molecular structure of polyvinylidene fluoride (PVDF).

1.2 Structure and properties of PVDF

PVDF exhibits a well-known polymorphism based on the crystal orientation of different phases such as alpha (α), beta (β), gamma (γ), delta (δ), and epsilon (ε), and are shown in Fig. 13.2 [8]. The nonpolar alpha (α) phase is an eminent, kinetically favoured, and

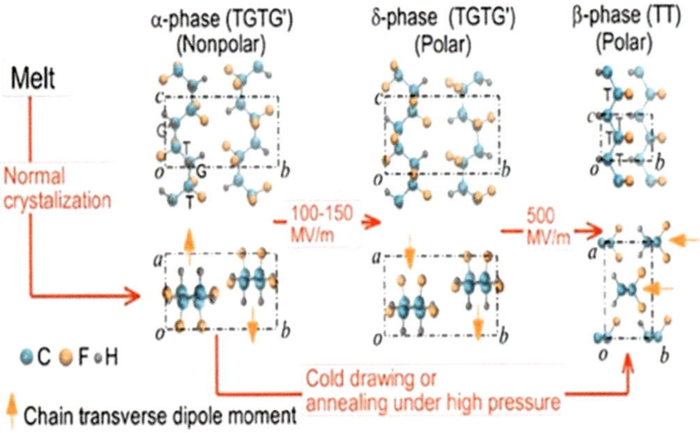

Figure 13.2
Electric field-induced phase transitions of PVDF. The unit cells are shown as viewed from the ab and bc planes. The transverse dipole moment of each polymer chain is shown using the orange arrow that points from the negatively charged fluorine atoms to the positively charged hydrogen atoms [8].

stable phase, that is formed by the effective cooling of molten PVDF [9,10]. The beta phase is the polar and thermodynamically stable phase, which is formed by the effective stretching straining, or quenching of the polymer. The crystal β phase has a planar zig-zag conformation with all-trans (TT) bonds remains in the same plane as the carbon backbone [6] i.e., the fluorine atoms on one side and hydrogen atoms on the other side of the chain [11]. The intermediate polar gamma phase has a trans-trans-trans-gauche (TTTG) conformation and it is formed under moderative stress of polymer or its reaction with certain chemicals. Gamma phase is acquired from ultra-high molecular weight polyvinylidene fluoride [12,13]. Delta and epsilon phases are the two other polymorphs of PVDF, they are still under debate. However, some scientists discovered that the δ crystals were generated by the distortion of one of the other crystalline forms [14]. A complexity in the morphology of alpha and gamma structures happens and overwhelms the spherulitic qualities of the PVDF when the melted polymer is permitted to harden at temperatures close to the melting point [15]. Amorphous PVDF regions have a density of 1.68 g/cm^3 and the polymorphs such as alpha, beta and gamma exhibit 1.93, 1.97, and 1.93 g/cm^3 respectively [16]. The dissolve thickness of PVDF homopolymers and VDF/HFP-copolymers is around 1.45−1.48 g/cm^3 at 230°C and 1 bar, i.e., a volumetric shrinkage factor of practically 20% happen in cooling from the melt to the solid. The glass transition temperature (T_g) of PVDF was observed in the range of −40 to−30°C and its melting temperature (Tm) is in the range of 155−192°C. The most important feature of PVDF is nontoxicity and also it is not hygroscopic, retains under 0.05% of water at room temperature. PVDF tends to release hydrogen fluoride (HF) which makes it vulnerable to attack from nucleophiles such as strong bases. It is soluble in polar solvents such as esters, acetone, and tetrahydrofuran and this soluble nature help for film casting. which, when all is said in done, likewise decreases the end-use temperature evaluations. Most commonly used copolymers of PVDF are poly(vinylidene fluoride-co-trifluoroethylene)[P(VDF-TrFE)] and poly(vinylidene fluoride-co-hexafluoropropylene) [P(VDF-HFP)] [24]. The properties and the application of PVDF may vary according to its phases. The alpha and beta forms of PVDF are extensively used in energy-related applications such as supercapacitor, lithium-ion batteries, fuel cells, sensors. In beta polymorph, molecular dipoles are spontaneously undergo polarization by the application of electrical field, temperature variation, magnetic field or interaction with nanoparticles, resulting in the polymer's piezo-, pyro-, and ferro- electrical reactions [8].

2. Synthesis and characterization PVDF nanocomposites for energy storage and harvesting applications

Portable electronic gadgets, stationary power systems, and hybrid electric cars are all driving demand for low-cost, small, and high-performance electrical energy harvesting and storage technologies. Batteries, fuel cells, sensors, supercapacitors, self-charging photo-power cell, solar cell and piezoelectric nanogenerators are some of the trending energy storage and harvesting devices due to their fast electrical-energy storage and charging-discharging capability [17]. PVDF and its nanocomposites play different roles in the fabrication of these devices including a dielectric separator, binder for holding active material, gel electrolytes, piezoelectric source, and so forth. There are so many methods for synthesizing polymer nanocomposites, and each of the methods has its advantages and disadvantages. The synthesizing method has a crucial role in the property enhancement of polymer nanocomposites. Among them, solvent casting, in situ synthesis, melt intercalation, electrospinning methods generally used in the fabrication of polymer nanocomposites used in the energy harvesting, and storing devices. PVDF is an engineering plastic, possess toughness, hardness and other mechanical properties [7]. However, to amplify the differentiated usefulness and utility of this polymer whereas keeping flexibility, diverse nanofillers have been incorporated. The ability of producing multifunctional materials with special physicochemical properties has fuelled the investigation on the PVDF nanocomposite. The enhancement in the property of these multiphase nanoscale systems comes about from an increment within the interfacial surface area resulted due to reduction in the filler size and quantum confinement effects induced by a nanoscale dimension. Another factor for using the maximum potential impact of nanoscale fillers on composite properties is the degree of interaction between the filler surface and matrix [7]. Nanoparticles with a negatively charged surface can interact with the positively charged $-CH_2$ groups of PVDF which helps the formation of the beta phase and thereby increase the piezoelectric property. Some of the generally used nanofillers are clay, silica, ceramic fillers such as barium titanate ($BaTiO_3$) and lead zirconium titanate (PZT); carbonaceous materials such as carbon nanotube (CNT), graphene, carbon fibres; metal nanoparticles based on Pd, Ag, Pt; and metal oxides such as CuO, ZrO_2, TiO_2, RuO_2, MnO_2, Fe_3O_4. The nucleating efficiency, supercooling and the interfacial interactions with the polymer matrix are important parameters of nanofillers that determines the structure and properties of PVDF nanocomposites [8]. The synthesis of PVDF nanocomposites is purely depending on the final application. The resultant properties of nanocomposites depend not only on the properties of the individual parents but also on their morphology, interfacial characteristics, and the method of synthesis. Innovative synthetic methods are underway to gain control over the internal morphology these polymer nanocomposites [18].

Khatun et al. prepared simple, cost-effective and efficient prototype self-charged hybrid photo-power cell (HPPC) using a large dielectric BaF_2/PVDF thin film as a storage unit in a single power pack. The PVDF/BaF_2 dielectric thin film was synthesized via in situ process followed by solution casting procedure as shown in Fig. 13.3.

The field emission scanning electron microscopy (FESEM) images of the modified BaF_2 nanoparticle (NP)/PVDF thin film is shown in Fig. 13.4. The SEM images clearly indicate the growth of spherical shaped BaF_2 NPs with a diameter of ~50–80 nm within the PVDF matrix. The authors found the agglomeration of NPs after 19.4 vol % loading.

They also examined the effect of BaF_2 NPs on the dielectric properties of the modified BaF_2/NPs PVDF thin film (Fig. 13.5) and found that the values of the dielectric constant and tan δ were increased sharply up to 19.4 vol % BaF_2 NP loading, thereafter a decreasing tendency was observed at 20Hz. At higher frequency (2 MHz) those parameters are increased slowly with NPs content.

The solid interconnection between BaF_2 NPs with the PVDF matrix was resulted by the good mixing of the contents and it was achieved by in situ synthesis. The strong interconnection diminishes the intermolecular space and improves the number of charge carriers, for example, dipoles which have instigated longer planar zig-zag conformation into the polymer matrix bringing about a significantly huge dielectric consistent of the PBF0.15 at this loading of the NPS. The NPs are well isolated from each other at lower doping of the NPs in the PVDF matrix, so there is no chance of successful interaction

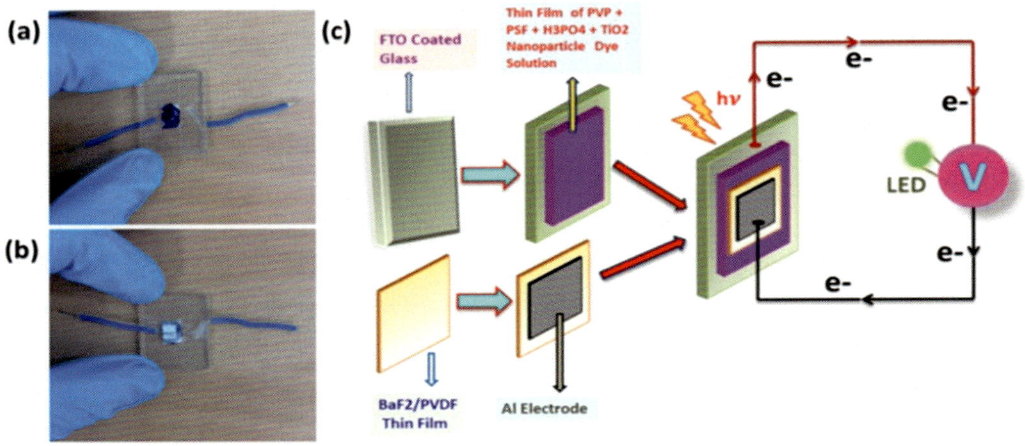

Figure 13.3
(A, B) Digital image of the HPPC, (C) Schematic representation of the fabrication of the HPPC [2].

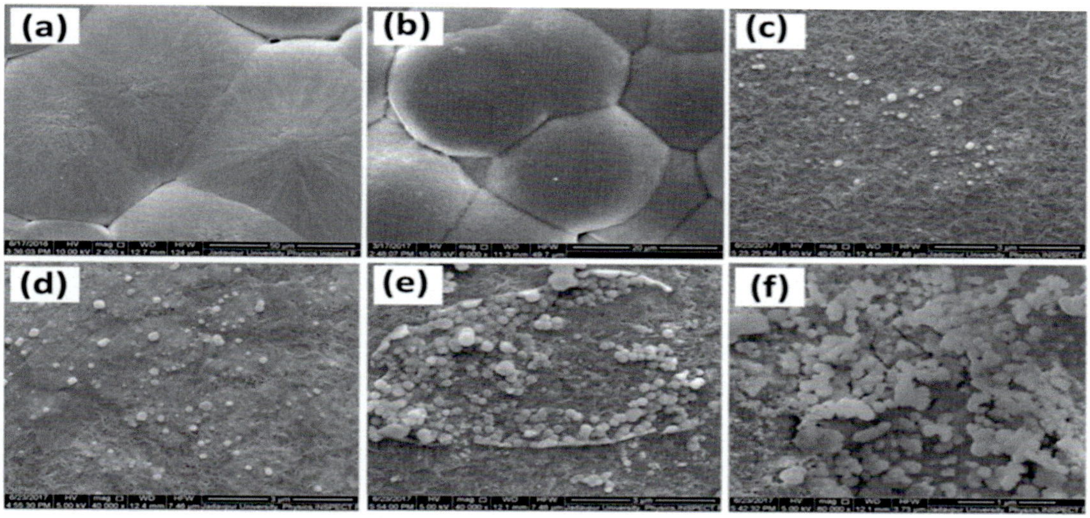

Figure 13.4
FESEM images of (A) pure PVDF, (B) PBF0.01, (C) PBF0.05, (D) PBF0.10, (E) PBF0.15, and (F) PBF0.2 [2].

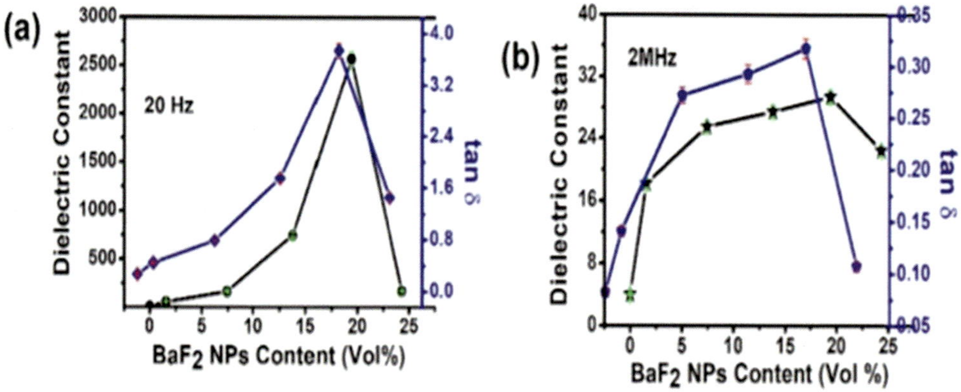

Figure 13.5
BaF_2 NPs content (volume %) dependency of (A) dielectric constant and (B) tangent loss [2].

between the NPs. At the point when the NPS content has been increased, the interfacial area per unit volume between the polymer and the NPs is also increased leading to the reduction in the interparticle distances. In this way, the normal interfacial polarization has expanded effectively up to the doping concentration of NPs when it has reached to a percolation level esteem that is 19.4vol%. Final HPPC produces a 1.3V photovoltaic voltage within 65s under visible light exposure (almost 1110 mW/cm^2) with a higher energy storage capacity of 79%. The device has the capacity of storing charge ~1200 F/m^2 with power density ~18.7 W/m^2 and charge density ~1500 C/m^2 [2].

The effective dispersion of nanofillers in the polymer matrix is supposed to be a big task for scientists those who are working in this area of polymer nanocomposites. They always engaged in the development of novel methods for the better dispersion of the nanofillers in the matrix. Surface modification of nanoparticles is one such innovative method that facilitates better dispersion of nanofillers into the polymer matrix. Niu et al. developed a barium titanate (BT)/PVDF nanocomposites by incorporating surface-modified barium titanate nanoparticles [19]. In this study they used the ball-milling technique followed by thermal treatment for the preparation of modified BT/PVDF nanocomposites is shown in Fig. 13.6.

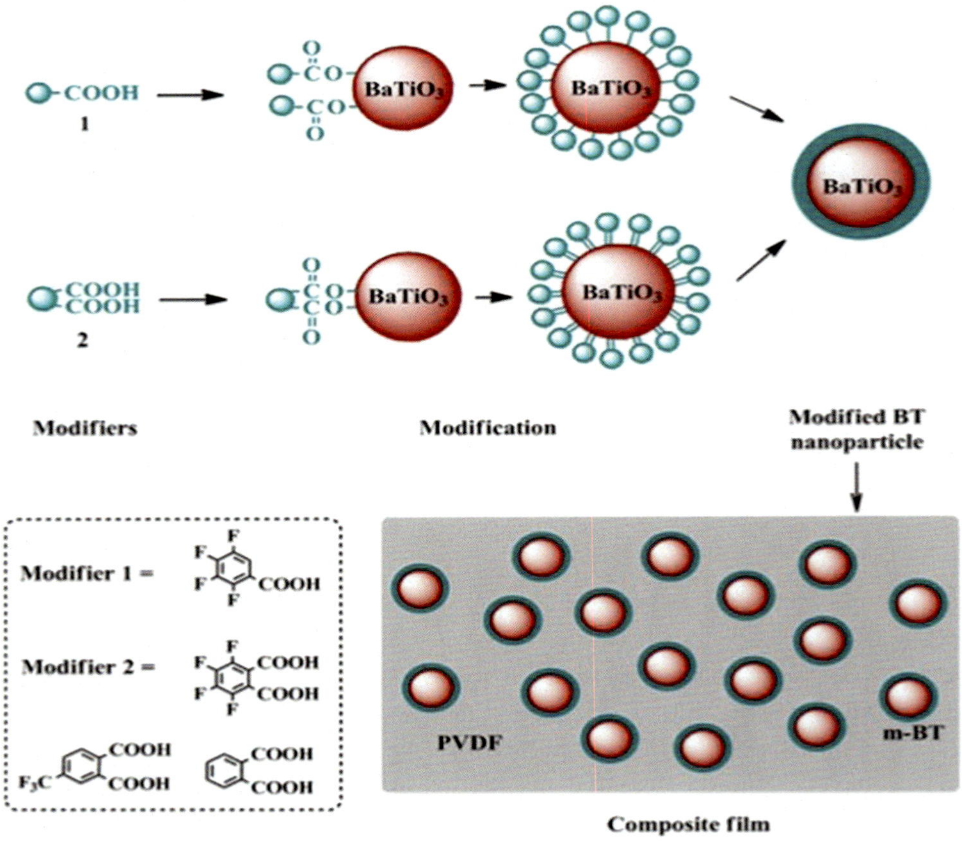

Figure 13.6
Process of nanoparticle modification and incorporation into the polymer matrix [19].

Mainly four types of novel small-molecule surface modifiers such as 2,3,4,5-tetrafluorobenzoic acid, 4(trifluoromethyl)phthalic acid, tetrafluorophthalic acid, and phthalic acid were used to modify BT particles. Theses modifiers are belonging to the carboxylic acid family and have low molecule weight and simple structure. The effect of modification of the BT nanoparticles was evaluated using FTIR spectroscopy and shown in Fig. 13.7.

The SEM images also confirmed the uniform distribution NPs in the polymer matrix without any phase separation. So, the ball-milling technique highly useful for the dispersion of nanofillers in the PVDF matrix, but one of the disadvantages of this method is its poor compatibility between the nanofiller and PVDF matrix. Except for the one treated with 4-(trifluoromethyl)phthalic acid the discharged energy density of nanocomposites filled with m-BT nanoparticles is higher than that of nanocomposites BT/PVDF.

Fig. 13.8 implied the discharged energy density of the nanocomposites filled with BT and m-BT nanoparticles. The discharged energy densities of the nanocomposites modified with tetrafluorophthalic acid and phthalic acid are 6.5 and 6.7 J/cm^3 respectively, it is 35.7% and 37.7% higher than that of BT/PVDF nanocomposites [19].

Satapathy et al. developed a novel high Q-factor PVDF nanocomposite for high Q-capacitor applications [20]. PVDF/V_2O_5/GNP nanocomposite films were prepared by colloidal blending was shown in Fig. 13.9. Colloidal blending was a traditionally used method for the fabrication of nanocomposite films.

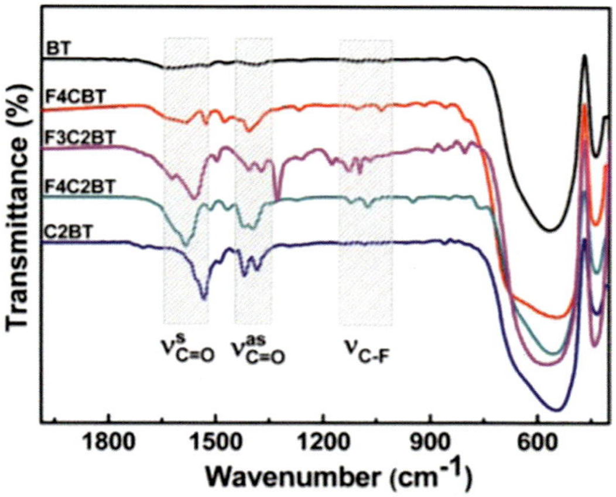

Figure 13.7
FTIR spectra of BT and the surface modified BT with modifiers [19].

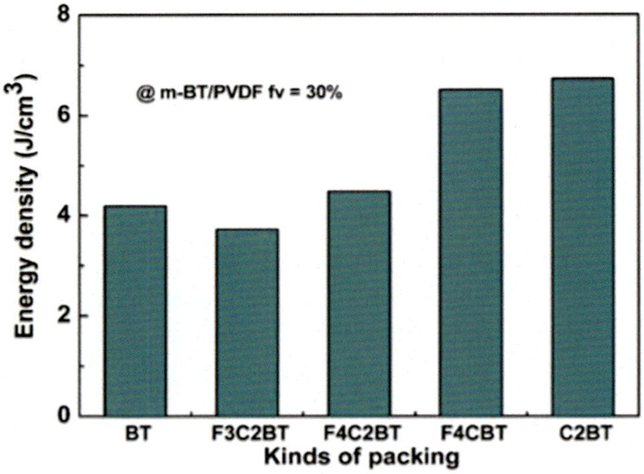

Figure 13.8
Discharged energy density of the nanocomposites filled with BT and m-BT nanoparticles [19].

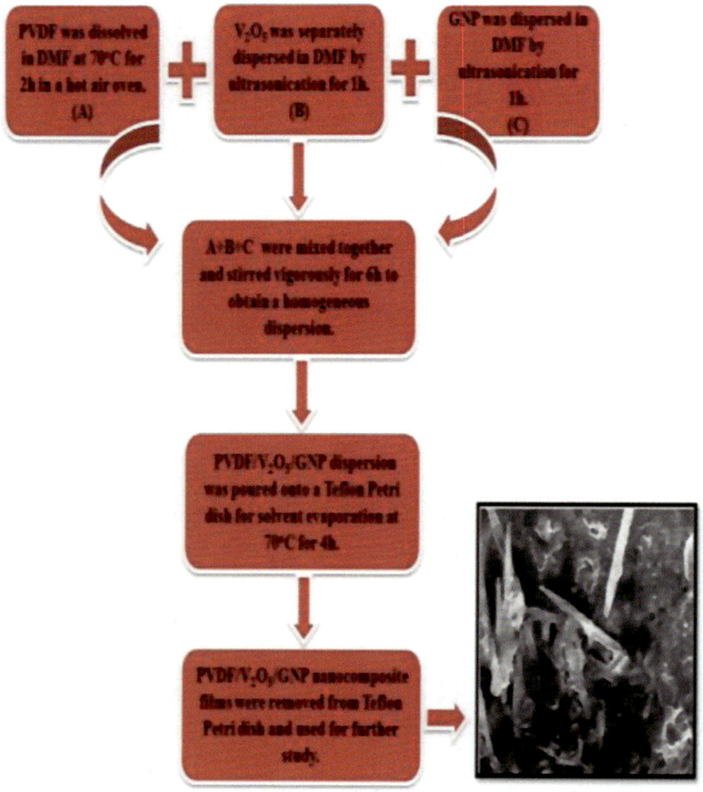

Figure 13.9
Protocol for the synthesis of PVDF/V_2O_5/GNP nanocomposite films [20].

The thermal properties of PVD/V_2O_5/GNP nanocomposite films and individual materials were analyzed using TGA (Fig. 13.10) and result showed that the GNP has better thermal stability than V_2O_5 and the thermal stability of nanocomposite was improved due to the good interaction between the nanofillers and the polymer matrix at higher nanofillers loadings.

Optical microscopy is one of the most useful techniques for determining the nanofillers dispersion state and it helps to understand the reinforcing effect of fillers in the polymer matrix. It also provides information on the morphology and agglomeration range from several micrometers to millimeters. To understand the reinforcing effect of V_2O_5 and GNP in the PVDF matrix, PVDF/V_2O_5/GNP nanocomposite films were analyzed using optical microscopy (Fig. 13.11).

The semiconductor behavior of as-prepared nanocomposite was analyzed using current-voltage characteristics (Fig. 13.12), and the results showed that the increase in GNP loading gradually increases the conducting behavior of the PVDF/V_2O_5/GNP nanocomposite films concerning the applied current.

The Q-factor is a dimensionless parameter, which defines how the system (oscillator or resonator) oscillates with the amplitude which gradually decreases to zero and the feasibility of the high Q-capacitor highly depend on the Q-factor of nanocomposites. The Q-factor of the nanocomposite can be studied directly from the electrical property measurement of nanocomposites. The results showed the addition of V_2O_5 enhances the Q-factor of PVDF/V_2O_5/GNP nanocomposite films. Furthermore, the low loading of GNP

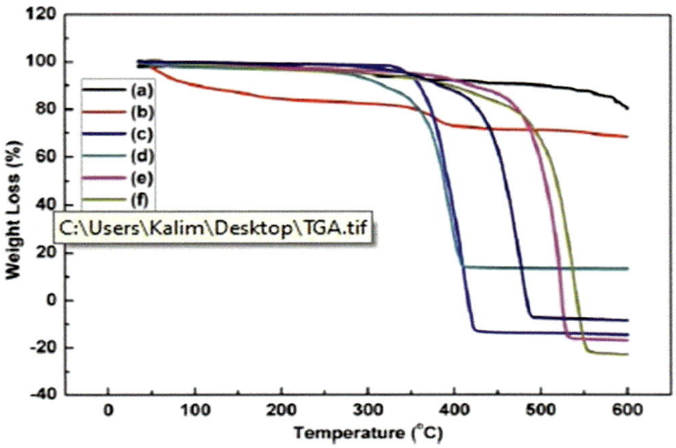

Figure 13.10
TGA thermograms of PVDF/V_2O_5/GNP nanocomposites (A) GNP, (B) V_2O_5, (C) PVDF, (D) 0.3wt% GNP, (E) 0.5wt% GNP, (F) 0.7wt% GNP, and (G) 1wt% GNP [20].

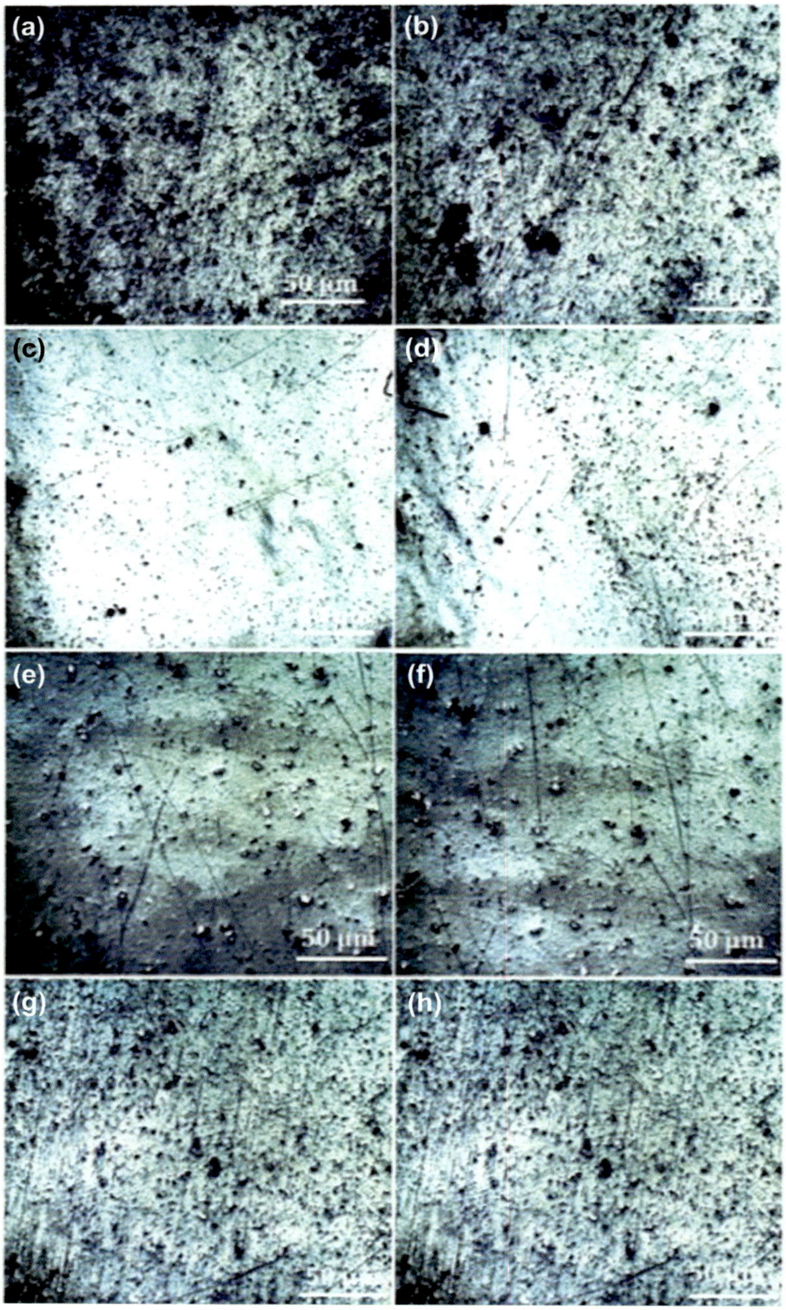

Figure 13.11
Optical microscopy images of PVDF/V_2O_5/GNP nanocomposite films (A,B) 0.3wt% GNP, (C,D) 0.5wt% GNP, (E,F) 0.7wt% GNP, (G,H) 1wt% GNP [20].

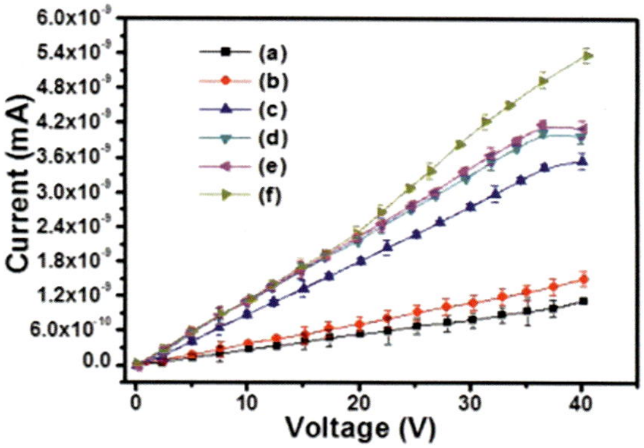

Figure 13.12
I–V curves of PVDF and PVDF/V_2O_5/GNP nanocomposites (A) PVDF (B) 5wt% V_2O_5 (C) 0.3wt% GNP (D) 0.5wt% GNP (E) 0.7wt% GNP (F) 1wt% GNP [20].

shows an essentially high Q-factor esteem demonstrating the low energy loss in the nanocomposites. The nanocomposites with 4.7wt% V_2O_5 and 0.3wt% GNP loading demonstrated Q-factor estimation of about 1099.04. Thus, the overall results proved that colloidal blending is an effective method for the fabrication of high-quality factor nanocomposite films [20].

There is a lot of preparation techniques used for the fabrication of polymer nanocomposite was already mentioned in the previous section. Among those methods, solvent casting is the most efficient and commonly used method for the fabrication of polymer nanocomposites for energy-related applications. The high efficiency of this method is due to the dispersion of nanofillers in a solvent in which the polymer is soluble. The most important advantage is that the apparatus required for the method was very simple. A large number of studies have been reported in the field of both energy storage and energy harvesting applications [21].

A novel flexible pressure sensor was fabricated using PVDF/rGO-titania nanolayers (TNL) nanocomposite via solvent casting. The hybrid composite consists of about 2.5 Wt. % TNL and reduced graphene oxide (rGO) each. The morphology of the nanocomposite was studied using SEM. Owing to the functional group interaction between the two nanomaterials the nanolayers are homogeneously distributed. The oxygen content of the functional groups on the rGO surface interact with the titanium nanolayers through hydrogen bonding and van Der Waals forces, resulting in good dispersion of the fillers within the PVDF medium. For the further substantiation of the SEM results, the authors took the TEM images of the surface of PVDF, PVDF/TNL, PVDF/rGO, and PVDF/rGO-TNL nanocomposite (Fig. 13.13). Apart from a small amount of agglomeration at high resolution, it was found that the TEM results are in agreement with the SEM results.

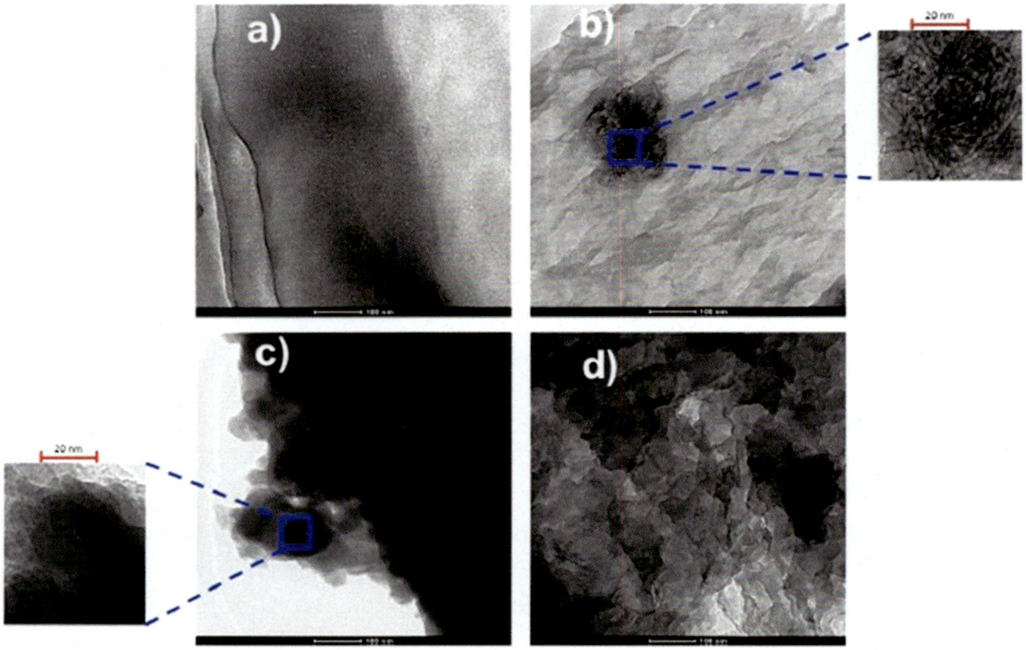

Figure 13.13
TEM images of surface of: (A) PVDF; (B) PVDF/TNL (C) PVDF/rGO and (D) PVDF/rGO-TNL [22].

The addition of hybrid nanofiller decreased the roughness of the PVDF (5.773 nm) to 3.45 nm was shown in the AFM images (Fig. 13.14). The surface of the hybrid composite possesses a smooth texture compared to the PVDF/TNL (6.405 nm) and PVDF/rGO (6.185 nm) composites. The morphological analysis of the hybrid composites proved the better dispersion of the TNL and rGO in the PVDF matrix.

The mechanical properties of the hybrid composite were examined by conducting tensile test and the results were given in Table 13.1. The synergistic effect offered by the hybrid filler (rGO and TNL) has increased the mechanical properties of the composites. The tensile strength and Young's modulus of the hybrid PVDF/rGO-TNL composite were two and 3.6 times higher than that of the neat PVDF. The decrease in the elongation at break of the hybrid composite compared neat PVDF indicates its brittle behavior [22].

The sensing ability is important for a good sensor, for novel PVDF/rGO-TNL composite sensor it was evaluated by measuring the relative resistance of the composite under different applied pressures and the results are shown in Fig. 13.15.

The hybrid composite possesses resistance relatively higher compared to other composites at all the three applied pressures. The flexible sensor showed strong sensing

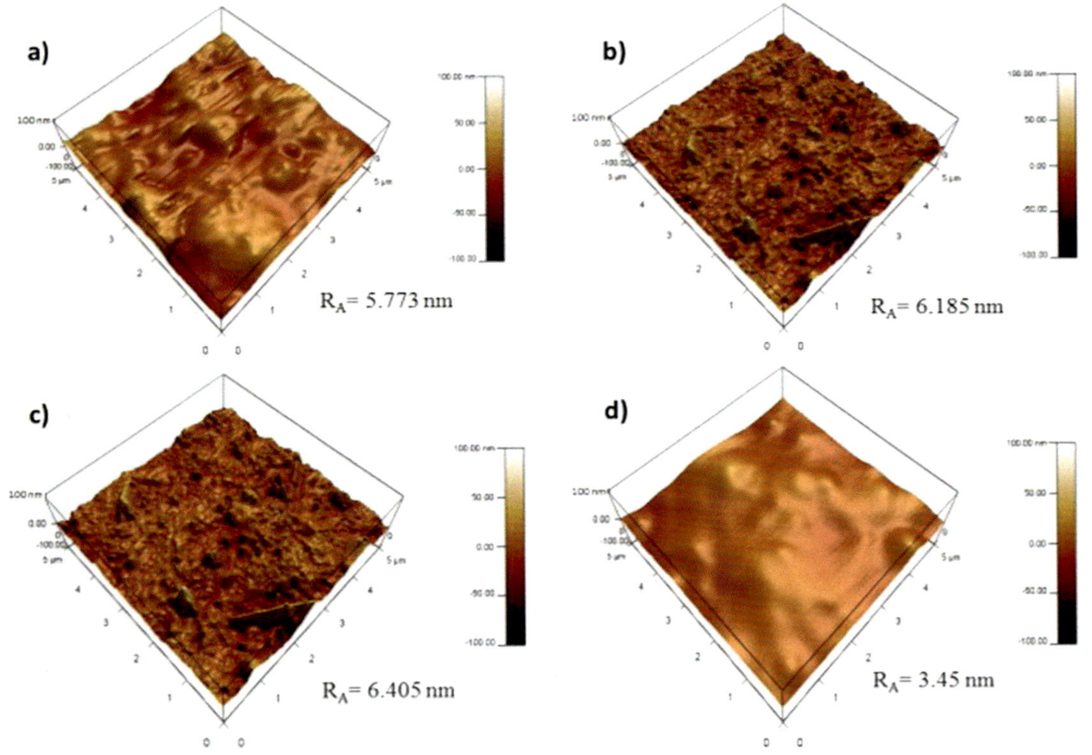

Figure 13.14
AFM images of surface of (A) PVDF (B) PVDF/rGO (C) PVDF/TNL and (D) PVDF/rGO-TNL [22].

Table 13.1: Tensile test results of PVDF/rGO-TNL, PVDF-TNL, PVD-rGO, and PVDF [22].

Samples	Tensile strength (MPa)	Young's modulus (MPa)	Elongation at break (%)
PVDF	21.85 ± 1.93	1365.50 ± 101.23	16.22 ± 1.45
PVDF/rGO	22.927 ± 1.197	2969 ± 380.6	7.352 ± 0.66
PVDF/TNL	41.53 ± 1.58	3112.7 ± 173.60	5.40 ± 0.56
PVDF/rGO-TNL	46.91 ± 0.99	5010.65 ± 243.35	4.01 ± 0.49

properties implying the better dispersion of nanofillers in the PVDF matrix, leading to the improved wettability and di-electrical properties of the sensor. The piezoelectric and dielectric properties of the polymer nanocomposite are known to be important parameters to consider during the fabrication of energy storage and harvesting equipment. Ma et al. demonstrated that soluble casting method can be used for the development of nanocomposite of Li-ion modified Montmorillonite (MMT Li) and the copolymer of

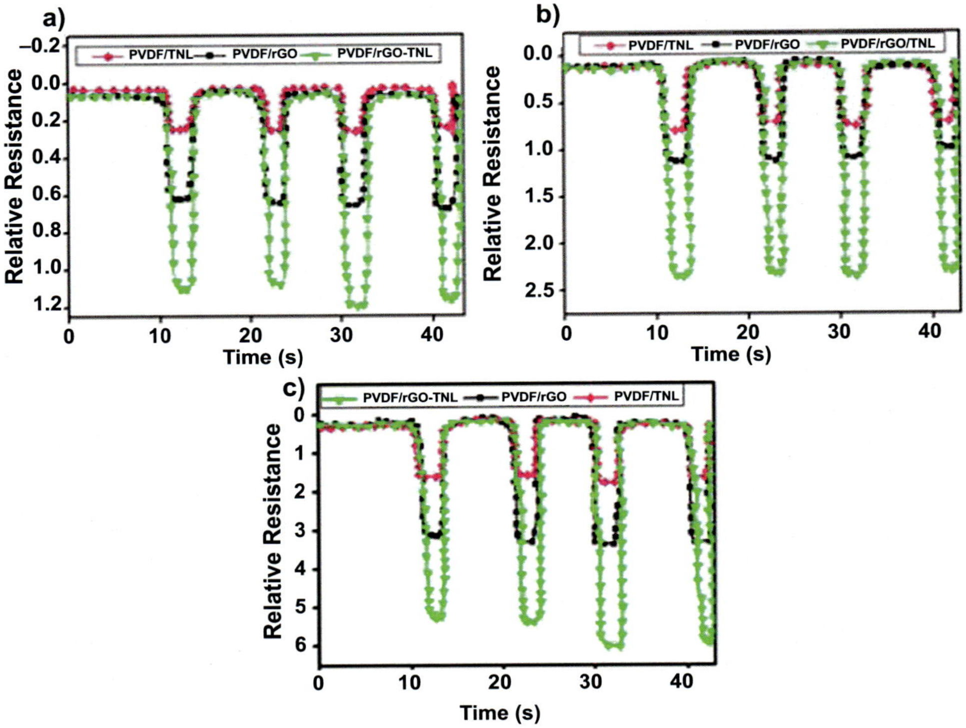

Figure 13.15
Sensing responses of PVDF/rGO, PVDF/TNL, and PVDF/rGO-TNL composites to: (A) 5 kPa; (B) 10.7 kPa; and (C) 17.6 kPa [22].

PVDF i.e., PVDF-HFP. Taking into account the function and content of adsorbed water, the phase transformations, as well as dielectric and piezoelectric properties of MMT(Li)/PVDF-HFP composite films, were investigated. MMT(Li)/PVDF-HFP composite film with 15wt% MMT and 1.72wt% H_2O displayed the great energy conversion and storage efficiency with high dielectric constant due to the Li+ polarization and β-stage nucleation by hydrogen bonds [23]. Chen et al. prepared a high energy density composite film using electrospun BaTiO3@Ag nanofiller in the PVDF matrix using solvent casting followed by hot compressing molding. The characterization results of the BT@Ag NFs/PVDF nanocomposite film showed superior properties and energy density 7.75Jcm^{-3} at 3500 kV/m [24]. Some of the researchers found that solvent casting is an efficient method for improving the breakdown strength of PVDF. They found that by inducing a small amount of functionalized MgO nanoparticles increased the break strength of PVDF nanocomposites. The breakdown strength of PVDF/MgO nanocomposite was estimated from the Weibull plots (Fig. 13.16).

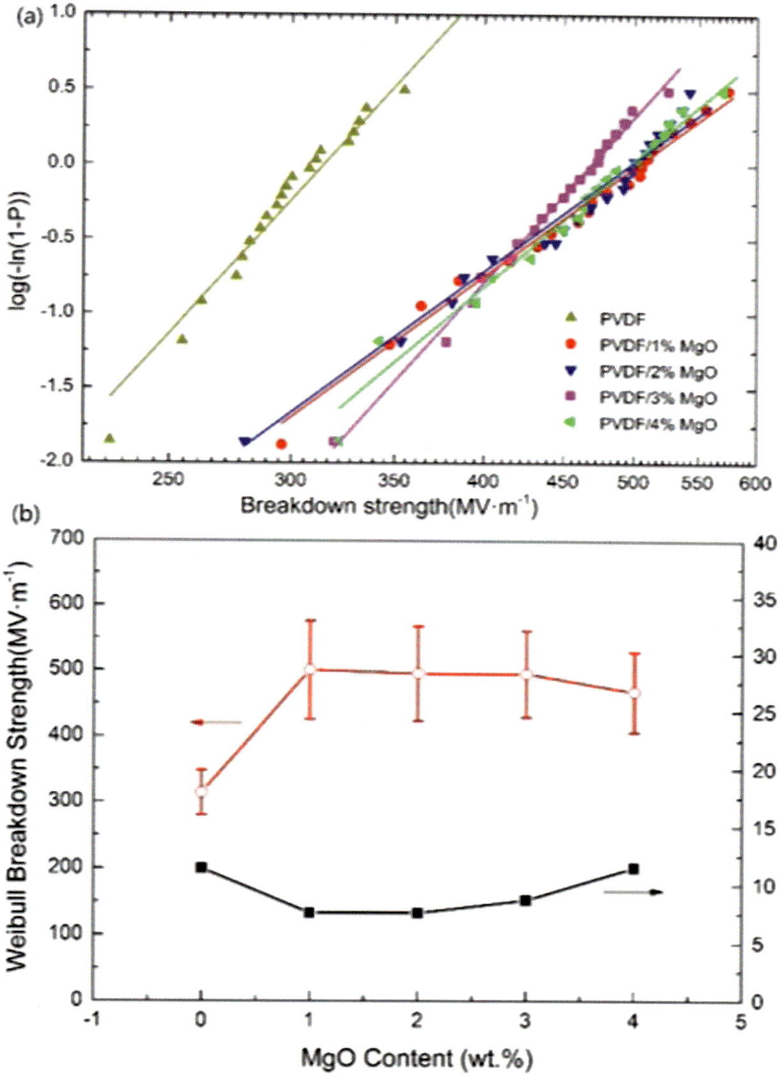

Figure 13.16
The Weibull breakdown strength of PVDF and PVDF/MgO nanocomposites [25].

It is evident from the study that the electrical breakdown strength of the PVDF/MgO nanocomposite was higher than the pristine PVDF films. The highest value of 501.4 MV/m is obtained for 1wt% MgO incorporated composite film which is 59% higher than pristine PVDF (314.5 MV/m). In conclusion, the existence of functionalized MgO nanoparticles can enormously increase the breakdown strength of the PVDF nanocomposites [25]. Recently, a breathable and flexible nanogenerator was developed by Kim et al. using a low-temperature hydrothermal method. The surface of electrospun PVDF nanofibers act as a substate for the growth of ZnO nanorods. Electrospinning is an efficient method for producing nanofibrous polymeric materials and favours the formation of beta phase or conversion of alpha to beta phase because of its remarkable in situ electrical poling and mechanical stretching impacts [25].The beta phase of the PVDF shows the piezo-, pyro- and ferroelectric property, and these properties made them an ideal polymer for energy-based applications. The piezoelectric polymers in the form of nanofibers possess superior properties compared to polymer films. The nanofibers exhibit high mechanical strength, high porosity, large specific area, high electrochemical stability, and dielectric strength. Therefore, the interaction between nanofillers and nanofiber shows a synergistic effect on the properties of the outcome. The successful growth of ZnO nanorods on the surface of the PVDF nanofibers was shown in Fig. 13.17. The ZnO nanorods and PVDF nanofibers showed an average diameter of 183 + 153 nm and 120 + 100 nm respectively.

The existence of the beta phase of PVDF in ZnO@PVDFnanofibrous composite was confirmed by analyzing the FTIR spectrum. The calculated percentage of beta phase was found to be higher for PVDF (83.8%) and ZnO@PVDF it is 80.2%. For measuring the piezoelectric response of nanogenerator they integrated the nanogenerator by comprising ZnO@PVDF nanofiber membrane sandwiched between two conductive knitted fabric electrodes as shown in Fig. 13.18.

Under all charging conditions, ZnO@PVDF nanogenerators generated significantly greater power output than PVDF nanogenerators (Fig. 13.19). Growing ZnO nanorods on the surface of PVDF nanofibers have increased open-circuit voltage by 231% and load currents by 244% and 210% respectively for 470 kW and 15 MW.

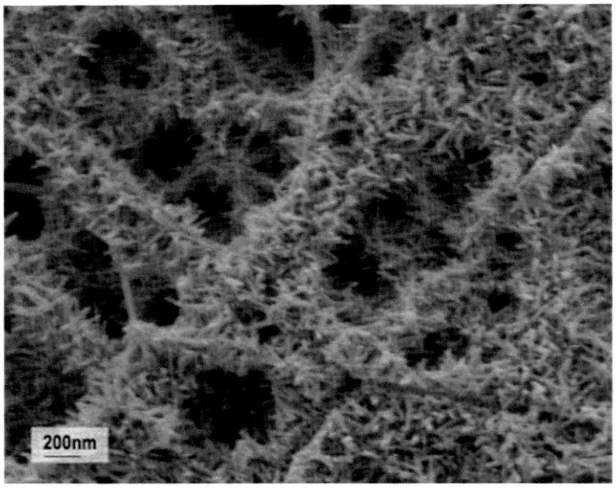

Figure 13.17
An FESEM image of the ZnO nanorods grown on the surface of the electrospun PVDF nanofiber [26].

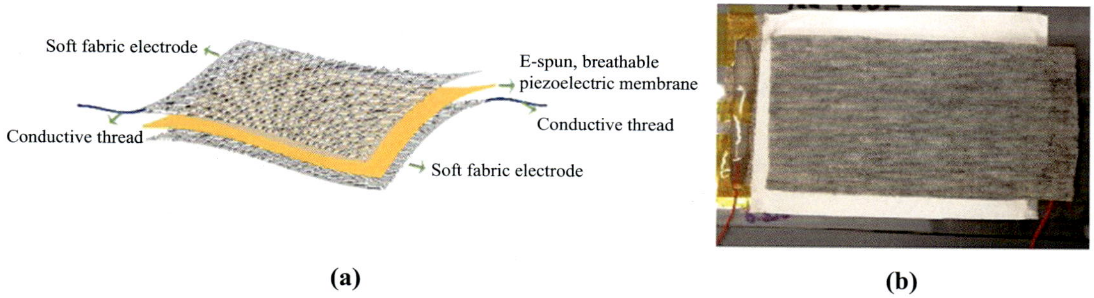

Figure 13.18
The breathable fibrous nanogenerator (A) schematic illustration and (B) photo [26].

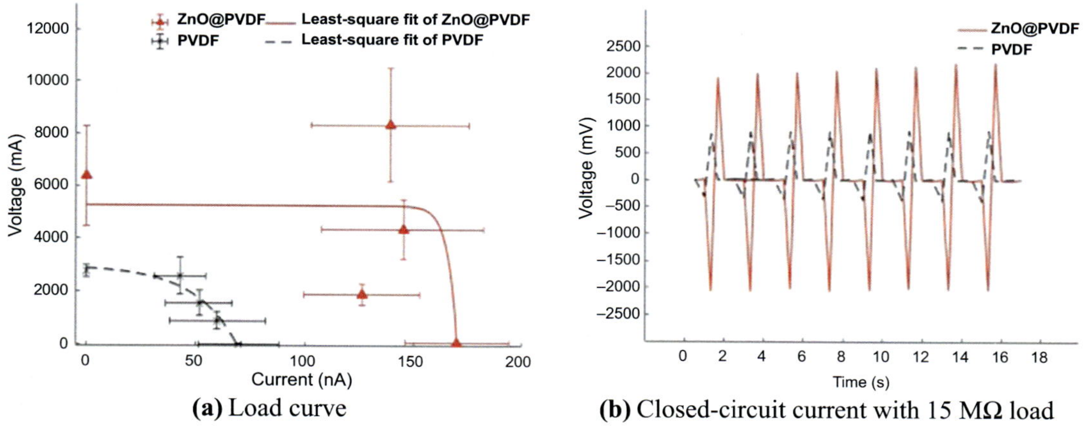

Figure 13.19
Open-circuit voltage measurements under 0.10 MPa impact of 1 Hz frequency: (A) conductive knitted fabric electrode nanogenerator; (B) aluminum foil electrode nanogenerator. The black lines are from PVDF, while the red lines from ZnO@PVDF [26].

The authors started this work to develop a breathable nanogenerator and breathability of piezoelectric fibrous nanogenerators was evaluated by measuring the water vapor permeability (WVP) of all the materials used in the nanogenerator preparation. The results showed that all the material had almost similar WVP value. They concluded that the novel ZnO@PVDF fibrous nanogenerator exhibit a significant level of piezoelectric response without compromising breathability and flexibility [26].

The use of versatile piezoelectric nanogenerator to harvest energy from the ambient mechanical energy is a pioneering step toward achieving a reliable and renewable energy source. Zhao et al. found a new method to enhance the piezoelectric property of PVDF by introducing ZnO nanorods into the PVDF matrix via a simple drop-casting method without doing any mechanical, electrical, or thermal treatment and scheme was in Fig. 13.20.

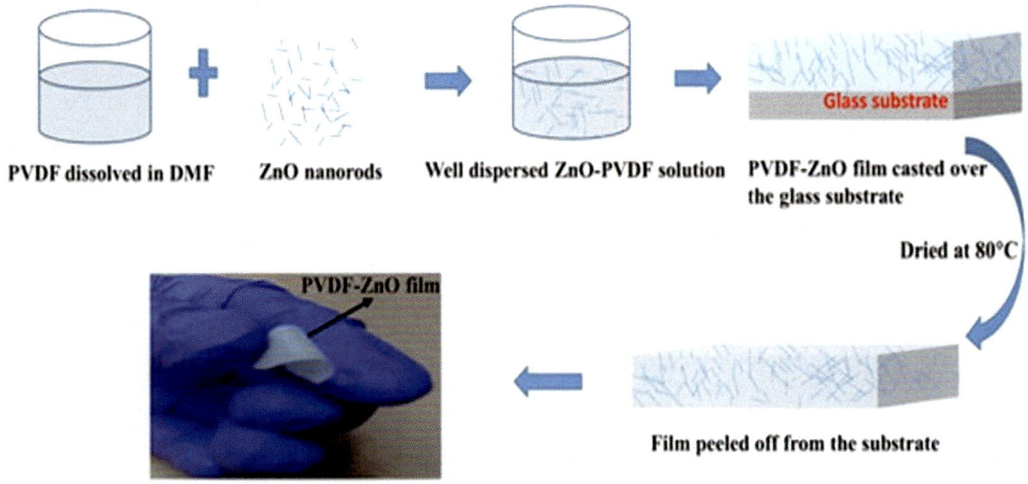

Figure 13.20
Schematic diagram showing the synthesis of flexible polyvinylidene fluoride (PVDF)-ZnO nanocomposite films [27].

The effect of ZnO loading on the structure of the PVDF film was examined using XRD spectrum Fig. 13.21. The diffractogram proved that the increasing filler loading diminishes the intensity of alpha peak and increased the intensity of beta phase.

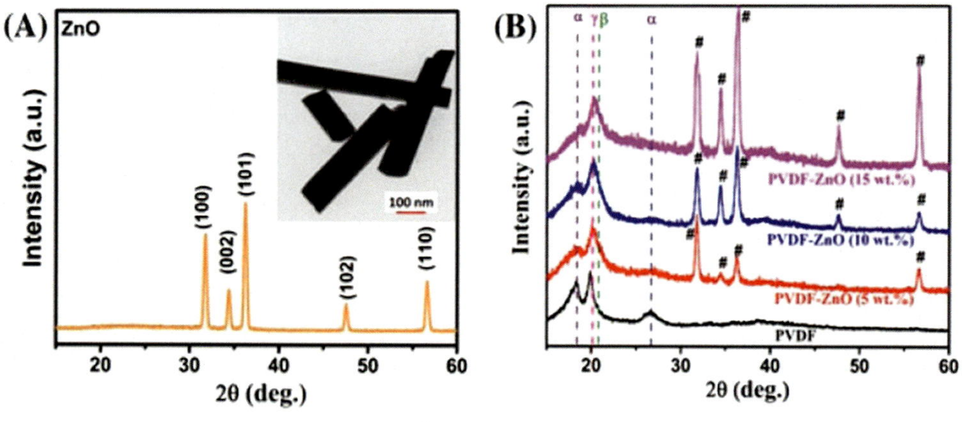

Figure 13.21
X-ray diffractometer (XRD) patterns of (A) ZnO nanorods and (B) polyvinylidene fluoride (PVDF) and PVDF-ZnO nanocomposite films with a different weight percentage of ZnO. The peaks marked with "#" correspond to ZnO. The inset of Fig. 2 shows a transmission electron microscopy image of ZnO nanorods [27].

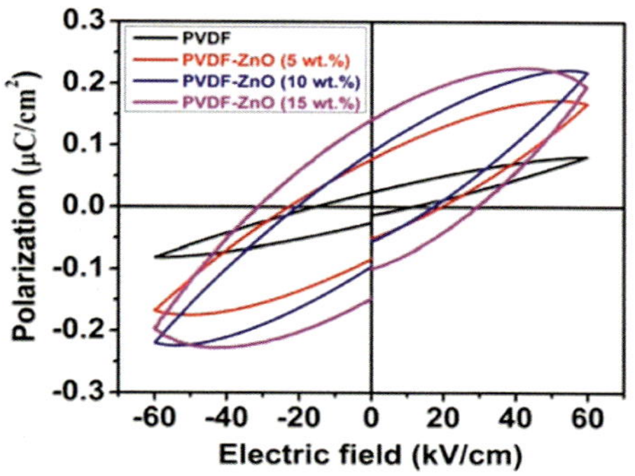

Figure 13.22
Polarization-electric field (P—E) hysteresis loops of polyvinylidene fluoride (PVDF) and ZnO-loaded PVDF nanocomposite films [27].

The P-E loop was used to represent the ferroelectric property of the PVDF/ZnO nanocomposite film and PVDF Fig. 13.22. The graphs clearly indicate that the addition ZnO nanorods have the power to increase the ferroelectric property of the film. The maximum power density was obtained for PVDF/ZnO nanogenerator with 15wt% of ZnO loading at the 7 MΩ load resistance [27].

In the fabrication of energy storage devices such as Li-ion batteries and supercapacitors, PVDF and its copolymer act as a multifunctional material. Because it acts as a binder, separator and gel electrolyte in the fabrication of these devices. Among these, the binding action of PVDF is more prominent. For improving the electrochemical performance of MnO_2 based supercapacitors various strategies have been adopted by many scientists and the problem was solved by using PVDF and its copolymers as binder [28]. Dong et al. introduced a simple chemical method for fabricating MnO_2 electrodes using PVDF as a binder (Fig. 13.23). They prepared a composite binder with graphene and PVDF. The specific capacitance of the chemically prepared electrodes (CBE) was compared with the electrode fabricated physically (PBE).

The electrochemical performance of the MnO_2/PVDF electrode was evaluated from the cyclic voltammetry and charge (CV) and galvanostatic charge/discharge (GCD) using a three-electrode system. The specific capacitance from CV was calculated using the equation [28],

$$C = \frac{\int IdV}{2vm\Delta V} \tag{13.1}$$

Where, C is the average specific capacitance ((F/g)), v is the potential scan rate (V/s), m is the mass of the active material (g), and ΔV is the width of the potential window (V).

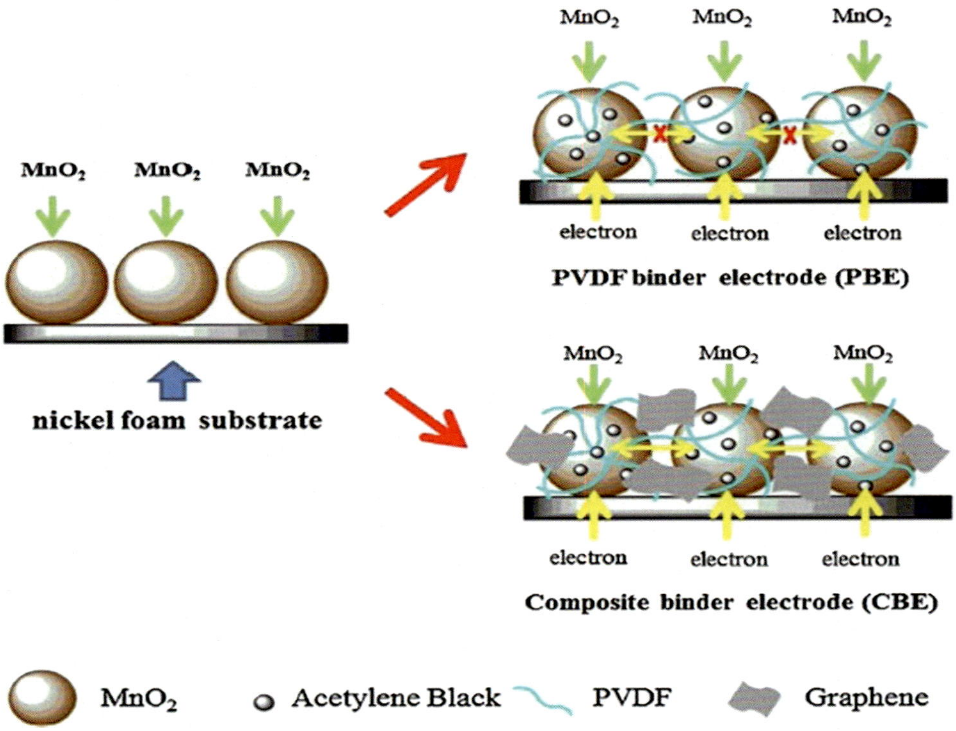

Figure 13.23
The schematic diagram for MnO_2 with a different binder and electron transfer [29].

The composite binders such 0% CBE, 1% CBE, 5% CBE, and 10% CBE possess a specific capacitance of 183, 190, 203, and 195 F/g, respectively, at a scan rate of 10 mV/s. The specific capacitance of composite binder with 5 wt% graphene showed higher specific capacitance. The GCD results of 5% CBE are in agreement with the CV results. PVDF's high adhesive properties hold the active material together, enhancing the working electrodes' specific capacitance. The PVDF/graphene composite binder filled the gap between the MnO_2 spheres, generating a network structure and increasing the working electrode's conductivity. CBE's cycling stability has improved, and it now has a specific capacitance of roughly 90.07 percent after 1000 cycles, whereas PBE's is only 83.3 percent [29].

3. Summary

The unreliable existence of renewable energy supplies, efficient energy storage systems is in urgent need for storing and generating electricity. Polymer nanocomposites have pulled in noteworthy research interests because of their promising potential for flexible applications extending from environmental remediation, energy storage, electromagnetic (EM) absorption, sensing and actuation, transport and safety, defense system, information industry, and so on. The incorporation of nanofillers into polymers will contribute to distinct properties of each component's synergistic effects in the PNCs. Polyvinylidene fluoride (PVDF) based nanocomposites have a significant role in the field of energy storage and energy-saving applications due to their excellent physicochemical property. The selection of the method for the synthesis of PVDF nanocomposites are purely depended on the properties are essential for the end-use application. This chapter concisely gives an overview of the properties of PVDF nanocomposites that are achieved from the different synthesizing techniques.

Acknowledgments

Authors gratefully acknowledge APJ Abdul Kalam Technological University, Thiruvananthapuram, Kerala for the financial support (KTU proceedings No. 4/1654/2019).

References

[1] B.C. Riggs, S. Adireddy, C.H. Rehm, V.S. Puli, R. Elupula, D.B. Chrisey, Polymer nanocomposites for energy storage applications, Mater. Today Proc. [Internet] 2 (6) (2015) 3853–3863. Available from: https://doi.org/10.1016/j.matpr.2015.08.004.
[2] F. Khatun, P. Thakur, N.A. Hoque, A. Kool, S. Roy, P. Biswas, et al., Nanocomposite fi lm for superior and highly durable self-charged hybrid, Energy Convers. Manag. [Internet] 171 (March 2018) 1083–1092. Available from: https://doi.org/10.1016/j.enconman.2018.06.050.
[3] Q. Wang, L. Zhu, Polymer nanocomposites for electrical energy storage, J. Polym. Sci., Part B: Polym. Phys. 49 (20) (2011) 1421–1429.
[4] C. Yang, H. Wei, L. Guan, J. Guo, Y. Wang, X. Yan, et al., Polymer nanocomposites for energy storage, energy saving, and anticorrosion, J. Mater. Chem. 3 (29) (2015) 14929–14941.
[5] S. Castagnet, J. Gacougnolle, P. Dang, Correlation between macroscopical viscoelastic behaviour and micromechanisms in strained a polyvinylidene fluoride (PVDF), Mater. Sci. & Eng. A 276 (2000) 152–159.
[6] L. Priya, J.P. Jog, Polymorphism in Intercalated Poly (Vinylidene Fluoride)/Clay Nanocomposites, March, 2002.
[7] D. Laboratories, Poly (Vinylidene Fluoride) Nanocomposites, 2017.
[8] C. Fang, Y. Tang, Polymer-Engineered Nanostructures for Advanced Energy Applications, 2017, pp. 31–51. Available from: http://link.springer.com/10.1007/978-3-319-57003-7.
[9] D.M. Esterly, D.M. Esterly, Manufacturing of Poly (Vinylidene Fluoride) and Evaluation of its Mechanical Properties Manufacturing of Poly (Vinylidene Fluoride) and Evaluation of its Mechanical Properties, 2002.
[10] Y.J. Yu, A.J.H. McGaughey, Energy barriers for dipole moment flipping in PVDF related ferroelectric polymers, J. Chem. Phys. 144 (1) (2016), 014901.
[11] P. Sajkiewicz, A. Wasiak, Z. Gocl, Phase transitions during stretching of poly (vinylidene uoride), Eur. Polym. J. 35 (1999) 423–429.
[12] I. Three, Oriented phase III poly (vinyl1dene fluoride), J. Polym. Sci. Polymer Lett. Edi. 17 (43) (1979) 585–589.
[13] Fluorine-containing Polymers, 1997.

[14] A.J. Lovinger, Annealing of poly(viny1idene fluoride) and formation of a fifth phase, Macromolecules 44 (6) (1982) 40–44.
[15] R. Gregorio, Effect of crystalline phase, orientation and temperature on the dielectric properties of poly (vinylidene fluoride) (PVDF), J. Mater. Sci. 4 (1999).
[16] N. Mekhilef, Viscoelastic and pressure–volume–temperature properties of poly(vinylidene fluoride) and poly(vinylidene fluoride)–hexafluoropropylene copolymers, J. Appl. Polymer (2001) 230–241.
[17] T. Kousksou, P. Bruel, A. Jamil, T. El Rhafiki, Y. Zeraouli, Energy storage: applications and challenges, Sol. Energy Mater. Sol. Cells 120 (Part A) (2014) 59–80.
[18] S.A. Hashmi, N. Yadav, M.K. Singh, Polymer Electrolytes for Supercapacitor and Challenges, 2020, pp. 231–297.
[19] Y. Niu, Y. Bai, K. Yu, Y. Wang, F. Xiang, H. Wang, Effect of the modifier structure on the performance of barium titanate/poly(vinylidene fluoride) nanocomposites for energy storage applications, ACS Appl. Mater. Interfaces 7 (43) (2015) 24168–24176.
[20] K. Digvijay Satapathy, K. Deshmukh, M. Basheer Ahamed, K. Kumar Sadasivuni, D. Ponnamma, K. Khadheer, S. Pasha, et al., High- quality factor poly (vinylidenefluoride) based novel nanocomposites filled with graphene nanoplatelets and vanadium pentoxide for high-Q capacitor applications, Adv. Mater. Lett. 8 (3) (2017) 288–294.
[21] G. Manjunatha, R. George, I. Hiremath, Functionalized graphene for epoxy composites with improved mechanical properties, Am. J. Mater. 6 (2016) 41–46.
[22] A. Al-Saygh, D. Ponnamma, M.A.A. AlMaadeed, P. Poornima Vijayan, A. Karim, M.K. Hassan, Flexible pressure sensor based on PVDF nanocomposites containing reduced graphene oxide-titania hybrid nanolayers, Polymers 9 (2) (2017) 33.
[23] Y. Ma, W. Tong, W. Wang, Q. An, Y. Zhang, Montmorillonite/PVDF-HFP-based energy conversion and storage films with enhanced piezoelectric and dielectric properties, Compos. Sci. Technol. 168 (2018) 397–403.
[24] J. Chen, X. Yu, F. Yang, Y. Fan, Y. Jiang, Y. Zhou, et al., Enhanced energy density of polymer composites filled with $BaTiO_3$@Ag nanofibers for pulse power application, J. Mater. Sci. Mater. Electron. 28 (11) (2017) 8043–8050.
[25] S.S. Chen, J. Hu, L. Gao, Y. Zhou, S.M. Peng, J.L. He, et al., Enhanced breakdown strength and energy density in PVDF nanocomposites with functionalized MgO nanoparticles, RSC Adv. 6 (40) (2016) 33599–33605.
[26] M. Kim, Y.S. Wu, E.C. Kan, J. Fan, Breathable and flexible piezoelectric ZnO@PVDF fibrous nanogenerator for wearable applications, Polymers 10 (7) (2018).
[27] H.H. Singh, N. Khare, Flexible ZnO-PVDF/PTFE based piezo-tribo hybrid nanogenerator, Nano Energy [Internet] 51 (2018) 216–222. Available from: https://doi.org/10.1016/j.nanoen.2018.06.055.
[28] N. Mahmood, C. Zhang, H. Yin, Y. Hou, Graphene-based nanocomposites for energy storage and conversion in lithium batteries, supercapacitors and fuel cells, J. Mater. Chem. 2 (1) (2014) 15–32.
[29] J. Dong, Z. Wang, X. Kang, Colloids and surfaces a: physicochemical and engineering aspects the synthesis of graphene/PVDF composite binder and its application in high performance MnO_2 supercapacitors, Colloids Surf. A Physicochem. Eng. Asp [Internet] 489 (2016) 282–288. Available from: https://doi.org/10.1016/j.colsurfa.2015.10.060.

CHAPTER 14

Synthesis and structural studies of superconducting perovskite $GdBa_2Ca_3Cu_4O_{10.5+\delta}$ nanosystems

V.S. Vinila, Jayakumari Isac

Centre for Condensed Matter, Department of Physics, CMS College, Kottayam, Kerala, India

Chapter Outline

1. Introduction

Superconducting ceramics are electronic ceramics with innumerable properties and potential applications [1]. They include superconducting power transmission wires, passive RF, powerful electromagnets, radars, superconducting quantum interference devices (SQUIDS), beam steering magnets in particle accelerators, magnetic field detectors and Josephson junctions, magnetic levitation, logic, and storage functions in computers, RSFQ logic, magnetic separation, integrated circuits, magnetic energy storage devices, Maglev trains, magnets for MRI and other imaging applications in medical field and ultrafast superconductive switches [2—4].Cuprates are widely studied ceramic superconductors, because of their high values of transition temperatures [5—7]. Gd based cuprate

Design, Fabrication, and Characterization of Multifunctional Nanomaterials. https://doi.org/10.1016/B978-0-12-820558-7.00022-4

superconductors attracted the attention of many researchers due to their immense superconducting, magnetic and thermoelectric applications [8–10]. Mössbauer studies on Gd based Bismuth ceramic superconductors [11], stabilization of the tetragonal phase and superconducting behavior of $RBa_2(Cu_{1-x} Fe_x)_3O_y$ (R = Y, Gd) [12], thermodynamic stabilities of $RCoO_3$ (R = Nd, Sm, Eu, Gd or Dy) using solid oxide–electrolyte emf technique [13] and grain refinement and enhancement of critical current density in the $V_{0.60}Ti_{0.40}$ alloy superconductor with Gd addition [14] have been studied. Electric field in Eu, Gd and Yb based high Tc superconductors have been analyzed [15]. Trapped field of 17.6T in melt-processed, bulk Gd-Ba-Cu-O reinforced with shrink-fit steel has been reported [16].

In this work, focus is on the designing, synthesis and structural studies of nanocrystalline, superconducting perovskite $GdBa_2Ca_3Cu_4O_{10.5+\delta}$ or GBCCO 1234 (as per the conventional nomenclature of ceramic superconductors) or simply GBCCO which is a type 2, high T_c superconductor (HTSC) which enters into its superconducting state at high values of temperature (materials showing superconductivity at high temperatures, approximately above 30 K are generally called "high temperature [high Tc] superconductors" or HTSC [17]). It belongs to the family of quasi two-dimensional cuprate superconductors named $Ba_2Ca_{(n-1)}Cu_nO_{(2n+4-\delta)}$ where n = 2,3,4 etc. [18]; "*n*" takes the value four in GBCCO. They share a two-dimensional layered perovskite of ABO3 structure [19,20] with superconductivity taking place in a copper-oxide plane [21]. Alternating layers of oxygen deficient copper-oxide planes produce superconductivity in oxy carbonate cuprates [22]. Element Gd which shows superconducting properties [23] is used, so as to enhance superconductivity in the final ceramic. As far as the available information, GBCCO is a new material.

2. Experimental section

2.1 Materials

Gadolinium Oxide (Gd_2O_3), Barium Carbonate ($BaCO_3$), Calcium Oxide (CaO) and Cupric Oxide (CuO) (reagents are of 99.99% purity, Fisher Scientific chemicals).

2.2 Synthesis

GBCCO is designed and synthesized via solid state reaction route following steps such as selection of precursor powders, weighing them according to molecular formula of the desired product, manual mixing, mechanical mixing or chemical mixing, calcination at optimum temperature, and regrinding. Sometimes calcination is followed by pelletization and sintering before subjecting to various characterization studies. Formation of ceramic phase by atomic diffusion of constituent materials during solid state reaction takes place when calcined at optimum temperature [24]. The calcination temperature decides the density of the ceramic product, on which electromechanical characteristics of the synthesized ceramic greatly depends [25,26]. The higher the calcination temperature, the higher the homogeneity and density of the final ceramic product and hence proper calcinations at the right temperature is necessary to obtain the best electrical and mechanical properties [27,28]. Feasibility and rate of a solid state reaction depend on reaction conditions, structural properties of the reactants, surface area of the solids, reactivity of the solids and thermodynamic free energy change associated with the reaction [29,30]. Solid state reaction technique is an environment friendly, solvent less technique with low cost of production (as ceramic crystals are not emerged in the solvent) [31,32]. More products are produced and waste elimination is not required at the end of the process due to the lack of solvent [33].

For the synthesis of GBCCO, Gadolinium Oxide (Gd_2O_3), Barium Carbonate ($BaCO_3$), Calcium Oxide (CaO), and Cupric Oxide (CuO) are accurately weighed according to the stoichiometric formula,

$$1/2Gd_2O_3 + 2BaCO_3 + 3CaO + 4CuO = GdBa_2Ca_3Cu_4O_{10.5+\delta} + 2CO_2 \quad (14.1)$$

Stoichiometry of GBCCO is given in Table 14.1.

Once precursor powders are accurately weighed, manual mixing is done in an agate mortar with pestle for 21 days to obtain ceramic sample. This is followed by mechanical mixing via ball milling of the sample with the aid of zirconium beads for two months along with daily sieving and mixing ("ball to powder weight" ratio is 40:1). After that attrition milling is done on the sample for 3 h. Time of attrition must not extend 12 h since changes of agglomeration predominates [34]. The ceramic sample is subjected to calcinations at temperatures 30, 500, 850, and 900°C in a furnace to form various GBCCO nanosystems. After each calcination cool oxygen is allowed into the furnace. Cooled sample is ground, sieved and mixed and subjected to structural studies. Optimum calcination temperature for the formation of GBCCO is found (by trial and error method) to be at 900°C.

Table 14.1: Stoichiometry of GBCCO.

Precursors	Molecular mass (in gm)	Stoichiometric mass (in gm)	Required mass for 40 gm of GBCCO (gm)	Measured mass for 40 gm of GBCCO (gm)
Gd_2O_3	362.4982	½ (362.4982) = 181.2491	7.44108	7.4411
$BaCO_3$,	197.3359	2(197.3359) = 394.6718	16.17544	16.1755
CaO	56.0774	3(56.0774) = 168.2322	6.90668	6.9065
CuO	79.5454	4(79.5454) = 318.1816	13.06276	13.0628

2.3 Characterization

Ceramic samples of GBCCO nanosystems, obtained after calcinations at 30, 500, 850, and 900°C are powdered and analyzed by X-ray diffraction patterns. The diffractometer used in this work for taking patterns is Bruker AXS D8 advance diffractometer. X-ray radiations of are of wavelength 1.54, and 184 Å. The specimen is scanned at a speed of 2 degree/min. Data obtained from XRD patterns (peak positions, 2θ values, intensities, and d spacings) is compared with JCPDS or ICDD database and applied XPERT PRO software to obtain lattice parameters and to analyze crystal system. XRD diffraction peak profile analysis methods used are Debye Scherrer's formula, Cauchy equation, and WH plot.

3. Results and discussion

3.1 XRD analysis of GBCCO

XRD patterns are obtained for GBCCO at four different calcination temperatures 30, 500, 850, and 900°C (GBCCO nanosystems), which is given in Fig. 14.1. Lattice parameters at final calcination temperature, 900°C, is, a = 24.6955 Å, b = 5.4667 Å, c = 5.1189 Å, $\alpha = \beta = \gamma = 90°$ with a unit cell volume (abc for orthorhombic) 691. 0669 (Å) [3], hence confirming the orthorhombic crystal structure ($a \neq b \neq c$, $\alpha = \beta = \gamma = 90°$) of GBCCO [35].

XRD data can be used to calculate the interplanar distance, d, according to Bragg's law

$$n\lambda = 2d\sin\theta \tag{14.2}$$

λ is the wavelength of the X-ray used (1.54,184 Å), d is the lattice spacing, n is the order of diffraction and θ is the Bragg's angle [36]. The observed preferred peaks for studies are high intensity peaks with (hkl) values—(110), (111), (321), and (521)—and the data of which at different calcination temperatures are provided in Table 14.2.

Analyzing XRD data of preferred peaks of GBCCO when calcined at different elevated temperatures, it is observed that diffraction angle (2θ) of the mentioned peaks increases, interplanar distance (d) decreases, FWHM (β) decreases, and intensity increases. This reflects the enhancement of crystallite size [37], decreasing in the concentration of lattice imperfections (structural defects) and consequently to the increasing of the crystallite size and ceramic quality with calcination temperature [38]. Gradual increase in the intensity indicates the crystallite growth of the GBCCO nanoparticles [39]. Crystallinity is accelerated by the calcination process [40].

Peak intensity denotes total scattering amplitude from crystal planes within the crystal lattice [41]. It depends on the structure and composition of the crystal. The intensity of diffracted beam in a particular direction is proportional to the modulus of square of

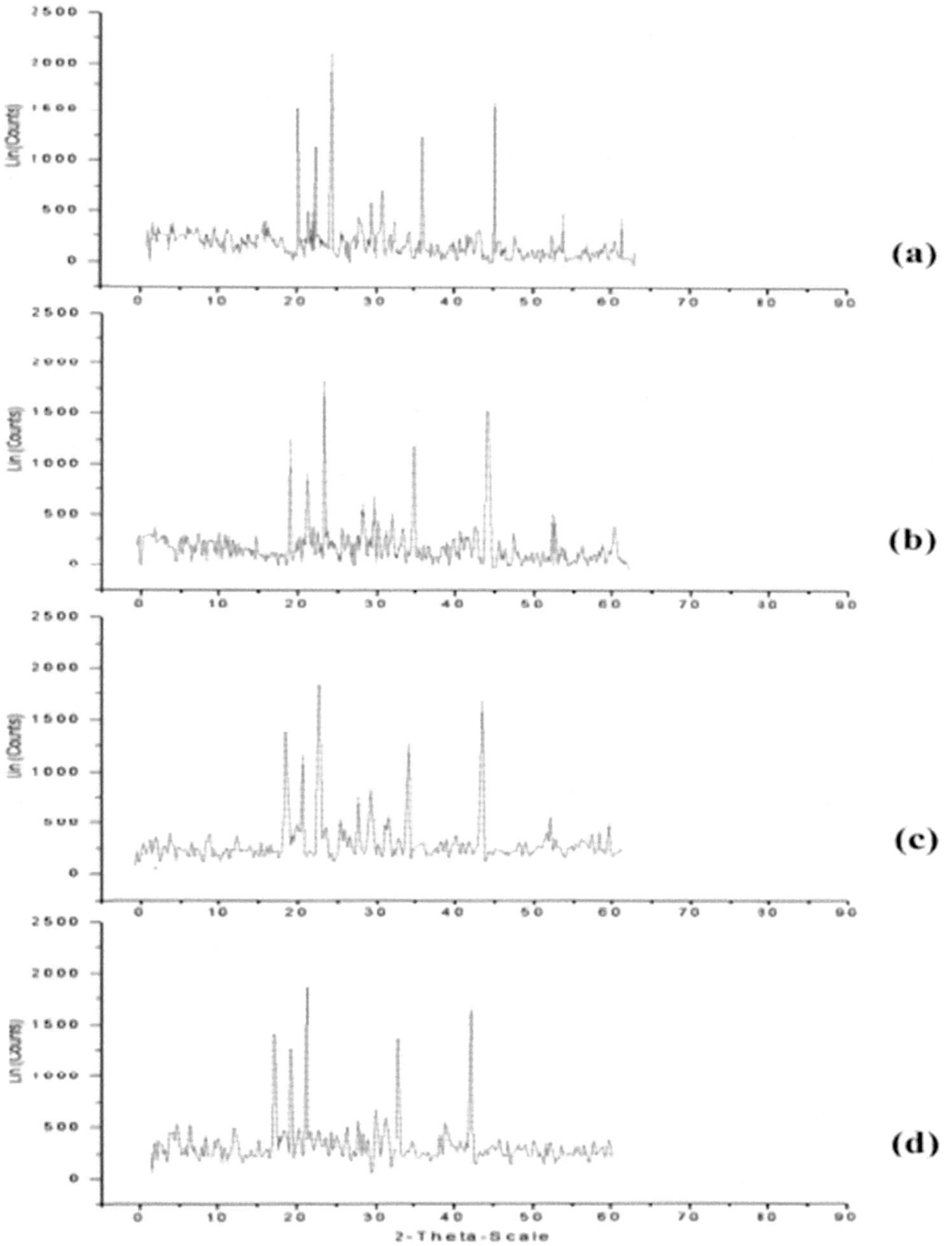

Figure 14.1
X-ray diffractograms of GBCCO at calcination temperatures (A) 900°C, (B) 850°C, (C) 500°C, and (D) 30°C.

Table 14.2: Data of high intensity peaks of GBCCO at different calcination temperatures.

Calcination temperature of GBCCO (°C)	(hkl) values of high intensity peaks	Diffraction angle 2θ°	β (radian × 10^{-3})	d in (Å) obtained	d in (Å) calculated
30	(110)	17.827	9.3676	4.689	4.972
	(111)	22.937	9.0362	3.764	3.875
	(321)	33.179	6.4195	2.659	2.698
	(521)	43.439	6.0183	2.144	2.182
500	(110)	18.053	9.1234	4.531	4.910
	(111)	23.389	8.8966	3.304	3.801
	(321)	34.786	6.2974	2.643	2.677
	(521)	43.985	5.9136	2.045	2.057
850	(110)	19.953	8.8792	3.884	4.446
	(111)	23.511	8.6175	2.928	3.781
	(321)	34.935	6.2625	2.521	2.566
	(521)	44.895	5.7217	1.943	2.017
900	(110)	20.780	8.4954	3.704	4.272
	(111)	24.603	7.9023	2.743	3.616
	(321)	35.235	5.9962	2.344	2.545
	(521)	45.009	5.2507	1.692	2.013

geometrical structure factor (F_{hkl}), which depends on the occupancy of atoms in different Wyckoff positions of a unit cell and then crystal [42,43]. Factors which influence the diffracted beam intensity for a polycrystalline ceramic are polarization, multiplicity factor, Lorentz factor, thermal vibration of atoms and absorption factor [44]. As crystallinity increases, number of planes orientated in a particular direction increases; hence multiplicity increases and causes increase in intensity [45]. Similarly, if all the Wyckoff positions in a crystal are occupied by the required number of atoms, then the structure factor gives high value, which will lead to high intensity [46].

3.2 Particle size of GBCCO

3.2.1 Using Debye Scherrer equation

Debye Scherrer equation for the calculation of particle size is

$$D = \frac{K\lambda}{\beta \cos\theta} \tag{14.3}$$

where K is the Scherrer constant, λ is wave length of the X-ray beam used (1.54,184 Å), β is the Full width at half maximum (FWHM) of the peak and θ is the Bragg angle [47]. Scherrer constant denotes the shape of the particle and its value is most commonly taken as 0.9 [48]. Scherrer equation accounts for broadening solely due to crystallite size [49]. Particle size of the high intensity peaks of GBCCO sample at different calcination

Table 14.3: Particle size of GBCCO at different calcination temperatures using Scherrer equation for peak (110).

Calcination temperature of GBCCO (°C)	Bragg angle θ (°)	β (radian × 10^{-3})	Particle size for peak (110) (D) (nm)
30	8.9136	9.3676	14.479
500	9.0265	9.1234	14.871
850	9.9765	8.8792	15.323
900	10.3901	8.4954	16.036

temperatures are extracted using Debye Scherrer equation and are shown in Tables 14.3–14.6. Particle sizes of preferred peaks are plotted as functions of calcination temperature, and are given in Figs. 14.2–14.5.

Particle size of GBCCO increases with increase in calcination temperature. Heat treatment causes the particles to anneal and form larger grains [50]. Migration of grain boundaries occurs at higher calcination temperatures resulting in coalescence of small grains and formation of large grains [51]. Increasing calcination temperature provides the system with sufficient kinetics to permit further growth of GBCCO grains [52]. The growth of GBCCO grains with temperature can be ascribed by nucleation–aggregation–agglomeration growth mechanism [53]. The increase in particle size with temperature takes place by the aggregation of these agglomerated particles to form large particle [54]. Having larger grains means more number of planes orientated in a particular direction and hence more crystallinity [55,56]. Obtained particle size of GBCCO for preferred high intensity peaks confirm its nanostructure property.

The dislocation density, δ of the calcined ceramic is defined as the length of dislocation lines per unit volume and can be evaluated from the relation [57].

$$\delta = \frac{1}{D^2} \tag{14.4}$$

Table 14.4: Particle size of GBCCO at different calcination temperatures using Scherrer equation for peak (111).

Calcination temperature of GBCCO (°C)	Bragg angle θ (°)	β (radian × 10^{-3})	Particle size for peak (111) (D) (nm)
30	11.4687	9.0362	15.131
500	11.6947	8.8966	15.381
850	11.7556	8.6175	15.882
900	12.3017	7.9023	17.355

Table 14.5: Particle size of GBCCO at different calcination temperatures using Scherrer equation for peak (321).

Calcination temperature of GBCCO (°C)	Bragg angle θ (°)	β (radian × 10^{-3})	Particle size for peak (321) (D) (nm)
30	16.5892	6.4195	21.779
500	17.3932	6.2974	22.297
850	17.4675	6.2625	22.430
900	17.6174	5.9962	23.844

Table 14.6: Particle size of GBCCO at different calcination temperatures using Scherrer equation for peak (521).

Calcination temperature of GBCCO (°C)	Bragg angle θ (°)	β (radian × 10^{-3})	Particle size for peak (110) (D) (nm)
30	21.7195	6.0183	23.965
500	21.9925	5.9136	24.436
850	22.4475	5.7217	25.337
900	22.5046	5.2507	27.621

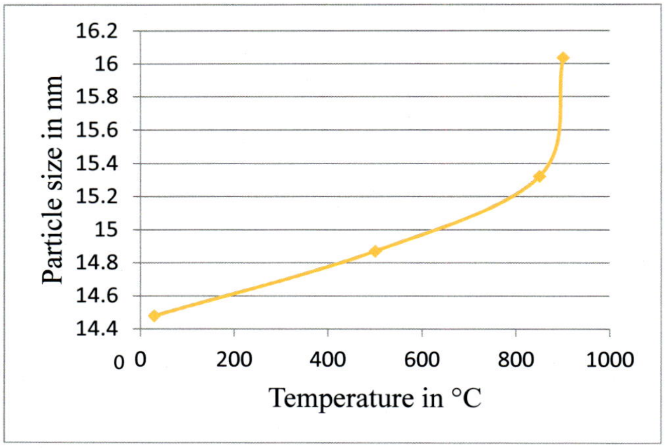

Figure 14.2
Effect of calcination temperature on particle size of GBCCO (peak 110).

The number of particles per unit surface area, N can be determined according to [58]

$$N = \frac{d}{D^3} \tag{14.5}$$

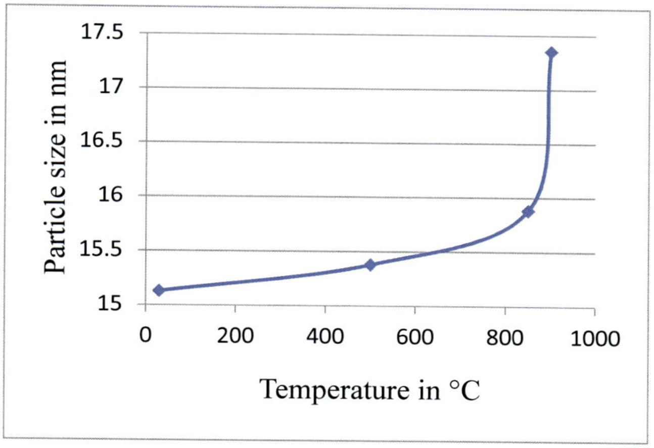

Figure 14.3
Effect of calcination temperature on particle size of GBCCO (peak 111).

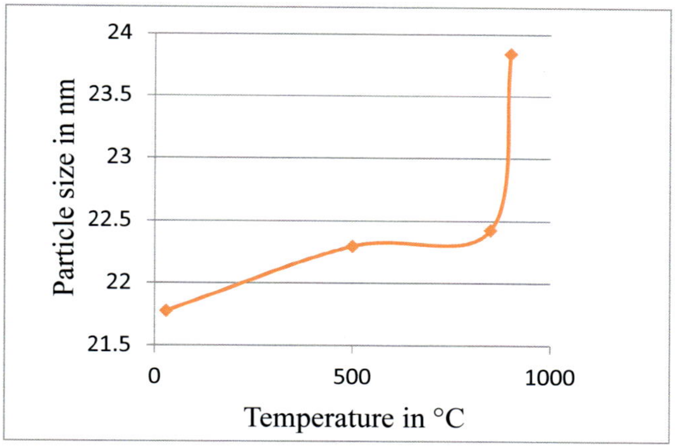

Figure 14.4
Effect of calcination temperature on particle size of GBCCO (peak 321).

Theoretically a dislocation is a crystallographic irregularity or a defect formed within the crystal and it is found that crystals with larger dislocation density were harder [59]. Using particle size (D), dislocation density (δ) and the number of particles per unit surface area (N) of GBCCO for mentioned peaks are calculated, which are given in Tables 14.7–14.10. Graphs plotting the effect of calcination temperature on δ and N are given in Figs. 14.6–14.13.

While considering dislocation density (δ) and number of particles per unit surface area (N) of GBCCO, for preferred high intensity peaks, as functions of calcination temperature, it is noticed that both decreases with increase in calcination temperature. The thermal

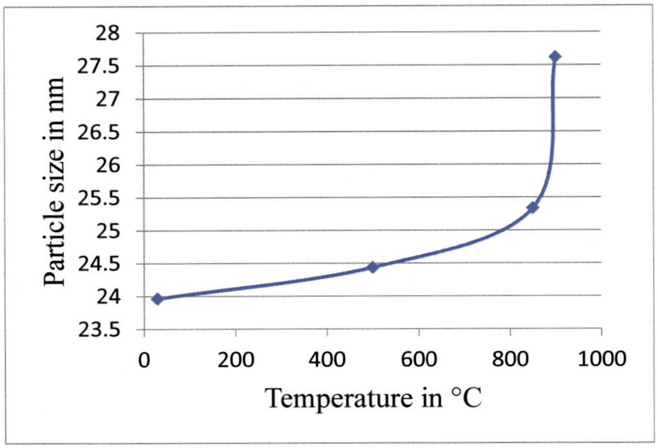

Figure 14.5
Effect of calcination temperature on particle size of GBCCO (peak 521).

Table 14.7: Dislocation density and number of particles per unit surface area of GBCCO at various calcination temperatures for peak (110).

Calcination temperature of GBCCO (°C)	Dislocation density (δ) (m)	No. of particles per unit surface area (N)
30	47.69×10^{14}	154×10^{12}
500	45.22×10^{14}	138×10^{12}
850	42.59×10^{14}	108×10^{12}
900	38.88×10^{14}	89×10^{12}

Table 14.8: Dislocation density and number of particles per unit surface area of GBCCO at various calcination temperatures for peak (111).

Calcination temperature of GBCCO (°C)	Dislocation density (δ) (m)	No. of particles per unit surface area (N)
30	43.68×10^{14}	109×10^{12}
500	42.27×10^{14}	90.8×10^{12}
850	39.64×10^{14}	73.1×10^{12}
900	33.2×10^{14}	52.5×10^{12}

Table 14.9: Dislocation density and number of particles per unit surface area of GBCCO at various calcination temperatures for peak (321).

Calcination temperature of GBCCO (°C)	Dislocation density (δ) (m)	No. of particles per unit surface area (N)
30	21.08×10^{14}	25.7×10^{12}
500	20.11×10^{14}	23.8×10^{12}
850	19.88×10^{14}	22.3×10^{12}
900	17.59×10^{14}	17.3×10^{12}

Table 14.10: Dislocation density and number of particles per unit surface area of GBCCO at various calcination temperatures for peak (521).

Calcination temperature of GBCCO (°C)	Dislocation density (δ) (m)	No. of particles per unit surface area (N)
30	17.41×10^{14}	15.6×10^{12}
500	16.74×10^{14}	14×10^{12}
850	15.58×10^{14}	11.9×10^{12}
900	13.1×10^{14}	8.03×10^{12}

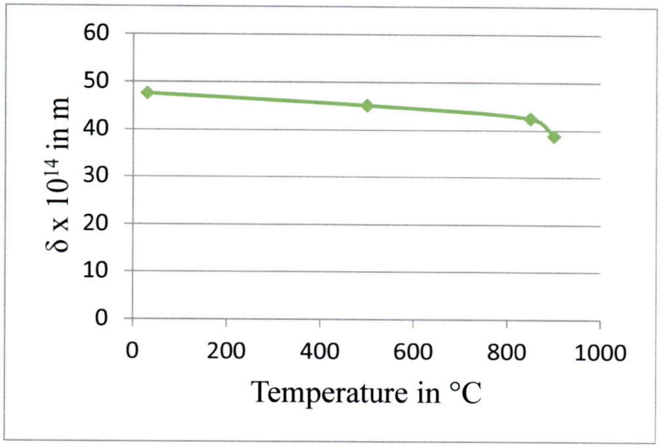

Figure 14.6
Effect of calcination temperature on δ of GBCCO (peak 110).

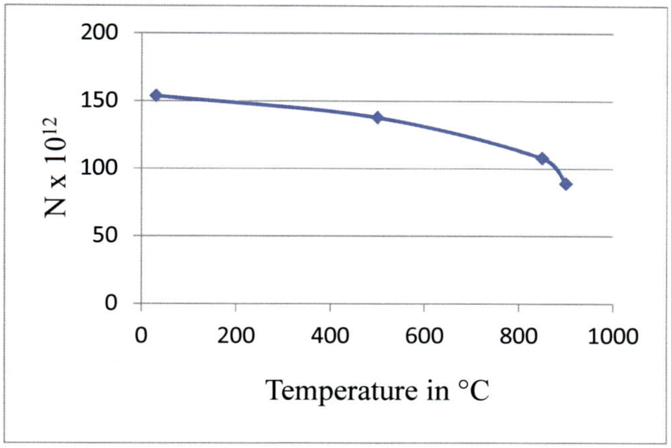

Figure 14.7
Effect of calcination temperature on N of GBCCO (peak 110).

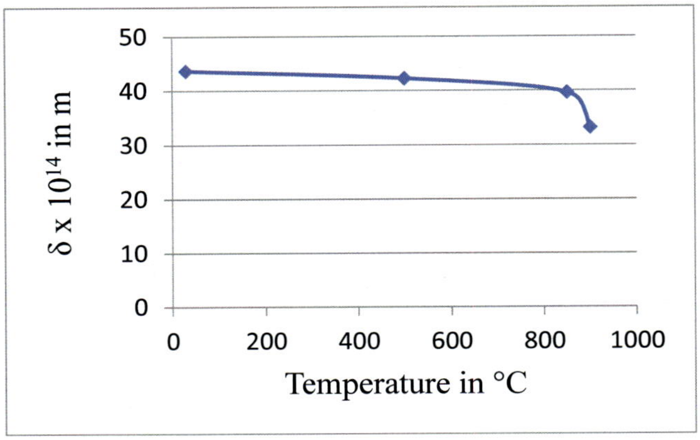

Figure 14.8
Effect of calcination temperature on δ of GBCCO (peak 111).

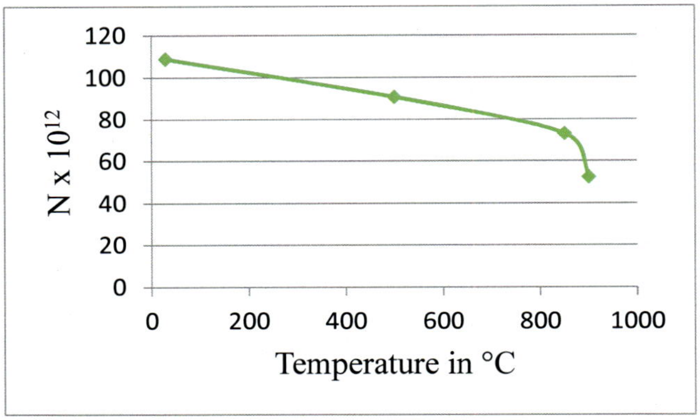

Figure 14.9
Effect of calcination temperature on N of GBCCO (peak 111).

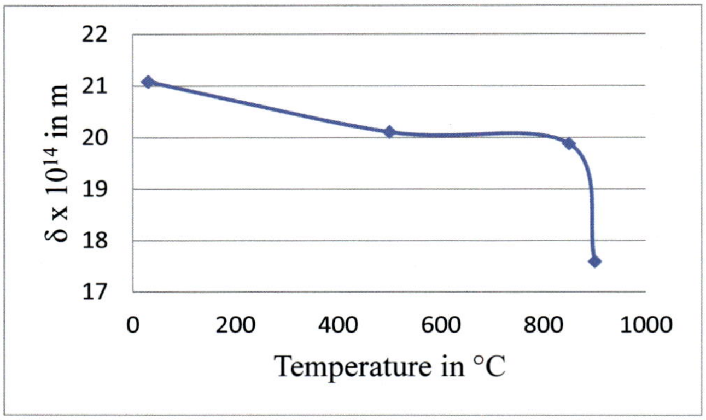

Figure 14.10
Effect of calcination temperature on δ of GBCCO (peak 321).

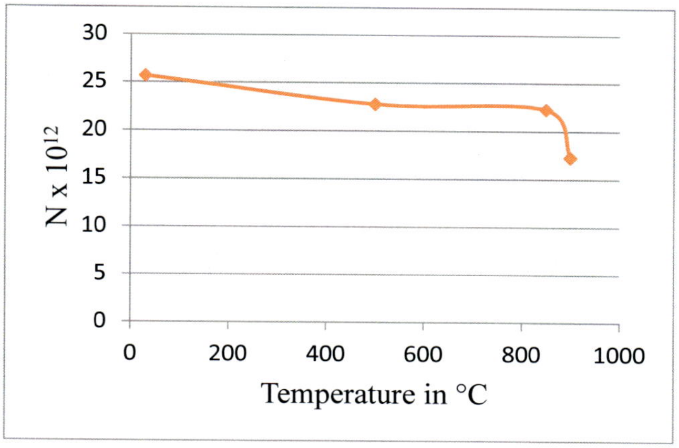

Figure 14.11
Effect of calcination temperature on N of GBCCO (peak 321).

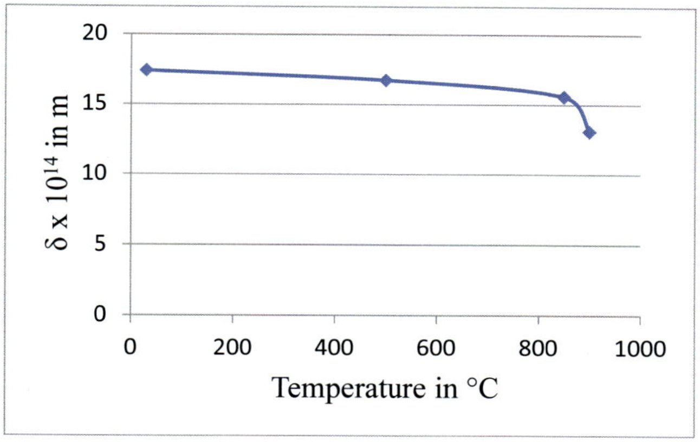

Figure 14.12
Effect of calcination temperature on δ of GBCCO (peak 521).

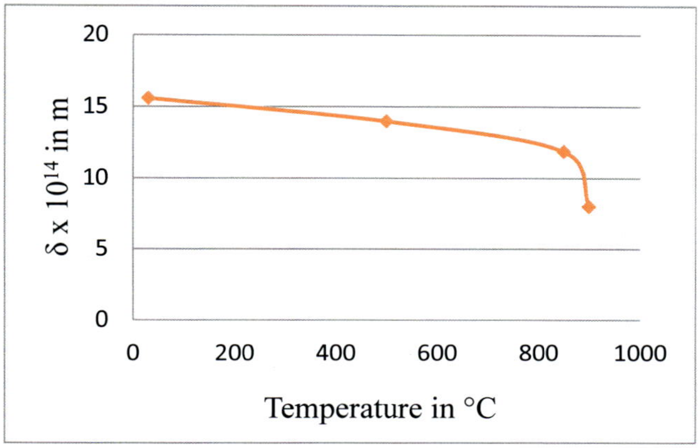

Figure 14.13
Effect of calcination temperature on N of GBCCO (peak 521).

calcination greatly reduces, but does not completely eliminate, the defect density [60]. Decrease in number of particles per unit surface area (N) with calcination temperature reflects, higher the calcination temperatures, higher the particle size [61].

3.2.2 Using Williamson Hall plot

XRD instrumental broadening due to sample is given by Ref. [62].

$$\mathrm{FW(s)} \times \cos\theta = \frac{K\lambda}{\mathrm{size}} + 4 \times \mathrm{strain} \times \sin\theta \tag{14.6}$$

Williamson and Hall proposed a method for obtaining size and strain broadening by considering peak width as a function of 2θ. By plotting $\mathrm{FW(s)} \times \cos\theta$ on y axis against sinθ on x axis, we get the strain component from the slope and the particle size component from the y intercept [63]. Using Williamson Hall plot, the particle size and strain of GBCCO at various calcination temperatures, considering high intensity peaks (111) and (521) are calculated. WH plots are given in Figs. 14.14 and 14.15. Particle size (D) and strain (e) of GBCCO for mentioned peaks at different calcination temperatures are given in Tables 14.11 and 14.12.

Effect of calcination temperature on particle size of GBCCO size (determined from WH plot) for observed high intensity peaks are plotted in Figs. 14.16 and 14.17.

As calcined at elevated temperatures, particle size increases and strain decreases. Grain growth with calcination temperature could be confirmed from this [64]. Decreasing strain value with treating temperature reflects reduced concentration of lattice imperfections or structural defects at high calcination temperatures [65].

Cauchy equation for calculating particle size using strain value (e) is given by [66].

$$\beta \cos\theta = \frac{\lambda}{D} + 4e \sin\theta \tag{14.7}$$

Particle size of GBCCO for preferred high intensity peaks (111) and (521) at different calcination temperatures are theoretically calculated using Cauchy equation, using corresponding strain values obtained from WH plots. Particle size by Cauchy equation is provided in Tables 14.13 and 14.14. Effect of calcination temperature on particle size is plotted in Figs. 14.18 and 14.19.

A comparison of particle size of GBCCO nanosystems, for observed peaks (111) and (521) obtained by different methods are given in Tables 14.15 and 14.16.

The particle size GBCCO nanosystems of measured by Scherrer method does not account for the strain induced and the instrumental factors affecting the broadening, hence the value is less than those obtained from Cauchy method and WH method [67].

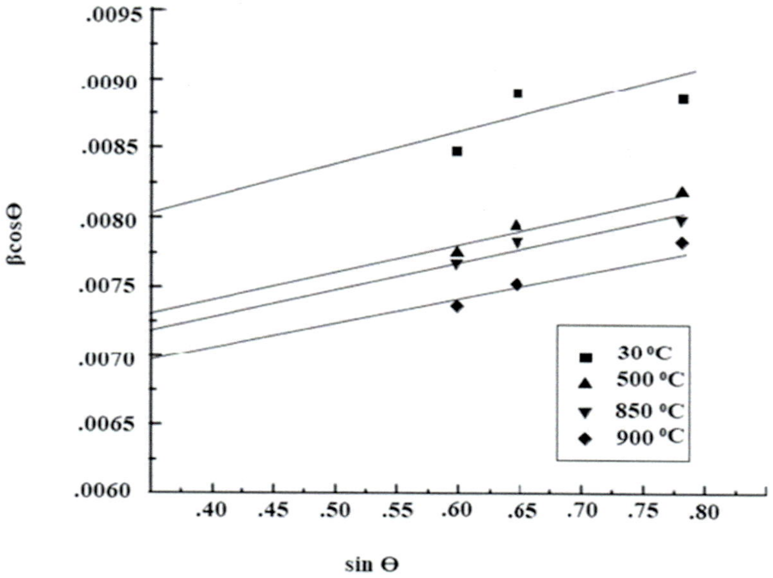

Figure 14.14
Williamson Hall plot of GBCCO (peak 111).

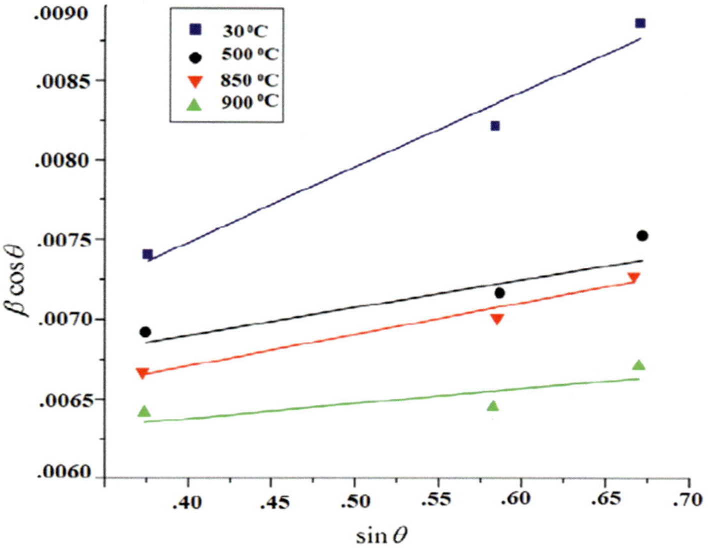

Figure 14.15
Williamson Hall plot of GBCCO (peak 521).

Table 14.11: Particle size and strain of GBCCO (peak 111) at different calcination temperatures using WH plot.

Calcination temperature of GBCCO (°C)	Particle size (D) (nm)	Strain (e)
30	16.832	0.01
500	17.131	0.0072
850	17.981	0.0062
900	18.551	0.0054

Table 14.12: Particle size and strain of GBCCO (peak 521) at different calcination temperatures using WH plot.

Calcination temperature of GBCCO (°C)	Particle size (D) (nm)	Strain (e)
30	24.40	0.0081
500	25.938	0.0077
850	26.502	0.0066
900	28.676	0.0061

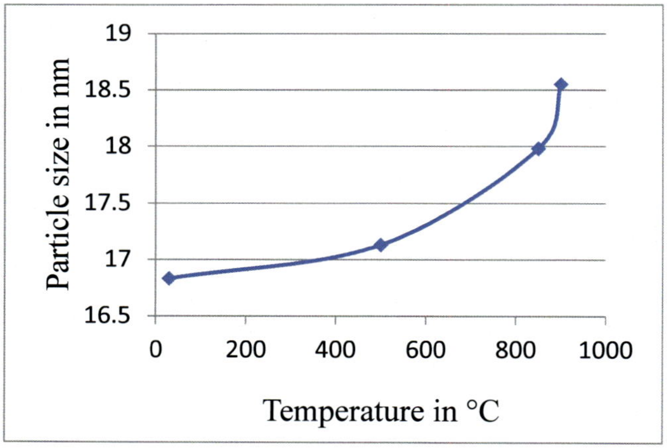

Figure 14.16
Calcination temperature and particle size (determined from WH plot) of GBCCO (peak 111).

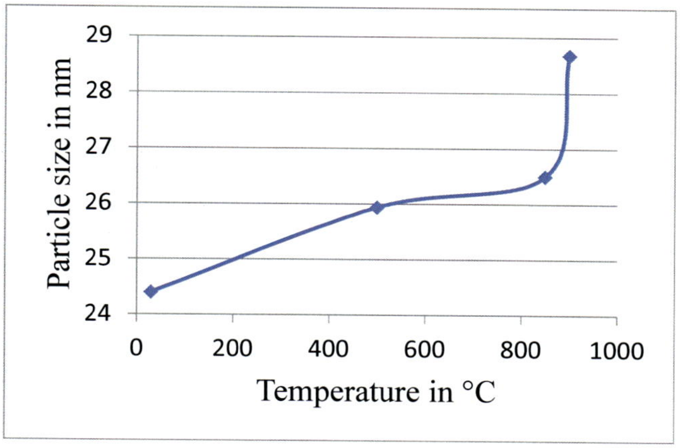

Figure 14.17
Calcination temperature and particle size (determined from WH plot) of GBCCO (peak 521).

Table 14.13: Particle size of GBCCO (peak 111) by Cauchy equation at different calcination temperatures.

Calcination temperature of GBCCO (°C)	Particle size (D) (nm)
30	17.55
500	18.51
850	18.92
900	20.83

Table 14.14: Particle size of GBCCO (peak 521) by Cauchy equation at different calcination temperatures.

Calcination temperature of GBCCO (°C)	Particle size (D) (nm)
30	24.9
500	26.81
850	27.223
900	29.943

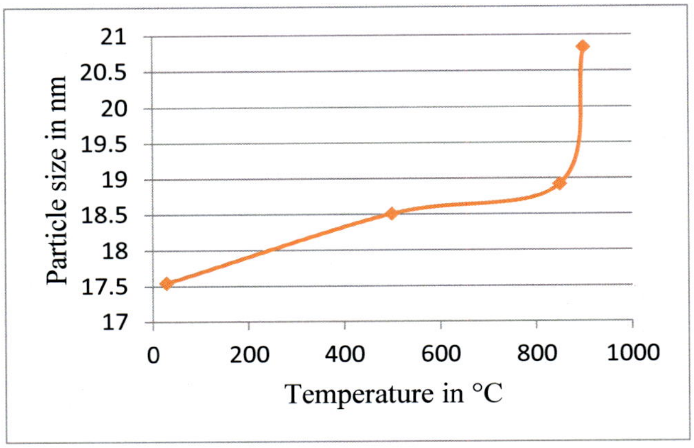

Figure 14.18
Calcination temperature and particle size (by Cauchy equation) of GBCCO (peak 111).

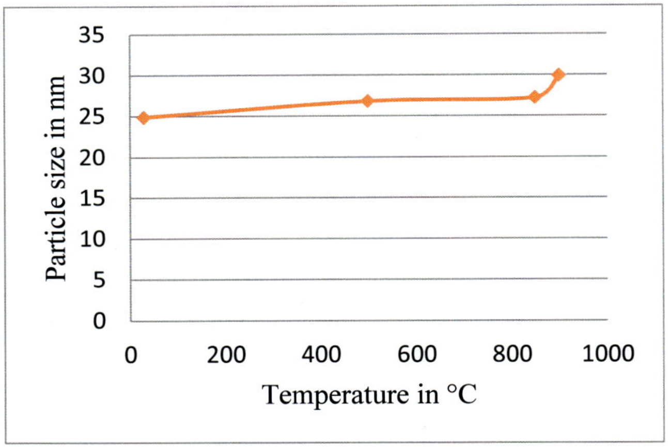

Figure 14.19
Calcination temperature and particle size (by Cauchy equation) of GBCCO (peak 521).

Table 14.15: Particle size of GBCCO (peak 111) by different methods.

Calcination temperature of GBCCO (°C)	Particle size by Scherrer eqn. (nm)	Particle size by WH plot (nm)	Particle size by Cauchy eqn. (nm)
30	15.131	16.832	17.55
500	15.381	17.131	18.51
850	15.882	17.981	18.92
900	17.355	18.551	20.83

Table 14.16: Particle size of GBCCO (peak 521) by different methods.

Calcination temperature of GBBCO (°C)	Particle size by Scherrer eqn. (nm)	Particle size by WH plot (nm)	Particle size by Cauchy eqn (nm)
30	23.965	24.40	24.9
500	24.436	25.938	26.81
850	25.337	26.502	27.223
900	27.621	28.676	29.943

4. Conclusion

Superconducting perovskite GBCCO nanosystems are designed and successfully synthesized by solid state reaction technique. Structural studies are done by XRD analysis. GBCCO is found to be in orthorhombic crystal system. XRD diffractograms taken at different calcination temperatures revealed that intensity, of the preferred high intensity peaks, (110), (111), (321), and (521), increases with increase in calcination temperature. FWHM and d spacing (interplanar distance) decreases with increase in calcination temperature. Increase in peak intensity and reduction in β indicate reduction of lattice defects in the crystal, increasing particle size and crystallinity. Calcination at high temperature improves the quality of the ceramic. Particle size of GBCCO is obtained at various calcination temperatures using Scherrer method and they are found to be increasing with the increase in calcination temperature. Higher calcination temperatures results in coalescence of small particles and formation of large particles. Increased thermal energy accelerates further grain growth in ceramics which can be ascribed by nucleation–aggregation–agglomeration growth mechanism. Dislocation density and number of particles per unit surface area of GBCCO nanosystems are determined as functions of calcination temperature. Both decreases with increase in calcination temperature. Thermal calcination reduces defect density. Decrease in number of particles per unit surface area (N) with calcination temperature indicates increasing particle size with calcination temperatures. Particle sizes and strains are extracted from WH plots of the samples. Using strain values theoretical calculation is done using Cauchy equation. Particle size determined by WH plots and Cauchy equation of GBCCO nanosystems are in agreement with those calculated using Debye Scherrer equation. Lattice strain decreases with elevated calcination temperatures. It can be concluded that calcination temperature greatly affects structural parameters. Crystallinity and ceramic quality increases and defect density decreases with increase in calcination temperature.

Acknowledgments

The authors acknowledge the Sophisticated Analytical Instruments Facility (SAIF) Kochi for providing instrumental data and the Principal, CMS College, Kottayam, Kerala for providing research facilities.

References

[1] R. Jacob, J. Isac, Morphological, thermal and optical studies of jute reinforced PbSrCaCuO polypropylene composite, Mod. Phys. Lett. B 30 (31) (2016) 1650379.

[2] W. Tabis, Charge order and its connection with fermi-liquid charge transport in a pristine high-Tc cuprate, Nat. Commun. 5 (2014) 5875.

[3] J. Prokleska, J. Pospisil, J.V. Poltierova, V. Sechovsky, J. Sebek, Low temperature AC susceptibility of UCoGe crystals, J. Phys. Conf. Series 200 (2010) 012161.

[4] M.I. Youssif, A.A. Bahgat, I.A. Ali, AC magnetic susceptibility technique for the characterization of high temperature superconductors, Egypt J. Sol. 23 (2) (2000) 231–250.

[5] A.S. Nair, R. Jacob, S. Issac, S. Rajan, V.S. Vinila, D.J. Satheesh, J. Issac, Band gap analysis of nanocrystalline L0.1ZY0.9BCCO ceramics, J. Adv. Chem. 11 (7) (2015) 3742–3750.

[6] A. Schilling, M. Cantoni, J.D. Guo, H.R. Ott, Superconductivity above 130 K in the Hg-Ba-Ca-Cu-O system, Nature 363 (1993) 56–58.

[7] V.S. Vinila, J. Isac, AC magnetic susceptibility study of superconducting ceramic GBCCO 1234, Int. J. Sci. Res. & Rev. 8 (5) (2019) 388–402.

[8] E.P. Khlybov, O.E. Omelyanovsky, A.J. Zaleski, A.V. Sadakov, D.R. Gizatulin, L.F. Kulikovaa, I. Kostyleva, V.M. Pudalov, Magnetic and superconducting properties of Fe As-based high-Tc superconductors with Gd, JETP Lett. (Engl. Transl.) 90 (5) (2009) 387–390.

[9] F. Inanir, S. Yildiz, K. Ozturk, S. Celebi, Magnetization of Gd diffused $YBa_2Cu_3O_{7-x}$ superconductor: experiment and theory, Chin. Phys. B 22 (7) (2013) 077402.

[10] A. Rao, Geetha, S.O. Manjunath, B. Christopher, U. Deka, High temperature superconductors $REBa_2Cu_3O_7$ (RE=Y, Gd and Eu) for possible thermo-electric applications, Rev. Appl. Phys. (RAP) 4 (2015) 1.

[11] E.V. Sampathkumaran, G. Wortmann, G. Kaindl, Mossbauer studies of Gd-doped bismuth-based ceramic superconductors, Bull. Mater. Sci. 14 (1991) 703–707. Springer.

[12] T.J. Kistenmacher, W.A. Brydan, J.S. Morgan, K. Moorjani, Stabilization of the tetragonal phase and superconducting behaviour in $RBa_2(Cu_{1-x}\ Fe_x)_3O_y$ ($R = Y$, Gd; $0 \le x \le 0.15$), Phys. Rev. B 36 (1987) 16.

[13] R. Subasri, R. Pankajavalli, O.M. Sreedharan, High temperature thermodynamic stabilities of $RCoO_3$ (R = Nd, Sm, Eu, Gd or Dy) using solid oxide – electrolyte emf technique, J. Alloys Compd. 269 (1–2) (1998) 71–74. Elsevier.

[14] S. Paul, S.K. Ramjan, R. Venkatesh, L.S. Sharath Chandra, M.K. Chattopadhyay, Grain refinement and enhancement of critical current density in the $V_{0.60}Ti_{0.40}$ alloy superconductor with Gd addition, Condensed matter superconductivity, in: Applied Superconductivity Conference, 2019, 1912.04507.

[15] D. Reotier, P. Vulliet, A. Yaouanc, P. Chaudouet, S. Garcon, J.P. Senateur, F. Weiss, L. Asch, G.M. Kalvius, Rare-earth valency and electric field in Eu, Gd and Yb based high Tc superconductors, Phys. C Supercond. 153–155 (3) (1988) 1543–1544. Elsevier.

[16] J.H. Durrell, A.R. Dennis, J. Jaroszynski, M.D. Ainslie1, K.G.B. Palmer, Y.H. Shi, A.M. Campbell, J. Hull, M. Strasik, E.E. Hellstrom, D.A. Cardwell1, A trapped field of 17.6 T in melt-processed, bulk Gd-Ba-Cu-O reinforced with shrink-fit steel, IOP Science, Supercond. Sci. & Technol. 27 (8) (2014) 082001.

[17] J.A.F. Livas, L. Boeri, A. Sanna, G. Profeta, R. Arita, M. Eremets, A Perspective on Conventional High-Temperature Superconductors at High Pressure: Methods and Materials, Physics Reports, vol. 856, Elsevier, 2020, pp. 1–78.

[18] M. Mumtaz, L. Ali, S. Azeem, S. Ullah, G. Hussain, M.W. Rabbani, A. Jabbar, K. Nadeem, Dielectric properties of Zn_x/CuTl – 1223 nanoparticle superconductor composites, J. Adv. Ceram. 5 (2) (2016) 159–166. Springer, 2226 4108.

[19] A.S. Bhalla, R. Gou, R. Roy, The perovskite structure – a review of its role in ceramic science and technology, Mater. Res. Innov. 4 (1) (2000) 3–26. Springer-Verlag.

[20] J. Wu, Perovskite lead-free piezoelectric ceramics, J. Appl. Phys. 127 (19) (2020) 190901.
[21] A. Younis, N.A. Khan, N.U. Bajwa, Dielectric properties of $Cu_{0.5}Tl_{0.5}Ba_2Ca_3Cu_{4-y}Zn_yO_{12-\delta}$ (y = 0, 3) superconductors, J. Korean Phys. Soc. 57 (2010) 1437–1443.
[22] B. Dalla Piazza, M. Mourigal, N.B. Christensen, G.J. Nilsen, P. RegennaT-Piggott, T.G. Perring, M. Enderle, D.F. McMorrow, D.A. Ivanov, H.M. Rønnow, Fractional excitations in the square-lattice quantum antiferromagnet, Nat. Phys. 11 (2014) 62–68.
[23] Y. Shi, G.B. Andreis, A.R. Dennis, J.H. Durrell, D.A. Cardwell, The growth and superconducting properties of RE–Ba–Cu–O single grains with combined RE elements (RE = Gd and Y), Supercond. Sci. Technol. 33 (2020) 3. https://iopscience.iop.org/article/10.1088/1361-6668/ab6ee4.
[24] M. Cernea, F. Vasiliu, M.C. Bartha, C. Plapcianu, I. Mercioniu, Characterization of ferromagnetic double perovskite Sr_2FeMoO_6 prepared by various methods, Ceram. Int. 40 (2014) 8.
[25] T. He, W. Sun, H. Huo, O. Kononova, Z. Rong, V. Tshitoyan, T. Botari, G. Ceder, Similarity of precursors in solid-state synthesis as text-mined from scientific literature, Chem. Mater. 32 (18) (2020) 7861–7873.
[26] M. Cernea, F. Vasiliu, C. Plapcianu, C. Valsangiacom, I. Mercioniu, I. Pasuk, R. Lowndes, R. Trusca, G. Aldica, L. Pintilie, Preparation by sol-gel and solid state reaction methods and properties investigation of double perovskite Sr_2FeMoO_6, J. Eur. Ceram. Soc. 33 (13–14) (2013) 2483–2490.
[27] V.S. Vinila, J. Isac, Temperature and frequency dependence of dielectric properties of superconducting ceramic $GdBa_2Ca_3Cu_4O_{10\cdot5}$, Int. J. Sci. Res. 7 (8) (2016) 2319–7064.
[28] L.T.T. Vien, N. Tu, M.T. Tran, N. Van Du, D.H. Nguyena, D.X. Viet, N.V. Quang, D.Q. Trung, P.T. Huy, A new far-red emission from Zn_2SnO_4 powder synthesized by modified solid state reaction method, Opt. Mater. 100 (2020) 109670. Elsevier.
[29] K. Huang, J. Yu, L. Zhang, J. Xu, P. Li, Z. Yang, C. Liu, W. Wang, X. Kan, Synthesis and characterizations of magnesium and titanium doped M-type barium calcium hexaferrites by a solid state reaction method, J. Alloys Compd. 825 (2020) 154072. Elsevier.
[30] R. Jacob, H.G. Nair, J. Isac, Impedance spectroscopy and dielectric studies of nanocrystalline iron doped barium strontium titanate ceramics, Process. & Appl. Ceram. 9 (2) (2015) 73–79.
[31] K. Ariyoshi, K. Yuzawa, Y. Yamada, Reaction mechanism and kinetic analysis of the solid-state reaction to synthesize single-phase $Li_2Co_2O_4$ spinel, J. Phys. Chem. C 124 (15) (2020) 8170–8177.
[32] J.L. Clabel, I.T. Awan, A.H. Pinto, I.C. Nogueira, V.D.N. Bezzon, E.R. Leite, D.T. Balogh, V.R. Mastelaro, S.O. Ferreira, E. Marega, Insights on the mechanism of solid state reaction between TiO_2 and $BaCO_3$ to produce $BaTiO_3$ powders: the role of calcination, milling, and mixing solvent, Ceram. Int. 46 (3) (2020) 2987–3001. Elsevier.
[33] H.R. Javadinejad, R.E. Kahrizsangi, Thermal and kinetic study of hydroxyapatite formation by solid-state reaction, Int. J. Chem. Kinet. (2020) 1–13.
[34] A. Mony, J. Isac, Fabrication and analysis of nanocrystalline superconductor YSrBiCuO at different calcination temperatures, Int. J. Sci. Res. 3 (12) (2012) 2319–7064.
[35] V.S. Vinila, R. Jacob, S. Rajan, A.S. Nair, D.J. Satheesh, J. Isac, Investigation of nano crystalline ceramic superconductor GdBaCaCuO: structural and morphological aspects, J. Phys. Photon 220–228 (2015), 5218-3718.
[36] M. Dongol, A. El-Denglawey, A.F. Elhady, A.A. Abuelwafa, Structural properties of nano 5, 10, 15, 20-Tetraphenyl-21H, 23H-porphine nickel (II) thin films, Curr. Appl. Phys. 12 (5) (2012) 1334–1339. Elsevier.
[37] R. Hari Krishna, B.M. Nagabhushana, Nagabhushana, N. SuriyaMurthy, S.C. Sharma, C. Shivakumara, R. Chakradhar, Effect of calcination temperature on structural, photoluminescence, and thermoluminescence properties of $Y_2O_3:Eu^{3+}$ nanophosphor, December 2012, J. Phys. Chem. C 117 (4) (2013) 1915–1924.
[38] S. Lalitha, S.Z. Karazhanov, P. Ravindran, S. Senthilarasu, R. Sathyamoorthy, J. Janabergenov, Electronic structure, structural and optical properties of thermally evaporated CdTe thin films, Phys. B Condens. Matter 387 (2007) 227–238. Elsevier.

[39] R. Sathyamoorthy, S. Senthilarasu, S. Lalitha, A. Subbarayan, K. Natarajan, X. Mathew, Electrical conduction properties of flash evaporated ZincPhthalocyanine (ZnPc) thin films, Solar Energy Mater. & Solar cells 82 (1 - 2) (2004) 69−177.
[40] E.R. Shaaban, N. Afify, A. EL-Taher, Effect of film thickness on microstructure parameters and optical constants of CdTe thin films, J. Alloys Compd. 482 (1−2) (2009) 400−404.
[41] Y. Ahmed, S.M. El-Sheikh, Z.I. Zaki, Changes in hydroxyapatite powder properties via heat treatment, Bull. Mater. Sci. 38 (2015) 7.
[42] S.A. Aly, N.Z. El Sayed, M.A. Kaid, Effect of annealing on the optical properties of thermally evaporated ZnO films, Vacuum 61 (1) (2001) 1−7.
[43] R.P.P. Singh, I.S. Hudiara, S.B. Rana, Effect of calcination temperature on the structural, optical and magnetic properties of pure and Fe-doped ZnO nanoparticles, Mater. Sci.-Poland 34 (2) (2016) 451−459.
[44] V.K. Pecharsky, P.Y. Zavalij, Fundamentals of Powder Diffraction and Structural Characterization of Materials, Springer, New York, 2003.
[45] A. Habiba, R. Haubnerb, N. Stelzer, Effect of temperature, time and particle size of Ti precursor on hydrothermal synthesis of barium titanate, Mater. Sci. Eng. B 152 (2008) 60−65.
[46] M. Dongol, A. El-Denglawey, M.S. Abd El Sadek, I.S. Yahia, Thermal annealing effect on the structural and the optical properties of nano CdTe films, Optik 126 (2015) 1352−1357. Elsevier.
[47] A. Sundaresan, C.N.R. Rao, Ferromagnetism as a universal feature of inorganic nano particles, Nano Today 4 (1) (2009) 96−106.
[48] H.I. Hsing, C.S. Hsib, C.C. Huang, S.L. Fu, Low temperature sintering and dielectric properties of $BaTiO_3$ with glass addition, Mater. Chem. Phys. 113 (2009) 658−663. Elsevier.
[49] M. Turker, Effect of production parameters on the structure and morphology of Ag nanopowders produced by inert gas condensation, Mater. Sci. & Eng. A 367 (2004) 74−81. Elsevier.
[50] J. Klinkaewnarong, S. Maensiri, Nanocrystalline hydroxyapatite powders by a polymerized complex method, Chiang Mai J. Sci. 37 (2) (2010) 243−251.
[51] J. Klinkaewnarong, S. Utara, Ultrasonic-assisted conversion of limestone into needle-like hydroxyapatite nanoparticles, Ultrason. Sonochem. 46 (2018) 18−25. Elsevier.
[52] I.R. Gibson, I. Rehman, S.M. Best, W. Bonfield, Characterization of the transformation from calcium deficient apatite to β-tricalcium phosphate, J. Mater. Sci. Mater. Med. 11 (2000) 799−804. Springer.
[53] W.W. So, J.S. Jang, Y.W. Rhee, K.J. Kim, S.J. Moon, Preparation of nanosized crystalline Cds particles by the hydro - thermal treatment, J. Colloid Interface Sci. 237 (2001) 136−141.
[54] L. He, Y.W. Mai, Z.Z. Chen, Fabrication and characterization of nanometer CaP(aggregate)/Al_2O_3 composite coating on Titanium, Mater. Sci. & Eng. A 367 (1−2) (2004) 51−56. Elsevier.
[55] M. Dongol, M.A. Zied, G.A. Gamal, A. EI-Denglawey, The effects of composition and heat treatment on the structural and optical properties of $Ge_{15}Te_{85-x}Cu_x$ thin films, Phys. B Condens. Matter 353 (3−4) (2004) 169−175. Elsevier.
[56] M.M. El-Nahassam, M.H. Ali, A. El-Denglawey, Structural and optical properties of nano-spin coated sol−gel porous TiO_2 films, Trans. Nonferrous Metals Soc. China 22 (12) (2012) 3003−3011. Elsevier.
[57] M. Dongol, M. Saad, E. Mohamed, A.A. Abuelwafa, Effect of thermal annealing on the optical properties of $Ge_{20}Se_{65}S_{15}$ thin films, Indian J. Phys. 94 (2020) 7.
[58] S.P. Singh, X.R. Huang, P. Kumar, M.E. Kassner, Assessment of the dislocation density using X − Ray topography in Al single crystals annealed for long times near the melting temperature, Materials Today Commun. 21 (2019) 100613.
[59] J.E. Ayers, The measurement of threading dislocation densities in semiconductor crystals by X- ray diffraction, J. Cryst. Growth 135 (1-2) (1994) 71−77. Elsevier.
[60] S. Neretina, R.A. Hughes, J.F. Britten, N.V. Sochinskii, J.S. Preston, P. Mascher, The role of substrate surface termination in the deposition of (111) CdTe on (0001) sapphire, Appl. Phys. A 96 (2009) 429−433.

[61] M. Figueiredo, A. Fernado, G. Martins, J. Freitas, F. Judas, H. Figueiredo, Effect of calcination temperature on the composition and microstructure of hydroxyapatite derived from human and animal bone, Ceram. Int. 36 (2010) 2383–2393. Elsevier.
[62] A. Khorsand Zak, W.H. Abd Majid, M.E. Abrishami, R. Yousefi, X - ray analysis of ZnO nanoparticles by Williamson Hall and size strain plot methods, Solid State Sci. 13 (2011) 251–256.
[63] Y.T. Prabhu, K.V. Rao, S. Sai Kumar, B. Siva Kumari, X – ray analysis by Williamson Hall and size strain plot methods of ZnO nanoparticles with fuel variation, World J. Nano Sci. Eng. 4 (2014) 21–28.
[64] D.J. Satheesh, N.V. Nair, J. Isac, Identification of a super paramagnetic phase of titanium substituted Mn-Zn ferrites and its magnetic and mossbauer studies, J. Sci. Res. & Rep. 9 (6) (2015) 1–8.
[65] R. Jacob, H.G. Nair, J. Isac, Structural and morphological studies of nanocrystalline ceramic $BaSr_{0.9}Fe_{0.1}TiO_4$, Int. Lett. Chem. Phys. & Astron. 44 (2015) 95–107. Sci Press Ltd.
[66] K. Venkateswarlu, M. Sandhyarani, A. Thangavelu, N. Rameshbabu, Estimation of crystallite size, lattice strain and dislocstion density of nanocrystalline carbonate substituted hydroxy apapite by X ray peak variance analysis, Procedia Mater. Sci. 5 (2014) 212–221.
[67] J.P. Borah, J. Barman, K.S. Sarma, Structural and optical properties of properties ZnS nanoparticles, Chalcogenide Lett. 5 (9) (2008) 201–208.

SECTION C

Development of multiferroic nanoparticles

CHAPTER 15

Design of multifunctional magnetoelectric particulate nanocomposites by combining piezoelectric and ferrite phases

J. Philip[1], R. Rakhikrishna[2]
[1]*Amal Jyothi College of Engineering, Kanjirappalli, Kerala, India;* [2]*Department of Physics, NSS College, Changanassery, Kerala, India*

Chapter Outline

1. Introduction

A general definition of magnetoelectric (ME) effect delineates the coupling between electric polarization and magnetization in a material directly or indirectly. The most important feature of ME materials is their ability to change electron spins by electric fields and electric dipoles by magnetic fields. Accordingly, both magnetization and polarization of such materials are functions of external electric and magnetic fields respectively [1,2].

In general, ME materials form a subclass of multiferroic materials, which possess more than one ferroic order like ferroelectricity, ferroelasticity, ferromagnetism, ferrotoroidism, etc. simultaneously. The relationship between ME and multiferroic couplings are well documented in literature [3]. ME composites can be prepared by a proper selection and mixing of electric and magnetic counterparts having intercoupling properties. All ME materials need not be multiferroic, but are necessarily both electrically and magnetically polarizable. Some of the multiferroic materials are ME as well. The coupling interaction

Design, Fabrication, and Characterization of Multifunctional Nanomaterials. https://doi.org/10.1016/B978-0-12-820558-7.00018-2

between electric field, magnetic field and strain in ME materials are well documented in literature [4,5]. The intercoupling between ferroelectric, ferroelastic, and ferromagnetic properties of multiferroic material and how a new ME property emerges from these interactions are also understood [5]. It is now known that ME property originates from cross-coupling of piezoelectric and piezomagnetic properties.

In general, ME materials can be classified into following two, based on their structure and origin of the property:

(1) Single-phase materials in which both magnetic and electric properties exist in the same phase.
(2) Composite materials in which electric and magnetic counterparts exist as separate phases, and their coupling give rise to ME phenomenon.

Usually single-phase materials show ME effect at very low temperatures and high fields. Ferro/antiferro electric to paraelectric and ferro/ferri/antiferro magnetic to paramagnetic transitions occur in single-phase materials. The conditions for the occurrence of single-phase ME effect are [6]

(1) Presence of adequate structural building blocks permitting ferroelectric type ionic movements.
(2) Magnetic interaction pathways, usually of symmetric or antisymmetric superexchange type.
(3) Fulfillment of symmetry conditions.
(4) Dipolar interactions and Zeeman energy.

Existence of these conditions at microscopic level is the foundation of ME performance in single-phase materials [7]. But it is difficult to have both ferroelectricity and magnetism existing in a single material because these two are mutually contradictory in their origin. Ferroelectricity is the result of relative shift of positive and negative ions and preferable condition for this is empty d orbitals, while magnetism is ordering of electron spins in incomplete ionic shells, and it requires partially filled d orbitals. Because of these incompatible conditions, single-phase MEs are very few in number [8]. $BiFeO_3$ is the first single-phase material which is found to exhibit room temperature ME property [9,10].

The weak or low temperature ME effect found in single-phase materials make them unsuitable for room temperature applications. So, researchers turned to development of artificial ME materials, which are composites. It is possible to prepare room temperature multiphase ME materials in composite form, with high ME property, following different preparation techniques. Individual phases in a composite need not be ME in nature [11], but there should be coupling between phases, which shall give rise to ME interaction. This can be achieved by combining piezoelectric and magnetostrictive phases in which the coupling arises from striction induced in the material through external electric or magnetic field.

The polarization/magnetization induced in the material in the presence of magnetic/electric field is due to the strain developed at the interface of phases. ME property in a composite follows a product tensor property [11]. In composites the material exhibits phase transitions in individual phases; from ferroelectric to paraelectric in piezoelectric phase and from ferro/ferri/antiferro magnetic to paramagnetic in magnetostrictive phase. Composite materials have the advantage of high design flexibility and multifunctionality at room temperature [12].

Mixing rules of properties are different for different composites prepared from two or more phases. Composition dependent properties of composites can be divided into three: sum properties, product properties and combination properties. Sum property like density and resistivity is the average contribution of the properties of individual phases. The ME effect is a known product tensor property [13]. It can be explained as: if one phase has X−Y effect and other has Y−Z effect in a composite system and the system is configured in such a way that the response of Y due to X is coupled or get transferred to the second phase, then the composite will exhibit X−Z effect [14]. In ME systems this property explains the cross-interaction of electric and magnetic phases in the composite. The ME interaction is generally mediated by strain in such systems. This means that the ME property is the product of magnetostriction in the magnetic phase and piezoelectricity in piezoelectric phase [15]. Following this, the direct and converse ME effect can be represented as

$$\text{Direct ME effect} = \left(\frac{\text{magnetic}}{\text{mechanical}}\right) \times \left(\frac{\text{mechanical}}{\text{electric}}\right) \tag{15.1}$$

$$\text{Converse ME effect} = \left(\frac{\text{electric}}{\text{mechanical}}\right) \times \left(\frac{\text{mechanical}}{\text{magnetic}}\right) \tag{15.2}$$

Thus, ME effect of a composite is an extrinsic property. The strain developed in the magnetostrictive phase, on applying a magnetic field, exerts a stress in the piezoelectric phase through the interface. This results in an induced polarization on the strained piezoelectric phase. Converse mechanism is also possible. From this concept, the ME effect and thus the ME coefficient of a composite is high only when the combining phases have high piezoelectric coefficient d_{33} and magnetostrictive coefficient λ_{11}. However, ME effect also depends on the composite micro/macrostructure and mechanical coupling across magnetic-piezoelectric interfaces. If the average magnitude of the property in a composite exceeds the value of the end components, then it is an example for a combination property.

2. Nanocomposite ME materials

In a composite system, if one or more phases are constituted by nanomaterials, then it is said to be a nanocomposite. A nanocomposite is usually a bulk material with reinforced nanofillers to enhance certain physical, chemical, or even biological properties. Properties of nanocomposites differ from their bulk counterparts and also from their component phases. Some nanocomposites have properties better by several orders of magnitude than their component phases due to large surface to volume ratio of embedded nanoparticles. By controlling the size and shape of nanomaterials the properties of nanocomposites can be controlled even with low volume fractions compared to bulk [16]. Some new properties also arise through the synergism between different phases, which are also influenced by size effects. Interfacial area and associated effects as well as mechanical coupling between phases are critical parameters in determining the efficiency of ME composites [17]. So such materials made with nanomaterials are interesting from application point of view [18].

In ME nanocomposites, ferroelectric/piezoelectric or magnetostrictive phases could be tuned and controlled at nanoscales, accomplishing new scales for exploring ME coupling mechanisms. Interfacial effects play a great role in determining the coupling efficiency of such materials. Losses at the interface on combining two phases in bulk ME composites can be reduced by reducing the particle size and combining them at atomic levels [11]. ME coupling at the atomic scales can be studied by combining different phases with similar crystal lattices and thereby designing ME nanocomposites. Such multiferroic nanostructures are of great interest.

Reduction in particle size in a nanocomposite affects the piezoelectric phase more than the magnetic phase, leading to dilution of ME property and efficiency [19,20]. Isolated ultrafine magnetic nanoparticles in ferroelectric host are expected to improve the coupling between piezoelectric and magnetostrictive phases as a result of increased area of contact [21].

Detailed thermodynamic theory of ME composites has been worked out, and is available in literature [22]. From these theoretical developments it can be inferred that ME property of composites depends on the dielectric constant (ε), permeability (μ), piezoelectric coefficient (d), and piezomagnetic or magnetostrictive coefficient (q) of individual phases. The relevant relations show that ME property of composites are also influenced by mechanical properties and coupling between individual phases.

3. Magnetoelectric coupling in composites

ME coupling signifies the efficiency of an ME material as transducer that functions between electric and magnetic energies. ME materials can perform bidirectional functions, viz. magnetic to electric and electric to magnetic couplings [23–25]. These can be expressed based on the following equations.

$$\Delta P = \alpha \Delta H \text{(direct ME effect)} \tag{15.3}$$

$$\Delta E = \alpha_E \Delta H \tag{15.4}$$

where α and α_E are the ME and ME voltage coefficients, respectively. Converse ME effect simultaneously occurs in an ME material as

$$\Delta M = \alpha_M \Delta E(\text{converse ME effect}) \tag{15.5}$$

where α_M is the ME magnetization coefficient.

ME effect can be measured under two distinctly different conditions, which refer to electric polarization by an applied magnetic field or to magnetization by an external electric field. Theoretically, ME output in a composite would be maximum for a material with 50% ferrite (magnetic) and 50% ferroelectric phases, provided there is no depolarizing field.

Experiments with PZT/$NiFe_2O_4$ composite system show that α_E declines after a peak value with increase in volume fraction of the ferrite [26]. The high concentration of ferrite phase makes it difficult to polarize the composite. This results in a low piezoelectric coefficient d_{33}. The charge developed on piezo phase leaks through the low resistance ferrite paths [27]. So there is an optimum concentration at which ME property of a composite is maximum.

4. Synthesis and properties of piezoelectric-ferrite particulate nanocomposites

The first ME composite to be prepared in laboratory was $BaTiO_3-CoFe_2O_4$ following a unidirectional solidification method [28–30]. The ME voltage coefficient obtained for this composite was 0.13 V/cm-Oe at room temperature. The maximum ME effect obtained in composites is about 100 times larger than that for single-phase materials [31]. The method of unidirectional solidification requires critical control over composition and processing. The complexity of material preparation makes this method unattractive for the synthesis of ME composites.

Particulate ceramic composites of ferrites and $BaTiO_3$ or $Pb(ZrTi)O_3$, by cost effective sintering process has been reported [32]. This method is rather easy, but the composite obtained has low ME effect compared to earlier techniques. However, this was a great step forward for synthesis of ME composites. Sintered ME composites have many advantages, like easy processing and low cost, compared to in situ composites. Moreover, molar ratio of phases, grain size of each phase, and sintering temperature are easily controllable [33]. Additionally, they provide opportunity to combine phases with widely different crystal structures.

The effect of sintering temperature on various electrical parameters of ME composites have been reported [33]. It is reported that the density of the material increases with sintering temperature, reaches a maximum and then decreases. This is explained as due to the interdiffusion between two phases and pores formed by stress caused by heterogeneities of sintering rate between two phases. Sintering influences dielectric properties, permeability, quality factor, *ac* conductivity, and ME properties. Similar results have been reported on Mg− Cu− Zn ferrites also [34].

Scientists are keen on preparing lead-free smart materials that perform well compared to lead based ones. Lead-fee materials are preferred for development of ME composites due to toxicity of lead. One of the phases of an ME composite ought to be piezoelectric. It is well known that piezoelectric ceramics having large piezoelectricity comes under the class of lead zirconate titanate (PZT) with perovskite (ABO_3) structure. Several researchers have tried to replace lead with nontoxic materials to develop efficient piezoelectric ceramics. Materials coming under the class of perovskite sodium potassium niobate (NKN) partially fulfills the above requirement and plays the main role in lead-free piezoelectric ceramics [35]. So, by combining these lead-free piezoceramics with magnetic ferrites, we can prepare ecofriendly ME composites. In the following sections we report our work and results on the development of two such materials.

5. Synthesis and properties of NKLN—(N/C) FO nanocomposites

The group of Sodium Potassium Niobate (NKN), with ABO_3 perovskite structure, similar to PZT, is found to be a good alternative to lead based ceramics [36]. Many members of this group have high piezoelectric coefficients. In this group, Sodium Potassium Lithium Niobate has good piezoelectric property with reported piezoelectric coefficient d_{33} in the range 200−235 pC/N [37]. We have chosen the composition $(Na_{0.5}K_{0.5})_{0.94}Li_{0.06}NbO_3$ (NKLN), which has the largest value for piezoelectric d_{33} coefficient, in our work. We have analyzed composite materials prepared by mixing piezoelectric NKLN and magnetostrictive nickel ferrite (NFO), or cobalt ferrite (CFO) nanoparticles followed by sintering. NFO is an important inverse spinel ferrite. Ferrimagnetic NFO has high spin polarization and high Curie temperature [38]. Cobalt Ferrites, with good magnetostriction properties, are hard magnetic materials with comparatively high coercivity and saturation magnetization [39]. Value of its magnetostriction coefficient, reported in literature, is −167 ppm [40].

The magnetic nanoparticles have been synthesized following simple coprecipitation route and the piezoelectric matrix by conventional solid-state reaction route. Sintering has the advantage of providing greater flexibility for combining the phases of widely different crystalline structures, and controllability on grain growth and other physical parameters [31].

The general formula of the composites prepared in this work is x NKLN-(1-x) MFO, where M stands for Nickel or Cobalt. Four different compositions of each with molar weight fractions $x = 0.7$, 0.75, 0.8, and 0.85 have been subjected to investigations. However, it is found that samples with $x = 0.7$ and 0.75 are highly porous and do not acquire the required mass density (>90% theoretical density) for further experimentation. Hence detailed investigations have been limited to samples with $x = 0.8$ and 0.85. The dense samples are polled electrically in an electric field of 1 kV/mm following contact electric poling at a temperature of 150°C for half an hour, and polled magnetically by placing it in a *dc* magnetic field of strength 1T for 30 min.

The samples have been characterized for their structure by powder XRD and morphology by TEM analysis. The chemical bonds in the samples have been identified by FTIR analysis. The dielectric, magnetic, magnetodielectric and ME properties of both groups of samples have been measured. The ME efficiency of a sample can be determined by measuring the ME voltage coefficient. A direct method to measure the ME coefficient is with an ME measurement set up [41]. In this method, ME coefficient is determined by measuring the induced voltage on applying an *ac* magnetic field across the sample. The sample pellet is mounted parallel in the thickness direction between the pole pieces of a constant *dc* bias magnetic field. An *ac* magnetic field is then applied parallel to the bias field and the variation of induced voltage is measured with variation in magnetic field strength [42].

The ME voltage coefficient, α, is given by

$$\alpha = \left(\frac{dE}{dH_{ac}}\right)_{H_{dc}}$$

From the slope of the variation of output voltage with applied magnetic field, ME voltage coefficient α is obtained as

$$\alpha = \frac{V}{(txH_{ac})}$$

where t is the thickness of sample, V is the output voltage and H_{ac} is the applied *ac* magnetic field in a constant *dc* magnetic field [43,44].

6. Results and discussion

From the powder XRD spectra of NKLN-NFO, the peaks corresponding to NKLN and NFO have been identified and compared with the data available in literature. The peaks of individual phases are clearly visible and do not contain any additional peaks, indicating that there is no chemical reaction between individual phases. The same is true with NKLN-CFO samples. The FTIR spectra contain vibrational peaks of metal oxides, corresponding to stretching vibrations of tetrahedral and octahedral metal-oxygen bonds.

Theoretical densities of the composites are calculated using rule of mixtures. The percentage densities of the pellets are found to be greatly influenced by the composition. The percentage density is found to increase with increase in NKLN percentage and reach maximum for the pellet with 80% NKLN and 20% NFO or CFO ($x = 0.80$). The density of the sintered ceramics depends on various factors like porosity, interdiffusion of phases and internal cracks. In such composites pores are formed by stress caused by heterogeneities of sintering rate between NKLN and NFO/CFO phases. So decrease in density with high ferrite concentration may be the result of increased porosity. The samples with $x = 0.85$ and 0.80 have achieved densities well above 90% of the respective theoretical densities. Correspondingly, polling and further investigations have been limited to these samples.

Magnetic properties of the samples have been investigated by recording their M-H curves. The thin hysteresis loops exemplify the soft magnetic behavior of nickel ferrite phase in the NKLN-NFO composite. The soft and hard magnetic behaviors of NFO and CFO based composites respectively are evident from their magnetic properties like saturation magnetization, coercivity, areas of M-H hysteresis loops etc. The piezoelectric d_{33} coefficients obtained for the samples are tabulated in Table 15.1. As expected, the piezoelectric coefficient increases with increase in NKLN content (or decrease in ferrite content). Addition of even small amount of ferrite is found to reduce the piezoelectric property significantly. One reason for this is the reduction in effective poling on adding low resistive ferrite to the piezoelectric matrix. Higher ferrite concentration makes connected chains of ferrite particles, which form leakage paths, leading to reduction in polarization and consequently the piezoelectric coefficients.

Dielectric properties (dielectric constant and loss) of ME composites depend on frequency, microstructure, temperature as well as magnetic field. Frequency dependences of room temperature dielectric properties, from 100 kHz to 1 MHz, of dense NKLN-NFO sample with $x = 0.85$, before and after poling, is shown in Fig. 15.1. Upon polling alignment of dipoles occur, thereby achieving domain growth in the ferroelectric phase. This results in the observed decrease in dielectric constant and dielectric loss in the low frequency region, after polling. Polled samples show resonance/relaxation in dielectric constant at selected frequencies in the measurement range, and dielectric loss curve shows corresponding sharp peaks at same frequencies. Occurrence of multiple resonances in dielectric curve after poling is due to the activation of different piezoelectric vibration modes in the sample. This comes from particulate microstructural features and random orientation of crystallites

Table 15.1: Piezoelectric, dielectric and ME properties of *x NKLN-(1-x) MFO* Nanocomposites.

Sample	*x*	Piezoelectric coefficient *pC/N*	DC dielectric constant (after poling)	ME coupling constant $x10^{-2}$ *V/cm-Oe*
x NKLN-(*1*−*x*)NFO	0.80	22.5 ± 0.5	202 ± 5	2.32 ± 0.15
	0.85	31.5 ± 0.5	264 ± 5	2.43 ± 0.15
x NKLN-(*1*-*x*)CFO	0.80	16.5 ± 0.5	281 ± 5	1.35 ± 0.10
	0.85	24.2 ± 0.5	338 ± 5	1.62 ± 0.10

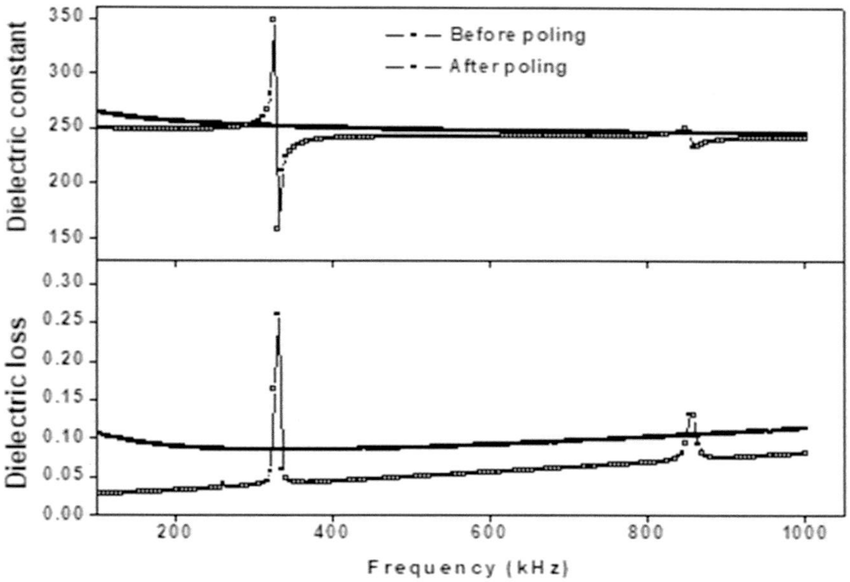

Figure 15.1
Frequency dependences of dielectric constant and dielectric loss, before and after polling, for *x NKLN-(1-x) NFO* nanocomposite with $x = 0.85$.

with respect to the applied field [45,46]. The other dense sample with $x = 0.80$ also shows a similar behavior with shift in resonance frequencies. Resonance is found to get shifted to higher frequencies with increase in piezoelectric phase. Similar results have been obtained for NKLN-CFO samples.

The frequency dependence of *ac* conductivity for both groups of composites also exhibit resonance peaks. In general, *ac* conductivity increases with frequency, which confirms that hopping conduction between nearest neighbor sites occur due to the formation of polarons, following localization of charge carriers.

It is found that dielectric properties have strong dependence on temperature. Dielectric constant, in general, increases with increase in temperature. Polarization has contributions from dipolar, ionic, electronic and interfacial polarizations, but at low frequencies dipolar and interfacial polarizations contribute maximum to overall polarization. It has been reported that in ME composites the process of thermally activated charge hopping enhances dielectric polarization resulting in an increase of dielectric constant with temperature [47]. In addition to this, increase in temperature enhances number of charge carriers within different phases of the material. This results in an increase of space charge carriers, which in turn enhances the dielectric properties [48]. All samples exhibit similar variations with temperature.

The ME voltage coefficient is a complex parameter that depends on the mole fractions of component phases, piezoelectric coefficient, magnetostriction coefficient, resistivity of phases and mechanical coupling between them. Porosity in the sample and increase in volume fraction of low resistive ferrite phase make effective poling difficult for composites. Porosity reduces mechanical coupling between phases and increase in ferrite phase provides a leakage path. So, efficient ME conversion will be obtained only for composites with high enough density and required ferrite content.

In Fig. 15.2 we plot the induced electric field (output voltage per unit thickness) across the sample against strength of the applied *ac* magnetic field, keeping the bias *dc* magnetic field at 1000 Oe, for the two polled samples of NKLN-NFO with $x = 0.85$ and $x = 0.80$. The values of ME voltage coefficient α obtained from these variations are also tabulated in Table 15.1. The highest value reported so far for ME coefficient for a sample prepared by simple solid-state reaction method is 1.3×10^{-2} *V/cm-Oe* for the sample $0.5Ni_{0.3}Zn_{0.62}Cu_{0.08}Fe_2O_4$ + $0.5\ Pb(Fe_{0.5}Nb_{0.5})\ O_3$ [49], while we get a value of 2.43×10^{-2} *V/cm-Oe.*

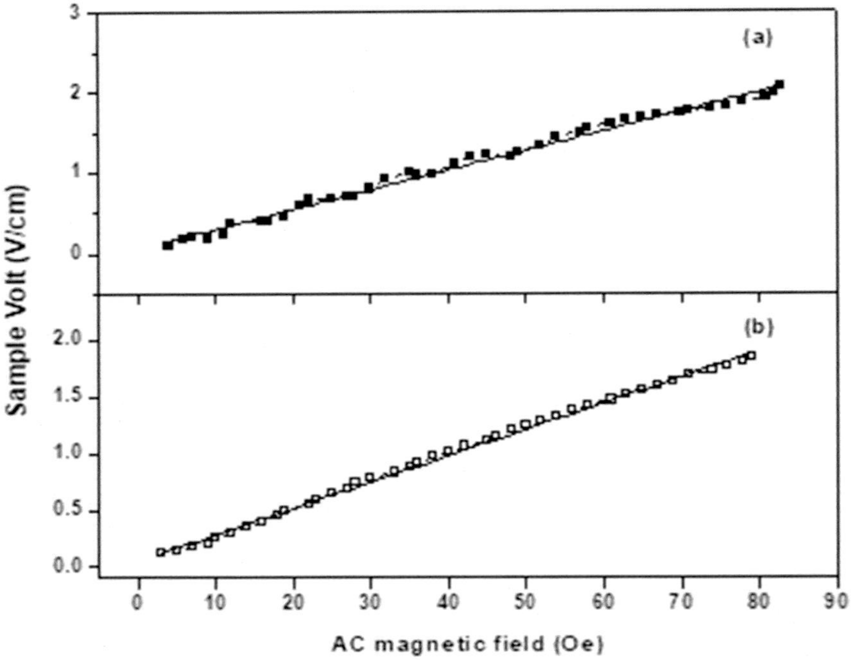

Figure 15.2
Variations of induced electric field with applied *ac* magnetic field in a *dc* bias field of 1000 Oe for *x NKLN-(1-x) NFO* samples with (A) $x = 0.85$ and (B) $x = 0.80$. ME coupling coefficients have been determined from these.

It is evident from Table 15.1 that we get a higher ME coupling coefficient for the sample with $x = 0.85$, which is the sample with lower ferrite content. However, samples with still lower ferrite content fail to meet the desired physical properties.

The values of ME coefficient α obtained for CFO based samples are lower than the corresponding values for NFO based samples. Though the magnetostriction coefficient of NFO is lower than CFO [50], we measure a lower ME coefficient for CFO based sample. ME property of a composite is controlled by the magnetic properties of the magnetic phase. CFO is a magnetically harder material than NFO and there is a limit for domain wall rotation compared to softer magnetic materials like NFO. This may the reason for the lower ME response of CFO based composite than NFO based one. From this one can conclude that magnetically softer ferrite phase provides better ME coupling in multiphase composites.

7. Conclusions

Four different molar concentrations of *x NKLN−(1-x) MFO* nanocomposites have been synthesized and their ME properties investigated. The resonance peaks appearing in the dielectric properties provide evidence for the evolution of piezoelectric modes in polled samples. The frequency response of *AC* conductivity indicate that conduction is mediated by small polarons in the sample. Piezoelectric and ME properties indicate that samples with $x = 0.85$ show higher values for both. The ME coefficients of CFO based samples are lower than the corresponding values for NFO based samples, and this is interpreted as due to the magnetic hardness of CFO compared to NFO.

References

[1] T. Lottermoser, T. Lonkai, U. Amann, D. Hohlwein, J. Ihringer, M. Fiebig, Magnetic phase control by an electric field, Nature 430 (2004) 541−544.

[2] T. Kimura, T. Goto, H. Shintani, K. Ishizaka, T. Arima, Y. Tokura, Magnetic control of ferroelectric polarization, Nature 426 (2003) 55−58.

[3] W. Eerenstein, N.D. Mathur, J.F. Scott, Multiferroic and magneto electric materials, Nature 442 (2006) 759−765.

[4] L.W. Martin, S. P Crane, Y.-H. Chu, M. B Holcomb, M. Gajek, M. Huijben, C.-H. Yang, N. Balke, R. Ramesh, Multiferroics and magneto electrics: thin films and nanostructures, J. Phys. Condens. Matter 20 (2008) 1−13.

[5] J.P. Velev, S.S. Jaswal, E.Y. Tsymbal, Multi-ferroic and magneto electric materials and interfaces, Phil. Trans. R. Soc. A 369 (2011) 3069−3097.

[6] J. Ryu, S. Priya, K. Uchiono, H.-E. Kim, Magneto electric effect in composites of magnetostrictive and piezoelectric materials, J. Electroceram. 8 (2002) 107−119.

[7] N.A. Hill, Why are there so few magnetic ferroelectrics? J. Phys. Chem. B 104 (2000) 6694−6709.

[8] M. Fiebig, Revival of the magneto electric effect, J. Phys. D Appl. Phys. 38 (2005) 123−152.

[9] G. Catalan, J.F. Scott, Physics and applications of bismuth ferrite, Adv. Mater. 21 (2009) 2463−2485.

[10] M.K. Singh, R.S. Katiyar, J.F. Scott, New magnetic phase transitions in $BiFeO_3$, J. Phys. Condens. Matter 20 (2008) 1−4.

[11] Y. Wang, J. Hu, Y. Lin, C.W. Nan, Multiferroic magneto electric composite nanostructures, NPG Asia Mater. 2 (2010) 61–68.
[12] H. Palneedi, V. Annapureddy, S. Priya, J. Ryu, Status and perspectives of multiferroic magneto electric composite materials and applications, Actuators 5 (2016) 1–31.
[13] J. Van Suchtelen, Product properties: a new application of composite materials, Philips Res. Rep. 27 (1972) 28–37.
[14] V. Wadhawan, Introduction to Ferroic Materials, Gordon and Breach Science Publishers, India, 2000, pp. 517–524.
[15] C.W. Nan, Magneto electric effect in composites of piezoelectric and piezomagnetic phases, Phys. Rev. B 50 (1994) 6082–6088.
[16] B. Singh, S. Chauhan, G. Verma, Nanocomposites – a review, J. Chem. Cheml. Sci. 5 (2015) 506–510.
[17] K. Sadhana, S.R. Murthy, S. Jie, Y. Xie, Y. Liu, Q. Zhan, R.W. Li, Magnetic field induced polarization and magneto electric effect of $Ba_{0.8}Ca_{0.2}TiO_3$-$Ni_{0.2}Cu_{0.3}Zn_{0.5}Fe_2O_4$ nanomultiferroic, J. Appl. Phys. 113 (2013) 1–2.
[18] C.C. Okpala, The benefits and applications of nanocomposites, Int. J. Adv. Eng. Technol. 5 (2014) 12–18.
[19] T. Shet, R. Bhimireddi, K.B.R. Varma, Grain size-dependent dielectric, piezoelectric and ferroelectric properties of $Sr_2Bi_4Ti_5O_{18}$ ceramics, J. Mater. Sci. 51 (2016) 9253–9266.
[20] M. George, A.M. John, S.S. Nair, P.A. Joy, M.R. Anantharaman, Finite size effects on the structural and magnetic properties of sol–gel synthesized $NiFe_2O_4$ powders, J. Magn. Magn Mater. 302 (2006) 190–195.
[21] Y. Guo, Y. Liu, J. Wang, R.L. Withers, H. Chen, L. Jin, P. Smith, Giant magnetodielectric effect in 0-3 $Ni_{0.5}Zn_{0.5}Fe_2O_4$-poly(vinylidene-fluoride) nanocomposite films, J. Phys. Chem. C 114 (2010) 13861–13866.
[22] M.I. Bichurin, V.M. Petrov, Modeling of magneto electric interaction in magnetostrictive-piezoelectric composites, Adv. Condens. Matter Phys. (2012) 1–12, 2011.
[23] K.F. Wang, J.M. Liu, Z.F. Ren, Multiferroicity: the coupling between magnetic and polarization orders, Adv. Phys. 58 (2009) 321–448.
[24] S.V. Suryanarayana, Magneto electric interaction phenomena in materials, Bull. Mater. Sci. 17 (1994) 1259–1270.
[25] C.A.F. Vaz, J. Hoffman, C.H. Ahn, R. Ramesh, Magneto electric coupling effects in multiferroic complex oxide composite structures, Adv. Mater. 22 (2010) 2900–2918.
[26] J. Zhai, N. Cai, Z. Shi, Y. Lin, C.W. Nan, Magnetic-dielectric properties of $NiFe_2O_4$/PZT particulate composites, J. Phys. D Appl. Phys. 37 (2004) 823–827.
[27] D.R. Patil, A.D. Sheikh, C.A. Watve, B.K. Chougule, Magneto electric properties of ME particulate composites, J. Mater. Sci. 43 (2008) 2708–2712.
[28] J. Van den Boomgaard, A.M.J.G. Van Run, J. Van Suchetelene, Magnetoelectricity in piezoelectric-magnetostrictive composites, Ferroelectrics 10 (1976) 295–298.
[29] J. van den Boomgaard, R.A.J. Born, A sintered magneto electric composite material $BaTiO_3$-Ni (Co, Mn) Fe_2O_4, J. Mater. Sci. 13 (1978) 1538–1548.
[30] J. van den Boomgaard, D.R. Terrell, R.A.J. Born, H.F.J.I. Giller, An in situ grown eutectic magneto electric composite material, J. Mater. Sci. 9 (1974) 1705–1709.
[31] J. Ryu, A.V. Carazo, K. Uchino, H.-E. Kim, Piezoelectric and magneto electric properties of lead Zirconate Titanate/Ni-Ferrite particulate composites, J. Electroceram. 7 (2001) 17–24.
[32] G. Harshe, J.P. Dougherty, R.E. Newnham, Theoretical modeling of 3-0, 0-3 magneto electric composites, Int. J. Appl. Electromagn. Mater. 4 (1993) 161–172.
[33] M.A. Rahman, M.A. Gafur, A.K.M. Akther Hossain, Structural, magnetic and transport properties of magneto electric composites, J. Magn. Magn Mater. 345 (2013) 89–95.
[34] M.M. Haque, M. Huq, M.A. Hakim, Influence of CuO and sintering temperature on the microstructure and magnetic properties of Mg-Cu-Zn ferrites, J. Magn. Mater. 320 (2008) 2792–2799.

[35] P.K. Panda, Review: environmental friendly lead-free piezoelectric materials, J. Mater. Sci. 44 (2009) 5049–5062.
[36] P.K. Panda, Environmental friendly lead-free piezoelectric materials, J. Mater. Sci. 44 (2009) 5049–5062.
[37] Y. Guo, K. Kakimoto, H. Ohsato, Phase transitional behavior and piezoelectric properties of $(Na_{0.5}K_{0.5})NbO_3-LiNbO_3$ ceramics, Appl. Phys. Lett. 85 (2004) 4121–4123.
[38] B. Sarkar, B. Dalal, V.D. Ashok, K. Chakrabarti, A. Mitra, S.K. De, Magnetic properties of mixed spinel $BaTiO_3$-$NiFe_2O_4$ composites, J. Appl. Phys. 115 (2014) 123908 1–12390810.
[39] K. Maaz, S. Karim, A. Mashiatullah, J. Liu, M.D. Hou, Y.M. Sun, J.L. Duan, H.J. Yao, D. Mo, Y.F. Chen, Structural analysis of nickel doped cobalt ferrite nanoparticles prepared by co-precipitation route, Physica B 404 (2009) 3947–3951.
[40] O. Caltun, H. Chiriac, N. Lupu, I. Dumitru, B.P. Rao, High magnetostrictive doped cobalt ferrite, J. Opt elect. and Adv. Mate. 9 (2007) 1158–1160.
[41] M. Mahesh Kumar, A. Srinivas, S.V. Suryanarayana, G.S. Kumar, T. Bhimasankaram, An experimental setup for dynamic measurement of magneto electric effect, Bull. Mater. Sci. 21 (1998) 251–255.
[42] S.S. Chougule, B.K. Chougule, Studies on electrical properties and the magneto electric effect on ferroelectric-rich $(x)Ni_{0.8}Zn_{0.2}Fe_2O_4+(1-x)$ PZT ME composites, Smart Mater. Struct. 16 (2007) 493–498.
[43] J. Kulawik, D. Szwagierczak, P. Guzdek, Magnetic, magneto electric and dielectric behavior of $CoFe_2O_4-Pb(Fe_{1/2}Nb_{1/2})O_3$ particulate and layered composites, J. Magnetism & Mag. Mater. 324 (2012) 3052–3057.
[44] R. Grossinger, G.V. Duong, R.S. Turtellia, The physics of magneto electric composites, J. Magnetism & Mag. Mater. 320 (2008) 1972–1977.
[45] J.P. Zhou, L. Lv, Q. Liu, Y.X. Zhang, P. Liu, Hydrothermal synthesis and properties of $NiFe_2O_4$@$BaTiO_3$ composites with well-matched interface, Sci. Technol. Adv. Mater. 13 (2012) 045001 1–04500112.
[46] C.E. Cimoga, I. Dumitru, L. Mitoseriu, C. Galassi, A.R. Iordan, M. Airimioaei, M.N. Palamaru, Magneto electric ceramic composites with double-resonant permittivity and permeability in GHz range: a route towards isotropic metamaterials, Scripta Mater. 62 (2010) 610–612.
[47] B.K. Bammannavar, L.R. Naik, R.K. Kotnala, Study of electrical properties and the magneto electric effect in $Ni_{0.2}Co_{0.8}Fe_2O_4 + PbZr_{0.8}Ti_{0.2}O_3$ particulate composites, Smart Mater. Struct. 20 (2011) 045005 1–7.
[48] B.K. Bammannavar, L.R. Naik, Electrical properties and magneto electric effect in $(x)Ni_{0.5}Zn_{0.5}Fe_2O_4+(1-x)$ BPZT composites, Smart Mater. Struct. 18 (2009) 085013 1–9.
[49] P. Guzdek, M. Sikora, L. Góra, C.Z. Kapusta, Magnetic and magneto electric properties of nickel ferrite–lead iron niobate relaxor composites, J. Eur. Ceram. Soc. 32 (2012) 2007–2011.
[50] M. Atif, M. Nadeem, R. Grossinger, R.S. Turtelli, Studies on the magnetic, magnetostrictive and electrical properties of sol–gel synthesized Zn doped nickel ferrite, J. Alloys Compd. 509 (2011) 5720–5724.

SECTION D

Green synthesis of nanomaterials

CHAPTER 16

Green synthesis of MN (M= Fe, Ni − N= Co) alloy nanoparticles: characterization and application

Amirsadegh Rezazadeh Nochehdehi[1], Neerish Revaprasadu[2], Sabu Thomas[3,4], Fulufhelo Nemavhola[1]

[1]Biomechanics Research Group (BRG), Mechanical and Industrial Engineering Department (DMIE), School of Engineering, College of Science, Engineering and Technology (CSET), University of South Africa, Pretoria, South Africa; [2]SARCHI Chair in Nanotechnology, Department of Chemistry, Faculty of Science and Agriculture, University of Zululand, KwaZulu-Natal, Zululand, South Africa; [3]International and Inter University Centre for Nanoscience and Nanotechnology, Mahatma Gandhi University, Kottayam, Kerala, India; [4]School of Energy Materials, Mahatma Gandhi University, Kottayam, Kerala, India

Chapter Outline

Design, Fabrication, and Characterization of Multifunctional Nanomaterials. https://doi.org/10.1016/B978-0-12-820558-7.00005-4

1. Introduction

Metallic alloy nanoparticles will initiate important and remarkable progress in nanotechnology due to their unique chemical and physical properties [1]. Magnetometallic nanoalloys represent a particular interesting class of nanomaterial, which will be used in various applications from industry to medicine [2,3]. In fact, nanoalloying technology is a purely interdisciplinary knowledge, which is related to material science and engineering, medical science, pharmacy and drug design, veterinary medicine, biology, applied physics, molecular chemistry, and even mechanical engineering, electrical engineering, and chemical engineering [4–6]. Analysts believe that nanotechnology, biotechnology, and information technology are three realms that form the third industrial revolution [7–10]. Generally, this has expanded the frontiers of knowledge and helps scientists to solve the technological problems which has led to the emergence of new applications in various sciences.

One of the most commonly used fields of nanotechnology, which has attracted attention by many researchers around the world, is medicinal application of nanotechnology [11]. In fact, nanomedicine is a multifunctional applications of nanotechnology products in medicine fields [12]. The purpose of medical nanotechnology to the prevention, diagnosis, care, and treatment of diseases defined. In this regard, magnetic nanoparticles have been developed by many researchers around the world for various applications in the field of medicine [13,14]. Scientists are developing smart nanocarriers for drug delivery to cells, which can circulate in the blood vessels, pass through the immune system, sticking to defective cells and his deadly drug load without side effects which is available in treatments such as chemotherapy [15].

In this study, special applications of magnetic nanoparticles have been considered [16]. We selected the iron, nickel, and cobalt alloy nanoparticles according to the Slater-Pauling curve, which has considered two main factors like high saturation magnetization and their good biocompatibility [17–20]. In addition, the synthesis procedure is the most important factor to produce a safe and biocompatible nanomaterial [21–23]. There are various methods to synthesize the alloy nanoparticles recognized by various scientists around the world. Coprecipitation route is the most widely used process to synthesis magnetic nanocrystalline particles as a simple method with controllable conditions. In fact, control of size and shape of nanoparticles are the most important factors, which can be controlled in this process [24]. Microemulsion technique is generally capable method to control the particle size distribution with very fine, shape-controlled, and highly crystalline, as compared to other synthesis procedure [25]. Contrary to the conventional coprecipitation method, nanoparticles synthesis by microemulsion technique, they precipitate inside the micelles [26]. Sol-gel method is the wet chemical technique known as a cheapest and lowest reaction temperature for synthesizing metal-oxide nanoparticles [27–29].

The materials produced by this method can be used in optical, electronic, and energy applications [30]. However, this method has many advantages, including synthesis at low temperatures, high purity and efficiency, synthesis of porous material, and high chemical reactivity of precursors [31]. Chemical reduction is one of the most commonly used method to synthesis of stable metallic nanoparticles with a large scale around 300–400 nm [32]. Different reduction agent like glucose, sodium borohydried, sodium citrate, ethanol, and ethylene glycol can be used in this process for stabilize and optimize the particle size and distribution [33,34].

Today, the synthesis of nanoparticles with the use of microwave heating is one of the safest, fastest, easiest, and most affordable methods for the synthesis of various nanomaterials with a specific size and structure, and it has attracted the attention of scientists [35]. This technique has various advantages such as uniform heat distribution, superheating, and selective heating, increasing the speed of reactions, and also subsequently reducing the time and energy needed to perform the synthesis of nanomaterials. For this reason this method is increasingly used in synthesis of different nanoparticles such as gold-silver-palladium nanostructures, metal-oxide, selenide and sulfide-base nanoparticles, and also porous nanostructures [36].

Chemical vapor deposition has been developed as an effective way to build a wide range of components and products in several industrial sectors, including semiconductors, the ceramic industry, and so on. This method is most often used in the synthesis of hard materials or coatings on products [37]. In addition, it has significant advantages, such as the ability to precipitate a wide range of materials, sediment deposition materials with very high purity, and high deposition rate. There are also disadvantages like safety issues related to explosion of the precursors, it has little safety. Extremely high costs are needed to produce high purity layers [38]. In recent decades, many methods for the synthesis of nanoparticles have emerged. One of the most important methods for the synthesis of uniform nanostructure materials, which has been able to generalize the synthesis of nanoparticles, is the pulsed laser ablation deposition technique.

To recognize the impact of parameters in this process on the properties of the synthesized nanomaterials, research should be done with goal to achieve a narrow distribution size of nanoparticles by scientists around the world [39]. After various study about synthesis procedures, we selected the polyol method, which is a liquid-phase synthesis method in multibasic alcohols with a high boiling point that has led to produce nanoparticles [40].

Polyol has a high degree of biodegradability and biocompatibility as far as there are some productions which has approved by US Food and Drug Administration (FDA) and others are known as green solvents. In generally, polyol provide a variety of advantages for the synthesis of nanoparticles [41]. The useful features of polyol can be related to polyol's degradation properties that allows the synthesis of metal nanoparticles directly and the easy removal of polyols from the surface of the nanoparticles after completion of synthesis, which has led to their higher purity, the ability to scale, and usability in continuous flow synthesis [41,42]. Ultimately, we synthesized the FeCo and NiCo nanoalloys then perform their characterization like SEM, TEM, XRD, and VSM techniques.

2. Experimental procedure

2.1 Substance materials

Pure Metallic salts like iron chloride ($FeCl_2.4H_2O$) and cobalt acetate ($CoAc_2.4H_2O$), nickel acetate ($(Ch_3COO)_2Ni.4H_2O$) polyvinyl pyrrolidone (PVP) as a surfactant, NaOH, polyethylene glycol (PEG), ethylene glycol (EG), ethanol, and acetone, which is produced by MERCK and also Sigma—Aldrich Co.

2.2 Synthesize procedure of FeCo magnetic nanoparticles

According to review the green process of synthesize various nanoparticles, the reduction polyline method was selected to synthesize iron-cobalt (FeCo) magnetic nanoparticles. Specified of weight ratio of an iron and cobalt salt (60:40) used in this procedure. At the first step, 30 mmol of $FeCl_2.4H_2O$ dissolved in 50 mL of polyethylene glycol (PEG) through using magnetic stirrer at the first step. After complete dissolution of iron salts, the color of solution is changed to green. Then, 6 mmol $CoAc_2.4H_2O$ added to the solution slowly. After complete dissolution of cobalt salts, the solution color is changed to purple.

The second step of synthesize the FeCo nanoparticles in order to achieve the PH at the range of 10—11, was added 0.5 g of PVP solid white powder to the compound, After complete dissolution. Finally, 5 g of sodium hydroxide (NaOH) was added to the compound after complete grinding. At this time, the solution color is changed to black. Thus, the solvent temperature increased to 453 K. The reaction vessel was refluxed for 2 h while stirring under nitrogen gas. Finally the suspension was cooled down to room temperature. The products was recovered from polyol and washed with ethanol and acetone several times. Then, the resulting powder was dried at 373 K for 1 h. Ultimately, the morphology, compositions, structure, and magnetic properties were analyzed.

2.3 Synthesize procedure of NiCo magnetic nanoparticles

According to review the green process of synthesize various nanoparticles, the reduction polyline method was selected to synthesize nickel-cobalt (FeCo) magnetic alloy nanoparticles. Specified of weight ratio of an iron and cobalt salt (50:50) used in this procedure. At the first step, 3.5 gr of $NiAc_2.4H_2O$ dissolved in 35 mL of EG through using magnetic stirrer, at the first step. After complete dissolution of nickel salts, the color of solution is changed to green. Then, 6 3.5 gr $CoAc_2.4H_2O$ added to the solution slowly. After complete dissolution of cobalt salts, the solution color is changed to purple.

The second step of synthesize the FeCo nanoparticles to achieve the PH at the range of 10−11, was added 1.1 g of PVP solid white powder to the compound, After complete dissolution. At this time, the solution color is changed to black one. Thus, the solvent temperature increased to 413 K. The reaction vessel was refluxed for 2 h while stirring under nitrogen gas. Finally the suspension was cooled down to room temperature. The products was recovered from polyol and washed with ethanol and acetone several times. Then, the resulting powder was dried at 400 K for 1.5 h. Ultimately, the morphology, compositions, structure, and magnetic properties were analyzed.

3. Results and discussion

3.1 Structure analysis

The structure of an MN alloy nanoparticles were performed by X-ray diffraction (XRD) technique. Fig. 16.1 Shows the XRD pattern of FeCo and NiCo alloy nanoparticles which were synthesized by the green method called "polyol process." According to the XRD spectrum presented of the FeCo nanoparticles at angles of $2\theta = 30.29$, 35.73, 45.98, 54.75, 57.49, 63.5, 65.25, 74.42, 82.66, respectively, related to the crystal plates 200, 311, 400,

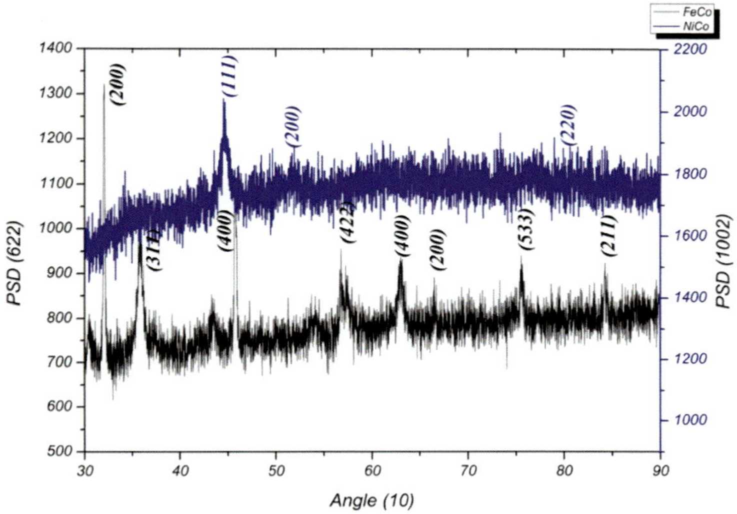

Figure 16.1
XRD pattern of FeCo (Black pattern) and NiCo (Blue pattern) nanoparticles synthesized by polyol reduction process.

511, 422, 400, 200, 533, and 211. It looks like it conforms to the standard model CJCPDS 98-006-0802. According to the XRD spectrum presented of the NiCo nanoparticles at angles of $2\theta = 44.6, 51.6, 77.7$ respectively related to the crystal plates 111, 200, 222 which were represented face centered cubic structure of NiCo nanoalloy. Therefore, it is determined that the black deposit of iron-cobalt and nickel-cobalt magnetic nanoparticles is obtained. The XRD pattern confirms that the magnetic nanoparticles are FeCo pure and have a spinal structure.

3.2 Microstructure analysis

The morphology of MN alloy nanoparticles were determined by transmitted electron microscopy (TEM). The TEM images shown in Fig. 16.2A and B indicated that the FeCo nanoparticles are composed of cubic-shape nanoparticles well dispersed with particle size under 10 nm. It shows nanoparticles rather in good crystallinity. In addition, Fig. 16.2C and D indicated that the NiCo alloy nanoparticles are composed of spherical shape which are in an agglomerated form with diameter around 20 nm.

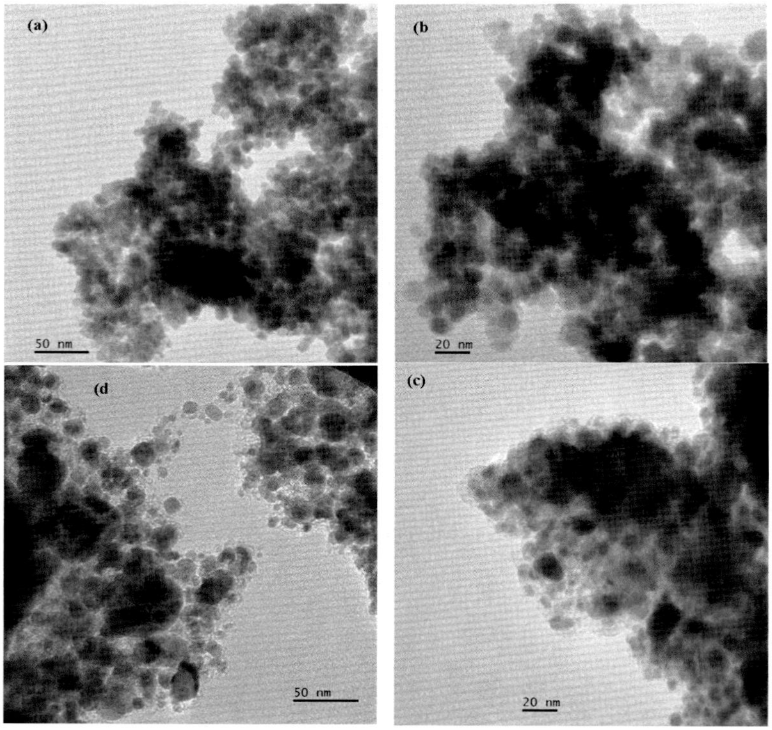

Figure 16.2
Transmitted Electron Microscopy (TEM): (A, B) FeCO alloy NPs; (C, D) NiCo alloy NPs which were synthesized by polyol method.

3.3 Elemental analysis

The elemental identification of an iron-cobalt nanoparticles determined by scanning electron microscopy (SEM) and an energy dispersive X-ray analyzer (EDX). The SEM images shown in Fig. 16.3 confirmed that the FeCo nanoparticles has a cubical shape and also the particles size is under 10 nm.

Figure 16.3

Scanning electron microscopy (SEM) of FeCO nanoparticles synthesized by polyol reduction method.

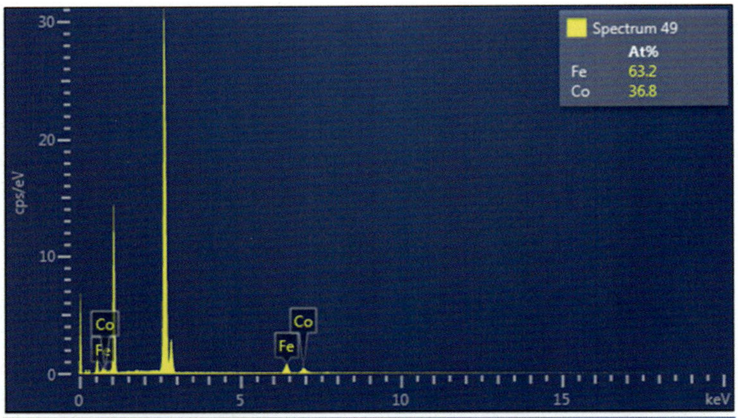

Element	Line Type	Apparent Concentration	k Ratio	Wt%	Wt% Sigma	Atomic %	Standard Label	Factory Standard
Fe	K series	5.18	0.05183	61.99	1.78	63.25	Fe	Yes
Co	K series	3.14	0.03141	38.01	1.78	36.75	Co	Yes
Total				100.00		100.00		

Figure 16.4

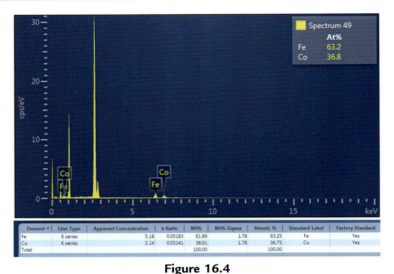

Element analysis of FeCo nanoparticles synthesized by polyol method by EDX technique.

In addition, the result of an EDX shown in Fig. 16.4 indicated that, the pure FeCo nanoparticles synthesized in this procedure. The FeCo nanoparticles completely formed in presence of 63.2% and 36.8% atomic ratio of iron and cobalt.

3.4 Magnetic analysis

Magnetorheological studies were performed by vibrating samples magnetometer (VSM) technique. The hysteresis loop of FeCo and NiCo alloy nanoparticles at 78.264 Oe field were shown in Fig. 16.5. The saturation magnetization (Ms), coercivity (Hc) of FeCo nanoparticles were calculated as 39.856 emu/g and 500.96 G, respectively. The magnetic saturation (Ms) and Coercivity (Hc) of NiCo nanoparticles were also calculated as 29.503 emu/g and 97.692 G, respectively. Both the particles show a soft ferromagnetic behavior with negligible coercivity and unsaturated magnetization values typical of superparamagnetic behavior near room temperature. This is expected because the magnetic nanostructures are by themselves ferromagnetic, but are so dispersed that their overall magnetization is weak. This may be due to the additional coating of PVP polymer which prevents the agglomeration of the particles.

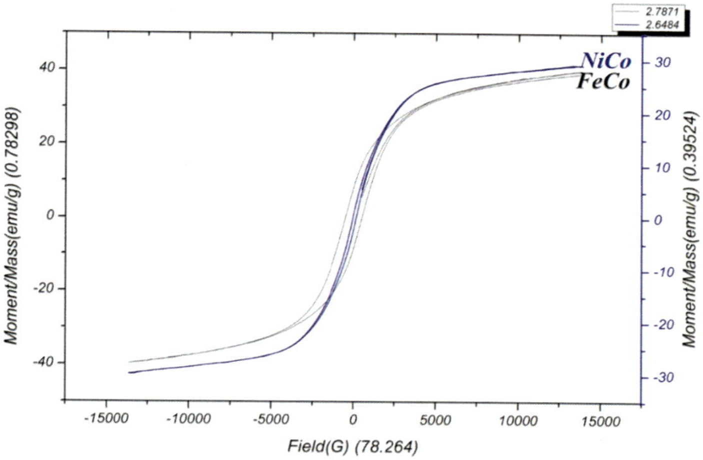

Figure 16.5
Magnetic saturation and hysteresis curve measured at room temperature for FeCo and NiCo alloy nanoparticles prepared using polyol process.

4. Conclusions

The iron-cobalt (FeCo) and nickel—cobalt (NiCo) nanoalloy were synthesized by the cheap and economical, easy, green, and environmentally friendly process called polyol method. According to the spectrum obtained from the XRD pattern showed that use of the FeCo and NiCo nanoparticles synthesized successfully and has a spinal structure. The morphology of FeCo and NiCo nanoparticles were cubic and spherical shape and the size was under 10 and 20 nm, respectively. The EDX results confirmed the pure FeCo and NiCo nanoparticles was formed in this procedure. Magnetization against field of these NPs confirmed the soft magnetic nature of these samples. It shown that magnetization of NiCo is lower than FeCo, it is completely according to Slater-Pauling Curve. As a result, because the FeCo and NiCo nanoparticles can show unique and specified properties, there will be various applications from industry to medicine. For example, it can be used as a microwave absorbance, or date storage, water purification, or catalysts. Recently, has been developed various application of magnetic nanoparticles on medicine. In the most important case, it can be used in medicinal applications like drug or gene delivery, as a contrast agent in MRI or heat therapy (hyperthermia) of various diseases.

In the next step, we can suggest that to evaluate the effect of reaction time and temperature, amount of surfactant, and solvent in size, shape, and magnetic saturation of dual phase and ternary alloy nanoparticles. In addition, we suggest that to measure the biocompatibility of the FeCo and NiCo synthesized nanoparticles according to the standard protocol ASTM756 and also ISO 10993.

Acknowledgments

Authors are grateful to Dr. Thomas Moyo and Mr. Patrick Itegbeyogene, magnetic analyzer laboratory, Durban University of Technology (DUT), South Africa, for permitting to carry out this experimental analysis.

References

[1] D. Alloyeau, C. Mottet, C. Ricolleau, Nanoalloys: Synthesis, Structure and Properties, Springer-Verlag London, 2012.
[2] K.H. Bae, H.J. Chung, T.G. Park, Nanomaterial for cancer therapy and imaging, Mol. Cell. 31 (4) (2011) 295–302.
[3] G. Oberdörster, et al., Principles for characterizing the potential human health effects from exposure to nanomaterial: elements of a screening strategy, Particle & Fiber Toxicology 2 (1) (2005) 1.
[4] S. Nie, et al., Nanotechnology applications in cancer, Annu. Rev. Biomed. Eng. 9 (2007) 257–288.
[5] J. Weiss, P. Takhistov, D.J. McClements, Functional materials in food nanotechnology, J. Food Sci. 71 (9) (2006) R107–R116.
[6] Z. Karimi, Nano-magnetic particles used in biomedicine: core and coating materials, J. Mater. Sci. Eng. C 33 (2013) 2465–2475.
[7] A. Jacob, K. Chakravarthy, Engineering magnetic nanoparticles for thermo-ablation and drug delivery in neurological cancers, Cureus 6 (4) (April 6, 2014).
[8] G. Baronzio, D. Hager, Hyperthermia in Cancer Treatment, Springer, Landes Bioscience, 2006.
[9] B. Minev, Cancer Management in Man: Chemotherapy, Biological Therapy, Hyperthermia and Supporting Measures, Springer, 2011.
[10] P. Del Pino, B. Pelaz, Hyperthermia Using Inorganic Nanoparticles, Elsevier: Science and Technology, 2012.
[11] S. Gubin, Magnetic Nanoparticles, Wiley-VCH, 2007.
[12] C. Binns, Medical applications of magnetic nanoparticles, J. Front. Nanosci. 6 (2014) 217–258.
[13] L.A. Thomas, Nanoparticle Synthesis for Magnetic Hyperthermia, PhD Thesis, Chemistry Department, University College London, 2010.
[14] A.R. Nochehdehi, S. Thomas, N. Revaprasadu, N. Kalarikkal, Biomedical applications of iron-and cobalt-based biomagnetic alloy nanoparticles, Nanosci. Med. 1 (2020) 333–371.
[15] V. Virendra Mohite, Self-Controlled Magnetic Hyperthermia. Master of Science Dissertation, Department of Mechanical Engineering, The Florida State University, 2004.
[16] E.P. Wohlfarth, Handbook of Magnetic Materials and Nanotechnology, North-Holland Publishing company, 1980.
[17] J.M.D. Coey, Magnetism and Magnetic Materials, 2009. Cambridge.
[18] M. Zamanpour, Cobalt-based Magnetic Nanoparticles: Design, Synthesis and Characterization, PhD Thesis, College of Engineering, Northeastern University, 2014.
[19] O.D. Jayakumar, H.G. Salunke, A.K. Tyagi, Synthesis and characterization of stoichiometric NiCo nano particles dispersible in both aqueous and non-aqueous media, J. Solid State Commun. 149 (2009) 1769–1771.
[20] I. Arief, P.K. Mukhopadhyay, Synthesis of dimorphic MR fluid containing NiCo nanoflowers by the polymer assisted polyol method and study of its magnetorheological properties, J. Phys. (Paris) B 448 (2014) 73–76.

[21] J.M. Soares, O.L.A. Conceicao, F.L.A. Machado, Magnetic coupling in $CoFe_2O_4$/FeCo − FeO core-shell nanoparticles, J. Magn. Magn Mater. 374 (2014) 192−196.
[22] T.L. Kline, Y.H. Xu, Y. Jing, J.P. Wang, Biocompatible high-moment FeCo-Au magnetic nanoparticles for magnetic hyperthermia treatment optimization, J. Magn. Magn Mater. 321 (10) (2011) 1525−1528.
[23] A.H. Lu, E.L. Salabas, F. Schüth, Magnetic nanoparticles: synthesis, protection, functionalization, and application, Angew. Chem. Int. Ed. 46 (8) (2007) 1222−1244.
[24] G.S. Chaubey, C. Barcena, P. Narayan, C. Rong, J. Gao, S. Sun, J.P. Liu, Synthesis and stabilization of FeCo nanoparticles, J. Am. Chem. Soc. 129 (23) (2007) 7214−7215.
[25] V. Tzitzios, G. Basina, D. Niarchos, W. Li, G. Hadjipanayis, Synthesis of air stable FeCo nanoparticles, J. Appl. Phys. 109 (2011) 07A313.
[26] A. Shokuhfar, S.S.S. Afghahi, Size controlled synthesis of FeCo alloy nanoparticles and study of the particle size and distribution effects on magnetic properties", J. Adv. Mater. Sci. & Eng. (2014) 1−10, https://doi.org/10.1155/2014/295390.
[27] S.J. Shin, Y.H. Kim, C.W. Kim, H.G. Cha, Y.J. Kim, Y.S. Kang, Preparation of magnetic FeCo nanoparticles by co-precipitation route, J. Curr. Appl. Phys. 7 (4) (2011) 404−408.
[28] I. Sharifi, H. Shokrollahi, M.M. Doroodmand, R. Safi, Magnetic and structural studies on $CoFe_2O$ nanoparticles synthesized by co-precipitation, normal micelles and reverse micelles methods, J. Magn. Magn Mater. 324 (10) (2012) 1854−1861.
[29] T.P. Braga, D. Felix Dias, M. Falcãode Sousa, J. MariaSoares, J. MarcosSasaki, Synthesis of air stable FeCo alloy nanocrystallite by proteic sol−gel method using a rotary oven, J. Alloys Compd. 622 (2015) 408−417.
[30] L. Simon Lobo, S. Kalainathan, A. RubanKumar, Investigation of electrical studies of spinel $FeCo_2O_4$ synthesized by sol−gel method, J. Superlattices & Microstruc. 88 (2015) 116−126.
[31] P. Nautiyal, M. Motin Seikh, O.I. Lebedev, A.K. Kundu, Sol−gel synthesis of Fe−Co nanoparticles and magnetization study, J. Magn. Magn Mater. 377 (2015) 402−405.
[32] A.R. Nochehdehi, S. Thomas, M. Sadri, S.S.S. Afghahi, S.M. Hadavi, Iron oxide biomagnetic nanoparticles (IO-BMNPs); synthesis, characterization and biomedical application−a review, J. Nanomed. Nanotechnol. 8 (1) (2017) 1−9.
[33] B. Kandapallil, R.E. Colborn, P.J. Bonitatibus, F. Johnson, Synthesis of high magnetization Fe and FeCo nanoparticles by high temperature chemical reduction, J. Magn. Magn Mater. 378 (2015) 535−538.
[34] K. Chokprasombat, P. Harding, S. Pinitsoontorn, S. Maensiri, Morphological alteration and exceptional magnetic properties of air-stable FeCo nanocubes prepared by a chemical reduction method, J. Magn. Magn Mater. 369 (2014) 228−233.
[35] Z. Qi, G. Jiang, J.M. Riggs, Synthesis of magnetically exchange coupled $SrFe_{12}O_{19}$/FeCo core/shell particles through microwave-polyol process, J. Mater. Lett. 163 (2016) 270−273.
[36] K.S. Steen Jensen, H. Sun, R.M.L. Werchmeister, K. Mølhave, J. Zhang, Microwave synthesis of metal nanocatalysts for the electrochemical oxidation of small biomolecules, J. Curr. Opin. Electrochem. 4 (1) (2017) 124−132, https://doi.org/10.1016/j.coelec.2017.08.014.
[37] S. Hosseininasab, N. Faucheux, G. Soucy, J.R. Tavares, Full range of wettability through surface modification of single-wall carbon nanotubes by photo-initiated chemical vapour deposition, J. Chem. Eng. J. 325 (2017) 101−113.
[38] P.L. Ong, S. Mahmood, T. Zhang, J.J. Lin, R.V. Ramanujan, P. Lee, R.S. Rawat, Synthesis of FeCo nanoparticles by pulsed laser deposition in a diffusion cloud chamber, J. Appl. Surf. Sci. 254 (7) (2008) 1909−1914.
[39] Happy, S.R. Mohanty, P. Lee, T.L. Tan, S.V. Springham, A. Patran, R.V. Ramanujan, R.S. Rawat, Effect of deposition parameters on morphology and size of FeCo nanoparticles synthesized by pulsed laser ablation deposition, J. Appl. Surf. Sci. 252 (8) (2006) 2806−2816.

[40] D. Thapa, et al., Properties of magnetite nanoparticles synthesized through a novel chemical route, Mater. Lett. 58 (21) (2004) 2692–2694.
[41] C.P. Poole, I.J. Owens, Introduction to Nanotechnology, John Wiley & Sons, Inc., NY. USA, 2003.
[42] A. Rezazadeh Nochehdehi, S. Thomas, M. Sadri, S.M. Hadavi, Y. Grohens, Fe, Co based bio-magnetic nanoparticles (BMNPs); synthesis, characterization and biomedical application, in: Sabu Thomas (Ed.), Chapter 6, Recent Trends in Nanomedicine and Tissue Engineering, 1st 8, River Publisher, 2017.

CHAPTER 17

Green synthesis of nanomaterials for photocatalytic application

S. Padhiari[1], M. Mishra[2], Garudadhwaj Hota[1]
[1]Department of Chemistry, National Institute of Technology, Rourkela, Odisha, India; [2]Department of Life Science, National Institute of Technology, Rourkela, Odisha, India

Chapter Outline

1. Introduction

Environmental pollution and energy crisis are the most two important global topics that remain a great challenge for modern-day researchers [1]. Since the extraordinary work of photoelectrochemical water splitting on TiO_2 electrode by Fujishima and Honda, the use of semiconducting nanomaterials in the field of photocatalysis for the production of clean hydrogen by solar water splitting, reduction of CO_2 into useful hydrocarbon fuels, and degradation of water and air pollutants over the surface of the catalyst has attracted tremendous research attention [2]. Various metal oxides (e.g., ZnO, Fe_2O_3, $BiVO_4$, WO_3, SnO_2, TiO_2) [3–8], metal sulfides (e.g., Cu_xS, CdS, SnS) [9–11], carbon-based materials (carbon nanotubes, carbon dots, GO) [12–14], organic semiconducting materials (e.g., g-C_3N_4) [15], and their nanocomposites (e.g., TiO_2/g-C_3N_4, ZnO/CdS, SnO_2/rGO) [16–18] are extensively studied in the field of photocatalysis. For instance, Hao et al. have manufactured high surface area g-C_3N_4/TiO_2 heterojunction photocatalysts that showed extraordinary high RhB degradation performance than its parent g-C_3N_4 and TiO_2 samples [19]. Similarly, Zhang with his research group have

Design, Fabrication, and Characterization of Multifunctional Nanomaterials. https://doi.org/10.1016/B978-0-12-820558-7.00006-6

developed a magnetic core-shell nano γ-Fe_2O_3@ZnO photocatalyst. This photocatalyst is magnetically recyclable, and the controlled deposition of the 3D layer of ZnO over a γ-Fe_2O_3 surface leads to a dramatic increase in the photocatalytic performance of γ-Fe_2O_3@ZnO toward ciprofloxacin degradation understimulated sunlight [20]. Cheng et al. also reported the low-temperature synthesis of binary CdS/CuS, CdS/g-C_3N_4, CuS/g-C_3N_4, and ternary CdS/g-C_3N_4/CuS nanocomposites. They have found that among all the synthesized samples due to better synergistic interaction between CdS, g-C_3N_4, and CuS, the ternary nanocomposites show enhanced photocatalytic H_2 production activity under visible-light illumination than the binary nanocomposites [21]. In all these reports, many tedious techniques like chemical reduction process, hydrothermal method, chemical vapor deposition method, and chemical precipitation method has been adopted, which involves the use of highly toxic chemicals, complicated multiple steps, and higher temperature. Some of these steps rely on hazardous complex chemical substances to the environment that can cause serious health risks. Besides this, some of the chemical substances might cause contamination in the fabricated nanocomposites, which reduces their catalytic efficiency. Therefore, the development of environmental benign method that reduces the use of toxic organic solvents and hazardous chemical agents for the synthesis of nanomaterials is highly desirable.

Compared to traditional chemical methods, in the recent past, the green synthetic method that utilizes reagents like plant extracts, bacteria, fungi for the synthesis of nanomaterials has received considerable attention in the field of nanotechnology. These synthetic technique does not require the use of high pressure or temperature, hazardous chemicals and stabilizing agents, complicated reaction steps and long reaction time. Unlike vastly used chemical and physical methods, this method is simple, cost-effective and even also compatible with industrial applications. Taking the above advantages into consideration there are several kinds of literature are available in which the green synthetic method has been adopted for the synthesis nanocomposites for their potential applications in various fields. For example, Atarod et al. have successfully synthesized Ag/TiO_2 nanocomposites by using *Euphorbia heterophylla* leaf extract. This catalyst is found to be effective for the reduction of methyl orange (MO), Congo red (CR), methylene blue (MB) dyes, and 4-nitrophenol (4-NP) at room temperature [22]. Similarly, Fu et al. also reported the Au NPs/chitosan (CTS)/activated coke (AC) nanocomposites, where CTS without the assistance of any extra chemical reagents serves as a linker, stabilizer, and reductant during the synthesis of the catalyst [23]. The catalytic activity of the synthesized catalyst was found to be effective toward the hydrogenation of 4-nitrophenol and azo dyes. Apart from this, Jayaprakash and coworkers with the assistance of microwaves have developed a green synthetic procedure to synthesize silver nanoparticles (Ag NPs) by using *Tamarindus indica* fruit extract as a redactor and capping agent. The Ag NPs synthesized by this method are found to be highly stable and shows good antibacterial activities [24].

Thus considering the importance and effectiveness of the green synthetic method, the investigation of photocatalytic materials synthesized in a greener way can have significant importance from a practical point of view.

This book chapter provides a compressive knowledge about the green synthesis of metal oxides, metal sulfides, carbonaceous, and organic semiconducting material based on various photocatalytic materials. The advantages of the use of plant extract, bacteria, and fungi over the employment of toxic chemical substances in the synthetic process is well explained. The photocatalytic applications of the synthesized materials toward the degradation of various organic compounds from aqueous medium, CO_2 reduction and NO oxidation from the air, and also hydrogen production employing solar water splitting are being highlighted. The role of biomass and biopolymers in the synthesis of different types of photocatalytic materials are also thoroughly reviewed. This chapter is expected to be helpful for the readers in providing depth understanding about the factors and challenges that lies in the green synthesis and photocatalytic applications of various nanomaterials.

1.1 ZnO-based nanomaterials for photocatalytic application

ZnO is a type of semiconductor having a broad direct bandgap width (3.37 eV), large excitation binding energy and deep violet/borderline ultraviolet (UV) absorption at room temperature. It is an excellent semiconductor oxide that possesses favorable excellent electrical, mechanical, and optical properties, similar to TiO_2. ZnO not only has antifouling and antibacterial properties, but also good photocatalytic activity. Furthermore, as reported by Liang et al. the production cost of ZnO is up to 75% lower than that of TiO_2 nanoparticles. The photoactivity of a catalyst is governed by its ability to create photogenerated electron-hole pairs. The major constraint of ZnO as a photocatalyst, however, is the rapid recombination rate of photogenerated electron-hole pairs, which perturbs the photodegradation reaction. Additionally, it has also been noted that the solar energy conversion performance of ZnO is affected by its optical absorption ability, which has been associated with its large bandgap energy. Therefore, intense efforts have been made to improve the optical properties of ZnO by minimizing the bandgap energy and inhibiting the recombination of photogenerated electron-hole pairs by coupling the semiconductor nanoparticles with other materials, such as noble metals, other semiconductor nanoparticles, and carbanions materials [25]. For example, Hassanpour et al. synthesized magnetically separable and reusable Co_3O_4/ZnO heterostructured nanocomposite photocatalyst by using a green and environmentally friendly microwave-assisted method. Results show that Co_3O_4/ZnO nanocomposite is synthesized homogenously, without further impurity. They found that the synthesized nanocomposite Co_3O_4/ZnO shows 86% and 91% of MB and RB dyes removing ability under UV light [26]. Apart from heterostructure formation with metal oxides, Plasmonic photocatalysis has recently emerged as a promising technology for developing high-performance photocatalysts. The noble metal nanoparticles (NPs) have attracted considerable interest due to their surface plasmon resonance (SPR) effect, which accelerates the separation process of the photogenerated electrons and holes in the semiconductor catalyst under visible-light. Many researchers have demonstrated that the Ag NPs deposited on semiconductor show efficient plasmon resonance in the visible region. Thus coupling of ZnO nanostructures with various metallic nanocrystals, including Au, Pt, and Ag, was found to possess improved UV and visible-light photocatalytic activity. However, there are a smaller number of literatures available about biosynthesis of heterostructure photocatalyst based on ZnO and noble metal nano particles. Sohrabnezhad and coworkers in 2016 have reported the synthesis of Ag/ZnO in montmorillonite by using Urtica dioica leaf extract as reducing agent. The SEM and TEM images reveals that Ag/ZnO nanoparticles was successfully incorporated in MMT matrix by using the green synthesis strategy. They also proved that photocatalytic activity of ZnO increases to a great extent by utilizing the unique properties of MMT and silver nano particles. The results of photocatalytic degradation suggest that modification of ZnO-MMT nanocomposite with silver nanoparticles increased the percentage of degradation MB from 37.57% in Ag/ZnO nanocomposite to 82.5% [27]. But however, the synthesis of noble metal nanoparticles decorated on the surface of ZnO by using conventional synthetic methods such as electrochemical deposition, photoreduction, hydrothermal, and precipitation-deposition lead to the formation of undefined morphology and uneven distribution of noble metal nanoparticles on the ZnO matrix. In contrast, the use of presynthesized noble metal nanoparticles capped with bifunctional ligands for the preparation of noble metal nanoparticles functionalized ZnO offers several advantages like the size of noble metal nanoparticles can be independently controlled particularly, a very narrow size distribution can be obtained via well-established synthetic methods and additional steps to remove surplus noble metal nanoparticles and impurities are not necessary. Xiao et al. reported that a well-defined Au/ZnO hybrid nanostructure can be fabricated via a facile and green self-assembly strategy by taking advantage of the pH mediated surface charge characteristics of Au@(dithiolated diethylentriaminepentaacetic) DTDTPA and ZnO, in which DTDTPA plays a crucial role as a directing and stabilizing medium in the dispersion of Au uniformly on the ZnO support as represented in Fig. 17.1.

The results demonstrate that the GNP/ZnO nanocomposite thus obtained with the well-defined hybrid nanostructure exhibits enhanced photocatalytic performances as compared to that of ZnO only [28]. Again it is well known that integration/hybridization of carbon materials with inorganic components represents a robust strategy for developing new functionalities and improving their properties. Therefore, methodologies for constructing inorganic hybrid materials with carbanion materials integrated in well-defined architectures are of significant importance and may pave a way for screening new efficient

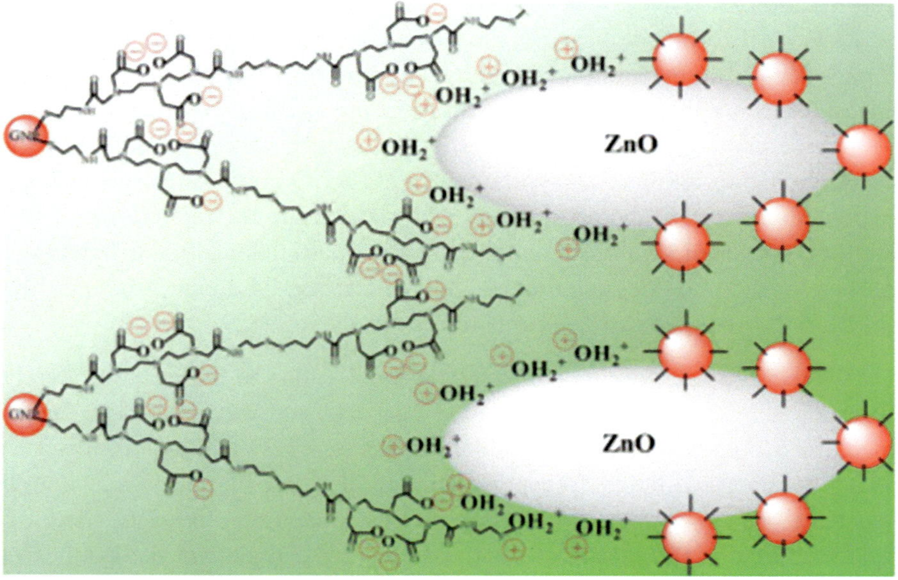

Figure 17.1
Schematic representation showing the self-assembly mechanism of Au@DTDTPA and ZnO.

photocatalytic materials. Among all the carbon-based nanomaterials, the lamellar structure of RGO bestows remarkable and unique properties, such as high specific surface area, large spring constants, high chemical, high mobility of charge carriers and electrochemical stability, which has been regarded as idea guest component for fabricating RGO-based functional materials. On account of its favorable properties, RGO provides new opportunities for photocatalytic carriers or promoters. Hence, great interests have been focused on the synthesis of RGO/semiconductors with GO as the raw material. As a support to deposit and stabilize semiconductors, RGO can also act as a synergist to promote the charge separation, restrain the e^-/h^+ recombination and improve the electron capture-storage-transport properties of the compound nanomaterials, which results in an enhanced photocatalytic activity of the synthesized composite material. Therefore, the coupling of RGO with ZnO can increase the catalytic efficiency of ZnO to a significant extent. For instance, Au-RGO, Bi_2WO_6-RGO, and TiO_2-RGO nanocomposites have been reported that display excellent performances in the photocatalytic degradation of dyes under visible-light. But the different types of synthesis methods that are developed for preparing reduced graphene oxide composites, such as sol-gel method, solvothermal method, UV-assisted photoreduction, and microwave irradiation suffer from various limitations, for example, in sol-gel and solvothermal process either a toxic reducing agent or deleterious solvents are needed, UV-assisted photocatalytic reduction of GO process is difficult to be scaled-up and microwave irradiation method is instantaneous and cannot be easily controlled. So the development of a convenient green method for the synthesis of RGO/semiconductor heterojunction is urgently needed. Li and coworkers synthesized ZnO-RGO nanocomposite by using a convenient, environmentally friendly one-step hydrothermal approach without using any extra reducing agent as represented in Fig. 17.2.When the photocatalytic activity of the synthesized nanocomposite was investigated they have found that the synthesized ZnO-RGO composites showed better photocatalytic activity toward the degradation of RhB dye and the reduction of CO_2 under the irradiation of simulated sunlight than the pure ZnO particles and the efficiency of the nanocomposites for the photodegradation of RhB was increased about 39% and the yield of methanol from the reduction of CO_2 was improved by 75% [29].

Moreover, apart from RGO, carbon quantum dots (CQDs) due to their ease of synthesis, functionalization, and outstanding properties such as size- and wavelength-dependent luminescence emission, resistance to photobleaching and good biocompatibility has been found application in various fields. Recently there are also few literatures are reported in which CQDs have been hybridized with different semiconductors, like TiO_2, SiO_2, Fe_2O_3, Ag_3PO_4, Cu_2O, and CdS. The results of photocatalytic experiments of the synthesized semiconductor/CQDs indicate that the nanocomposites exhibited a superior photocatalytic performance than pure semiconducting materials. For example, Bozetine et al. synthesized ZnO/CQDs nanocomposites by using a simple, one-pot method synthetic method. They presented that as the reaction for the synthesis of ZnO/CQDs takes place in aqueous solution so the synthetic procedure obeys the "green chemistry" principles. When the photocatalytic activity of the synthesized nanohybrid was evaluated by the photocatalytic degradation of rhodamine B (RhB), the result shows that the nanocomposite was active under visible-light irradiation and its efficiency was much higher than that of ZnO prepared under similar experimental conditions [30].

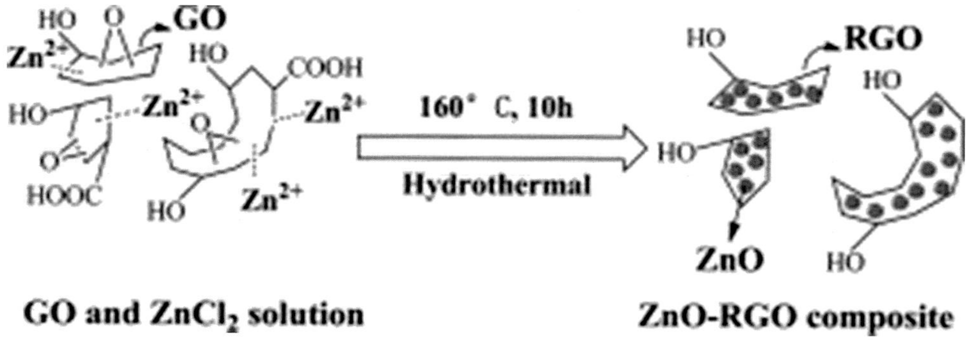

Figure 17.2
Schematic representation for the green synthesis of ZnO-RGO nanocomposite.

Besides the above modification strategy, the development of ZnO based photocatalyst in extremely small size (50 nm) to provide sufficient surface area for photocatalytic reaction remain a great challenge. Because the small nanosized particles readily undergo agglomeration, they cannot be recovered easily from the reaction system. One way to overcome these limitations is to assemble ZnO nanoparticles having a high surface area with materials that can assist electron-hole separation and can increase the recovery efficiency of the ZnO photocatalyst from the reaction system. For example, Pant et al. reported the synthesis of Ag—ZnO/reduced graphene oxide (Ag—ZnO/RGO) composite by using a green and facile one-step hydrothermal approach. The morphology of the synthesized nanocomposite suggests that Ag-doped ZnO flowers are homogeneously distributed over the surface of RGO. Furthermore, the mechanism for the degradation of the dye by the synthesized nanocomposite (Fig. 17.3) suggests that both Ag nanoparticles and RGO sheets act as good electron acceptors which favors the rapid separation of photogenerated electrons and holes and also provides the easiness of particles recovery from reaction system after the photodegradation reaction [31].

1.2 TiO_2-based nanomaterials for photocatalytic application

Among a large number of metal oxide-based semiconductors that have been reported to catalyze the photocatalytic reactions, titanium dioxide (TiO_2) has been extensively studied as an effective photocatalyst due to its excellent photocatalytic properties. Nanosized TiO_2-based photocatalytic treatment is a highly effective method for the degradation and detoxification of organic and inorganic pollutants from wastewater. Generality, the

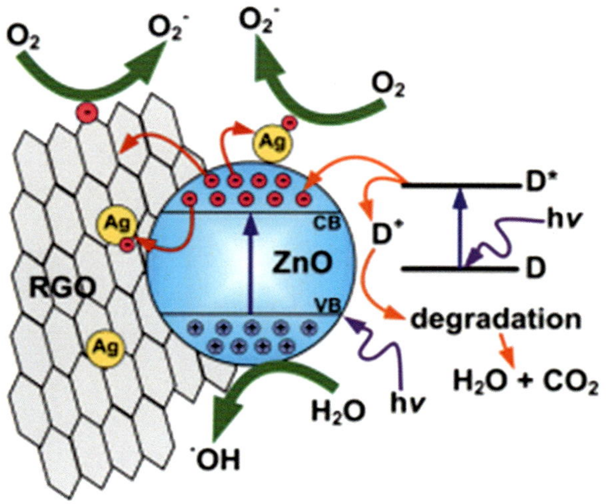

Figure 17.3
The schematic diagram of the mechanism of dye degradation by Ag—ZnO/RGO photocatalyst.

synthesis of TiO_2-based nanomaterials involves the utilization of highly expensive, toxic, and hazardous chemical substances, which when released to the environment creates serious ecological toxicity. So to avoid the adverse effect of the conventional chemical synthetic method, green synthetic root is adopted. Green synthesis is a simple, ecofriendly, and less toxic way of synthesizing nanoparticles from biodegradable materials like plant extracts, microbes, and enzymes. However, the synthesis of nanoparticles using plant extracts is the most useful strategy as it lessens the chance of associated contaminants and reduces the reaction time. Inspired by the above advantages Goutam et al. synthesized TiO_2 nanoparticles by using leaf extract of a petro plant, *Jatropha curcas* L using green chemistry approach as represented in Fig. 17.4. They have shown that the presence of various organic materials such as phenols and tannins present in the leaf extract act as a reducing as well as capping agents for metal ions. Furthermore, they also demonstrated that the green synthesized TiO_2 nanoparticles show remarkable photocatalytic west water treatment ability, which removes 82.26% of COD and 76.48% of Cr^{+6} simultaneously from tannery wastewater under solar light [32].

However, some intrinsic drawbacks have limited the wide application of TiO_2 based nanomaterials in some fields. On the one hand, wide bandgap makes TiO_2 (anatase: 3.2 eV, rutile: 3.0 eV) occupy only 3%–5% of the total solar spectrum. On the other hand, the fast recombination of photogenerated electron-hole pairs also leads to decreased

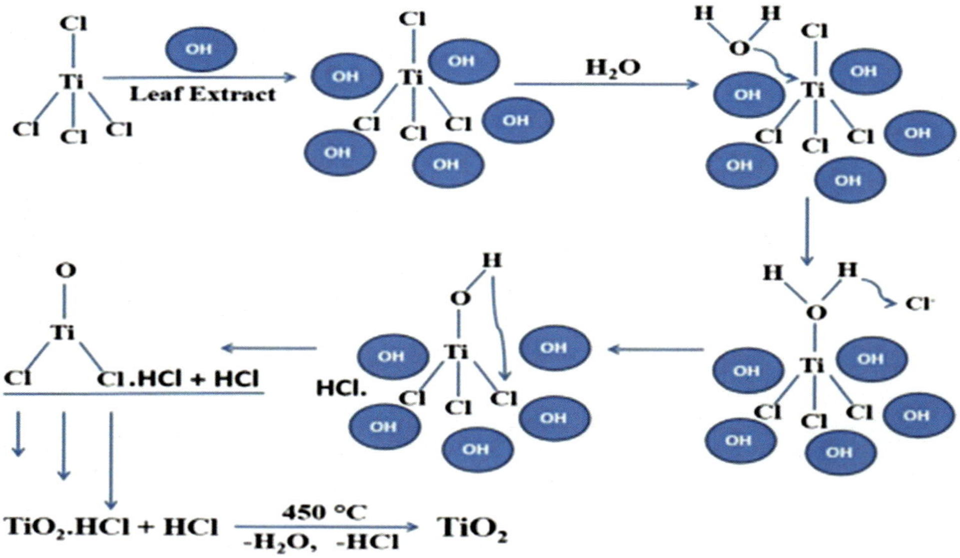

Figure 17.4
Possible reaction mechanism for the formation of TiO_2 nanoparticles in the presence of leaf extract of *Jatropha curcas* L.

efficiency in the photocatalytic activity. Over the last few years, modification with nonmetals have been considered to be an ideal an inexpensive choice for narrowing the bandgap of TiO_2 and increasing the visible-light-harvesting capacity of the synthesized hybrid TiO_2 nanocomposites. For example, Parayil et al. reported the synthesis of carbon modified TiO_2 composite materials by using a sucrose mediated green synthetic root. When the photocatalytic activity of the resultant materials was tested they have found that the C-modified TiO_2 exhibit higher hydrogen evolution ability understimulated sunlight than pure TiO_2. They have shown that optimal loading of mesoporous carbon causes efficient trapping of electrons in the shallow states which enables the effective separation of photogenerated electrons and holes as a result of which the photocatalytic hydrogen evolution activity of the synthesized nanohybrid increases to a great extent [33].

Apart from modification, an effective approach to tackle the limitations of TiO_2 as a photocatalyst is to dope TiO_2 with nonmetal elements, so that the light absorbance of TiO_2 can be extended into visible reason. Numerous contributions have been devoted recently to developing effective nonmetal-doped TiO_2 photocatalysts with visible-light activity. The preparation method is another crucial factor that should be considered for applications. Traditional methods to prepare C-doped TiO_2 include high-temperature sintering of a carbon-containing TiO_2 precursor, CVD or pyrolysis, and sol-gel methods. Usually, for these methods, high-temperature treatment (400−850°C) is required; expensive, toxic, or unstable precursors are used; undesirable gaseous byproducts are usually produced in the preparation process; and the procedures are somewhat tedious, which, in all, make the preparation cost high and large-scale application difficult. A green synthesis approach is efficient in scale-up and benign to human health and the environment, which is based on minimizing the number of preparation steps without utilizing either excessively harmful reagents or unstable precursors and without generating particularly toxic byproducts. For example, Dong et al. prepared mesoporous C-doped TiO_2 photocatalyst by a one-pot green synthetic approach using sucrose as a new carbon doping source for the first time. They also found that postthermal treatment is an effective, facile method to promote the visible-light activity of C-doped TiO_2. The thermal treatment of the as-prepared C-doped TiO_2 has little influence on the crystal size and could decrease the specific surface areas and pore volumes slightly. However, the optical and surface properties improved significantly, which is caused by the changes of the doped carbon element. For the sample treated at an optimum temperature of 200°C, the visible-light absorbance was broadened due to the increased content of doped carbon and the recombination of electron-hole pairs was suppressed due to the reduction of surface defects, which, in all, contributed to the promoted visible-light photocatalytic activity [34]. The same group of researchers also reported efficient visible-light photocatalytic C-doped TiO_2 with a mesoporous structure that was synthesized successfully by a one-step green synthetic approach using $Ti(SO_4)_2$ and glucose as precursors. $Ti(SO_4)_2$ hydrolysis, carbon doping, TiO_2 crystallization, and formation of the mesoporous structure have been simultaneously achieved during this single process. They have further also reported that this facile method is free of using expensive or unstable precursors and the production of undesirable byproducts during the synthesis process. Carbon is doped into the TiO_2 lattice in the form of substitutional carbon for oxygen sites. The observed new electronic states above the valence band edge are directly responsible for the electronic origin of bandgap narrowing and visible-light photoactivity of the C-doped TiO_2. The green synthesized C-doped TiO_2 exhibits efficient photocatalytic activity in degradation of toluene in the gas phase under visible-light irradiation concerning P25 and C-doped TiO_2 prepared by the solid-state method. This one-step green method provides an effective approach for future industrial applications in pollution control and solar energy conversion owing to its low cost and easy scaling up [35].

Like C, there are several literatures are available where N is doped with TiO_2 and the resulted N-doped TiO_2 is found to show exceptionally high photocatalytic activity than undoped TiO_2. For example, Parida et al. synthesized the first time N-doped TiO_2 by using a facile and environmentally friendly approach as represented in Fig. 17.5. They have shown that the synergetic combination of hierarchical mesoporous frame, bandgap

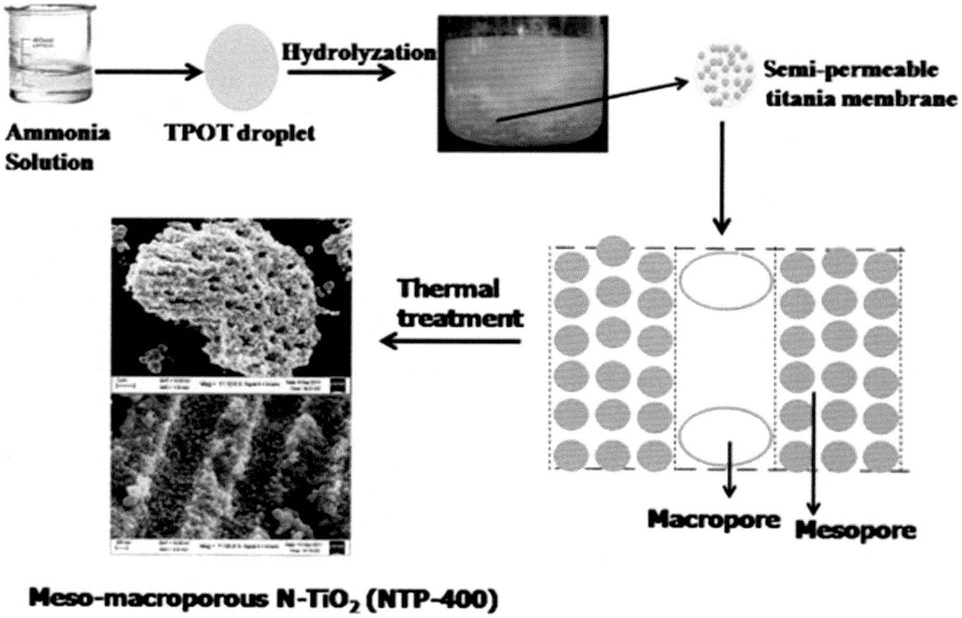

Figure 17.5
Preparation method of hierarchical meso-macroporous N-doped TiO_2.

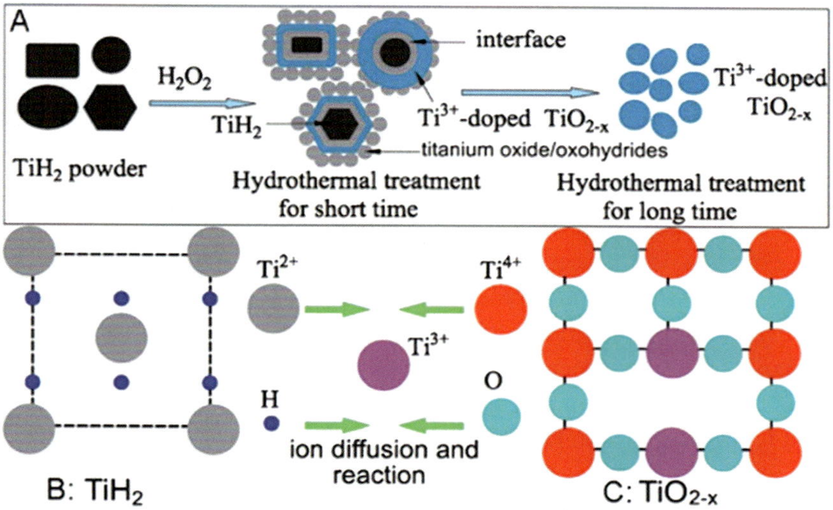

Figure 17.6

(A) Schematic of the formation mechanisms for the rice-shaped Ti^{3+} self-doped TiO_{2-x} nanoparticles. (B) and (C) The interface diffusion—redox diagram. The green arrows indicate ion diffusion.

reduction, and higher visible-light absorption as a result of N-doping makes the synthesized N-doped TiO_2 as an efficient photocatalyst for the evolution of H_2 under visible-light irradiation [36].

However, doping with heteroatoms can cause dopant induced charge recombination and/or traps, and dopant induced thermal instability. According to recent reports reduced $TiO_2(TiO_2-x)$, which contains the Ti^{3+} or oxygen vacancy, has emerged as an effective approach to improve visible photoactivity of TiO_2 nanocrystals. The presence of Ti^{3+} in nonstoichiometric TiO_2-x improves the wet stability, which is important for photocatalysis and superhydrophilic effects. However, theoretical work suggested that to achieve efficient visible-light activity, the concentration of Ti^{3+} must be high enough to induce a continuous vacancy band of electronic states just below the conduction band edge of TiO_2. Otherwise, a low Ti^{3+} doping concentration only creates localized oxygen vacancy states that deteriorate the electron mobility and exhibit a negligible visible photoactivity. Thus, it is exceptionally desirable to explore facile synthetic methodologies for high concentration Ti^{3+} doped TiO_2 that are active in the visible region it is exceptionally desirable to explore facile synthetic methodologies for high concentration Ti^{3+} doped TiO_2 that are active in the visible region. Several methods have been reported to produce Ti^{3+} and/or oxygen vacancy self-doped TiO_2-x. Although these methods have their own advantages, they were limited by high cost and complicated procedures, such as heating of TiO_2 in reducing (e.g., H_2 or CO) environments in high pressure, chemical vapor deposition, or bombardment of TiO_2 with high energy laser irradiation or Ar^+ ion sputtering. These methods are fairly impractical for large-scale production. Furthermore, in these approaches, the reduction occurs mainly in the surface region and the surface oxygen vacancies in TiO_2 are usually not thermally stable enough in the air because the Ti^{3+} is easily oxidized, even by the dissolved oxygen present in water. In contrast to the current reduction-based method, the development of the method that can produce a high concentration of Ti^{3+} throughout the surface and the bulk of TiO_2 nanoparticles is much more essential. For instance, Liu et al. synthesized rice-shaped Ti^{3+} self-doped TiO_2-x nanoparticles by mild hydrothermal treatment of TiH_2 in H_2O_2 aqueous solution as represented in Fig. 17.6. Furthermore, they have shown that the formation of Ti^{3+} self-doped TiO_2-x nanocomposite proceeds through "surface oxide—interface diffusion—redox" reaction mechanism i.e., initially the surface of TiH_2 powder gets oxidized by H_2O_2 forming an intermediate product which is composed of unreacted TiH_2 as the core and titanium oxides and Oxo hydrides as the shell. Under hydrothermal condition with an increase in reaction time a solid interface diffusion—redox reaction occurs, which results in the formation of phase-pure rice-shaped Ti^{3+} self-doped TiO_2-x nanoparticles. Under visible-light irradiation, the samples exhibit higher photocatalytic activity for hydrogen evolution and photo-oxidation of MB than that of the commercial P25 TiO_2 nanoparticles [37].

Apart from the doping strategy, in recent past composites of graphene and TiO_2 have been found to show advantageous enhancement of photocatalytic activity because graphene can facilitate charge separation in the composite materials. For instance, Zhang et al. prepared chemically bonded TiO_2/graphene sheets (GS) nanocomposites using a green and facile one-step hydrothermal method where $TiCl_4$ and graphene oxide (GO) are used as the precursors for TiO_2 and GS. From the results of XRD analysis, they have shown that in the synthesized nanocomposite the existence of GS suppresses the crystal growth and promotes the formation of the rutile phase of TiO_2. Compared to that of P25, a significant improvement for photocatalytic hydrogen evolution was observed for the hydrothermally prepared TiO_2/GS composites and comparable to that of the sole-gel obtained TiO_2/GS composites reported by the same group in their previous work [38]. Similarly, Yang et al. successfully synthesized P25/graphene nanocomposite photocatalysts with P25 and different ratios of graphene oxide through a green and facile one-step microwave-assisted method. This synthetic method produces a P25/graphene hybrid composite with strong interactions between the two components. The resulting hybrid materials have been found to show significant advancement over bare P25 in the photodegradation of MB, which is attributed to increased absorptivity of dyes, extended light absorption range, and efficient charge separation properties of the synthesized nanohybrid [39]. Muthirulan et al. also reported the synthesis of TiO_2—graphene nanocomposite by using an environmentally friendly and cost-effective strategy. Photodegradation studies revealed that TiO_2—graphene nanocomposite possess greater degradation efficiency for Acid Orange 7 dye compared to TiO_2 under solar irradiation, indicating that the electron transfer between TiO_2 and graphene will greatly retard the recombination of photoinduced charge carriers and prolong electron lifetime, which contributes to the enhancement of photocatalytic performance [40]. Notably, in most of the above-discussed methods, the P25 nanoparticles were used as a countercomponent to the graphene in the composites. However, this tactic severely limits the tunability of the composite properties since one species is already fixed in the binary composites. In other cases, titanium alkoxides are widely employed as a TiO_2 precursor. Owing to the high reactivity of alkoxides. However, for achieving improved photocatalytic activity sophisticated control over reaction conditions is essential during solution-based synthesis. Shah et al. reported a facile and robust one-step synthesis of TiO_2–graphene composite as represented in Fig. 17.7 and its performance in photocatalytic applications. The most notable aspect of this approach was the use of an aqueous solution and lack of harsh chemicals (such as hydrazine) during synthesis. The reduction of graphite oxide, hydrolysis of $TiCl_4$, and the crystallization of the produced TiO_2 are concurrently carried out in a single-step reaction by hydrothermal treatment. So they effectively exclude the use of titanium alkoxides as highly reactive reagent or hydrazine as a toxic reducing agent, which have been encountered in synthetic conditions in conventional systems in addition, in the synthesized catalysts, TiO_2 was present in both anatase and rutile phases, whereas the reported methods have mainly dealt with a single phase. The catalysts show enhanced photocatalytic activity toward the degradation of the RhB dye and benzoic acid under visible-light irradiation [41].

Up to now, we have discussed the method of synthesis of TiO_2–graphene nanocomposite that mainly relies on the anchoring method, namely, the formation of nanocrystals of TiO_2 in the presence of pristine on functionalized graphene nanosheets. The biggest limitation of these hybrid materials is that the ultrafine TiO_2 nanoparticles tend to agglomerate,

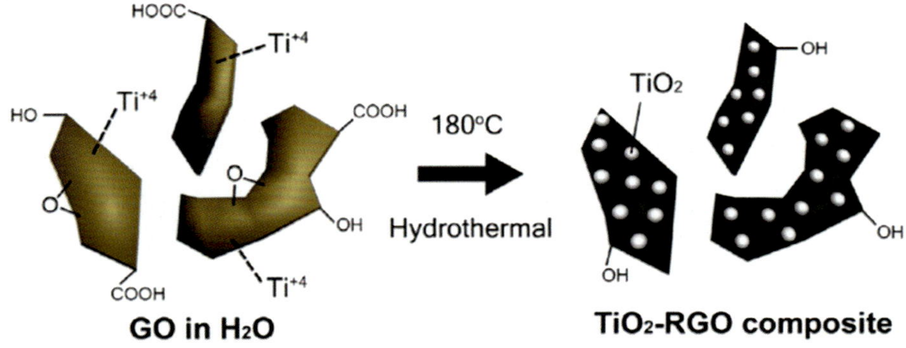

Figure 17.7
Reaction scheme of synthesizing TiO_2–reduced graphene oxide nanocomposites.

which will lead to a decrease in the effective surface area of both TiO_2 and graphene during cyclic utilization of the synthesized nanohybrid in the photocatalytic degradation process. One way to overcome these limitations is to fabricate graphene wrapped TiO_2 nanostructure. For instance, Liu et al. reported a green and direct protocol to fabricate TiO_2@GO core/shell structures through a layer-by-layer assembly of the stable GO nanosheets onto TiO_2 microspheres without any modified agents as represented in Fig. 17.8. The photocatalytic activity of the as-synthesized materials were evaluated by decolorization of an aqueous solution of RhB dye under both UV and visible-light irradiation. The photocatalytic degradation result indicates that the TiO_2@GO core/shell structures exhibit excellent photocatalytic activity and recyclability property under the irradiation of both UV and visible-light [42].

However, more recently one-dimensional TiO_2 nanotube has been receiving immense research interest as they possess the combine properties of zero-dimensional TiO_2 nanoparticles (e.g., chemically stable, inexpensive, nontoxic, and abundant in nature) and 1-D nanostructure (e.g., large specific surface area, good electron conductivity and high aspect ratio). But like TiO_2 nanoparticles, pure TNT also suffers from rapid recombination of photoinduced electron/hole pairs and a narrow light-response range, which results in low efficiency in the utilization of solar energy. Hybridizing TiO_2 nanotube with graphene

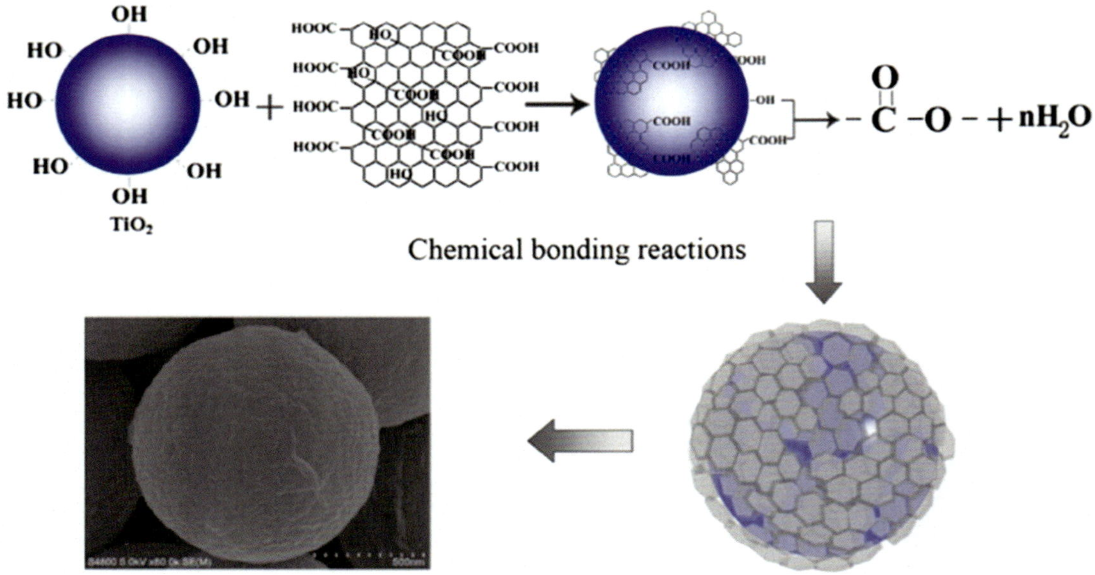

Figure 17.8
Schematic representation of the preparation of graphene oxide encapsulated TiO_2@GO core/shell structure.

can significantly enhance the photocatalytic activity of pure TiO_2 nanotube. For example, Dang et al. reported a green and facile one-step alkali assisted hydrothermal process to prepare titanate nanotube/graphene (TNT/GR) photocatalysts. The photocatalytic performance of the synthesized TNT/GR nanocomposite was evaluated in detail by H_2 generation from water splitting under visible-light irradiation. They have shown with the introduction of GR, the separation efficiency of photoinduced electron/hole pairs was improved efficiency and the reaction space for water reduction was enlarged greatly (not only on the surface of semiconductor catalysts but also on GR sheets) consequently enhancing photocatalytic H_2 production efficiency. Furthermore, above the optimal level of GR content, an excessive amount of GR sheets covers the active sites on the surface of TNTs and also suppress the contact of the sacrificial agents with TNTs, which decrease the H_2 evolution rate of TNT/GR nanocomposite [43].

In recent years, polymers with extended pi-conjugated electron systems have attracted considerable attention because of their high absorption coefficients in the visible region and improved conductivity, which allows high mobility of charge carriers. Among conductive polymers, polyaniline (PANI) has been widely used to improve electronic conductivity as well as solar energy transfer and photocatalytic activity of semiconducting materials, due to its easiness of preparation, comparatively low cost, and excellent environmental stability. In the available literatures, different ways to produce PANI/semiconductor heterojunction photocatalyst have been demonstrated, however, aniline is classically chosen as the starting monomer in most of the toxic work. Thus the development of environmentally friendly synthesis techniques for preparation of PANI/semiconductor heterojunction is highly required. Hidalgo et al. the first time reported the synthesis of PANI—TiO_2 nanocomposite mesoporous films by using a simple, safe, low-cost, and environmentally friendly procedure. The result of the photocatalytic activity of the synthesized PANI—TiO_2 mesoporous films indicates that they show improved water-splitting reactions under simulated solar light [44]. Semiconductor QDs have many unique properties such as quantum size effect and multiple exciton generation effect comparing to their bulk counterparts. Among the semiconductor QDs, CdSe QDs has attracted great attention in the past decades. To our knowledge, CdSe is an important group semiconductor with a direct energy bandgap of 1.75 eV for the bulk phase. Particularly, CdSe has a higher conduction band edge than that of TiO_2, which is beneficial for the injection of excited electrons from CdSe to TiO_2. Wang and coworkers have developed a facile and green method was employed to synthesize CdSe quantum dots (QDs)-modified TiO_2. It exhibited much higher photocatalytic activities than both TiO2 and bulk CdSe-modified TiO_2 on degrading malachite green (MG) aqueous solution under visible-light irradiation. The vital role that quantum size effect of CdSe QDs plays in enhancing photocatalytic properties of CdSe QDs-modified TiO_2 is discussed in detail for the first time. Due to the more negative conduction band potentials of CdSe QDs than that of bulk CdSe, the electron transfer rate from CdSe QDs to TiO_2 is much faster than that of bulk CdSe to TiO_2, which would be beneficial for an efficient charge carrier's separation. As a result, more quantities of active species ($O_2{\bullet}^-$, H_2O_2, •OH) are generated during the photocatalytic reaction in CdSe QDs-modified TiO_2 system. This work could not only provide a facile and green strategy to fabricate semiconductor QDs-modified TiO_2 photocatalyst but also contribute to a deeper understanding of the specific role of semiconductor QDs during the photocatalytic reaction process [45]. The noble metal plays a dual role in the photocatalytic process: on one hand, the SPR caused by noble metal can be utilized to harvest the visible-light on the other hand, the formation of Schottky barrier between the semiconductor and noble metal is obviously beneficial for the separation of electron–hole pairs. Until now, several strategies including chemical reduction, UV irradiation, sol–gel, and hydrothermal method have been developed to synthesize the Ag/TiO_2 heterojunction. However, these methods usually need either rigorous synthetic condition or poisonous reducer, and Ag nanoparticles in the prepared Ag/TiO_2 heterojunction hardly have high crystallinity and uniform size. Thus, it is desirable to develop a facile and generic method to synthesize the Ag/TiO_2 heterojunction with well-defined Ag nanocrystals. Atarod et al. showed an environmentally friendly method to prepare stable Ag/TiO_2 nanocomposite employing *E. heterophylla* leaf extract. The flavonoids present in the extract of leaves of *E. heterophylla* act as both reducing and capping/stabilizing. Also, the catalytic activity of Ag/TiO_2 nanocomposite for the reduction of a variety of dyes in water was also studied. Also, this methodology offers the competitiveness of recyclability of the catalyst without significant loss of catalytic activity, and the catalyst could be easily recovered and reused several times, thus making this procedure environmentally more acceptable [22]. Similarly, Yang et al. reported a green and facile bioinspired method assisted by a small organic molecule, 3-(3,4-dihydroxyphenyl) propionic acid (diHPP) for in situ preparation of Ag/TNT heterojunctions under benign conditions (Fig. 17.9). As prepared in the TEM images of Ag/TNT nanocomposites, hemispherical Ag nanocrystals about 3.8 nm in diameter are dispersed uniformly on the TNT surface and form the heterojunction structure with TNT. During the bioinspired synthetic process, the diHPP molecules serve as the dual functions, i.e., the reducing agent of Ag^+ and the capping agent to stabilize Ag nanocrystals and hinder their agglomeration. The Ag/TNT heterojunctions show much higher photocatalytic activity than pure TNT and P25 under visible-light, which is attributed to the SPR effect and the inhibiting effect for photogenerated carriers of Ag nanocrystals [46].

Apart from the formation of Ag/TiO_2 heterojunction, combining TiO_2, Ag and GO is a promising strategy to simultaneously introduce excellent absorption, conduction, and visible-light-harvesting capacity in the resultant composite, which could facilitate effective degradation of pollutants. Nasrollahzadeh et al. have developed an efficient, simple, inexpensive and environmentally benign method for preparation of the Ag/RGO/TiO_2

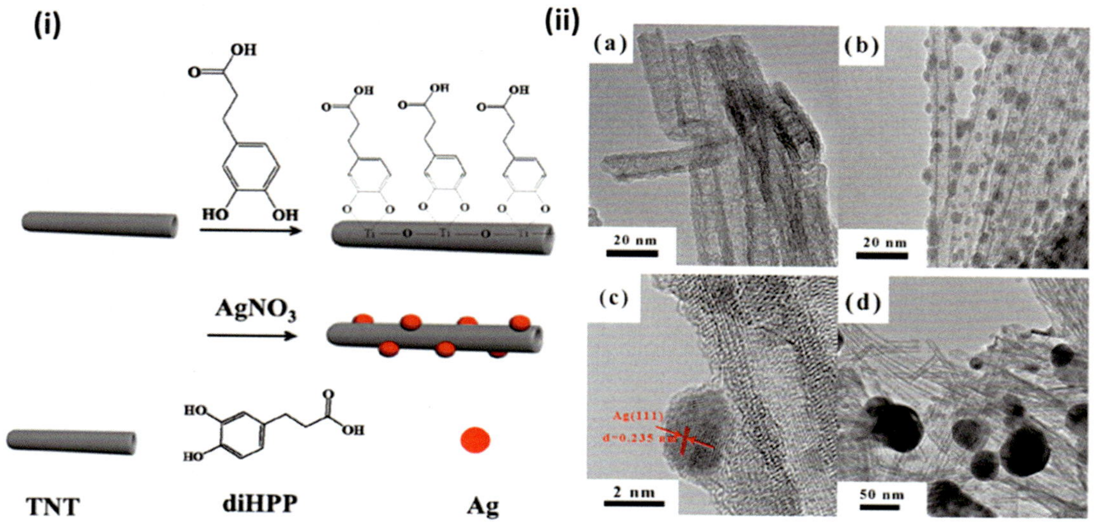

Figure 17.9
(1) schematic representation of synthesis of Ag/TiO_2 heterojunction. (2) TEM images of TNT (a) Ag/TNT heterojunction (b), and Ag/TNT nanocomposites prepared without diHPP (c); and HRTEM image (d).

nanocomposite using *E. helioscopia* L. leaf extract as a stabilizing and reducing agent. The advantages of this environmentally benign and safe protocol include the elimination of toxic, expensive and harmful chemicals, the use of water as the solvent, easy-synthesis of the Ag/RGO/TiO_2 nanocomposite under environmentally friendly conditions, easy separation of the catalyst and experimental ease. Moreover, the Ag/RGO/TiO_2 nanocomposite exhibited high catalytic activity for the reduction of 4-NP, CR and MB in water at room temperature. Furthermore, the catalyst can be recovered and reused for several consecutive runs with no significant loss of catalytic activity [47]. Moreover, Liu et al. also synthesized ternary TiO_2–SiO_2–Ag nanocomposite by utilizing lysozyme as both inducing agents of TiO_2 and reducing agents of Ag^+. The resultant TiO_2–SiO_2–Ag nanocomposites exhibit high, stable, and tunable photocatalytic activity in catalyzing the decomposition of RhB under the visible-light radiation, which can be attributed to the synergistic effect of the enhanced light-harvesting ability from plasmonic Ag nanoparticles and enhanced adsorption capacity of RhB from SiO_2 moiety. Naik et al. also reported a simple, ecofriendly and low-cost approach for the preparation of stable Au/TiO_2 nanocomposites with a green synthetic method using aqueous extract of *C. tamala* leaves as the reducing and capping agent. The nanocomposites prepared by this green method exhibited comparable or higher visible-light activities for degradation of MO than the similar catalysts prepared by other chemical methods. This green synthesis approach is a rapid and better alternative to chemical synthesis and also effective for the large-scale synthesis of Au/TiO_2 nanocomposites [48].

It is not industrially practical to recycle the powdered TiO_2 photocatalysts (e.g., nanoparticles and nanosheets) from aqueous solution after photocatalytic reactions. Nanocrystalline TiO_2 immobilized on supporting materials such as glass, sand, zeolite, silica, cotton fiber and ceramic can improve the separation efficiency. However, the nanocrystals deposition is generally nonuniform and easily detach from the support, which frequently leads to a substantial decrease in photocatalytic activity and selectivity. Light-harvesting is inefficient due to the high content of inactive supports. Therefore, the synthesis of a multifunctional photocatalyst that is both highly photoactive and effectively recyclable is a great challenge. Therefore, the immobilization of nanostructured TiO_2 on suitable supports endowed with reduced bandgap and improved pollutant adsorption capacity is desirable for the large-scale practical applications. Shabani et al. reported the synthesis of Fe_3O_4 and TiO_2 nanoparticles in the presence of pomegranate juice. Multifunctional cotton fabrics were coated with Fe_3O_4/TiO_2 nanocomposite, using the pad-dry-cure method. The effect of concentration and precipitating agent was investigated on the morphology and particle size of the products. The photocatalytic behavior of Fe_3O_4–TiO_2 nanocomposite was evaluated using the degradation of acid black, acid brown, and CR under UV light irradiation. The results show that pomegranate juice is a suitable green capping agent for the preparation of Fe_3O_4–TiO_2 nanocomposites [49].

1.3 Green synthesis of metal sulfide-based nanomaterials for photocatalytic application

However, till now most of our discussions have been focused on the progress of green synthesis and photocatalytic applications of metal oxide-based nanomaterials for energy and environmental pollutant remediation. But the most commonly investigated metal oxides like TiO_2, ZnO due to their wide bandgap can only respond to UV light below 387 nm. Since the spectrum of solar light contains 5% UV light (<380 nm), 45% Vis light (380–780 nm), and 50% infrared (IR) light (>780 nm) so they cannot utilize solar energy efficiently. As an alternative of metal oxides, the metal sulfide photocatalysts possess high conduction band potentials originating from the overlapping of d and sp orbitals while the valence bands consist of S 3p orbitals, which are much more negative than O 2p orbitals, thus resulting in conduction band positions negative enough to reduce H_2O to H_2 and narrowband gaps with a suitable response to the solar spectrum [2]. However, up to now, many methods available for the fabrication of metal sulfide-based composites utilize Na_2S, thiourea, and dimethyl sulfoxide as the sulfur source, which can cause various health and environmental risks. Such issues can be overcome by using the biomolecules assisted synthesis method. For example, Zhang et al. developed an environment-friendly strategy toward the one-pot synthesis of CuS/reduced graphene oxide (CuS/rGO) nanocomposites with the use of L-cysteine as a reducing agent, sulfur donor, and linker to anchor CuS nanoparticles onto the surface of rGO. The formation of CuS/rGO nanocomposites occurs in a single-step process, carried out by hydrothermal treatment of aqueous dispersion of GO and $CuCl_2$ in the presence of L-cysteine as represented in Fig. 17.10. Furthermore, they have also suggested that the CuS/rGO nanocomposites show significant photocurrent response under visible-light irradiation ($\lambda > 400$ nm) and good photocatalytic activity in the degradation of MB, which could be ascribed to the efficient charge transport of rGO sheets and hence reduced recombination rate of excited carriers [50].

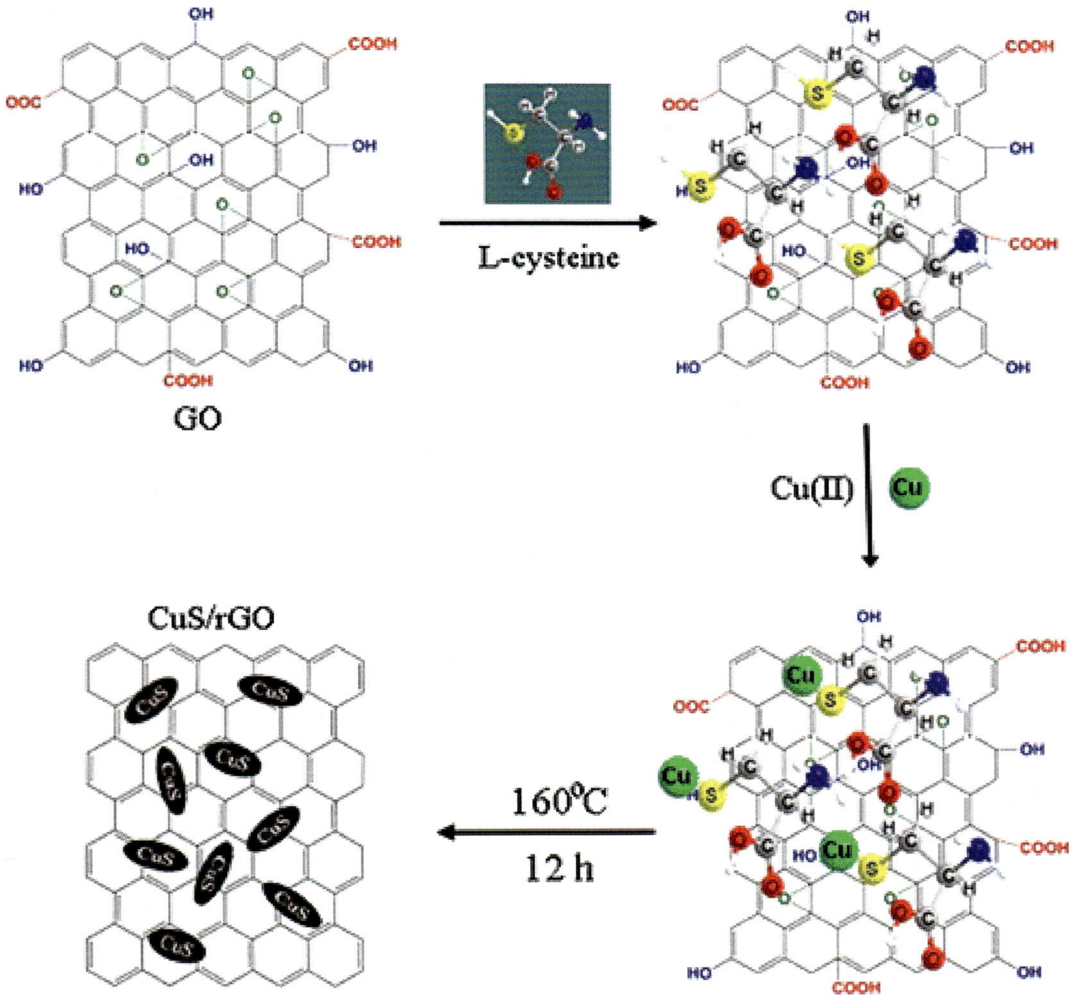

Figure 17.10
Schematic diagram to illustrate the preparation process of the CuS/rGO nanocomposites with the use of L-Cysteine.

Similarly, Vadivel et al. reported a simple, facile and one-pot synthesis of Bi_2S_3 microflowers hierarchical structure—RGO composite by a microwave-induced solvothermal process using $BiNO_35H_2O$, D-penicillamine, and GO as precursors. This method is environmentally benign because the toxic Na_2S, thiourea, thioacetamide were avoided and D-penicillamine is used as a sulfur source. The photocatalytic properties of as-synthesized Bi_2S_3 microflower/RGO composites were examined using the photocatalytic degradation of crystal violet (CV) a triphenylmethane dye under visible-light irradiation. This novel hybrid material shows superior photocatalytic activity than pure Bi_2S_3 microflowers under ambient conditions [51]. Apart from CuS and Bi_2S_3, cadmium sulfide (CdS) with suitable bandgap (Eg = 2.4 eV) has been extensively used in the visible-light-driven photocatalytic field such as degradation of organic pollutants, photocatalytic H_2 production, CO_2 reduction and photoelectric conversion on solar cells and so on. For example, Li and coworkers have synthesized CeO_2/CdS-DETA composites by using a two-step hydrothermal method. They have found that the photocatalytic H_2 performance has been greatly improved compared to pure CeO_2 and CdS-DETA, and the H_2 production efficiency of CdS-DETA loaded with 7% of CeO_2 reached 14.84 mmol/g h under visible-light irradiation, which is twice more than that of CdS-DETA. The excellent photocatalytic performance of CeO_2/CdS-DETA is attributed due to the presence of highly active sites, oxygen vacancies in CeO_2, and excellent electron transfer efficiency from CdS-DETA to CeO_2 [52]. But, as cadmium is highly toxic in nature and its long term impact over living organism is unknown so concern has been raised over the use of CdS as photocatalyst. Therefore, cadmium-free sulfides, such as zinc sulfide compounds, are very promising alternatives due to their environmental abundance and natural presence in the human body. Mansur et al. reported the development of a novel nano-photocatalyst based on semiconductor ZnS QDs conjugated with chitosan via a single-step "green chemistry" aqueous colloidal process. The synthesis of ZnS QDs was performed using a chemical precipitation method with chitosan as the biocompatible capping ligand. The prepared ZnS conjugates were assayed as nano-photocatalysts for the photodegradation of MB and MO, which were used as "model" pollutant [53]. Similarly, Yuan et al. for the first also reported the synthesis of SnS_2 QDs grown on RGO composite and their applications as "green" photocatalysts for visible-light-driven Cr(VI) reduction reaction. The consequences of photocatalytic activity indicate that the catalytic efficiency of SnS_2 QDs can be significantly enhanced by the loading of graphene due to the improvement in electron transport. The 1.5% graphene/SnS_2 photocatalyst showed almost three times higher activity compared with the pure SnS_2 QDs under similar experimental conditions [54].

1.4 Green synthesis of g-C_3N_4-based nanomaterials for photocatalytic application

Apart from metal oxides and metal sulfides, since the first report on photocatalytic H_2 evolution over g-C_3N_4 by Wang et al. in 2009, this two dimensional organic semiconducting material due to its fascinating merits, such as moderate bandgap ($\approx$2.7 eV), proper electronic band structure, nontoxicity, low cost, good stability, and easy preparation has received worldwide attention. The pure bulk g-C_3N_4 prepared by thermal polycondensation of low-cost nitrogen-containing organic precursors suffers from several shortcomings, including low specific surface area, insufficient visible-light utilization, and, particularly, rapid recombination of photogenerated charge carriers. Accordingly, the exploration of efficient g-C_3N_4 based photocatalysts with better separation/transfer efficiency of photogenerated electron-hole pairs has become an important research direction. Among the latest advances in g$-C_3N_4-$related photocatalytic systems, constructing g-C_3N_4 based heterostructures is a research hotspot, due to its feasibility and effectiveness for the spatial separation of photogenerated electron-hole pairs. Thus, many approaches for the preparation of g-C_3N_4 based heterostructure nanocomposite have been attempted, such as chemical precipitation processes, physical vapor deposition, and hydrothermal techniques. Although, in these chemical and physical methods, g-C_3N_4 based heterostructure with relatively high density and surface ratios can be successfully fabricated. But, these processes involve the use of highly toxic chemicals or high temperatures, and complicated manipulations. Some of them might cause contamination in the nanocomposites, thereby reducing the electron transfer capacity and photocatalytic activity. Accordingly, there has been considerable interest and significance in developing a facile, green and environment-friendly strategy for the synthesis of g-C_3N_4 based nanocomposite with improved photocatalytic performance [55]. For example, Jiang et al. reported a novel synthetic approach employing bamboo leaves as sources of both the C/N dopant and reductant for the formation of C/N codoped TiO_2 modified with Ag and g-C_3N_4 (Ag/CN$-TiO_2$@g-C_3N_4). In this case, the ternary composite has a hierarchical structure and a large surface area, which increases the contact area of reactants. Degradation of RhB and hydrogen generation was carried out to evaluate the photocatalytic activity of as-prepared samples under visible-light irradiation. It is found that with respect to single and binary catalysts, the Ag/CN$-TiO_2$@g-C_3N_4 ternary composite shows the highest photocatalytic activity (degradation of RhB, H_2 evolution from water splitting) as a result of the fast generation, separation, and transportation of the photogenerated carriers [56]. Similarly, Zhang and coworkers also reported a facile and green method for the synthesis of PgC_3N_4/Ag_3PO_4 without any surfactant as represented in Fig. 17.11. And they have found that the as-prepared PgC_3N_4/Ag_3PO_4 composite shows higher photocatalytic property than pure Pg-C_3N_4 and Ag_3PO_4 under visible-light irradiation. The numerous active edges and diffusion channels of Pg-C_3N_4 nanosheets can greatly speed up mass transfer and the spread of photogenerated charge carriers, consequently resulting in enhancement photocatalytic activity [57].

Chen et al. successfully prepared Cr-doped $SrTiO_3$/g-C_3N_4 hybrid nanocomposites by a green and facile method. The structural characterizations show that Cr-doped $SrTiO_3$

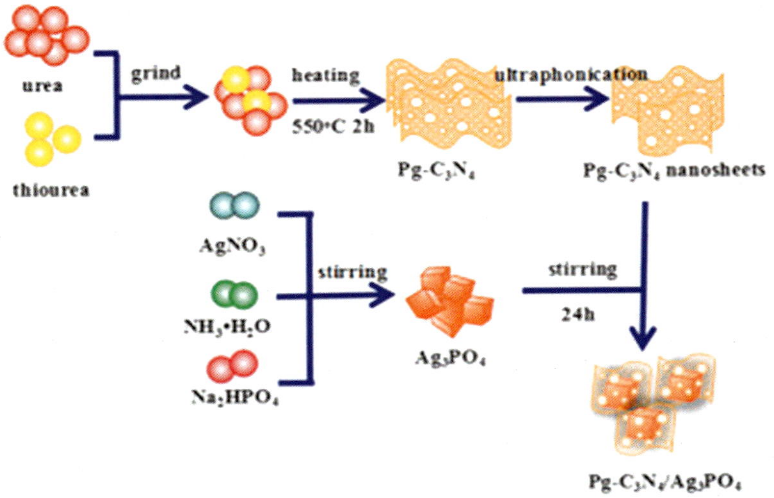

Figure 17.11
Schematic route for the synthesis of Pg-C_3N_4/Ag_3PO_4 photocatalysts.

spheres have been successfully loaded on the g-C_3N_4 nanosheets. Furthermore, the as-prepared Cr-doped $SrTiO_3$/g-C_3N_4 hybrid nanocomposites exhibit considerable stability and significantly enhanced photocatalytic activity for the degradation of RhB under visible-light irradiation. The photocatalytic activity of $SrTiO_3$/g-C_3N_4 -70% is almost 4.5 times higher than that of pure g-C_3N_4 and 3.5 times higher than that of pure Cr-doped $SrTiO_3$, respectively [58].

2. Conclusion

Due to their ultrasmall size, nanoparticles offer completely different properties in comparison to its bulk form and thus create a new existing opportunity in the field of nanoscience and nanotechnology. Green synthetic root being an environmentally benign, inexpensive, and easily scaled-up method, has received considerable research attention in recent past. Nanoparticles synthesized by using various kinds of natural components like plant extract, fungi, bacteria, and yeast are free from toxic contaminants that are required in therapeutic applications. Furthermore, the green synthetic method has also been proven to provide better control over shape, size, morphology, and other specific features of the sized nanocomposites. This book chapter was organized to encompass the “state of the art” research on the “green” synthesis of nanocomposites and their use in photocatalytic applications. Updated literature survey based on the mechanism of the synthetic of various nanocomposites and their application in the photocatalytic field has been provided to encounter the existing problem in green synthesis and to extend the laboratory-based work to an industrial scale by addressing the existing health and environmental issues.

Acknowledgment

The authors would like to acknowledge NIT Rourkela, India (Odisha) for providing the research facility and funding to carry out this work.

References

[1] T. Di, Q. Xu, W.K. Ho, H. Tang, Q. Xiang, J. Yu, Review on metal sulphide-based Z-scheme photocatalysts, ChemCatChem 11 (2019) 1394–1411.
[2] K. Zhang, L. Guo, Metal sulphide semiconductors for photocatalytic hydrogen production, Catal. Sci. Technol. 3 (2013) 1672–1690.
[3] G. Wang, Y. Ling, D.A. Wheeler, K.E.N. George, K. Horsley, C. Heske, J.Z. Zhang, Y. Li, Facile synthesis of highly photoactive α-Fe_2O_3-based films for water oxidation, Nano Lett. 11 (2011) 3503–3509.
[4] R. Kumar, S. Anandan, K. Hembram, T. Narasinga Rao, Efficient ZnO-based visible-light-driven photocatalyst for antibacterial applications, ACS Appl. Mater. Interfaces 6 (2014) 13138–13148.
[5] S.G. Kumar, K.S.R.K. Rao, Tungsten-based nanomaterials (WO_3 & Bi_2WO_6): modifications related to charge carrier transfer mechanisms and photocatalytic applications, Appl. Surf. Sci. 355 (2015) 939–958.
[6] H. Zhou, Z. Li, X. Niu, X. Xia, Q. Wei, The enhanced gas-sensing and photocatalytic performance of hollow and hollow core-shell SnO_2-based nanofibers induced by the Kirkendall effect, Ceram. Int. 42 (2016) 1817–1826.
[7] M.F.R. Samsudin, S. Sufian, B.H. Hameed, Epigrammatic progress and perspective on the photocatalytic properties of $BiVO_4$-based photocatalyst in photocatalytic water treatment technology: a review, J. Mol. Liq. 268 (2018) 438–459.
[8] S. Leong, A. Razmjou, K. Wang, K. Hapgood, X. Zhang, H. Wang, TiO_2 based photocatalytic membranes: a review, J. Membr. Sci. 472 (2014) 167–184.
[9] M. Tanveer, C. Cao, Z. Ali, I. Aslam, F. Idrees, W.S. Khan, F.K. But, M. Tahir, N. Mahmood, Template free synthesis of CuS nanosheet-based hierarchical microspheres: an efficient natural light driven photocatalyst, CrystEngComm 16 (2014) 5290–5300.
[10] Y.J. Yuan, D. Chen, Z.T. Yu, Z.G. Zou, Cadmium sulfide-based nanomaterials for photocatalytic hydrogen production, J. Mater. Chem. A. 6 (2018) 11606–11630.
[11] Z. Wu, Y. Xue, Y. Zhang, J. Li, T. Chen, SnS_2 nanosheet-based microstructures with high adsorption capabilities and visible light photocatalytic activities, RSC Adv. 5 (2015) 24640–24648.
[12] M. Oveisi, M. Alinia Asli, N.M. Mahmoodi, Carbon nanotube based metal-organic framework nanocomposites: synthesis and their photocatalytic activity for decolorization of colored wastewater, Inorg. Chim. Acta. 487 (2019) 169–176.
[13] D. Xu, B. Cheng, S. Cao, J. Yu, Enhanced photocatalytic activity and stability of Z-scheme Ag_2CrO_4-GO composite photocatalysts for organic pollutant degradation, Appl. Catal. B Environ. 164 (2015) 380–388.
[14] J. Wang, M. Gao, G.W. Ho, Bidentate-complex-derived TiO_2/carbon dot photocatalysts: in situ synthesis, versatile heterostructures, and enhanced H_2 evolution, J. Mater. Chem. A. 2 (2014) 5703–5709.
[15] S. Cao, J. Yu, g-C_3N_4-Based photocatalysts for hydrogen generation, J. Phys. Chem. Lett. 5 (2014) 2101–2107.
[16] X. Wang, G. Liu, Z.G. Chen, F. Li, L. Wang, G.Q. Lu, H.M. Cheng, Enhanced photocatalytic hydrogen evolution by prolonging the lifetime of carriers in ZnO/CdS heterostructures, Chem. Commun. (2009) 3452–3454.
[17] K. Li, S. Gao, Q. Wang, H. Xu, Z. Wang, B. Huang, Y. Dai, J. Lu, In-situ-reduced synthesis of Ti^{3+} self-doped TiO_2/g-C_3N_4 heterojunctions with high photocatalytic performance under LED light irradiation, ACS Appl. Mater. Interfaces 7 (2015) 9023–9030.

[18] H. Shen, X. Zhao, L. Duan, R. Liu, H. Wu, T. Hou, X. Jiang, H. Gao, Influence of interface combination of RGO-photosensitized SnO_2 @RGO core-shell structures on their photocatalytic performance, Appl. Surf. Sci. 391 (2017) 627–634.
[19] R. Hao, G. Wang, C. Jiang, H. Tang, Q. Xu, In situ hydrothermal synthesis of g-C_3N_4/TiO_2 heterojunction photocatalysts with high specific surface area for Rhodamine B degradation, Appl. Surf. Sci. 411 (2017) 400–410.
[20] N. Li, J. Zhang, Y. Tian, J. Zhao, J. Zhang, W. Zuo, Precisely controlled fabrication of magnetic 3D γ-Fe_2O_3@ZnO core-shell photocatalyst with enhanced activity: ciprofloxacin degradation and mechanism insight, Chem. Eng. J. 308 (2017) 377–385.
[21] F. Cheng, H. Yin, Q. Xiang, Low-temperature solid-state preparation of ternary CdS/g-C_3N_4/CuS nanocomposites for enhanced visible-light photocatalytic H_2 -production activity, Appl. Surf. Sci. 391 (2017) 432–439.
[22] M. Atarod, M. Nasrollahzadeh, S. Mohammad Sajadi, Euphorbia heterophylla leaf extract mediated green synthesis of Ag/TiO_2 nanocomposite and investigation of its excellent catalytic activity for reduction of variety of dyes in water, J. Colloid Interface Sci. 462 (2016) 272–279.
[23] Y. Fu, L. Qin, D. Huang, G. Zeng, C. Lai, B. Li, J. He, H. Yi, M. Zhang, M. Cheng, X. Wen, Chitosan functionalized activated coke for Au nanoparticles anchoring: green synthesis and catalytic activities in hydrogenation of nitrophenols and azo dyes, Appl. Catal. B Environ. 255 (2019) 117740.
[24] N. Jayaprakash, J.J. Vijaya, K. Kaviyarasu, K. Kombaiah, L.J. Kennedy, R.J. Ramalingam, M.A. Munusamy, H.A. Al-Lohedan, Green synthesis of Ag nanoparticles using Tamarind fruit extract for the antibacterial studies, J. Photochem. Photobiol. B Biol. 169 (2017) 178–185.
[25] C.B. Ong, L.Y. Ng, A.W. Mohammad, A review of ZnO nanoparticles as solar photocatalysts: synthesis, mechanisms and applications, Renew. Sustain. Energy Rev. 81 (2018) 536–551.
[26] M. Hassanpour, H. Safardoust-Hojaghan, M. Salavati-Niasari, Degradation of methylene blue and Rhodamine B as water pollutants via green synthesized Co_3O_4/ZnO nanocomposite, J. Mol. Liq. 229 (2017) 293–299.
[27] S. Sohrabnezhad, A. Seifi, The green synthesis of Ag/ZnO in montmorillonite with enhanced photocatalytic activity, Appl. Surf. Sci. 386 (2016) 33–40.
[28] F. Xiao, F. Wang, X. Fu, Y. Zheng, A green and facile self-assembly preparation of gold nanoparticles/ZnO nanocomposite for photocatalytic and photoelectrochemical applications, J. Mater. Chem. 22 (2012) 2868–2877.
[29] X. Li, Q. Wang, Y. Zhao, W. Wu, J. Chen, H. Meng, Green synthesis and photo-catalytic performances for ZnO-reduced graphene oxide nanocomposites, J. Colloid Interface Sci. 411 (2013) 69–75.
[30] H. Bozetine, Q. Wang, A. Barras, M. Li, T. Hadjersi, S. Szunerits, R. Boukherroub, Green chemistry approach for the synthesis of ZnO-carbon dots nanocomposites with good photocatalytic properties under visible light, J. Colloid Interface Sci. 465 (2016) 286–294.
[31] H. Raj Pant, B. Pant, H. Joo Kim, A. Amarjargal, C. Hee Park, L.D. Tijing, E. Kyo Kim, C. Sang Kim, A green and facile one-pot synthesis of Ag-ZnO/RGO nanocomposite with effective photocatalytic activity for removal of organic pollutants, Ceram. Int. 39 (2013) 5083–5091.
[32] S.P. Goutam, G. Saxena, V. Singh, A.K. Yadav, R.N. Bharagava, K.B. Thapa, Green synthesis of TiO_2 nanoparticles using leaf extract of *Jatropha curcas* L. for photocatalytic degradation of tannery wastewater, Chem. Eng. J. 336 (2018) 386–396.
[33] S.K. Parayil, H.S. Kibombo, C.M. Wu, R. Peng, J. Baltrusaitis, R.T. Koodali, Enhanced photocatalytic water splitting activity of carbon-modified TiO_2 composite materials synthesized by a green synthetic approach, Int. J. Hydrogen Energy 37 (2012) 8257–8267.
[34] F. Dong, S. Guo, H. Wang, X. Li, Z. Wu, Enhancement of the visible light photocatalytic activity of C-doped Tio_2 nanomaterials prepared by a green synthetic approach, J. Phys. Chem. C 115 (2011) 13285–13292.
[35] F. Dong, H. Wang, Z. Wu, One-step "Green" synthetic approach for mesoporous C-doped titanium dioxide with efficient visible light photocatalytic activity, J. Phys. Chem. C 113 (2009) 16717–16723.

[36] K.M. Parida, S. Pany, B. Naik, Green synthesis of fibrous hierarchical meso-macroporous N doped TiO_2 nanophotocatalyst with enhanced photocatalytic H_2 production, Int. J. Hydrogen Energy 38 (2013) 3545–3553.

[37] X. Liu, S. Gao, H. Xu, Z. Lou, W. Wang, B. Huang, Y. Dai, Green synthetic approach for Ti^{3+} self-doped TiO_2-x nanoparticles with efficient visible light photocatalytic activity, Nanoscale 5 (2013) 1870–1875.

[38] X. Zhang, Y. Sun, X. Cui, Z. Jiang, A green and facile synthesis of TiO_2/graphene nanocomposites and their photocatalytic activity for hydrogen evolution, Int. J. Hydrogen Energy 37 (2012) 811–815.

[39] Y. Yang, E. Liu, J. Fan, X. Hu, W. Hou, F. Wu, Y. Ma, Green and facile microwave-assisted synthesis of TiO_2/graphene nanocomposite and their photocatalytic activity for methylene blue degradation, Russ. J. Phys. Chem. A. 88 (2014) 478–483.

[40] P. Muthirulan, C. Nirmala Devi, M. Meenakshi Sundaram, A green approach to the fabrication of titania-graphene nanocomposites: insights relevant to efficient photodegradation of Acid Orange 7 dye under solar irradiation, Mater. Sci. Semicond. Process. 25 (2014) 219–230.

[41] M.S.A. Sher Shah, A.R. Park, K. Zhang, J.H. Park, P.J. Yoo, Green synthesis of biphasic TiO_2-reduced graphene oxide nanocomposites with highly enhanced photocatalytic activity, ACS Appl. Mater. Interfaces 4 (2012) 3893–3901.

[42] H. Liu, X. Dong, X. Wang, C. Sun, J. Li, Z. Zhu, A green and direct synthesis of graphene oxide encapsulated TiO_2 core/shell structures with enhanced photoactivity, Chem. Eng. J. 230 (2013) 79–85.

[43] H. Dang, X. Dong, Y. Dong, J. Huang, Facile and green synthesis of titanate nanotube/graphene nanocomposites for photocatalytic H_2 generation from water, Int. J. Hydrogen Energy 38 (2013) 9178–9185.

[44] D. Hidalgo, S. Bocchini, M. Fontana, G. Saracco, S. Hernández, Green and low-cost synthesis of PANI-TiO_2 nanocomposite mesoporous films for photoelectrochemical water splitting, RSC Adv. 5 (2015) 49429–49438.

[45] P. Wang, D. Li, J. Chen, X. Zhang, J. Xian, X. Yang, X. Zheng, X. Li, Y. Shao, A novel and green method to synthesize CdSe quantum dots-modified TiO_2 and its enhanced visible light photocatalytic activity, Appl. Catal. B Environ. 160–161 (2014) 217–226.

[46] D. Yang, Y. Sun, Z. Tong, Y. Tian, Y. Li, Z. Jiang, Synthesis of Ag/TiO_2 nanotube heterojunction with improved visible-light photocatalytic performance inspired by bioadhesion, J. Phys. Chem. C 119 (2015) 5827–5835.

[47] M. Nasrollahzadeh, M. Atarod, B. Jaleh, M. Gandomirouzbahani, In situ green synthesis of Ag nanoparticles on graphene oxide/TiO_2 nanocomposite and their catalytic activity for the reduction of 4-nitrophenol, Congo red and methylene blue, Ceram. Int. 42 (2016) 8587–8596.

[48] C. Liu, D. Yang, Y. Jiao, Y. Tian, Y. Wang, Z. Jiang, Biomimetic synthesis of TiO_2-SiO_2-Ag nanocomposites with enhanced visible-light photocatalytic activity, ACS Appl. Mater. Interfaces 5 (2013) 3824–3832.

[49] A. Shabani, G. Nabiyouni, J. Saffari, D. Ghanbari, Photo-catalyst Fe_3O_4/TiO_2 nanocomposites: green synthesis and investigation of magnetic nanoparticles coated on cotton, J. Mater. Sci. Mater. Electron. 27 (2016) 8661–8669.

[50] Y. Zhang, J. Tian, H. Li, L. Wang, X. Qin, A.M. Asiri, A.O. Al-Youbi, X. Sun, Biomolecule-assisted, environmentally friendly, one-pot synthesis of CuS/reduced graphene oxide nanocomposites with enhanced photocatalytic performance, Langmuir 28 (2012) 12893–12900.

[51] S. Vadivel, V.P. Kamalakannan, Keerthi, N. Balasubramanian, D-pencillamine assisted microwave synthesis of Bi_2S_3 microflowers/RGO composites for photocatalytic degradation - a facile green approach, Ceram. Int. 40 (2014) 14051–14060.

[52] Z. Li, J. Zhang, J. Lv, L. Lu, C. Liang, K. Dai, Sustainable synthesis of CeO_2/CdS-diethylenetriamine composites for enhanced photocatalytic hydrogen evolution under visible light, J. Alloys Compd. 758 (2018) 162–170.

[53] A.A.P. Mansur, H.S. Mansur, F.P. Ramanery, L.C. Oliveira, P.P. Souza, "Green" colloidal ZnS quantum dots/chitosan nano-photocatalysts for advanced oxidation processes: study of the photodegradation of organic dye pollutants, Appl. Catal. B Environ. 158–159 (2014) 269–279.
[54] Y.J. Yuan, D.Q. Chen, X.F. Shi, J.R. Tu, B. Hu, L.X. Yang, Z.T. Yu, Z.G. Zou, Facile fabrication of "green" SnS_2 quantum dots/reduced graphene oxide composites with enhanced photocatalytic performance, Chem. Eng. J. 313 (2017) 1438–1446.
[55] N. Cheng, J. Tian, Q. Liu, C. Ge, A.H. Qusti, A.M. Asiri, A.O. Al-youbi, X. Sun, Au-nanoparticle-loaded graphitic carbon nitride nanosheets: Green photocatalytic synthesis and application toward the degradation of organic pollutants, ACS Appl. Mater. Interfaces 5 (2013) 6815–6819.
[56] Z. Jiang, D. Liu, D. Jiang, W. Wei, K. Qian, M. Chen, J. Xie, Bamboo leaf-assisted formation of carbon/nitrogen co-doped anatase TiO_2 modified with silver and graphitic carbon nitride: novel and green synthesis and cooperative photocatalytic activity, Dalton Trans. 43 (2014) 13792–13802.
[57] J. Zhang, J. Lv, K. Dai, Q. Liu, C. Liang, G. Zhu, Facile and green synthesis of novel porous g-C_3N_4/Ag_3PO_4 composite with enhanced visible light photocatalysis, Ceram. Int. 43 (2017) 1522–1529.
[58] X. Chen, P. Tan, B. Zhou, H. Dong, J. Pan, X. Xiong, A green and facile strategy for preparation of novel and stable Cr-doped $SrTiO_3$/g-C_3N_4 hybrid nanocomposites with enhanced visible light photocatalytic activity, J. Alloys Compd. 647 (2015) 456–462.

SECTION E

Development of metal phthalocyanine nanostructures

CHAPTER 18

Metal phthalocyanines and their composites with carbon nanostructures for applications in energy generation and storage

K. Priya Madhuri, Neena S. John
Centre for Nano and Soft Matter Sciences, Bengaluru, Karnataka, India

Chapter Outline

Design, Fabrication, and Characterization of Multifunctional Nanomaterials. https://doi.org/10.1016/B978-0-12-820558-7.00009-1

1. Introduction

The present-day information-oriented society is being significantly driven by inorganic semiconductors for applications such as transistor-based computers, light emitting diodes, semiconductor lasers, and optical communication devices such as waveguides. However, the recent advances made in the field of organic materials opened up avenues for next generation devices toward applications involving organic electronics, organic optoelectronics, and organic photonics. The newly emerging fields are mainly concerned with solution processing to obtain novel thin films coatings for flexible device manufacture [1,2]. Devices based on organic materials possess several advantages over inorganic semiconductor based devices such as flexibility, light-weight and cost-effectiveness [3,4]. Organic photovoltaic devices (OPVs), organic light emitting diodes (OLEDs) and organic field effect transistors (OFETs) have become the central research topics for the development in the fields of electronics and optoelectronics [5,6]. Organic materials for use in optoelectronic devices are often referred to as photo and electroactive molecules as they are characterized by the capability of light absorption and emission at the desired wavelength. They can also photo-generate as well as transport charge carriers and participate in the injection of charge carriers upon application of an electrical bias. Electroactive organic molecules based on π-electron systems include small molecules and oligomers with well-defined structures that are crystalline in nature and are termed as organic semiconductors. The representative classes of electroactive molecular materials are hydrogen phthalocyanine (H_2Pc) and its derivatives, metal phthalocyanines (MPc), porphyrins, perylene bisdiimides, and polycondensed aromatic hydrocarbons such as anthracene, pentacene, and fullerenes [7,8]. These molecules can exist as molecular crystals or can be amorphous. Chemical structures of a few organic molecules are shown in Fig. 18.1.

A substantial amount of research has been dedicated to metal-free phthalocyanine, H_2Pc and other MPcs because of the inherent versatile properties and prospective applications that could be exploited [9]. The possibilities of varying the metal center and functionalization of the surrounding four isoindoline ligands to tune the chemical and electronic properties render MPcs an exciting class of electroactive molecules. MPcs are probably the most studied class of functional molecular materials, structurally similar to porphyrins and are also known as porphyrazanes [7]. This comes from the fact that the MPcs possess fascinating structural diversity and exceptional stability. These materials are two-dimensional (2D) and hold 18 electrons in their system to form a delocalized aromatic cloud over an arrangement of alternate carbons and nitrogens rendering them as *p*-type semiconducting donors. The delicate balance between the contributions from the central metal cation and the π-electrons of the macrocyclic ligand can lead to many exciting functionalities. They are also well known for their excellent thermal and chemical

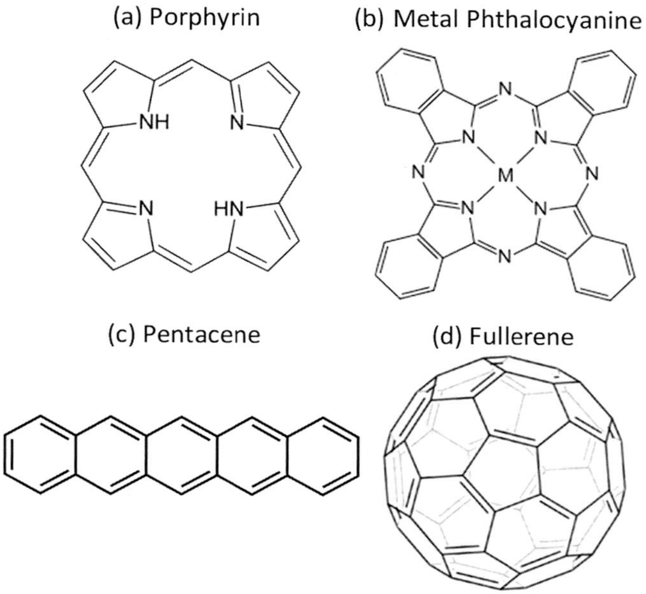

Figure 18.1
Structures of some small organic electroactive molecules.

resistance. The stability may arise due to extensive conjugation in the system. They can be heated to temperatures as high as 500°C without any chemical decomposition, and their chemical structure remains unaltered even when exposed to oxidative acids and bases [10–13].

A few exceptional properties of MPcs in optical, electrical and electrochemical realms play an important role in serving emerging applications like energy storage and conversion. The MPc nanostructures or thin films exhibit characteristic intense absorption bands in visible and near-infrared (NIR) regions of the solar spectrum [14]. Further, the anisotropic structural arrangements in MPcs offer maneuvering of specific absorption bands. For example, promoting the growth of the triclinic PbPc polymorph increases the NIR sensitivity for improved solar cell efficiencies [15]. Due to their visible-NIR responsiveness, photoinduced water splitting reactions are also increasingly explored using MPcs as one of the layers forming a *p*/*n* heterojunction [16,17]. On the other hand, MPcs behave as molecular semiconductors and their electrical conductance ranges widely from 10^{-4} to as low as 10^{-12} S/cm [14]. The electrical properties are also structure-dependent and thus witness contrasting current responses. In OPVs and OLEDs, a layer of MPc is sandwiched between the electrically conducting substrates and the heterojunction with the electron transport layer is formed, while MPc acts as the hole transport layer [18,19]. One of the advanced applications include the combination of OFET and OLED to create a fully flexible device structure [20,21]. Apart from these, MPcs possess interesting electrochemical properties as they bear redox active pyrrolic nitrogen moieties along with the centrally located metal cations. These redox processes become important for catalyzing and identifying a variety of biomolecules and other industrial effluents such as CO_2, oxides of nitrogen, known as electrochemical detection [22,23]. They also find utmost importance in energy generation applications such as hydrogen evolution and oxygen reduction reactions [24]. Thus, MPcs are the molecular materials that possess excellent stability and are useful candidates for numerous device-specific applications. The detailed physical properties of the MPcs are discussed in the following sections. Although the conductivity of the native MPcs is found to be poor, they are often found to be combined with conductive carbon matrices such as graphene, reduced graphene oxide, fullerenes, carbon nanotubes and mesoporous carbon [25–28]. Improved electronic conductivities as a result of the synergy between the two sp^2 hybridized systems result in improved device responses for a plethora of applications such as OPVs, OFETs, gas sensors, energy storage and energy generation [29]. Further, the combination of MPcs with carbon based nanomaterials gives rise to enhanced electrochemical supercapacitance and other electrochemical applications [30]. The combination of MPcs contributing to pseudocapacitance and carbon materials contributing to electrochemical double layer capacitance is an excellent example for the synergy of the composites.

2. Properties of metal phthalocyanines

2.1 Structural characteristics

Depending on the type of metal cation that is accommodated inside a Pc ring, the size and shape of the corresponding MPc can be varied. The size of a metal-free Pc, i.e., H_2Pc is found to be 3.93 Å and the largest diameter can go up to 3.96 Å when the incorporated metal ion is Zn (II) or Pt (II) with ionic radii of 0.74 Å. With nearly every transition metal cation that can be incorporated, the planar configuration of the molecule is retained with square planar geometry bearing coordination number 4 (increased D_{4h} symmetry from D_{2h} in metal-free H_2Pc) as shown in Fig. 18.2A. Further, depending on the oxidation state of the metal cation held in the central core, one or even two metal ions can be incorporated into the Pc ring with planarity retained. When the heavy metal ion is substituted (Pb and Sn with ionic radii of 1.19 and 1.18 Å), the planarity of the complex is lost due to the protrusion of the metal ion out of the planar conjugated ring, giving rise to shuttlecock-like structure (Fig. 18.2B). With higher coordination number, pyramidal, tetrahedral, and octahedral structures arise with one or two axial ligands attached to central metal ion (M = Ti, V, Ru, In) as represented in Fig. 18.2C. Heavy f-block elements (actinides and lanthanides) are exceptionally large and give rise to sandwich complexes with metal located between two Pc rings. These structures are also termed as bis-phthalocyanines (Fig. 18.2D) [10,31–33].

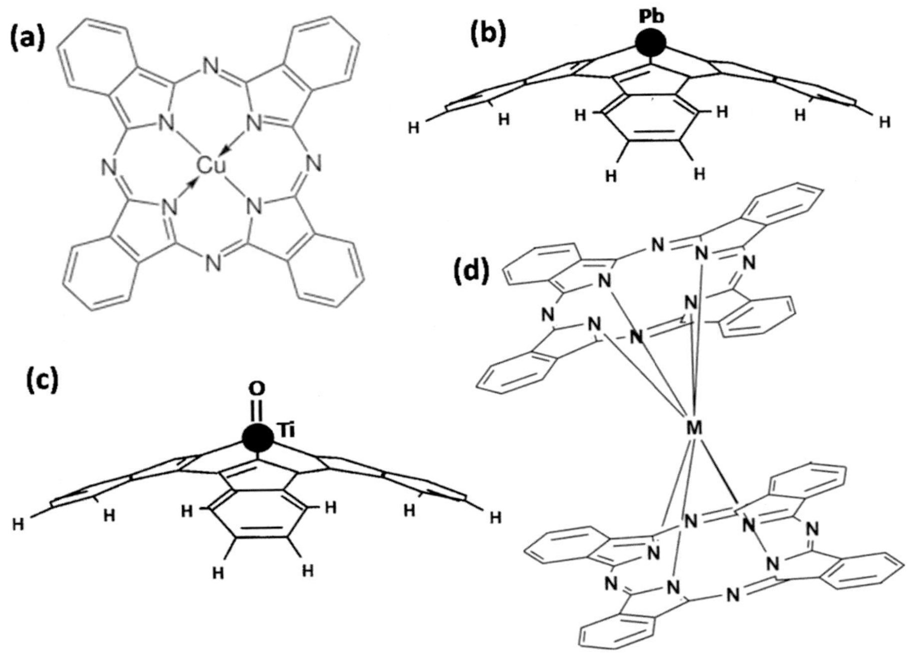

Figure 18.2
Schematic of (A) planar MPc, (B) shuttlecock MPc, (C) shuttlecock MPc with axial ligand and (D) sandwich-type MPc structures.

One of the most interesting features of MPcs is that the strong π-π and van der Waals interactions give rise to different spatial arrangements or polymorphic forms leading to exciting anisotropic properties. The polymorphs have various stacking columnar structures which differ in tilting angles between their molecular column and stacking axes. The interplanar distances between the successive macrocycles also vary considerably along the columnar structure. Almost all the MPcs are crystalline in nature and are known to stack in a herringbone-type fashion. The most common types of polymorphic crystals exhibited by planar unsubstituted MPcs are of two types denoted as α (P1) and β (P21/a). The stabilities of the two phases are dependent on temperature with α phase being metastable (formed at lower temperatures), and β phase is thermodynamically stable (formed at elevated temperatures, $> 100°C$). The two polymorphs differ in their tilt angle (δ), with α form of crystal exhibiting $\delta = 65°$ and β form, $\delta = 45°$. A schematic of α and β crystalline phases showing different molecular packing is represented in Fig. 18.3A [34,35]. However the packing in nonplanar MPcs is different compared to its planar counterpart. The loss of ring planarity with the accommodation of bulky metal cation (Pb or Sn), imparts different spatial arrangements. Certain other MPcs like VOPc and

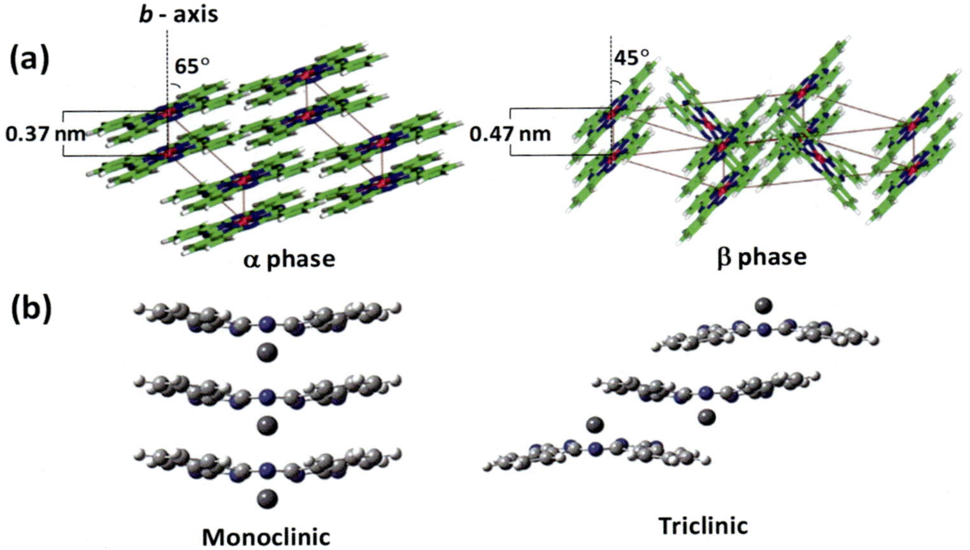

Figure 18.3

(A) Schematic showing the molecular packing in α and β crystalline forms of a planar MPc [34]. (B) Polymorphism in nonplanar MPc with monoclinic and triclinic forms on flat substrate surfaces [32]. *(A) Reproduced by permission of The Royal Society of Chemistry. (B) Reproduced from Ref. N. Hamamoto, H. Sonoda, M. Sumimoto, K. Hori, H. Fujimoto, Theoretical study on crystal polymorphism and electronic structure of lead (II) phthalocyanine using model dimers, RSC Adv. 7 (2017) 8646–8653. Published by The Royal Society of Chemistry.*

TiOPc have an oxygen atom sticking out of the planar conjugated ring, which along with its dipolar character imparted to the molecule gives rise to specific polymorphism [36]. The nomenclature for polymorphs among nonplanar MPcs is usually different and is represented as monoclinic (*P21/b*) and triclinic (*P*1). The monoclinic form is also known as *H*-aggregate in which the macrocycles stack in a head-to-head fashion while the triclinic form, known as *J*-aggregate has a stacking in a head-to-tail configuration (Fig. 18.3B) [37].

In simple terms, the monoclinic arrangement gives rise to linearly stacked column of MPc macrocycles, while the triclinic offers an arrangement such that molecules stack with adjacent overlapping of the macrocyclic rings resulting in interesting anisotropic properties. When deposited on flat substrates, these two phases can take the forms of face-on and edge-on orientations with macrocycles lying either parallel or perpendicular to the plane of the substrate [15]. The differences in nature of stacking of MPc molecules can give rise to structurally dependent physical properties that becomes crucial for the fabrication of MPc based devices. The molecules assembled in long-range ordered (1D) networks have a significant impact on optical, electrical and magnetic properties.

The nature of the substrate and the templating layers exert a large impact on the orientation of the incoming MPc molecules, which in turn lead to the formation of different polymorphic architectures. The face-on molecular orientation of the triclinic polymorph, is particularly useful for OPV applications resulting in significantly improved solar cell performance [18,38]. The end-to-end overlaps of MPc moieties in this configuration can extend π-delocalization laterally, thereby providing more photon harvest in NIR region [39]. In a slightly different configuration, the face-on arrangement of monoclinic polymorph gives rise to enhanced out-of-plane conduction owing to π-electron delocalization in a vertical direction, which is utilized for OLEDs and gas sensors. On the other hand, edge-on orientation of the monoclinic form provides charge delocalization along the direction, parallel to the gate dielectric, which is exploited for OFET applications [40,41].

2.2 Optical properties

Since MPcs are extensively conjugated aromatic chromophoric systems, they exhibit intense bands in their absorption spectra and are brilliantly blue colored complexes. Their absorption ranges from ultraviolet to visible and NIR regions with a high molar absorptivity of ~ 10^5 L/mol cm. The characteristic B (Soret band) and Q bands appearing around 300−450 and 650−750 nm, respectively, are a hallmark to the recognition of any given MPc. While MPcs exhibit a single Q band, the metal-free Pc has a split Q band with equal intensity, due to lowered D_{2h} symmetry. Incorporation of metal inside the macrocycle leads to changes in electronic structure, which results in shifting of the absorption bands as a function of metal size, coordination and oxidation states [1,14]. The appearance of B and Q bands are a result of various electronic transitions occurring between the HOMO and LUMO of the macrocycle. The B band is caused due to weaker transitions from a_{2u} and b_{2u} to e_g* and the Q bands arise as a result of a_{1u} to e_g* (Fig. 18.4A). Other electronic transitions include metal to ligand (d-π*) and from ligand to metal (π-d) in the ultraviolet regions of 250−350 nm. MPcs often aggregate in the form of H and J with characteristic absorption features. In PbPc, the H-aggregate with monomers stacking in head-to-head fashion exhibits blueshift and J-aggregate with head-to-tail configuration exhibits red-shift. The monomer units also are identified in the spectrum lying between H- and J-aggregates (Fig. 18.4B). This enables them as useful candidates for the applications such as optical sensors, photoreceptors, photosensitizers, solar cells and OLEDs [42−44].

2.3 Electrical properties

The electrical properties of organic molecules are different from the inorganic counterparts in which the bonding in the former case occurs mainly through weak van der Waals forces

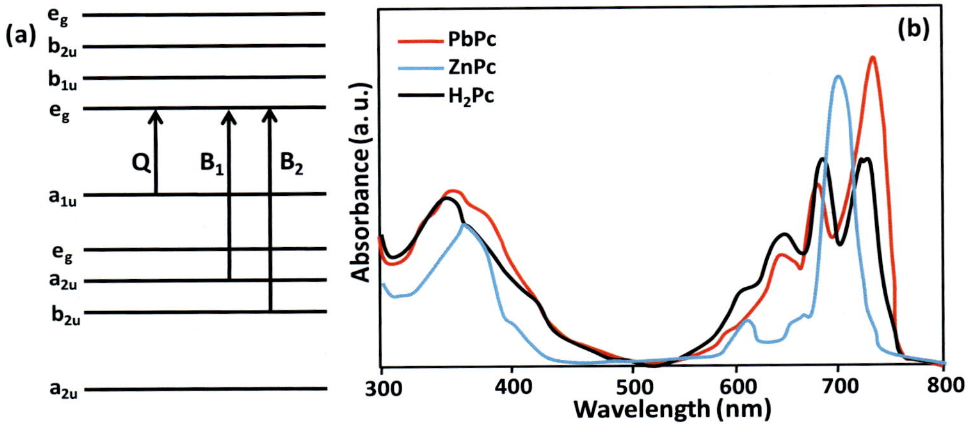

Figure 18.4
(A) Molecular orbital level diagram of MPc showing electronic transitions giving rise to specific absorption bands (B) UV-Vis absorption spectra of various MPcs.

with minimum orbital overlapping. The latter case is characterized by strong covalent and ionic interactions through atomic orbitals overlapping. Thus, distinct valence and conduction bands transform to HOMO and LUMO in organic molecules. In highly delocalized MPc systems, the sp^2 hybridized carbon atoms with p_z orbitals from adjacent carbon atoms form an electron charge cloud above and below the molecular plane of the MPc macrocycle [45–47]. These π-electrons are mobile, which confer conductivity in MPcs and open up avenues for various applications as electronic devices in gas sensors, transistors, OLEDs, OPVs, and so forth. In general, MPcs are *p*-type semiconductors in which the conductivity is either due to the intrinsic nature or the type of organization of MPc molecules at a supramolecular level. This type of assembly has an extended π-orbital overlap promoting the conductivity pathway in a specific direction depending on the type of molecular packing, as discussed in polymorphism. The p-type intrinsic conductivity, although on the insulator borderline (10^{-12} S/cm), can be enhanced tremendously by the hole and electron doping, making them promising materials for organo-electronic applications, with thermal activation energy lower than 2 eV [14]. The out-of-plane protrusion of a heavy element as in PbPc, is considered as a representative one in its monoclinic configuration possessing highest D.C. conductivity of 10^{-4} S/cm due to its exceptional stacking arrangement. This has also been referred to be an organic 1-D conductor [48]. Fig. 18.5 shows the schematic of the dependence of the MPc molecular organization on electrical properties which can be utilized for different applications. The desired molecular organization can be achieved by using templating layers beneath the MPc thin films. The templating layer interacts with the film to offer specific orientation of molecules which differ in their electrical properties. Native substates such as graphite and Si themselves offer different kind of substrate-molecule interactions in which the former

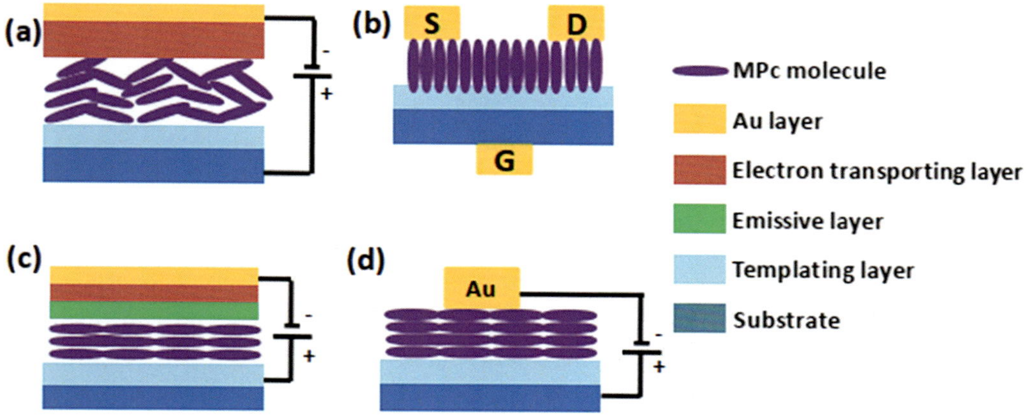

Figure 18.5
Schematic showing the dependence of MPc molecular organization on different applications (A) OPV, (B) OFET, (C) OLED, and (D) gas sensors.

induces face-on stacking due to π-π interactions while the latter favors edge-on orientation due to van der Waals and hydrogen bonding. Madhuri et al. demonstrated in detail the enhanced out-of-plane electrical conductivity arising from face-on stacked PbPc thin film deposited on graphite substrate owing to π-electron delocalization in the vertical axis [49]. Fig. 18.5A shows the triclinic phase with end-to-end overlap of macrocycles. This property has been particularly beneficial for improved solar performance providing extensive π delocalization laterally with more NIR photon harvest. Fig. 18.5B shows the edge-on orientation of MPc moieties, useful for OFET applications in which the charge delocalization along the lateral direction lies parallel to the gate dielectric. The face-on stacked orientation of MPc moieties (Fig. 18.5C and D) exhibit enhanced out-of-plane conductivity owing to π-electron delocalization in vertical direction, beneficial for OLEDs and gas sensors.

Halogenation of MPcs is popularly adopted to enhance their conductivity in which electron acceptors such as halogens, particularly iodine, will promote the hole charge carriers. Iodinated MPc crystals are very well established and are known to exhibit temperature dependent conductivity akin to metals, with conductivity as high as 10^4 S/cm. Thus iodinated MPcs are also known as metallic conductors. The established conductivity is due to the presence of iodine in the form of triiodide chains partially oxidizing the metal macrocycle. Iodine doping can be performed in MPc thin films as well, altering MPc packing and improving charge transport across the MPc domains by hole doping [50,51]. The doping concept has been utilized in plethora of applications such as lithium-ion batteries, oxygen reduction reactions, enhanced NIR absorption for photovoltaic devices, and gas sensors. It has also given rise to interesting magnetic properties indicating coupling of local spins with itinerant electrons [50].

2.4 Magnetic properties

The recent interest in magnetism lies with molecular magnets, wherein the organic molecules possessing metal atoms with unpaired spins behave as molecular magnets due to several coupled magnetic atoms. Transition MPcs exhibit unique properties that primarily depend on the electronic ground state configuration of the type of the metal-substituted in the macrocycle. The *d*-electronic configuration of the occupied central metal atom, the columnar stacking of the macrocycles with differing tilting angles and different interplanar distances also lead to tremendous effect on the magnetic properties. Like electrical properties, the molecular packing in MPcs also affect the magnetic properties giving rise to ferromagnetic and antiferromagnetic interactions. The resultant of the magnetic coupling is the sum of various superexchange interactions between the metal and the macrocycle orbitals. Applications of these systems include a wide range from high-capacity storage devices to quantum computers. Magnetic interactions between the metal ions of MPcs adsorbed on ferromagnetic surfaces lead to essential applications such as chemical switches and giant magnetoresistance [52–54].

2.5 Electrochemical properties

MPcs are metal-N_4 complexes that are recognized for their excellent physico-chemical and redox properties. The complexes are known for various energy applications such as electrocatalysis, electrochemical sensing, photoelectrocatalysis and supercapacitors. The various metal ions amidst the macrocycle ring of MPc complexes possess redox properties that can be utilized for electrochemical applications. When the substituted metal ion is the transition metal, the redox reactions occur at lower potentials compared to macrocyclic redox processes. The *d*-orbitals are positioned between the π (HOMO) and π^* (LUMO) orbitals of the Pc ligands. Thus the redox potentials observed for the metal ions may occur at the potentials close to the macrocyclic ligands and are reversible in nature [31,55]. Table 18.1 gives the oxidation and reduction potentials of different MPcs obtained in non-aqueous media against standard calomel electrode (SCE).

Table 18.1: List of oxidation and reduction potentials of different MPcs (M = Mn, Fe, Co, Ni, Cu, Zn and H_2) [56,57].

MPc	Oxidation potential (V vs. SCE)	Reduction potential (V vs. SCE)	Type of electron transfer
FePc	0.19	–	$M^{2+} \leftrightarrow M^{3+}$
CoPc	0.77	−0.37	$M^{2+} \leftrightarrow M^{3+}$
NiPc	1.05	−0.85	Ligand redox
CuPc	0.98	−0.84	Ligand redox
ZnPc	0.68	−0.89	Ligand redox
H_2Pc	1.10	−0.66	Ligand redox

Further, the oxidation potentials in MPcs can be correlated to the electronic spectra (UV-Vis) and the charge transfer bands (NIR) [31]. Since MPcs possess dual nature of π-donor and acceptor functionalities, they have been found to demonstrate excellent electrocatalytic activity for numerous technologically significant redox reactions. The electrochemical performance of the MPc electrodes substantially depend on factors like applied potential, pH of the electrolyte and the nature of central metal ion. Applications include electroreduction of CO_2, harmful chemicals like SO_2, chlorate ions, thionyl chloride, electrooxidation of hydrazine, SO_2, organic acids, and biomolecules [58–60]. Initially, the metal cation undergoes oxidation, which further aids in oxidizing adsorbed biomolecules. Water electrolysis is another exciting application seeking the development of electrocatalysts with lowered overpotentials [58]. Due to their visible-NIR responsiveness, photoinduced water splitting reactions are increasingly explored using MPc composites or *p*/*n* heterojunctions [16,17]. Recently, MPcs have been reported as outstanding candidates for charge storage applications in which they serve as pseudocapacitors, i.e., materials which possess charges as a virtue of their redox properties and contribute toward electrochemical supercapacitance. This is also referred to as faradaic capacitance. Further, the presence of pyrrolic nitrogens in the Pc complex aid in promoting the pseudocapacitance by undergoing faradaic reactions [30,61] Thus the MPcs with the metal cation in the macrocycle along with pyrrolic nitrogen atoms act as active redox centers with specific and sharp redox transitions [62].

3. Preparation of metal phthalocyanine-carbon nanocomposites

Metal phthalocyanine-carbon nanocomposites can be prepared using simple synthetic approaches in which the morphology and the size aspect can be tailor-made to suite a specific application. The general synthetic procedures involved for the synthesis of MPc—carbon composites can be broadly categorized into physical and chemical methods. In physical method, preformed constituents such as MPc and carbon materials (graphene, rGO or CNTs) are used as precursors with physical mixing of the two components by sonication or heating to yield a composite material having a good physical interface. The chemical method involves a certain chemical reaction that occurs between the two individual precursors often associated with covalent interactions to give chemically attached MPcs on the surface of carbon matrix. The following discussion gives synthesis methods using different carbon nanostructures for MPc composites.

3.1 Metal phthalocyanine-reduced graphene oxide nanocomposites

Graphene oxide (GO) and reduced graphene oxide (rGO) are most commonly prepared adopting modified Hummers' method using natural graphite powder. Unlike graphene, GO offers a wide range of chemical methods for the attachment of different functional groups on to its surface thereby allowing control on their optical transparency, electrical, and thermal conductance. Studies show that the GO has a basal plane of sp^3 hybridized carbon atoms with a sheet-like structure and contain numerous oxygen containing chemically bonded hydroxyl and epoxy groups on the basal plane and carboxyl, lactol rings, ketone, and ester groups at the edges. The reduced form of GO is known as rGO in which the oxidized areas are restored to sp^2 bonded carbon network. Studies reveal that there is a remarkable amount of topological defects after the reduction process and these defects induce in-plane and out-of-plane strain of rGO surface. The increased concentrations of the defects in the basal plane during oxidation and remnant functional groups render them inferior to pure graphene. However, due to the removal of majority of the oxygen containing functional groups that limit the electron transfer pathway, rGO possesses better electrical conductance over GO. The typical thickness of the GO and rGO layers are about 1.0 ± 0.2 nm and 0.3 ± 0.2 nm, with interlayer distance of 0.34 nm in rGO similar to that of a graphene. Thus GO and rGO offer wide variety of composite formation by tuning or functionalizing their surface to selectively bind to certain molecules, which aid in specific applications. The delocalized π electrons in sp^2 hybridized carbon atoms participate in offering synergy to the incoming organic molecules [63,64].

3.1.1 Physical mixing

The general strategy adopted for the preparation of MPc—GO or rGO composite through physical mixing involves addition of GO or rGO to MPc with heating or ultrasonication. In situ formation of rGO using GO precursor in the presence of MPc can be achieved by the addition of a reducing agent that facilitates better interaction between the two components. MPc nanostructures can be obtained from phthalonitrile and the corresponding metal salt followed by heating the mixture in a high boiling solvent like ethylene glycol. The following description is enumerated with literature examples.

The nanocomposite of tetrasulfonate salt of CuPc (TSCuPc) with rGO has been fabricated through the reduction of GO in the presence of water soluble TSCuPc using hydrazine hydrate as a reducing agent at 90°C. The resulting dark green solution was stable for more than three months and contained rGO monolayer sheets grafted with TSCuPc [26]. In another report, the GO obtained by modified Hummers' method was dispersed in N,N-Dimethylformamide (DMF) and treated with NiPc nanofibers dispersed in trifluoroacetic acid and heated to 100°C in a hydrothermal vessel for 24 h. The NiPc nanofibers were first synthesized by heating a mixture of phthalonitrile and corresponding metal salt in the ratio 1:4 in presence of ammonium heptamolybdate catalyst suspended in ethylene glycol solvent at 100°C for 8 h [30]. The NiPc nanofibers were then mixed with GO suspended in DMF and treated hydrothermally at a temperature of 120°C. Madhuri et al. has demonstrated the preparation of various MPc nanostructures (MnPc, FePc, CoPc, NiPc, CuPc, ZnPc) dispersible in polar solvents [65]. The obtained nanostructures are elongated in shape and are designated as respective MPc nanofibers (MPc NF) as shown in Fig. 18.6A. The morphology of the obtained MPc NFs were studied using scanning electron microscope (SEM) and transmission electron microscope (TEM). Fig. 18.6B and C show the SEM and TEM images of CuPc NF. The diameter of the fiber was found to be ~100 nm and length >12 μm. The TEM image in the inset of Fig. 18.6C shows clear lattice fringes that match the (100) plane of β-CuPc arising at $2\theta = 7.06°$. Wang et al. dissolved the synthesized CuPc structures in DMF containing rGO suspension which was obtained by reducing GO using hydrazine hydrate. The resulting mixture is vigorously stirred for about 2 h to yield a rGO-CuPc nanocomposite [66]. Mukherjee et al. prepared rGO-CuPc nanocomposite using GO previously synthesized using modified Hummers' method and CuPc nanotubes obtained through the method described systematically in their report. The precursors were dispersed in DMF followed by the addition of hydrazine hydrate and ammonia solution [67]. In a different report by Mukherjee et al. rGO-ZnPc nanocomposite was prepared using GO, previously obtained using modified Hummers'

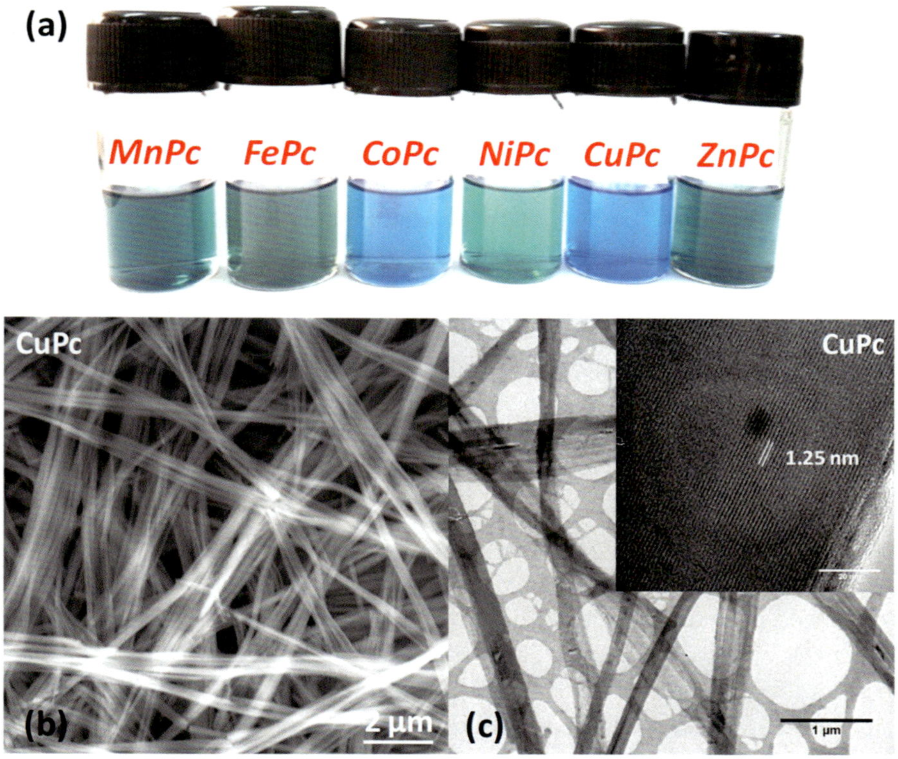

Figure 18.6
(A) Photograph of various transition metal MPc nanofibers dispersed in ethanol medium. (B) SEM and (C) TEM images of synthesized CuPc NF [65]. *Reproduced by permission of John Wiley and Sons.*

method and ZnPc synthesized by suspending phthalonitrile, zinc acetate and ammonium heptamolybdate in ethylene glycol contained in an autoclave and subjected to 180°C for 24 h. Finally an equal amount of GO and ZnPc were mixed in a DMF solution followed by the addition of hydrazine hydrate and ammonia and heating at 90°C for 2 h [68].

3.1.2 Chemical synthesis

Unlike physical mixing method, the chemical synthetic route provides covalent bonding between the carbon matrix and the MPc nanostructures. The covalent bonding is easily achieved by functionalizing the respective components with desired chemical groups. The synthesis of rGO-ZnPc hybrid is based on the covalent linkage of ZnPc to GO by an esterification reaction and subsequent in situ reduction of GO moieties to rGO during mild thermal treatment using DMF solvent as reported by Song et al. [69] A simple and elegant method of concomitant electropolymerization of tetraamino functionalized CoPc (TACoPc) at electrochemically reduced GO to prepare highly stable nanocomposite has been reported by Mani et al. This method bridges excellent physicochemical property of rGO along with redox rich chemistry of TACoPc [70]. Zhang et al. prepared tetranitro-CoPc (TNCoPc) and rGO composite using GO and TNCoPc dispersed in ethylene glycol as a reducing agent. The resulting mixture was subjected to hydrothermal treatment at 160°C for 24 h. TNCoPc nanostructure was fabricated using 4-nitrophthalonitrile, $Co(OAc)_2$, and ammonium heptamolybdate in a high boiling solvent like ethylene glycol [71]. On the other hand, metal phthalocyanine has been combined with just GO to maximize the surface functionalization and chemical interactions apart from having π-π interactions in a few cases. Yang et al. adapts a facile wet-chemical route to synthesize CoPc-GO nanostructures involving the intercalation and adsorption of CoPc onto the layers of GO through π-stacking. GO was first prepared by modified Hummers' method. For the preparation of CoPc-GO, a defined amount of CoPc in acetone was vigorously stirred for 2 h and ultrasonicated for 1 h followed by the addition of GO. This mixture was put into the oil bath at 60°C until the solvent evaporated. The obtained powder was washed with mixture of ethanol, water and ammonia and dried [72]. In another simple method reported by Zhang et al., 2,11,20,29-tetra-tert-butyl-2,3-naphthalocyanine in chloroform was mixed with GO chloroform solution which was obtained previously from modified Hummers' method. The resulting mixture was sonicated for 15 min followed by centrifugation [73]. In a different report by Zhang et al. developed a facile method for the synthesis of FePc and nitrogen doped graphene nanocomposite. The details of the liquid phase approach adapted for the synthesis of nitrogen doped graphene using GO and ammonia solution has been reported elsewhere [74].

The preparation methods of MPc-rGO composite can be presented as the schematic given in Fig. 18.7. The figure shows the initial preparation of MPc nanostructures by adding phthalonitrile, corresponding metal salt and ammonium heptamolybdate in catalytic

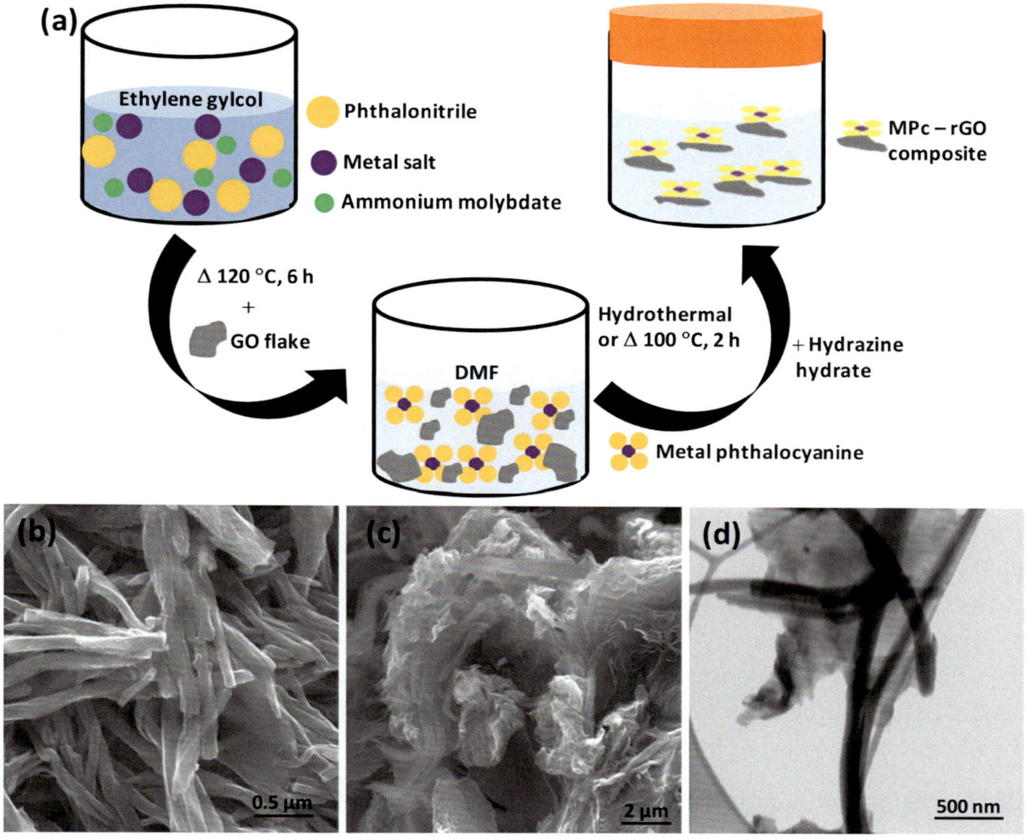

Figure 18.7

(A) Schematic showing the synthesis procedure for the formation of MPc—GO/rGO composite. (B) and (C) SEM images of NiPc NF and NiPc NF - rGO composite. (D) TEM image of NiPc NF - rGO composite. *Reprinted from Ref. K.P. Madhuri, N.S. John, Supercapacitor application of nickel phthalocyanine nanofibers and its composite with reduced GO, Appl. Surf. Sci. 449 (2018) 528—536. Copyright (2018) with permission from Elsevier.*

amounts and heating the mixture in a high boiling solvent like ethylene glycol. In the next step, MPc nanostructures can be combined with GO, which is earlier prepared by modified Hummers' method and further by adding a suitable reducing agent, rGO can be obtained. To further this, the rGO can also be functionalized to harbor functionalized MPcs structures through covalent attachment. The schematic of the preparation method can be represented as shown in Fig. 18.7A. Fig. 18.7B and C shows the typical morphology obtained for NiPc NF and NiPc NF—rGO composite. It is seen that the NiPc NF has a width of ~150 nm and length of ~ 5 µm. Fig. 18.7C and D show the SEM and TEM image of the composite in which the NiPc NF is found to be interspersed with rGO matrix. This indicates that the obtained MPc—rGO composite through this method offers MPcs to adhere and have a good physical interface with rGO surface [30].

3.2 Metal phthalocyanine-carbon nanotube nanocomposites

Carbon nanotubes (CNTs) can be simply described as the sheets of graphite rolled up into a tubular-like structure. CNTs are considered to be 1-D structures due to their high aspect ratio (length to diameter ratio). The bonding in CNTs is sp^2 hybridization and consists of honeycomb lattices with seamless structures. The most important structures of CNTs are single-walled carbon nanotubes (SWCNTs) and multiwalled carbon nanotubes (MWCNTs) in which SWCNTs are made of only single-wrapped graphene sheet while MWCNTs are similar to the collection of concentric cylinders of SWCNTs. The length of CNTs can run into several micrometers with diameters as small as 0.7 nm in SWCNTs up to 15 nm in MWCNTs. In addition to two different basic structures, CNTs can also be classified into three types depending on how the graphene sheet is rolled up during the synthesis process; armchair, zigzag, and chiral. This structural configuration has a direct consequence on the nanotube's electrical properties. CNTs are also known for their excellent mechanical strength and thermal conductivity. Similar to rGO, CNTs also provide scope for functionalizing their surface to selectively graft certain molecules forming nanocomposites. Further, the sp^2 hybridized carbon with delocalized π-electrons can harbor organic molecules with a better synergy [75—77].

Similar to the case of composite formation with rGO, CNT dispersion may be mechanically mixed with MPc solution to form CNT-MPc composite. In an alternate route, MPcs can also be in situ synthesized or polymerized in the presence of CNTs to obtain the composite promoting a better physical interface between CNTs and MPcs rather than just physical mixing. Zhang et al. reported the preparation of MWCNTs and CuPc composite by dissolving CuPc in chloroform by sonicating for 15 min followed by the addition of 10 mg of purified MWCNT and further sonication for 2 h. The resultant solution is centrifuged and the retained supernatant fraction was washed over PVDF membrane using chloroform. The retained composite over the filter was suspended back in chloroform and dried [78]. In a recent article by Jiang et al., MPc-CNT hybrid was prepared by dispersing purified CNTs (20 mg) in DMF followed by sonication. Then MPc (1.5 mg) contained in DMF was added to the CNT suspension. The mixture was further sonicated for 30 min and stirred overnight at room temperature. The solid was filtered and washed with DMF, ethanol, and water, and finally vacuum dried to yield final product [79]. Morozan prepared the composite mixture by first dissolving MPc powders in THF and then a portion of purified CNTs powder was added to this solution followed by sonication for 20 min to form a homogenous dispersion. THF solvent was evaporated under a stream of nitrogen to yield a composite powder [80].

In chemical synthesis, MPcs may be covalently grafted on to CNTs having functional groups. Cao et al. synthesized the composite of FePc and SWCNTs using covalent functionalization of SWCNTs. In this method, the pyridyl groups were anchored on the surface of CNTs through diazonium reaction. FePc moieties were then coordinated to Py-CNTs through the bond formation between the nitrogen of pyridyl group and iron center of FePc moiety [81]. Another report by Cao et al. MWCNTs were functionalized with dodecylamine and suspended in dichloroethane containing TiOPc dispersed in polyvinyl butyral (PVB) and sonicated for 2 h. The functionalized MWCNTs/TiOPc composite was found to have good dispersibility in PVB matrix and the obtained modified CNT/TiOPc composite were fabricated using dip coating technique [82]. In a different report by Wang et al. MWCNTs were modified with dodecyl chains by first converting MWCNTs to carboxy terminated chains and then treating with dodecyl bromide and tetra-n-octyl ammonium bromide in a phase-transfer reaction. Similarly, the carboxy-substituted CuPc was reacted with dodecyl alcohol in benzene mixed with small quantity of pyridine at 100°C for 40 h. The good compatibility between MWCNTs and CuPc was achieved due to π-π interactions and dodecyl long chain interactions apart from improving their solubility in organic compounds. Finally, the composite film was obtained by mixing the solutions of two components prepared by cast-coating method [83]. Verma et al. functionalized CNTs using corona electrostatic discharge to graft −OH groups on the walls of CNTs which facilitate noncovalent interactions. To prepare a composite material, about 10 mg of CNTs were dispersed in toluene and sonicated for 30 min. A weighed quantity of MPc (CuPc, VPc, CoPc) was added to sonicated CNTs and magnetically stirred for 20 h at 60°C followed by ultrasonification [84]. Chidembo et al. prepared a composite containing acid-functionalized MWCNT and tetraamino substituted NiPc by grinding the two components in an equal ratio dissolved in DMF, ultrasonicated for 30 min and dried [61]. The covalent attachment of tetraamino-NiPc onto the −COOH functional groups of MWCNTs offered enhanced supercapacitive properties. In a report by Asedegbega et al. CNTs synthesized by catalytic vapor decomposition method were oxidized with HNO_3 and further treated with ethylenediamine in hexane to give amine-functionalized CNTs. This was further treated with tetrasulfonic acid−substituted FePc and stirred at room temperature for 17 h [85]. The preparation of ZnPc-SWCNT composite has been reported by Ballesteros et al. [28] Initially, carboxylated SWCNTs was prepared by treating SWCNTs with N-octylglycine and 4-formyl-benzoic acid. Highly substituted ZnPc with benzyl alcohol groups were treated with acid-functionalized SWCNTs to yield esterified product. The solution was filtered and the residue was washed with several portions of water, acetone, methanol, and so forth to finally obtain a purified composite.

A general route for the preparation of MPc−CNT composite can be described as the initial preparation of MPc nanostructures from phthalonitrile precursor and can also be subjected to functionalization of the phenyl rings of the macrocycle. This may be combined with

SWCNTs or MWCNTs obtained through CVD technique or commercially procured by sonication. Additionally CNTs can also be functionalized with acid, acid chloride, amine, halogenation, and so forth to harbor functionalized MPc structures via covalent attachment. A schematic of the preparation method of the composite via chemical functionalization can be represented as shown in Fig. 18.8. According to Fig. 18.8, acidified MWCNTs are treated with thionyl chloride to convert the carboxyl groups to acid chloride groups in the process, evaporating the excess of chloride. The product was dispersed in 1-methyl-2-pyrrolidinone to which amine substituted CuPc was added. The resulting mixture was stirred for 24 h, filtered, washed, and finally dried to yield CuPc-MWCNTs composite [86].

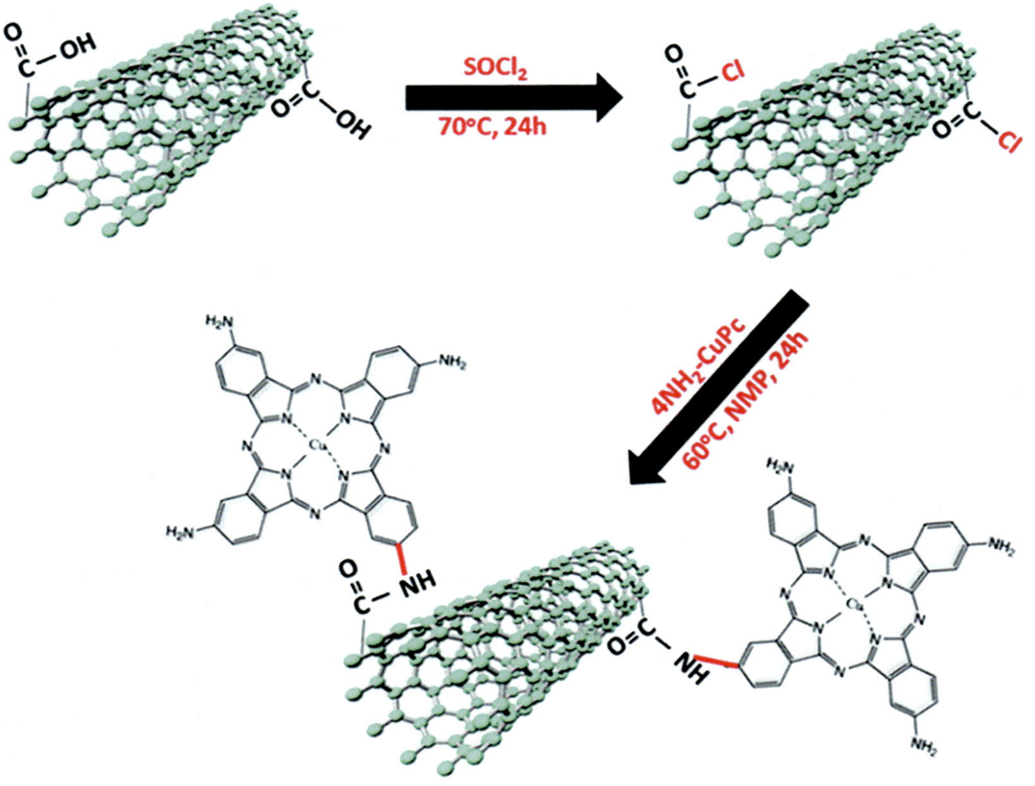

Figure 18.8

Schematic showing the synthesis procedure for the formation of MPc – MWCNT composite via covalent attachment. *Reproduced from Ref. X. Li, W. Xu, Y. Zhang, D. Xu, G. Wang, Z. Jiang, Chemical grafting of multi-walled carbon nanotubes on metal phthalocyanines for the preparation of nanocomposites with high dielectric constant and low dielectric loss for energy storage application, RSC Adv. 5 (2015) 51542–51548 – Published by The Royal Society of Chemistry.*

3.3 Metal phthalocyanine - porous carbon nanocomposites

Porous carbon materials, also called activated carbon materials, are 3D, and their composites have attracted a huge attention as advanced electrode materials in supercapacitors, batteries, and hybrid energy storage systems. Their high electrical conductivity, porosity, diverse structures, superior chemical stability, and low cost offer other potential applications such as electrochemical catalysis, water treatment, air/gas purification, food processing, etc. Porous carbon materials are classified into different types based on the size of their pores; microporous <2 nm, mesoporous <50 nm, and macroporous >50 nm. The structure of porous carbon is a continuous network with sp^2 hybridized carbon atoms designed to retain the porosity. This structure possesses excellent mechanical stability and can provide an efficient pathway for the transport of electrons or ions with accessible active centers, which favor high electrochemical performance. Porous carbon material exhibits a high specific surface area and is more resistant to structural changes imposed by hydrolytic effects in aqueous environments. Moreover, the structure of 3D porous carbon can be a good support to integrate other electrochemical active materials through covalent functionalization or doping of heteroatoms to realize better performance of the composite materials. This includes oxidation, sulfonation, halogenation, grafting and attachment of nanoparticles and surface coating with active nanomaterials [87–90].

Similar to rGO and CNTs, porous carbon materials can also be used in pristine form or can be functionalized to harbor functionalized form of MPcs structures through covalent attachment. Li et al. reported the preparation of FePc–ordered mesoporous carbon composite (OMC) in which OMC (10 mg) was firstly dispersed in DMF followed by 30 min ultrasonication. FePc (20 mg) was further added to this suspension and was vehemently stirred for 48 h at room temperature. Subsequently, the resulting suspension was centrifuged to obtain the powder. A series of FePc–OMC composites were prepared containing different mass ratios [91]. Magdesieva et al. reported the synthesis of transition MPc and porphyrins complex supported on activated carbon fibers (ACF). A weighed amount of catalysts were suspended in CH_2Cl_2 or acetone and appropriate amount of ACF was also added to the same solution. The resulting mixture was magnetically stirred for three to four days at room temperature followed by vacuum filtration. The ACF containing adsorbed active materials was dried in a vacuum oven at 100°C for 1 h. Similarly, MPc has been combined with another important 2D material such as carbon nitride [92]. Lee et al. prepared ZnPc–mesoporous carbon nitride composite by impregnation method. A certain amount of ZnPc was dissolved in THF and mesoporous carbon nitride was added to the solution at room temperature with continuous stirring. The solvent was completely evaporated to yield a composite powder. The procedure for the synthesis of mesoporous carbon nitride is elaborated elsewhere [93].

Li et al. reported the preparation of CoTAPc and OMC composite [94]. OMC (0.1 g) was treated with potassium persulfate in water whose pH was adjusted to 12. The solution was stirred at 85°C for 3 h. The solution was cooled and filtered until neutral. The carboxyl-containing OMC (0.1 g) and CoTAPc (0.005 g) were dispersed and ultrasonicated in DMSO. N,N′-dimethylpyridin-4-amine and N,N′-dicyclohexylcarbodiimide were added and the reaction was stirred for 24 h at room temperature. The formation of amide bond between OMC and CoTAPc aids in successful grafting of CoTAPc onto OMC. The product was washed with DMF to remove ungrafted CoTAPc and finally with water and ethanol to yield CoTAPc-OMC composite. Takanabe et al. gives a detailed preparation of mesoporous graphitic carbon nitride (mpg-C_3N_4) in which the Pt nanoparticles were impregnated. This was further utilized to make a composite with MgPc offering π-π interactions. This was simply achieved by dissolving 0.01−0.5 wt% of the photocatalyst MgPc in 1 mL of THF and 0.5 g of Pt-mpg-C_3N_4 powder. The suspension was continuously stirred at room temperature until THF evaporated completely. The obtained powder was treated in a muffle furnace at 100°C for 1 h [95].

A schematic of the general route adopted for the preparation of MPc−porous carbon composite can be presented as shown in Fig. 18.9. The pores of the mesoporous silica template are initially impregnated with carbon source such as sucrose, followed by polymerization and carbonization through pyrolysis [96,97]. This yields corresponding mesoporous carbon replica, which can be further used to anchor the MPc nanostructures through covalent or noncovalent interactions as described in the synthesis section. Fig. 18.9C and D show the high resolution TEM images of porous carbon matrix doped with FePc particles. Fig. 18.9E shows high-angle annular dark field−scanning transmission electron microscopy (HAADF−STEM) image of the porous carbon-FePc composite. The obtained carbon matrix is 3D and is a highly porous network containing FePc particles.

4. Applications of metal phthalocyanines and their composites in energy generation, conversion, and storage

As discussed previously, MPcs are the molecular materials that are photo and electro-responsive. The aromatic chromophoric system allows for an intense absorption bands in visible and NIR regions with high molar absorptivity. The particular stacking among MPc units causes broadening in Q band absorption providing vectorial charge transport in high-performance solar cells. Further, the presence of central metal cation along with pyrrolic nitrogen groups imparts excellent electrochemical properties which are useful in energy conversion and energy storage applications. These constituents participate in reduction and oxidation reactions and facilitate the transport of electrons thereby

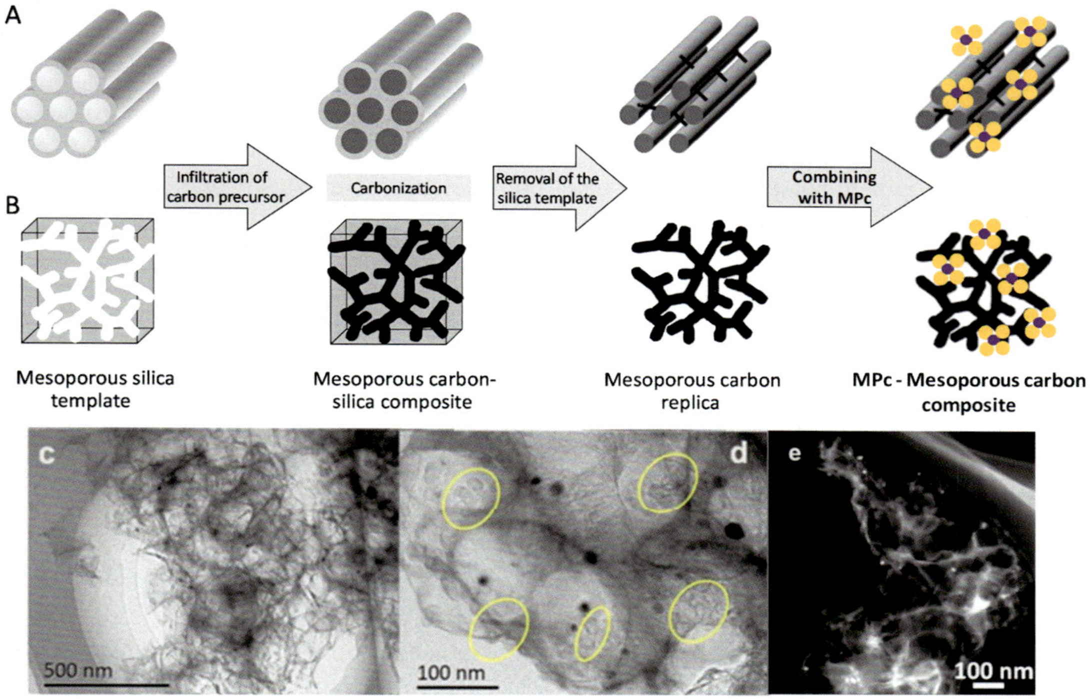

Figure 18.9

Schematic showing the synthesis procedure for the formation of (A) MPc—porous carbon vesicles and (B) MPc—mesoporous carbon composite. (C) and (D) High resolution TEM images of porous carbon matrix doped with FePc particles. (E) HAADF-STEM image of porous carbon matrix doped with FePc particles. *(A) and (B) Adapted from A. Walcarius, Recent trends on electrochemical sensors based on ordered mesoporous carbon, Sensors 17 (2017) 1863 and published by MDPI. (C—E) Reproduced from Ref. Y. Jiang, Y. Xie, X. Jin, Q. Hu, L. Chen, L. Xu, J. Huang, Highly efficient iron phthalocyanine based porous carbon electrocatalysts for the oxygen reduction reaction, RSC Adv. 6 (2016) 78737—78742 — Published by The Royal Society of Chemistry.*

catalyzing many important chemical reactions. This particular property of MPcs can be exploited for energy applications such as generation, conversion, and storage. Applications in these three categories are detailed in the following sections.

4.1 Solar cells

Photovoltaic technology is recognized as one of the essential components to cater the future global energy needs. The potential applications of solar cells based on small π-conjugated organic molecules lie primarily in their low-cost processing methods on large-area substrates. MPcs are highly distinguished for their extraordinary light-harvesting properties in the NIR spectral region along with their thermal robustness. MPcs have played a significant role in the success of dye-sensitized solar cells. An enormous amount of research has been dedicated for the synthetic variations on main Pc scaffold. Urbani et al. gives a detailed description of the various structural aspects that influence the performance of solar cells, considerably [98]. Vasseur et al. demonstrated the use of a nonplanar MPc, PbPc layer/C60 in a planar heterojunction solar cell with high NIR sensitivity. Optimized solar cell with a power conversion efficiency (PCE) of 2.6% and external quantum efficiencies (EQE) above 11% in the range of 320−900 nm was achieved [15]. In a slightly different experiment carried out by Vasseur et al., CuI layers were inserted to induce the formation of strongly textured and crystalline PbPc layers in which the heterojunction device yielded an EQE of 34% at 900 nm with PCE of about 2.9%. The growth of NIR polymorph of PbPc was induced by the underlying CuI layers [18]. Fig. 18.10 gives the schematic representation of a dye sensitized solar cell employing PbPc layers laid on different templating layers with different molecular orientations as a function of quantum efficiency. Similarly, another nonplanar MPc, TiOPc was explored by Vasseur et al. in which the structure of TiOPc thin films is correlated to the photovoltaic performance of a planar heterojunction with C60 film. Here, the introduction of 1,1,2H,2H-perfluorodecyltrichlorosilane (FDTS) templating layer induces the formation of edge-on TiOPc with EQE of above 27% and PCE of 2.6% [99]. The introduction of CuI layers for inducing NIR active polymorph of PbPc was also investigated by Kim et al. in which the PCE was improved from 1.3% to 2.5% [100]. These studies reveal that when the underlying templating layers induce the formation of triclinic PbPc phase which is more NIR active, the EQE of the device also increases. Other parameters include thin film deposition conditions, film crystallinity, and relevant electrodes for achieving better solar performance.

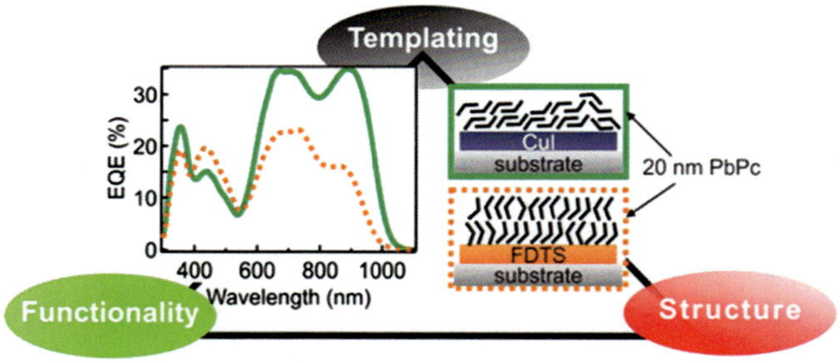

Figure 18.10

Schematic representation of PbPc thin film structure-dependent on templating layer determining the solar efficiency. *Reprinted with permission from Ref. K. Vasseur, K. Broch, A.L. Ayzner, B.P. Rand, D. Cheyns, C. Frank, F. Schreiber, M.F. Toney, L. Froyen, P. Heremans, Controlling the texture and crystallinity of evaporated lead phthalocyanine thin films for near-infrared sensitive solar cells, ACS Appl. Mater. Interfaces 5 (2013) 8505−8515. Copyright (2013) American Chemical Society.*

Transparent electrodes required for organic solar cells based on aqueous solution of TSCuPc–SWCNTs composite was investigated by Raissi et al. [101] The composite ink was highly stable exhibiting good conductivity with a PCE of 7.4% when poly(3-hexylthiophene-2,5-diyl) (P3HT)–indene-C60 (ICBA) bisadduct as active front layer and phenyl C61-butyric acid methyl ester (PCBM) - poly[N-9′-heptadecanyl-2,7-carbazole-alt-5,5-(4′,7′-di-2-thienyl-2′,1′,3′-benzothiadiazole)] (PCDTBT) active back layers. Ince et al. developed a novel NIR ZnPc bearing a donor-chromophore-acceptor anchoring groups, which was used as a dye sensitizer for the investigation of the solar cell performance [102]. The new ZnPc with an extended π-conjugation produces red-shift in the absorption spectra giving a modest PCE of about 3.3%. In a different report by Nouri et al. functional perovskite solar cell was prepared using an inexpensive, stable and soluble tetra-n-butyl-substituted CuPc (CuBuPc) as a hole transporting layer with TiO_2/rGO as electron acceptor and transport mediators [103]. The device exhibited an optimum PCE of 22% and higher stability when compared to pristine TiO_2 based devices. Bartelmess et al. presents a new family of pyrene (Py) substituted MPc i.e., ZnPc-Py and H_2Pc-Py in combination with SWCNTs. The advantage of pyrene substitution is that the units provide non-covalent functionalization of SWCNTs via additional π-π interactions in which the carbon skeleton is not impacted by chemical modification. These electrodes have been integrated into photoactive electrodes revealing stable and reproducible photocurrent with monochromatic PCE of about 15% and 23% without and with +0.1 V applied bias [104]. Hatton et al. utilized a hybrid layer of TSCuPc–MWCNTs at ITO-PCBM:P3HT interface having a PCE of about 25% [105]. The report also provides experimental evidence for TSCuPc phase enhancing light absorption and MWCNTs facilitating conduction of holes to the underlying ITO electrode. Gomis et al. summarize the evolution of the MPc sensitized solar cells over time with various photovoltaic parameters [106]. The article also highlights recent progress in MPc molecules covering various strategies employed to increase their photovoltaic performance.

4.2 Electrocatalysis and photocatalysis for energy generation and energy conversion

The various metal ions in the macrocyclic ring of MPc compounds possess redox properties and can be utilized for electrochemical applications. When the substituted metal is a transition group element, it can undergo redox reactions at lower potentials compared to the macrocyclic redox processes. Since MPcs possess dual nature of π donor and acceptor functionalities, they have been found to demonstrate excellent electrocatalytic activity for numerous technologically significant redox reactions [53]. The oxidation and reduction potential of MPcs can be explored using electrochemical techniques such as cyclic voltammetry. The electronic structure of MPcs has been studied using DFT, and it is shown that HOMO level of certain MPcs (M = Fe, Co, Cu) is 3*d* centered, while the

HOMO in few MPcs could be ring centered. The LUMO levels of all the MPcs are ring centered [31]. The oxidation potential in MPcs can be correlated to the electronic spectra (UV-Vis) and charge transfer bands (NIR). In brief, the metal-centered and ring/ligand centered redox processes are reversible and can be represented as shown,

$$\left[M^{2+}Pc^{2-}\right]^{0} \leftrightarrow \left[M^{3+}Pc^{2-}\right]^{+1} + e^{-}$$
$$\left[M^{2+}Pc^{2-}\right]^{0} \leftrightarrow \left[M^{2+}Pc^{1-}\right]^{+1} + e^{-}$$

4.2.1 Electrochemical catalysis

The drive to replace the scanty and expensive Pt based catalysts has led to the development of many transition metal ions based organo-electrocatalysts for various relevant energy conversion reactions such as oxygen reduction reaction (ORR) and hydrogen evolution reaction (HER). ORR is one of the most important reactions in energy conversion systems such as fuel cells or batteries. The utilization of noble metals, high overpotential and degradation of catalysts are a few issues that impair the practicality of ORR. Electrocatalytic HER is another fundamental electrochemical reaction for the production of sustainable source of hydrogen fuel that can be used in a zero-emission combustion engine. Recent developments include the exploration of transition metal macrocycles such as MPcs for understanding the mechanism of ORR and HER achieving high energetic efficiency.

Zhang et al. synthesized phthalocyanine tethered FePc (Pc-FePc) on graphitized carbon black (GCB) for exploring its electrocatalytic nature toward ORR in O_2 saturated 0.1 M KOH solution. The Pc-FePc/GCB electrode presents a positive potential of 0.88 V versus reversible hydrogen electrode (RHE) with a highest cathodic current density of 3.02 mA/cm^2. The ORR performance was further illustrated by obtaining a Tafel slope of 42 mV/decade suggesting higher transfer coefficient for ORR kinetics [107]. The mechanism behind ORR is the initial adduct formation between O_2 and metal cation, followed by the sequential addition of $2H^+$ from the electrolyte along with $2e^-$ transfers to produce H_2O_2. The subsequent transfer of $2H^+$ and $2e^-$ cleave H_2O_2 to produce H_2O. The production of H_2O_2 during ORR is not desirable as it may lead to poisoning of the catalyst and thus deactivate it. The same group carried out a systematic study of the transition metal (Fe, Co, Ni, Cu) phthalocyanines as electrocatalysts toward ORR in O_2 saturated 0.1 M KOH solution [62]. In particular, the FePc electrode exhibited the most positive potential of 0.89 V versus RHE with highest cathodic current density of 1.36 mA/cm^2 compared to other MPcs. The Tafel slope for FePc of about 52 mV/decade was obtained, comparable to that of commercial Pt/C revealing that FePc as electrocatalyst share similar rate determining step in the lower potential regime. In a different report by Zhang et al. FePc and nitrogen doped graphene composite was utilized as a novel nonprecious catalysts toward ORR in O_2 saturated 0.1 M KOH [74]. The ORR by N_2 doped graphene and FePc electrodes occurred at −0.08 and −0.06 V, whereas the onset potential of the composite commenced at −0.01 V versus Ag/AgCl, which was as good as a commercial Pt/C. The catalytic current density of the composite was also found to be much larger compared to individual components. Such excellent performance was attributed to higher electrical conductivity and synergistic π-π interactions between FePc and nitrogen doped graphene. Taniguchi et al. studied FePc self-assembled on GO as a hybrid electrocatalyst affording ultrafast ORR in saturated 0.1 M KOH solution [108]. It was demonstrated that the FePc - rGO hybrid offered a reduction peak at −0.08 V versus Ag/AgCl, while 20% Pt/C exhibited a peak around −0.14 V. Furthermore, the hybrid displays a very small Tafel slope of about 33 mV/decade, which was less than half of that obtained for Pt/C (88 mV/decade). Cao et al. developed a bio-inspired composite consisting of FePc anchored on the walls of CNTs via pyridyl groups (FePc-Py-CNTs) for the electrocatalytic reduction of O_2 [81]. The electrochemical evaluation of the composite was tested in O_2 saturated 0.1 M KOH and compared with 20% Pt/C. the composite exhibits exceptionally high ORR with positive half-wave potential of 0.915 V versus RHE with the Tafel slope of 27 mV/decade. This value was quite low when to the Pt/C with the Tafel slope of 64 mV/decade.

Mukherjee et al. solvothermally synthesized graphene-supported ZnPc composite and their mechanism for ORR was tested in O_2 saturated neutral phosphate buffer solution using rotating disk electrode technique [68]. The onset potential for the composite was observed to be −0.04 V versus Ag/AgCl. A strong cathodic ORR peak appears at −0.35 V versus Ag/AgCl with cathodic peak current of 0.23 mA/cm^2. Similar studies in 0.1 M KOH solution indicated the onset potential of 0.14 V with peak current density of 0.20 mA/cm^2 with a strong ORR peak appearing at −0.28 V versus Ag/AgCl. The same group explored the electrocatalytic activity of MnPc toward ORR in O_2 saturated 0.1 M KOH solution [109]. A pronounced ORR peak appeared at −0.15 V versus SHE (or 0.56 V vs. RHE) with high current density of 2.84 mA/cm^2. Liu et al. studied the catalytic activity of electrochemically reduced GO (ERGO) supported with FePc toward ORR and compared with FePc coupled with graphene in alkaline medium (0.1 M KOH) [110]. In contrast to poor catalytic activity of ERGO, FePc loaded composite showed significantly higher catalytic activity for ORR with peak potential of −0.15 V versus SCE and onset potential of −0.015 V versus SCE. This potential is positively shifted by 45 mV in comparison with commercial Pt/C catalyst. Fig. 18.11A shows the linear sweep voltammograms of differently modified electrodes indicating the superior catalytic activity of FePc−ERGO composite with lower onset potential compared to other electrodes. Fig. 18.11B shows the graph for calculated number of electron taken part in ORR. The FePc−ERGO composite predominantly follows $4e^-$ pathway with minimum H_2O_2 production observed in Fig. 18.11C.

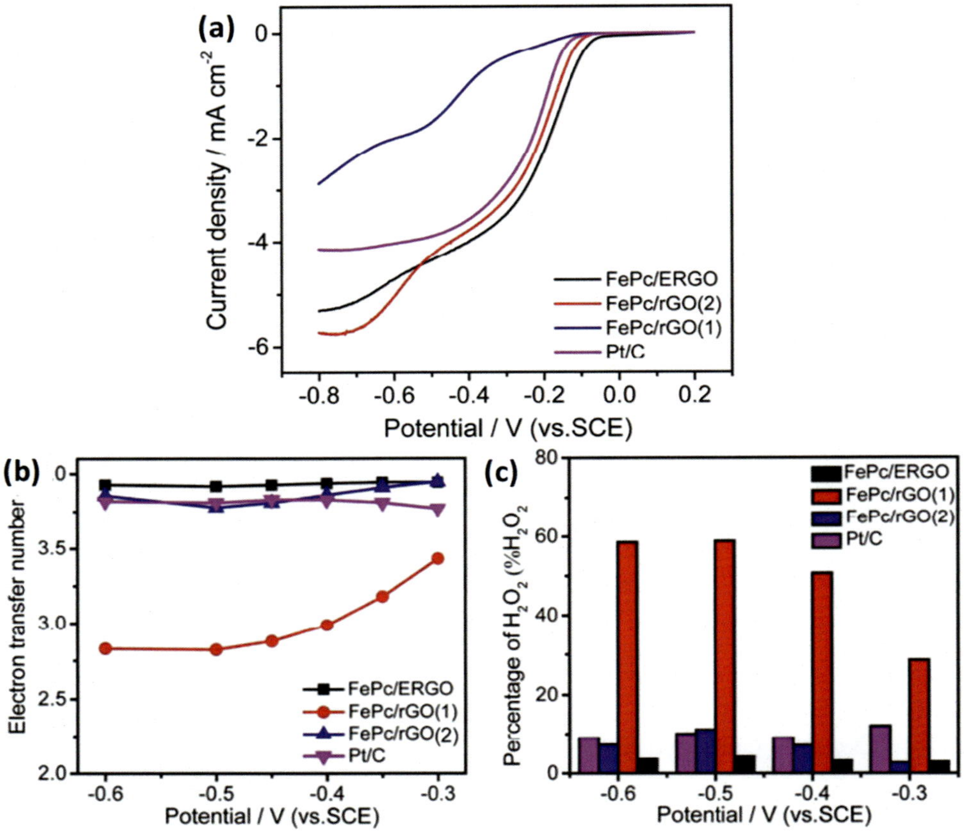

Figure 18.11

(A) Linear sweep voltammograms of differently modified electrodes. (B) A graph showing calculated electron transfer numbers and (C) Percentage of H_2O_2 production during ORR. *Reprinted with permission from Ref. D. Liu, Y.T. Long, Superior catalytic activity of electrochemically reduced GO supported iron phthalocyanines toward oxygen reduction reaction, ACS Appl. Mater. Interfaces 7 (2015) 24063–24068. Copyright (2015) American Chemical Society.*

In a different report by Liu et al. FePc has been covalently functionalized on graphene in 0.1 M NaOH [111]. The hybrid catalyst exhibited an ORR peak potential at −0.12 V versus SCE which is more positive when compared to individual components with a Tafel slope of 41 mV/decade. Li et al. investigated the comparative ORR electrocatalytic nature of FePc loaded on rGO (FePc-rGO), mesoporous carbon vesicle (FePc-MCV) and ordered mesoporous carbon (FePc-OMC) in O_2 saturated 0.1 M KOH solution [91]. The ORR peak potential for FePc-rGO, FePc-MCV, and FePc-OMC were found to be 0.787, 0.813, and 0.863 V versus RHE with concomitant increase in current density of −1.47, −2.21 and −2.89 mA/cm^2. This indicates that the OMC is a better choice to support FePc molecules with high dispersion and abundant active sites. Tafel analysis was also performed for three composites, with 40.42, 46.45 and 53.65 for FePc-OMC, FePc-MCV and FePc-rGO respectively. Jiang et al. demonstrated the enhanced electrocatalytic performance of Pt-free FePc supported on graphene for efficient ORR in O_2 saturated 0.1 M KOH. The composite exhibits a pronounced cathodic ORR peak at 0.9 V versus Ag/AgCl, which is more positive compared to Pt/C and individual components [25]. The obtained current density was also twice as large compared to pristine FePc and graphene catalysts. Cui et al. reported tetracumyl phenoxy-substituted FePc supported on graphene (FePc(CP)$_4$-Gr) as a hybrid catalyst toward ORR in O_2 saturated 0.1 M NaOH [112]. The hybrid exhibits ORR peak potential at −0.195 V and an onset potential of −0.07 V versus SCE, which is more positive when compared to Pt/C and individual constituents. The derived Tafel slope of 31 mV/decade was obtained for the hybrid. The same group presents a novel FePc tetra-substituted with pyridine groups (FeTPPc) decorated on graphene sheets (FeTPPc-Gr) synthesized through solvothermal method [113]. The ORR activity was checked in O_2 saturated 0.1 M NaOH solution and it was found that the FeTPPc-Gr exhibits a remarkable reduction peak at −0.118 V versus SCE with a Tafel slope of −41 mV/decade. These values are quite low when compared to Pt/C and individual components. The following Table 18.2 gives the list of the MPc nanocomposites with experimental details regarding the ORR electrocatalysis.

MPcs also find applications in water splitting by catalyzing HER from water. Electrocatalytic HER of alkyne substituted CoPc decorated on rGO was explored by Aykuz et al. [116] The electrocatalytic responses were measured in 0.1 M $LiClO_4$ electrolyte and the highest catalytic was observed for the hybrid with an overpotential of −0.38 V versus Ag/AgCl, Tafel slope of 137 mV/decade and current density ratio (ratio of current density on modified to bare electrode) of 13.3 mA/cm^2. The same group investigated electrocatalytic HER using metal phthalocyanines (CoPc and ZnPc) bearing terminal alykyne groups in composite with polyaniline in dimethyl sulfoxide (DMSO) and tetrabutylammonium perchlorate (TBAP) electrolyte [117]. The modified CoPc and ZnPc exhibited an overpotential of 138 and 72 mV versus Ag/AgCl with Tafel slope of 211 and 274 mV/decade. Ozcesmeci et al. performed electrocatalytic HER using supramolecular CoPc carrying cobaloxime moieties in phosphate buffer solution [118]. An onset potential of −0.36 V versus Ag/AgCl with a Tafel slope of 79 mV/decade was observed. Monama et al. incorporated CuPc nanostructures in metal organic framework (CuPc-MOF) for HER application in DMSO—TBAP solvent mixture containing H_2SO_4 [119]. The onset potential observed for the HER was about +0.3 V and the potential at which the current density increased was about −0.68 V versus Ag/AgCl and a Tafel slope of 176 mV/decade was obtained. The same group explored tetranitro substituted CuPc grown on MOF and found that the onset potential observed was −0.71 V vas Ag/AgCl, with a Tafel slope of 147 mV/decade [120].

Table 18.2: List of electrochemical parameters for the ORR activity of different MPc nanocomposites.

Sl. No.	Materials	Electrolyte	Onset potential	ORR potential	Current density mA/cm^2	Tafel slope mV/decade	Refs.
1	Pc-FePc - graphitized carbon black	0.1 M KOH	–	0.88 V versus RHE	3.02	42	[107]
2	FePc - graphitized carbon black	0.1 M KOH	0.98 V versus RHE	0.89 V versus RHE	1.36	52	[62]
3	FePc - nitrogen doped graphene	0.1 M KOH	−0.01 V versus Ag/AgCl	−0.14 V versus Ag/AgCl	–	–	[74]
4	FePc - rGO	0.1 M KOH	–	−0.08 V versus Ag/AgCl	–	33	[108]
5	ZnPc - rGO	0.1 M KOH	−0.14 V versus Ag/AgCl	−0.28 V versus Ag/AgCl	0.20	–	[68]
6	ZnPc - rGO	0.1 M PBS	−0.04 V versus Ag/AgCl	−0.35 V versus Ag/AgCl	0.23	–	[68]
7	MnPc	0.1 M KOH	0.83 V versus RHE	−0.15 V versus SHE or 0.56 V versus RHE	2.84	–	[109]
8	Tetra-tert-butyl substituted CoPc - MWCNTs	0.1 M NaOH	−0.094 V versus SCE	−0.26 V versus SCE	3.5	–	[80]
9	FePc - MWCNTs	0.1 M NaOH	−0.192 V versus SCE	−0.12 V versus SCE	4.5	–	[80]
10	FePc - electrochemically reduced GO	0.1 M KOH	−0.015 V versus SCE	–	–	–	[110]
11	FePc - graphene	0.1 M NaOH	–	−0.12 V versus SCE	–	41	[111]
12	FePc - rGO	0.1 M KOH	0.89 V versus RHE	0.78 V versus RHE	1.47	53.65	[91]
13	FePc - mesoporous carbon vesicle	0.1 M KOH	0.94 V versus RHE	0.81 V versus RHE	2.21	46.45	[91]
14	FePc - ordered mesoporous carbon	0.1 M KOH	0.95 V versus RHE	0.86 V versus RHE	2.89	40.42	[91]
15	FePc - ordered mesoporous carbon	0.5 M H_2SO_4	0.725 V versus RHE	0.49 V versus RHE	0.87	–	[91]

16	FePc - graphene	0.5 M H_2SO_4	−0.02 V versus SCE	—	2.7	—	[114]
17	Binuclear FePc - graphene	0.5 M H_2SO_4	0.12 V versus SCE	—	4.2	—	[114]
18	FePc - graphene	0.1 M KOH	—	0.9 V versus Ag/AgCl	—	—	[25]
19	Tetra-cumylphenoxy substituted FePc - graphene	0.1 M NaOH	−0.07 V versus SCE	−0.195 V versus SCE	—	31	[112]
20	Tetra-pyridyloxy substituted FePc - graphene	0.1 M NaOH	—	−0.118 V versus SCE	—	41	[113]
21	FePc-pyridine-CNTs	0.1 M KOH	—	Half-wave potential of 0.915 V versus RHE	—	27	[81]
22	Iron polyphthalocyanine - MWCNTs	0.5 M H_2SO_4	—	Half-wave potential of 0.8 V versus RHE	—	—	[115]

Madhuri et al. demonstrated the use of pristine CuPc and CoPc nanofibers as electrode materials for electrochemical HER [65]. The glassy carbon electrode modified CuPc and CoPc exhibited an onset potential of −250 mV and −290 mV with a Tafel slope of 120 and 107 mV/decade respectively.

MPcs have been employed for other important electrocatalytic reaction such as CO_2 reduction. Electrochemical reduction of CO_2 signifies a possible means of generating fuels and organic feedstocks such as ethanol, methane, CO, formic acid, etc. The redox potential for this reaction is akin to hydrogen evolution in aqueous solutions. Zhang et al. developed CoPc-CNTs hybrid structures as active electrocatalysts toward CO_2 reduction in CO_2 saturated 0.1 M $KHCO_3$ neutral solution [121]. The hybrid also exhibits an excellent catalytic activity with a current density of 15 mA/cm^2 at an overpotential of 0.52 V versus RHE. Magdesieva et al. investigated electrochemical reduction of CO_2 with phthalocyanine metal complex adsorbed on activated carbon fiber nanoporous material in presence of CO_2 saturated 0.1 M $KHCO_3$ neutral solution [92]. An optimal potential range of −1.3 to −1.4 V versus SCE was found and current densities for CO production were as high as 50–70 mA/cm^2. Jiang et al. revealed the hidden performance of MPcs hybridized with CNTs for electrocatalytic reduction of CO_2 in 0.1 M $KHCO_3$ saturated with CO_2 [79]. Both CoPc-CNT and FePc-CNT show significant cathodic current below −0.4 V versus Ag/AgCl with a current density of 10 mA/cm^2 obtained at about −0.8 V. Unlike their hybrid counterparts, it was found that the current density of FePc and CoPc was much smaller. Morlanes et al. developed perfluorinated CoPc as a bifunctional catalyst immobilized on carbon electrodes for the simultaneous CO_2 reduction and water splitting which is represented in Fig. 18.12A [122]. Fig. 18.12B shows the current-time profile of the bifunctional catalyst controlled between 2 and 3 V. A high faradaic efficiency of 90% was maintained for CO evolution with no reduction in current densities and the evolved gas was estimated using a gas chromatograph. The reduction of CO_2 becomes apparent at −0.5 V versus RHE in 0.5 M $NaHCO_3$. Water oxidation using the same catalyst was also investigated in potassium phosphate buffer solution at pH 7. It was indicated that the onset of water oxidation occurs at 1.7 V versus RHE with current densities reaching 6 mA/cm^2. In addition to this, Vasudevan et al. investigated different MPcs as electrocatalysts for various reduction reactions [55]. Use of MPcs results in reduction of CO_2 at a lesser negative potential with increased current densities when compared to precious metal electrodes such as Pd, In, Sn, etc. Electroreduction of N_2 to ammonia at ambient pressure and temperature can be achieved using FePc. Current efficiency and stability of the electrocatalytic process depend on the type of central metal ion in MPc. Vasudevan et al. also demonstrated electroreduction of nitrogen oxides, sulfur oxides, thionyl chloride and chlorate ions. The same group also demonstrated the use of MPcs as catalysts toward electrochemical oxidation of hydrazine, glucose, hydroxylamine and other organic compounds [55].

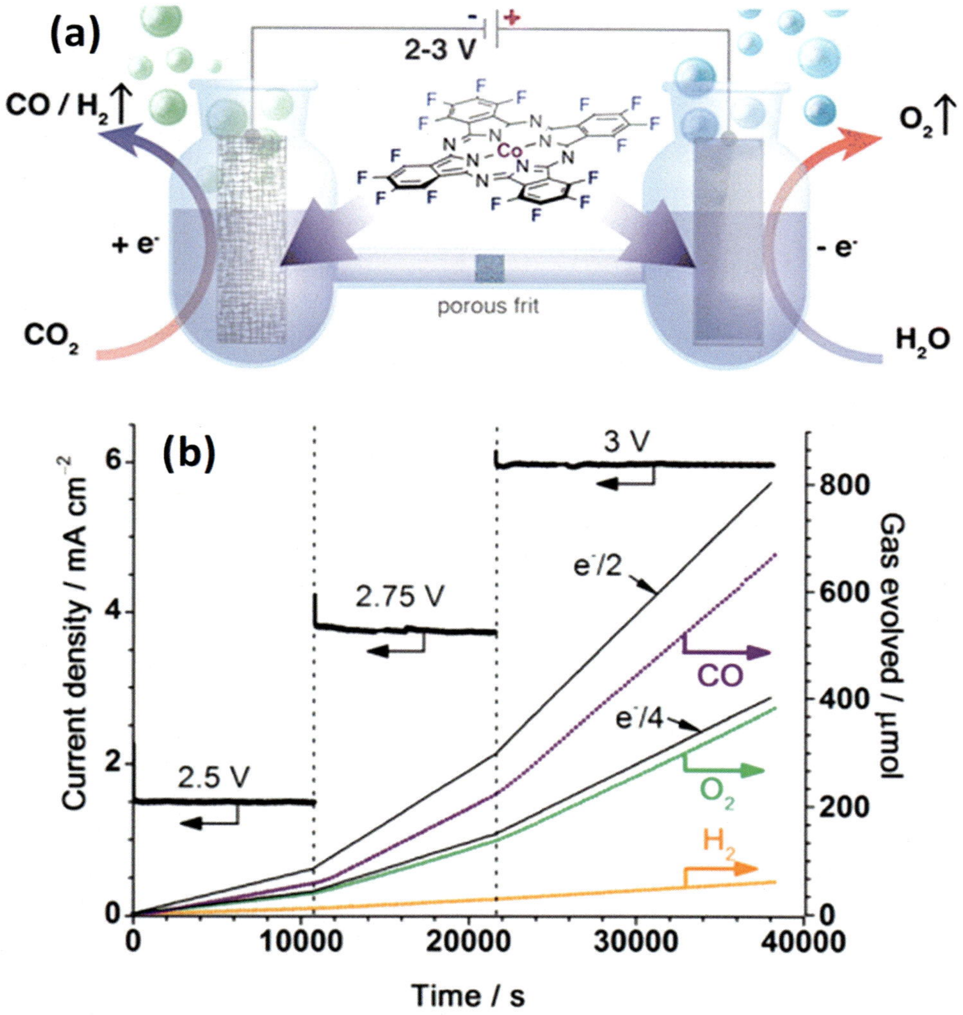

Figure 18.12

(A) Schematic illustration of the setup for simultaneous reduction of CO_2 and oxygen evolution using perfluorinated CoPc as a bifunctional catalyst immobilized on carbon electrode. (B) Performance of perfluorinated CoPc electrode operating between 2 and 3 V versus RHE showing the evolution of CO, H_2 and O_2. *Reprinted with permission from Ref. N. Morlanés, K. Takanabe, V. Rodionov, Simultaneous reduction of CO2 and splitting of H_2O by a single immobilized cobalt phthalocyanine electrocatalyst, ACS Catal. 6 (2016) 3092–3095. Copyright (2016) American Chemical Society.*

4.2.2 Photocatalysis

Abe et al. examined a water splitting system comprising of TiO_2 and ZnPc/C60 bilayer and compared to conventional TiO_2 and Pt based systems [16]. The proposed setup demonstrated the water splitting without any external bias. The production of H_2 and O_2

has also been compared with photoelectrochemical system under a potential applied which is discussed under photoelectrochemical setup. Hagiwara et al. studied the photocatalytic activity of $KTaO_3$ modified with different MPcs and cocatalyzed with Pt nanoparticles [123]. It was found that the most effective dye for improving water splitting was CrPc with a redox potential of −0.8 V versus SCE. For understanding central cation effects, relationship between redox properties of Pc and photocatalytic H_2 formation rate was studied. A strong dependency was observed between redox potential and H_2 formation. Higher rate of H_2 formation was observed for a MPc with redox potential around −0.8 V versus SCE. Luo et al. developed a solar driven hydrogen evolution system using CoPc as a water reduction catalyst along with fluorescein as photosensitizer and triethylamine (TEA) as sacrificial electron donor in ethanol water mixture [124]. The maximum H_2 evolution was observed at a pH of 12. Takanabe et al. developed a mesoporous graphite carbon nitride sensitized with MgPc for photocatalytic hydrogen evolution in presence of triethanolamine as sacrificial reagent even under irradiation longer than 600 nm [95]. With 3 wt% of Pt as cocatalyst, the photocatalytic H_2 formation was 50 μmol/h and exhibited a quantum efficiency of 5.6% at 420 nm. Terao et al. designed water soluble tetrasulfonate-substituted CuPc as catalyst toward photochemical water oxidation [125]. The onset potential attributed to water oxidation is located at 0.99 V versus SCE in 0.1 M, pH 9.5 borate buffer. Tiwari et al. studied the performance of porous TiO_2 functionalized with ZnPc dye for photocatalytic hydrogen production under simulated solar light with triethanolamine as sacrificial agent [126]. The photocatalyst exhibits best performance for H_2 evolution of 2260 μmol for 5 h. A homogenous molecular system consisting of CoPc molecular catalyst paired with $[Ru^{II}(bpy)_3]Cl_2$ as photosensitizer and TEA as sacrificial reagent was explored for photogeneration of H_2 from water by Xie et al. [127] More than 300 μmol of H_2 was produced when the mixture of ethanol and water in 4:1 ratio was used as a solvent. Zhang et al. studied photocatalytic production of H_2 using ZnPc sensitized carbon nitride using ascorbic acid as a sacrificial layer [128]. The H_2 production in the case of the composite was much higher compared to pristine carbon nitride at wavelength from 400 nm to > 800 nm with the photoactivity of 125 μmol/h. When the organic photoelectrode is irradiated with visible light, a series of photophysical events such as light absorption, formation of electron-hole pair, and their transfer to *p*/*n* interface and conduction of electrons and holes into respective layers occur. MPcs particularly help in the effective absorption of the visible light radiation, which forms the crucial step in photocatalysis. An organic p/n bilayer can provide oxidizing and reducing potentials at the MPc interface being *p*-type and fullerene as *n*-type surfaces, respectively [129].

Yuan et al. studied neutral NiPc as a stable catalyst for visible light-driven hydrogen evolution from water [130]. Iridium complex and TEA were used as sensitizer and sacrificial agent in acetone-water mixture under wavelength > 420 nm. The maximum rate of H_2 production of about 14.5 μmol/h was observed at an initial pH value of 10.

To improve the photocatalytic activity of graphene based catalysts, Huang et al. covalently functionalized SiPc onto nitrogen doped rGO along with Pt as cocatalyst and demonstrated good H_2 production performance under both UV and visible light radiation [131]. The amount of H_2 produced by pristine nitrogen doped rGO is only 0.63 μmol/mg which doubles to 1.5 μmol/mg on sensitizing it with SiPc. Watanabe projected the fundamental principles and recent progress in the production of H_2 employing photocatalytic water splitting using organic-inorganic composite photocatalyts [132]. The direct photoconversion of greenhouse gases such as CO_2 and methane using CoPc sensitized TiO_2 was reported by Yazdanpur et al. [133] The compositions of the resulting products were identified using gas chromatography. Components such as oxalic acid, acetaldehyde, water, acetic acid and carbon monoxide were formed. The photoconversion reaction was not observed until the TiO_2 surface was sensitized with CuPc. In another report by Pal et al., ordered mesoporous organosilica spheres with Au nanoparticles and ZnPc were explored to photocatalyze the reduction of oxygen to form H_2O_2 [134]. The mechanism of photocatalytic reduction of O_2 is due to photon absorption by ZnPc according to its band gap photoexcitation energy and an electron is promoted from valence to conduction band to create electron-hole pairs. The photogenerated electrons interact with O_2 and reduce it to H_2O_2. Kumar et al. presents visible light-driven photocatalytic oxidation of thiols to disulfides using FePc functionalized GO as a photocatalyst [135]. The quantitative identification of photooxidized products was done using gas chromatography. The work also demonstrated the reusable nature of the catalyst without any loss in the activity.

4.2.3 Photoelectrochemical catalysis

Photoelectrochemical catalysis is photocatalysis assisted by a small applied potential. This can be achieved by application of an external bias or employing a photovoltaic cell. When a photoelectrode is irradiated with visible light, a series of photophysical events result in conduction of electrons and holes into respective layers generating electrical energy, which can be further utilized to directly cause a chemical catalysis. MPcs particularly help in the effective absorption of the UV and visible light radiation which forms the crucial step. An organic *p*/*n* bilayer can provide oxidizing and reducing potentials at the MPc interface being *p*-type and fullerene as *n*-type surfaces, respectively. To create an efficient visible light-responsive interface, Abe et al. has investigated a great deal on a solid/water interface that have potential photoelectrode characteristics especially utilizing an organic layer [136]. An organic bilayer consisting of H_2Pc and fullerene as p/n heterojunction was found to act as a photoelectrode in a water phase system which induced photoreduction of $Fe^{III}(CN)_6^{3-}$. This experiment demonstrated the successful photoinduced electron transfer reaction through the conduction of photogenerated charge carriers. This led to the development of another p/n heterojunction using PTCBI and CoPc layers for photochemical electrolysis of water. The stoichiometric photoelectrochemical water splitting into constituent molecular oxygen and hydrogen occurred at an applied bias potential of +0.38 V versus Ag/AgCl for O_2/OH^- couple, with O_2 generated at CoPc photoanode [137]. The same group reported another novel photodevice responsive to the entire visible-light energy range <750 nm for H_2 evolution unlike the previous photodevice that was not responsive beyond 500 nm [129]. The photodevice consists of $H_2Pc/C60{-}Pt$ in the water phase and the photoelectrochemical water splitting to form H_2 and O_2 occurred at a potential of −0.32 V versus Ag/AgCl for H^+/H_2 couple. The faradaic efficiency for H_2 evolution was calculated to be 90%. Similarly another photodevice consisting of an organic p/n layer composed of ZnPc and C60, in which C60 is loaded with Pt nanoparticles, has been used for the photoelectrochemical H_2 production [17]. It was demonstrated that the efficient photo-assisted H_2 evolution occurred at the photocathode at a potential of −0.32 V versus Ag/AgCl. In a different report by Abe et al. an organophotoelectrode based on TiO_2 and ZnPc/C60 bilayer has been employed for simultaneous water oxidation and H^+ reduction [16]. The H_2 and O_2 evolution occurred at voltages as low as 0.25 V. This indicates effective charge transfer between the two electrodes and suppression of charge carrier recombination. The same group also extended the effective 2-mercaptoethanol oxidation using aluminum phthalocyanine chloride/C60 composite photoelectrode [138]. Fig. 18.13A shows the schematic illustration of photocatalytic water splitting using organophotoelectrodes, $H_2Pc/C60$ heterojunction with electron conduction pathway. Fig. 18.13B gives the cyclic voltammograms of $H_2Pc/C60$ photoelectrode. The obtained current density under illumination is appreciable compared to the dark current. Fig. 18.13C shows the plot of increasing H_2 evolution with time [129].

In another report by Cheng et al. dye-sensitized photoelectrochemical water splitting was reported by anchoring ZnPc sensitized or functionalized on CNTs [139]. The experiment was carried out in alkaline conditions under UV and visible light radiation. The photoelectrochemical activity shows distinctive curves with photocurrent of 0.32 mA/cm at 1.2 V versus RHE for ZnPc functionalized CNTs. On the other hand, the generated photocurrents were negligible on pristine CNTs less than 0.005 mA/cm^2 under identical experimental conditions. Li et al. reported a photoelectrochemical device by an interfacial self-assembly of CuPc on the surface of TiO_2 nanorod arrays [140]. The generated photocurrent reaches 2.40 mA/cm^2 at 1.23 V versus RHE. The obtained photocurrent is 2.4 times higher than the pristine TiO_2. The device also shows good stability and the photocurrent density does not decline even after 8 h of continuous operation. The mechanism involved here is that the holes from the valence band of the TiO_2 are consumed in the process in which Cu^{2+} is oxidized to Cu^{3+} and Cu^{4+}, thereby oxidizing water to produce O_2. Vishwanath et al. reported the electrochemical growth of nanostructured CuPc thin film on Cu substrate and explored for the photoelectrochemical hydrogen evolution [141]. The CuPc electrode showed an enhanced activity over bare Cu and ITO electrode posited with CuPc.

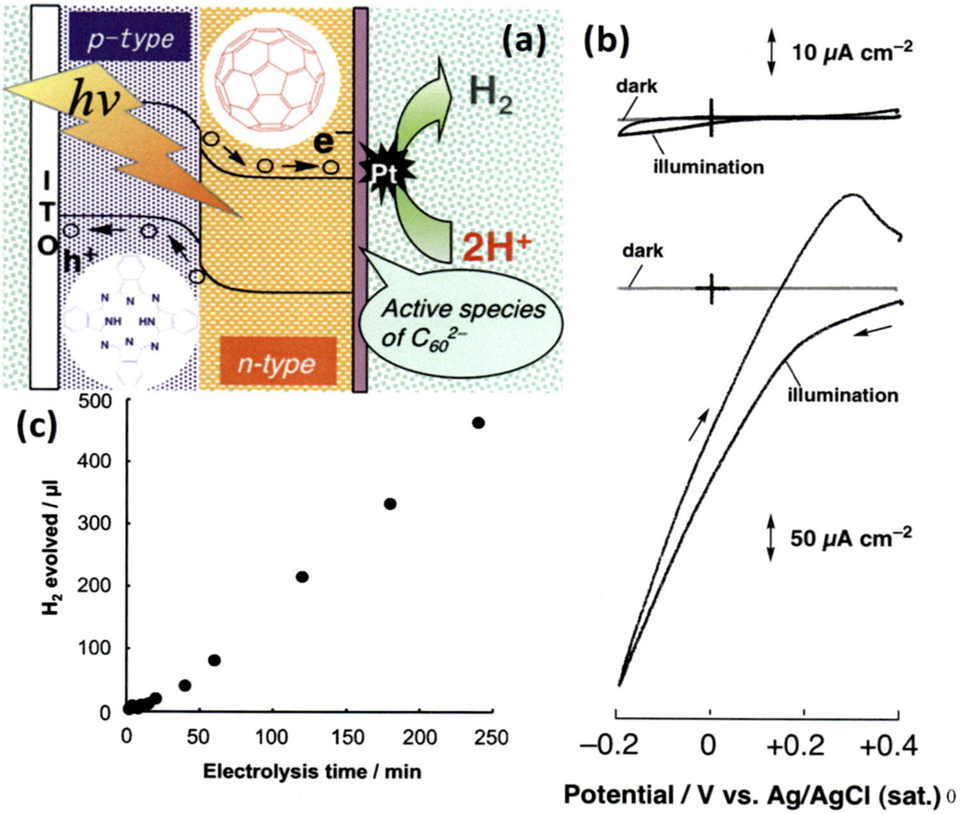

Figure 18.13

(A) An illustration showing the photocatalytic water splitting employing an organic bilayer electrode, $H_2Pc/C60$ heterojuntion. (B) Cyclic voltammograms of ITO/$H_2Pc/C60$ in phosphoric acid electrolyte and (C) Time course of H_2 production at ITO/$H_2Pc/C60$ heterojunction. *Reprinted with permission from Ref. T. Abe, S. Tobinai, N. Taira, J. Chiba, T. Itoh, K. Nagai, Molecular hydrogen evolution by organic p/n bilayer film of phthalocyanine/fullerene in the entire visible-light energy region, J. Phys. Chem. C 115 (2011) 7701–7705. Copyright (2011) American Chemical Society.*

Zheng et al. built ZnPc/carbon nitride heterojunction for visible light photoelectrocatalytic conversion of CO_2 to methanol that occurred at -1.0 V [142]. This process is also compared to photocatalytic and electrocatalytic CO_2 reduction reaction. The superior efficiency of photoelectrocatalytic process is due to the synergistic function and the energy band matching between ZnPc and carbon nitride resulting in faster electron transfer with less probability for electron-hole pair recombination. Zaini et al. reported the photoelectrochemical properties of $BiVO_4/WO_3$ and $BiVO_4/WO_3$ functionalized with CoPc transparent photoanodes toward oxygen evolution reaction [143]. The former photoanode presented an onset potential of $+0.5$ V versus RHE with a photocurrent density of 1.1 mA/cm^2 at 1.8 V, while the latter photoanode presented an onset potential of 0.1 V versus RHE with a photocurrent of 2.4 mA/cm^2 at 1.8 V. This has been attributed to effective visible light absorption and the charge transfer at the interface.

4.3 Electrochemical energy storage

Electrochemical supercapacitors are also known as power capacitors and ultracapacitors, and they serve as an intermediate bridging the gap between batteries (high energy) and dielectric capacitors (high power). There are two types of electrochemical capacitors, electrical double layer capacitors (EDLC) and pseudocapacitors. The EDLC systems are the ones that store charges at the electrode and electrolyte interface, called as nonfaradaic process, while the pseudocapacitors are the materials that possess charges as a virtue of their redox properties, also referred to as a faradaic process. Electrochemical supercapacitance is another field in which MPcs are applied due to their rich redox chemistry. In MPcs, the metal cation in the macrocyclic center along with the pyrrolic nitrogen atoms act as an active redox center giving rise to pseudocapacitance with specific and sharp redox transitions arising from the metal cation. The combination of EDLC from carbon support and pseudocapacitors (MPc) can provide an overwhelming performance in terms of higher specific capacitance and excellent cycling stability. The combination of MPcs with conductive carbons also aids in enhanced surface area and conductivity and increased cycling stability.

Supercapacitive behavior of a novel functional material, nickel (II) octa [(3,5-biscarboxylate)-phenoxy] phthalocyanine (NiOBCPPc) upon covalent integration with functionalized SWCNT-phenylamine has been reported by Agboola et al. [144] The supercapacitive behavior of the composite material was investigated by galvanostatic charge-discharge method in which the hybrid exhibited a superior geometrical capacitance of 186 mF/cm^2 at a current density of 138 μA/cm^2 when compared to the individual counterparts. The NiOBCPPc and SWCNT-phenylamine exhibited 54 and 74 mF/cm^2 respectively. The composite exhibits an excellent stability over 1000 charge-discharge cycles. In a different report by Chidembo et al. the supercapactive property of NiTAPc in composite with MWCNTs has been investigated and it was found that the composite possess a maximum specific capacitance of 981 F/g with an excellent stability over 1500 continuous charge-discharge cycles [61]. The specific power density and energy density of the composite is found to be 700 W/kg and 134 W h/kg. It is speculated that the NiPc functionalized with tetraamino groups (containing nitrogen groups) possibly contribute to superior specific capacitance. Similarly, a nitrogen enriched MPc compound, cobalt tetrapyrazinoporphyrazine (CoTPyzPz) has been conjugated with acid modified and phenylamine modified MWCNTs. Both the composites exhibited a superior capacitance of 1984 and 2028 F/g, attributed to high protonation and deprotonation rate in 1 M H_2SO_4 [145]. The composite was found to be stable for over 1000 continuous cycling with specific energy and specific power density of 250 W h/kg and 952 W/kg, respectively. The same group reports the composite structure of CoTPyzPz with GO in aqueous neutral Na_2SO_4 with a specific capacitance of 500 F/g having long cycle life and excellent short response time [146]. Ramachandran et al. reported the supercapacitor application of octamethyl substituted CoPc in combination with GO. The composite exhibited a specific capacitance of 291 F/g at 0.5 A/g with a cycling stability of about 100% over 500 cycles [147].

Madhuri et al. investigated the supercapacitance of rGO combined with unsubstituted NiPc nanofibers in acidic medium [30]. Fig. 18.14A and B show the typical cyclic voltammograms and galvanostatic charge-discharge curves of NiPc nanofiber (NF)-rGO modified electrode in comparison with the constituent components. The composite materials exhibit superior specific capacitance of 223 F/g at 1 A/g, which is fourfold higher compared to individual components with 48 and 86 F/g estimated for NiPc nanofibers and rGO, respectively. It is also shown that the composite was quite stable even for 1000 continuous galvanostatic cycling. A highly uniform 3-D flower-like tetranitrophthalocyanine iron (TNFePc) hierarchical nanostructures have been synthesized through solvothermal technique and explored as supercapacitor device [148]. The specific capacitance of the TNFePc was calculated to be 63 F/g at 0.3 A/g. A capacitance of 45 F/g was still retained even at a higher current density of 2 A/g with the

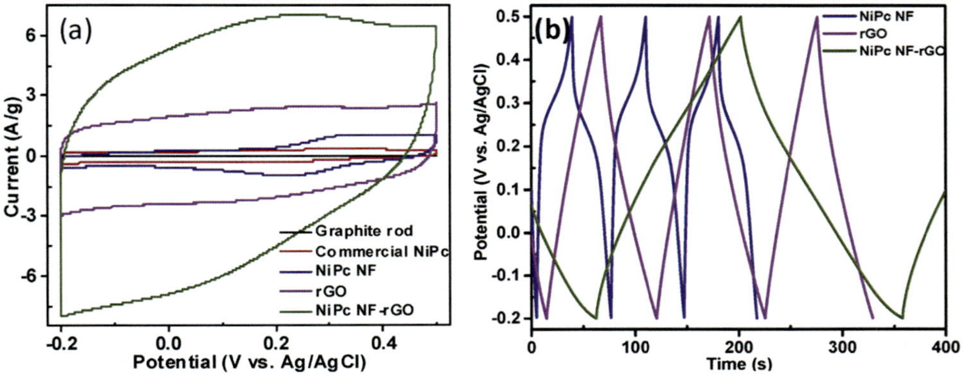

Figure 18.14

(A) Cyclic voltammograms of different modified electrodes scanned at 100 mV/s in 1 M H_2SO_4 aqueous electrolyte. (B) Typical chronopotentiometry charge-discharge cycles for differently modified electrodes at a current density of 1 A/g. *Reprinted from Ref. K.P. Madhuri, N.S. John, Supercapacitor application of nickel phthalocyanine nanofibers and its composite with reduced GO, Appl. Surf. Sci. 449 (2018) 528–536. Copyright (2018) with permission from Elsevier.*

maintenance ratio of more than 71%. Further, the specific capacitance retention was over 95% even after subjecting to 700 continuous charge-discharge cycling. In a different report by Samanta et al. CuPc nanowires were obtained from a top-down approach using bulk CuPc powder [149]. The CuPc nanowire electrode exhibits 260 F/g capacitance at 0.3 A/g and possess long cycling stability up to 3000 cycles at 6 A/g. Table 18.3 gives a list of various reported MPc based composite nanomaterials and their experimental energy storage parameters.

The incorporation of MPcs into batteries to improve the qualities of devices has been a topic of research recently. It has been reported that the organic cathode materials are advantageous over inorganic counterparts in having high theoretical specific capacity, abundance, environmental benignity, fast reaction kinetics, etc. A conjugated system like MPc with its inherent high stability, excellent electrochemical functionality are used to achieve certain goals like introduction of multi-active sites on MPc ring, poor solubility of MPc to achieve better cyclic stability, and a MPc moiety possessing good conductivity [150]. The latter part is achieved by either iodine doping or compositing with conductive carbon matrices. Chen et al. reported a novel molecular model of carboxyl substituted NiPc doped with I_2 used as a cathode material in a lithium-ion battery [150]. The device retained a capacity of about 300 and 500 mA h/g at 0.4 mA/g for tetra-carboxyl and octa-carboxyl substituted MPcs with the cyclic stability of 50 and columbic efficiency of 70%. The same group explored tetra-carboxyl substituted NiPc by replacing —H of the carboxyl with Li as an organic electrode material (t-CAL-Pc) [151]. Tetradiacyl substituted NiPc was also studied for battery performance (t-DA-Pc). It was found that the cyclic stability was improved with slight decrease in irreversible capacity. A capacity of 169 mA h/g after 100 cycles and 241 mA h/g at 0.4 mA/g after 200 cycles was still retained for t-CAL-Pc and t-DA-Pc, exhibiting columbic efficiency of 94% and 97% respectively. In all the cases, the I_2 doped product showed significantly smaller charge transfer resistance and impedance compared to the undoped product which verified the effect of I_2 doping on the conductivity of the cathode material. Wang et al. developed a fully conjugated CuPc metal organic framework in composite with I_2 serving as cathode material for a Na—I_2 battery [152]. The battery exhibits a stable specific capacity of 150 mA h/g at 0.3 mA/g even after 3200 cycles. Very little has been explored on the utilization of MPc-carbon composites for battery applications. He et al. developed FePc-rGO composite as a cathode material for lithium-ion battery application [153]. A discharge capacity of 186 mA h/g at 300 mA/g is maintained after 100 cycles. The study indicated that the self-assembled FePc nanospheres on the surface of rGO form a layer of active lithium intercalating sites providing conducting network.

Table 18.3: List of experimental energy storage parameters for different MPc composite nanomaterials.

Sl. No.	Materials	Specific/areal capacitance at a current density	Specific energy and power density	Cycling stability	Electrolyte	Refs.
1	NiOBCPPc-SWCNTs-phenylamine	186 mF/cm^2 at 138 μA/cm^2	—	1000 cycles with energy deliverable efficiency of 118%	1 M Na_2SO_4	[144]
2	NiTAPc-MWCNTs	981 F/g, 200 mF/cm^2 at 1 A/g	134 W h/kg and 701 W/kg	1500 cycles with energy deliverable efficiency of 121%	1 M H_2SO_4	[61]
3	CoTPyzPz-GO	500 F/g at 1 A/g	44 W h/kg and 31 kW/kg	1000 cycles with energy deliverable efficiency of 99%	1 M Na_2SO_4	[146]
4	Octamethyl substituted CuPc-GO	291 F/g at 0.5 A/g	3.38 W h/kg and 0.138 kW/kg	5000 cycles	1 M H_2SO_4	[147]
5	NiPc nanofibers-rGO	223 F/g at 1 A/g	15.19 W h/kg and 350 W/kg	1000, cycles with energy deliverable efficiency of 88.24%	1 M H_2SO_4	[30]
6	3-D flower-like TNFePc	63 F/g at 0.3 A/g	—	700 cycles	Ethanol and trifluoroacetate acid mixture	[148]
7	CuPc nanowires	260 F/g at 0.3 A/g	—	3000 cycles with energy deliverable efficiency of 99%	1 M NaOH	[149]
8	CoTPyzPz-MWCNTaf	1984 F/g at 1 A/g	250 W h/kg and 952 W/kg	1000 cycles	1 M H_2SO_4	[145]

5. Conclusions

Electroactive metal phthalocyanine nanostructures are highly promising and inexpensive molecular materials to realize various applications in energy landscape as discussed in this chapter. The insights derived from various experimental works elucidate the properties of MPcs in detail opening up a broad spectrum of applications. The semiconducting MPc macrocycles associate with the sp^2 network of graphene or CNTs through π-interactions enhancing the synergy between the two components for better electronic conduction. This has been enormously advantageous for engineering photovoltaic devices. The promise of using MPcs shows that a broad array of research can be executed by varying metal centers and organic functionalization when anchored on different conducting carbon matrices offering excellent physical interface. The applications of MPcs as organocatalysts in electrochemical applications of hydrogen evolution and oxygen reduction reactions have great potential. Characteristic features such as sharp redox transitions, active, stability/durability and offering low potential for hydrogen evolution/oxygen reduction reactions enable MPcs to be employed widely for electrochemical catalytic applications. Because of their NIR responsiveness, they have also been successfully incorporated for photovoltaic devices in solar water splitting reactions. MPcs are also active materials for the energy storage implications such as supercapacitors and batteries. The prospective implementation of MPc-carbon composites lie in the designing of flexible energy storage and conversion devices such as solar cells, fuel cells, supercapacitors, and batteries. The development of preferable cell configurations and appropriate structural designs for leading a fully flexible, portable, and wearable electronic device is a topic desired for intensive and focused research towards realization of its beneficial outcomes.

References

[1] S.A. Jenekhe, The special issue on organic electronics, Chem. Mater. 16 (2004) 4381–4382.

[2] R.H. Friend, R.W. Gymer, A.B. Holmes, J.H. Burroughes, R.N. Marks, C. Taliani, D.D.C. Bradley, D.A. Dos Santos, J.L. Bredas, M. Logdlund, W.R. Salaneck, Electroluminescence in conjugated polymers, Nature 397 (1999) 121–128.

[3] Y. Yang, J. Ouyang, L. Ma, R.H. Tseng, C.W. Chu, Electrical switching and bistability in organic/polymeric thin films and memory devices, Adv. Funct. Mater. 16 (2006) 1001–1014.

[4] Y.H. Chen, D.G. Ma, H.D. Sun, J.S. Chen, Q.X. Guo, Q. Wang, Y.B. Zhao, Organic semiconductor heterojunctions: electrode-independent charge injectors for high-performance organic light-emitting diodes, Light: Sci. Appl. 5 (2016) 16042.

[5] S.R. Forrest, Ultrathin organic films grown by organic molecular beam deposition and related techniques, Chem. Rev. 97 (1997) 1793–1896.

[6] O. Ostroverkhova (Ed.), Handbook of Organic Materials for Optical and (Opto) Electronic Devices: Properties and Applications, Elsevier, 2013.

[7] K.M. Kadish, K.M. Smith, R. Guilard (Eds.), The Porphyrin Handbook: Inorganic, Organometallic and Coordination Chemistry, vol. 3, Elsevier, 2000.

[8] H.S. Nalwa (Ed.), Supramolecular Photosensitive and Electroactive Materials, Gulf Professional Publishing, 2001.
[9] Phthalocyanine, Wikipedia, The Free Encyclopedia. https://en.wikipedia.org/w/index.php?title=Phthalocyanine&oldid=950674252.
[10] J. Jianzhuang, Functional Phthalocyanine Molecular Materials: Structure and Bonding, Springer-Verlag Berlin Heidelberg, 2010.
[11] C.C. Leznoff, A.B.P. Lever (Eds.), Phthalocyanines: Properties and Applications, vols. 1–4, VCH Publishers Cambridge, 1996.
[12] N.B. McKeown (Ed.), Phthalocyanine Materials: Synthesis, Structure and Function, Cambridge University Press Cambridge, 1998.
[13] J. Jiang (Ed.), Functional Phthalocyanine Molecular Materials, vol. 135, Springer, 2010.
[14] C.G. Claessens, U. Hahn, T. Torres, Phthalocyanines: from outstanding electronic properties to emerging applications, Chem. Rec 8 (2008) 75–97.
[15] K. Vasseur, B.P. Rand, D. Cheyns, L. Froyen, P. Heremans, Structural evolution of evaporated lead phthalocyanine thin film for near-infrared sensitive solar cells, Chem. Mater. 23 (2011) 886–895.
[16] T. Abe, K. Fukui, Y. Kawai, K. Nagai, H. Kato, A water splitting system using an organo-photocathode and titanium dioxide photoanode capable of bias-free H_2 and O_2 evolution, Chem. Commun. 52 (2016) 7735–7737.
[17] T. Abe, Y. Hiyama, K. Fukui, K. Sahashi, K. Nagai, Efficient *p*-zinc phthalocyanine/*n*-fullerene organic bilayer electrode for molecular hydrogen evolution induced by the full visible-light energy, Int. J. Hydrogen Energy 40 (2015) 9165–9170.
[18] K. Vasseur, K. Broch, A.L. Ayzner, B.P. Rand, D. Cheyns, C. Frank, F. Schreiber, M.F. Toney, L. Froyen, P. Heremans, Controlling the texture and crystallinity of evaporated lead phthalocyanine thin films for near-infrared sensitive solar cells, ACS Appl. Mater. Interfaces 5 (2013) 8505–8515.
[19] J.M. Mativetsky, H. Wang, S.S. Lee, L. Whittaker-Brooks, Y.L. Loo, Face-on stacking and enhanced out-of-plane hole mobility in graphene-templated copper phthalocyanine, Chem. Commun. 50 (2014) 5319–5321.
[20] O.A. Melville, B.H. Lessard, T.P. Bender, Phthalocyanine-based organic thin-film transistors: a review of recent advances, ACS Appl. Mater. Interfaces 7 (2015) 13105–13118.
[21] E. Zysman-Colman, S.S. Ghosh, G. Xie, S. Varghese, M. Chowdhury, N. Sharma, D.B. Cordes, A.M. Slawin, I.D. Samuel, Solution-processable silicon phthalocyanines in electroluminescent and photovoltaic devices, ACS Appl. Mater. Interfaces 8 (2016) 9247–9253.
[22] M. Wang, K. Torbensen, D. Salvatore, S. Ren, D. Joulie, F. Dumoulin, D. Mendoza, B. Lassalle-Kaiser, U. Isci, C.P. Berlinguette, M. Robert, CO_2 electrochemical catalytic reduction with a highly active cobalt phthalocyanine, Nat. Commu. 10 (2019) 3602.
[23] M.A.H. Khan, M.V. Rao, Q. Li, Recent advances in electrochemical sensors for detecting toxic gases: NO_2, SO_2 and H_2S, Sensors 19 (2019) 905.
[24] A. Koca, A. Kalkan, Z.A. Bayir, Electrocatalytic oxygen reduction and hydrogen evolution reactions on phthalocyanine modified electrodes: electrochemical, in situ spectroelectrochemical, and in situ electrocolorimetric monitoring, Electrochim. Acta 56 (2011) 5513–5525.
[25] Y. Jiang, Y. Lu, X. Lv, D. Han, Q. Zhang, L. Niu, W. Chen, Enhanced catalytic performance of Pt-free iron phthalocyanine by graphene support for efficient oxygen reduction reaction, ACS Catal 3 (6) (2013) 1263–1271.
[26] A. Chunder, T. Pal, S.I. Khondaker, L. Zhai, Reduced graphene oxide/copper phthalocyanine composite and its optoelectrical properties, J. Phys. Chem. B 114 (2010) 15129–15135.
[27] F. Roth, C. Lupulescu, T. Arion, E. Darlatt, A. Gottwald, W. Eberhardt, Electronic properties and morphology of Cu-phthalocyanine - C60 composite mixtures, J. Appl. Phys. 115 (2014) 033705.

[28] B. Ballesteros, G. de la Torre, C. Ehli, G.M.A. Rahman, F. Agullo-Rueda, D.M. Guldi, T. Torres, Single-wall carbon nanotubes bearing covalently linked phthalocyanines - photoinduced electron transfer, J. Am. Chem. Soc. 129 (2007) 5061–5068.
[29] Z. Yin, J. Zhu, Q. He, X. Cao, C. Tan, H. Chen, Q. Yan, H. Zhang, Graphene-based materials for solar cell applications, Adv. Energy. Mater. 4 (2013) 1–19.
[30] K.P. Madhuri, N.S. John, Supercapacitor application of nickel phthalocyanine nanofibers and its composite with reduced graphene oxide, Appl. Surf. Sci. 449 (2018) 528–536.
[31] M.S. Liao, S. Scheiner, Electronic structure and bonding in metal phthalocyanines, metal = Fe, Co, Ni, Cu, Zn, Mg, J. Chem. Phys. 114 (2001) 9780–9791.
[32] N. Hamamoto, H. Sonoda, M. Sumimoto, K. Hori, H. Fujimoto, Theoretical study on crystal polymorphism and electronic structure of lead (II) phthalocyanine using model dimers, RSC Adv. 7 (2017) 8646–8653.
[33] F. Przyborowski, C. Hamann, The monoclinic modification of lead phthalocyanine - a quasi-one-dimensional metal with characteristic order of molecular dipoles, Cryst. Res. Tech. 17 (1982) 1041–1045.
[34] P. Erk, H. Hengelsberg, M.F. Haddow, R. van Gelder, The innovative momentum of crystal engineering, CrystEngComm 6 (2004) 475–483.
[35] S. Heutz, S.M. Bayliss, R.L. Middleton, G. Rumbles, T.S. Jones, Polymorphism in phthalocyanine thin films: mechanism of the $\alpha \rightarrow \beta$ transition, J. Phys. Chem. B 104 (2000) 7124–7129.
[36] M. Brinkmann, J.,-C. Wittmann, M. Barthel, M. Hanack, C. Chaumont, Highly ordered titanyl phthalocyanine films grown by directional crystallization on oriented poly(tetrafluoroethylene) substrate, Chem. Mater. 14 (2002) 904–914.
[37] H. Yonehara, K. Ogawa, H. Etori, C. Pac, Vapor deposition of oxotitanium (IV) phthalocyanine on surface-modified substrates: effects of organic surfaces on molecular alignment, Langmuir 18 (2002) 7557–7563.
[38] H.-S. Shim, H.J. Kim, J.W. Kim, S.-Y. Kim, W.-I. Jeong, T.-M. Kim, J.-J. Kim, Enhancing photovoltaic response of organic solar cells using a crystalline molecular template, J. Mater. Chem. 22 (2012) 9077–9081.
[39] M. Hiramoto, K. Kitada, K. Iketaki, T. Kaji, Near infrared light driven organic p-i-n solar cells incorporating phthalocyanine J-aggregate, Appl. Phys. Lett. 98 (2011) 023302.
[40] T. Sakurai, T. Ohashi, H. Kitazume, M. Kubota, T. Suemasu, K. Akimoto, Structural control of organic solar cells based on nonplanar metallophthalocyanine/C_{60} heterojunctions using organic buffer layers, Org. Electron. 12 (2011) 966–973.
[41] K. Xiao, W. Deng, J.K. Keum, M. Yoon, I.V. Vlassiouk, K.W. Clark, A.P. Li, I.I. Kravchenko, G. Gu, E.A. Payzant, B.G. Sumpter, Surface-induced orientation control of CuPc molecules for the epitaxial growth of highly ordered organic crystals on graphene, J. Am. Chem. Soc. 135 (2013) 3680–3687.
[42] J. Mack, M.J. Stillman, Transition assignments in the ultraviolet - visible absorption and magnetic circular dichroism spectra of phthalocyanines, Inorg. Chem. 40 (2001) 812–814.
[43] D. Roy, N.M. Das, N. Shakti, P.S. Gupta, Comparative study of optical, structural and electrical properties of zinc phthalocyanine Langmuir-blodgett thin film on annealing, RSC Adv. 4 (2014) 42514–42522.
[44] T.C. Rosenow, K. Walzer, K. Leo, Near-Infrared organic light emitting diodes based on heavy metal phthalocyanines, J. Appl. Phys. 103 (2008) 043105.
[45] K.C. Kao, W. Hwang, Electrical transport in solids with particular reference to organic semiconductors, Int. Ser. Sci. Solid State, Pergamon Press, New York, 14 (1981).
[46] H. Ishii, K. Sugiyama, E. Ito, K. Seki, Energy level alignment and interfacial electronic structures at organic/metal and organic/organic interfaces, Adv. Mater. 11 (1999) 605–625.
[47] F. Gutmann, L.E. Lyons, Organic Semiconductors, Wiley, New York, 1967.

[48] L. Ottaviano, L. Lozzi, A.R. Phani, A. Ciattoni, S. Santucci, S. Di Nardo, Thermally induced phase transition in crystalline lead phthalocyanine films investigated by XRD and atomic force microscopy, Appl. Surf. Sci. 136 (1998) 81–86.
[49] K.P. Madhuri, P. Kaur, M.E. Ali, N.S. John, Nanoscale conductance in lead phthalocyanine thin films: influence of molecular packing and humidity, J. Phys. Chem. C 121 (2017) 9249–9259.
[50] K.P. Madhuri, N.S. John, S. Angappane, P.K. Santra, F. Bertram, Influence of iodine doping on the structure, morphology, and physical properties of manganese phthalocyanine thin films, J. Phys. Chem. C 122 (2018) 28075–28084.
[51] K.P. Madhuri, P.K. Santra, F. Bertram, N.S. John, Current mapping of lead phthalocyanine thin films in the presence of gaseous dopants, Phys. Chem. Chem. Phys. 21 (2019) 22955–22965.
[52] H. Yamada, T. Shimada, A. Koma, Preparation and magnetic properties of manganese (II) phthalocyanine thin films, J. Chem. Phys. 108 (1998) 10256–10261.
[53] Z. Wang, M.S. Seehra, Ising-like chain magnetism, arrhenius magnetic relaxation, and case against 3D magnetic ordering in β-manganese phthalocyanine ($C_{32}H_{16}MnN_8$), J. Phys. Condens. Matter 28 (2016) 136002.
[54] S. Schmaus, A. Bagrets, Y. Nahas, T.K. Yamada, A. Bork, M. Bowen, E. Beaurepaire, F. Evers, W. Wulfhekel, Giant magnetoresistance through a single molecule, Nat. Nanotechnol. 6 (2011) 185.
[55] P. Vasudevan, N. Phougat, A.K. Shukla, Metal phthalocyanines as electrocatalysts for redox reactions, Appl. Organomet. Chem. 10 (1996) 591–604.
[56] A. Wolberg, J. Manassen, Electrochemical and electron paramagnetic resonance studies of metalloporphyrins and their electrochemical oxidation products, J. Am. Chem. Soc. 92 (1970) 2982–2991.
[57] D.W. Clack, N.S. Hush, I.S. Woolsey, Reduction potentials of some metal phthalocyanines, Inorg. Chim. Acta 19 (1976) 129–132.
[58] N. Chebotareva, T. Nyokong, First-row transition metal phthalocyanines as catalysts for water electrolysis: a comparative study, Electrochim. Acta 42 (1997) 3519–3524.
[59] J. Oni, P. Westbroek, T. Nyokong, Electrochemical behavior and detection of dopamine and ascorbic acid at an iron (II) tetrasulfophthalocyanine modified carbon paste microelectrode, Electroanalysis 15 (2003) 847–854.
[60] K.P. Madhuri, N.S. John, Metallophthalocyanine-nanofibre-based electrodes for electrochemical sensing of biomolecules, Bull. Mater. Sci. 41 (2018) 118.
[61] A.T. Chidembo, K.I. Ozoemena, B.O. Agboola, V. Gupta, G.G. Wildgoose, R.G. Compton, Nickel (II) tetra-aminophthalocyanine modified MWCNTs as potential nanocomposite materials for the development of supercapacitors, Energy Environ. Sci. 3 (2010) 228–236.
[62] Z. Zhang, S. Yang, M. Dou, H. Liu, L. Gu, F. Wang, Systematic study of transition-metal (Fe, Co, Ni, Cu) phthalocyanines as electrocatalysts for oxygen reduction and their evaluation by DFT, RSC Adv. 6 (2016) 67049–67056.
[63] A.T. Dideikin, A.Y. Vul, Graphene oxide and derivatives: the place in graphene family, Front. Phys. 6 (2019) 149.
[64] C. Gomez-Navarro, J.C. Meyer, R.S. Sundaram, A. Chuvilin, S. Kurasch, M. Burghard, K. Kern, U. Kaiser, Atomic structure of reduced graphene oxide, Nano Lett. 10 (2010) 1144–1148.
[65] K.P. Madhuri, N.S. John, Solution processable transition metal phthalocyanine nanofibers, ChemistrySelect 4 (2019) 7292–7299.
[66] Z. Wang, R. Wei, X. Liu, Dielectric properties of reduced graphene oxide/copper phthalocyanine nanocomposites fabricated through π-π interaction, J. Electron. Mater. 46 (2017) 488–496.
[67] M. Mukherjee, U.K. Ghorai, M. Samanta, A. Santra, G.P. Das, K.K. Chattopadhyay, Graphene wrapped copper phthalocyanine nanotube: enhanced photocatalytic activity for industrial waste water treatment, Appl. Surf. Sci. 418 (2017) 156–162.
[68] M. Mukherjee, M. Samanta, G.P. Das, K.K. Chattopadhyay, Investigation of ORR performances on graphene/phthalocyanine nanocomposite in neutral medium, Microsc. Microanal. 25 (2019) 1416–1421.

[69] W. Song, C. He, W. Zhang, Y. Gao, Y. Yang, Y. Wu, Z. Chen, X. Li, Y. Dong, Synthesis and nonlinear optical properties of reduced graphene oxide hybrid material covalently functionalized with zinc phthalocyanine, Carbon 77 (2014) 1020–1030.
[70] V. Mani, S.T. Huang, R. Devasenathipathy, T.C. Yang, Electropolymerization of cobalt tetraamino-phthalocyanine at reduced graphene oxide for electrochemical determination of cysteine and hydrazine, RSC Adv. 6 (2016) 38463–38469.
[71] L. Zhang, H. Li, Q. Fu, Z. Xu, K. Li, J. Wei, One-step solvothermal synthesis of tetranitro-cobalt phthalocyanine/reduced graphene oxide composite, Nano 10 (2015) 1550045.
[72] J.H. Yang, Y. Gao, W. Zhang, P. Tang, J. Tan, A.H. Lu, D. Ma, Cobalt phthalocyanine–graphene oxide nanocomposite: complicated mutual electronic interaction, J. Phys. Chem. C 117 (2013) 3785–3788.
[73] X. Zhang, Y. Feng, S. Tang, W. Feng, Preparation of a graphene oxide–phthalocyanine hybrid through strong $\pi-\pi$ interactions, Carbon 48 (2010) 211–216.
[74] C. Zhang, R. Hao, H. Yin, F. Liu, Y. Hou, Iron phthalocyanine and nitrogen-doped graphene composite as a novel non-precious catalyst for the oxygen reduction reaction, Nanoscale 4 (2012) 7326–7329.
[75] D. Tasis, N. Tagmatarchis, A. Bianco, M. Prato, Chemistry of carbon nanotubes, Chem. Rev. 106 (2006) 1105–1136.
[76] C. Scoville, R. Cole, J. Hogg, O. Farooque, A. Russell, Carbon nanotubes, Notes Lecture (1991) 1–11.
[77] A. Aqel, K.M.A. El-Nour, R.A. Ammar, A. Al-Warthan, Carbon nanotubes, science and technology part (I) structure, synthesis and characterisation, Arab. J. Chem. 5 (2012) 1–23.
[78] L. Zhang, H. Yu, L. Liu, L. Wang, Study on the preparation of multi-walled carbon nanotube/ phthalocyanine composites and their optical limiting effects, J. Compos. Mater. 48 (2014) 959–967.
[79] Z. Jiang, Y. Wang, X. Zhang, H. Zheng, X. Wang, Y. Liang, Revealing the hidden performance of metal phthalocyanines for CO_2 reduction electrocatalysis by hybridization with carbon nanotubes, Nano Res. 12 (2019) 2330–2334.
[80] A. Morozan, S. Campidelli, A. Filoramo, B. Jousselme, S. Palacin, Catalytic activity of cobalt and iron phthalocyanines or porphyrins supported on different carbon nanotubes towards oxygen reduction reaction, Carbon 49 (2011) 4839–4847.
[81] R. Cao, R. Thapa, H. Kim, X. Xu, M.G. Kim, Q. Li, N. Park, M. Liu, J. Cho, Promotion of oxygen reduction by a bio-inspired tethered iron phthalocyanine carbon nanotube-based catalyst, Nat. Commu. 4 (2013) 2076.
[82] L. Cao, H. Chen, M. Wang, J. Sun, X. Zhang, F. Kong, Photoconductivity study of modified carbon nanotube/oxotitanium phthalocyanine composites, J. Phys. Chem. B 106 (2002) 8971–8975.
[83] Y. Wang, H.Z. Chen, H.Y. Li, M. Wang, Fabrication of carbon nanotubes/copper phthalocyanine composites with improved compatibility, Mater. Sci. Eng. B 117 (2005) 296–301.
[84] A.L. Verma, S. Saxena, G.S.S. Saini, V. Gaur, V.K. Jain, Hydrogen peroxide vapor sensor using metal-phthalocyanine functionalized carbon nanotubes, Thin Solid Films 519 (2011) 8144–8148.
[85] E. Asedegbega-Nieto, M. Pérez-Cadenas, J. Carter, J.A. Anderson, A. Guerrero-Ruiz, Preparation and surface functionalization of MWCNTs: study of the composite materials produced by the interaction with an iron phthalocyanine complex, Nanoscale Res. Lett. 6 (2011) 353.
[86] X. Li, W. Xu, Y. Zhang, D. Xu, G. Wang, Z. Jiang, Chemical grafting of multi-walled carbon nanotubes on metal phthalocyanines for the preparation of nanocomposites with high dielectric constant and low dielectric loss for energy storage application, RSC Adv. 5 (2015) 51542–51548.
[87] X. Deng, J. Li, L. Ma, J. Sha, N. Zhao, Three-dimensional porous carbon materials and their composites as electrodes for electrochemical energy storage systems, Mater. Chem. Front. 3 (2019) 2221–2245.
[88] E. Pérez-Mayoral, I. Matos, M. Bernardo, I.M. Fonseca, New and advanced porous carbon materials in fine chemical synthesis. Emerging precursors of porous carbons, Catalysts 9 (2019) 133.
[89] J. Lee, J. Kim, T. Hyeon, Recent progress in the synthesis of porous carbon materials, Adv. Mater. 18 (2006) 2073–2094.
[90] A. Stein, Z. Wang, M.A. Fierke, Functionalization of porous carbon materials with designed pore architecture, Adv. Mater. 21 (2009) 265–293.

[91] M. Li, X. Bo, Y. Zhang, C. Han, L. Guo, Comparative study on the oxygen reduction reaction electrocatalytic activities of iron phthalocyanines supported on reduced graphene oxide, mesoporous carbon vesicle, and ordered mesoporous carbon, J. Power Sources 264 (2014) 114–122.
[92] T.V. Magdesieva, T. Yamamoto, D.A. Tryk, A. Fujishima, Electrochemical reduction of CO_2 with transition metal phthalocyanine and porphyrin complexes supported on activated carbon fibers, J. Electrochem. Soc. 149 (2002) 89–95.
[93] S.C. Lee, H.O. Lintang, L. Yuliati, Photocatalytic removal of phenol under visible light irradiation on zinc phthalocyanine/mesoporous carbon nitride nanocomposites, J. Exp. Nanosci. 9 (2014) 78–86.
[94] N. Li, W. Lu, K. Pei, Y. Yao, W. Chen, Ordered-mesoporous-carbon-bonded cobalt phthalocyanine: a bioinspired catalytic system for controllable hydrogen peroxide activation, ACS Appl. Mater. Interfaces 6 (2014) 5869–5876.
[95] K. Takanabe, K. Kamata, X. Wang, M. Antonietti, J. Kubota, K. Domen, Photocatalytic hydrogen evolution on dye-sensitized mesoporous carbon nitride photocatalyst with magnesium phthalocyanine, Phys. Chem. Chem. Phys. 12 (2010) 13020–13025.
[96] A. Walcarius, Recent trends on electrochemical sensors based on ordered mesoporous carbon, Sensors 17 (2017) 1863.
[97] Y. Jiang, Y. Xie, X. Jin, Q. Hu, L. Chen, L. Xu, J. Huang, Highly efficient iron phthalocyanine based porous carbon electrocatalysts for the oxygen reduction reaction, RSC Adv. 6 (2016) 78737–78742.
[98] M. Urbani, M.E. Ragoussi, M.K. Nazeeruddin, T. Torres, Phthalocyanines for dye-sensitized solar cells, Coord. Chem. Rev. 381 (2019) 1–64.
[99] K. Vasseur, B.P. Rand, D. Cheyns, K. Temst, L. Froyen, P. Heremans, Correlating the polymorphism of titanyl phthalocyanine thin films with solar cell performance, J. Phys. Chem. Lett. 3 (2012) 2395–2400.
[100] H.J. Kim, H.-S. Shim, J.W. Kim, H.H. Lee, J.-J. Kim, CuI interlayers in lead phthalocyanine thin films enhance near-infrared light absorption, App. Phys. Lett. 100 (2012) 134.
[101] M. Raïssi, L. Vignau, E. Cloutet, B. Ratier, Soluble carbon nanotubes/phthalocyanines transparent electrode and interconnection layers for flexible inverted polymer tandem solar cells, Org. Electron. 21 (2015) 86–91.
[102] M. Ince, J.H. Yum, Y. Kim, S. Mathew, M. Grätzel, T. Torres, M.K. Nazeeruddin, Molecular engineering of phthalocyanine sensitizers for dye-sensitized solar cells, J. Phys. Chem. C 118 (2014) 17166–17170.
[103] E. Nouri, M.R. Mohammadi, Z.X. Xu, V. Dracopoulos, P. Lianos, Improvement of the photovoltaic parameters of perovskite solar cells using a reduced-graphene-oxide-modified titania layer and soluble copper phthalocyanine as a hole transporter, Phys. Chem. Chem. Phys. 20 (2018) 2388–2395.
[104] J. Bartelmess, B. Ballesteros, G. de la Torre, D. Kiessling, S. Campidelli, M. Prato, T. Torres, D.M. Guldi, Phthalocyanine – pyrene conjugates: a powerful approach toward carbon nanotube solar cells, J. Am. Chem. Soc. 132 (2010) 16202–16211.
[105] R.A. Hatton, N.P. Blanchard, V. Stolojan, A.J. Miller, S.R.P. Silva, Nanostructured copper phthalocyanine-sensitized multiwall carbon nanotube films, Langmuir 23 (2007) 6424–6430.
[106] L. Martín-Gomis, F. Fernández-Lázaro, Á. Sastre-Santos, Advances in phthalocyanine-sensitized solar cells (PcSSCs), J. Mater. Chem. A 2 (2014) 15672–15682.
[107] Z. Zhang, M. Dou, J. Ji, F. Wang, Phthalocyanine tethered iron phthalocyanine on graphitized carbon black as superior electrocatalyst for oxygen reduction reaction, Nano Energy 34 (2017) 338–343.
[108] T. Taniguchi, H. Tateishi, S. Miyamoto, K. Hatakeyama, C. Ogata, A. Funatsu, S. Hayami, Y. Makinose, N. Matsushita, M. Koinuma, Y. Matsumoto, A self-assembly route to an iron phthalocyanine/reduced graphene oxide hybrid electrocatalyst affording an ultrafast oxygen reduction reaction, Part. Part. Syst. Char. 30 (2013) 1063–1070.
[109] M. Mukherjee, M. Samanta, P. Banerjee, K.K. Chattopadhyay, G.P. Das, Endorsement of manganese phthalocyanine microstructures as electrocatalyst in orr: experimental and computational study, Electrochim. Acta 296 (2019) 528–534.

[110] D. Liu, Y.T. Long, Superior catalytic activity of electrochemically reduced graphene oxide supported iron phthalocyanines toward oxygen reduction reaction, ACS Appl. Mater. Interfaces 7 (2015) 24063–24068.
[111] Y. Liu, Y.Y. Wu, G.J. Lv, T. Pu, X.Q. He, L.L. Cui, Iron (II) phthalocyanine covalently functionalized graphene as a highly efficient non-precious-metal catalyst for the oxygen reduction reaction in alkaline media, Electrochim. Acta 112 (2013) 269–278.
[112] L. Cui, G. Lv, Z. Dou, X. He, Fabrication of iron phthalocyanine/graphene micro/nanocomposite by solvothermally assisted $\pi-\pi$ assembling method and its application for oxygen reduction reaction, Electrochim. Acta 106 (2013) 272–278.
[113] L. Cui, G. Lv, X. He, Enhanced oxygen reduction performance by novel pyridine substituent groups of iron (II) phthalocyanine with graphene composite, J. Power Sources 282 (2015) 9–18.
[114] T. Li, Y. Peng, K. Li, R. Zhang, L. Zheng, D. Xia, X. Zuo, Enhanced activity and stability of binuclear iron (III) phthalocyanine on graphene nanosheets for electrocatalytic oxygen reduction in acid, J. Power Sources 293 (2015) 511–518.
[115] X. Wang, B. Wang, J. Zhong, F. Zhao, N. Han, W. Huang, M. Zeng, J. Fan, Y. Li, Iron polyphthalocyanine sheathed multiwalled carbon nanotubes: a high-performance electrocatalyst for oxygen reduction reaction, Nano Res. 9 (2016) 1497–1506.
[116] D. Akyüz, B. Keskin, U. Şahintürk, A. Koca, Electrocatalytic hydrogen evolution reaction on reduced graphene oxide electrode decorated with cobaltphthalocyanine, Appl. Catal. B 188 (2016) 217–226.
[117] D. Akyüz, H. Dinçer, A.R. Özkaya, A. Koca, Electrocatalytic hydrogen evolution reaction with metallophthalocyanines modified with click electrochemistry, Int. J. Hydrogen Energ. 40 (2015) 12973–12984.
[118] İ. Özçeşmeci, A. Demir, D. Akyüz, A. Koca, A. Gül, Electrocatalytic hydrogen evolution reaction with a supramolecular cobalt (II) phthalocyanine carrying four cobaloxime moieties, Inorganica Chim. Acta 466 (2017) 591–598.
[119] G.R. Monama, K.D. Modibane, K.E. Ramohlola, K.M. Molapo, M.J. Hato, M.D. Makhafola, G. Mashao, S.B. Mdluli, E.I. Iwuoha, Copper (II) phthalocyanine/metal organic framework electrocatalyst for hydrogen evolution reaction application, Int. J. Hydrogen Energ. 44 (2019) 18891–18902.
[120] G.R. Monama, M.J. Hato, K.E. Ramohlola, T.C. Maponya, S.B. Mdluli, K.M. Molapo, K.D. Modibane, E.I. Iwuoha, K. Makgopa, M.D. Teffu, Hierarchiral 4-tetranitro copper (II) phthalocyanine based metal organic framework hybrid composite with improved electrocatalytic efficiency towards hydrogen evolution reaction, Results Phys 15 (2019) 102564.
[121] X. Zhang, Z. Wu, X. Zhang, L. Li, Y. Li, H. Xu, X. Li, X. Yu, Z. Zhang, Y. Liang, H. Wang, Highly selective and active CO_2 reduction electrocatalysts based on cobalt phthalocyanine/carbon nanotube hybrid structures, Nat. Commun. 8 (2017) 14675.
[122] N. Morlanés, K. Takanabe, V. Rodionov, Simultaneous reduction of CO_2 and splitting of H_2O by a single immobilized cobalt phthalocyanine electrocatalyst, ACS Catal. 6 (2016) 3092–3095.
[123] H. Hagiwara, N. Ono, T. Ishihara, Effects of redox potential of metallophthalocyanine dye on photocatalytic activity of $KTa(Zr)O_3$ for water splitting, Chem. Lett. 39 (2010) 178–179.
[124] G.G. Luo, X.C. Li, J.H. Wang, Visible light-driven hydrogen evolution from aqueous solution in a noble-metal-free system catalyzed by a cobalt phthalocyanine, Chem. Select 1 (2016) 425–429.
[125] R. Terao, T. Nakazono, A.R. Parent, K. Sakai, Photochemical water oxidation catalyzed by a water-soluble copper phthalocyanine complex, ChemPlusChem 81 (2016) 1064–1067.
[126] A. Tiwari, N.V. Krishna, L. Giribabu, U. Pal, Hierarchical porous TiO_2 embedded unsymmetrical zinc – phthalocyanine sensitizer for visible-light-Induced photocatalytic H_2 production, J. Phys. Chem. C 122 (2018) 495–502.
[127] A. Xie, X.L. Liu, Y.C. Xiang, G.G. Luo, A homogeneous molecular system for the photogeneration of hydrogen from water based on a $[Ru^{II}(bpy)_3]^{2+}$ photosensitizer and a phthalycyanine cobalt catalyst, J. Alloys & Compound. 717 (2017) 226–231.

[128] X. Zhang, L. Yu, C. Zhuang, T. Peng, R. Li, X. Li, Highly asymmetric phthalocyanine as a sensitizer of graphitic carbon nitride for extremely efficient photocatalytic H_2 production under near-infrared light, ACS Catal. 4 (2014) 162–170.
[129] T. Abe, S. Tobinai, N. Taira, J. Chiba, T. Itoh, K. Nagai, Molecular hydrogen evolution by organic p/n bilayer film of phthalocyanine/fullerene in the entire visible-light energy region, J. Phys. Chem. C 115 (2011) 7701–7705.
[130] Y.J. Yuan, J.R. Tu, H.W. Lu, Z.T. Yu, X.X. Fan, Z.G. Zou, Neutral nickel (II) phthalocyanine as a stable catalyst for visible-light-driven hydrogen evolution from water, Dalton Trans. 45 (2016) 1359–1363.
[131] J. Huang, Y. Wu, D. Wang, Y. Ma, Z. Yue, Y. Lu, M. Zhang, Z. Zhang, P. Yang, Silicon phthalocyanine covalently functionalized N-doped ultrasmall reduced graphene oxide decorated with Pt nanoparticles for hydrogen evolution from water, ACS Appl. Mater. Interfaces 7 (2015) 3732–3741.
[132] M. Watanabe, Dye-sensitized photocatalyst for effective water splitting catalyst, Sci. Technol. Adv. Mat. 18 (2017) 705–723.
[133] N. Yazdanpour, S. Sharifnia, Photocatalytic conversion of greenhouse gases (CO_2 and CH_4) using copper phthalocyanine modified TiO_2, Sol. Energy Mater. Sol. Cells 118 (2013) 1–8.
[134] M. Pal, V. Ganesan, U.P. Azad, Photochemical oxygen reduction by zinc phthalocyanine and silver/gold nanoparticle incorporated silica thin films, Thin Solid Films 525 (2015) 172–176.
[135] P. Kumar, G. Singh, D. Tripathi, S.L. Jain, Visible light driven photocatalytic oxidation of thiols to disulfides using iron phthalocyanine immobilized on graphene oxide as a catalyst under alkali free conditions, RSC Adv. 4 (2014) 50331–50337.
[136] T. Abe, K. Nagai, K. Sekimoto, A. Tajiri, T. Norimatsu, Novel characteristics at a fullerene/water interface in an organic bilayer photoelectrode of phthalocyanine/fullerene, Electrochem. Commun. 7 (2005) 1129–1132.
[137] T. Abe, K. Nagai, S. Kabutomori, M. Kaneko, A. Tajiri, T. Norimatsu, An organic photoelectrode working in the water phase: visible-light-induced dioxygen evolution by a perylene derivative/cobalt phthalocyanine bilayer, Angew. Chem. 45 (2006) 2778–2781.
[138] S. Zhang, T. Abe, T. Iyoda, K. Nagai, Photoelectrode characteristics of partially hydrolyzed aluminum phthalocyanine chloride/fullerene C60 composite nanoparticles working in a water phase, Molecules 17 (2012) 10801–10815.
[139] Y. Cheng, A. Memar, M. Saunders, J. Pan, C. Liu, J.D. Gale, R. Demichelis, P.K. Shen, Dye functionalized carbon nanotubes for photoelectrochemical water splitting – role of inner tubes, J. Mater. Chem. A 4 (2016) 2473–2483.
[140] Y. Li, M. Yang, Z. Tian, N. Luo, Y. Li, H. Zhang, A. Zhou, S. Xiong, Assembly of copper phthalocyanine on TiO_2 nanorod arrays as co-catalyst for enhanced photoelectrochemical water splitting, Front. Chem. 7 (2019) 334.
[141] R.S. Vishwanath, S. Kandaiah, Facile electrochemical growth of nanostructured copper phthalocyanine thin film via simultaneous anodic oxidation of copper and dilithium phthalocyanine for photoelectrochemical hydrogen evolution, J. Solid State Electr. 20 (2016) 767–773.
[142] J. Zheng, X. Li, Y. Qin, S. Zhang, M. Sun, X. Duan, H. Sun, P. Li, S. Wang, Zn phthalocyanine/carbon nitride heterojunction for visible light photoelectrocatalytic conversion of CO_2 to methanol, J. Catal. 371 (2019) 214–223.
[143] A. Ziani, T. Shinagawa, L. Stegenburga, K. Takanabe, Generation of transparent oxygen evolution electrode consisting of regularly ordered nanoparticles from self-assembly cobalt phthalocyanine as a template, ACS Appl. Mater. Interfaces 8 (2016) 32376–32384.
[144] B.O. Agboola, K.I. Ozoemena, Synergistic enhancement of supercapacitance upon integration of nickel (II) octa [(3, 5-biscarboxylate)-phenoxy] phthalocyanine with SWCNT-phenylamine, J. Power Sources 195 (2010) 3841–3848.
[145] J. Lekitima, K.I. Ozoemena, N. Kobayashi, Electrochemical capacitors based on nitrogen-enriched cobalt (II) phthalocyanine/multi-walled carbon nanotube nanocomposites, ECS Trans. 50 (2013) 125.

[146] J.N. Lekitima, K.I. Ozoemena, C.J. Jafta, N. Kobayashi, Y. Song, D. Tong, S. Chen, M. Oyama, High-performance aqueous asymmetric electrochemical capacitors based on graphene oxide/cobalt (II)-tetrapyrazinoporphyrazine hybrids, J. Mater. Chem. A 1 (2013) 2821–2826.
[147] R. Ramachandran, Q. Hu, F. Wang, Z.X. Xu, Synthesis of N-$CuMe_2Pc$ nanorods/graphene oxide nanocomposite for symmetric supercapacitor electrode with excellent cyclic stability, Electrochim. Acta 298 (2019) 770–777.
[148] J. Mu, C. Shao, Z. Guo, M. Zhang, Z. Zhang, P. Zhang, B. Chen, Y. Liu, Solvothermal synthesis and electrochemical properties of 3D flower-like iron phthalocyanine hierarchical nanostructure, Nanoscale 3 (2011) 5126–5131.
[149] M. Samanta, P. Howli, U.K. Ghorai, M. Mukherjee, C. Bose, K.K. Chattopadhyay, Solution processed copper phthalocyanine nanowires: a promising supercapacitor anode material, Physica E Low Dimens. Syst. Nanostruct. 114 (2019) 113654.
[150] J. Chen, Q. Zhang, M. Zeng, N. Ding, Z. Li, S. Zhong, T. Zhang, S. Wang, G. Yang, Carboxyl-conjugated phthalocyanines used as novel electrode materials with high specific capacity for lithium-ion batteries, J. Solid State Electr. 20 (2016) 1285–1294.
[151] J. Chen, J. Guo, T. Zhang, C. Wang, N. Ding, Q. Zhang, H. Yang, X. Liu, D. Li, Z. Li, S. Zhong, Electrochemical properties of carbonyl substituted phthalocyanines as electrode materials for lithium-ion batteries, RSC Adv. 6 (2016) 52850–52853.
[152] F. Wang, Z. Liu, C. Yang, H. Zhong, G. Nam, P. Zhang, R. Dong, Y. Wu, J. Cho, J. Zhang, X. Feng, Fully conjugated phthalocyanine copper metal–organic frameworks for sodium–iodine batteries with long-time-cycling durability, Adv. Mater. 32 (2020) 1905361.
[153] D. He, W. Xue, R. Zhao, W. Hu, A.J. Marsden, M.A. Bissett, Reduced graphene oxide/Fe-phthalocyanine nanosphere cathodes for lithium-ion batteries, J. Mater. Sci. 53 (2018) 9170–9179.

CHAPTER 19

Fabrication of nanostructures with excellent self-cleaning properties

Ajit Behera[1], Dipen Kumar Rajak[2], K. Jeyasubramanian[3]

[1]*Department of Metallurgical & Materials Engineering, National Institute of Technology, Rourkela, Odisha, India;* [2]*Sandip Institute of Technology and Research Centre, Nashik, Maharashtra, India;* [3]*Mepco Schlenk Engineering College, Sivakasi, Tamil Nadu, India*

Chapter Outline

Design, Fabrication, and Characterization of Multifunctional Nanomaterials. https://doi.org/10.1016/B978-0-12-820558-7.00014-5

1. What is a self-cleaning property of materials?

Everyone wants a dirt-free clean surface. In current decade, self-cleaning material is going to fulfill this dream. Self-cleaning surfaces are a special category of materials having ability to washout any surface contaminants (dirt, pollutants, and bacteria) which leads to labor-saving and offers a healthy environment [1,2]. This self-cleaning quality on the surfaces is mainly mimicked from natural phenomena detected in lotus leaves, gecko feet, and water striders and in many other creatures in this world. Self-cleaning characteristics can be introduced into any kind of synthetic surface by controlling the wettability property. Self-cleaning activity on the surfaces preferentially applied in many industries, domestic, agriculture, and military sector [3,4]. The first attempt to fabricate the self-cleaning surface was taken in 1995 by Paz et al., who synthesized a transparent TiO_2-film on a glass substrate and investigated on self-cleaning activity [5]. Gradual research on this topic give rise the first commercial production in 2001 that is Pilkington glass coated surface [6]. After the commercial production of self-cleaning glass, TiO_2 nanoparticles are incorporated into other material surfaces to facilitate the self-cleaning property [7].

According to the surface activities, the self-cleaning surfaces are categorized mainly into three groups: (1) hydrophobic and superhydrophobic surface, (2) hydrophilic and superhydrophilic surface, and (3) photocatalytic.

Fig. 19.1 showing hydrophobic, superhydrophobic, hydrophilic, superhydrophilic, and photocatalytic materials from nature as well as from manufacture. The main effective factor for categorization of these materials, i.e., wettability, is given in the figure mentioning with their angle condition.

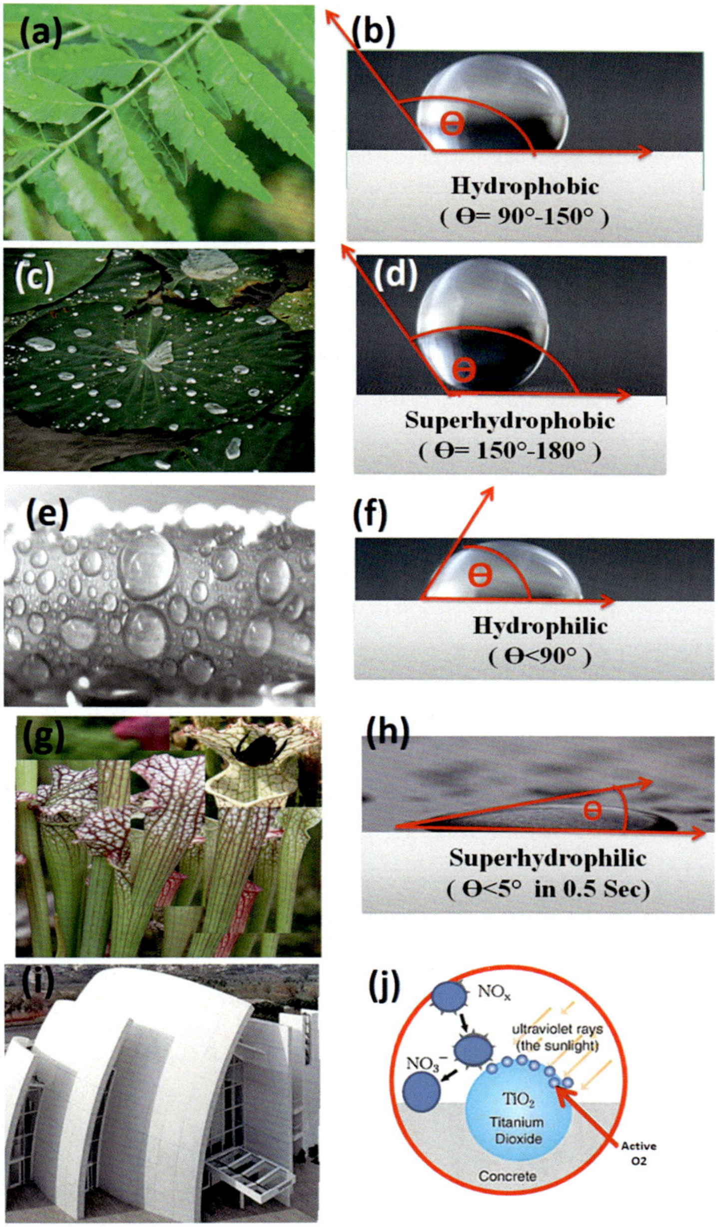

Figure 19.1

(A) Hydrophobic nature in neem plant leave, (B) schematic representation of hydrophobic angle, (C) superhydrophobic nature in lotus leaf, (D) schematic representation of superhydrophobic angle, (E) hydrophilic materials surface, (F) schematic representation of hydrophilic angle, (G) superhydrophilic nature in pitcher plant, (H) schematic representation of superhydrophilic angle, and (I) photocatalytic concrete material, and (J) photocatalytic TO_2 effect.

2. Market size of self-cleaning structure

Superhydrophobic coatings can be implemented in various surfaces irrespective of their complex geometry and composition. Furthermore, technological developments have favored the production of highly transparent coatings with high optical clarity, which increasingly find application in self-cleaning and fog-free screens and materials. The global hydrophobic coatings market size was USD 1.34 billion in 2015 and is expected to experience significant rise due to its increasing reach in the automotive, aerospace and construction industries [8].Hydrophobic coatings for glass components provide high resistance to water in heavy rain. As a result, high demand for automotive window and window manufacturing is expected to increase during the planning period. The aerospace industry is expected to gain ground due to the increasing use of anticorrosion and antiice or antiwetting coatings. Demand for the product is expected to increase significantly in the medical industry due to its increasing use in the manufacture of surgical tools and the use of biocompatible stainless steel and other materials. Furthermore, the increasing application of coatings in the optical industry in the manufacture of glasses and spectacles should have a positive impact on the growth of the industry in the next 9 years. The global self-cleaning glass market was valued at USD 118.0 million in 2018 and is projected to reach USD 162.9 million by 2026, at a compound annual rate of 4.1% [9,10]. Self-cleaning glass is a specialized category of glass that has also hydrophilic and photolytic properties and can keep its surface free of dirt and grime. Self-cleaning glass is applied in areas such as glass facades, outdoor store fountains, display cases, terraces, balconies, suspended glazing, windows, and doors. Fig. 19.2 illustrates the significant growth rates of self-cleaning materials in the automotive, aerospace, medical, construction, optical, and other sectors [11–13].

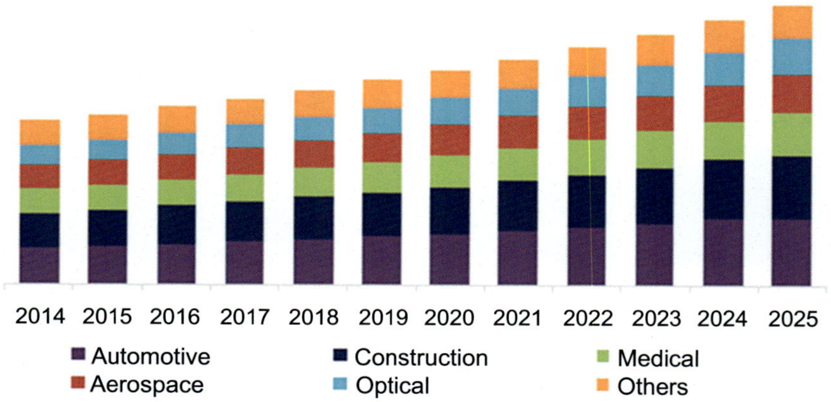

Figure 19.2
Year-wise growth rate of self-cleaning materials in different industries in the running market.

North America and Europe account for more than 50% of the total market share due to high production volumes from the end-user segments, in particular automobiles, aircraft, and medical devices. The Middle East and Africa are expected to experience significant growth due to growth in the automotive sector in the region. The main players in the world market for self-cleaning glass are Saint-Gobain Glass, Nippon Sheet Glass Co., Ltd., Cardinal Glass Industries, Inc., Asahi Glass Co., Atis Group, Guardian Industries, Roof-Maker Limited, Wuxi Yaopi Glass Engineering Co., Viridian Glass, Kneer-Südfenster, Clear Glass Solutions, Dongguan City of East Pearl River Glass Company Limited, Foshan Qunli Glass Company Limited, Gevergel (FoShan) Engineering Glass Company Limited, ITC International Trading and Consulting Pty Limited, PPG Industries, Inc., Saint-Gobain SA., Shanghai HuZheng Nanotechnology Company Limited, and others [14,15].

3. Surface characteristics of self-cleaning materials

The material capability to self-cleaning generally depends on the hydrophobicity or hydrophilicity of the surface. Whether you clean an aqueous or organic material from surface, water perform principal part in the self-cleaning activity. More specifically, the water contact angle on the materials surface plays an important role that helps to determine the self-cleaning ability of a material. This angle is also influenced by the micro-/nanoroughness of the surface that are well described in the following models with respect to the wettability (or adhesion) of a self-cleaning surface.

3.1 Wettability

The contact angle (θ) of the liquid drop on the material surface measures the wettability. The co-relationship of θ with wettability is illustrated in Fig. 19.3A. The Thomas Young's equation is the basis of all wetting phenomena [16]:

3.1.1 Young's model of wetting

Young's model of wetting is used to describe the relationship of water droplet angle Θ with the surface energies of the water, the surface, and the surrounding air (Fig. 19.3A). This model is generally used to explain the self-cleaning on an ideally flat surface as seen in lotus leaves mechanism. Young's model is presented in Eqs. (19.1) and (19.2) as follows [17]:

$$\gamma_{SG} - \gamma_{SL} = \gamma_{LG}\ Cos\Theta \tag{19.1}$$

Where.

Θ = Contact angle of water on the surface.

γ_{SG} = Surface energy of the surface-air interface.

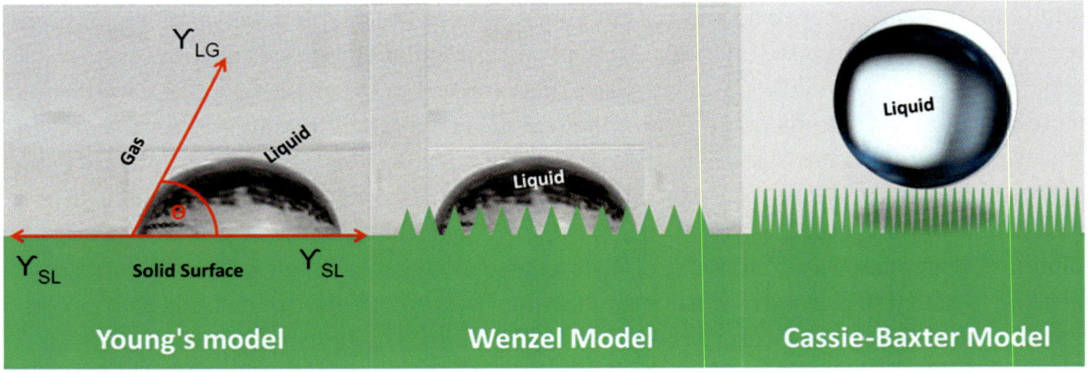

Figure 19.3
Schematic of (A) Young's model, (B) Wenzel, and (C) Cassie-Baxter models for wettability.

γ_{SL} = Surface energy of surface-liquid interface.

γ_{LG} = Surface energy of liquid-air interface.

Eq (19.1) simplified as,

$$Cos(\Theta_0) = (\gamma_{SA} - \gamma_{SL}\gamma_{LA}) = Cos(\Theta) = \frac{\gamma_{SG} - \gamma_{SL}}{\gamma_{LG}} \tag{19.2}$$

This equation is only reasonable for an ideal solid surface that is surface is smooth, inert, chemically homogeneous, and rigid.

3.1.2 Wenzel's model of wetting

When a drop of water is on a nonflat surface and the topographic characteristics of the surface lead to a larger surface area than that of a perfectly flat version of the same surface, the Wenzel model is a more valid predictor of wettability than the Young's model (Fig. 19.3B). Wenzel's model is presented in Eq. (19.3) [18]:

$$\text{Cos } (\Theta) = R_f Cos(\Theta_0) \tag{19.3}$$

Where

Θ = Contact angle of water predicted by Wenzel's model.

R_f = Ratio of surface area of rough surface to the surface area of a flat projection of the same surface.

Thus, the Wenzel's model of wetting is describing the interface between a water droplet and a micro/nano-rough surface.

3.1.3 Cassie-Baxter's model of wetting

This model is a modification of Wenzel's model and is applicable for complex systems that are representative of water-surface interactions in nature. This model is showing the water droplet that traps air in micro/nano-rough surface just below it (Fig. 19.3C). The Cassie-Baxter model is presented in Eq. (19.4) [19]:

$$\mathrm{Cos}(\Theta_{CB}) = R_f \mathrm{Cos}(\Theta_0) - f_{LA}\left(R_f \mathrm{Cos}(\Theta_0) + 1\right) \tag{19.4}$$

Where

Θ_{CB} = Contact angle of water predicted by Cassie-Baxter's model.

f_{LA} = Liquid-air fraction, the fraction of the liquid droplet that is in contact with air.

Cassie Baxter's model of wetting is used to describe the interface between a water droplet and a surface when the water droplet creates air pockets between itself and the surface topographical features on the surface.

3.2 Drag reduction

As the Cassie-Baxter model shows, superhydrophobic surfaces have a highly convex microstructure, which maintains a large amount of air in the groove between the solid surface, i.e., the actual contact surface between the liquid and the superhydrophobic surface consists of two interfaces: liquid-solid interface and liquid-gas interface [20]. On a superhydrophobic surface, the droplet can only come into contact with the separated convex microstructure, which greatly reduces the contact surface of water and the solid surface, and the coefficient of friction of the liquid-gas interface is much lower than that of the liquid-solid interface, so that the superhydrophobic surface will achieve the goal of reducing fluid resistance. Ke et al. determined that the static contact angle had little influence in reducing resistance, and the dynamic contact angle was an important factor in reducing fluid resistance [21]. Applying a superhydrophobic surface to water pipes and pipelines can reduce the friction resistance of the transport medium in the pipeline, thus reducing the cost of pipeline transport. And making a superhydrophobic surface on the surface of the boat can reduce the resistance of the ship to fluids, save energy and reduce consumption.

4. Self-cleaning surfaces

The different kind of self-cleaning surfaces such as superhydrophobic, superhydrophilic, and photocatalytic are discussed below:

4.1 Hydrophobic and superhydrophobic surfaces

In day to day life, the droplets of car window, glasses, tall-building glass surface slide to take away the dirt, but with the evaporation of droplets, the dirt will stick on the surface. The solution of this problem is found out by some researcher that the droplets on the flat surface can remove the dirt by sliding. This removal is due to the microstructure of superhydrophobic surface for which the droplets could roll to remove dirt with a small tilt [22]. If the contact angle lies in between 90 degrees and 150 degrees, then that surface is called hydrophobic surface. The word hydrophobic comes from the Greek word (hydro means water, and phobia means fearing or hating). If the contact angle is greater than 150 degrees, and the sliding angle is less than 10 degrees, then the water droplets will slide easily with the dirt (Fig. 19.4A–B). This property mimicked from lotus leaves (*Nelumbo nucifera*) since it is water repellent and adhesive resistance which keep them free from contamination even being rinse in polluted water [23]. The leave cells have papillae or microstructured asperities results very rough surface. Just above the microscale roughness, the papillae surface is overlying with nanoscale asperities contains epicuticular waxes (hydrophobic hydrocarbons). Basically, the plant cuticle is a composite material consists of spread cut in with low surface energy waxes, growing at different hierarchical levels [24]. The different leveled surfaces of the lotus leaves are made up of convex cells (which look

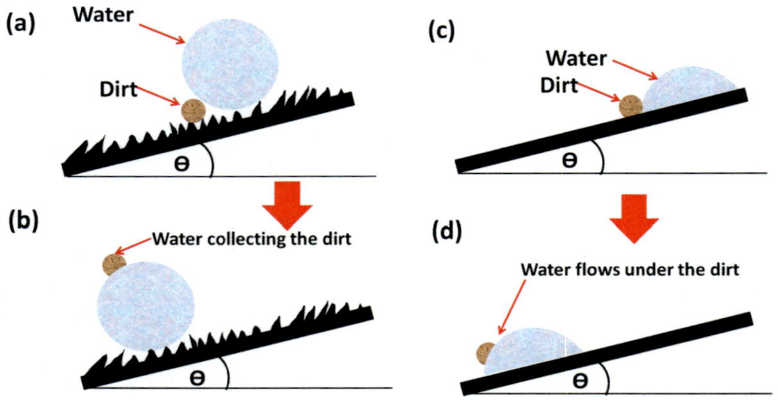

Figure 19.4
Superhydrophobic surface at (A) point of touching of water droplet with the dirt, and (B) dirt collecting by the water droplet; superhydrophilic surface at (C) point of touching of water droplet with the dirt, and (D) dirt collecting by the water droplet.

like bumps) and a much smaller layer of waxy tubules. The water pearls on the leaves of the plants rest on top of the nanofunctions, because air is trapped in the valley of the convex cells, which minimizes the contact area of the water droplets. Therefore, the lotus leaf indicate a remarkable superhydrophobicity. The static contact angle and the hysteresis of the contact angle of the lotus leaf are determined, respectively, around 164 degrees and 3 degrees [25]. At small tilt angles, the water droplets on the sheet roll up and cause dirt or contaminants, leading to self-cleaning. The ability of the droplets to form and roll depends not only on hydrophobicity, but also on contact angle hysteresis. In the plant world, the other natural superhydrophobic surfaces are taro leaves (*Colocasia esculenta*), Indian canna leaves (*Cannageneralis bailey*), and rice leaves (whatever the type of rice) also represent superhydrophobicity. In the animal world, some examples are the butterfly wings, the leg of the mosquitoes (Gerrisremigis), and the feet of geckos [26–28]. SEM micrographs of butterfly wings show a hierarchy, resulting from aligned microgrooves, covered with fine strip-stacked nanorays. The legs of spiders (Gerrisremigis) are highly water repellent due to their hierarchical morphology [29]. They are made up of hydrophobic waxy microhairs, and each hair is covered with nanogrooves. As a result, air is trapped between micro- and nanohairs, which repel water. Gecko's strong grip when studying the surface characteristics of the toes. There is a hierarchical morphology of each foot that is made up of millions of tiny hairs called bristles. In addition, each dental floss is made up of a smaller pile and each pile is covered with a flat spatula. In addition to strong adhesion to adhere the surface, the gecko foot has a unique self-cleaning property that does not require water like the lotus leaf [30].

4.2 Hydrophilic and super-hydrophilic self-cleaning surfaces

Hydrophilic, meaning to love water (comes from the Greek words: hydro means water and philos means love) is generally defined as having a contact angle of less than 90° [31]. Superhydrophilicity allows surfaces to clean a wide variety of dirt or debris and here the contact angle is less than 5 degrees. On such a surface, water droplets spread very rapidly and water flows from the surface at a considerable speed. For self-cleaning superhydrophilic surfaces, cleaning occurs because the water on the surface can be act to a great extent (extremely small contact angle with water) to position itself between contaminating debris and the surface to be washed (Fig. 19.4C–D). From nature, pitcher plant and shark skin are examples of self-cleaning superhydrophilic surfaces. Pitcher plant is a carnivorous plant. Insects caught on the edge of the petal jar (peristome) fall prey to the bottom of the jar. One of the most important characteristics of the jar to catch insects is its slippery surface. The surface of the peristome shows superhydrophilicity due to microtopography of the surface and the secretion of hygroscopic nectar. As a result, stable water films are formed in wet conditions. This film of water causes insects that walk above the rim to slide down from the jar and repelling the oils on their feet. Another example of antifouling, self-cleaning, and low-adhesion surfaces is shark skin. This hydrophilic surface allows sharks to quickly maneuver in the water. Shark skin is made up of periodically arranged diamond-shaped dermal denticles superimposed on triangular triangle riblets. The wettability of these surfaces is normally high, so the angle of contact with water tends to be approximately 0°. Despite such cleaning ability, it is still in a mature state with respect to hydrophobic coatings, and research in this field is still underway to innovate effective cleaning ability for such coatings by varying material compositions. These are high surface energy substrates that attract water and allow surface wetting. Many surfaces tend to be more water friendly, including glass, steel, or stainless steel and many coatings and paints. If the droplets extend almost flat, with a contact angle of less than about 5 degrees, they will show superhydrophilicity [32].

4.3 Photocatalysis self-cleaning materials

One of the most widely used self-cleaning products, TiO_2, uses a unique self-cleaning mechanism that combines an initial photocatalytic step and subsequent superhydrophilicity. TiO_2 is widely used because of its nontoxicity, availability, cost effectiveness, chemical stability, and favorable physical and chemical properties. A TiO_2 layer, usually on glass windows, when exposed to UV light, will generate free electrons that will interact with oxygen and water in the air to create free radicals. These free radicals in turn decompose any embedded organic matter deposited on the glass surface. TiO_2 also transforms normally hydrophobic glass into a superhydrophilic surface (contact angle of approximately 0°) [33]. Therefore, when precipitation occurs, instead of the water dripping onto the window surface and instantly falling onto the glass, the raindrops spread quickly on the hydrophilic surface. The water will descend along the surface of the window, in the form of a film rather than a drop, essentially acting as a squeegee to remove debris from the surface. Studies have shown that the self-cleaning effect of TiO_2 can be enhanced by water flow or precipitation. Therefore, one of the best applications of TiO_2 self-cleaning surfaces is construction materials, as these could be exposed to abundant sunlight and precipitation. Later, in the 1990s, a wide range of self-cleaning hydrophilic coatings such as cement, tile, marquee materials, glass, aluminum cladding, and plastic films was commercialized. HydrotechTM, a photo-induced superhydrophilic technology introduced by TOTO Ltd. (a Japanese company). This technology uses sunlight to break down washable dirt/contaminants. It could be effectively used in building materials, coatings and construction paints. High processing temperatures are required to manufacture TiO_2-coated, self-cleaning photocatalytic glazing products in commercial applications. There are several forms of TiO_2, among which the primary phases are the anatase, rutile, and brookite phases [34]. The most common form of TiO_2 is the rutile phase, which is densely packed and used in pigments such as sunscreens and paints. The anatase phase is rare and has an open crystal structure, making it highly photocatalytic.

The anatase and rutile phases have a tetragonal structure. The brookite phase is orthorhombic, it is extremely rare. The TiO_2 anatase phase when heated to over 400°C becomes the rutile phase. Photocatalysis can generally be classified into two kinds of process. The process in which the adsorbate molecule is photoexcited and interacts with the catalyst substrate in the ground state is known as catalyzed photoreaction. Furthermore, if the initial photoexcitation takes place on the catalytic substrate and transfers an electron or energy to a molecule in the ground state, the process is called sensitized photoreaction. The quantum yield (number of events that occur per absorbed photon) determines the efficiency of a photocatalytic process. By analyzing all possible paths for electrons and holes, quantum efficiency or yield is calculated. As TiO_2 is a semiconductor, during the absorption of light greater than or equal to its prohibited band, it is excited to produce electrons and holes. Most of these charge carriers undergo recombination and some migrate to the surface. The electrons produced move from the valence band to the conduction band where it reacts with atmospheric oxygen to produce superoxide radicals. These superoxide radicals are very energetic, breaking down organic dirt into carbon dioxide and water, which is called the cold combustion process. The decomposition of stearic acid into carbon dioxide and water vapor occurs in the presence of atmospheric oxygen. TiO_2 is even used in paints and cosmetics as a pigment and as a food additive. The material is also used in antipollution applications and for water purification (the membrane is covered with TiO_2 that kills bacteria in the water) [35,36].

5. Low surface energy material for hydrophobic surface

There are two basic parameters required to achieve the self-cleaning surface, one is the surface energy of the material combined with the second one that is the surface roughness. Some low surface energy materials are discussed below.

5.1 Silicones

PDMS (polydimethylsiloxane) belongs to a group of organosilicon compounds, commonly called silicones. The intrinsic deformability and hydrophobic properties of PDMS make it a very suitable material for the production of superhydrophobic surfaces. Various methods are practiced to produce superhydrophobic surfaces using PDMS. The modification of the surface in the PDMS can be performed using a pulsed CO_2 laser as the excitation source to introduce peroxide groups on the surface of the PDMS. These peroxides are capable of initiating graft polymerization of 2-hydroxyethyl methacrylate in PDMS. The water contact angle of the treated PDMS was measured at 175 degrees. The reason for such an increase in the angle of contact with water was due to the porosity and order of the chains on the PDMS surface. The PDMS elastomer containing microcomposite and nanocomposite structures is also used to produce superhydrophobic surfaces by laser engraving to induce roughness on the PDMS surface. The surface produced by this technique had a water contact angle of up to 160 degrees and a slip angle of less than 5 degrees. Electrospinning technique used to produce superhydrophobic membranes. Here, the electrospun fibers formed by a PS-PDMS block mixed with a PS homopolymer (polystyrene) have reached a water contact angle of approximately 163 degrees. The large angle of contact with water is due to the combined effect of the enrichment of the fiber surfaces by the PDMS component and the roughness of the surface due to the small diameter of the fiber (150–400 nm). The superhydrophobic surface can be synthesized using a casting technique. In the casting process, a PS-PDMS micellar solution in the presence of moist air gave a superhydrophobic surface with a water contact angle of about 163 degrees [37].

5.2 Fluorocarbons

Fluoropolymers are attracting a lot of interest these days due to their extremely low surface energies. Curing these polymers will result in superhydrophobic surfaces. Superhydrophobicity can be obtained by stretching a Teflon film (polytetrafluoroethylene). The superhydrophobic property obtained is due to the presence of fibrous crystals with large fractions of space on the surface. The rough surface produced by treatment with oxygen plasma on the Teflon surface and showing a superhydrophobic nature with a water contact angle of 168 degrees. Due to limited solubility, many fluorinated materials have not been used directly, but have been bonded with other raw materials to create superhydrophobic surfaces. A microporous transparent honeycomb polymer film with microsized pores can be produced by pouring a polymer solution under wet conditions. In this process, the fluorinated glass substrate was placed on a substrate support. A metal blade was attached perpendicular to the substrate and the space between the blade and the substrate was adjusted. A fluorinated copolymer solution was provided between the slide and the substrate. Moist air (relative humidity of approximately 60% at room temperature) was supplied to the surface of the solution at a flow rate of 10 per min. Transparent honeycomb-patterned films produced by this method exhibited superhydrophobic properties with a WCA of approximately 160 degrees [38].

5.3 Organic materials

Although silicones and fluorocarbons are widely used to produce superhydrophobic surfaces, recent research has shown that hydrophobicity can also be achieved using paraffinic hydrocarbons. A highly porous low-density polyethylene superhydrophobic surface can be produced by controlling the crystallization time and nucleation rate and the angle of contact with water is 173 degrees. The superhydrophobic film can be obtained by electrostatic spinning and spraying with a solution of PS in dimethylformamide.

The obtained surface was composed of porous microparticles and nanofibers. Recent research has shown that alkyl ketene, polycarbonate, and polyamide also have superhydrophobic properties [39].

5.4 Inorganic materials

Superhydrophobic properties have also been demonstrated by some inorganic materials. Recent research carried out on materials such as ZnO and TiO_2 has resulted in the production of films with reversibly switchable wettability. When the ZnO film was exposed to UV radiation, electron pairs were produced, resulting in adsorption of a hydroxyl group on the ZnO surface. Therefore, the superhydrophobic property of the film becomes superhydrophilic. Storing the UV irradiated film in the dark for a week makes it superhydrophobic again [40].

6. Fabrication of superhydrophobic materials

Generally, there are two-step fabrication processes seen in the fabrication of hydrophobic surfaces, where (1) the rough material is constructed and (2) the material is chemically modified to meet the desired characteristics. A great number of fabrication techniques have been reported, such as layer-by-layer assembly, laser process, the solution-immersion method, sol-gel techniques, chemical etching, and Hummers' method, lithography, templating, femtosecond laser pulsing, and etching.

6.1 Electrospinning technique

With the development of nanotechnology, electrospinning has become important in the synthesizing of continuous nanofibers with high surface roughness. Electrospinning is a dominant technique for the synthesis of fine nanofibers. This technique is widely used by various research groups to provide sufficient surface roughness to induce superhydrophobicity. An electrospinning setup consists of a high voltage power supply, a grounded collector, a syringe loaded with a precursor solution, and a pump to regulate the flow of the precursor solution. When the high voltage applied to the tip of the syringe exceeds the surface tension of the precursor solution, a Taylor cone is produced, which first forms a jet. For the formation of the Taylor cone, the applied voltage must exceed a critical voltage. The unstable jet accumulates on the collector surface where the fibers are obtained. The parameters that govern electrospinning are the surface tension of the precursor solution, the concentration of the precursor solution, the viscosity of the precursor solution, the applied voltage, the flow rate of the solution and the distance between the tip of the syringe and the collector. If the viscosity of the solution is high, the risk of clogging is high, and if it is low, electric spraying instead of electro-spinning occurs. Pearl formation in the resulting fibers could be controlled by optimizing the viscosity of the solution. For electrical spin to occur, a sufficient voltage (5–30 kV) must be applied, otherwise sparking will occur. The distance parameter is critical because it must be optimized to facilitate evaporation of the solvent before collecting the fibers in the collector. Today, there is a controllable moisture electrolytic spinning configuration that remedies pearl formation in the resulting fibers. Hydrophobic surfaces are necessary for applications such as anticorrosion, antifreeze, and water-repellent systems. In industry, they are widely used for ultradry and surface applications. By applying the coating, air molecules come into contact between the coating and the substrate, which increases the water contact angle. Superhydrophobic coatings improve energy efficiency in the marine industry by reducing skin friction scratches that occur on ship hulls. Such a coating increases the speed of the boat and also acts as anticorrosive systems, preventing any organic pollutant or marine microorganism from coming into contact with the hulls of the boats. In vehicles, superhydrophobic coatings are applied to the lenses to prevent them from snagging, helping to clean the car. Furthermore, such superhydrophobic membranes are used in water desalination plants for the efficient production of freshwater. A well-known superhydrophobic coating used on small boats is the "HullKote Speed Polisher" which provides protection of the surface of the boats and is easy to drag [41].

6.2 Wet chemical reaction and hydrothermal reaction

The wet chemical reaction is a simple technique that can effectively control the dimensionality and morphology of the nanostructures (nanoparticles, nanowires, and mesoporous inorganics) produced. This method has been widely used in the manufacture of superhydrophobic surfaces on metal substrates such as Cu, Al, and steel. A chemical composition process used to produce a superhydrophobic surface on a Cu substrate by immersing it in a solution of tetradecanoic acid for approximately one week, resulting in a change in the surface of the substrate. For example, an acidic or basic fluoroalkyl silane etchant solution can be applied to polycrystalline metals to form a superhydrophobic surface. Superhydrophobic surfaces on Ni substrates have been created by a wet chemical process in which monoalkylphosponic acid reacts with Ni to produce flowering microstructures that constitute a continuous covering. A stable superhydrophobic surface is produced on a copper substrate using oxalic acid as the reaction reagent, then chemical modification is performed using vinyl terminated poly(dimethylsiloxane) (PDMSVT). A layer of interconnected $Cu(OH)_2$ nanowires were generated in a Cu plate by immersing it in the mixture of NaOH and $K_2S_2O_8$ solution. After chemical modification with dodecanoic acid, the surface showed superhydrophobicity [42,43].

6.3 Electrochemical deposition

Electrochemical deposition is widely used to develop superhydrophobic surfaces as it is a versatile technique for preparing structures on the microscopic and nanometric scale. A galvanic deposition technique on metals can be used to deposit a solution of metal salts, which has resulted in the formation of a superhydrophobic surface with a water contact angle of approximately 173 degrees. The produced surface can float effortlessly on a water surface similar to that of pond skaters. Electrochemical deposition method used to induce long-chain fatty acids to produce a copper mesh with hierarchical micro- and nanometric structure that exhibited superhydrophobicity. Superhydrophobic 3D porous copper films have been manufactured using hydrogen bubbles as a dynamic model for metal coating. Since the films have been electrodeposited and grown in the interstitial spaces between the hydrogen bubbles, the pore diameter and the wall thickness of the porous copper films have been successfully adapted by adjusting the electrolyte concentration. The porous structure with numerous dendrites in different directions forming a solid film [44].

6.4 Spraying and physical method

Spraying is a simple and inexpensive method of manufacturing superhydrophobic surfaces, and can be used to repair mechanical surface damage by local spraying. Although this method relies heavily on the adhesion of the coating to achieve stability. Different concentrations of functionalized silica nanoparticles and different polymers were adopted to carry out a spraying method at low cost and in one step to achieve a uniform superhydrophobic coating without special equipment. When the functionalized silica concentration was 2.5% by weight and no polymer was added, the maximum contact angle was 162 degrees and the water slip angle was less than 5 degrees. The composite coating has shown great potential and could be used for various industrial substrates such as glass, plastic, and metal. Due to the ability of dopamine to spontaneously polymerize in an alkaline aqueous solution, the superhydrophobic porous polymeric membrane can be manufactured by spraying a poly (fluorinated ester) polydimethylsiloxane block copolymer solution. The method was simple and included only three steps: spraying the coating, drying and peeling. The physical method is very simple and requires no chemical reagents or expensive equipment. The superhydrophobic surface of the polyolefin could be made by abrasion machines and could also be applied to other hydrophobic polymeric materials to obtain a superhydrophobic property. Due to its ease of use and low cost, this method will have a very practical application [45,46].

6.5 Lithography

Lithography is a conventional technique used to create micro- and nanopatterns. The different lithographic techniques practiced in practice are photolithography, electron beam lithography, X-ray lithography, soft lithography, and nanosphere lithography, etc. continues [47]. For example, the photocatalytic lithography technique used for Au composite surfaces to make superhydrophilic and superhydrophobic patterns. Electron beam lithography and plasma etching are also used to produce a series of nanowells and nanopiles on the surface of the material. When this surface was hydrophobized with octadecyl trichlorosilane, it exhibited superhydrophobic properties with an angel in contact with water that reached up to 164 degrees [48].

6.6 Sol-gel method and polymerization reaction

The sol-gel method can be used in the manufacture of superhydrophobic surfaces on all types of solid substrates. The sol-gene technique can be divided into several stages where (a) the sample will be washed, (b) the clean sample will be immersed in a solution, (c) the coated sample will be washed again and (d) the sample will be will dry. For example, superhydrophobic surfaces prepared in a Cu alloy using hexamethylenetetramine and ethylene glycol, a powerful bidentate chelating agent for Cu^{2+} and Fe^{2+} ions with high stability constant, as a styling reagent. Soil immersion method used to produce ordered pore indium oxide matrix films using colloidal polystyrene monolayers. It is observed that the superhydrophobic properties exhibited by the film can be controlled by increasing the size of the pores in the film [49].

6.7 Laser process

Laser treatment has proven to be a promising manufacturing technique due to three main factors: (a) its excellent control of nanosurface roughness on a microscopic scale, (b) it is one-step processing under conditions adequate, and (c) the ability to work with different types of materials. Laser treatment can be used to increase the contact angle of static water droplets by increasing the roughness of the material surface. Lasers are generally used to apply complex structures to the material, where the material is often covered with a fluorine-based material afterward to create a hydrophobic surface. The fluorine-based material and the surface roughness lead to a superhydrophobic material [50].

6.8 Flame treatment

Flame treatment is a special and simple method of making a superhydrophobic surface. Based on the difference in thermal properties of organic and inorganic materials, the micro-/nanostructure and morphology of the superhydrophobic biomimetic surfaces were fabricated. An organic-inorganic compound consisting of polydimethylsiloxane particles and SiO_2 can be used as a powerful hydrophobic material. In this compound, these two components showed different thermal responses in the flame treatment process and then built the expected superhydrophobic micro-/nanostructure. The roughness of the superhydrophobic surface could be controlled by modifying the flame treatment time, to obtain an ultralow slip angle that reaches a limit value of 1 degree and considerably eliminates anisotropic wettability. Compared to other conventional methods, this method has the advantages of simplicity, high efficiency, and low cost. These characteristics will give it wide application possibilities [51].

6.8.1 Plasma treatment

Plasma treatment of surfaces is essentially a dry etching of the surface. This is accomplished by filling a chamber with gases, such as oxygen, fluorine, or chlorine, and by accelerating ion species from an ion source through plasma. Ionic acceleration to the surface forms deep grooves within the surface. In addition to topography, plasma treatment can also provide surface functionalization by using different gases to deposit different elements on surfaces. The roughness of the surface depends on the duration of the plasma etching [52].

6.9 Self-assembly and layer-by-layer methods

Self-assembly and layer-by-layer assembly (LBL) techniques are based on the sequential adsorption of a substrate in solutions of compounds of opposite charges. These techniques remain the most popular and well-established methods of forming multilayer thin films. Self-assembly and LBL deposition are inexpensive techniques in which micro- and nanometric superhydrophobic structures can be easily manufactured with finely controlled surface morphologies. For example, a superhydrophobic rambutan-type surface with hollow aniline spheres can be prepared by a self-assembly technique in the presence of perfluorooctane sulfonic acid. Carbon nanotube arrays are built on the cotton substrate to reproduce the structure of the lotus leaves. To control the assembly of carbon nanotubes in cotton fibers, these fibers are modified by carbon nanotubes treated as macroinitiators [53,54].

6.10 Chemical etching

Chemical etching is a cheap and easy method to control since the reaction conditions require a specific temperature and immersion time. This method is generally used in the manufacture of superhydrophobic surfaces on metal substrates. Chemical etching is a type of surface treatment commonly applied to metals. The engraving process consists of several stages in which (a) the sample is cleaned, often by ultrasound, and dried, (b) the material is engraved in an aqueous solution, which often contains copper, (c) the material is rinsed in water and often with ethanol, and (d) the material will dry. For example, combining chemical etching and oxidation technology to make a superhydrophobic surface of steel plate in a low concentration acid solution; The process contained hydrochloric acid and potassium chloride in an oxygen-enriched environment, to prepare a superhydrophobic surface on steel substrates. When the contact angle of the superhydrophobic surface was 166 degrees, and the slip angle was less than 2 degrees, it showed excellent hydrophobicity; Furthermore, the superhydrophobic surface had long-term chemical stability. These properties are favorable for the possible applications of steel materials for the field of reduction of resistance [55,56].

6.11 Hummers' method

Hummers' method is to use a solution of $NaNO_3$ and $KMnO_4$ dissolved in H_2SO_4 to oxidize the graphite to graphite oxide. The Hummers' method was a very fast means of oxidation. The Hummers' method is very simple and quick to implement, it is now widely used. For example, Hummers' method of producing hydrophobic aerogel. Following the Hummers' method, various applications of solutions and drying processes were applied, ending with a reaction with 1H, 2H, 2H-perfluorodecanothiol. Although the product is highly hydrophobic, the reaction requires much more effort and time than the other aforementioned methods. In addition to the six manufacturing techniques mentioned, many other techniques that can be applied to make a hydrophobic coating. Each described method has its advantages and disadvantages, where some methods are very easy and fast, but they can only be applied to specific types of materials and others are complex, but they allow obtaining high-quality results in many types of materials [57].

6.11.1 Nanocasting

Nanocasting is a method based on smooth lithography that uses elastomeric molds to produce nanostructured surfaces. For example, polydimethylsiloxane was poured onto the lotus leaf and used to create a negative PDMS matrix. The PDMS was then coated with a trimethylchlorosilane nonstick monolayer and used to make a positive PDMS matrix from the former. Since the natural structure of the lotus leaf allows a pronounced self-cleaning ability, this modeling technique was able to reproduce the nanostructure, resulting in surface wettability similar to that of the lotus leaf. Furthermore, the ease of this methodology allows for the translation into massive replication of nanostructured surfaces [58].

6.12 3D printing

This is a rapid prototyping technique. The nanostructured superhydrophobic surface can be made on a powder bed using laser/electron source operated by computer-aided designing programming. There are several categories of 3D printing technology such as selective laser melting, selective laser sintering, and electron-beam melting mainly tried to fabricate the superhydrophobic materials [59].

7. Fabrication of hydrophilic materials

In general, improving the hydrophilicity of surfaces can be accomplished in two ways: by depositing a molecular film of new material more hydrophilic than the substrate, or by modifying the chemistry of the substrate surface [60]. Deposition of surface coatings is the most common method for inorganic substrates. However, modification of surface chemistry is used in the case of polymeric materials. Several techniques for synthesizing surface hydrophilicity are discussed in the following sections.

7.1 Deposited molecular structures

Monolayers can be formed from organic molecules that adsorb to the solid surface. Organic molecules can come from a solution or a vapor phase. This ultimately changes the wetting characteristics of the surface. Much research has been done on densely compact molecular structures used in metals. Alkanethiols have been generally used in Au, Ag, Cu, and Pd. Chlorosilanes have been used in silicon oxide, Al, and Ti. Phosphoric acids can also be used in Ti and Al. The Langmuir–Blodgett film technique is used to mechanically deposit monolayers and multilayers. However, a major disadvantage of using this method is that it suffers from poor stability when multiple layers are contacted with liquids. The hydrophilic surface is obtained if the end of the deposited organic layers is polar. If a saturated hydrocarbon group or a fluorinated group is at the end of the layer, the water will not be attracted to the surface, which will cause hydrophobic conditions. The hydrogen bond is how the water molecules will be attracted to the surface if the surface has chemical groups like −OH, −COOH, and POOH. A zero-contact angle has never been recorded. In addition to establishing self-assembled monolayers of short functional molecules chemically bonded to inorganic surfaces, he has recently focused on coating materials with macromolecules and biomacromolecules. These are popular in modifying polymers that come into contact with biofluids, such as blood. Biomacromolecules, such as albumin and heparin, has been widely used in the health sector to provide hydrophilic characteristics that complement the body's needs.

Synthetic polymers such as polyethylene glycol and phospholipid macromolecules have been extensively studied. The presence of a hydrophilic coating in bioengineering applications can cause the adsorption of proteins on the surface, which is highly undesirable. Therefore, having such protective coatings prevents protein adsorption when the materials come into contact with biological fluids [61,62].

7.2 Modification of surface chemistry

The modification of surface chemistry has been studied in the last decades. Plasma, corona, flame, photons, electrons, ions, X-rays, UV-rays, and ozone are methods that have been studied to modify the chemistry of polymer surfaces without affecting their volume properties. Oxidation of polymer surfaces can be accomplished by plasma treatment in an air or oxygen, corona and flame environment. In plasma and corona treatments, the electrons are accelerated and then bombard the polymer with energies two to three times higher than necessary to break the molecular bonds, producing free radicals. This generates cross-links and reacts with the surrounding oxygen to produce oxygen-based functionality. Hydroxyl, peroxy, carbonyl, ether, and ester groups are typical polar groups that are created on the surface. In flame treatment, the surface combustion of the polymer takes place with the formation of radicals such as hydroperoxide and hydroxyl. The oxidation depth by flame treatment is approximately 5–10 nm. This increases to more than 10 nm for air plasma treatment. Plasma, corona and flame treatments cause significant surface oxidation and ultimately highly wettable surfaces. Due to environmental chemistry, the polar groups produced during surface oxidation tend to be buried in the mass when in contact with air for an extended period. However, its presence remains on the surface in contact with water or any other polar environment. Many polymers undergo oxidation and degradation under UV light. For example, outdoor consumer products made from polymers require the incorporation of UV absorbers when exposed to the sun to prevent discoloration, cracking, and discoloration. The wavelength of light varies from 10 to 400 nm, the incident photons have enough energy to break the intermolecular bonds of most polymers. This allows for structural and chemical changes in the macromolecules. Exposure of the polymer to UV radiation has the result of making the surface more hydrophilic due to chain cleavage, crosslinking, and increased density of oxygen-based polar groups. Surface hydrophilicity can be improved by alkaline treatment of polymers at elevated temperatures. The surface of the polymer, such as polyethylene terephthalate, contains groups such as hydroxyl and carboxyl groups that contribute to the hydrophilicity of the interaction during etching with concentrated bases. Oxide conducting surfaces can be electrochemically treated using an anode potential to control their wetting characteristics [63,64].

8. Applications

Self-cleaning surfaces have gained tremendous interest, due to their numerous applications and outstanding properties, such as anticorrosion, antiicing, and antibacteria. The growth of this self-cleaning surface materials in many sectors such as automobile industries, aero-industries, chemical industries, textiles industries, and electronics industries have already shown. Some of the applications are discussed in the following sections.

8.1 Automobile industries

The self-cleaning coating was developed at the Nissan Technical Center in the UK in conjunction with nanotechnology firm, nanolabs, in the hope that Nissan owners would never have to wash their cars again. Nissan is the first carmaker to apply this technology, called "Ultra-Ever Dry" [65], on automotive bodywork. It has responded well to common use cases including rain, spray, frost, sleet and standing water. Fig. 19.5A showing the effect of mud on the car body after and before coated with the self-cleaning materials. Self-cleaned materials also used on the front glass, and mirror as shown in Fig. 19.5(B). Self-cleaning water resistance plastics are also used in automotive industries (Fig. 19.5C).

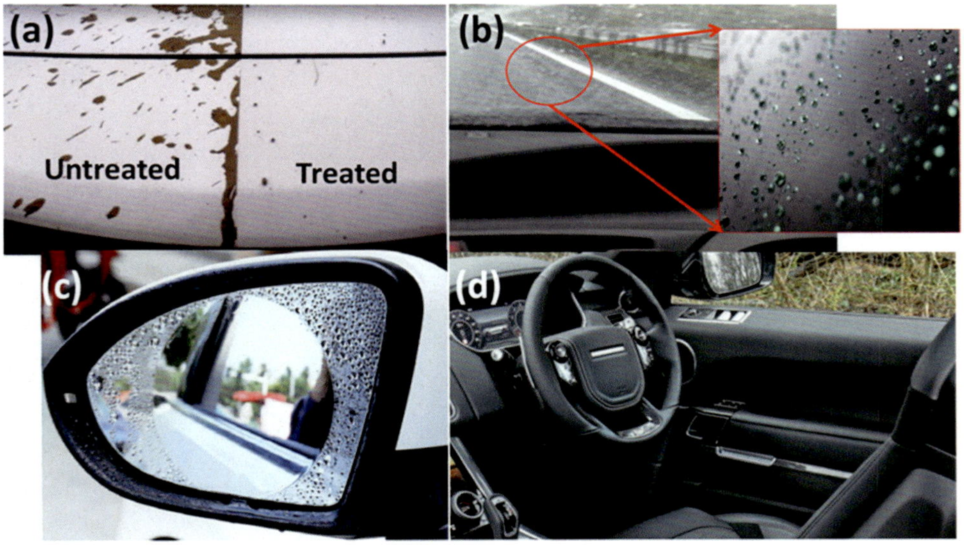

Figure 19.5
Self-cleaned surface at (A) car body, (B) front glass, (C) mirror, and (D) cabin polymer.

8.2 Aeroindustries

Between 2000 and 2005 there is a focus on fixed-wing aircraft by hydrophobic and self-cleaning structural aircraft parts. Coating for deicing an aircraft, which has other potential applications in the aerospace industry, such as coating of helicopter rotor blades and an aircraft maintenance system. Self-cleaning superhydrophobic coating for aircraft structures (especially Ti or Ti-based alloy structures) that are sensitive to the accumulation of ice, water, and other contaminants. Hard, wear-resistant, and phobic coating for an aerodynamic surface to improve deicing properties through low-pressure plasma vapor deposition technologies, such as plasma-assisted chemical vapor deposition, vapor phase chemical deposition, physical vapor deposition (sputtering') and reactive spraying. The liner includes a functional top layer that is harder than the surface of the airfoil and has a high angle of contact with water. Hydrophobic coatings with anticorrosive and self-cleaning properties for spacecraft survival systems and self-cleaning of exterior walls [66]. Self-cleaning coating with antimicrobial properties for use in aircraft cabins. Fig. 19.6 showing the self-cleaning exterior surface used on airplanes and satellites.

8.3 Electronic industries

Hydrophobic electronics treatments can be used to increase reliability, improve performance and add value to consumer electronics by repelling oil and water and allowing easy cleaning of fingerprint-resistant surfaces. Hydrophobic applications include antiwetting applications, inkjet printer nozzles, microfluidic channels, inkjet repellency, hard drives, stainless steel components, needles, and syringes, and many others. Hydrophobic electronic repellency treatments for electronics are compatible with the following types of materials: metals, polymers, glass/ceramics, particles, semiconductors,

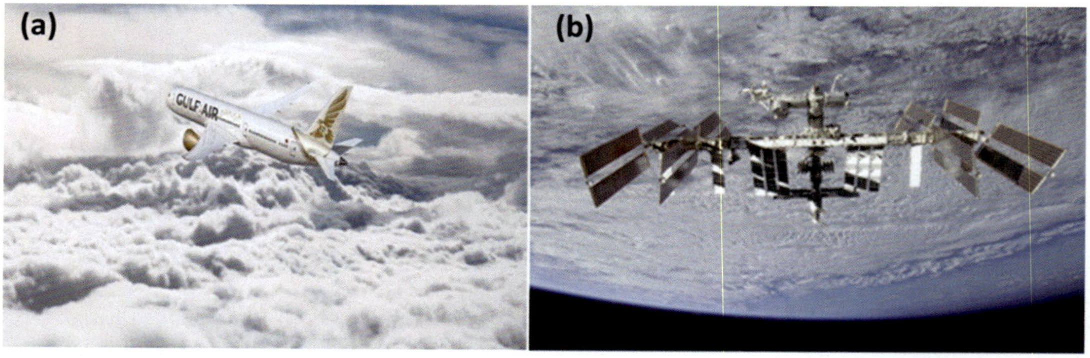

Figure 19.6
Self-cleaning coating is used in (A) airplane outer surface to avoid the galvanic activities, (B) Satellite to avoid the clogging effect.

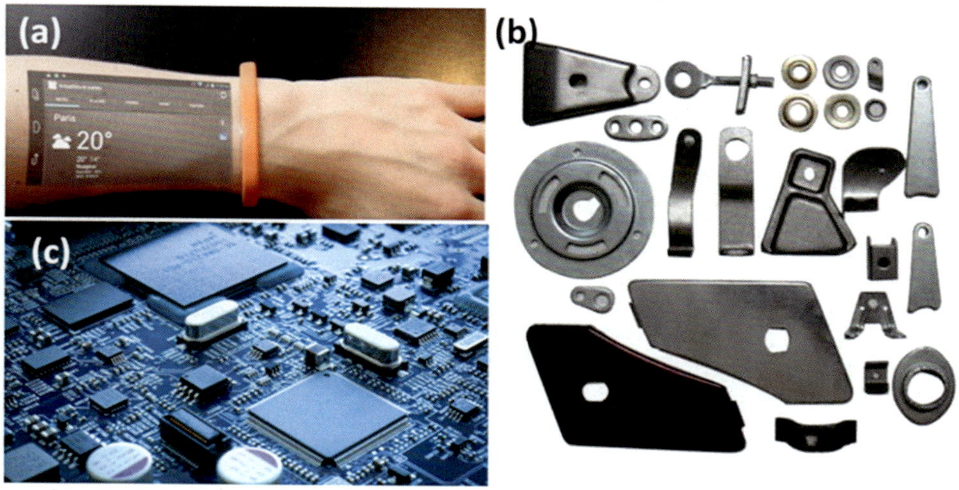

Figure 19.7
(A) Inflexible and the wearable electronic device, and (B) stainless steel components.

fabrics/others, and many others. Fig. 19.7 showing some self-cleaning surface coating observed in a wearable electronic device, various stainless steel electronics components. High sensitive superhydrophobic self-cleaning infrared nanosensor is also synthesized by carbon nanoparticles or TiO_2 nanotube layers or ZnO nanorod. Now, most of the glass materials in electronics have a high demand for transparent or colored superhydrophobic material coating [67,68].

8.4 Medical industries

In medical establishments, it is necessary to equip materials and surfaces with a high level of hygiene using antimicrobial agents to protect them against bacteria and other microorganisms to prevent infections caused by bacteria and contribute significantly to reducing health costs. Thanks to innovative hydrophilic coatings, invasive operations that were previously impossible can now be carried out safely through the cardiovascular, circulatory, neurological, urological and vascular systems, allowing for faster recoveries and fewer complications. This growing and exciting field stimulate growth and demand for new tools and innovations. Our hydrophilic coatings are formulated to attract and retain water, giving them lubricating qualities that reduce potential trauma and improve performance during these invasive procedures. Hydromer medical device coatings provide lubricity, some of which are low-particle coatings, and help improve the performance of these medical devices [69]. Additional self-cleaning products included reusable tissue box covers, mousepads, reception mats, and place mats that provide a cleaner rest area for personal effects and medical equipment. And the newest product the

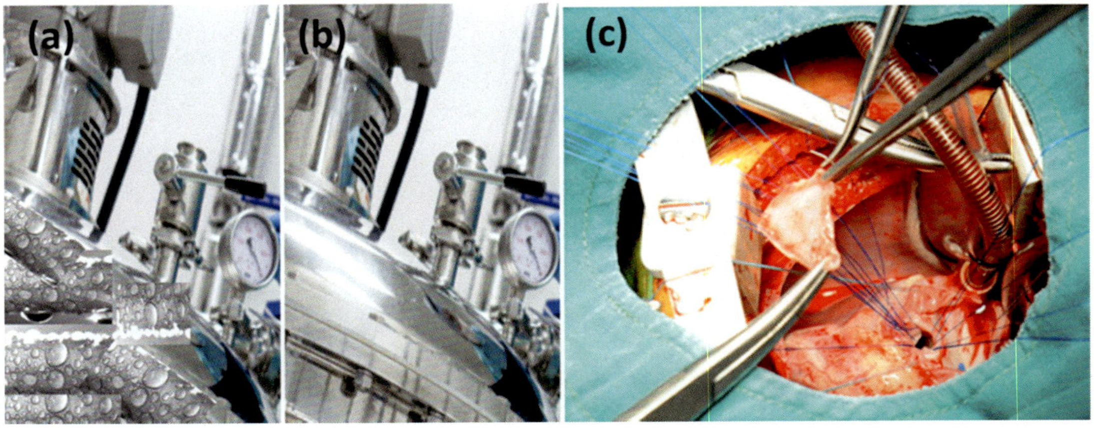

Figure 19.8
Liquid nitrogen equipped cylinder (A) without self-cleaning coating, (B) with self-cleaning coating, and (C) self-cleaning material coated surgical instruments for easily visible surgery.

team is about to launch is a transparent, self-cleaning film for kiosks and tablets [70]. Fig. 19.8A shows the nitrogen cylinder that is equipped in medical industries which clearly showing the effect of the self-cleaning material coating. Fig. 19.8B shows the surgical instruments that are coated with the self-cleaning materials to avoid any contamination in precise surgery.

8.5 Textile industries

Today, the textile industry has more than one technique for developing self-cleaning textiles. Another important application of TiO_2 is the self-cleaning of fabrics. Coffee and red wine stains on cotton fabrics can be easily removed with TiO_2 nanosols under UV radiation. Furthermore, the TiO_2 nanosol coating has demonstrated excellent UV protection against the cotton fabric. Fabrics coated with cellulose acetate and TiO_2 could retain their self-cleaning activity after several stains and washes. The TiO_2-SiO_2 coated fabrics have been tested for their self-cleaning ability against fading of red wine stains. TiO_2-SiO_2 with a molar ratio of 30:70 has shown optimal efficacy in stain removal. Fabrics coated with porous $Au/TiO_2/SiO_2$ nanocomposites manufactured to enhance self-cleaning performance assisted by visible light. Red wine and brown stains on the nanocomposite coated tissues were completely removed after 20 h of visible light irradiation [71]. Fig. 19.9 shows some self-cleaning coated (TiO_2, TiO_2-SiO_2) fabric for wearing purpose and for the umbrella.

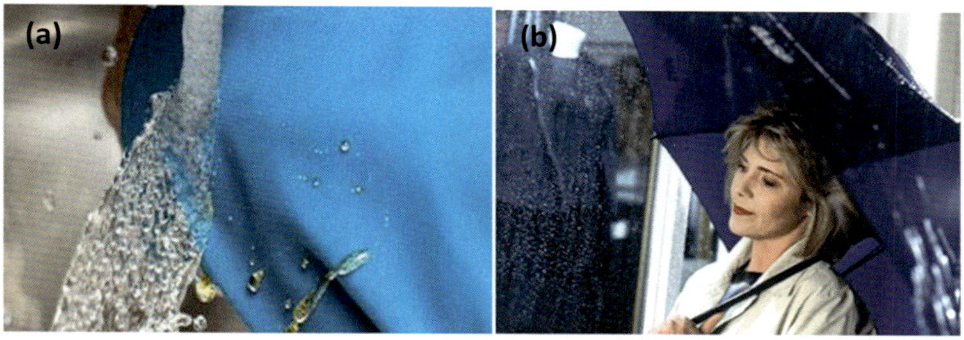

Figure 19.9
Self-cleaning coated materials in (A) wear materials, and (B) umbrella.

8.6 Other industries

Self-cleaning materials are growing to increase their importance from larger industries to smaller retailer industries. In energy harvesting equipment like solar cell losses more than 30% efficiency due to accumulation of dust particles and other organic pollutants. This problem can be rectified by the application of self-cleaning materials on the upper surface of the solar panel [72]. A typical solar cell (silicon micromorph) can be designed on a glass substrate via plasma-enhanced chemical vapor deposition method to develop the self-cleaning and antireflective properties [73]. Superhydrophobic coatings are used in ultradry surface applications. The coating causes an almost imperceptibly thin layer of air to form on a surface. The coating can be sprayed on objects to make them waterproof. The spray is anticorrosive and antifreeze; it has cleaning capabilities, and can be used to protect circuits and networks [74]. Superhydrophobic coatings have important applications in the marine industry. They can reduce drag from skin friction to ship hulls, thereby increasing energy efficiency. Such a coating would allow ships to increase speed or range while reducing fuel costs [75]. They can also reduce corrosion and prevent marine organisms from growing in the hull of a ship. In addition to these industrial applications, superhydrophobic coatings have potential uses on vehicle windshields to prevent raindrops from sticking to the glass. The coatings also remove salt deposits without using freshwater. Additionally, superhydrophobic coatings can easily collect other minerals from the seawater brine. Hydrophobic materials are also used for oil and water separation media [76].

9. Summary

Due to the high demand of the market for self-cleaning materials, the scientific community is focusing on the basic concept behind these materials which is represented here. Surface characteristics of self-cleaning materials such as wettability, drag reduction are well described here. The concept of surface cleaning of surface can easily understand by the wettability modes from Young's model to Cassie-Baxter's model of wetting. The three type of self-cleaning surfaces that are superhydrophobic, superhydrophilic and photocatalysis surfaces well described here. Different parameters required for self-cleaning surface are presented with various fabrication procedures of the surface. Current application in automobile industries, aero-industries, electronic industries, medical industries, textile industries, and others are given in this chapter.

References

[1] P. Ragesh, V.A. Ganesh, S.V. Nair, A.S. Nair, A review on self-cleaning and multifunctional materials, J. Mater. Chem. 2 (2014) 14773–14797, https://doi.org/10.1039/C4TA02542C.

[2] Q. Xu, W. Zhang, C. Dong, T.S. Sreeprasad, Z. Xia, Biomimetic self-cleaning surfaces: synthesis, mechanism and applications, J. R. Soc. Interface 13 (122) (2016) 20160300, https://doi.org/10.1098/rsif.2016.0300.

[3] M. Eseev, A. Goshev, S. Kapustin, Y. Tsykareva, Creation of superhydrophobic coatings based on MWCNTs xerogel, Nanomaterials 9 (11) (2019) 1584, https://doi.org/10.3390/nano9111584.

[4] M.P. Madeira, A.O. Lobo, B.C. Viana, E.C. Silva Filho, J.A. Osajima, Systems developed for application as self-cleaning surfaces and/or antimicrobial properties: a short review on materials and production methods, Cerâmica 65 (375) (2019), https://doi.org/10.1590/0366-69132019653752693.

[5] Y. Paz, Z. Luo, L. Rabenberg, A. Heller, Photooxidative Self-Cleaning Transparent Titanium Dioxide Films on Glass, J. Mater. Res. vol. 10 (Issue 11) (1995) 2842–2848, https://doi.org/10.1557/JMR.1995.2842, 357.

[6] A. Mills, A. Lepre, N. Elliott, S. Bhopal, I.P. Parkin, S.A. O' Neill, Characterisation of the photocatalyst Pilkington Activ™: a reference film photocatalyst? J. Photochem. Photobiol. Chem. 160 (Issue 3) (2003) 213–224.

[7] H. Dylla, M.M. Hassan, L.N. Mohammad, T. Rupnow, E. Wright, Evaluation of environmental effectiveness of titanium dioxide photocatalyst coating for concrete pavement, J. Transp. Res. Rec. 2164 (Issue 1) (2010) 46–51, https://doi.org/10.3141/2164-06.

[8] See: https://www.infoholicresearch.com/press-release/self-cleaning-coating-market-to-touch-5-34-billion-by-2023/, Dt: 18/04/2020.

[9] See: https://www.transparencymarketresearch.com/self-cleaning-glass-market.html, Dt: 18/04/2020.

[10] See: https://www.verifiedmarketresearch.com/product/self-cleaning-glass-market/, Dt: 18/04/2020.

[11] See: https://www.techsciresearch.com/report/self-cleaning-glass-market/4096.html, Dt: 18/04/2020.

[12] See: https://www.reportsanddata.com/report-detail/self-cleaning-glass-market, Dt: 18/04/2020.

[13] See: https://medium.com/@rithwik.mavuluri/self-cleaning-glass-market-size-share-and-forecast-2026-c35c1b1a10b8, Dt: 18/04/2020.

[14] See: https://menafn.com/1099853441/Global-Self-Cleaning-Photocatalytic-Coatings-Market-growth-analysis-research-trends-demand-and-market-size-2020Advanced-Materials-JTJ-SRO-Cristal-Eoxolit-Fumin-etc, Dt: 18/04/2020.

[15] See: https://marketresearch.biz/report/self-cleaning-glass-market/, Dt: 18/04/2020.

[16] W. Song, D.D. Veiga, C.A. Custódio, J.F. Mano, Bioinspired degradable substrates with extreme wettability properties, Adv. Mater. Comm. 21 (Issue 18) (2009) 1830–1834, https://doi.org/10.1002/adma.200803680.

[17] Q. Xu, W. Zhang, C. Dong, T.S. Sreeprasad, Z. Xia, Biomimetic self-cleaning surfaces: synthesis, mechanism and applications, J. R. Soc. Interface 13 (Issue 122) (2016), https://doi.org/10.1098/rsif.2016.0300.

[18] D. Byun, J. Hong, Saputra, J.H. Ko, Y.J. Lee, H.C. Park, B.K. Byun, J.R. Lukes, Wetting characteristics of insect wing surfaces, JBE 6 (2009) 63–70.

[19] E. Bormashenko, Why does the Cassie–Baxter equation apply? Colloid. Surface. Physicochem. Eng. Aspect. 324 (Issues 1–3) (2008) 47–50, https://doi.org/10.1016/j.colsurfa.2008.03.025.

[20] E.J. Lobaton, T.R. Salamon, Computation of constant mean curvature surfaces: application to the gas-liquid interface of a pressurized fluid on a superhydrophobic surface, J. Colloid Interface Sci. 314 (Issue 1) (2007) 184–198, https://doi.org/10.1016/j.jcis.2007.05.059.

[21] H. Dong, M. Cheng, Y. Zhang, H. Wei, F. Shi, Extraordinary drag-reducing effect of a superhydrophobic coating on a macroscopic model ship at high speed, J. Mater. Chem. 1 (2013) 5886–5891, https://doi.org/10.1039/C3TA10225D.

[22] T. Bharathidasan, S.V. Kumar, M.S. Bobji, R.P.S. Chakradhar, B.J. Basu, Effect of wettability and surface roughness on ice-adhesion strength of hydrophilic, hydrophobic and superhydrophobic surfaces, Appl. Surf. Sci. 314 (2014) 241–250, https://doi.org/10.1016/j.apsusc.2014.06.101.

[23] H.R. Allcock, L.B. Steely, A. Singh, Hydrophobic and superhydrophobic surfaces from polyphosphazenes, Polym. Int. vol. 55 (Issue 6) (2006) 621–625, https://doi.org/10.1002/pi.2030.

[24] S. Dash, S.V. Garimella, Droplet evaporation on heated hydrophobic and superhydrophobic surfaces, Phys. Rev. 89 (Issue 4) (2014), https://doi.org/10.1103/PhysRevE.89.042402.

[25] M.J. Niedziółka, F. Lapierre, Y. Coffinier, S.J. Parry, F. Zoueshtiagh, T. Foat, V. Thomy, R. Boukherroub, EWOD driven cleaning of bioparticles on hydrophobic and superhydrophobic surfaces, Lab Chip 11 (2011) 490–496, https://doi.org/10.1039/C0LC00203H.

[26] M.H. Kim, H. Kim, K.S. Lee, D.R. Kim, Frosting characteristics on hydrophobic and superhydrophobic surfaces: a review, Energy Convers. Manag. 138 (2017) 1–11, https://doi.org/10.1016/j.enconman.2017.01.067.

[27] B.M. Mognetti, H. Kusumaatmaja, J.M. Yeomans, Drop dynamics on hydrophobic and superhydrophobic surfaces, Faraday Discuss. 146 (2010) 153–165, https://doi.org/10.1039/B926373J.

[28] Y. Ofir, B. Samanta, P. Arumugam, V. M. Rotello, Controlled fluorination of FePt nanoparticles: hydrophobic to superhydrophobic surfaces, Adv. Mater. 19 (Issue 22) (2007) 4075–4079, https://doi.org/10.1002/adma.200700169.

[29] C. Antonini, F. Villa, M. Marengo, Oblique impacts of water drops onto hydrophobic and superhydrophobic surfaces: outcomes, timing, and rebound maps, Exp. Fluid 55 (2014), https://doi.org/10.1007/s00348-014-1713-9. Article number: 1713.

[30] Y. Wang, Atomic force microscopy measurement of boundary slip on hydrophilic, hydrophobic, and superhydrophobic surfaces, J. Vac. Sci. Technol. 27 (2009) 754, https://doi.org/10.1116/1.3086637.

[31] J. Son, S. Kundu, L.K. Verma, M.S. Aaron, J. Danner, C.S. Bhatia, H. Yang, A practical superhydrophilicself cleaning and antireflective surface for outdoor photovoltaic applications, Sol. Energy Mater. Sol. Cell. 98 (2012) 46–51, https://doi.org/10.1016/j.solmat.2011.10.011.

[32] S.K. Sethi, G. Manik, Recent progress in super hydrophobic/hydrophilic self-cleaning surfaces for various industrial applications: a review, Polym. Plast. Technol. Eng. 57 (Issue 18) (2018) 1932–1952, https://doi.org/10.1080/03602559.2018.1447128.

[33] T. Adachi, S.S. Latthe, S.W. Gosavi, N. Roy, N. Suzuki, H. Ikari, K. Kato, K. Katsumata, K. Nakata, M. Furudate, T. Inoue, T. Kondo, M. Yuasa, A. Fujishima, C. Terashima, Photocatalytic, superhydrophilic, self-cleaning TiO_2 coating on cheap, light-weight, flexible polycarbonate substrates, Appl. Surf. Sci. 458 (2018) 917–923, https://doi.org/10.1016/j.apsusc.2018.07.172.

[34] T. Kamegawa, K. Irikawa, H. Yamashita, Multifunctional surface designed by nanocomposite coating of polytetrafluoroethylene and TiO_2 photocatalyst: self-cleaning and superhydrophobicity, Sci. Rep. 7 (2017), https://doi.org/10.1038/s41598-017-14058-9. Article number: 13628.

[35] S. Afzal, W.A. Daoud, S.J. Langford, Superhydrophobic and photocatalytic self-cleaning cotton, J. Mater. Chem. 2 (2014) 18005–18011, https://doi.org/10.1039/C4TA02764G.

[36] W.A. Daoud, Self-Cleaning Materials and Surfaces: A Nanotechnology Approach, 2013, ISBN 978-1-119-99177-9, p. 366.

[37] D. Cai, A. Neyer, R. Kuckuk, H.M. Heise, Raman, mid-infrared, near-infrared and ultraviolet-visible spectroscopy of PDMS silicone rubber for characterization of polymer optical waveguide materials, J. Mol. Struct. 976 (Issues 1–3) (2010) 274–281, https://doi.org/10.1016/j.molstruc.2010.03.054.

[38] Y. Zhou, M. Li, X. Zhong, Z. Zhu, P. Deng, H. Liu, Hydrophobic composite coatings with photocatalytic self-cleaning properties by micro/nanoparticles mixed with fluorocarbon resin, Ceram. Int. 41 (Issue 4) (2015) 5341–5347, https://doi.org/10.1016/j.ceramint.2014.12.090.

[39] Y. Li, X. Liu, X. Nie, W. Yang, Y. Wang, R. Yu, J. Shui, Multifunctional organic–inorganic hybrid aerogel for self-cleaning, heat-insulating, and highly efficient microwave absorbing, Materials 29 (Issue 10) (2019) 1807624, https://doi.org/10.1002/adfm.201807624.

[40] I.P. Parkin, R.G. Palgrave, Self-cleaning coatings, J. Mater. Chem. 15 (2005) 1689–1695, https://doi.org/10.1039/B412803F.

[41] R. Menini, M. Farzaneh, Production of superhydrophobic polymer fibers with embedded particles using the electrospinning technique, Polym. Int. 57 (Issue 1) (2008) 77–84, https://doi.org/10.1002/pi.2315.

[42] B. Xu, Z. Cai, Fabrication of a superhydrophobic ZnO nanorod array film on cotton fabrics via a wet chemical route and hydrophobic modification, Appl. Surf. Sci. 254 (Issue 18) (2008) 5899–5904, https://doi.org/10.1016/j.apsusc.2008.03.160.

[43] H.Y. Zhu, Y. Lan, X.P. Gao, S.P. Ringer, Z.F. Zheng, D.Y. Song, J.C. Zhao, Phase transition between nanostructures of titanate and titanium dioxides via simple wet-chemical reactions, J. Am. Chem. Soc. 127 (18) (2005) 6730–6736, https://doi.org/10.1021/ja044689+.

[44] X. Zhang, F. Shi, X. Yu, H. Liu, Y. Fu, Z. Wang, L. Jiang, X. Li, Polyelectrolyte multilayer as matrix for electrochemical deposition of gold clusters: toward super-hydrophobic surface, J. Am. Chem. Soc. 126 (10) (2004) 3064–3065, https://doi.org/10.1021/ja0398722.

[45] I.D. Alvim, F.S. Souza, I.P. Koury, T. Jurt, F.B.H. Dantas, Use of the spray chilling method to deliver hydrophobic components: physical characterization of microparticles, Cienc. Tecnol. Aliment. 33 (2013), https://doi.org/10.1590/S0101-20612013000500006 supl. 1 Campinas.

[46] B. Liu, Y. He, Y. Fan, X. Wang, Fabricating super-hydrophobic lotus-leaf-like surfaces through soft-lithographic imprinting, Macromol. Rapid Commun. 27 (Issue 21) (2006) 1859–1864, https://doi.org/10.1002/marc.200600492.

[47] M. Luo, W. Tang, J. Zhao, C. Pu, Hydrophilic modification of poly(ether sulfone) used TiO_2 nanoparticles by a sol-gel process, J. Mater. Process. Technol. 172 (Issue 3) (2006) 431–436, https://doi.org/10.1016/j.jmatprotec.2005.11.004.

[48] J.M. Calvert, Lithographic patterning of self-assembled films, J. Vac. Sci. Technol. B 11 (Issue 6) (1998), https://doi.org/10.1116/1.586449.

[49] T. Textor, B. Mahltig, A sol-gel based surface treatment for preparation of water repellent antistatic textiles, Appl. Surf. Sci. 256 (Issue 6) (2010) 1668–1674, https://doi.org/10.1016/j.apsusc.2009.09.091.

[50] A. Kietzig, M.N. Mirvakili, S. Kamal, P. Englezos, S.G. Hatzikiriakos, Laser-patterned super-hydrophobic pure metallic substrates: Cassie to Wenzel wetting transitions, J. Adhes. Sci. Technol. 25 (Issue 20) (2011) 2789–2809.

[51] I.S. Bayer, A.J. Davis, A. Biswas, Robust superhydrophobic surfaces from small diffusion flame treatment of hydrophobic polymers, RSC Adv. 4 (2014) 264–268, https://doi.org/10.1039/C3RA44169E.

[52] S.M. Mukhopadhyay, P. Joshi, S. Datta, J.G. Zhao, P. France, Plasma assisted hydrophobic coatings on porous materials: influence of plasma parameters, J. Phys. D Appl. Phys. 35 (1927), https://doi.org/10.1088/0022-3727/35/16/305.

[53] P.K. Deshmukh, K.P. Ramani, S.S. Singh, A.R. Tekade, V.K. Chatap, G.B. Patil, S.B. Bari, Stimuli-sensitive layer-by-layer (LbL) self-assembly systems: targeting and biosensory applications, J. Contr. Release 166 (Issue 3) (2013) 294–306, https://doi.org/10.1016/j.jconrel.2012.12.033.

[54] Y. Tsuge, J. Kim, Y. Sone, O. Kuwaki, S. Shiratori, Fabrication of transparent TiO_2 film with high adhesion by using self-assembly methods: application to super-hydrophilic film, Thin Solid Films 516 (Issue 9) (2008) 2463–2468, https://doi.org/10.1016/j.tsf.2007.04.084.

[55] Y. Liu, X. Yin, J. Zhang, Y. Wang, Z. Han, L. Ren, Biomimetic hydrophobic surface fabricated by chemical etching method from hierarchically structured magnesium alloy substrate, Appl. Surf. Sci. 280 (2013) 845–849, https://doi.org/10.1016/j.apsusc.2013.05.072.

[56] C. Dong, Y. Gu, M. Zhong, L. Li, K. Sezer, M. Ma, W. Liu, Fabrication of superhydrophobic Cu surfaces with tunable regular micro and random nano-scale structures by hybrid laser texture and chemical etching, J. Mater. Process. Technol. 211 (Issue 7) (2011) 1234–1240, https://doi.org/10.1016/j.jmatprotec.2011.02.007.

[57] T. Chen, B. Zeng, J.L. Liu, J.H. Dong, X.Q. Liu, Z. Wu, X.Z. Yang, Z.M. Li, High throughput exfoliation of graphene oxide from expanded graphite with assistance of strong oxidant in modified Hummers method, J. Phys.: Conf. Ser. 188 (2009) 012051, https://doi.org/10.1088/1742-6596/188/1/012051.

[58] F. Ilhan, E.F. Fabrizio, L.M. Corkle, D.A. Scheiman, A. Dass, A. Palczer, M.A.B. Meador, J.C. Johnston, N. Leventis, Hydrophobic monolithic aerogels by nanocasting polystyrene on amine-modified silica, J. Mater. Chem. 16 (2006) 3046–3054, https://doi.org/10.1039/B604323B.

[59] Y. Yang, X. Li, X. Zheng, Z. Chen, Q. Zhou, Y. Chen, 3D-printed biomimetic super-hydrophobic structure for microdroplet manipulation and oil/water separation, Adv. Mater. vol. 30 (Issue 9) (2018) 1704912, https://doi.org/10.1002/adma.201704912.

[60] L. Cao, H. Hu, D. Gao, Design and fabrication of micro-textures for inducing a superhydrophobic behavior on hydrophilic materials, Langmuir 23 (8) (2007) 4310–4314, https://doi.org/10.1021/la063572r.

[61] See: http://www1.lsbu.ac.uk/water/hydrophobic_hydration.html, 18/04/2020.

[62] D. Ahmad, I.V.D. Boogaert, J. Miller, R. Presswell, H. Jouhara, Hydrophilic and hydrophobic materials and their applications, Energy Sources, Part A Recovery, Util. Environ. Eff. 40 (Issue 22) (2018) 2686–2725, https://doi.org/10.1080/15567036.2018.1511642.

[63] H.M. Shang, Y. Wang, S.J. Limmer, T.P. Chou, K. Takahashi, G.Z. Cao, Optically transparent superhydrophobic silica-based films, Thin Solid Films 472 (Issues 1–2) (2005) 24 37–43, https://doi.org/10.1016/j.tsf.2004.06.087.

[64] T. Ishizaki, N. Saito, O. Takai, Correlation of cell adhesive behaviors on superhydrophobic, superhydrophilic, and micropatterned superhydrophobic/superhydrophilic surfaces to their surface chemistry, Langmuir 26 (11) (2010) 8147–8154, https://doi.org/10.1021/la904447c.

[65] See: https://interestingengineering.com/this-self-cleaning-car-paint-could-put-car-washes-out-of-business, 19/04/2020.

[66] See: https://www.lexology.com/library/detail.aspx?g=e508721d-0a6c-43c5-b436-218c43ef855f, 19/04/2020.

[67] See: https://www.nanowerk.com/spotlight/spotid=19644.php, 19/04/2020.

[68] F. Geyer, M.D. Acunzi, A.S. Aghili, A. Saal, N. Gao, A. Kaltbeitzel, T.F. Sloo, R. Berger, H.J. Butt, D. Vollmer, When and how self-cleaning of superhydrophobic surfaces works, Sci. Adv. 6 (3) (2020) eaaw9727, https://doi.org/10.1126/sciadv.aaw9727.

[69] See: https://www.futuremarketsinc.com/nanocoatings-in-the-medical-industry/, 19/04/2020.

[70] See: https://www.prnewswire.com/news-releases/self-cleaning-surfaces-in-healthcare-improve-patient-experience-and-create-cleaner-facilities-300641876.html, 19/04/2020.

[71] See: https://www.technicaltextile.net/articles/self-cleaning-textile-an-overview-2646, 19/04/2020.

[72] See: https://www.asme.org/topics-resources/content/self-cleaning-solar-panels-maximize-efficiency, 19/04/2020.

[73] G. He, C. Zhou, Z. Li, Review of self-cleaning method for solar cell array, Procedia Eng. 16 (2011) 640–645, https://doi.org/10.1016/j.proeng.2011.08.1135.

[74] A.A. Raad, K.K. Ganesh, S.B. Alexandru, Fabrication of Transparent Superhydrophobic Polytetrafluoroethylene Coating, Appl. Surf. Sci. 444 (2018) 208–215, https://doi.org/10.1016/j.apsusc.2018.02.206.
[75] M.S. Selim, S.A. El-Safty, M.A. Shenashen, S.A. Higazy, A. Elmarakbi, Progress in biomimetic leverages for marine antifouling using nanocomposite coatings, J. Mater. Chem. B 8 (2020) 3701–3732, https://doi.org/10.1039/C9TB02119A.
[76] Z. Xue, Y. Cao, N. Liu, L. Feng, L. Jiang, Special wettable materials for oil/water separation, J. Mater. Chem. 2 (2014) 2445–2460, https://doi.org/10.1039/C3TA13397D.

SECTION F

Development of carbon-based nanoparticles

CHAPTER 20

Low-dimensional carbon-based nanomaterials: synthesis and application in polymer nanocomposites

Nidhin Divakaran[1], Manoj B. Kale[2], Lixin Wu[3]

[1]School for Advanced Research in Polymers (SARP) — LARPM, Central Institute of Plastics Engineering & Technology (CIPET) — IPT, Patia, Bhubaneswar, Odisha, India; [2]CAS Key Laboratory of Design and Assembly of Functional Nanostructures, and Fujian Provincial Key Laboratory of Nanomaterials, State Key Laboratory of Structural Chemistry, Key Laboratory of Optoelectronic Materials Chemistry and Physics, Fujian Institute of Research on the Structure of Matter, Chinese Academy of Sciences, Fuzhou, Fujian, China; [3]CAS Key Laboratory of Design and Assembly of Functional Nanostructures, Fujian Key Laboratory of Nanomaterials, Fujian Institute of Research on the Structure of Matter, Chinese Academy of Sciences, Fuzhou, Fujian, China

Chapter Outline

1. Introduction

The role of carbon in the proliferation of modern science and technology is immense. The discovery of fullerene commenced the revolution of carbon-based materials and it was

Design, Fabrication, and Characterization of Multifunctional Nanomaterials. https://doi.org/10.1016/B978-0-12-820558-7.00011-X

followed by the highly efficient carbon nanotubes and graphene. The need for the researchers to enhance the comprehensive properties thrusted the existence of carbon-based materials, having luminescent property. The discovery of carbon nanodots in the year 2006 satisfied this requirement. Carbon nanodots are also termed as carbon quantum dots (CQDs) or carbon dots (CDs). It can be coined as nanoparticles with the dimension being less than 10 nm. This supplemented the versatility of carboniferous materials and provided impetus in its commercial applications. Carbon, being black in color, cannot emit light but the discovery of carbon nanodots changed its notion. The invention of the carbon nanodots could be drawn back to the research on carbon nanotubes (SWNT) fluorescent materials in the year 2004. Sun et al. [1] termed these carbon nanotubes as carbon nanodots (CND). He devised the enhanced fluorescent properties within the CND by surface passivation. The discovery of carbon nanodots was a revolution in technological innovation with the advantages such as low toxicity, chemical stability, water solubility, high luminescence, and less photobleaching. CNDs are similar to semiconductor quantum dots and luminous organic dyes. The application of CNDs lies in the commercial use in various fields such as photo voltaic devices, optoelectronics, photocatalysis, biosensing devices [2]. The CND structure can be described in the form of carbogenic core consisting of crystalline and amorphous structure with the functional groups attached to the surface [3–5]. The structure is also analogous with the graphene configuration due to the presence of sp^2 carbon segments. The core carbon nanodots also possess the amorphous carbon structure graphene/graphitic structure or the diamond like structure. Graphene quantum dots could be regarded as one of the CNDs as they possess the similar arrangement of oxygen terminated functional groups on their surface [6]. The molecular orbital theory can define the structure of the CNDs. The CNDs exhibit $n\text{->}\pi$ and $\pi\text{-} > n$ transition, owing to the difference in the energy levels. The π states signifies the sp^2 hybridization in the core, while the n states specify the oxygen functional group attached to the sp^2 hybridized carbon. The functional group could be carbonyl, amine, thiol, and amides. CNDs are hydrophobic or hydrophilic depending on the structure of the surface [7–9]. The fluorescence properties of the CNDs can be described on the basis of the synthesis procedure to develop it. The synthesis technique defines the capability of CNDs to emit fluorescence at different wavelength. The fluorescence properties are commonly tunable with it having the ability of emitting blue, green or red light, independent of the excitation wavelength [10].

2. Synthesis of carbon nanodots

Xu et al. accidently discovered the fluorescent carbon nanodots while undergoing the purifying procedure for the single walled carbon nanotubes using arc-discharged soot. A plethora of research has been conducted to synthesize carbon nanodots which includes the basic preparation of nanomaterials like top-down approach and the bottom approach [11]. Fig. 20.1 displays the schematic setup of the preparation technique to synthesize carbon

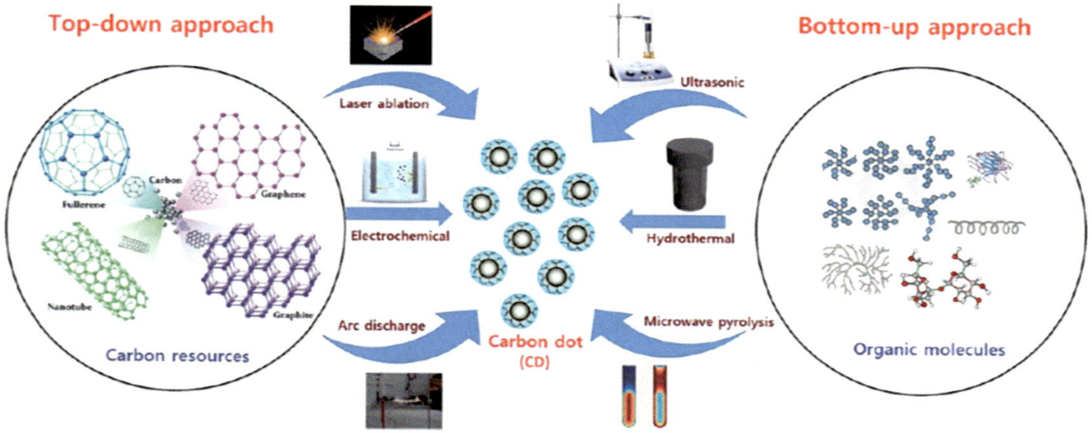

Figure 20.1

Synthesis procedure adopted to fabricate carbon nanodots. *Reproduced with permission from Ref. B. De, N. Karak, Recent progress in carbon dot-metal based nanohybrids for photochemical and electrochemical applications, J. Mater. Chem. A. 5 (2017) 1826–1859. https://doi.org/10.1039/C6TA10220D. Copyright 2017, Royal Science of Chemistry.*

nanodots. The top-down approach involves the synthesis of the nanomaterials by fragmenting bulk carbon nanomaterials having dimensions higher than CNDs for example graphite or carbon nanotubes. The procedure adopted to synthesize the CNDs involves electrochemical technique, microwave synthesis, laser ablation, and hydrothermal methods, and CNDs are obtained as a product from these methods. The CNDs substantially obtained are not fluorescent [13]. They can be modified into highly fluorescent CNDs by optimizing the reactivity on their surface with various polar moieties after the synthesis procedure [14,15]. The bottom-up synthesis technique includes the carbonization of various molecular precursors such as citric acid or sucrose. The formation of carbon nanodots is carried out by carbonization these compounds at a very low melting point at comparatively less temperature [16]. The other alternative method involves mixing the carbon sources with various precursors such as urea and thiourea [17].

3. Carbon nanodots based polymer composites

The comprehensive properties of carbon nanodots provides scope for a thorough study of its interaction with polymer. The hybrid structure of CND/polymer composites tends to extend the versatility of polymers in various commercial applications [18]. There exists the interrelation between the chemistry involved in the interaction of CNDs with the polymer and inclusive property optimization of the polymer and hence lots of research has been conducted to explore the interactive effect of CNDs in various polymers [19].

3.1 Polyaniline composite of carbon nanodots

Polyaniline (PANI) is a type of the conducting polymers which displays the conductivity on redox with excellent environmental stability and ease of synthesis. It comprises of a low-cost monomer having wide range of application in the development of electronic devices such as sensor, corrosion protection, and radar absorbing materials [20,21]. The rigid backbone of PANI plays a major role in its intimidating process and it is due to its high degree of conjugation. The conjugated structure also fetches conductivity in PANI and its reasonable cost propels PANI as a versatile polymer. A few researches have been conducted on the CND-based PANI composites. Ge et al. developed the PANI/cobalt sulfide (CoS)/CND composites via in situ polymerization and attempted to study the magnetic properties of the composite. The presence of enhanced magnetocapacitances in the composite played a vital role in displaying electromagnetic wave absorbing capability [22]. A thorough analysis was conducted on the influence of magnetic field on the electrochemical and electromagnetic behavior of the composite. It was observed that applying magnetic field of 0.5 T has displayed larger magnetocapacitance and electromagnetic wave absorption capabilities in the composite.

The PANI/CoS/CND composites display a uniform morphology as displayed in Fig. 20.2 and possess the navellike holes. The obvious difference in morphology in the figure is observed due to the presence of magnetic fields and it has contributed to the development of PANI nanotubes instead of nanofibers. The CND plays a very important role in the formation of a thoroughly dispersed phase within PANI. Ghosh et al. devised a facile approach in the construction of controlled grafting of PANI chains using fluorescent spherical CND. CNDs are easily available by the oxidation process from the candle soot [23].

Fig. 20.3 displays the schematic representation of controlled growth of CND on PANI chains. The covalently attached carbon nanodots has improved the specific capacitance of PANI considerably and has contributed to the excellent electrochemical performance. The increased conductivity and the more compact nano structure formation due to the Π-Π stacking interactions has led to the significant development in electrochemical properties. This has accorded the development of supercapacitors using CND.

3.2 Epoxy composite of carbon nanodots

Epoxy constitute the class of reactive prepolymers and polymers which consist of epoxide groups. The cross-linking mechanism of epoxy resins is through catalytic homopolymerization and also in the presence of extensive range of coreactants including polyfunctional amines, phenols, alcohols, and thiols [24]. They are also known as hardeners or curing agent. Epoxy resins are extensively used in coating, adhesives, and composite materials [25–27]. The properties of epoxy can be modified with several

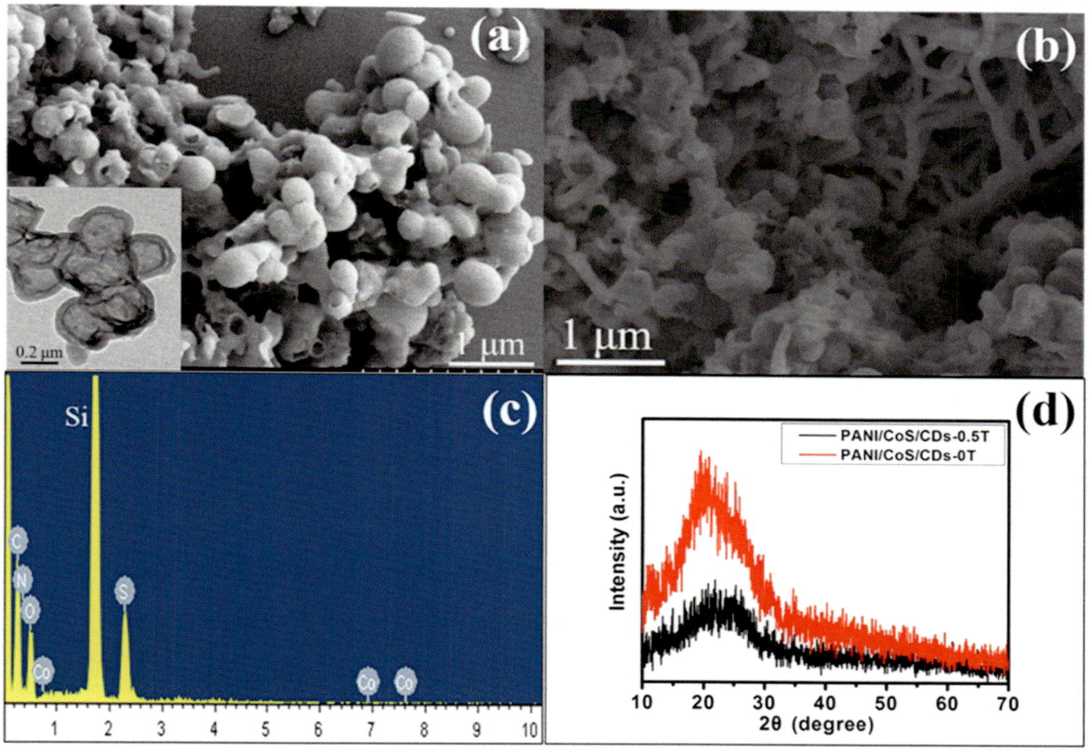

Figure 20.2

SEM images of (A) PANI/CoS/CNDs-0.5T (inset refers to the TEM image) and (B) PANI/CoS/CNDs-0T (C) EDX spectrum of PANI/CoS/CNDs-0.5T (D) XRD spectrum of PANI/CoS/CD-0.5T and PANI/CoS/CD-0T. *Reproduced with permission from Ref. C. Ge, X. Zhang, J. Liu, F. Jin, J. Liu, H. Bi, Hollow-spherical composites of polyaniline/cobalt sulfide/carbon nanodots with enhanced magneto-capacitance and electromagnetic wave absorption capabilities, Appl. Surf. Sci. 378 (2016) 49–56. https://doi.org/10.1016/j.apsusc.2016.03.210. Copyright 2016, Elsevier.*

nanofillers reinforcement and their inclusive properties can be enhanced. CNDs tends to play an important role in the property enhancement of epoxy. Zhang et al. [28] devised epoxy polymerized CND luminescent thin films and analyzed their properties. The water-soluble CNDs were synthesized and were constituted within epoxy matrix under different wt.% to fabricate the thin film. The tensile modulus of the epoxy films with CNDs displayed consistent augmentation in the values elucidating the inference that CNDs display superior mechanical strength and flexibility on interaction with epoxy. Fig. 20.4 exhibits the tensile strength enhancement of epoxy resin with the addition of CNDs.

The quantum yield tests carried out on the CNDs constituted epoxy films displayed better photo stability behavior of the films due to the presence of CND. The decaying of CNDs is at a lower rate and they act as a fluorophore for thin films. CNDs are very compatible

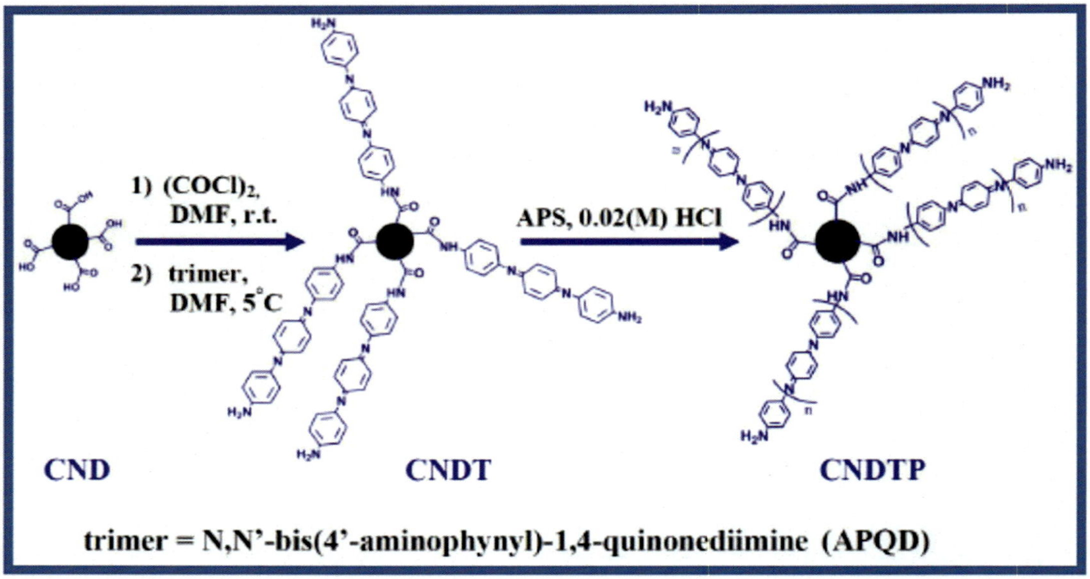

Figure 20.3

Controlled grafting of PANI chains on CND surface following a "grafting from" approach. *Reproduced with permission from Ref. T. Ghosh, R. Ghosh, U. Basak, S. Majumdar, R. Ball, D. Mandal, A.K. Nandi, D.P. Chatterjee, Candle soot derived carbon nanodot/polyaniline hybrid materials through controlled grafting of polyaniline chains for supercapacitors, J. Mater. Chem. A. 6 (2018) 6476–6492. https://doi.org/10.1039/c7ta11050b. Copyright 2018, Royal Science of Chemistry.*

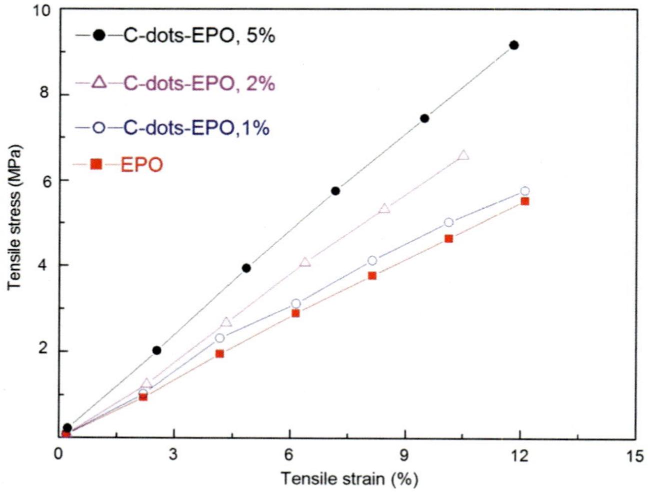

Figure 20.4

Tensile strength analysis of epoxy films with different % of CNDs. *Reproduced with permission from Ref. C. Zhang, L. Du, C. Liu, Y. Li, Z.Z. Yang, Y.C. Cao, Photostable epoxy polymerized carbon quantum dots luminescent thin films and the performance study, Results Phys. 6 (2016) 767–771. https://doi.org/10.1016/j.rinp.2016.10.013. Copyright 2016, Elsevier.*

with the epoxy resin and have improved their fluorescence properties. Chen et al. [29] studied the influence of amido-functionalized CNDs into the epoxy matrix and an attempt was made to develop them into transparent and luminescent composites. The functionalized CND provided immaculate dispersion within epoxy due to its covalent bonding interface. The homogenously dispersed CND decreases the light scattering in the polymer and boosted the transparency, eight-fold enhanced luminescence to flare up the potential of epoxy as an encapsulating material in white light-emitting diodes. The CNDs exhibited highest photoluminescence intensity due to the proper compatibility with epoxy matrix and reduced aggregation on account of similar polarity with epoxy. The white light was obtained from the epoxy/CNDs composite encapsulated InGaN chip, thereby satisfying the potential application in LED. The photoluminescence spectra of the white LED culminates the application of epoxy/CNDs composite as an encapsulating material in LEDs and an improved photoluminescence characteristic.

3.3 Polyvinyl butyral composite of carbon nanodots

Polyvinyl butyral (PVB), synthesized from polyvinyl alcohol reacting with butylaldehyde, is widely used for applications involving optical properties, strong bonding, toughness, and flexibility [30,31]. It is also the general 3D printer filament due to its strong heat resistant properties. PVB is regarded as the acetal due to its formation from the reaction between aldehyde and alcohol [32]. PVB has the widespread application in the form of laminated glass. The good dispersibility and excellent fluorescent luminescence property of CND tends to play a major role in the property enhancement of PVB [33,34]. Arthisree et al. [35] investigated the influence of CNDs dispersed into PVB matrix and studied the photoluminescence properties. The strong hydrogen bonding between CND and PVB was observed and confirmed using FTIR. The presence of CND into polymer matrix has resulted the decrement of the band gap, as inferred from UV-Vis spectroscopy analysis. The Raman analysis of the PVB/CND composite display diminution in the values of I_d/I_g due to stacking-like arrangement of CND in the PVB matrix.

The morphological analysis affirmed the excellent reinforcement of CND into PVB with the bright region representing the polymer network and the dark cluster region denoting the random orientation, aggregation, and entrapment of CND in the PVB matrix. The phase inversion, hydrophilic nature, and high aspect ratio of CND has resulted in an increase in surface roughness of PVB, as confirmed by AFM. The presence of CND also enhanced the electrical conductivity of PVB matrix, thereby augmenting its prospects for application as photoluminescent material.

3.4 Chitosan-based composite of carbon nanodots

Chitosan is a linear polysaccharide consisting of randomly arranged β-(1−4) connected D-glucosamine and N- acetyl-D-glucosamine. It contains of several functional groups such as hydroxyl and carboxyl [36]. It retains the property of hydrophilicity, biocompatibility, chemical inertness, and excellent film-forming capacity [37—39]. It possesses an extensive commercial application such as food processing, cosmetics, fabrics, and water treatment. It is also used in drug delivery to the patients [40]. The amalgamation of CND and chitosan can improve the morphology of the polymer and the dispersion of CND in polymer can be maintained. Sheng et al. [41] devised the chitosan-CND film and used it as an immobilization matrix to trap protein, hemoglobin for electrochemistry, and bioelectrocatalysis. The construction of electrodes using CND-chitosan composites was carried out and the direct electron transfer (DET) between electrode and protein was observed. The DET between electrode and protein is very important to study the protein-mediated biological redox reactions such as photosynthesis, respiration, and bioenergetics metabolism. The presence of CND has facilitated the DET and has enhanced the reputation of favorable electrode materials. SEM images validates the structure of chitosan-CND composite with the CNDs entrapped in the 3D chitosan network and pose as a conductor due to its superior intrinsic conductivity. CND also enhance the active surface area of electrode to entrap hemoglobin for electrochemical analysis. Fig. 20.5 displays the schematic illustration of Chitosan-CND systems. The entrapped hemoglobin retains its structural intactness, high bioactivity, and good affinity with H_2O_2. CNDs can act as a proficient electrical wiring the redox centers of proteins to the electrode and the combination of CND and chitosan will generate greater opportunities for developing novel bioelectrocatalytic systems.

Qi et al. fabricated a quartz crystal microbalance (QCM) humidity sensor using CND-chitosan film with high sensitivity, fast response-recovery, and tiny hysteresis [42].

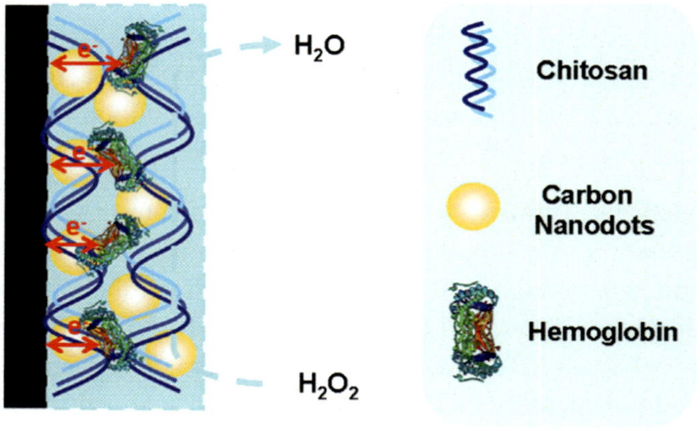

Figure 20.5
Schematic diagram of hemoglobin-CND-chitosan system. *Reproduced with permission from Ref. M. Sheng, Y. Gao, J. Sun, F. Gao, Carbon nanodots-chitosan composite film: a platform for protein immobilization, direct electrochemistry and bioelectrocatalysis, Biosens. Bioelectron. 58 (2014) 351—358. https://doi.org/10.1016/j.bios.2014.03.005, Copyright 2014, Elsevier.*

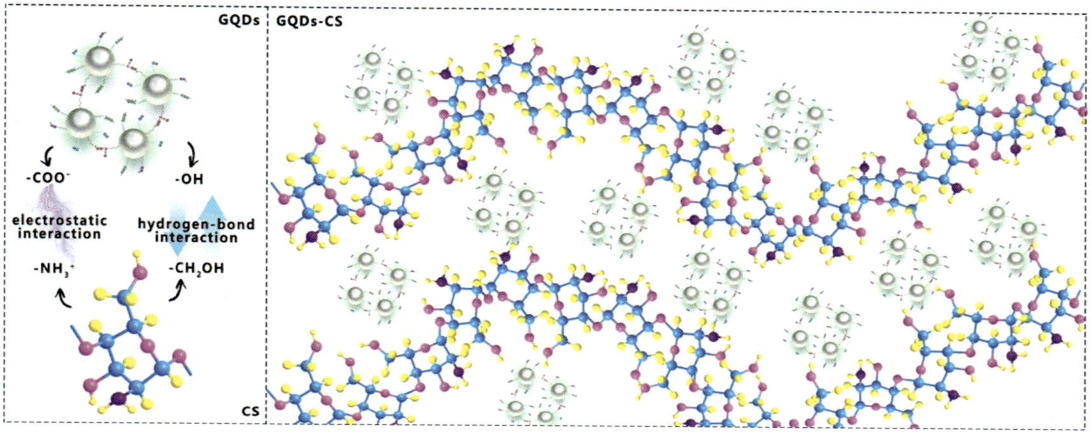

Figure 20.6
Schematic diagram of graphene quantum dots/chitosan composites. *Reproduced with permission from Ref. P. Qi, T. Zhang, J. Shao, B. Yang, T. Fei, R. Wang, A QCM humidity sensor constructed by graphene quantum dots and chitosan composites, Sensor. Actuator. A Phys. 287 (2019) 93–101. https://doi.org/10.1016/j.sna.2019.01.009. Copyright 2019, Elsevier.*

Fig. 20.6 shows the interaction mechanism of CND with chitosan. The morphological analysis using SEM and TEM specifies the formation of cellular and highly porous random 3D structure of chitosan-CND. The presence of CND can change the overall morphology of chitosan by dispersing thoroughly in the polymer. The optimization analysis of QCM humidity sensor using CND-chitosan films implies the logic of appropriate mass proportion of CND and chitosan could make the charge balance between two materials and can develop into homogenous, compact sensitive membrane. The humidity sensing characteristics of chitosan-CND films displays excellent reproducibility and reversibility. The hysteresis curve of the QCM sensors modified by chitosan-CND represent tiny hysteresis and it has enhanced the performance of the sensor. The intrinsic hydrophilicity of CND and chitosan has provided more water absorption sites and this sensor is capable of monitoring water vapor content in wide humidity range.

3.5 Polypyrole based composite of carbon nanodots

Polypyrole (PPy) is an organic polymer synthesized from the polymerization of polypyrole, which on oxidation becomes a conducting polymer [43]. It is the most-used polymer among the conducting polymers due to its ease in synthesis, superior electrical conductivity, and redox properties. PPy loses its conductivity at overoxidation [44,45]. The excellent intrinsic properties of PPy has prompted the application in supercapacitors, batteries, biosensors, antistatic coatings, EMI shielding, and drug delivery applications [46]. PPy tends to be excellent in stimulus responsive properties, and hence it could be developed into a very smart biomaterial [47]. The good conducting property of CND will play a major role in the development of PPy into the first-grade sensors and supercapacitors. Zhang et al. [48] implemented the concept of using graphene oxide/CND/PPy as a ternary composite in electrode active material for supercapacitors. The presence of CND in the electrode active material promotes the electron transportation in the ternary composite, diminishing the internal resistance, and charge transfer resistance of the electrode. Fig. 20.7 specifies the role of CND in the electron transfer within the composites and enhancement of the dielectric constant.

Morphological analysis using TEM substantiate the sandwiching of CND between GO film and PPy layer in the ternary composite of GO/CND/PPy and the UV-Vis spectra, photo luminescence emission spectra reveal the existence of electron transfer between GO/CND and CND/PPy in the composite system. The thorough reinforcement of CND within the GO/PPy matrix has eliminated the drawback of the conducting polymer by reducing the poor cycle stability and increasing the capacitance. The composite possesses large specific capacitance, higher energy density, and excellent capacitance retention after 5000 cycles. This infers the role of CND in enhancing the performance of hybrid electric double layer capacitor. Jian et al. [49] constructed a CND reinforced PPy nanowires via electrostatic

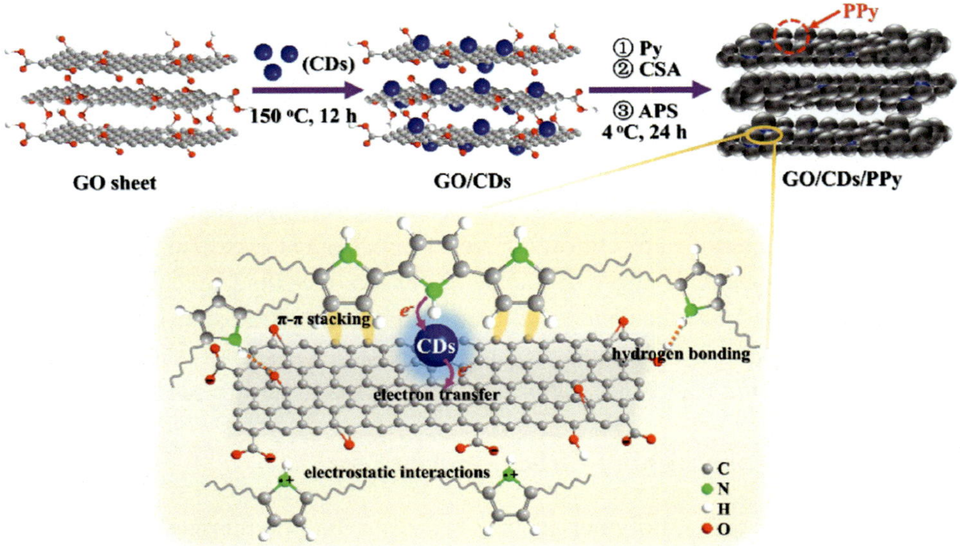

Figure 20.7
Schematic diagram of synthesis process of GO/CND/polypyrole composite. *Reproduced with permission from Ref. X. Zhang, J. Wang, J. Liu, J. Wu, H. Chen, H. Bi, Design and preparation of a ternary composite of graphene oxide/carbon dots/polypyrrole for supercapacitor application: importance and unique role of carbon dots, Carbon N. Y. 115 (2017) 134–146. https://doi.org/10.1016/j.carbon.2017.01.005. Copyright 2017, Elsevier.*

self-assembly strategy as high-performance energy storage material. The dotted line structure of the composite has prompted a high-performance capacitance and the fabricated supercapacitor offers high areal capacitance and good cycling capability (85.2% capacitance retention after 5000 cycles). Fig. 20.8 displays the TEM images of PPy nanowire and CND/PPy nanowires affirming the covalent attachment of CND on the surface of PPy nanowires. This is due to the adsorption of CND on the surface of PPy via electrostatic self-assembly strategy.

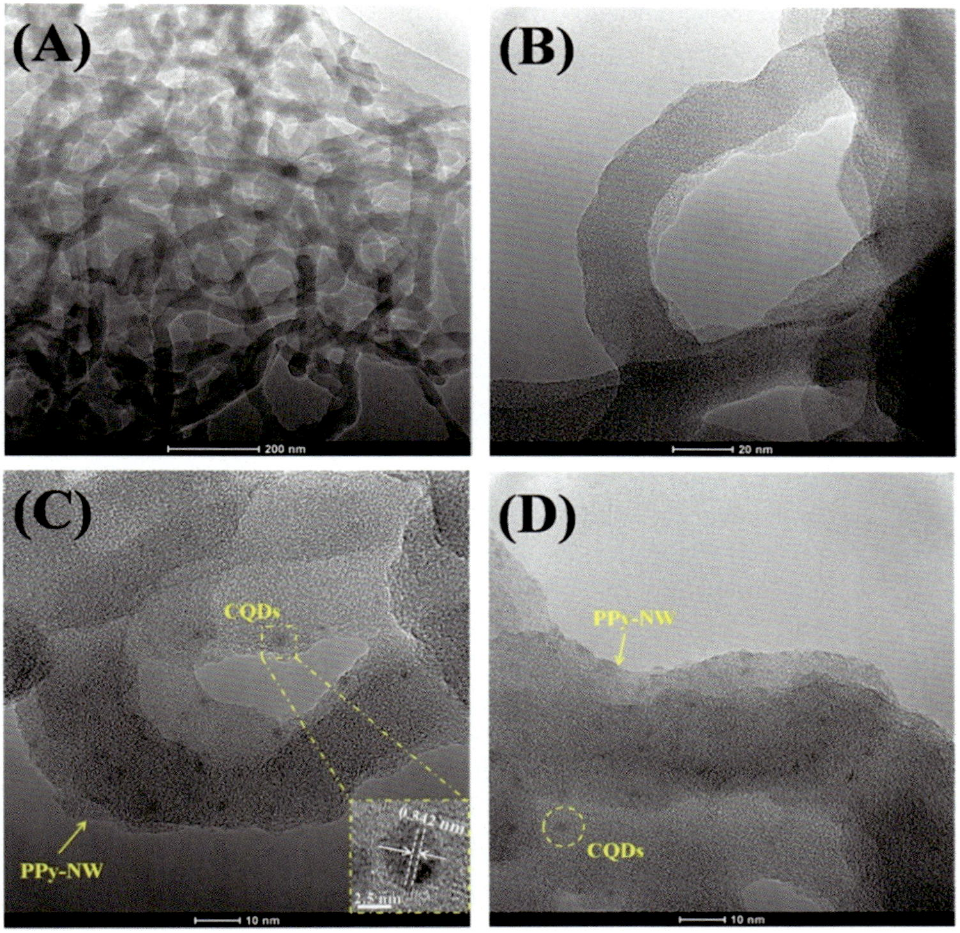

Figure 20.8

TEM images of pure PPy nanowires (A) and (B). Different magnification TEM images of CND/ PPy nanowires (C) and (D). *Reproduced with permission from Ref. X. Jian, J.G. Li, H.M. Yang, L.L. Cao, E.H. Zhang, Z.H. Liang, Carbon quantum dots reinforced polypyrrole nanowire via electrostatic self-assembly strategy for high-performance supercapacitors, Carbon N. Y. 114 (2017) 533–543. https://doi.org/10.1016/j.carbon.2016.12.033. Copyright 2017, Elsevier.*

The electrochemical impedance spectroscopy validates the excellent performance of CND/PPy nanowires due to the reduction of charge transfer resistance by CND and this can ensure fast charge transfer. The presence of CND also contributed to the development of PPy nanowires with larger surface area, more active edges, and superior conductive paths. This advances the nanowires as a promising electrode material for high-performance electrochemical storage devices.

3.6 Polyurethane based composite of carbon nanodots

Polyurethane (PU) is a thermosetting polymer comprised of urethane links. They are synthesized by reacting triisocynate with polyol. PU is a versatile polymer used in extensive applications. It is strong, moldable and has been the popular polymer in the biomedical industry [50–53]. The comprehensive properties such as toughness, durability, biocompatibility, and degradation rate can be tailored based on the application of PU. PU is generalized by the presence of urethane linkage in their repeating unit [54]. Various types of PU such as thermosetting, thermoplastic, hard, soft, and flexible can be synthesized depending on the nature and length of the isocynate and hydroxyl molecules [55,56]. The main connecting linkage within PU polymer is −NHCOO− [57]. The incorporation of CND within PU could be progressive toward the development of green environmental friendly polymers. Tian et al. [58] developed this green low-CNDs reinforced PU composites for strain sensing application. The mechanism of the composites relies on the instant response of photoluminescence intensity to the external strain applied over a large range (up to 250% strain). Green chemistry is adopted here by dispersing CNDs in the aqueous solution of PU and these composites were fabricated toward the strain sensors. FTIR and XPS survey results conclude the functionalization of CND using different surface groups doped with nitrogen. The photoluminescence intensity of CND/PU composite film varied with the different concentration addition of CND. The strain sensing mechanism displayed the recovering of original shape of the CND-PU film within the elastic range of 40%. The as fabricated strain sensors using CND/PU film is highly biocompatible, cost effective, corrosion resistant and can be supplemented to wide-ranging substrate system. The photoluminescence controlling mechanism using the CND-based strain sensors pave the way for widespread commercial application. Yao et al. [59] synthesized the silane functionalized CND using one-pot hydrothermal synthesis and used them as a novel compatibilizers with thermoplastic polyurethane (TPU) polymer. The compatabilization efficiency of the silanized CND resulted in preventing the cohesiveness of polymer chains by rigid nanoparticle core. It boosted the interfacial interaction due to the molecular entanglement and hydrogen bonding. The resultant TPU/CND composite also augmented the thermal stability, mechanical property and tunable fluorescent properties of the polymer. Fig. 20.9 displays

Figure 20.9

Optical micrograph of polyurethane blends with various loadings of Si-CNDs under visible light (left) and UV light (right). *Reproduced with permission Ref. N. Yao, H. Wang, L. Zhang, M. Tian, D. Yue, One-pot hydrothermal synthesis of silane-functionalized carbon, Appl. Surf. Sci. (2020) 147124. https://doi.org/10.1016/j.apsusc.2020.147124. Copyright 2020, Elsevier.*

the tunable fluorescence properties of TPU/CND polymer with the blend, displaying blue fluorescent at 1 wt.% and green fluorescent at 5 wt.%. The fluorescent properties are attributed to the nature of CNDs and its compatibilization efficiency.

4. Polyvinyl alcohol composites of carbon nanodots

Polyvinyl alcohol (PVA) is commonly used water-soluble synthetic polymer, a hydrolysis product of polyvinyl acetate [60]. There are fully hydrolyzed and partially hydrolyzed PVA depending on the hydrolysis condition. The solubility of PVA in water depend on the degree of hydrolysis. It is insoluble in organic solvent, and possesses solubility pattern and easy degradability [61–63]. Hence it is also termed as green polymer. PVA displays excellent compatibility with number of polymers [64]. So it has a wide range of application especially as water-soluble polymer in biomedical field. PVA is an exquisite host polymer matrix for composite films [65,66]. The advantageous properties of PVA and CND can be combined and novel composite materials for versatile applications can be developed. Liu et al. [67] devised a highly sensitive electroluminescence (ECL) sensor using Tris(bipyridine)ruthenium(II) chloride $Ru(bpy)_3^{2+}$/CND as a modifier. CND possess the ability to enhance the ECL intensity and this property played a catalytic role in the development of sensors. The constructed sensor could be proposed for sophoridine analysis using CND/PVA nanocomposite. The presence of CNDs in the nanocomposites has improved the conductivity and the electron transmission rate, CND/PVA nanocomposites display promising fabrication reproducibility and stability. CNDs function as an electron tunnel enhancing the ECL signal and provide stability to the sensor.

The sophoridine analysis in human serum by the ECL sensors provided gratifying results, thrusting the sensors toward potential commercial applications. Hoang et al. explored the effects of interaction between CNDs and PVA on the changes in optical properties or physiochemical properties of PVA. The embedding mechanism of CND into PVA is very crucial and proper synthesis procedure need to be adopted [68]. The unique combination of CNDs having low-lying states and insulating PVA could develop the emergence of charge trapping ability of resultant composites. The hydrogen bonding formation and the condensation reaction between CND and PVA increase the water-resistant property of the composites. Fig. 20.10 displays the cross-linking-induced photoluminescence (PL) quenching mechanism of CND within PVA. The presence of acidic groups and basic groups onto the surface of CND prompt the reaction with PVA and external strain experienced by capping ligands were reported to displace the surface atoms over CND from their relaxed position resulting new energy states within the CND band gap. This imparts recombination pathways to quench PL.

The chemical bonding between CQD and PVA also produced redshift of the absorption band from surface fluorophore of CNDs. The interaction between CND and PVA polymer matrix plays a very important role in optimizing the optical properties of the composites.

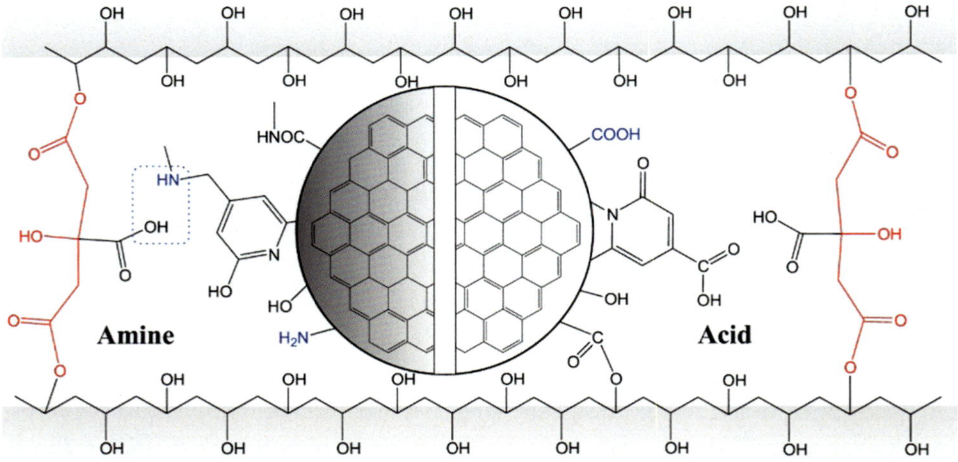

Figure 20.10

Cross-linking induced photoluminescence quenching mechanism in PVA/CND composite. *Reproduced with permission from Ref. Q.B. Hoang, V.T. Mai, D.K. Nguyen, D.Q. Truong, X.D. Mai, Cross-linking induced photoluminescence quenching in polyvinyl alcohol-carbon quantum dot composite, Mater. Today Chem. 12 (2019) 166–172. https://doi.org/10.1016/j.mtchem.2019.01.003. Copyright 2019, Elsevier.*

5. Conclusion

This chapter focused on the application of low-dimensional carbon nanomaterials, i.e., carbon nanodots (CND) in various polymer nanocomposites. CNDs possess distinguishing characteristics for meticulous interaction with the polymer and enhance their properties. CNDs are chemically functionalized with various organic modifiers for superior interaction with the polymer. CNDs possess excellent tuneable optoelectronic properties with high photoluminescence and fluorescence properties. The versatile properties of CNDs prove to be an ideal foil within the polymer to boost their inclusive properties. The exemplary conductive properties of CNDs have enhanced the electron transfer mechanism in insulating polymers. Their reinforcement in polymers have pertained their applications in sensors and supercapacitors. The optoelectronic features of CNDs have thrusted the application of polymers in biomedical field and drug delivery. The contemporary development of CND-based polymer nanocomposites has provided impetus in harnessing innovative materials and vindicated the versatility of carbon-based nanomaterials. This may provide an insight in exploring the further application of CNDs in nearly all available polymers to fabricate a hybrid composite.

Acknowledgments

Nidhin Divakaran and Manoj B Kale contributed equally in this book chapter.

References

[1] Y.P. Sun, B. Zhou, Y. Lin, W. Wang, K.A.S. Fernando, P. Pathak, M.J. Meziani, B.A. Harruff, X. Wang, H. Wang, P.G. Luo, H. Yang, M.E. Kose, B. Chen, L.M. Veca, S.Y. Xie, Quantum-sized carbon dots for bright and colorful photoluminescence, J. Am. Chem. Soc. 128 (2006) 7756–7757, https://doi.org/10.1021/ja062677d.
[2] H. Li, Z. Kang, Y. Liu, S.T. Lee, Carbon nanodots: synthesis, properties and applications, J. Mater. Chem. 22 (2012) 24230–24253, https://doi.org/10.1039/c2jm34690g.
[3] S.Y. Song, K.K. Liu, J.Y. Wei, Q. Lou, Y. Shang, C.X. Shan, Deep-ultraviolet emissive carbon nanodots, Nano Lett. 19 (2019) 5553–5561, https://doi.org/10.1021/acs.nanolett.9b02093.
[4] R. Wang, K.Q. Lu, Z.R. Tang, Y.J. Xu, Recent progress in carbon quantum dots: synthesis, properties and applications in photocatalysis, J. Mater. Chem. A. 5 (2017) 3717–3734, https://doi.org/10.1039/c6ta08660h.
[5] L. Xiao, H. Sun, Novel properties and applications of carbon nanodots, Nanoscale Horizons 3 (2018) 565–597, https://doi.org/10.1039/c8nh00106e.
[6] M. Li, T. Chen, J.J. Gooding, J. Liu, Review of carbon and graphene quantum dots for sensing, ACS Sens. 4 (2019) 1732–1748, https://doi.org/10.1021/acssensors.9b00514.
[7] A. Sharma, T. Gadly, A. Gupta, A. Ballal, S.K. Ghosh, M. Kumbhakar, Origin of excitation dependent fluorescence in carbon nanodots, J. Phys. Chem. Lett. 7 (2016) 3695–3702, https://doi.org/10.1021/acs.jpclett.6b01791.

[8] F. Arcudi, L. Đorđević, M. Prato, Rationally designed carbon nanodots towards pure white-light emission, Angew. Chem. Int. Ed. 56 (2017) 4170–4173, https://doi.org/10.1002/anie.201612160.
[9] N. Papaioannou, A. Marinovic, N. Yoshizawa, A.E. Goode, M. Fay, A. Khlobystov, M.M. Titirici, A. Sapelkin, Structure and solvents effects on the optical properties of sugar-derived carbon nanodots, Sci. Rep. 8 (2018) 1, https://doi.org/10.1038/s41598-018-25012-8.
[10] A. Cadranel, J.T. Margraf, V. Strauss, T. Clark, D.M. Guldi, Carbon nanodots for charge-transfer processes, Acc. Chem. Res. 52 (2019) 955–963, https://doi.org/10.1021/acs.accounts.8b00673.
[11] Z.A. Qiao, Y. Wang, Y. Gao, H. Li, T. Dai, Y. Liu, Q. Huo, Commercially activated carbon as the source for producing multicolor photoluminescent carbon dots by chemical oxidation, Chem. Commun. 46 (2010) 8812–8814, https://doi.org/10.1039/c0cc02724c.
[12] B. De, N. Karak, Recent progress in carbon dot-metal based nanohybrids for photochemical and electrochemical applications, J. Mater. Chem. A. 5 (2017) 1826–1859, https://doi.org/10.1039/C6TA10220D.
[13] L. Sui, W. Jin, S. Li, D. Liu, Y. Jiang, A. Chen, H. Liu, Y. Shi, D. Ding, M. Jin, Ultrafast carrier dynamics of carbon nanodots in different pH environments, Phys. Chem. Chem. Phys. 18 (2016) 3838–3845, https://doi.org/10.1039/c5cp07558k.
[14] A. Sciortino, E. Marino, B. Van Dam, P. Schall, M. Cannas, F. Messina, Solvatochromism unravels the emission mechanism of carbon nanodots, J. Phys. Chem. Lett. 7 (2016) 3419–3423, https://doi.org/10.1021/acs.jpclett.6b01590.
[15] L. Wang, Y. Bi, J. Gao, Y. Li, H. Ding, L. Ding, Carbon dots based turn-on fluorescent probes for the sensitive determination of glyphosate in environmental water samples, RSC Adv. 6 (2016) 85820–85828, https://doi.org/10.1039/c6ra10115a.
[16] H. Li, X. He, Z. Kang, H. Huang, Y. Liu, J. Liu, S. Lian, C.H.A. Tsang, X. Yang, S.T. Lee, Water-soluble fluorescent carbon quantum dots and photocatalyst design, Angew. Chem. Int. Ed. 49 (2010) 4430–4434, https://doi.org/10.1002/anie.200906154.
[17] L. Bao, Z.L. Zhang, Z.Q. Tian, L. Zhang, C. Liu, Y. Lin, B. Qi, D.W. Pang, Electrochemical tuning of luminescent carbon nanodots: from preparation to luminescence mechanism, Adv. Mater. 23 (2011) 5801–5806, https://doi.org/10.1002/adma.201102866.
[18] B. Demir, M.M. Lemberger, M. Panagiotopoulou, P.X. Medina Rangel, S. Timur, T. Hirsch, B. Tse Sum Bui, J. Wegener, K. Haupt, Tracking hyaluronan: molecularly imprinted polymer coated carbon dots for cancer cell targeting and imaging, ACS Appl. Mater. Interfaces 10 (2018) 3305–3313, https://doi.org/10.1021/acsami.7b16225.
[19] S. Wu, M. Qiu, Z. Tang, J. Liu, B. Guo, Carbon nanodots as high-functionality cross-linkers for bioinspired engineering of multiple sacrificial units toward strong yet tough elastomers, Macromolecules 50 (2017) 3244–3253, https://doi.org/10.1021/acs.macromol.7b00483.
[20] P. Liu, J. Yan, Z. Guang, Y. Huang, X. Li, W. Huang, Recent advancements of polyaniline-based nanocomposites for supercapacitors, J. Power Sources 424 (2019) 108–130, https://doi.org/10.1016/j.jpowsour.2019.03.094.
[21] C.O. Baker, X. Huang, W. Nelson, R.B. Kaner, Polyaniline nanofibers: broadening applications for conducting polymers, Chem. Soc. Rev. 46 (2017) 1510–1525, https://doi.org/10.1039/c6cs00555a.
[22] C. Ge, X. Zhang, J. Liu, F. Jin, J. Liu, H. Bi, Hollow-spherical composites of polyaniline/cobalt sulfide/carbon nanodots with enhanced magnetocapacitance and electromagnetic wave absorption capabilities, Appl. Surf. Sci. 378 (2016) 49–56, https://doi.org/10.1016/j.apsusc.2016.03.210.
[23] T. Ghosh, R. Ghosh, U. Basak, S. Majumdar, R. Ball, D. Mandal, A.K. Nandi, D.P. Chatterjee, Candle soot derived carbon nanodot/polyaniline hybrid materials through controlled grafting of polyaniline chains for supercapacitors, J. Mater. Chem. A. 6 (2018) 6476–6492, https://doi.org/10.1039/c7ta11050b.
[24] Y. Wang, S. Chen, X. Chen, Y. Lu, M. Miao, D. Zhang, Controllability of epoxy equivalent weight and performance of hyperbranched epoxy resins, Compos. B Eng. 160 (2019) 615–625, https://doi.org/10.1016/j.compositesb.2018.12.103.

[25] M. Jouyandeh, N. Rahmati, E. Movahedifar, B.S. Hadavand, Z. Karami, M. Ghaffari, P. Taheri, E. Bakhshandeh, H. Vahabi, M.R. Ganjali, K. Formela, M.R. Saeb, Properties of nano-Fe_3O_4 incorporated epoxy coatings from Cure Index perspective, Prog. Org. Coating 133 (2019) 220–228, https://doi.org/10.1016/j.porgcoat.2019.04.034.
[26] P. Song, H. Qiu, L. Wang, X. Liu, Y. Zhang, J. Zhang, J. Kong, J. Gu, Honeycomb structural rGO-MXene/epoxy nanocomposites for superior electromagnetic interference shielding performance, Sustain. Mater. Technol. 24 (2020) e00153, https://doi.org/10.1016/j.susmat.2020.e00153.
[27] L. Wang, L. Chen, P. Song, C. Liang, Y. Lu, H. Qiu, Y. Zhang, J. Kong, J. Gu, Fabrication on the annealed $Ti_3C_2T_x$ MXene/Epoxy nanocomposites for electromagnetic interference shielding application, Compos. B Eng. 171 (2019) 111–118, https://doi.org/10.1016/j.compositesb.2019.04.050.
[28] C. Zhang, L. Du, C. Liu, Y. Li, Z.Z. Yang, Y.C. Cao, Photostable epoxy polymerized carbon quantum dots luminescent thin films and the performance study, Results Phys 6 (2016) 767–771, https://doi.org/10.1016/j.rinp.2016.10.013.
[29] L. Chen, C. Zhang, Z. Du, H. Li, L. Zhang, W. Zou, Fabrication of amido group functionalized carbon quantum dots and its transparent luminescent epoxy matrix composites, J. Appl. Polym. Sci. 132 (2015) 1–7, https://doi.org/10.1002/app.42667.
[30] Z. Lei, Z. Chen, Y. Zhou, Y. Liu, J. Xu, D. Wang, Y. Shen, W. Feng, Z. Zhang, H. Chen, Novel electrically conductive composite filaments based on Ag/saturated polyester/polyvinyl butyral for 3D-printing circuits, Compos. Sci. Technol. 180 (2019) 44–50, https://doi.org/10.1016/j.compscitech.2019.05.003.
[31] B.Y. Wen, X.J. Wang, Y. Zhang, Ultrathin and anisotropic polyvinyl butyral/Ni-graphite/short-cut carbon fibre film with high electromagnetic shielding performance, Compos. Sci. Technol. 169 (2019) 127–134, https://doi.org/10.1016/j.compscitech.2018.11.013.
[32] L. Hao, K. Zhu, G. Lv, D. Yu, A comparative study of nanoscale poly N-(vinyl) pyrrole in polyvinyl butyral coatings for the anti-corrosion property of zinc: nanotubes vs nanoparticles, Prog. Org. Coating 136 (2019) 105251, https://doi.org/10.1016/j.porgcoat.2019.105251.
[33] J. Ma, T. Xiao, N. Long, X. Yang, The role of polyvinyl butyral additive in forming desirable pore structure for thin film composite forward osmosis membrane, Separ. Purif. Technol. 242 (2020) 116798, https://doi.org/10.1016/j.seppur.2020.116798.
[34] Y. Hou, Y. Xu, H. Li, Y. Li, Q.J. Niu, Polyvinyl butyral/modified SiO_2 nanoparticle membrane for gasoline desulfurization by pervaporation, Chem. Eng. Technol. 42 (2019) 65–72, https://doi.org/10.1002/ceat.201800327.
[35] D.L. Arthisree, R.R. Sumathi, G. Joshi, Effect of graphene quantum dots on photoluminescence property of polyvinyl butyral nanocomposite, Polym. Adv. Technol. 30 (2019) 790–798, https://doi.org/10.1002/pat.4516.
[36] P. Zimet, Á.W. Mombrú, D. Mombrú, A. Castro, J.P. Villanueva, H. Pardo, C. Rufo, Physico-chemical and antilisterial properties of nisin-incorporated chitosan/carboxymethyl chitosan films, Carbohydr. Polym. 219 (2019) 334–343, https://doi.org/10.1016/j.carbpol.2019.05.013.
[37] E. Avcu, F.E. Baştan, H.Z. Abdullah, M.A.U. Rehman, Y.Y. Avcu, A.R. Boccaccini, Electrophoretic deposition of chitosan-based composite coatings for biomedical applications: a review, Prog. Mater. Sci. 103 (2019) 69–108, https://doi.org/10.1016/j.pmatsci.2019.01.001.
[38] M. Mujtaba, R.E. Morsi, G. Kerch, M.Z. Elsabee, M. Kaya, J. Labidi, K.M. Khawar, Current advancements in chitosan-based film production for food technology; a review, Int. J. Biol. Macromol. 121 (2019) 889–904, https://doi.org/10.1016/j.ijbiomac.2018.10.109.
[39] J. Uranga, A.I. Puertas, A. Etxabide, M.T. Dueñas, P. Guerrero, K. de la Caba, Citric acid-incorporated fish gelatin/chitosan composite films, Food Hydrocoll. 86 (2019) 95–103, https://doi.org/10.1016/j.foodhyd.2018.02.018.
[40] Á. Molnár, The use of chitosan-based metal catalysts in organic transformations, Coord. Chem. Rev. 388 (2019) 126–171, https://doi.org/10.1016/j.ccr.2019.02.018.

[41] M. Sheng, Y. Gao, J. Sun, F. Gao, Carbon nanodots-chitosan composite film: a platform for protein immobilization, direct electrochemistry and bioelectrocatalysis, Biosens. Bioelectron. 58 (2014) 351–358, https://doi.org/10.1016/j.bios.2014.03.005.
[42] P. Qi, T. Zhang, J. Shao, B. Yang, T. Fei, R. Wang, A QCM humidity sensor constructed by graphene quantum dots and chitosan composites, Sensor. Actuator. A Phys. 287 (2019) 93–101, https://doi.org/10.1016/j.sna.2019.01.009.
[43] Y. Xie, Electrochemical performance of transition metal-coordinated polypyrrole: a mini review, Chem. Rec. 19 (2019) 2370–2384, https://doi.org/10.1002/tcr.201800192.
[44] Z. Wang, X. Xu, J. Kim, V. Malgras, R. Mo, C. Li, Y. Lin, H. Tan, J. Tang, L. Pan, Y. Bando, T. Yang, Y. Yamauchi, Nanoarchitectured metal-organic framework/polypyrrole hybrids for brackish water desalination using capacitive deionization, Mater. Horizons. 6 (2019) 1433–1437, https://doi.org/10.1039/c9mh00306a.
[45] K. Yamani, R. Berenguer, A. Benyoucef, E. Morallón, Preparation of polypyrrole (PPy)-derived polymer/ZrO_2 nanocomposites: effects of nanoparticles interface and polymer structure, J. Therm. Anal. Calorim. 135 (2019) 2089–2100, https://doi.org/10.1007/s10973-018-7347-z.
[46] H. Li, J. Yin, Y. Meng, S. Liu, T. Jiao, Nickel/cobalt-containing polypyrrole hydrogel-derived approach for efficient ORR electrocatalyst, Colloids Surfaces A Physicochem. Eng. Asp. 586 (2020) 124221, https://doi.org/10.1016/j.colsurfa.2019.124221.
[47] M. Maruthapandi, A.P. Nagvenkar, I. Perelshtein, A. Gedanken, Carbon-dot initiated synthesis of polypyrrole and polypyrrole@CuO micro/nanoparticles with enhanced antibacterial activity, ACS Appl. Polym. Mater. 1 (2019) 1181–1186, https://doi.org/10.1021/acsapm.9b00194.
[48] X. Zhang, J. Wang, J. Liu, J. Wu, H. Chen, H. Bi, Design and preparation of a ternary composite of graphene oxide/carbon dots/polypyrrole for supercapacitor application: importance and unique role of carbon dots, Carbon N. Y. 115 (2017) 134–146, https://doi.org/10.1016/j.carbon.2017.01.005.
[49] X. Jian, J.G. Li, H.M. Yang, L.L. Cao, E.H. Zhang, Z.H. Liang, Carbon quantum dots reinforced polypyrrole nanowire via electrostatic self-assembly strategy for high-performance supercapacitors, Carbon N. Y. 114 (2017) 533–543, https://doi.org/10.1016/j.carbon.2016.12.033.
[50] J. Yang, W. Yang, X. Wang, M. Dong, H. Liu, E.K. Wujcik, Q. Shao, S. Wu, T. Ding, Z. Guo, Synergistically toughening polyoxymethylene by methyl methacrylate–butadiene–styrene copolymer and thermoplastic polyurethane, Macromol. Chem. Phys. 220 (2019) 1–10, https://doi.org/10.1002/macp.201800567.
[51] M. Dong, C. Wang, H. Liu, C. Liu, C. Shen, J. Zhang, C. Jia, T. Ding, Z. Guo, Enhanced solid particle erosion properties of thermoplastic polyurethane-carbon nanotube nanocomposites, Macromol. Mater. Eng. 304 (2019) 1–11, https://doi.org/10.1002/mame.201900010.
[52] K. Chang, H. Jia, S.Y. Gu, A transparent, highly stretchable, self-healing polyurethane based on disulfide bonds, Eur. Polym. J. 112 (2019) 822–831, https://doi.org/10.1016/j.eurpolymj.2018.11.005.
[53] S. Bahrami, A. Solouk, H. Mirzadeh, A.M. Seifalian, Electroconductive polyurethane/graphene nanocomposite for biomedical applications, Compos. B Eng. 168 (2019) 421–431, https://doi.org/10.1016/j.compositesb.2019.03.044.
[54] A.B. Baker, S.R.G. Bates, T.M. Llewellyn-Jones, L.P.B. Valori, M.P.M. Dicker, R.S. Trask, 4D printing with robust thermoplastic polyurethane hydrogel-elastomer trilayers, Mater. Des. 163 (2019) 107544, https://doi.org/10.1016/j.matdes.2018.107544.
[55] J.W. Su, W. Gao, K. Trinh, S.M. Kenderes, E. Tekin Pulatsu, C. Zhang, A. Whittington, M. Lin, J. Lin, 4D printing of polyurethane paint-based composites, Int. J. Smart Nano Mater. 10 (2019) 237–248, https://doi.org/10.1080/19475411.2019.1618409.
[56] H. Zhao, W. She, D. Shi, W. Wu, Q. chao Zhang, R.K.Y. Li, Polyurethane/POSS nanocomposites for superior hydrophobicity and high ductility, Compos. B Eng. 177 (2019) 107441, https://doi.org/10.1016/j.compositesb.2019.107441.

[57] G.J. Zhu, P.G. Ren, H. Guo, Y.L. Jin, D.X. Yan, Z.M. Li, Highly sensitive and stretchable polyurethane fiber strain sensors with embedded silver nanowires, ACS Appl. Mater. Interfaces 11 (2019) 23649–23658, https://doi.org/10.1021/acsami.9b08611.

[58] Y. Tian, Y. Zhao, F. Kang, F. Lyu, Z. Li, J. Lu, Y. Yang, Green low-cost carbon nanodots-polyurethane composites with novel anisotropic anti-quenching mechanism for strain sensing, arXiv, Appl. Phys. A (2019).

[59] N. Yao, H. Wang, L. Zhang, M. Tian, D. Yue, One-pot hydrothermal synthesis of silane-functionalized carbon, Appl. Surf. Sci. (2020) 147124, https://doi.org/10.1016/j.apsusc.2020.147124.

[60] K. Sun, Z. Wang, J. Xin, Z. Wang, P. Xie, G. Fan, V. Murugadoss, R. Fan, J. Fan, Z. Guo, Hydrosoluble graphene/polyvinyl alcohol membranous composites with negative permittivity behavior, Macromol. Mater. Eng. 305 (2020) 1–8, https://doi.org/10.1002/mame.201900709.

[61] J. Shen, P. Zhang, L. Song, J. Li, B. Ji, J. Li, L. Chen, Polyethylene glycol supported by phosphorylated polyvinyl alcohol/graphene aerogel as a high thermal stability phase change material, Compos. B Eng. 179 (2019) 107545, https://doi.org/10.1016/j.compositesb.2019.107545.

[62] N. Yang, P. Qi, J. Ren, H. Yu, S. Liu, J. Li, W. Chen, D.L. Kaplan, S. Ling, Polyvinyl alcohol/silk fibroin/borax hydrogel ionotronics: a highly stretchable, self-healable, and biocompatible sensing platform, ACS Appl. Mater. Interfaces 11 (2019) 23632–23638, https://doi.org/10.1021/acsami.9b06920.

[63] J. Zhang, L. Wan, Y. Gao, X. Fang, T. Lu, L. Pan, F. Xuan, Highly stretchable and self-healable MXene/polyvinyl alcohol hydrogel electrode for wearable capacitive electronic skin, Adv. Electron. Mater. 5 (2019) 1–10, https://doi.org/10.1002/aelm.201900285.

[64] L.Y. Jun, N.M. Mubarak, L.S. Yon, C.H. Bing, M. Khalid, P. Jagadish, E.C. Abdullah, Immobilization of peroxidase on functionalized MWCNTs-buckypaper/polyvinyl alcohol nanocomposite membrane, Sci. Rep. 9 (2019) 1, https://doi.org/10.1038/s41598-019-39621-4.

[65] M. Liao, H. Liao, J. Ye, P. Wan, L. Zhang, Polyvinyl alcohol-stabilized liquid metal hydrogel for wearable transient epidermal sensors, ACS Appl. Mater. Interfaces 11 (2019) 47358–47364, https://doi.org/10.1021/acsami.9b16675.

[66] Y. Zhao, D. Wang, Y. Gao, T. Chen, Q. Huang, D. Wang, Stable Li metal anode by a polyvinyl alcohol protection layer via modifying solid-electrolyte interphase layer, Nanomater. Energy 64 (2019) 103893, https://doi.org/10.1016/j.nanoen.2019.103893.

[67] Z. Liu, X. Zhang, L. Cui, K. Wang, H. Zhan, Development of a highly sensitive electrochemiluminescence sophoridine sensor using $Ru(bpy)_3^{2+}$ integrated carbon quantum dots – polyvinyl alcohol composite film, Sensor. Actuator. B Chem. 248 (2017) 402–410, https://doi.org/10.1016/j.snb.2017.03.115.

[68] Q.B. Hoang, V.T. Mai, D.K. Nguyen, D.Q. Truong, X.D. Mai, Crosslinking induced photoluminescence quenching in polyvinyl alcohol-carbon quantum dot composite, Mater. Today Chem. 12 (2019) 166–172, https://doi.org/10.1016/j.mtchem.2019.01.003.

SECTION G

Development of nanofibers

CHAPTER 21

Electrospun polymer composites and ceramics nanofibers: synthesis and environmental remediation applications

Shabna Patel[1], Garudadhwaj Hota[2]
[1]*Department of Mathematics & Science, UGIE, Rourkela, Odisha, India;* [2]*Department of Chemistry, National Institute of Technology, Rourkela, Odisha, India*

Chapter Outline

1. Introduction

Production of clean and toxic-free water in more economic route is the major concern of the current research. Due to rapid growth of population, industrialization, and modernized society water recourses are being polluted though out the globe. Apart from this, water bodies are also being contaminated with various microbes that causes different waterborne diseases, and leads to around deaths of 1.8 million annually [1]. The large-scale production and discharge of industrial effluents that includes toxic organic and inorganic pollutants from different manufacturing industries like leather, textile, paper, cosmetic, and pharmaceuticals to the nearby water reservoir is highly hazardous to public health and

Design, Fabrication, and Characterization of Multifunctional Nanomaterials. https://doi.org/10.1016/B978-0-12-820558-7.00019-4

living animals [2]. Therefore, elimination of these chemical pollutants from waste water in a more effective and convenient process is an urgent and highly challenging for researchers in this era [3]. A variety of decontamination processes have been applied to eliminate these pollutants. Some of them are adsorption, photocatalytic degradation, chemical oxidation, flocculation, membrane separation, and electrooxidation [4]. However, adsorption is found to be the most promising and effective approach for the removal of these contaminants from aqueous media arising out of its easy operation, cost-effectiveness, and high adsorption efficiency [5]. Hence, it is very important to develop efficient and novel adsorbent materials for the removal of toxic dyes and inorganic metal ions with high sorption capacity, good compatibility, and reusability [6].

In comparison to bulk materials, nanosized adsorbent materials draw more research attention, because of their larger specific surface area, higher surface energy, and uniform size distribution. Mahapatra and coworkers have reported on the synthesis of hybrid iron oxide-alumina composite nanoadsorbent and studied the adsorptive removal of CR dye from aqueous media [7]. However; the separation of these nanoadsorbent materials from the aqueous system after their use was difficult because of the extremely small nano size dimension, high surface energy and high dispersibilty, which restrict their practical applications [8,9].

Owing to their porous structures, high specific surface area, and interconnected fibers channel, one-dimensional (1D) electrospun nanofibers have been used as support materials for immobilization of reactive nanoparticles [10]. Recent research focused on the synthesis of organic-inorganic composite nanofibers, where different metal, metal oxides nanoparticles are embedded inside the polymer nanofibers matrix. The incorporation of these nanosized particles into the polymer nanofibers matrix has been significantly improved the mechanical, thermal properties, chemical stability, and also enhance the application potential of the composite nanofibers membrane in the energy and environmental field [11,12]. Electrospinning is found to be a facile and cost effective method for the fabrication of organic-inorganic composite nanofibers and hence widely used for this purpose in recent years. By using this electrospinning method, it is easier to incorporate/functionalize various nanomaterials into the polymer nanofibers matrix and as a result of which the physicochemical properties of the polymeric nanofibrous matrix has been improved [13]. Kim and coworkers have synthesized ZnO- nylon six composite nanofibers via electrospinning method and used as catalyst for photocatalytic degradation of methylene blue (MB) dye from aqueous media [11]. Nevertheless, among the different nanomaterials, nanosized iron oxide materials are extensively used for decontamination of toxic inorganic (metal ions) and organic pollutants arising out of their nontoxicity, cost-effectiveness, easy availability, high chemical stability, and adsorption efficiency. In recent years, Khosravi and coworkers, have synthesized magnetic iron oxide nanospheres by solvothermal method and have used for adsorption of reactive yellow and reactive orange anionic dyes from water media [14]. Likewise, Li and coworkers have also fabricated iron oxide functionalized nylon-6 composite nanofibers electrospun membrane and have examined its efficiency for the adsorptive removal of Cr (VI) ions form the aqueous media [9].

2. Synthesis of nanofibers by electrospinning

Electrospinning is a simple, convenient and cost effective method for large-scale production of continuous submicron-sized fibers by using a wide variety of polymeric and inorganic salt precursors materials [15]. Polymer nanofibers prepared by electrospinning technique have a diameter in the order of few nanometers to over 1 micrometer and exhibit unique characteristics, such as extraordinary high surface area to volume ratio, tunable porosity, excellent structural and mechanical properties, low basis weight, and high axial strength combined with extreme flexibility in surface functionalization [16,17].

A standard electrospinning setup consists of four primary components: DC high voltage power supply, grounded electrically conductive collector, spinneret (syringe with a metallic needle), and a syringe pump. In a typical electrospinning process, the positive charge of a high voltage power supply is connected to a metallic needle fitted with a syringe and negative electrode is connected to the grounded metallic collector. The polymer solution is pushed to the needle tip with the help of a syringe pump. On the application of a high voltage DC power to the polymer solution passing through the metallic needle, a pendant-shaped droplet forms at the tip of the needle. Then the electrostatic force of repulsion exists among the polymeric molecules that overcome the surface tension of the fluid. As a result of which the polymeric droplet formed at the tip of the needle will deform into a shape of conical droplet, which is known as Taylor cone [18]. As the electrostatic force of repulsion overcomes the surface tension of the conical droplet, an electrified jet of polymer solution is ejected from the needle tip. The interaction between the electric field and the surface tension of the polymeric fluid stretches the charged polymeric jet stream and provides a whipping motion, followed by the evaporation of the solvent. This helps the electrified charged jet stream to be continuously elongated as a long and thin cylindrical filament and then this cylindrical filament solidifies and gets deposited onto the grounded conductive collector, which results in the formation of a uniform nanofibers mat [19].

2.1 Process parameters of the electrospinning method

The size and morphology of the electrospun nanofibers are depended on a number of electrospinning parameters, which are commonly divided into three main categories. They are solution parameters, process parameters, and ambient parameters. The solution parameters include viscosity of the solution/concentration, conductivity, and surface tension. Similarly, the important process parameters include applied voltage, feed rate, and distance between the needle tip to collector and also the ambient parameters like humidity and temperature can be tuned in the electrospinning process to obtain the desired nanofibers matrix [19].

2.1.1 Solution concentration/viscosity

Viscosity of polymer is one of the most significant parameter, which influences the fiber diameter in electrospinning method. Higher is the viscosity of the solution, larger is the fiber diameter of the resultant fibers. Furthermore, the viscosity of the polymer solution in a solvent is depended on the polymer concentration and its molecular weight. In the year 2001, Deitzel et al. have reported that the fiber diameter increases with increasing polymer concentration according to a power law relationship [20]. Furthermore, beaded fibers are obtained when the solution concentration is decreased to the entanglement concentration (C_e). Again with increase in solution concentration above C_e prevents the formation of beaded fibers and smooth fibers formation are observed when C_e increases up to 2–2.5 times [21]. Fong et al. have reported that higher concentration of the polymer solution resulted the formation of fewer beads. They have used 1–4.5 wt% of PEO polymer solution for electrospinning and the resulting fiber membranes were analyzed by using SEM. Fig. 21.1 represents the SEM images of electrospun PEO nanofibers with increase in polymer solution viscosity. The viscosity of 1 and 4.5 wt% PEO solutions were found to be 13 and 1250 cP respectively. It was observed that with increase in viscosity the beads formation completely disappeared and the shape of the beads changed from spherical to spindle like when the polymer concentration varied from to lower to higher levels [22].

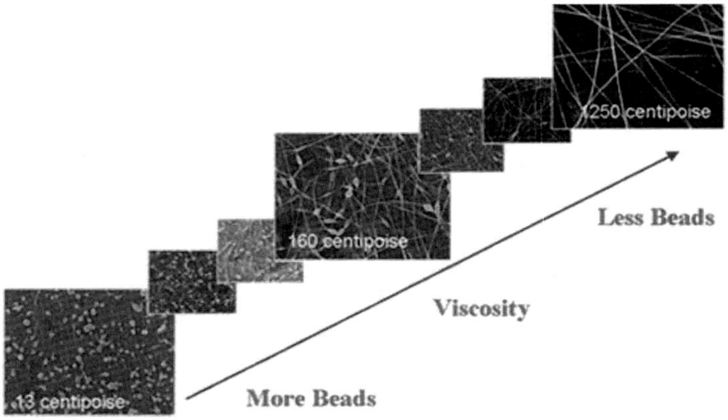

Figure 21.1
SEM images of electrospun nanofibers from different polymer concentration solution [22].

2.1.2 Applied voltage

In the electrospinning process, applied voltage is another crucial parameter that is used to control the fiber diameter and morphology. When the applied voltage overcomes the threshold voltage, fiber formation occurs. This induces the necessary charge on the solution along with the electric field and initiates the electrospinning process. With increases in the applied voltage, the electrostatic force of repulsion among polymer molecules on the fluid jet increases, which ultimately favors the decrease in the fiber diameter. Therefore, higher applied voltage causes greater stretching of the solution due to the greater columbic repulsive forces in the jet as well as a stronger electric field. These effects lead to the narrowing of fibers diameter followed by rapid evaporation of solvent from the fibers. The fibers have smaller diameter and smooth surface morphology at an optimum applied voltage. However, with further increase in applied voltage there is also greater probability of formation of beads as shown in Fig. 21.2 [18].

2.1.3 Flow rate/feed rate

The flow rate/feed rate is the rate of flow of the polymer solution from the syringe which can be controlled by using a syringe pump. This is another important electrospinning process parameter. It influences the velocity of charged fluid jet and transfer rate of the polymeric material. The flow rate of the electrospinning solution should always be kept minimum so that the solvent will get sufficient time for evaporation [23]. It was realized that in case of e-spun polystyrene polymer nanofibers, the fibers diameter and interfibers porosity increases with increase in the flow rate of polymer solution. Furthermore, by tuning the flow rate of polymer solution, one can also be able to change the morphological

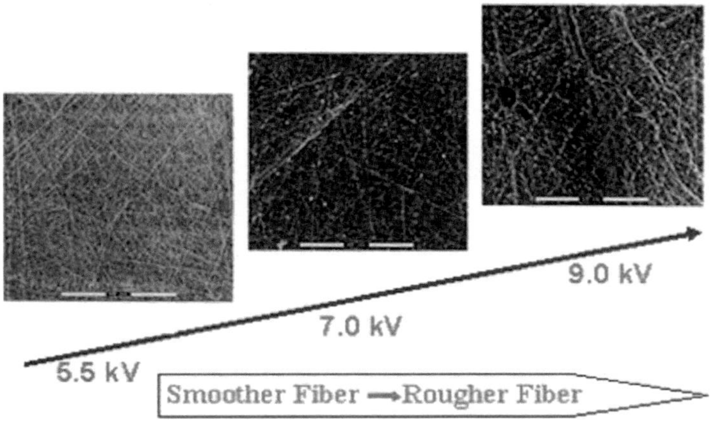

Figure 21.2
SEM images of PEO nanofibers electrospun with increase in applied voltage [18].

structure of nanofibers. It was also observed that the high flow rates result in the formation of beaded fibers due to unavailability of proper drying time prior to reach the grounded collector [16].

Zhang et al. have reported on the electrospinning mediated synthesis of PAN nanofibers. They have optimized the electrospinning condition to produce uniform and bead-free PAN nanofibers. PAN polymer solution in DMF with concentrations 4, 6.5, and 10 wt% were taken for electrospinning. It was observed that polymer solution with both 4 and 6.5 wt% produced nanofibers with beads. However, straight, uniform, bead-free nanofibers with a smooth surface morphology were formed, when the polymer concentration was increased to 10.0 wt% as shown in Fig. 21.3. Furthermore, the fiber quality was improved when the feed rate decreased from 10 to 8 μL/min as the lower feed rate will allow the solvent to have enough time for evaporation [24].

2.2 Synthesis of inorganic-organic composite nanofibers

Nirmala et al. have synthesized polyacrylonitrile (PAN) nanofibers and then decorated with CoS, CdS, and ZnS semiconductor nanoparticles on the PAN polymer nanofibers surface by electrospinning followed by dip coating method. The advantage of using the dip coating method for functionalization of these sulfide nanoparticles on the electrospun nanofibers surface is to control the concentration of nanoparticles on PAN nanofibers mat surface. For electrospinning of PAN nanofibers, PAN-DMF solution (10 wt%) was prepared by dissolving and electrospun with an applied electric voltage of 17 kV. Cadmium acetate salt precursor was dissolved in DMF solvent with 1:5 ratios. After that 0.5 mL of aqueous ammonium sulfide solution was added dropwise and very slowly to the cadmium acetate solution, with magnetic stirring to prepare CdS nanoparticles. Using the similar method, CoS and ZnS nanoparticles were also prepared. To prepare the electrospun PAN-metal sulfide composite nanofibers membrane, the previously prepared colloidal

Figure 21.3
SEM images of pure PAN nanofibers with a PAN loading (A) 4 wt%, (B) 6.5 wt%, and (C) 10 wt% [24].

nanoparticles solutions were gently poured into PAN nanofibers taken in a Petri dish and kept for 12 h. After that the composite nanofibers mats were washed with ethanol and then dried at 25°C under vacuum condition [25].

It was observed that the SEM images of pristine e-spun PAN nanofibers exhibited smooth and bead-free cylindrical surfaces having uniform diameters (200–250 nm) with highly aligned and bundle-like morphology. However, the SEM images of composite nanofibers represent a significant morphological difference from that of the pristine nanofibers. The aligned morphology was disturbed after dipping the PAN nanofibers into the colloidal nanoparticles solution. The diameter of the resulting composite nanofibers was slightly increased (about 250–400 nm) than that of pristine PAN nanofibers. It can also be observed that the semiconductor nanoparticles having diameter in the range of 50–100 nm were uniformly decorated on the e-spun PAN nanofibers surface after the dip coating process [25].

In case of organic-inorganic electrospun composite nanofibers adsorbents synthesis, metal oxides nanoparticles are commonly used as inorganic components, which show high adsorption efficiency for cationic and anionic pollutants due to their many exceptional physicochemical properties [26]. For example, nanosized Fe_3O_4 particles have been successfully used as adsorbent for the remediation of various metal ions from the aqueous solution due to their relatively large surface area, excellent magnetic response and ease of surface modification [27]. The synthesis of iron oxide–based inorganic-organic hybrid electrospun nanofibers mats have been reported by various research groups.

Recently, Liu et al. have reported on the synthesis of novel 1D nanostructured Fe_3O_4/PAN composite nanofibers by a simple two-step process, an electrospinning followed by solvothermal method (Fig. 21.4). They have used electrospinning method to prepare PAN nanofibers membrane from PAN/DMF/DMAc solution by applying a DC high voltage of 12 kV. The distance between the needle tip and the collector drum was maintained at 15 cm and the feed rate of the electrospinning solution was 0.9 mL/h. For the preparation of PAN/Fe_3O_4 composite nanofibers the previously prepared PAN nanofibers membrane was solvothermally treated with $FeCl_3$/DEG solution at a temperature of 200°C for 14 h. It was observed from the SEM image, that the PAN nanofibers turned into wavier and their surface become rough due to the formation of Fe_3O_4 nanoparticles. TEM micrograph revealed the presence of a uniform Fe_3O_4 nanoparticles coating layer of thickness 20 nm on the surface of the PAN nanofiber. The diameter of the nanofibers was around 500 nm and the coated Fe_3O_4 nanoparticles size were found to be around 60 nm from TEM studies [28].

More recently, Zhao et al. have fabricated branched polyethylenimine (b-PEI) grafted onto electrospun magnetic iron oxide (Fe_3O_4)/polyacrylonitrile composite nanofiber mat. They have prepared the inorganic-organic hybrid nanofibers membrane by using both

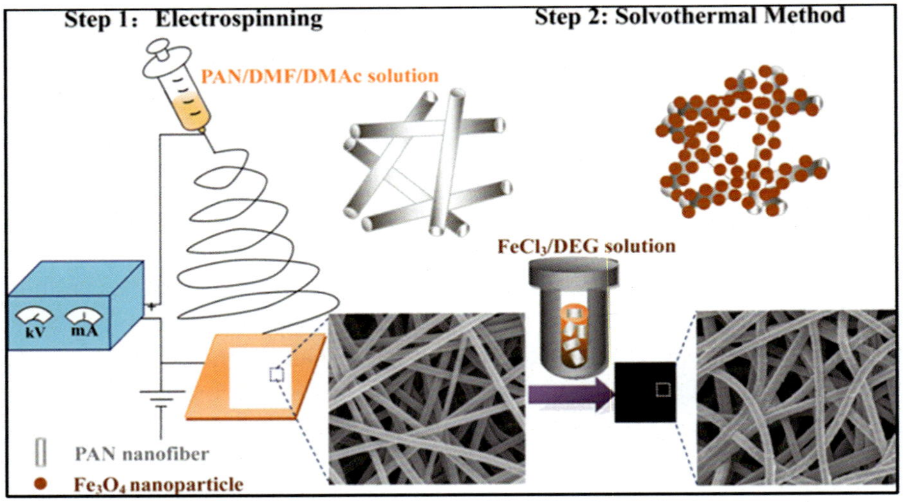

Figure 21.4
Schematic representations for the preparation of Fe_3O_4/PAN composite nanofibers [10].

electrospinning method and one step hydrothermal treatment process. In the first step, e-spun PAN nanofibers membrane was prepared by using 8wt% PAN/DMF solution and 15 kV applied voltage. Then the previously prepared PAN nanofibers membrane and Ferrous Chloride ($FeCl_2.4H_2O$)/b-PEI salt precursors were taken in an autoclave and hydrothermally treated at 140°C for 8 h. Then the composite membrane was rinsed with distilled water and ethanol and then dried at 60°C for 12 h [29].

Xu and coworkers have prepared hierarchically TPEE/Iron oxide nanofibers membrane by electrospinning thermal plastic elastomer ester (TPEE) polymer and iron alkoxide sol precursors along with hydrothermal strategy. The synthesis procedure involves two steps. Firstly, they have prepared iron alkoxide sol and added to TPEE polymer solution followed by electrospinning to from TPEE/iron alkoxide (IA) composite nanofibers. Then, with the growth of iron oxide nanostructure in situ on TPEE nanofibers, the prepared TPEE/IA composite nanofibers membrane was immersed into 100 mL deionized water at 100°C for different time interval (for 20, 40, 60, and 80 min). The plausible mechanism for the formation of TPEE/iron oxide composite nanofibers membrane is shown in Fig. 21.5. During the hydrothermal process the IA of composite electrospun TPEE/IA nanofibers get hydrolyzed into iron hydroxide, which then transformed into iron oxides through the decomposition reaction. As a result of which IO nanoparticles were grown in situ on TPEE nanofibers template to from PTEE/IO nanocomposite fibers [8,9]. From the SEM images, it can be observed that when the hydrothermal time was 20 min, it exhibited a conglutination phenomenon, which was supposed to be the hydrolysis of IA.

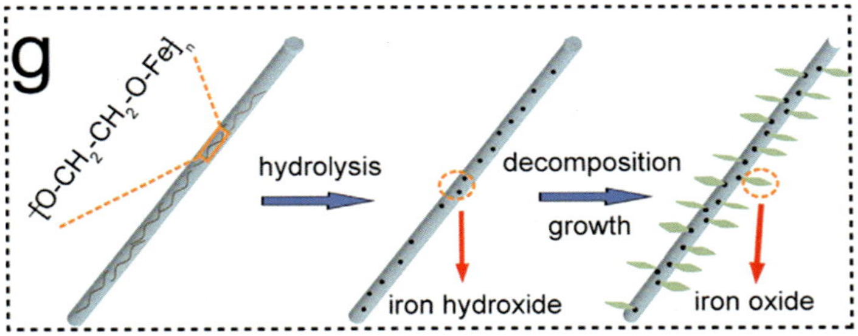

Figure 21.5
Schematic representations for the preparation of TPEE/IO composite nanofibers membrane [8].

After 40 min, hierarchical structures could be seen on the surface of nanofibers indicating the formation of iron oxides. With increasing hydrothermal time the density of iron oxides become higher [9].

Mahapatra et al. have used electrospinning method to synthesize PAN/Ag composite nanofibers. First they have prepared e-spun PAN/$AgNO_3$ composite nanofibers membrane using PAN/DMF/$AgNO_3$ solution. After that, the silver ions were reduced into metallic silver nanoparticles on PAN nanofibers surface by three different wet chemical techniques. They are chemical reduction of Ag^+ ions using $NaBH_4$, heat treatment, and refluxing the PAN/DMF solution containing $AgNO_3$ at 80°C for 2 h followed by electrospinning (Fig. 21.6). The TEM image of PAN/Ag composite nanofibers prepared by refluxed method is shown in Fig. 21.7A. It can be seen that uniform and spherical Ag nanoparticles of size 7–10 nm are formed on and within the PAN nanofibers surface. Fig. 21.7B represents the TEM image of PAN/Ag composite nanofibers synthesized by $NaBH_4$

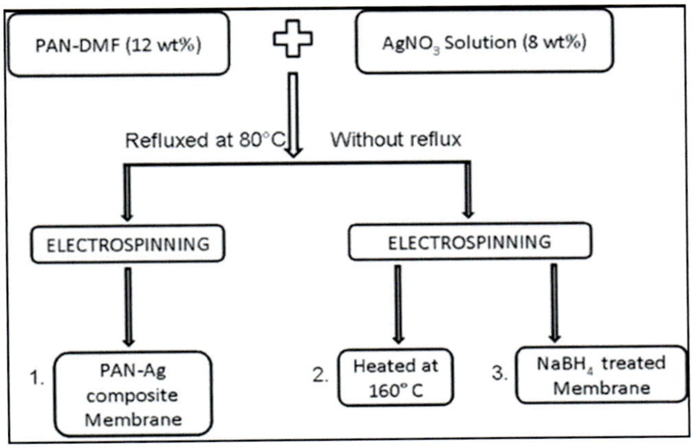

Figure 21.6
Schematic representations for the synthesis of PAN/Ag composite nanofibers [30].

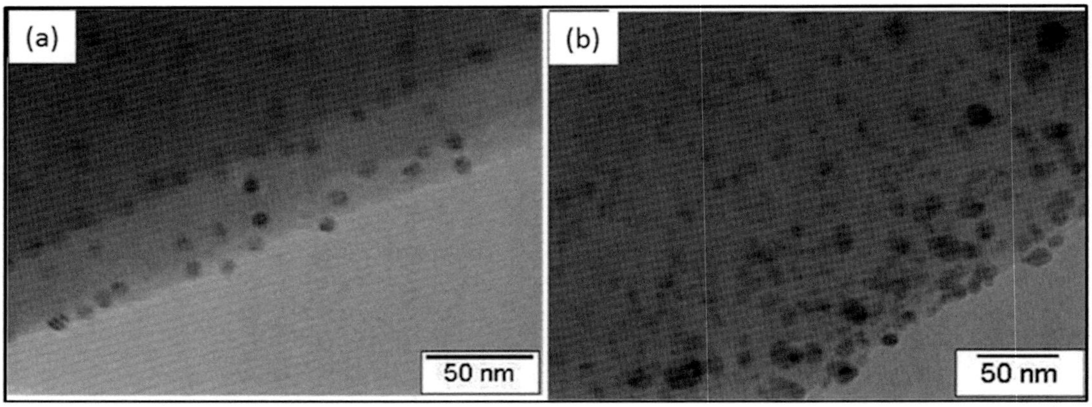

Figure 21.7
TEM micrographs of PAN-Ag composite nanofibers prepared by (A) refluxed method and (B) $NaBH_4$ reduction method [30].

reduction method. In such case also formation of spherical shape polydispersed Ag nanoparticles are observed on the nanofiber surface [30].

More recently, Panthi et al. have studied on the formation of Ag_3O_4/PAN composite nanofibers by electrospinning method. First, they have functionalized the PAN nanofibers surface with amidoxime (-C(CN_2) = NOH) groups, which can facilitate the attachment of Ag^+ ions through coordinate bonding. Then, the Ag + ions coordinated with PAN nanofibers surface were allowed to react with PO_4^{3-} ions to prepare Ag_3O_4/PAN composite nanofibers. The TEM characterization of e-spun Ag_3O_4/PAN composite nanofibers indicates that Ag_3O_4 nanoparticles were well dispersed and formed on the surface of PAN nanofibers. Such functionalization of inorganic nanoparticles in nonaggregated state onto the surface of e-spun polymer nanofibers possess large surface area and surface active sites and hence could be beneficial to exhibit exceptional physical and chemical properties [31].

Kim et al. have studied on formation of ZnO/nylon-6 composite nanofibers by using both electrospinning and hydrothermal synthetic techniques. They have noticed that ZnO rods with unique mop-brush-like morphology were formed on the surface of nylon-6 nanofibers. The schematic representation is presented in Fig. 21.8. They have prepared the composite nanofibers membrane in two steps. At first, they have blended ZnO nanoseeds in nylon-6 polymer solution and then electrospun to prepare ZnO/nylon-6 composite membrane. In the second step, mop-brush-like morphology of ZnO nanorods were grown on the surface of nylon-6 nanofibers via hydrothermal treatment of ZnO/nylon-6 hybrid nanofibers membrane with ZnO precursor solution by using an autoclave heated at 110°C for 4 h. From the SEM and FE-SEM studies of ZnO/nylon-6 composite nanofibers, it could be seen that large amount of ZnO nanorods having mop-brush like morphology are decorated on

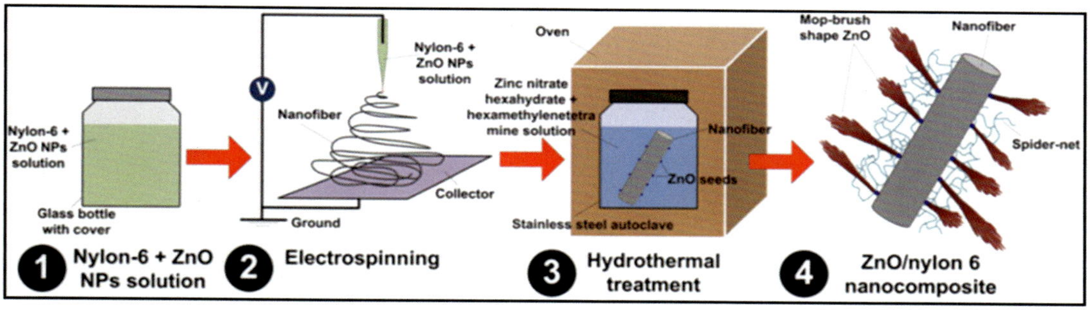

Figure 21.8

Schematic representations for the fabrication of mop-brush-shaped ZnO rods on the surface of electrospun nylon-6 nanofiber [11].

the surface of nanofibers. Kim et al. have also observed that ZnO nanoseeds content was playing an important role by providing the nucleation site for the development and growth of long ZnO mop-brushes on nanofibers surface during hydrothermal treatment [11].

Hassan et al. have reported on the fabrication of electrospun polyurethane (PU) nanofibers functionalized with bimetallic ZnO/Ag nanoparticles. They have prepared ZnO/Ag composite nanoparticles by sol-gel process and then added to the PU/THF/DMF polymer solution (10 wt%) with constant stirring. Then the ZnO/Ag nanoparticles embedded composite PU polymer solution was electrospun by applying DC voltage of 15 kV and maintaining 15 cm distance between the needle tip and the metallic collector. The schematic representation of formation of ZnO/Ag nanocomposite embedded PU nanofibers membrane is shown in Fig. 21.9. The fibers diameters of pristine and doped PU nanofibers are found to be in the range of 200–300 nm with unique interconnected spider-net-like structures [32].

Similarly, Son and coworkers have reported on the synthesis of inorganic-organic functional composite nanofibers via electrospinning method. They have used poly(vinyl alcohol), PVA polymer containing hydroxyl functional groups, that facilitate the surface functionalization of PVA. First they have chemically attached the 3,4-dihydroxyhydrocinnamic acid with the hydroxyl groups of PVA molecule. After that they have grafted catechol (1,2 dihydroxy benezene) molecules onto PVA surface by performing a series of chemical treatment to produce catechol-grafted PVA (PVA-g-ct). To obtain PVA-g-ct nanofibers they have prepared the PVA-g-ct polymer solution (3 wt%) in hexafluoroisopropanol (HFIP) solvent and electrospun into nanofibers by tuning the various electrospinning process parameters. The low reduction potential of catechol molecules help in reduction of noble metal ions into metal nanoparticles, and Son and coworkers have used the catechol-grafted PVA nanofibers as a template for formation and stabilization of metallic nanoparticles. They have prepared metal-polymer hybrid

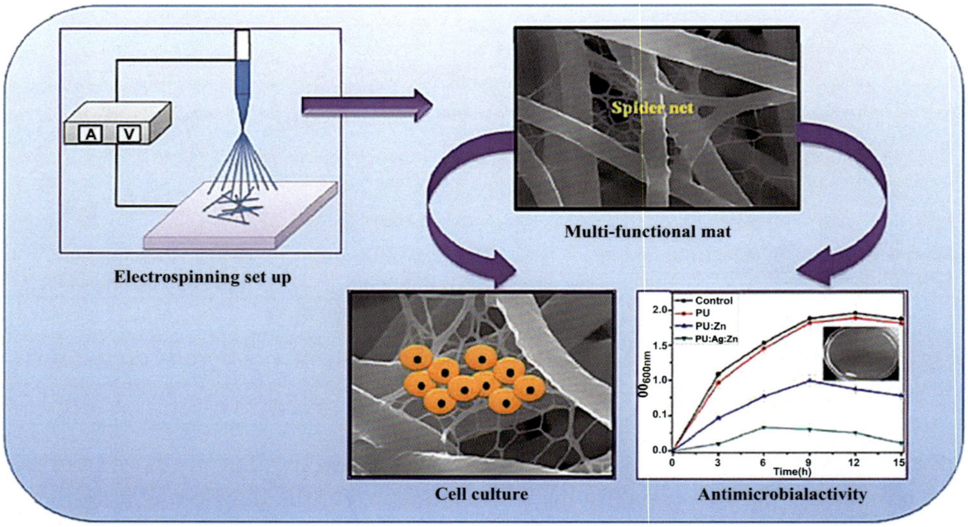

Figure 21.9
Schematic representation for the steps involved in the fabrication of composite nanofibers by electrospinning, antibacterial test, and proliferation of cells on the nanocomposite [32].

nanofibers by immersing PVA-g-ct nanofibers into salt precursor solution of Ag^{+}, Au^{3+}, and Pt^{4+} at an ambient temperature as a result of which metal ions form complexes with catechol, followed by in situ reduction into metallic nanoparticles. They have observed that Ag nanoparticles have spontaneously formed and attached to the PVA-g-ct nanofibers surface at ambient temperature. However, for gold and platinum noble metals it was observed that the ions form complexes with the nanofiber template but partially get reduced and, hence an additional heat treatment process was required for complete reduction and formation of Au and Pt nanoparticles on nanofibers surface [33].

2.3 Synthesis of ceramics/metal oxide nanofibers

Recently, electrospinning technique has been explored as an effective method and used for preparation of different 1D metal oxide and ceramics nanofibers. To synthesize ceramic nanofibers inorganic/organometallic salt precursor's solutions has been prepared and then mixed with a suitable polymer solution. Then the mixed composite solutions are electrospun into composite nanofibers followed by calcination at higher temperature to form pure ceramic nanofibers. The fabrication of ceramic nanofibers using electrospinning method follows these three major steps:

(a) Preparation of an electrospinning composite solution containing a polymer and salt precursor to form the ceramic materials.

(b) Electrospinning of the composite solution to generate composite nanofibers containing an inorganic precursor materials and polymer.
(c) Calcination of the composite nanofibers to form the desired ceramic nanofibers at an elevated temperature, by removal of polymer components from the precursor fibers.

Various research groups have prepared various metal oxides nanofibers by electrospinning method using different precursors-polymer solutions followed by calcination at higher temperature [31]. Zhang et al. have synthesized porous CeO_2 nanofibers with diameter 100–140 nm by electrospinning PVP and cerium nitrate ($Ce(NO_3)_2.6H_2O$) salt precursors solution followed by calcination at high temperature. It was observed that the solution and process parameters affecting the spinnability and nanofibers morphology [34]. Eu-doped SnO_2 nanofibers were prepared by Jiang et al. by electrospinning method. The prepared nanofibers are composed of crystallite grains with an average size of about 10 nm and Eu^{+3} ions are successfully doped into the SnO_2 lattice [35]. Similarly, pure α-Fe_2O_3 and Pt nanoparticles functionalized α-Fe_2O_3 nanowire/nanofibers were synthesized by a single step electrospinning method. The diameters of the nanofibers were found to be 100 nm. Pt doping showed remarkable effect on the morphology of the nanofibers [36]. α - Fe_2O_3 nanofibers were successfully synthesized by Ghasemi et al. by electrospinning a PVA and ferric nitrate composite solution followed by calcination in air. The formation, structure and morphology of the fibers were characterized by FT-IR, XRD, and SEM analytical techniques. The diameter of the α-Fe_2O_3 nanofibers were found to be in the range of 50–90 nm [37]. Fig. 21.10 represents the SEM micrographs of electrospun composite and α-Fe_2O_3 nanofibers.

The e-spun PVA/$Fe(NO_3)_3$ composite fibers were found to be uniform in size in the range of 150–250 nm, smooth surface with no significant bead formation. After sintering at 800°C, the PVA molecules evaporated leads to formation of α-Fe_2O_3 nanofibers. After

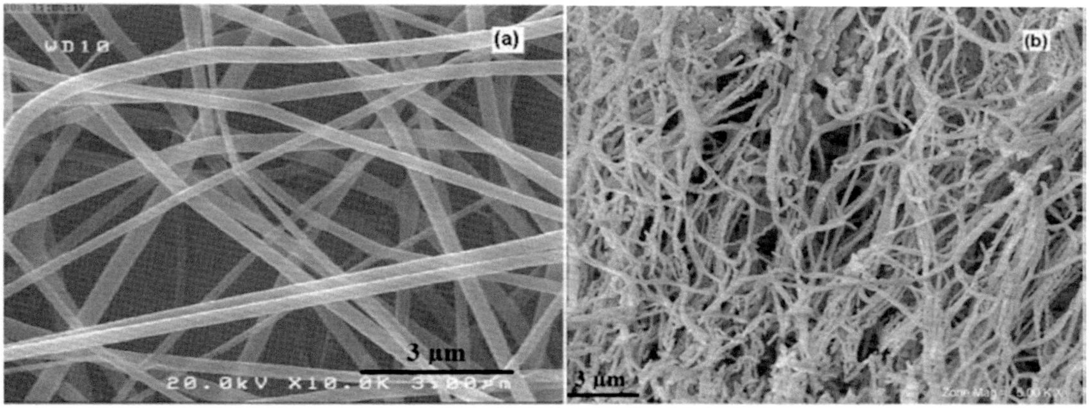

Figure 21.10
SEM images of (A) electrospun PVA/$Fe(NO_3)_3.9H_2O$ fibers and (B) α-Fe_2O_3 fibers [37].

sintering, it was observed that the fibers become bent and morphology become rough along with rupture of continuous fibers structure in axial direction. Apart from this, a reduction in fiber diameter was observed. Fe^{+3}-doped TiO_2 nanofibers were prepared by combination of sol-gel and electrospinning methods followed by heat treatment. XRD and XPS studied confirmed the formation of doped TiO_2 nanofibers and diameter of the nanofibers were found to be around 100 nm [38]. Similarly, $CdTiO_3$, $ZnWO_4$, $BiPO_4$, and ZnO/Bi_2WO_6 nanofibers were prepared by electrospinning the polymer-salt precursor's solutions followed by calcination at higher temperature. The obtained nanofibers were found to be highly porous structure and excellent photocatalyst materials [39–42]. Recently, Luo et al. have synthesized zirconia (ZrO_2) embedded carbon nanowires using electrospinning method. First they have prepared the $PVP/ZrOCl_2$ composite nanofibers using electrospinning method. Then the obtained composite nanofibers were sintered at 600°C in N_2 atmosphere to form ZrO_2 embedded carbon nanowires. The structure, formation, composition and morphology of the nanofibers were characterized by x-ray diffraction (XRD), fourier transform infrared (FTIR), Raman, x-ray photoelectron spectroscopy (XPS) and scanning electron microscope (SEM) analytical techniques [43]. In recent years, materials scientists have focused their study on the development of binary oxides nanofibers by electrospinning technique and have used for environmental remediation applications. For example, NiO–ZnO, $NiO–Al_2O_3$, ZnO–CuO, $ZnO–SnO_2$, and SnO_2–CuO heterostructured nanofibers were synthesized by electrospinning the polymer and mixed oxide salt precursors solutions followed by heat treatment. The analytical techniques such as XRD, FESEM, TEM, EDS, and XPS were used to confirm the formation of heterostructure nanofibers [44–47]. Fig. 21.11 represents the FE-SEM and EDS of NiO-γ Al_2O_3 composite ceramics nanofibers. It was observed from this figure that after calcination of $NiAC_2/AlAC_2$/PVP composite nanofibers at 1000°C, pure $NiO–Al_2O_3$ composite ceramics nanofibers were formed. The EDS study confirmed the formation composite ceramics nanofibers.

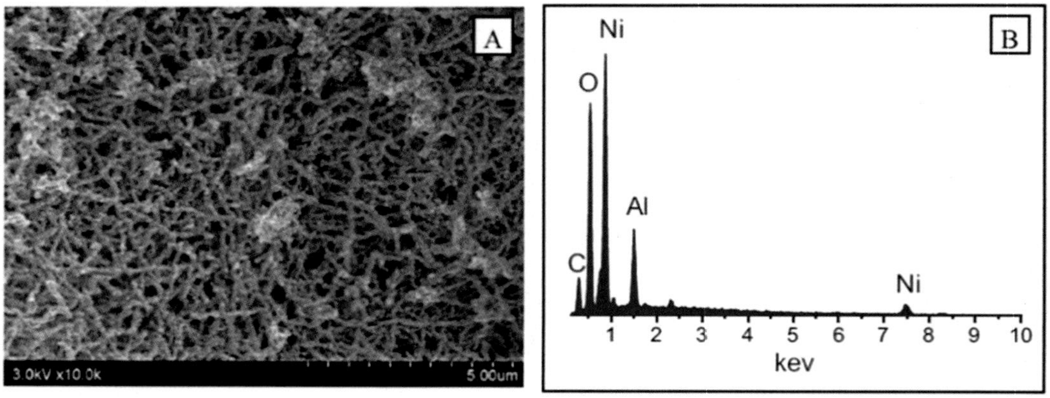

Figure 21.11
The FE-SEM image and EDS of NiO-γ Al_2O_3 composite nanofibers [45].

2.4 Environmental remediation applications of nanofibers materials

Among different traditional methods that are being used for remediation of organic and inorganic pollutants from aqueous stream, adsorption is found to be a simple and efficient method that is more widely used for environmental applications. In this regard, recent research focused on the synthesis of novel nanoadsorbents materials for efficient environmental remediation. Among different nanoadsorbents, highly porous fibers like nanostructured 1D materials have attracted much research attention due to their unique physical and chemical properties. In adsorption experimental studies batch adsorption experiments are most widely performed by varying different adsorption parameters such as effect of pH, adsorbent dose, time, initial concentration of pollutants etc. The amount of pollutant (q) adsorbed per unit mass of adsorbent and R, the dye removal efficiency can be calculated as per Eqs. (21.1) and (21.2), respectively,

$$q_e = \frac{(C_0 - C_e)V}{m} \tag{21.1}$$

$$\%R = \frac{(C_0 - C_t)}{C_0} \times 100 \tag{21.2}$$

Where, C_0 and C_e represent initial and equilibrium concentration of pollutant in solution (mg/L), V is the volume of the solution in liter (L) and C_t is the concentration at time t, m is the mass of the adsorbent (g).

2.4.1 Adsorption of inorganic pollutants

More recently, Malik et al. have developed PAN/magnetite composite nanofibers and have used as adsorbent for the removal of highly toxic lead (Pb^{2+}) ions from aqueous system. Most commonly, lead enter into the water bodies from discharges of industrial effluents by many industrial activities which include lead smelting, lead battery plants, lead paint pigments and the mining activities. The major health problems caused due to the lead contamination are kidney failure, damage to central nervous system, malfunctioning of other organs including the brain, and also effect on cellular processes. The PAN/magnetite composite nanofibers mat was synthesized by simple electrospinning process followed chemical precipitation of Fe_3O_4. The obtained composite nanofibers membrane was examined for the removal of Pb^{2+} ions from water media and found to be an efficient nanofibrous composite adsorbent with a maximum adsorption capacity of 156.25 mg/g [48].

Similarly, Li et al. have reported on the fabrication of magnetic nonwoven composite nanofibers by blending Fe_3O_4 nanoparticles and halloysite (Hal) nanotubes into polyethylene oxide/chitosan (PEO/CS) polymer solution via electrospinning method. Then the prepared magnetic nonwoven composite nanofibers have been used as adsorbent for the removal of heavy metal ions such as Cr(VI), Cd(II), Cu(II), and Pb(II) from the aqueous system. They have found that the composite nanofibers revealed high removal efficiency toward removal of different heavy metal ions such as cadmium, copper, lead, and chromium, and the adsorption capacity was observed to be in the order of Cr(VI) < Cd(II) < Cu(II) < Pb(II). The effect of counterions and reusability of composite membranes nanoadsorbent for the removal of different heavy metal ions were also determined, which indicated that the electrospun hybrid nanofibers adsorbents were widely adaptable and reusable [49].

Luo et al. have synthesized ZrO_2 embedded carbon nanowires by electrospinning method followed by calcination under N_2 atmosphere. The prepared electrospun ZrO_2-embedded nanowires (ZCNs) were used as adsorbent for adsorption of both arsenite {As(III)} and arsenate {As(V)} ions from aqueous media. The maximum adsorption capacity value of As(III) and As(V) onto ZCNs were found to be 28.6 and 106.56 mg/g at 40°C. Furthermore, it was observed that ZCNs catalyze the oxidation of As(III) to As (V) using dissolved oxygen [43]. A high adsorption capacity was also achieved in a wide pH range of 5−11 for As (III) and pH range 5 to 9 for As (V) ions. Adsorption of As ions onto ZCNs was found to be endothermic in nature and followed chemisorption mechanism. They have also studied the reusability and stability of ZCNs adsorbent materials. It was observed that the ZrO_2 embedded electrospun nanowires can be reused up to five successive cycles without major loss in its capacity.

Recently, Qiu et al. have synthesized flower shaped MoS_2—polyaniline (PANI) core-shell nanocomposite immobilized on PAN nanofibers by hydrothermal, in situ polymerization process, and electrospinning method. The prepared nanocomposite membrane was utilized as nanoadsorbent for removal of Cr(VI) ions from water. It was observed that the MoS_2@PANI/PAN nanocomposite fibers showed a high adsorption capacity of 6.57 mmol/g with good recyclability. The mechanism of Cr(VI) adsorption by composite MoS_2@PANI/PAN electrospun nanofibers was due to electrostatic force of attraction, ion exchange, and reduction of Cr(VI) to Cr(III) ions and is represented in Fig. 21.12. It was found that the surface charge of the MoS_2 nanoparticles and the surface functional group of e-spun nanofibers play important role in remediation of Cr(VI) ions. Furthermore, the prepared e-spun composite nanofibers can be recycled and reused successfully for more than six cycles without major loss in its adsorption capacity. The development of electrospun flower shaped MoS_2@PANI core-shell nanocomposite on PAN nanofibers surface was found to be a green, economic and convenient process and was used as adsorbent for effective removal toxic Cr(VI) ions from aqueous environment [50].

More recently, Min et al. have fabricated novel iron doped chitosan e-spun nanofibers mat (Fe@CTS) by electrospinning method. They have blended Fe^{+3} ions to chitosan/PEO solution prior to electrospinning. Then the as-spun composite nanofibers mat were exposed

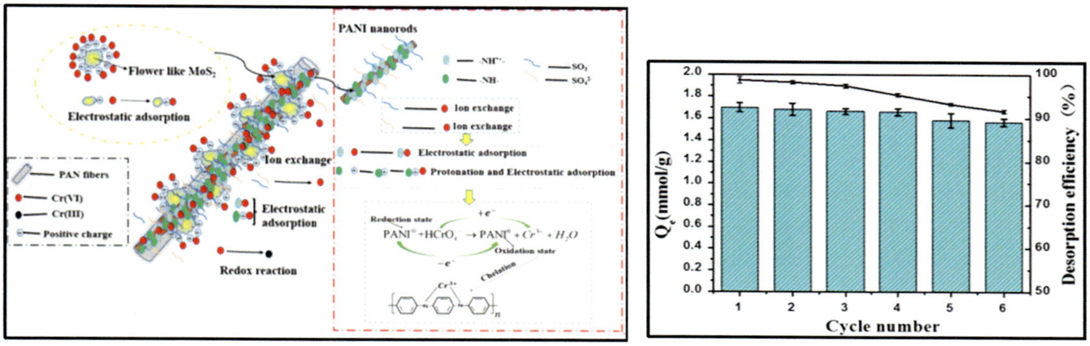

Figure 21.12
Adsorption mechanism of Cr(VI) and reusability of MoS_2@PANI/PAN nanocomposite fibers [50].

to ammonia vapor followed by washing thoroughly with deionized water to remove water soluble PEO polymer which leads to form Fe@CTS nanofibers mat. The Fe@CTS nanofibers mat was found to be very effective toward adsorption of arsenite As(III) ions from water. Batch adsorption studies were conducted to examine the removal efficiency of As(III) ions by tuning various process parameters. The maximum uptake capacity was found to be 36.1 mg/g. XPS studies revealed that C—OH, C—O—C, C—N, and Fe—O groups of Fe@CTS nanofibers mat play an important role in As(III) adsorption. It was also noticed that the prepared Fe doped chitosan nanofibers mat was very efficient, low cost and reusable As(III) adsorbent material [51]. Similarly, Zhao, et al. have synthesized inorganic/organic composite nanofibers and have investigated extensively as adsorbent for waste water treatment. They have reported on the synthesis of b-PEI grafted PAN nanofibers mounted with Fe_3O_4 nanoparticles (b-PEIFePAN) by electrospinning method and have explored the removal of hexavalent chromium ions. The b-PEI embedded PAN nanofiber, not only act as a flexible support to mount Fe_3O_4 nanoparticles but also help in enhancing the adsorption efficiency. The maximum adsorption capacity of the electrospun inorganic-organic nanofibers mat toward Cr(VI) removal was found to be 685 mg/g and was observed to be higher than that of other reported adsorbents [29]. Both electrostatic attraction and reduction actions were found to play important role toward the mechanism of Cr(VI) adsorption. Beheshti and coworkers also synthesized chitosan/MWCNT/Fe_3O_4 composite electrospun nanofibers by electrospinning technique and have been utilized as nanoadsorbent for removal of Cr(VI) from aqueous media. The obtained results showed that chitosan/MWCNT/Fe_3O_4 has high potential toward decontamination of Cr(VI) from water and wastewater system [52].

2.4.2 Adsorption of organic pollutants

Recently, Patel et al. have used three different methods to synthesize PAN/iron oxide composite nanofibers membrane and studied their adsorption behavior toward the removal of CR dye from aqueous media. In the first method, they have prepared PAN/iron acetylacetonate solution using DMF solvent and then electrospun into composite nanofibers. For in situ formation of iron oxide nanoparticles on the of PAN nanofibers surface the prepared e-spun composite nanofibers membrane was hydrothermally treated in an autoclave at 150°C for 12 h. The obtained PAN-iron oxide composite nanofibers membrane was designated as PAN-IO (H). In the second method, iron oxide nanoparticles have been prepared previously and then mixed with PAN-DMF homogeneous polymer solution followed by electrospinning to prepare nanocomposite fibers. The obtained composite nanofibers membrane was represented as PAN-IO (B). In the last method, in order to decorate iron oxide nanoparticles on the PAN nanofibers surface, the previously prepared e-spun PAN nanofibers mats were immersed into IA solution for 2 h, then transfer into an autoclave and treated hydrothermally at 80°C for 1 h. This iron oxide decorated PAN polymer nanofibers membrane was designated as PAN-IO (A). The prepared different functional nanofibers mats were examined as hybrid nanoadsorbent for adsorption of CR dye from water media and their adsorption capabilities have been evaluated.

The adsorption capabilities of functional PAN nanofibers mats prepared by hydrothermal and alkoxide methods {PAN-IO (H) and PAN-IO (A)} were observed to be twice than that of functional nanofibers mat prepared by blended method {PAN-IO (B)}. The proposed mechanism of in situ formation and development of nanosized iron oxide particles on the PAN nanofibers surface for PAN-IO (H) and PAN-IO (A) is shown in Fig. 21.13. The high equilibrium adsorption capacities of PAN-IO (H) and PAN-IO (A) composite nanofibers mats adsorbents toward the remediation of CR dye can be explained as below. The point of zero charge (PZC) value of prepared iron oxide nanoparticles with particle size 12−100 nm is 7.8−8.8. Therefore, at pH = 7, pH below the PZC, the surfaces of nanosized iron oxide particles are positively charged and hence electrostatic force of attraction exist between the positively charged iron oxide surface and negatively charged CR dye molecules. Therefore, the adsorption efficacies of functional PAN-IO (H) and PAN-IO (A) composite nanofibers mats are higher as compared to that of the iron oxide blended PAN-IO (B) hybrid nanofibers and unfunctionalized PAN nanofibers mat [53].

Very recently, Fard et al. have prepared hydroxylated α-Fe_2O_3 nanofibers in the combination of electrospinning technique along with in-situ polymerization process. First α-Fe_2O_3 nanofibers were prepared. Then vinyl acetate monomer was used to form polyvinylacetate (PVAc) polymer on the surface of α-Fe_2O_3 nanofibers. After that the PVAc polymer was hydrolyzed in an alkaline solution to convert the acetate functional groups into hydroxyl groups. The effectiveness of the prepared nanofibers membrane was examined for the adsorption studies of basic dyes such as BR 18, BR 46, and BB 41. It was observed that the maximum sorption capacities of these basic dyes occurred at pH = 8.5. The adsorption capacity of the functional α-Fe_2O_3 nanofibers toward basic dyes

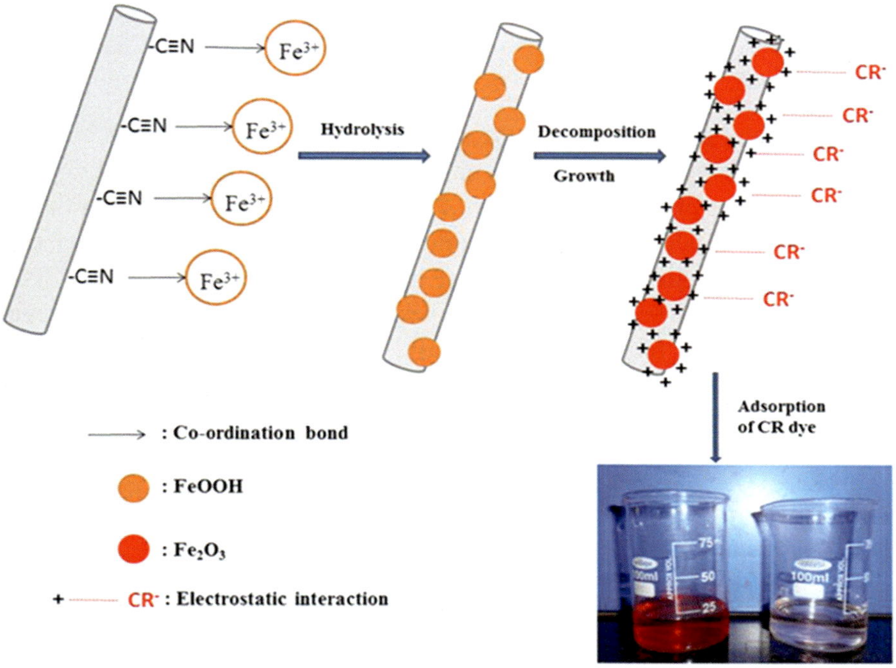

Figure 21.13
Schematic representations of iron oxide nanoparticles grown on the surface of PAN nanofibers and mechanism of CR dye removal [53].

removal significantly increases from 5.05 to 765.51, 10.28 to 825.34 and 6.85–789.42 mg/g for BR 18, BR 46, and BB 41, respectively, with the increase in pH value from 2.1 to 8.5. At alkaline pH the large number of hydroxyl groups present on the surface of hydroxylated α-Fe_2O_3 nanofibers turns to negatively charged surface. Therefore, strong electrostatic interaction occurred among the negatively charged surface of the functionalized α-Fe_2O_3 nanofibers adsorbent and the positively charged basic dye molecules [54].

Similarly, Peng et al. have prepared hollow and γ-Al_2O_3 nanofibers by the electrospinning of aluminum nitrate ($Al(NO_3)_3$)/PAN precursor solution, followed by calcination at 800°C for 2h with a heating rate of 5°C/min. Then the obtained γ-Al_2O_3 hollow nanofibers were used as adsorbent to eradicate three model dyes pollutants such as Congo red, Methyl blue, and acid fuchsine from aqueous solution. The porous hollow γ-Al_2O_3 nanofibers exhibited excellent dye adsorption efficiency with more than 90% removal. They have also carried out a comparative study performing the same experimental condition toward the removal of CR dye by using γ-Al_2O_3 powder. The efficiency of γ-Al_2O_3 powder for CR dye adsorption was much lower than that of the prepared e-spun porous and hollow γ-Al_2O_3 nanofibers. After 60 min of contact time, only about 12.3% of CR dye was removed from the water with an equilibrium adsorption capacity of 2.55 mg/g [55].

In a recent report, Tang et al. 2018 have synthesized SiO_2@ZrO_2 nanofibers mats by electrospinning technique in combination with the impregnation method followed by calcination in vacuum or mixture of hydrogen and nitrogen atmosphere. Then the prepared nanofiber membrane has been used as an adsorbent for titan yellow dye removal. It was seen that 99.9% of titan yellow solution was adsorbed by positively charged SiO_2@ZrO_2 nanofibers membrane at pH less than 6.7. The maximum adsorption capacity of coaxial SiO_2@ZrO_2 nanofibers was found to be 63.27 mg/g [56]. Zhang et al. 2016 have studied the fabrication of e-spun porous CeO_2 nanofibers by using the $Ce(NO_3)_3.6H_2O$/PVP precursor solution for electrospinning and then sintered at high temperature. The obtained porous CeO_2 nanofibers have been used as an adsorbent for the removal of MO. From the adsorption experiment it is seen that MO can be removed by the porous nanofibers rapidly and effectively. The rapid adsorption ability is ascribed due to their large surface area [34]. In 2015, Liu and coworkers have prepared Fe_3O_4 incorporated PAN nanofibers mat (Fe-NFM) by electrospinning followed by solvothermal treatment and targeted to remove tetracycline (TC), a typical class of antibiotic from aqueous solution. The adsorption data fitted well with Langmuir isotherm model with a maximum adsorption capacity of 315.31 mg/g. Electrostatic interaction and complexation between TC and Fe-NFM both played important roles in adsorption of TC [28].

3. Concluding remarks

This chapter summarizes the recent development of electrospun composite polymer nanofibers particularly in situ functionalization of nanomaterials in polymer nanofibers matrix and also development ceramics/metal oxide nanofibers. These nanofiber materials have been prepared by using simple and more convenient electrospinning method. It is also possible for large-scale generation of novel inorganic nanoparticles functionalized nanofibrous materials using this technique. The adsorption behavior of electrospun polymer composite and ceramics nanofibers toward removal of inorganic pollutants such as As, Cr, Cd, Hg ions, and organic pollutants such as synthetic dyes like CR, MO, MG, and antibiotics from water has been highlighted. The functionalized nanofibrous materials have a great application potential for development of highly porous, low cost and high surface area novel adsorbent materials. Therefore, extensive studies are highly required to generate the novel and highly efficient nanoporous functional adsorbent materials which can be used as next generation nanoadsorbents to solve the modern growing environmental problems.

Acknowledgment

The authors would like to acknowledge NIT Rourkela, Odisha for providing necessary research and lab facilties.

References

[1] World Health Organization (WHO), Guidelines for Drinking-Water Quality: Incorporating 1st and 2nd Addenda, third ed., WHO Press, Geneva, 2008.
[2] R.K. Sadasivam, S. Mohiyuddin, G. Packirisamy, Electrospun polyacrylonitrile (PAN) templated 2D nanofibrous mats: a platform towards practical application for dye removal and bacterial disinfection, ACS Omega 2 (2017) 6556–6569.
[3] R. Xing, W. Wang, T. Jiao, K. Ma, Q. Zhang, W. Hong, H. Qiu, J. Zhou, L. Zhang, P. Qiuming, Bioinspired polydopamine sheathed nanofibers containing carboxylate graphene oxide nanosheet for high-efficient dyes scavenger, ACS Sustain. Chem. Eng. 5 (2017) 4948–4956.
[4] G. Ma, Y.M. Zhu, Z.Q. Zhang, L.C. Li, Preparation and characterization of multiwalled carbon nanotube/ TiO_2 composites: decontamination of organic pollutant in water, Appl. Surf. Sci. 21 (2014) 811–817.
[5] S. Wang, E. Ariyanto, Competitive adsorption of malachite green and Pb ions on natural zeolite, J. Colloid Interface Sci. 314 (2007) 25–31.
[6] Y. Xie, B. Yan, H. Xu, J. Chen, Q. Liu, Y. Deng, H. Zeng, Highly regenerable mussel-inspired Fe_3O_4@polydopamine-ag core-shell microspheres as catalyst and adsorbent for methylene blue removal, ACS Appl. Mater. Interfaces 6 (2014) 8845–8852.
[7] A. Mahapatra, B.G. Mishra, G. Hota, Adsorptive removal of Congo red dye from wastewater by mixed iron oxide – alumina nanocomposites, Ceram. Int. 39 (2013) 5443–5451.
[8] G.R. Xu, J.N. Wang, C.J. Li, Preparation of hierarchically nanofibrous membrane and its high adaptability in hexavalent chromium removal from water, Chem. Eng. J. 198–199 (2012) 310–317.
[9] C.J. Li, Y.J. Li, J.N. Wang, J. Cheng, PA6@Fe_xO_y nanofibrous membrane preparation and its strong Cr (VI)-removal performance, Chem. Eng. J. 220 (2013) 294–301.
[10] Q. Liu, Y. Zheng, L. Zhong, X. Cheng, Removal of tetracycline from aqueous solution by Fe_3O_4 incorporated PAN electrospun nanofiber mat, J. Environ. Sci. 28 (2015) 29–36.
[11] H.J. Kim, H.R. Pant, C.H. Park, L.D. Tijing, N.J. Choi, C.S. Kim, Hydrothermal growth of mop-brush-shaped ZnO rods on the surface of electrospun nylon-6 nanofibers, Ceram. Int. 39 (2013) 3095–3102.
[12] F. Pan, Q. Cheng, H. Jia, Z. Jiang, Facile approach to polymer-inorganic nanocomposite membrane through a biomineralization-inspired process, J. Membr. Sci. 357 (2010) 171–177.
[13] J. Zhu, S. Wei, X. Chen, A.B. Karki, D. Rutman, D.P. Young, Z. Guo, Electrospun polyimide nanocomposite fibers reinforced with core-shell fe-feo nanoparticles, J. Phys. Chem. C 114 (2010) 8844–8850.
[14] M. Khosravi, S. Azizian, Adsorption of anionic dyes from aqueous solution by iron oxide nanospheres, J. Ind. Eng. Chem. 20 (2014) 2561–2567.
[15] I.S. Chronakis, Novel nanocomposites and nanoceramics based on polymer nanofibers using electrospinning process - a review, J. Mater. Process. Technol. 167 (2005) 283–293.
[16] N. Bhardwaj, S.C. Kundu, Electrospinning: a fascinating fiber fabrication technique, Biotechnol. Adv. 28 (2010) 325–347.
[17] S. Ramakrishna, K. Fujihara, W.E. Teo, T.C. Lim, Z. Ma, Introduction to Electrospinning and Nanofibers, 2005, https://doi.org/10.1142/5894.
[18] Z.M. Huang, Y.Z. Zhang, M. Kotaki, S. Ramakrishna, A review on polymer nanofibers by electrospinning and their applications in nanocomposites, Compos. Sci. Technol. 63 (2003) 2223–2253.
[19] X. Shi, W. Zhou, D. Ma, Q. Ma, D. Bridges, Y. Ma, A. Hu, Electrospinning of nanofibers and their applications for energy devices, J. Nanomater. 2015 (2015) 1–20.

[20] J. Deitzel, J. Kleinmeyer, D. Harris, N.B. Tan, The effect of processing variables on the morphology of electrospun nanofibers and textiles, Polymer 42 (2001) 261–272.
[21] M.G. McKee, G.L. Wilkes, R.H. Colby, T.E. Long, Correlations of solution rheology with electrospun fiber formation of linear and branched polyesters, Macromolecules 37 (2004) 1760–1767.
[22] H. Fong, I. Chun, D.H. Reneker, Beaded nanofibers formed during electrospinning, Polymer 40 (1999) 4585–4592.
[23] X.Y. Yuan, Y.Y. Zhang, C. Dong, J. Sheng, Morphology of ultrafine polysulfone fibers prepared by electrospinning, Polym. Int. 53 (2004) 1704–1710.
[24] D. Zhang, A.B. Karki, D. Rutman, D.P. Young, A. Wang, D. Cocke, T.H. Ho, Z. Guo, Electrospun polyacrylonitrile nanocomposite fibers reinforced with Fe_3O_4 nanoparticles: fabrication and property analysis, Polymer 50 (2009) 4189–4198.
[25] R. Nirmala, D. Kalpana, R. Navamathavan, Y.S. Lee, H.Y. Kim, Preparation and characterizations of silver incorporated polyurethane composite nanofibers via electrospinning for biomedical applications, J. Nanosci. Nanotechnol. 13 (2013) 4686–4693.
[26] Y.E. Miao, R. Wang, D. Chen, Z. Liu, T. Liu, Electrospun self-standing membrane of hierarchial SiO_2@γ-AlOOH (bohemite) core/sheath fibers for water remediation, ACS Appl. Mater. Interfaces 4 (2012) 5353–5359.
[27] P. Veerakumar, L.P. Muthuselvam, C.T. Hung, K.C. Lin, F.C. Chou, S.B. Liu, Biomass derived activated carbon supported Fe_3O_4 nanoparticles as recyclable catalysts for reduction of nitro arenes, ACS Sustain. Chem. Eng. 4 (2016) 6772–6782.
[28] Q. Liu, L. Bin Zhong, Q.B. Zhao, C. Frear, Y.M. Zheng, Synthesis of Fe_3O_4/polyacrylonitrile composite electrospun nanofiber mat for effective adsorption of tetracycline, ACS Appl. Mater. Interfaces 7 (2015) 14573–14583.
[29] R. Zhao, X. Li, Y. Li, Y. Li, B. Sun, N. Zhang, S. Chao, C. Wang, Functionalized magnetic iron oxide/polyacrylonitrile composite electrospun fibers as effective chromium (VI) adsorbents for water purification, J. Colloid and Interface Sci. 505 (2017) 1018–1030.
[30] A. Mahapatra, N. Garg, B.P. Nayak, B.G. Mishra, G. Hota, Studies on the synthesis of electrospun PAN-Ag composite nanofibers for antibacterial application, J. Appl. Polym. Sci. 124 (2012) 1178–1185.
[31] G. Panthi, S. Park, S. Chae, T. Kim, H. Chung, S. Hong, M. Park, H. Kim, Immobilization of Ag_3PO_4 nanoparticles on electrospun PAN nano fi bers via surface oximation: bifunctional composite membrane with enhanced photocatalytic and antimicrobial activities, J. Ind. Eng. Chem. 45 (2017) 277–286.
[32] M.S. Hassan, T. Amna, F.A. Sheikh, S.S. Al-Deyab, K. Eun Choi, I.H.H. wang, M.S. Khil, Bimetallic Zn/Ag doped polyurethane spider net composite nanofibers: a novel multipurpose electrospun mat, Ceram. Int. 39 (2013) 2503–2510.
[33] H.Y. Son, J.H. Ryu, H. Lee, Y.S. Nam, Bioinspired templating synthesis of metal-polymer hybrid nanostructures within 3D electrospun nanofibers, ACS Appl. Mater. Interfaces 5 (2013) 6381–6390.
[34] Y. Zhang, R. Shi, P. Yang, X. Song, Y. Zhu, Q. Ma, Fabrication of electrospun porous CeO_2 nanofibers with large surface area for pollutants removal, Ceram. Int. 42 (2016) 14028–14035.
[35] Z. Jiang, R. Zhao, B. Sun, G. Nie, H. Ji, J. Lei, C. Wang, Highly sensitive acetone sensor based on Eu-doped SnO_2 electrospun nanofibers, Ceram. Int. 42 (2016) 15881–15888.
[36] L. Guo, N. Xie, C. Wang, X. Kou, M. Ding, H. Zhang, Y. Sun, H. Song, Y. Wang, G. Lu, Enhanced hydrogen sulfide sensing properties of Pt-functionalized α-Fe_2O_3 nanowires prepared by one-step electrospinning, Sensor. Actuator. B 255 (2018) 1015–1023.
[37] E. Ghasemi, H. Ziyadi, A.M. Afshar, M. Sillanpää, Iron oxide nanofibers: a new magnetic catalyst for azo dyes degradation in aqueous solution, Chem. Eng. J. 264 (2015) 146–151.
[38] J. Shen, Y. Wu, L. Fu, B. Zhang, F. Li, Preparation of doped TiO2 nanofiber membranes through electrospinning and their application for photocatalytic degradation of malachite green, J. Mater. Sci. 49 (2014) 2303–2314.
[39] M.S. Hassan, T. Amna, M.S. Khil, Synthesis of high aspect ratio $CdTiO_3$ nanofibers via electrospinning: characterization and photocatalytic activity, Ceram. Int. 40 (2014) 423–427.

[40] J. Lu, M. Liu, S. Zhou, X. Zhou, Y. Yang, Electrospinning fabrication of $ZnWO_4$ nanofibers and photocatalytic performance for organic dye, Dyes Pigments 136 (2017) 1–7.
[41] G. Liu, S. Liu, Q. Lu, H. Sun, Z. Xiu, Synthesis of mesoporous $BiPO_4$ nanofibers by electrospinning with enhanced photocatalytic performances, Ind. Eng. Chem. Res. 53 (2014) 13023–13029.
[42] X. Liu, Q. Lu, J. Liu, Electrospinning preparation of one dimensional $ZnO/BiWO_6$ heterostructured submicrobelts with excellent photocatalytic performance, J. Alloys Compd. 662 (2016) 598–606.
[43] J. Luo, X. Luo, C. Hu, J.C. Crittenden, J. Qu, Zirconia (ZrO_2) embedded via electrospinning for efficient as removal from water combined with DFT studies, ACS Appl. Mater. Interfaces 8 (2016) 18912–18921.
[44] P.P. Dorneanu, A. Airinei, N. Olaru, M. Homocianu, V. Nica, F. Doroftei, Preparation and characterization of NiO, ZnO and NiO-ZnO composite nanofibers by electrospinning method, Mater. Chem. Phys. 148 (2014) 1029–1035.
[45] B. Li, H. Yuan, P. Yang, B. Yi, Y. Zhang, Fabrication of the composite nanofibers of NiO/γ-Al_2O_3 for potential application in photocatalysis, Ceram. Int. 42 (2016) 17405–17409.
[46] A. Naseri, M. Samadi, N.M. Mahmoodi, A. Pourjavadi, H. Mehdipour, A.Z. Moshfegh, Tuning composition of electrospun ZnO/CuO nanofibers: toward controllable and efficient solar photocatalytic degradation of organic pollutants, J. Phys. Chem. C 121 (2017) 3327–3338.
[47] P. Pascariu, A. Airinei, N. Olaru, L. Olaru, V. Nica, Photocatalytic degradation of Rhodamine B dye using ZnO-SnO_2 electrospun ceramic nanofibers, Ceram. Int. 42 (2016) 6775–6781.
[48] H. Malik, U.A. Quereshi, M. Muqeet, R.B. Mahar, F. Ahmed, Z. Khatri, Removal of lead from aqueous solution using polyacrylonitrile/magnetite nanofibers, Environ. Sci. Pollut. Res. 25 (2018) 3557–3564.
[49] L. Li, F. Wang, Y. Lv, J. Liu, D. Zhang, Z. Shao, Halloysite nanotubes and Fe_3O_4 nanoparticles enhanced adsorption removal of heavy metal using electrospun membranes, Appl. Clay Sci. 161 (2018) 225–234.
[50] J. Qiu, F. Liu, S. Cheng, L. Zong, C. Zhu, C. Ling, A. Li, Recyclable nanocomposite of flowerlike MoS_2@Hybrid acid-doped PANI immobilized on porous PAN nanofibers for the efficient removal of Cr(VI), ACS Sustain. Chem. Eng. 6 (2018) 447–456.
[51] L.-L. Min, L.M. Yang, R.X. Wu, L.B. Zhong, Z. Yuan, Y.M. Zheng, Enhanced removal of arsenite from aqueous solution by an iron doped electrospun chitosan nanofibers mat: preparation, characterization and performance, J. Colloid Interface Sci. 535 (2019) 255–264.
[52] H. Beheshti, M. Irani, L. Hosseini, A. Rahimi, M. Aliabadi, Removal of Cr(VI) from aqueous solutions using chitosan/MWCNT/Fe_3O_4 composite nanofibers-batch and column studies, Chem. Eng. J. 284 (2016) 557–564.
[53] S. Patel, G. Hota, Iron oxide nanoparticles-immobilized Pan nanofibers: synthesis and adsorption studies, RSC Adv. 6 (2016) 15402–15414.
[54] G.C. Fard, M. Mirjalili, F. Najafi, Hydroxylated α-Fe_2O_3 nanofibers: optimization of synthesis conditions, anionic dyes adsorption kinetic, isotherm and error analysis, J Taiwan Inst. Chem. Eng. 70 (2017) 188–199.
[55] C. Peng, J. Zhang, Z. Xiong, B. Zhao, P. Liu, Fabrication of porous hollow γ-Al_2O_3 nanofibers by facile electrospinning and its application for water remediation, Microporous Mesoporous Mater. 215 (2015) 133–142.
[56] Y. Tang, Z. Liu, K. Zhao, S. Fu, Positively charged and flexible Sio_2@ZrO_2 nanofibrous membranes and their application in adsorption and separation, RSC Adv. 8 (2018) 13018–13025.

Realization of relaxor PMN-PT thin films using pulsed laser ablation

Pius Augustine[1,2]
[1]Department of Physics, Sacred Heart College (Autonomous), Thevara, Kochi, India; [2]Material Research Centre, Indian Institute of Science, Bangalore, Karnataka, India

Chapter Outline

1. Introduction

> *"Career in the discovery of new science and new materials is a most rewarding and appreciated life work."*

A beautiful message from Prof. Mildred Dresselhous (late) of MIT, USA, a pioneer in material science research, is an inspiration for any student of material science. Fabrication and stabilization of potential materials, and their utilization for the progress of mankind to make the planet a better place to live in, is the prime responsibility of every material scientist.

Design, Fabrication, and Characterization of Multifunctional Nanomaterials. https://doi.org/10.1016/B978-0-12-820558-7.00012-1

Physics, the study of energy, has immensely contributed to make the life on earth easier. From time immemorial, man is searching for new and renewable energy resources. Generation of electricity from vibrations is an example for renewable energy resource, which opened a new arena in physics, i.e., piezoelectricity. The term piezoelectricity was coined from the word *piezein* meaning to squeeze, and electricity, and essentially means pressure electricity. After the discovery of piezoelectricity by Curie brothers in 1880, its avenues were analyzed and explored by scientists [1]. Natural piezoelectric materials were used in the World War I for sensing applications. But the acute shortage of natural piezoelectric materials encouraged scientist to search for artificial piezoelectric materials, which led to the development of a new branch of science, viz., "ferroelectrics," and further inquiry led to the discovery of a new class of ferroelectrics called relaxor ferroelectric materials, which exhibits giant piezoelectric and dielectric properties. It is the dielectric behavior that distinguishes relaxor ferroelectrics from normal ferroelectrics. While relaxors exhibit a temperature as well as field dependent dielectric spectrum, normal ferroelectric spectrum is only temperature dependent [2].

Exceptional ferroelectric and piezoelectric response of lead based piezoceramics, make them excellent candidates for piezoelectric applications like sensors, actuators, and so forth. Recent years witnessed a 9% annual growth rate in the multibillion dollar piezoelectric industry in which quartz and PZT command a major share [3]. Although, lead zirconate titanate (PZT) (1 mm plate) is having very high piezoelectric coefficient d_{33}, strain developed with application of 1000 V is less than 0.1%. The market demand for high strain piezoelectric actuators fueled intense research for identifying suitable alternatives for PZT, having high d_{33} coefficient. (1-x) $[Pb(Mg_{1/3}Nb_{2/3})O_3] - (x)[PbTiO_3]$ (PMN-PT), a relaxor ferroelectric material having exceptional dielectric and piezeoelectric response, was identified as a possible and better alternative to PZT for piezoelectric applications [4,5].

Though PMN-PT is a useful material in terms of its unusual electric response, even 60 years after its discovery, this material is not still widely utilized in the piezoelectric industry owing to its structural and polar disorders and stoichiometric complexity. Synthesis of pure perovskite PMN-PT without secondary paraelectric pyrochlore phase and inconsistency in properties of the synthesized ceramic repel the researchers from this material, which impedes its march ahead in the piezoelectric industry at large.

A high temperature stabilized and device worthy PMN-PT ceramic having excellent properties could be realized using a novel partial covering method combined with modulated heating. A detailed account of the synthesis difficulties and attempts made to overcome the constraints and to realize device worthy PMN-PT ceramic having excellent and consistent response is the major focus point of this chapter. Quality of the ceramic target is very important in the synthesis of device worthy PMN-PT thin films through pulsed laser ablation.

Realization of phase-pure PMN-PT ceramic is an art as it is science.

2. Hurdles in the synthesis of PMN-PT ceramic

One of the major difficulty with synthesis of PMN-PT like lead based relaxor systems is the formation of secondary paraelectric pyrochlore phase during the synthesis of PMN-PT which kills the expected electrical response by the system. Compounds having general formula $A_2B_2O_7$ (A and B metals) exhibit a structure similar to pyrochlore (NaCa) (NbTa) $O_6F/(OH)$. Because of their paraelectric character, their presence in the PMN-PT will deteriorate its relaxor characteristics [5].

Formation of three types of pyrochlore phases with different crystal symmetry have been reported in the synthesis of PMN-PT ceramic. They are

(1) Stoichiometric pyrochlore $Pb_2Nb_2O_7$.
(2) B-site deficient lead rich pyrochlore $Pb_3Nb_2O_8$ (P3N2, JCPDS-ICDD No. 83–1959).
(3) A-site (Pb) deficient cubic pyrochlore $Pb_3Nb_4O_{13}$ (P3N4, JCPDS-ICDD No. 0443).

B-site deficient lead rich pyrochlore, which is formed during calcination at lower temperature will react with PbO and be transformed to perovskite PMN-PT at higher calcination temperature. But A-site (Pb) deficient cubic pyrochlore is formed due to lead loss and will continue in the perovskite as parasitic secondary phase, which will deteriorate the quality of PMN-PT, and make it unsuitable for applications [6]. Assuming the direct reaction between PbO and Nb_2O_5 as the reason for the formation of paraelectric pyrochlore phase, for the synthesis of PMN based systems, a more general columbite B-site precursor method was put forth by Ref. [5]; which arrest the formation of parasitic and paraelectric pyrochlore phase.

Though the mechanism of pyrochlore phase formation in PMN-PT is not completely understood, various studies indicate that the parasitic pyrochlore phases are easily formed at low temperatures and need high temperature for their conversion to PMN. Prominent XRD peaks and their relative intensities of $Pb_3Nb_4O_{13}$ pyrochlore are shown in Fig. 22.1 (card number: 00-023-0352). Intense peak at around $2\theta = 29.2°$ is the major indication of its presence, if formed during the synthesis of PMN-PT ceramic and thin films.

3. Bulk ceramics synthesis: solid-state reaction

Stoichiometric proportion of the powder precursors may be finely mixed through mechanical grinding (Agate mortar and pestle) followed by high temperature heat treatment for the reaction to take place at an appreciable rate. Thermodynamics (free energy change involved) and kinetic factors (rate at which reaction occurs) decide the occurrence of the reaction. Thorough grinding ensures homogeneous mixing of very fine particles which enhance the proximity of the particles and rate of reaction. Enhanced surface to volume ratio realized through reduced particle size will favor the occurrence of

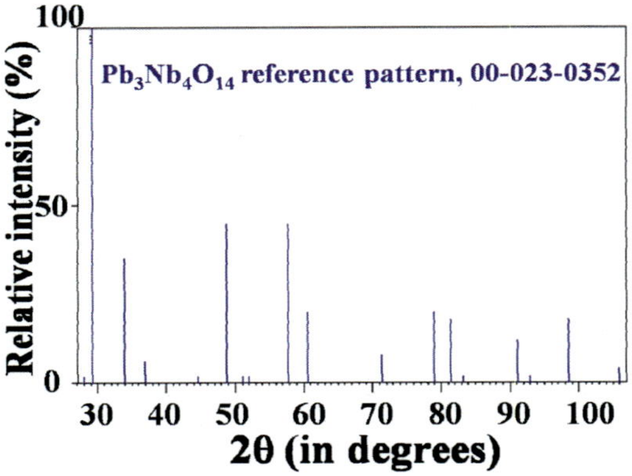

Figure 22.1
Depicts the prominent peak positions in the XRD pattern of cubic pyrochlore $Pb_3Nb_4O_{13}$ (card number: 00-023-0352) and their relative intensities.

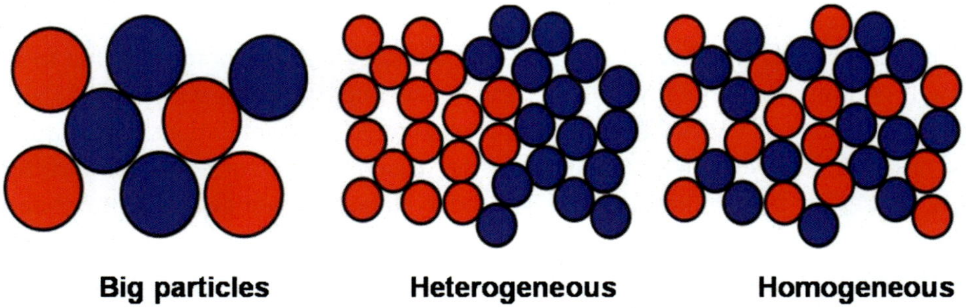

Figure 22.2
Kind of mixing during grinding.

the solid-state reaction in general; and preparation of lead based samples in particular. Solid-state reaction takes place at the interface of the solids and proceeds as the reactants diffuse from the bulk to the interface, which is favored at higher temperature. Reduced atomic diffusion distance due to fine particle size prevents fluctuation in stoichiometry during calcination and alleviates compositional nonhomogeneity prior to heat treatment (Fig. 22.2). As a rule of thumb, about two-thirds of the average melting temperatures of the solids and enhanced surface of contact expedite the solid-state reaction. Solid-state reaction proceeds in three steps: viz. (1) transport of matter to the interface, (2) reaction at the interface, and (3) transfer of matter away from the interface.

Sintering also called firing or densification is the process by which the compacted green body of the calcined and ground ceramic powder will be administered a high temperature

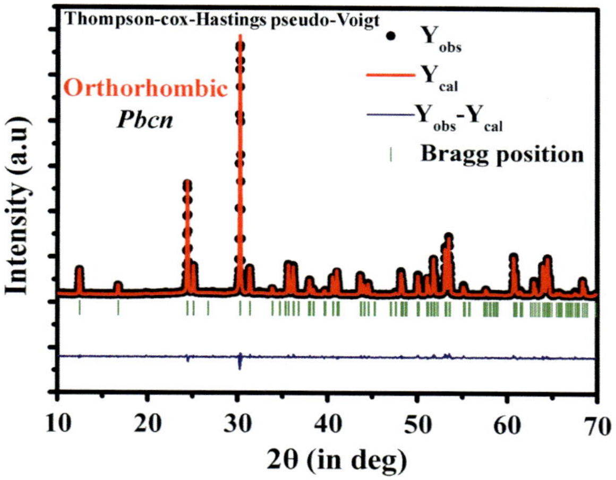

Figure 22.3
Rietveld refined powder XRD pattern of single phase columbite-like $MgNb_2O_6$ composition prepared with 2% excess MgO.

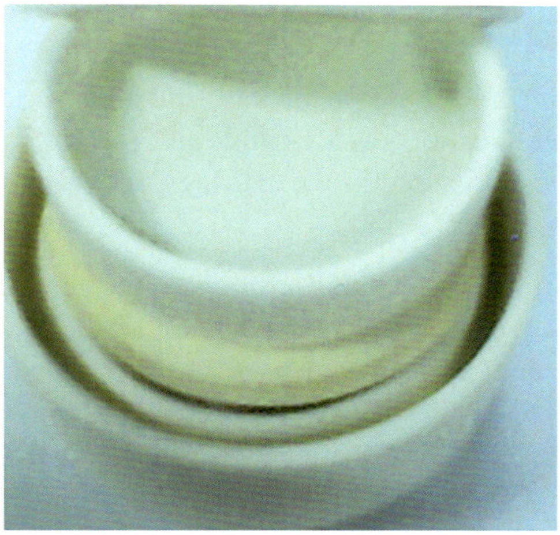

Figure 22.4
Schematic showing the partial covering method used for the synthesis of PMN-PT ceramic crucible arrangement (top dish used to close the set up is not shown).

treatment. This process eliminates the pores and results in the formation of dense pellet with approximately 95% theoretical density, which confers enhanced electrical properties to the ceramics. Standardization of sintering conditions is crucial as it determines the grain growth and directly influence the properties of the ceramic prepared.

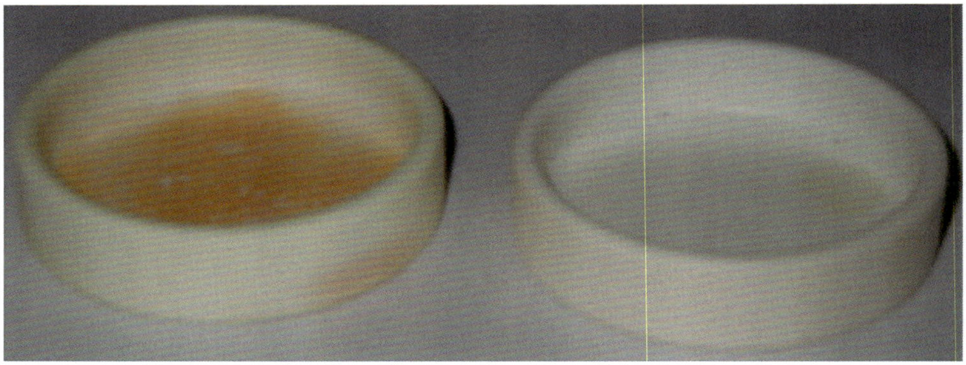

Figure 22.5
Shows the coloration in crucible in which sample with excess PbO was calcined, due to the removal of unused PbO through stabilization heating.

4. Synthesis of PMN-PT ceramics columbite B-stie precursor method

4.1 Single phase columbite precursor ($MgNb_2O_6$)

Chemistry of synthesis decides the physics of the material.

Columbite-like structure $MgNb_2O_6$ has perovskite like oxygen octahedra that makes it an ideal starting material for PMN and PMN-PT synthesis. The solid-state preparation in which columbite-like $MgNb_2O_6$ is used as the starting material is called columbite-precursor method [5].

Since the quality of the starting material has profound influence on the ferroelectric, dielectric, and piezoelectric properties of the final ceramic, the synthesis, characterization, and structural analysis of columbite-like $MgNb_2O_6$ phase plays a significant role in the synthesis of phase-pure PMN-PT through columbite precursor route. Unreacted Nb_2O_5 left in the precursor is considered as one of the reasons for the formation of stable paraelectric pyrochlore phase in PMN-PT. This parasitic phase in PMN-PT drastically reduces its ferroelectric property and makes it unsuitable for device applications [5].

Singh and Bajpai [7] have reported the formation of a corundum-like unwanted phase $Mg_4Nb_2O_9$ in the solid-state synthesis of columbite-like $MgNb_2O_6$, and the processing temperature is presented as the major reason for the same. Of the five possible product phases of MgO and Nb_2O_5 reaction, columbite-like phase $MgNb_2O_6$ and corundum-like phase $Mg_4Nb_2O_9$ are stable at the normal calcination temperature [8] and can be used as precursor for the synthesis of PMN-PT. But, a mixed phase of $MgNb_2O_6$ and $Mg_4Nb_2O_9$ is not ideal for the synthesis of phase-pure and device quality PMN-PT ceramic. The necessity of pure columbite-like phase was emphasized in earlier reports as well [9]. Swarts and Shrout [5] reported that, 3%–5% excess MgO in the initial precursor considerably improves the formation of perovskite PMN phase. Excess MgO assists the formation of pure PMN-PT perovskite phase in two ways, (1) it ensures the completion of Nb_2O_5 reaction, so that direct reaction of Nb_2O_5 with PbO can be checked; (2) cubic pyrochlore $Pb_3Nb_4O_{13}$ formed due to the reaction between PbO and Nb_2O_5 further reacts with PbO and forms rhombohedral pyrochlore which in turn reacts with MgO at elevated temperatures to form PMN [10]. The whole process is dependent on the modification in the MgO reactivity while encountering the pyrochlore phase.

Earlier studies have shown that even a small decrease in Mg/Nb ratio, will have significant negative impact on the formation of columbite-like phase $MgNb_2O_6$ [11]. $MgNb_2O_6$ and $Mg_4Nb_2O_9$ are formed due to the reaction between MgO and Nb_2O_5 in the molar ratio 1:1 and 4:1 respectively. In my study, when stoichiometric proportions of MgO and Nb_2O_5 precursors were used, corundum-like $Mg_4Nb_2O_9$ phase was also formed along with columbite-like phase, $MgNb_2O_6$. Though the formation of corundum-like phase needs higher temperature processing for a longer time for its predominance, a small percentage may be present along with columbite-like phase, $MgNb_2O_6$, during the reaction of MgO and Nb_2O_5 in the temperature of interest [8,12]. Spatial fluctuations of the composition (Mg/Nb ratio) could be responsible for the impurity phase formation in the synthesis of columbite-like $MgNb_2O_6$. It has been found that 2%–4% of excess MgO helps in the formation of pure columbite-like phase, $MgNb_2O_6$, while, a higher MgO excess (20%) favors corundum-like phase $Mg_4Nb_2O_9$ formation [12]. Hence, to get a phase-pure perovskite PMN-PT through columbite route, 2–4 mol% excess of MgO may be used in the synthesis of columbite-like $MgNb_2O_6$ (Fig. 22.3).

4.2 Synthesis of PMN-PT ceramics

The basic steps involved in the synthesis of PMN-PT ceramic.

1. $MgO + Nb_2O_5$ —— $MgNb_2O_6$
2. $MgNb_2O_6 + PbO + TiO_2$ —— $PMN - PT$

Chemicals used for synthesis were: PbO (Acros/Alfa Aesar 99.9+ %, MW- 223.19), MgO (Aldrich 99+ %, MW: 40.31), Nb_2O_5 (Alfa Aesar 99.8%, MW -265.81), TiO_2- Titanium (lV) oxide (Aldrich 99.8%, MW -79.9).

Fine and well mixed precursors is an essential prerequisite for solid-state reaction as it enables proximity of the chemicals during thermal treatments. Chemical homogeneity of B-site cations is very important for superior relaxor properties, which can be ensured by a thorough grinding of B-site precursors. Also, as the sintering kinetics greatly depend on the size of the particles; reduced particle size achieved by thorough grinding reduces the path length for the atomic diffusion which in turn will accelerate the densification process and/or attainment of chemical equilibrium at a comparatively lower temperature.

Stoichiometric amounts of the precursors: PbO, $MgNb_2O_6$, and TiO_2 are to be thoroughly (repeat thoroughly) ground prior to calcinations. Calcination is carried out in a specially designed alumina crucibles at strong lead atmosphere. Also certain percentage of excess PbO in the initial precursor to compensate the lead loss may be advisable. Through a partial covering combined with modulated heating quality PMN-PT ceramic could be realized even in stoichiometric composition. Although several studies suggested the use of excess PbO [6,13,14] to minimize pyrochlore formation, the exact amount of excess PbO required has not been reported. While it was presumed that unused PbO would evaporate at the calcination temperature ~850°C, reports suggest the formation of off-stoichiometric ceramics with the inclusion of unreacted PbO in the grain boundary, which would affect the electrical properties of the perovskite sample [15–19]. However, studies on the removal of unused PbO from the samples are lacking.

This section presents a novel method for synthesizing phase-pure and stoichiometric perovskite phase PMN-PT using single phase columbite precursor $MgNb_2O_6$ and calcination by a partial covering method combined with modulated heating for minimizing both Pb loss and pyrochlore formation during PMN-PT synthesis. A high temperature stabilization of the ceramic is given to maximize the perovskite phase and to strengthen the ceramic.

Significance of modulated heating can be understood from the following discussion. The cubic $Pb_3Nb_4O_{13}$ (P_3N_4) pyrochlore phase forms at 600°C, whereas the perovskite phase formed between 900 and 950°C [20]. Greater bonding strength of oxygen and cations, and lower formation temperature of pyrochlore (450–600°C) compared to perovskite (850–1050°C) [21], poses a challenge in the synthesis of PMN-PT. Although high temperature (1050°C) is ideal for maximum perovskite formation [21], the synthesis temperature of PMN-PT in most studies is restricted to 850°C [22] to minimize PbO evaporation and thereby suppressing the formation of pyrochlore phase.

4.3 Partial covering method and modulated heating: a novel approach

A partial covering method, designed with alumina dishes, could be used to provide strong Pb atmosphere during calcination. Alumina (Al_2O_3) dishes were selected against platinum to avoid the possible reaction of Pb melt with platinum. The size, shape and dimensions of the alumina dishes were so chosen as to minimize the air space around the sample and within the covering dishes (Fig. 22.4).

The process used flat alumina dishes to ensure uniform and steady heating during calcinations and sintering. Additional alumina lids were used to occupy the space inside the pair of covering crucibles and to increase the weight of the lids, so as to ensure perfect closure of the sample inside the dish. An alumina dish containing PbO was placed over it to provide Pb/PbO atmosphere during calcination and sintering. The complete set up was kept in two comparatively bigger alumina dishes arranged as a tight module (no alumina paste was used).

To minimize the synthesis cost of the ceramic preparation in all possible ways, only conventional furnace was used for calcination and sintering. A higher ramp of 10°C/min was administered during calcination and sintering in the range 400—600°C to avert the possible formation of pyrochlore and lower ramp of 2°C/min above 600°C (modulated heating) was applied for the gradual formation and stabilization of perovskite phase. It was observed that the formation of the cubic pyrochlore (P_3N_4) was prevented, when a rapid heating rate (10°C/min) was used initially up to 600°C. When the same was done with heating rates of 2 and 5°C/min in the lower temperature range up to 600°C, pyrochlore phase was seen along with the perovskite PMN-PT phase, even though all the other precautions were taken. Hence, a minimum heating rate of 10°C/min or even higher for calcination in the temperature range up to 600°C will be efficient in preventing pyrochlore phase formation. Subsequent slow heating rate of 2°C/min from 600 to 850°C may enhance the P_3N_2 pyrochlore phase formation which is an excellent precursor for perovskite PMN-PT. Calcination and stabilization temperatures were fixed at 850 and 1050°C for 6 and 4 h respectively. The same procedure can be followed for the PMN-PT preparation with and without excess PbO.

4.4 Significance of stabilization heating

Although PMN-PT perovskite will be formed during calcination at 850°C, probability of formation of pyrochlore phase during cooling of the sample is high, as the system goes through 600°C down to 450°C, which is the favorable range for the formation of cubic pyrochlore $Pb_3Nb_4O_{13}$. This can be averted by eliminating the unused PbO from the sample prior to cooling. Since the PMN-PT phase formation is expected to be close to 100% at around 1000°C [21] a slow heating up to 1050°C ensure the completion of perovskite formation. On the other hand if the mixed precursor is directly heated to 1050°C, Pb evaporation will be enhanced and will lead to the formation of PMN-PT ceramic with inferior properties, due to the formation of paraelectric and secondary pyrochlore phase in the composition. Also, the completion of the perovskite phase is expected to take place above 1000°C. Hence, a slow heating up to 1050°C ensure the completion of perovskite formation. Now the system (newly formed perovskite PMN-PT) will be retained at 1050°C for 4 h, which is called stabilization heating. Since PMN-PT phase is already formed, and its melting temperature is above 1280°C, stabilization heating will not affect the perovskite PMN-PT phase. However, it will maximize the perovskite phase and remove the unused PbO (melting point ~888°C), if at all left in the sample; and thus enable the formation of single phase perovskite PMN-PT ceramics [23,24].

To establish the effectiveness of the newly devised partial covering method and modulated heating, ceramics samples of the compositions $x = 0.3$, 0.325, 0.35, and 0.375 were prepared without using excess PbO in the initial precursor and analyzed. In all the cases phase-pure PMN-PT ceramic could be realized, which supports the effectiveness of partial covering method and modulated heating technique for realizing the single phase ceramic.

4.5 Removal of unused PbO and stabilization

After the phase stabilization, alumina dish in which the stoichiometric sample was calcined had no discoloration, while the dish with samples having excess PbO, had orange coloration indicating the PbO vaporization (Fig. 22.5). Absence of PbO deposit on the dish containing stoichiometric sample indicates complete utilization of PbO during phase formation as well as stability of the sample during stabilization heating. The stabilization process also helps to increase the tolerance factor [13], which in turn pushes the resultant perovskite PMN-PT phase to a noncentrosymmetric structure with improved properties. At stabilization temperatures above 1100°C, decomposition of the perovskite phase with release of PbO was observed even in samples without excess PbO, confirming decomposition of the sample at higher temperature, which may lead to the formation of pyrochlore phase during the cooling process. Stability of perovskite above 1100°C was tested in 0.675 PMN—0.325 PT also, which confirmed the degradation of samples at very high temperature.

4.6 Pellet compacting and sintering

Phase confirmed calcined powder samples were cold pressed after mixing thoroughly with 5% PVA (two drops for 5 g sample) for sintering, in tungsten carbide die set (top punch-tungsten carbide brazed with steel). A uniaxial pressure of 350 MPa was given for 5 min

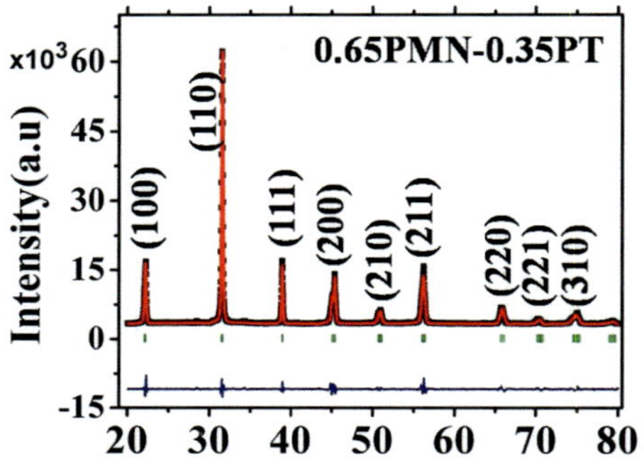

Figure 22.6
X-ray diffractogram of 0.65 PMN—0.35 PT ceramic prepare through partial covering combined with modulated heating.

and an optimized sintering temperature of 1150°C for 4 h was applied to the samples during densification. The partial covering method adopted for pellet sintering in which the pellets were covered with the same PMN-PT packing powder to prevent decomposition of the sample as well as to avoid Pb loss and associated pyrochlore formation on the surface. The pellets containing the binder PVA were given a dwell time of 3 h at 600°C during sintering to remove PVA. Modulated heating applied during calcination was adopted in the temperature range of 400–600°C, and the sintering was done at a temperature of 1150°C for 4 h.

5. Functional studies to test the quality of the ceramic

All the compositions (Fig. 22.6) showed improved ferroelectric, dielectric and piezoelectric properties. Piezoelectric coefficient of 475 pC/N could be realized and all samples prepared were showing substantially low dielectric loss which established device worthiness of the samples and the quality of the partial covering combined with modulated heating to realize high quality PMN-PT system. Hysteresis loop study (Fig. 22.7) shows the improved ferroelectric response exhibited by 0.65 PMN–0.35 PT stoichiometric system which was synthesized through columbite B-site precursor method, partial covering, and modulated cum stabilized heating.

6. Thin-film growth of PMN-PT using pulsed laser deposition

Dense ceramic pellets realized through modulated heating cum stabilization could be used for growing device worthy PMN-PT thin film through laser ablation. Ability to make stoichiometric transfer of material from multicomponent target to growing film made PLD an excellent technique for preparation of thin films of lead based systems like PMN-PT, as the functional response of PMN-PT like systems is dependent on the precision of the

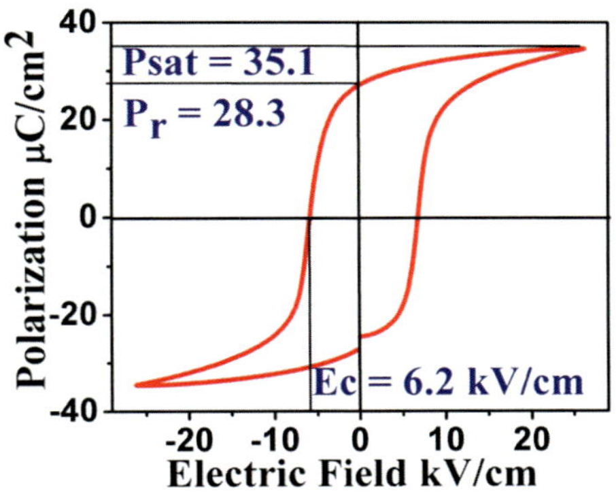

Figure 22.7
Ferroelectric hysteresis trace using PE loop tracer (Radiant Technologies, USA) by placing the unpoled disc into silicone oil.

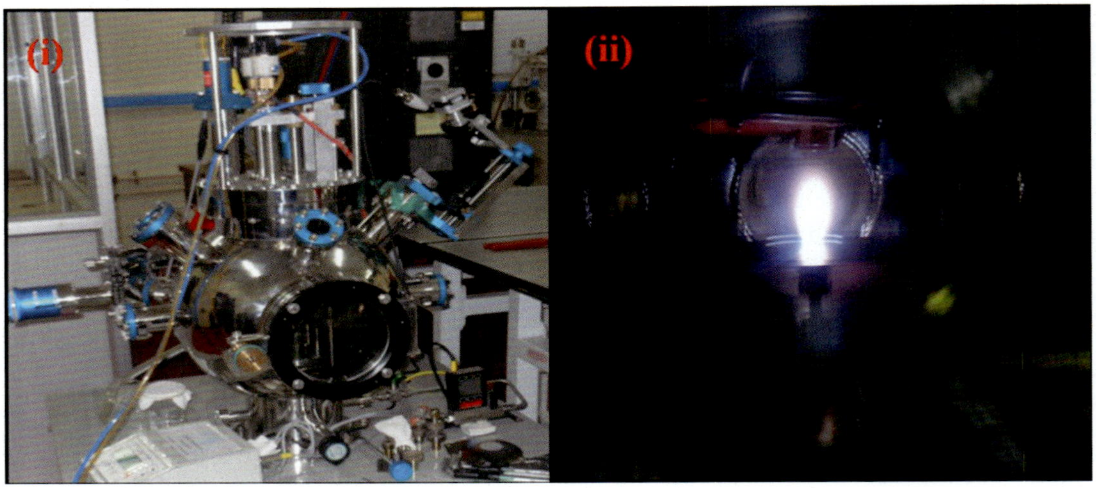

Figure 22.8
(1) Photograph of 14″ PLD chamber used for laser ablation. (2) Plume formed during ablation process a result of the vaporized constituents of the target. The emanated species move toward the substrate which is fixed on the extended construction of the substrate holder with heating assembly.

chemical composition [25,26]. Ever since its discovery in 1960, LASER has been a powerful tool in many technological applications. Narrow frequency bandwidth, coherence and high power density enables the production of highly intense beam capable of vaporizing even hardest and heat resistant materials. KrF excimer laser is ideal for ablation of PMN-PT system, as its wavelength (248 nm) matches the absorption spectrum of PMN-PT [27].

However, the challenges associated with thin-film research are many. Two major issues with thin films are, (1) clamping effect from substrate restricts the mechanical displacement, and (2) thin films are only nanometers in thickness. Therefore, for thin films, single grain (epitaxial) film is expected to show best response. If the film is polycrystalline, with multidomains and domain walls, there will be little space available for domain switching and motion, which considerably reduces the piezoelectric response in comparison with the bulk counterpart. Based on theoretical calculations, an average of one-third piezoelectric response can be expected in polycrystalline thin films compared to the bulk counterparts [28].

Synthesis of PMN-PT relaxor films invites special attention, as the electromechanical response exhibited by the polycrystalline film greatly depends on the thickness of the film. Miniaturization of electronic gadgets demand smaller actuators and sensors for which thick films of PMN-PT-like systems can be useful. PMN-PT films with thickness of one micron or more is required for optical applications as well [29]. The ferroelectric behavior of the PMN-PT thin films greatly depends on the orientation as well as crystalline quality of the film; and an increased thickness will favor domain rotation, which in turn will enhance the ferroelectric response of the developed films.

Increase in substrate temperature and a higher deposition rate favor the growth of thick ferroelectric films with improved response. Higher deposition rate with laser pulse ensures large mass transfer from the ablation plume, and minimize the loss of volatile materials during deposition at higher temperature. Nevertheless, the restrictions in the mechanical displacement of the thin film due to the clamping effect from the substrate; and the crack formation with increase in the thickness of the film are two major complementary challenges in the growth of PMN-PT thick film using PLD. Studies on the PMN-PT thin films, which are grown, using PLD are scanty due to the complexities involved in the PLD growth process (Fig. 22.8) [30–32]. Presence of large number of elemental species in PMN-PT composition poses additional complexity in the synthesis of crack free and device worthy thick films. An increased thickness warrants increased substrate temperature during deposition for better crystallinity and stabilization. But high temperature affects/interfere with the integration of piezoelectric film with other materials; and favor the formation of pyrochlore phase due to lead evaporation, which will upset the functional response of PMN-PT thin (or thick) films. Earlier reports suggest that, template layer is inevitable to have a complete elimination of pyrochlore on metal coated substrate like Pt/TiO_2/SiO_2/Si substrate [33,34]. Conducting perovskite oxide buffer layers like $La_{0.5}Sr_{0.5}CoO_3$ (LSCO) is reported to be ideal for preventing the formation paraelectric and conducting pyrochlore phase during the growth of PMN-PT thin films.

Using the high temperature stabilized targets thin films of PMN-PT having thickness from 100 nm to 1 μm could be realized. Film was grown on LSCO buffered Pt/TiO_2/SiO_2/Si substrate and the film so developed was crack free and exhibited improved functional response [35,36].

Deposition parameters for the realization of PMN-PT thin films are given below for reference (Table 22.1). However, substantial optimization would be very much essential to realize good quality PMN-PT thin films for device fabrication.

Table 22.1: Optimized deposition parameters for LSCO and PMN-PT thin-film growth.

Parameter	LSCO	PMN-PT
Substrate-target distance (cm)	3.5–4	3.5–4
O_2 partial pressure during deposition (mbar)	4–4.5	0.5–0.7
Energy fluence (J/cm^2)	2.5–2.9	2.5–2.9
Frequency (Hz)	4–6	8–10
Deposition temperature (°C)	550	675
Annealing time	30 min (550°C) and 30 min (675°C)	40 min (675°C)

7. Conclusion

This chapter described the novel partial covering combined with modulated heating to realize stoichiometric relaxor multicomponent system (1-x) PMN-x PT. Enhanced functional response exhibited by the ceramic pellets established the quality of the synthesis technique as well as the quality of the ceramic sample. The dense ceramic pellets synthesized using partial covering combined with modulated heating followed by stabilization could be ablated to realize device worthy PMN-PT thin films through pulsed laser ablation. This new technique may be extended with or without modifications for realizing lead based systems in general [35,36].

Acknowledgments

Pius is thankful to Science and Engineering Research Board (SERB), DST, Govt. of India for granting research award, Teacher Associate for Research Excellence (TARE—TAR/2020/000241), to carry out research in association with Prof. Karuna Kar Nanda, Material Research Center, Indian Institute of Science Bangalore. This book chapter is prepared as part of the initiation of the project at IISc Bangalore. Author also likes to acknowledge Prof. M. S. Ramachandra Rao, Material Science Research Center, Department of Physics IIT Madras who guided for the successful completion of the work.

References

[1] A.A. Vives, Piezoelectric Transducers and Applications, second ed., 2008.
[2] O. Noblanc, P. Gaucher, G. Calvarin, J. Appl. Phys. 79 (8) (1996) 4291.
[3] D. Udayan, K.R. Sahu, A. De, Solid State Phenom. 232 (2015) 235.
[4] E.M. Jayasingh, K. Prabhakaran, R. Sooraj, C. Durgaprasad, S.C. Sharma, Ceram. Int. 35 (2009) 591.
[5] S.L. Swarts, T.R. Shrout, Mat. Res. Bull. 17 (1982) 1245.
[6] P. Kumar, S. Sharma, O.P. Thakur, C. Prakash, T.C. Goel, Ceram. Int. 30 (2004) 585.
[7] K.N. Singh, P.K. Bajpai, Phys. B 405 (2010) 303.
[8] Y.C. You, H.L. Park, Y.G. Song, H.S. Moon, G.C. Kim, J. Mater. Sci. Lett. 13 (1994) 1487.
[9] E.R. Camargo, E. Longo, E.R. Leite, J. Sol. Gel Sci. Technol. 17 (2000) 111.
[10] M. Inada, Natl. Tech. Rep. 27 (1977) 95.
[11] P. Augustine, M.S.R. Rao, Ceram. Int. 41 (2015) 11984.
[12] K. Sreedhar, N.R. Pavaskar, Mater. Lett. 53 (2002) 452.
[13] S.Y. Chen, C.M. Wang, S.Y. Cheng, Mater. Chem. Phys. 49 (1997) 70.
[14] L.B. Kong, J. Ma, W. Zhu, O.K. Tan, J. Alloys Compd. 236 (2002) 242.
[15] L.E. Cross, Ferroelectrics 151 (1) (1994) 305.
[16] J.H. Ma, X.J. Meng, J.L. Sun, T. Lin, F.W. Shi, G.S. Wang, J.H. Chu, Appl. Surf. Sci. 240 (2005) 275.
[17] P. Ravindranathan, S. Komarneni, A.S. Bhalla, R. Roy, J. Am. Ceram. Soc. 74 (12) (1991) 2996.
[18] S. Shah, M.S.R. Rao, Appl. Phys. A 71 (2000) 65.
[19] H. Suzuki, T. Naoe, H.T. Miyazaki, T. Ota, J. Eur. Ceram. Soc. 27 (2007) 3769.
[20] Y.C. Liou, Y.C. Huang, C.T. Wu, MRS Proc. 848 (2004). FF9-2.
[21] H.C. Ling, M.F. Yan, B.I. Lee, E.J.A. Pope (Eds.), Chemical Processing of Ceramics, 1, Marcel Dekker Inc., New York, 1994, p. 397.
[22] C. Tantigate, J. Lee, A. Safari, Appl. Phys. Lett. 66 (1995) 1610.

[23] P. Augustine, S. Samanta, M. Rath, M. Miryala, K. Sethupathi, M. Murakami, M.R. Rao, Mater. Res. Bull. 95 (2017) 47–55.
[24] P. Augustine, M. Rath, M.R. Rao, Ceram. Int. 43 (12) (2017) 9408–9415.
[25] D.B. Chrisey, G.K. Hubler, Wiley-Interscience, Pulsed Laser Deposition of Thin Films, New York, 1994.
[26] S. Yokoyama, Y. Honda, H. Morioka, S. Okamoto, H. Funakubo, T. Iijima, H. Masuda, Y.T. Sito, H. Okino, O. Sakata, S. Kumura, J. Appl. Phys. 98 (2005) 094106.
[27] M. Ohring, The Materials Science of Thin Films, Academic, New York, 1991.
[28] T.M. Shaw, S.T. McKinstry, P.C. McIntyre, The properties of ferroelectric films at small annual review of materials, Science 30 (1) (2000) 263.
[29] K. Shinozaki, S. Hayashi, N. Wakiya, T. Kiguchi, J. Tanaka, N. Ishizawa, K. Sato, M. Kondo, K. Kurihara, IEEE Trans. Ultrason. Ferroelectr. & Freq. Contr. 55 (5) (2008).
[30] G.R. Bai, S.K. Streiffer, P.K. Baumann, O. Auceillo, K. Ghosh, S. Stemmer, A. Munkholm, C. Thompson, R. Rao, E.B. Eom, Appl. Phys. Lett. 76 (2000) 3106.
[31] Z. Kighelman, D. Damjanovic, N. Setter, J. Appl. Phys. 90 (2001) 4682.
[32] V. Nagarajan, C.S. Ganpule, B. Nagaraj, S. Aggarwal, E.D. Williams, A.L. Roytburd, R. Ramesh, Appl. Phys. Lett. 77 (2000) 438.
[33] M.C. Jiang, T.B. Wu, J. Mater. Res. 9 (1994) 1879.
[34] K.L. Saenger, R.A. Roy, D.B. Beach, K.F. Etzold, Mater. Res. Soc. Symp. Proc. 285 (1993) 421.
[35] P. Augustine, M. Rath, G.R. Haripriya, K. Sethupahi, M.R. Rao, Ceram. Int. 46 (17) (2020) 26767–26776.
[36] P. Augustine, M. Miryala, S. Samanta, S.P.K. Naik, K. Sethupathi, M. Murakami, M.R. Rao, 46 (5), (2020) 5658–5664.

Index

Note: 'Page numbers followed by "*f*" indicate figures and "*t*" indicate tables.'

H

K

L

M

N

T

U

V

Printed in the United States
by Baker & Taylor Publisher Services